101 Springer Series in Solid-State Sciences

Edited by Klaus von Klitzing

Springer Series in Solid-State Sciences

Editors: M. Cardona P. Fulde K. von Klitzing H.-J. Queisser

Managing Editor: H. K. V. Lotsch

Volumes 1–89 are listed at the end of the book

90 **Earlier and Recent Aspects of Superconductivity** Editors: J. G. Bednorz and K. A. Müller

91 **Electronic Properties of Conjugated Polymers III** Basic Models and Applications
Editors: H. Kuzmany, M. Mehring, and S. Roth

92 **Physics and Engineering Applications of Magnetism** Editors: Y. Ishikawa and N. Miura

93 **Quasicrystals** Editors: T. Fujiwara and T. Ogawa

94 **Electronic Conduction in Oxides** By N. Tsuda, K. Nasu, A. Yanase, and K. Siratori

95 **Electronic Materials** A New Era in Materials Science
Editors: J. R. Chelikowsky and A. Franciosi

96 **Electron Liquids** By A. Isihara

97 **Localization and Confinement of Electrons in Semiconductors**
Editors: F. Kuchar, H. Heinrich, and G. Bauer

98 **Magnetism and the Electronic Structure of Crystals**
By V. A. Gubanov, A. I. Liechtenstein, and A.V. Postnikov

99 **Electronic Properties of High-T_c Superconductors and Related Compounds**
Editors: H. Kuzmany, M. Mehring, and J. Fink

100 **Electron Correlations in Molecules and Solids** By P. Fulde

101 **High Magnetic Fields in Semiconductor Physics III**
Quantum Hall Effect, Transport and Optics By G. Landwehr

102 **The Physics of Conjugated Conducting Polymers** Editor: H. Kiess

103 **Molecular Dynamics Simulations** Editor: F. Yonezawa

104 **Products of Random Matrices in Statistical Physics**
By A. Crisanti, G. Paladin, and A. Vulpiani

105 **Self-Trapped Excitons** By K. S. Song and R. T. Williams

106 **Physics of High-Temperature Superconductors** Editors: S. Maekawa and M. Sato

107 **Electronic Properties of Polymers** Orientation and Dimensionality of Conjugated Systems
Editors: H. Kuzmany, M. Mehring, and S. Roth

108 **Site Symmetry in Crystals** Theory and Applications
By R. A. Evarestov and V. V. Smirnov

G. Landwehr (Ed.)

High Magnetic Fields in Semiconductor Physics III

Quantum Hall Effect, Transport and Optics

Proceedings of the International Conference,
Würzburg, Fed. Rep. of Germany,
July 30 – August 3, 1990

With 429 Figures

Springer-Verlag
Berlin Heidelberg New York
London Paris Tokyo
Hong Kong Barcelona
Budapest

Professor Dr. Gottfried Landwehr
Physikalisches Institut, Universität Würzburg, Am Hubland
W-8700 Würzburg, Fed. Rep. of Germany

ISBN-13: 978-3-642-84410-2 e-ISBN-13: 978-3-642-84408-9
DOI: 10.1007/978-3-642-84408-9

Library of Congress Cataloging-in-Publication Data. High magnetic fields in semiconductor physics III: quantum Hall effect, transport and optics: proceedings of the international conference, Würzburg, Fed. Rep. of Germany, July 30-August 3, 1990 / editor, G. Landwehr. p. cm. – (Springer series in solid-state sciences; 101). Includes bibliographical references and index. ISBN-13: 978-3-642-84410-2 1. Quantum Hall effect–Congresses. 2. Semiconductors–Junctions–Congresses. 3. Superlattices as materials–Congresses. I. Landwehr, G. (Gottfried), 1929-. II. Series. QC612.H3H54 1992 537.6'226–dc20 91-45269

Softcover reprint of the hardcover 1st edition 1992

Typesetting: Camera ready by authors
54/3140-543210 – Printed on acid-free paper

Preface

This volume contains contributions presented at the International Conference "The Application of High Magnetic Fields in Semiconductor Physics", which was held at the University of Würzburg from July 30 to August 3, 1990.

In the tradition of previous Würzburg conferences on the subject – the first conference was held in 1972 – only invited papers were presented orally. In order to give an overview of the subject, all lecturers were asked not only to review the latest results in their field but also to give a general introduction. Therefore this volume is different from the usual conference proceedings.

The scope of the conference has not essentially changed from previous ones. Already in 1988, contributions concerning two-dimensional electronic systems dominated. This trend has continued. In fact, there is special emphasis on the quantum Hall effect in this book. Almost half of the invited papers and a substantial number of the poster contributions deal with the integral and fractional quantum Hall effect. This subject was preferentially chosen because 1990 was the 10th anniversary of Klaus von Klitzing's discovery of the effect. A look at the content of this book shows that the effect is still being actively investigated. During the last two years considerable insight into the problem of edge currents has been gained. Also the influence of contacts is much better understood these days. Especially exciting seems the problem of Wigner condensation in the fractional quantum Hall regime. Whereas in previous years information was mainly deduced from transport experiments, optical methods have become competitive. Also progress has been made on the theoretical side as far as the spin-polarisation of the fractional quantum Hall states is concerned.

During the last two years the activities in the field of high magnetic field physics in conjunction with semiconductors have not declined. On the contrary, one has the impression that the application of high magnetic fields to semiconductors has grown in the recent past. For the elucidation of the electronic band structure of semiconductors high magnetic fields are still an indispensable tool. The investigation of two-dimensional electronic systems is frequently connected with the use of high magnetic fields.

This volume contains 53 contributions which were presented as posters. Unfortunately, not all contributions to the conference could be incorporated in the book for technical reasons. Together with the 37 printed invited papers the book gives a good impression of the present state of the field.

The organizing committee consisted of G. Landwehr (Chairman), K. von Klitzing, H. Hajdu, and W. Ossau.

The financial support of the conference by the following sponsors is gratefully acknowledged:

Deutsche Forschungsgemeinschaft
Bayerisches Staatsministerium für Wissenschaft und Kunst
Regionalverband Bayern der Deutschen Physikalischen Gesellschaft
Industrial sponsors: Siemens AG, München; IBM, Stuttgart;
Bruker Analytische Meßtechnik, Karlsruhe; Oxford Instruments, Wiesbaden; Cryophysics/Odelga Physics, Nürtingen; Vakuumschmelze, Hanau.

Würzburg, June 1991 *G. Landwehr*

Contents

Part I Integer Quantum Hall Effect, Localization

Part IV Magneto-Transport in 2D Systems

Part V Magneto-Transport in 3D Systems

Part VI Magneto-Optics, Mainly in Quantum Well Structures

Part I

Integer Quantum Hall Effect, Localization

Ten Years of the Quantum Hall Effect: Development and Present State of the Theory

J. Hajdu

Institut für Theoretische Physik, Universität zu Köln, Zülpicherstr. 77, W-5000 Köln 41, Fed. Rep. of Germany

Abstract. Shortly after the discovery of the Quantum Hall Effect had been announced /1/, a number of approaches were suggested to explain this surprising phenomenon:

The semi-phenomenological approach by Aoki and Ando /2/

The gauge argument by Laughlin /3/

The single point impurity model by Prange /4/

The Streda formula /5/

The lattice model by Thouless et al. /6/
(Preliminary form of the topological approach).

Semi-classical percolation models by Allen and Tsui /7/, Iordansky /8/ and Ono /9/.

In this survey I will try to sketch the development of these approaches and to show their place in the present state of the theory. Furthermore, I would like to make a few comments on some recent investigations and to mention a few still unsolved problems.

All of the approaches listed above refer to independent electrons (no Coulomb interaction). That a two-dimensional system of independent electrons moving in a random potential (impurities) and a strong perpendicular magnetic field indeed shows a QHE was confirmed by Ando in an early numerical analysis /10/. (A survey of the numerical work will be given by Professor Aoki in this conference).

1. Semi-phenomenological approach to the QHE

Aoki and Ando /2/ seem to be the first who fully realized that the QHE is due to a special type of localization-delocalization phenomenon occuring in a two-dimensional disordered electron system in the presence of a strong perpendicular magnetic field. Their approach can be formulated in the following program:

- Neglect the Coulomb interaction between the electrons
- Assume that (due to disorder) the Landau levels are broadened to (practically) non-overlapping bands centered around the Landau energies

$$\varepsilon_\nu = \hbar\omega_c(\nu + \frac{1}{2}) \tag{1.1}$$

$\nu = 0,1,2,\ldots$

- Define localized state $|\alpha\rangle$ by vanishing velocity expectation value,

$$\langle\alpha\,|\,\vec{v}\,|\,\alpha\rangle = 0 \tag{1.2}$$

Springer Series in Solid-State Sciences, Vol. 101
High Magnetic Fields in Semiconductor Physics III Editor: G. Landwehr

- Assume that in each Landau band delocalized states exist only in some small ranges around the band center, all other states being localized
- Use Kubo's conductivity formula to formally prove that at zero temperature the static longitudinal conductivity σ vanishes and the Hall conductivity takes the quantized values

$$\sigma_H = i\frac{e^2}{h} \tag{1.3}$$

as long as the Fermi energy ε_F varies in the ith mobility gap (localization regime).

The approach of Aoki and Ando is semi-phenomenological in the sense that localization is assumed rather than demonstrated to occur in the disordered system under consideration. To arrive at (1.3) for i = 1,2,... these authors added a further hypothesis: a sort of electron-hole symmetry in each Landau band. However, after Streda /5/ published a new general Hall conductivity formula for independent electrons it seemed to be possible to avoid any additional hypothesis. For T = 0 this formula reads

$$\sigma_H(\varepsilon_F) = ec\left(\frac{\partial N(\varepsilon_F)}{\partial B}\right)_{\varepsilon_F} + \sigma_H^I(\varepsilon_F) \tag{1.4}$$

where $N(\varepsilon)$ is the integrated density of states and $\sigma_H^I(\varepsilon_F)$ is a somewhat complicated expression which, however, vanishes if ε_F lies in a band gap. If this is the ith gap

$$N = \frac{i}{2\pi\ell^2} = i\frac{eB}{\hbar c} \tag{1.5}$$

holds and (1.4) yields the value (1.3). To extend this result to the ith mobility gap the Aoki-Ando localization description has to be implemented into the Streda formula. When elaborating on this it becomes apparent that some steps in the derivation of Streda's formula from the exact Kubo Hall conductivity formula are only justified if the extension of the system in the direction of the electric field (L_x) is finite. This corresponds to the Dirichlet boundary condition for the wave functions,

$$\psi_\alpha(x = \pm\frac{L_x}{2}, y) = 0 \tag{1.6}$$

In the perpendicular direction of the vector potential the periodic boundary condition

$$\psi_\alpha(x, y + L_y) = \psi_\alpha(x) \tag{1.7}$$

is appropriate, being compatible with a finite Hall current j_y. Due to the boundary condition (1.6) the energy spectrum of a free electron in zero electric field depends on the momentum $p_y = hk$, repspectively on the quantity $X = l^2k$, $l = \sqrt{ch/eB}$ which is for $L_x \to \infty$ the x coordinate of the cyclotron center. In other words the degeneracy of the Landau levels is lifted, $\varepsilon_\alpha = \varepsilon_n(X)$ being bent up at $X \simeq \pm(L_x/2-1)$ such that $\varepsilon_n(X) \to \infty$ for $X \to \pm\infty$. If $L_x >> 1$ a bulk regime exists in which the spectrum is up to exponentially small corrections identical to the degenerate Landau levels. Since the spectrum bends up no energy gaps exist. This remains essentially true also in the presence of disorder. Consequently, the Streda formula is not immediately conclusive in the case where it is strictly valid. In fact the high field conductivity formulas /11/ used in /2/ are only approximately equivalent to the exact Kubo formula and therefore not suitable to study fundamental problems such as the exact quantization of the Hall conductivity /12/.

In looking for an alternative version of the Kubo formula which is suitable for carrying out the Aoki-Ando program we run into a further problem: the periodic boundary condition (1.7) is not compatible with the usual formalism of quantum mechanics in which the operator $\hat{y}$ associated with the coordinate y is the multiplication by y. Indeed, if Ψ satisfies (1.7)

$$\psi' = \hat{y}\psi = y\psi \tag{1.8}$$

violates this condition. A possible way out of this dilemma is rolling down the cylinder in a plane /12/. This is equivalent to considering a periodic Hamiltonian

$$H(x, y + L_y) = H(x, y) \tag{1.9}$$

rather than a periodic wave function. For such a Hamiltonian the well-known Bloch theorem holds stating that the energy is a periodic function in the reciprocal lattice,

$$\varepsilon_n(k + g) = \varepsilon_n(k) \tag{1.10}$$

for any g = integer $(2\pi/Ly)$. Furthermore

$$\langle \alpha \mid v_y \mid \alpha \rangle = \frac{1}{\hbar}\frac{\partial \varepsilon_\alpha}{\partial k} \tag{1.11}$$

$\alpha = (n,k)$, $\varepsilon_\alpha = \varepsilon_n(k)$. (Due to the Dirichlet boundary condition (1.6) $\langle\alpha|v_x|\alpha\rangle \equiv 0$). Owing to the periodicity (1.10) k can be restricted to the first Brillouin zone. With $\theta = kL_y$

$$-\pi \leq \vartheta < \pi \tag{1.12}$$

In this Bloch representation the Kubo formula for the Hall conductivity takes the form

$$\sigma_H = -\frac{e^2}{\hbar}\sum_n \left(\frac{1}{L_x}\int_{-\pi}^{\pi} d\vartheta\, f(\varepsilon_n(\vartheta))\frac{\partial X_n(\vartheta)}{\partial\vartheta}\right) \tag{1.13}$$

where f is the Fermi distribution function and

$$X_n(\vartheta) = \langle \alpha \mid x \mid \alpha \rangle \tag{1.14}$$

is the x-component of the center of mass of the Bloch wave function $\Psi_\alpha = \Psi_{nk}$ /13/. A formula similar to (1.13) was previously found by Jonson /14/. Localized states are defined as before:

$$\langle \alpha \mid v_y \mid \alpha \rangle = 0 \leftrightarrow \frac{\partial \varepsilon_\alpha}{\partial \vartheta} = 0, \quad -\pi \leq \vartheta < \pi \tag{1.15}$$

This implies

$$\frac{\partial X_n(\vartheta)}{\partial \vartheta} = 0, \quad -\pi \leq \vartheta < \pi \tag{1.16}$$

i.e. in a localized state the center of mass cannot be shifted by changing θ. Obviously such states do not contribute to σ_H.

It is now not difficult to formulate general conditions under which (1.13) yields the quantized values (1.3) /13/:

1. It is possible to distinguish bulk and edge states according to the position of the center of mass $X_n(\theta)$.
2. The bulk states are delocalized near the band centers ($\varepsilon = \varepsilon_\nu = \hbar\omega_c(\nu + 1/2)$) and localized elsewhere. The edge states are all delocalized.
3. The Bloch bands $\varepsilon_n(\theta)$, $-\pi \leq \theta < \pi$ of delocalized states can be connected to form smooth energy branches $\varepsilon_\nu(\theta)$, $-\infty < \theta < \infty$.
4. The total variation of $X_\nu(\theta)$ with θ is independent of disorder, i.e. the same as for free electrons - in spite of the fact that the Brillouin zones of localized states dropped.

If, at $T = 0$, ε_F lies in the range of localized bulk states of the ν-th band having energies $\varepsilon > \varepsilon_\nu$ and $\theta_\nu^\pm$ are the two solutions of the equation $\varepsilon_\nu(\theta) = \varepsilon_F$ with $X_\nu^+ = X_\nu(\theta_\nu^-) > X_\nu^- = X_\nu(\theta_\nu^+)$ then

$$L_{x\nu} = X_\nu^+ - X_\nu^- \tag{1.17}$$

is the effective system size in the x-direction determined by the electron distribution. Consequently, at $T = 0$

$$\sigma_H = -\frac{e^2}{h}\sum_{\nu=0}^{i-1}\left(\frac{1}{L_x}\int_{\vartheta_\nu^-}^{\vartheta_\nu^+} d\vartheta \frac{\partial X_\nu(\vartheta)}{\partial\vartheta}\right) = \frac{e^2}{h}\sum_{\nu=0}^{i-1}\left(\frac{L_{x\nu}}{L_x}\right) \tag{1.18}$$

Since $L_{x\nu} < L_x$, the sum in (1.18) is smaller than i - even for free electrons with integer filling $\eta = 2\pi l^2 n = i$. The deviation of the result in (1.18) from (1.3) is due to the fact that the current across the x-axis is inhomogeneous. Therefore, the correct procedure is to calculate the average of the local current density $j_y(x)$ instead of that of the mean current density $J_y = -ev_y/L_xL_y$. But doing this is equivalent to replacing the geometrical size L_x by the physical size $L_{x\nu}$ in (1.18). Alternatively we may refer to the relation

$$\tilde{\varepsilon}_\nu(\vartheta_\nu^-) - \tilde{\varepsilon}_\nu(\vartheta_\nu^+) = -eU_H \tag{1.19}$$

where $\tilde{\varepsilon}_\nu(\theta)$ is the energy in the presence of the electric field and U_H is the Hall voltage (independent of ν). Thus, we arrive at

$$\sigma_H = i\frac{e^2}{h} \tag{1.20}$$

(for ε_F being in the ith bulk mobility gap). The crucial point in this derivation is the stability condition

$$X_\nu^- - X_\nu^+ = L_{x\nu} \qquad \text{independent of disorder} \tag{1.21}$$

This condition is indispensible to establish the Aoki-Ando program. No semi-phenomenological approach to the QHE can avoid, I believe, this or an equivalent requirement. In the topological approach (see below) the condition (1.21) is replaced by special properties of the geometry.

When going from the (Kubo) formula (1.13) to (1.18) we had to assume that the Bloch bands $\varepsilon_n(\theta)$ $\pi \leq \theta < \pi$ can uniquely be connected to smooth energy branches $\varepsilon_\nu(\theta)$, $-\infty < \theta < \infty$ (condition 3). As pointed out by Ohtsuki and Ono /15/ this assumption contradicts the von Neumann-Wigner no-crossing theorem according to which accidental degeneracy can (practically always) be excluded if the Hamiltonian depends only on a single continuous parameter (θ). Since in a random potential also no degeneracy due to symmetry can be expected, band gaps occur instead of band crossings. These are caused by mixing of states brought about by the random potential. Such mixing occurs between edge states on the same side and on different sides, as well as between edge states and bulk states of equal energy. Due to mixing localized and delocalized states can behave similarly with respect to θ. In both cases closed trajectories parametrized by θ occur in the (ε_n, X_n)-plane. Fortunately, the mixing of states - which is anyway much less important in the case of long-range impurities /16/ - does not affect the foregoing explanation of the QHE significantly. In fact, the conditions given above can be shown to be unnecessarily restrictive /17/.

The framework we developed to discuss the Aoki-Ando program is suitable to comment also on some other approaches to the QHE.

The Hamiltonian in the Bloch representation $H(\theta)$ (which is obtained by replacing p_y by $p_y + h\theta/L_y$ in the periodic Hamiltonian (1.9)) can be interpreted as the Hamiltonian of a system on the surface of a cylinder embedded in 3d space. The Bloch quantum number θ corresponds to an axial gauge field with flux $\phi = (hc/e)\theta$. Since, in a homogenous electric field in the x-direction, the electron energy in lowest order in the field is given by

$$\tilde{\varepsilon}_n(\vartheta) = \varepsilon(\vartheta) + eEX_n(\vartheta)$$

the Hall conductivity (1.13) multiplied by E, i.e. the Hall current density, can be interpreted as (-c times) the adiabatic derivative of the average electron energy with respect to ϕ, averaged over an elementary flux unit $\phi_o = h/e$. This is the Hall current expression used in the gauge argument /3/ which asserts that increasing ϕ by ϕ_o results in shifting an electron from one edge of the system to another. Due to the mixing of bulk and edge states resulting in "minigaps" mentioned above, a disordered electron system does not exhibit such a gauge symmetry. Therefore the gauge symmetry has to be interpreted as an assumption similar to the stability condition (1.17). Thus, the gauge argument does not hold like a strict mathematical theorem but is approximate and requires an appropriate physical interpretation.

The condition (1.21) has turned out to be necessary and sufficient to prove that, for integer filling, the current lost by localization is exactly counterbalanced by excess current carried in delocalized states /18/. This so-called compensation was first noticed by Ando et al. /19/ and later by Prange /4/.

So far the only system for which the conditions 1-4 have been shown to be rigorously satisfied consists of independent electrons and point impurities with vanishing density in the thermodynamic limit /20/. This and the foregoing remark establishes the connection between the phenomenological approach and the scattering theoretical approach to the QHE initiated by Prange /4/ and further developed by Chalker /21/, Joynt and Prange /22/, Brenig /23/ and Brenig and Wysokinski /24/.

The Aoki-Ando program can also be discussed for a system with periodic boundary conditions in both directions,

$$\begin{aligned}\psi_\alpha(x+L_x, y) &= \psi_\alpha(x,y)\\ \psi_\alpha(x, y+L_y) &= \psi_\alpha(x,y)\end{aligned} \qquad (1.22)$$

This is equivalent to considering a system on a torus. Again, to use usual quantum mechanics, we roll down the torus on a plane, i.e. we repeat the original rectangular system both in the x- and the y-direction. In other words we consider a double-periodic Hamiltonian

$$H(x+L_x, y+L_y) = H(x,y) \tag{1.23}$$

If the flux associated with B is, in units of ϕ_o, an integer, then the magnetic translations commute with each other and the Bloch theorem applies. In particular

$$\langle \alpha \mid \vec{v} \mid \alpha \rangle = \frac{1}{\hbar}\frac{d\varepsilon_n(\vec{k})}{d\vec{k}} \tag{1.24}$$

($\alpha = (n,\underline{k})$). Again, the Bloch wave numbers $\underline{k} = (k_x,k_y)$ can be restricted to the first Brillouin zone. With $\theta_1 = k_xL_x$, $\theta_2 = k_yL_y$

$$-\pi \leq \vartheta_1, \quad \vartheta_2 < \pi \tag{1.25}$$

In the Bloch representation the Kubo formula for the Hall conductivity can be written

$$\sigma_H = \frac{e^2}{h}\sum_n \frac{1}{2\pi i}\int_{-\pi}^{\pi} d\vartheta_1 \int_{-\pi}^{\pi} d\vartheta_2 f(\varepsilon_n(\vec{\vartheta}))Q_n(\vec{\vartheta}) \tag{1.26}$$

$$Q_n(\vec{\vartheta}) = \left[\left\langle \frac{\partial u_n}{\partial \vartheta_1} \mid \frac{\partial u_n}{\partial \vartheta_2}\right\rangle - \left\langle \frac{\partial u_n}{\partial \vartheta_2} \mid \frac{\partial u_n}{\partial \vartheta_1}\right\rangle\right] \tag{1.27}$$

Following now the Aoki-Ando program we define localized states as before by

$$\langle \alpha \mid \vec{v} \mid \alpha \rangle = 0, \quad \frac{d\varepsilon_\alpha}{d\vec{\vartheta}} = 0 \tag{1.28}$$

and assume that, in each Landau band, all states are localized except for those in the vicinity of the Landau energies ε_ν. Again, the localized states do not contribute to σ_H. For $T = 0$ ($f = \Theta(\varepsilon_F - \varepsilon_\alpha)$) two cases can be distinguished: if ε_F lies in a mobility gap

$$\sigma_H = \frac{e^2}{h}\sum_n{}' \left\{\frac{1}{2\pi i}\int_{-\pi}^{\pi} d\vartheta_1 \int_{-\pi}^{\pi} d\vartheta_2 Q_n(\vec{\vartheta})\right\} \tag{1.29}$$

where the sum goes over all occupied bands of extended states. Since the quantity in the curly bracket can be shown to be integer (Chern number) the possible values of σ_H are given by

$$\sigma_H = \frac{e^2}{h}k \tag{1.30}$$

where k is an <u>arbitrary</u> integer. Therefore we are not yet at our goal because explaining the observed quantization requires showing that k coincides with i, the number of occupied Landau energies, $\varepsilon_{i-1} \leq \varepsilon_F < \varepsilon_i$. Presumably this can be done by using perturbation theory. For the time being k = i is an additional assumption. If, on the other hand, ε_F lies in a region of delocalized states the curly bracket is in general not an integer.

The foregoing treatment of a 2d disordered system with torus geometry /25/ is in some respect similar to and in other respects substantially different from the topological approach to the QHE developed by Niu, Thouless and Wu /26/ and by Avron and Seiler /27/. The basic idea of the approach goes back to the work by Thouless et al. /6/. In the mathematically very elegant formulation by Avron and Seiler an adiabatic Hall conductivity $\sigma_H(\underline{\theta})$ is considered in the ground state of a many-particle Hamiltonian $\mathcal{H}(\underline{\theta})$ which depends on two parameters $\theta = (\theta_1, \theta_2)$. However, in contrast to our treatment of the torus geometry, these parameters are not Bloch quantum numbers but (up to some constant factors) fluxes of real physical fields. The measured macroscopic conductivity is identified with the average of $\sigma_H(\underline{\theta})$ over a domain which was chosen to be the torus (1.25). The condition for this average Hall conductivity $\langle\sigma_H\rangle$ to be quantized is now that the ground state of $\mathcal{H}(\underline{\theta})$ is non-degenerate and its energy is separated from the rest of the spectrum. For a single electron in a random potential these conditions may be satisfied and thus $\langle\sigma_H\rangle$ quantized - whereas σ_H, as given by (1.26), is not (because ε_F is not situated in a gap of H). Recently Riess /28/ pointed out that in a disordered system the $\underline{\theta}$-parameter space is possibly not a torus and therefore $\langle\sigma_H\rangle$ is not a topological invariant.

We may conclude that the semi-phenomenlogical approach (which assumes localization instead of proving it) explains the idealized QHE to occur in a 2d system of independent electrons in a random potential if certain conditions are satisfied. Since these conditions are by no means trivial and hard to be verified, the semi-phenomenological approach is rather unsatisfactory. Moreover this approach fails to predict any new measurable effects. The dissipative (longitudinal) conductivity vanishes identically in both the cylinder and the torus geometry.

2. Semi-Classical High Field Model

In the limit of a slowly varying random potential and very strong magnetic fields ($B \to \infty$), the motion of electrons can be described in terms of classical mechanics /29/. The remarkable fact about this high field limit is that it incorporates localization for $\varepsilon \approx \varepsilon_\nu$ (classical continuum percolation transition) and thus explains qualitatively the QHE /8,9,22,30/.

The natural way to get at the high field model is to decompose the electron motion in a slow drift motion of the cyclotron center (X,Y) in the random potential and a rapid cyclotron motion of the relative coordinates $\xi = v_y/\omega_c$, $\eta = -v_x/\omega_c$,

$$H = \frac{m\omega_c^2}{2}(\xi^2+\eta^2) + V(X+\xi, Y+\eta) \qquad (2.1)$$

(ξ,η) and (X,Y) are pairs of conjugate coordinates,

$$[\xi,\eta] = -\ell^2, \quad [X,Y] = \ell^2 \qquad (2.2a,b)$$

Passing to the high field limit involves two steps:

1. Neglecting in V the relative coordinates (which are of the order l)

$$V(X+\xi, Y+\eta) \to V(X,Y) \qquad (2.3)$$

and

2. going on the rhs of (2.2b) to the limit $l^2(\sim 1/B) \to 0$

$$[X,Y] \to 0 \qquad (2.4)$$

i.e. treating X and Y as classical quantities.
With reference to (2.2a) the eigenvalues of (2.1) (in infinite space) are given by

$$\varepsilon_\nu(X,Y) = \varepsilon_\nu + V(X,Y) \qquad (2.5)$$

They form identical equidistant bands of width $\Gamma = |V_{max} - V_{min}|$ and center spacing $\hbar\omega_c \gg \Gamma$. The classical Hamiltonian (2.5) describes reversible motion along equipotentials V(X,Y) = const. If the space average of the random potential V(x,y) is chosen to vanish all equipotentials with $\varepsilon \approx \varepsilon_\nu$ are closed. These lines correspond to localized states. Open equipotentials, which correspond to delocalized states exist only at V = 0, i.e. at the center $\varepsilon \approx \varepsilon_\nu$ of the bands (provided the system is sufficiently large).

The potential $e\phi(x)$ of a weak electric field added to V gives rise to a Hall current J_y. It is not difficult to show /8,9/ that the corresponding Hall conductivity is

$$\sigma_H = \frac{e^2}{h} \sum_{\nu=0} f(\varepsilon_\nu) \qquad (2.6)$$

For T = 0 and $\varepsilon_{i-1} \leq \varepsilon_F < \varepsilon_i$ (2.6) reduces to (1.3). Thus the high field limit furnishes the QHE. At finite field unfortunately not even corrections of the order 1/L ($\approx 10^{-4}$) can be excluded. This is a severe weakness of the approach since the observed accuracy of the quantiztaion is of the order 10^{-8}. Furthermore it is doubtful if the random potential is slowly varying in all samples showing QHE.

Eq. (2.6) can also be derived by linear response solution of the collision-free kinetic equation

$$\frac{\partial f}{\partial t} + \frac{\partial f}{\partial X}\dot{X} + \frac{\partial f}{\partial y}\dot{y} = 0 \qquad (2.7)$$

(or equivalently by working out the Kubo formula in the high field limit) /31/. Linear response solution of (2.7) has also been used to get expressions for the off-diagonal (Hall-type) thermo-electric transport coefficients defined by

$$\begin{aligned} j_y &= \sigma_H\left(E + \frac{1}{e}\frac{\partial \xi}{\partial x}\right) - \beta_H \frac{\partial T}{\partial x} \\ q_y &= \gamma_H\left(E + \frac{1}{e}\frac{\partial \xi}{\partial x}\right) - \kappa_H \frac{\partial T}{\partial x} \end{aligned} \qquad (2.8)$$

where $\underline{q}$ is the heat current and ζ is the chemical potential /32/. The coefficient $\beta_H(T,\zeta)(=\gamma_H/T)$ and $K_H(T,\zeta)$ show also QHE in the sense that they are entirely determined by the <u>free</u> electron density of states, similar to $\sigma_H(T,\zeta)$, given by (2.6). The broadening of the Landau levels to bands enters only when expressing ζ by the electron concentration $\zeta = \zeta(n,T,B)$.

Due to time reversal invariance all the dissipative transport coefficients vanish identically in the high field model. At finite temperatures dissipation may be brought about by variable range hopping /33,34/. For hopping between equipotentials /35/ the estimated value of the hopping reference temperature T_0 and its linear variation with B are in agreement with experiment /36/. For the thermopower $\underline{\underline{\alpha}} = \underline{\underline{\beta}}\underline{\underline{\sigma}}^{-1}$ the theory predicts $\alpha_{xy} = 0$ and $\alpha_{xx} = \text{const}$ in the $i = 0$ mobility gap and

$$\alpha_{xy} \propto \frac{1}{T} \exp\left(-\sqrt{\frac{T_0}{T}}\right), \quad \alpha_{xx} \propto \frac{1}{T^2} \exp\left(-2\sqrt{\frac{T_0}{T}}\right) \tag{2.9}$$

for $i \geq 1$. However, the validity of these results seems to be doubtful since coherent many-step processes are neglected.

As mentioned above the localization-delocalization transition (LDT) inherent in the high field model can be described in terms of continuum percolation theory /37/ (cf. also /22,29/). The most significant quantity is the correlation length ξ which diverges at a critical value of the occupation probability

$$p(\varepsilon) = \frac{\int dx \int dy \Theta(\varepsilon - V(x,y))}{\int dx \int dy} \tag{2.10}$$

with a certain exponent ν,

$$\xi \propto |p - p_c|^{-\nu} \tag{2.11}$$

For $B \to \infty$ $\xi(\varepsilon)$ limits the spatial extent of the wave function of energy ε. In two dimensions $\nu = 4/3$ and, for a symmetric potential, $p_c = p(\varepsilon_c) = 1/2$. For a finite system of size L two mobility edges exist. Finite size scaling hypothesis yields

$$|\varepsilon_{c2} - \varepsilon_{c1}| \propto L^{-\frac{1}{\nu}} \tag{2.12}$$

With the ansatz $l_{in} \sim T^{-1/2}$ for the inelastic mean free path the value $\nu = 1,2$ was extracted from experiments /38/.

In the semi-classical high field approach Planck's constant enters only via counting of states in phase space $(X, P = (eB/c)Y)$. Since all wave mechanical effects are neglected the approach fails in a regime $\Delta\varepsilon$ around ε_c where many saddle points of the potential and open equipotentials exist, and the probability of tunneling transitions between closed equipotentials separated by a saddle point takes it highest value. At finite magnetic fields such quantum corrections become significant. This complex problem has been attacked from different directions /39,40,41/. Milnikov and Sokolov /40/ estimated the Thouless localization length at a saddle point and obtained the critical behaviour (2.11) with exponent $\nu = \nu_p + 1$ where ν_p is the percolation exponent in the high field limit. With $\nu_p = 4/3$ one gets $\nu = 7/3$. Chalker and Coddington /41/ simplify the problem by considering a 2d network of trajectories. The scattering at the nodes is described by a transfer matrix and the links between the nodes are characterized by phase shifts. The localization length is now defined as the inverse of the smallest Lyapunov exponent associated with the transfer matrix and is calculated numerically by using finite size scaling. The result for the critical exponent is $\nu = 2.5 \pm 0.5$. Crucial for this result are back-scattering and interference. If these are excluded the network model does not show any delocalization transition - not even the classical one /42/. Therefore the very good agreement of ν with the value obtained in /40/ is somewhat embarrassing.

In summary we can conclude that the high field model and its various extensions proved to be fruitful. We can expect further progress although the inclusion of quantum corrections in a systematic way is presumably not much easier than developing a general microscopic theory.

3. Microscopic Theories of LDT and QHE

The intention of the microscopic theory is to determine $\sigma(n,T,B)$ and $\sigma_H(n,T,B)$, including the possibility of localization. This theory of localization in high magnetic fields is a very delicate subject and only little progress has been achieved so far. For a review of the development until 1985, see the contributions by Ando and by Ono in ref. /43/. For a recent survey of concepts and approaches, see /44/. I will restrict myself to mentioning a remarkable investigation dealing with the localization regime and some recent results concerning the transition regime.

As is well known, the density of states of a 2DEG in strong magnetic fields, calculated in the self-consistent Born approximation, consists of semi-ellipses centered around the Landau energies ε_ν /45/. In this approximation the $T = 0$ Hall conductivity plotted against the filling factor η is like a garland which touches the classical straight Hall line $\sigma_H^0 = \eta e^2/h$ from below at the points $\eta = i$ /46/. Efetov and Marikhin /47/ calculated now corrections in the so-called instanton approximation and found, as expected, exponentially small tails added to the density of states. However, $\sigma(\varepsilon_F)$ and $\sigma_H(\varepsilon_F)$ retain their previous gap values $\sigma(\varepsilon_F) = 0$ and $\sigma_H(\varepsilon_F) = i\ e^2/h$ in these tails. Consequently, a small but finite interval $\Delta\eta$ exists around each integer value $\eta = i$ in which $\sigma(\eta) = 0$ (localization) and $\sigma_H(\eta) = i\ e^2/h$ (quantum plateau formation); the instanton correction turns the slope of $\sigma_H(\eta)$ at $\eta = i$ into the horizontal position. This is a remarkable and promising result (in spite of some ambiguities in handling the Kubo Hall conductivity formula).

If the transition between localized and delocalized states is a critical phenomenon then we expect for the conductance G a scaling equation

$$\frac{d\ln G(L)}{d\ln L} = \beta(\ln G(L)) \tag{3.1}$$

At fixed points ($\beta = 0$) G(L) is scale invariant (characteristic for LDT) and the slope of β determines the critical exponent ν. According to Abrahams et al. /48/ in two dimensions $\beta \neq 0$, i.e. there is no LDT. This is obviously incorrect in high magnetic fields since the QHE requires both localized and delocalized states.

The task of the microscopic theory is to determine the β-function. The standard procedure is to set up an appropriate field theory characterized by a certain Lagrangian $\mathcal{L}$. As recognized by Pruisken /49/ (cf. also /50/) the axial vector character and the time reversal breaking property of the magnetic field require an additional term ($\mathcal{L}_2$) to appear in the Lagrangian,

$$\mathcal{L} = \sigma^0\mathcal{L}_1 + \sigma_H^0\mathcal{L}_2 \tag{3.2}$$

With appropriate boundary conditions for the fields $\mathcal{L}_1$ is a topological invariant. In (3.2) the coupling constants are the conductivities in the self-consistent Born approximation. As this theory indicates, the appropriate description of the high magnetic field case requires two-parameter scaling,

$$\frac{d\ln G}{d\ln L} = \beta(G, G_H), \quad \frac{d\ln G_H}{d\ln L} = \beta_H(G, G_H) \tag{3.3}$$

which is compatible with LDT in two dimensions. In fact, a two-parameter flow (3.3) was proposed by Khmelnitskii /51/ before the field theory was formulated. Unfortunately all standard methods of calculating β-functions are not suitable in the present case. The reason for this is the global orientational structure associated with the topological term. (For a simple model illustrating this structure cf. /52/). Although there is considerable experimental evidence for two-parameter scaling /53/ the details of the flow diagram are still contraversial /54/. Notice that at the LDT there is only one correlation length. Therefore finite size scaling calculations based on a one parameter flow (yielding critical exponents for the correlation length) can be quite compatible with a two-parameter flow in the conductances. This was recently confirmed by McKinnon /55/.

Chalker and Daniell /56/ studied the wave number and frequency dependent diffusion function $D(q,\omega) \sim \sigma(q,\omega)$ (Einstein relation) for small (mesoscopic) systems ($L < \xi$) and found the power law

$$D(\vec{q},\omega) \propto \left(\frac{q^2}{\omega}\right)^{-\frac{\eta}{2}} \tag{3.4}$$

with exponent

$$\eta = \begin{cases} 0 \\ 0.4 \pm 0.1 \end{cases} \quad \text{for} \quad \frac{Dq^2}{\omega} \begin{cases} < 1 \\ > 1 \end{cases} \tag{3.5}$$

where D is the diffusion constant (the exponent η should not be mixed up with the filling factor). The anomalous behaviour for $Dq^2/\omega > 1$ is connected with an algebraic decay of correlations,

$$S(\varepsilon,\vec{r},\vec{r}',\omega) = \left\langle \psi_\varepsilon^*(0)\psi_\varepsilon(\vec{r})\psi_{\varepsilon+\omega}^*(\vec{r}')\psi_{\varepsilon+\omega}(0)\right\rangle \propto |\,\vec{r}-\vec{r}'\,|^{-\eta} \tag{3.6}$$

Zirnbauer /57/ studied a system with spin-orbit coupling at $\underline{B} = 0$ and obtained $\nu = 2.0$ and an algebraic decay of S with spectrum of exponent η,

$$S \propto \int d\eta' \rho(\eta')\,|\,\vec{r}-\vec{r}'\,|^{-\eta'} \tag{3.7}$$

As pointed out by Pook and Janßen /58/ these "anomalies" are indications for multifractal character of the wave function at the LDT point. In a pioneer work Aoki /59/ suggested that $\psi(\underline{r})$ is there an ordinary fractal (i.e. self-similar on a number of length scales). The concept of multifractality seems to be the adequate tool both for describing the LDT and for organizing the numerical analysis /58/.

In conclusion we can say that the idea of critical behaviour at the Landau energies, with two-parameter scaling depending on a correlation length with exponent $\nu \simeq 7/3$ is very appealing /60/. Pruisken's model shows encouraging features - but it is too complicated to be solved and, perhaps, too simple to be realistic. The LDT ought to be described in terms of multifractality.

4. Local Potential and Current Distribution

The first attempts to determine the potential and the current distribution in a quantum Hall system self-consistently is due to MacDonald et al. /61/ and Riess /62/. For different improvements cf. /63/. All these investigations deal with an ideal electron gas (no disorder) and treat the Coulomb interaction in Hartree approximation. For this model Heinonen and Taylor /64/ found strong deviations from the classical behaviour $\sigma_H^\circ \sim \eta$. Disorder is taken into account in recent works by Ohtsuki and Ono /65/ and by Ando /66/. The crucial question is where the Hall current flows. Unfortunately this problem has not been settled. According to Ono and Ohtsuki /67/ the linear response current to a homogeneous field flows in the bulk of the system. Nonetheless it can be expressed in terms of edge states. Johnston and Schweitzer /68/ came to the conclusion that the current flows at the edges but for certain potential distributions it flows in the bulk. The Hall conductivity of a finite system calculated by using linear response theory is for $\eta > \sim 1/2$ always below the classical value σ_H° /67/. The relative deviation is of the order 1/L. For the Hall conductance the relative deviation is only of the order l^2/L^2 /69/. It has been conjectured that the Coulomb interaction which was neglected in /67/, /68/ would effectively reduce this deviation.

In the case of free electrons no such size effects occur (i.e. $\sigma_H = \sigma_H^\circ$) if the state of the system is described by the quasi-equilibrium distribution function

$$f = f(\frac{m}{2}v^2 + Cp_y) \tag{4.1}$$

/70/. This distribution is stationary because the momentum p_y perpendicular to the external electric field is conserved - in contrast to the "shifted" distribution

$$f = f\left(\frac{m}{2}(\vec{v} - \vec{v}_D)^2\right) \tag{4.2}$$

used in /64/. In (4.1) the constant C is determined by the total current. When taking into account the Coulomb potential, $mv^2/2$ is replaced by the Hartree energy $\varepsilon_\alpha(C)$. The resulting average local current density is large at the edges but the signs are opposite. The uncompensated part of the current flows in the bulk /70/. In the disorderd systems the quasi-equilibrium distribution is not justified. What is urgently asked for is a self-consistent local linear response theory for systems with disorder. As long as the transport problem is discussed in terms of global conductances (cf. e.g. /71/), the current distribution remains an open question - even if these conductances can be expressed in terms of edge states.

5. Outlook

It does not require any prophetic talents to predict that the future theoretical research regarding the QHE will mainly be concentrated on the following topics:

Localization theory

Interrelation between the integer and the fractional QHE

Relation to magnetotransport in mesoscopic systems, Contacts

Scattering between Landau levels (in the bulk and at the edges of the system)

Variable range hopping

Coulomb effects, local current and potential distribution

References

1. K. von Klitzing, G. Dorda, M. Pepper; Phys. Rev. Lett. 45, 494 (1980)
2. H. Aoki, T. Ando: Solid State Commun. 38, 1079 (1981)
3. R.B. Laughlin: Phys. Rev. B23, 5632 (1981)
4. R.E. Prange: Phys. Rev. B23, 4802 (1981)
5. P. Středa: J. Phys. C15, L717 (1982)
6. D.J. Thouless, M. Kohmoto, M.P. Nightingale, M. de Nijs: Phys. Rev. Lett. 49, 405 (1982)
7. D. Tsui, S.J. Allen: Phys. Rev. B24, 4082 (1981)
8. S.V. Iordansky: Solid State Commun. 48, 1 (1982)
9. Y. Ono: in Y. Nagaoka, H. Fukuyama (Eds.): Anderson Localization, Solid State Sciences 39 (Springer, Berlin, Heidelberg, New York 1982), p. 202
10. T. Ando, Ref. /9/, p. 176
11. R. Kubo, S.M. Mijake, N. Hashitsume: Solid State Physics 17, 269 (Academic, New York 1965)
12. M. Janßen: Diplomarbeit, Köln 1986
13. J. Hajdu, M. Janßen, O. Viehweger: Z. Physik B66, 433 (1987)
14. M. Jonson: Phys. Scr. 32, 435 (1985)
15. T. Ohtsuki, Y. Ono: Solid State Commun. 68, 787 (1988)
16. T. Ohtsuki, Y. Ono: J. Phys. Soc. Jpn. 58, 3863 (1989)
17. O. Viehweger, W. Pook, M. Janßen, J. Hajdu: Z. Physik B78, 11 (1990)
18. M. Janßen, J. Hajdu: Z. Physik B70, 461 (1988)
19. T. Ando, Y. Matsumoto, Y. Uemura, J. Phys. Soc. Jpn. 39, 279 (1975)
20. U. Gummich, M. Janßen: Z. Physik B79, 347 (1990)
21. J. Chalker: J. Phys. C16, 4297 (1983)
22. R. Joynt, R.E. Prange: Phys. Rev. B29, 3303 (1984)
23. W. Brenig: Z. Physik B50, 305 (1983)
24. W. Brenig, K.I. Wysokinski: Z. Physik 63, 149 (1986)
25. W. Pook, J. Hajdu: Z. Physik B66, 427 (1987)
26. Q. Niu, D.J. Thouless, Y.S. Wu: Phys. Rev. B31, 3372 (1985)
27. J. Avron, R. Seiler: Phys. Rev. Lett. 54, 259 (1985)
28. J. Riess: Europhys. Lett. 12, 253 (1990)
29. M. Tsukada: J. Phys. Soc. Jpn. 41, 1466 (1976)
30. S. Lury, R.F: Kazarinov: Phys. Rev. B27, 1386 (1983)
31. S.M. Apenko, Yu. E. Lozovik: JETP 62, 328 (1985)
32. A. Grunwald, J. Hajdu: Solid State Commun. 63, 289 (1987)
33. Y. Ono: J. Phys. Soc. Jpn. 51, 237 (1982)
34. K.I. Wysokinski, W. Brenig: Z. Physik B54, 11 (1983)
35. A. Grunwald, J. Hajdu: Z. Physik B78, 17 (1990)
36. G. Ebert, K. v. Klitzing, C. Probst, E. Schubert, K. Ploog, G. Weimann: Solid State Commun. 45, 625 (1983)
37. S.A. Trugman, Phys. Rev. B27, 7539 (1983)
38. H.P. Wei, D.C. Tsui, M.A. Paalen, A.M.M. Pruisken: in G. Landwehr (Ed.): High Magnetic Fields in Semiconductor Physics II, Solid State Sciences 87 (Springer, Berlin, Heidelberg, New York 1989) p. 2
39. H.A. Fertig, B.I. Halperin: Phys. Rev. B36, 7969 (1987); H.A. Fertig: Phys. Rev. B38, 996 (1988)
40. G.V. Mil'nikov, I.M. Sokolov: JEPT Lett. 48, 536 (1988)

41. J. Chalker, P.D. Coddington: J. Phys. C 21, 2665 (1988)
42. L. Jaeger; Diplomarbeit, Köln 1990
43. Y. Nagaoka (Ed): Anderson Localization, Progr. Theoret. Phys. Suppl. 84 (1985)
44. B. Kramer, Y. Ono, T. Ohtsuki: Ref. /38/, p. 24
45. T. Ando, Y. Uemura: J. Phys. Soc. Jpn. 36, 959 (1974)
46. T. Ando, Y. Matsumoto, Y. Uemura: Ref. /19/; A. Grunwald, Diplomarbeit, Köln 1984
47. K.B: Efetov, V.G. Marikhin: Phys. Rev. B40, 1226 (1989)
48. E. Abrahams, P.W. Anderson, D.C. Liccardello, T.V. Ramakrishnan: Phys. Rev. Lett. 42, 673 (1979)
49. A.M.M. Pruisken: Nucl. Phys. B235, 277 (1984)
50. H. Levine, S.B. Libby, A.M.M. Pruisken: Nucl. Phys. B240, 30, 49, 71 (1984)
51. D.E. Khmelnitskii: JETP Lett. 38, 553 (1983)
52. M. Janßen: Z. Physik B67, 227 (1987)
53. H.P. Wei, D.C. Tsui, A.M.M. Pruisken: Phys. Rev. B33, 1488 (1986); H.P. Wei, D.C. Tsui, M. Paalanen, A.M.M. Pruisken: Phys. Rev. Lett. 61, 1294 (1988)
54. T. Ando: in G. Landwehr (Ed.): High Magnetic Fields in Semiconductor Physics, Solid State Sciences 71 (Springer, Berlin, Heidelberg, New York 1987) p. 2; H. Aoki, T. Ando: ibid., p. 45; A. McKinnon; Ref. /38/, p. 10
55. A. McKinnon: J. Phys. Cond. Matter 1, 10407 (1989)
56. J.T. Chalker, G.J. Daniell: Phys. Rev. Lett. 61, 593 (1988)
57. M. Zirnbauer: to be published
58. W. Pook, M. Janßen: Z. Phys., in press
59. H. Aoki: J. Phys. C16, 1893; L205 (1983)
60. B. Huckestein, B. Kramer: Phys. Rev. Lett. 64, 1437 (1990)
61. A.H. MacDonald, T.M. Rice, W.F. Brinkman: Phys. Rev. B31, 8265 (1985)
62. J. Riess: J. Phys. C17, L849 (1984); Phys. Rev. B31, 8265 (1985)
63. W. Maas: Europhys. Lett. 2, 1 (1986); P.K. Yoegshwar, W. Brenig: Europhys. Lett. 7, 737 (1988); H. Nielsen: Ref. /38/, p. 50
64. O. Heinonen, P.L. Taylor; Phys. Rev. B28, 6119; B32, 633 (1985)
65. T. Ohtsuki, Y. Ono: J. Phys. Soc. Jpn. 58, 2482 (1989); 59 637 (1990)
66. T. Ando: J. Phys. Soc. Jpn. 58, 3771 (1989)
67. Y. Ono, T. Ohtsuki: Z. Physik B68, 445 (1987)
68. R. Johnston, L. Schweitzer: Z. Physik B68, 77 (1987)
69. W. Pook: private communication
70. D. Pfannkuche: Doktorarbeit, Köln 1990
71. M. Büttiker: Phys. Rev. B38, 9375 (1988)

An Overview of the Numerical Studies of the Quantum Hall Effect

H. Aoki

Department of Physics, University of Tokyo, Hongo,
Tokyo 113, Japan

The present status of the numerical study of the quantum Hall systems is reviewed.

1. Introduction

The last ten years have witnessed a fascinating development in the physics of the quantum Hall effect[1-5]. The quantum Hall effect has also inspired field theoretical approaches. Numerical studies have played an essential role as a first-principles approach to these developments. This is a natural consequence of the fact that the quantum Hall effect, like lattice gauge theories, demands non-perturbative, first-principles approaches. On the occasion of the tenth anniversary of the discovery of quantum Hall effect, I shall give an overview of the numerical studies and the development of ideas for the quantum Hall effect from the early stages to the latest results.

2. Wavefunction and Hall conductivity

2.1 Early studies

The numerical study of the two-dimensional(2D) disordered electron system in strong magnetic fields started when Aoki and Kamimura[6] first posed the localization problem for this system, which was followed by computer simulations (direct diagonalization[7] and real-space renormalization [8]) by Aoki in the mid 1970's. Aoki and Kamimura pointed out that the dynamics of electrons is governed by the quantum hopping of the centres of the cyclotron motion, which renders, strangely enough, the random potential a *dual* role of contributing to both the quantum transport of cyclotron motions and the localization of the states.

The range of the potential dominates the motion: while a slowly-varying potential (over the length scale of cyclotron radius) reduces the motion to a classical drift of the centre of the cyclotron motion along the equi-potential contours [9-12], short-range potentials give rise to maximal quantum hopping.

After the seminal discovery of the quantum Hall effect by von Klitzing, Dorda and Pepper[13] in 1980, which was followed by theoretical studies by Aoki and Ando[14], and by Laughlin[15], numerical studies also focussed their attention on this remarkable phenomenon. Computer simulations of the Hall conductivity plateaux for disordered systems were first done by Aoki[16] and

High Magnetic Fields in Semiconductor Physics III Editor: G. Landwehr

by Ando[17]. This was also extended to the Laughlin's geometry with extra magnetic flux, which twists the phase of the wavefunction at the periodic boundaries[10,19]. The result for the flux dependence may be regarded as an Aharonov-Bohm (AB) effect in finite quantum Hall systems[18]. The Hall conductivity, determined essentially by the phase of the wavefunction, is especially sensitive to external fluxes. Recent advances in fabricating nanometer semiconductor structures have enabled experimentalists to actually observe these interference effects [20].

The Kubo-formula Hall conductivity, when averaged over the AB fluxes, has been shown to be a topological invariant (an integer) in terms of a differential geometry for the dependence of the wavefunctions on the AB fluxes. This was first pointed out analytically by Thouless *et al* for periodic systems[21]. This idea has been extended to many-body systems[22] and disordered systems [23,24]. Numerical results for the latter systems have revealed the behaviour of the topological invariant, which reflects the dependence of the Hall conductivity on the degree of localization in finite systems.

2.2 Lattice systems

For lattice or periodic systems, the Landau level splits into p bands (called the Hofstadter butterfly) for $H = q/p$, where H is the magnetic flux in units of the flux quantum penetrating the unit cell of the lattice, and the quantum Hall effect becomes complicated accordingly[21].

Lattice systems differ from usual systems in a number of ways, including non-parabolicity of the energy band[25]. One manifestation is seen when $H = q/p$ with $q \neq 1$. The result for H=3/8 (Fig.1) shows that the behaviour of the

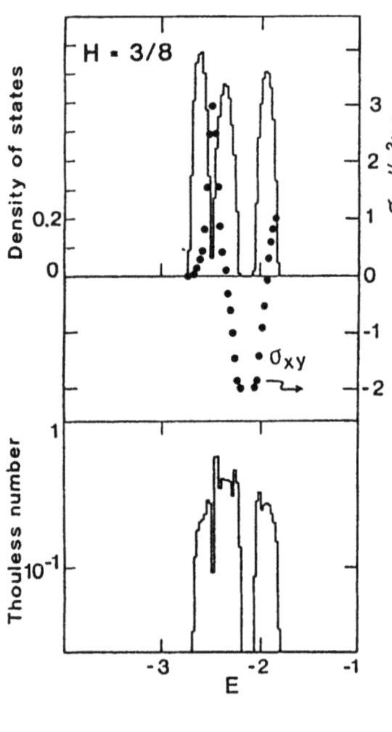

Fig. 1. Density of states, Hall conductivity (dots) and the Thouless number for 24×24 square lattice systems with H=3/8, W=0.5, where W is the fluctuation width of site energies in units of the transfer energy [25].

localization length against energy may not be monotonic within each Landau band[25]. Another effect of lattice structure appears in the honeycomb lattice, which contains a linear term in the energy dispersion[26].

2.3 Quantum interference and mesoscopic effects

Since the phase of wavefunctions dominates the quantum Hall effect, the problem of quantum fluctuations in the quantum Hall conduction[27] is interesting. The root-mean-square amplitude of the fluctuations in the Hall conductivity, $\delta\sigma_{xy}$, is numerically shown to vanish sharply as the average, $\langle\sigma_{xy}\rangle$, approaches a plateau[28]. This is because both $\delta\sigma_{xy}$ and $\langle\sigma_{xy}\rangle$ are dominated by the localization. One can thus construct a diagram for the $\delta\sigma_{xy}-\langle\sigma_{xy}\rangle$ trajectories. The study is extended to short/long range scatterers, dense/sparse scatterer densities to explore the universality of the diagram (Fig.2)[29]. Also of interest is the magneto-fingerprint effect for modulated H or single scatterer movement (Fig.3), which shows that σ_{xy} oscillates rapidly against the movement in finite systems[29].

Ohtsuki and Ono[30] have numerically studied the distribution of the Hall current induced by an electric field using the linear response theory(Fig.4). They conclude that, even for disordered systems, the total current ***distribution*** resembles what one would expect for a clean system, which may be thought

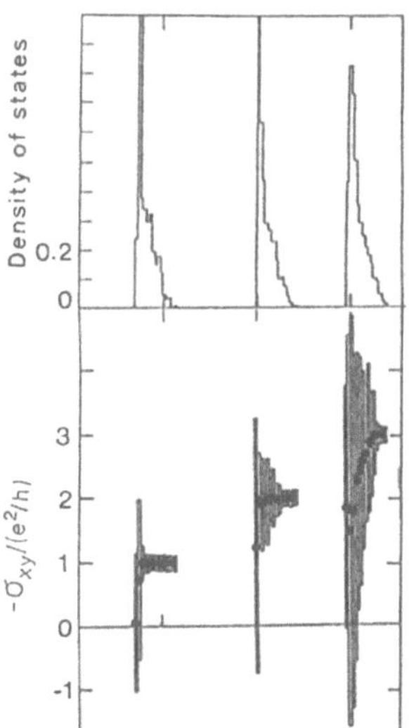

Fig. 2. Density of states, Hall conductivity with its average (dots) and rms fluctuation width for 24×24 square lattice systems with H=1/8 in which only a fraction (10%) of the sites deviate from zero in energy with a distribution of W=2.0 centred at +1.0(repulsive on average) [29].

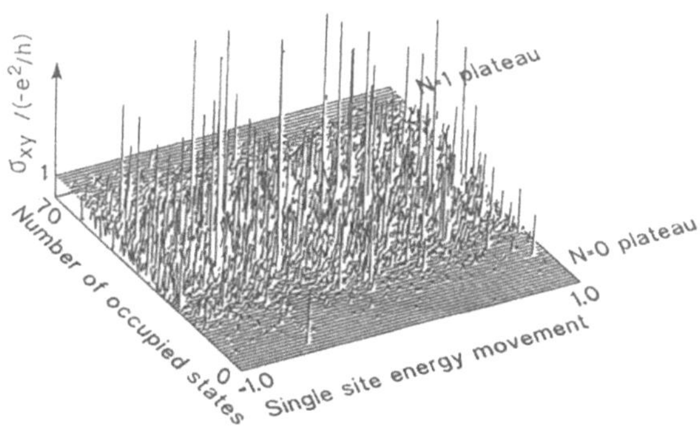

Fig. 3. Hall conductivity plotted against the shift of the energy of a single site for progressive number of occupied states in one sample of 24×24 square lattice with H=1/8 with W=1 [29].

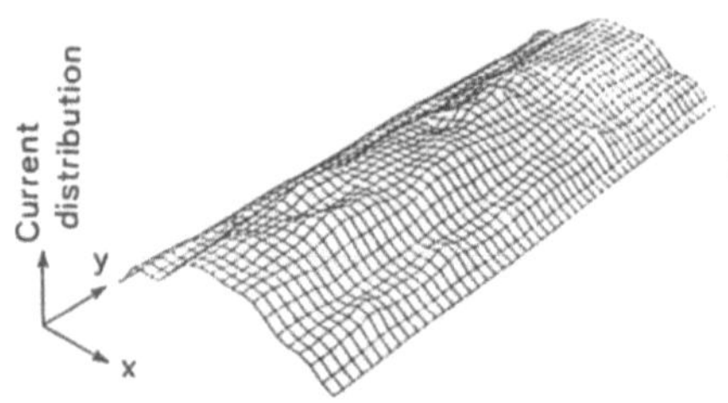

Fig. 4. An example of the distribution of total Hall current along y direction in a random system of rectangular shape with the Landau level filling of 17/22[30].

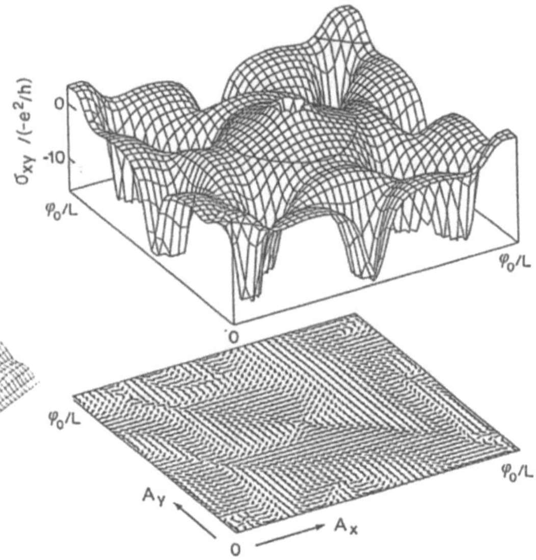

Fig. 5. Hall conductivity against the AB fluxes, (A_x, A_y), for a self-similar system (third generation Sierpinsky gasket). The lower panel shows the relative phase of the wavefunction on a site [33].

of as a generalization of the 'compensation theorem' of Aoki and Ando[14] for the global current. Later, this was analytically justified by Ando[31].

For systems confined by walls, we have to consider edge effects arising from the states localized along the edges, some of which may hybridize with bulk states. There is ample literature on this subject[32].

The self-interference of wavefunctions, which is the cause of all the quantum fluctuations and mesoscopic effects, occurs systematically in self-similar systems in which modulations of various length scales coexist. Numerical results (Fig. 5) indeed show that the quantum Hall effect in a self-similar system exhibits anomalously large quantum fluctuations, with a long-tailed auto-correlation in σ_{xy}[33].

3. Localization in quantum Hall systems

The year of the discovery of the quantum Hall effect almost coincided with that for the emergence of the scaling theory of localization[34]. The latter provides sound grounds to believe that 2D is really special, or marginal. The scaling hypothesis demands that the β function should only depend on universality classes, which may be orthogonal (H=0), unitary ($H \neq 0$)[35], or symplectic (spin-orbit interactions)[36]. A numerical test of the scaling hypothesis started when MacKinnon and Kramer[37] employed the finite-size scaling method to show that single-parameter scaling does indeed exist.

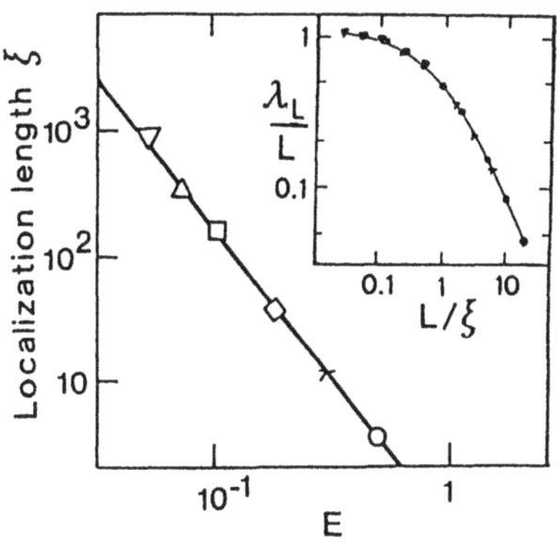

Fig. 6. Finite-size scaling result for the localization length against energy. The inset shows the finite-size scaling plot [40].

For the quantum Hall system, the quantum Hall theory indicates [14] that all the states *cannot* be localized in the quantum limit ($H \to \infty$), in sharp contrast to the scaling prediction for $H=0$. Aoki and Ando[38] have performed finite-size scaling studies of localization to elaborate on the Thouless-number study by Ando. The following results have emerged: (i)Single-parameter scaling is analytically shown to fail in the quantum Hall system for length scales of several cyclotron radii. (ii)Two-parameter scaling analysis of the numerical finite-size scaling results lead to a power-law critical behaviour for the inverse localization length, $\alpha(E)$, as a function of the energy, measured from the centre of each Landau level, with

$$\alpha(E) \sim |E|^s.$$

The critical exponent of localization, $s \sim 2$, is independent of the nature (short/long range) of the random potential, but is dependent on the Landau index in the practical range of sample sizes. The delocalized states at the centre of the Landau level is shown to have a critical (fractal) character[39].

Recently, Huckestein and Kramer[40] have employed a random matrix model(Fig.6). They conclude that (a)for sufficiently large sample sizes, single-parameter scaling seems to be restored, (b)the refined result, $s = 2.34 \pm 0.04$, for the critical exponent for N=0 Landau level is close to the result, $s = 7/3$, of the classical percolation combined with quantum tunneling for the infinitely slowly-varying random potential[41]. The result (b) confirms the range-independence of critical behaviour[38].

Ando[42] goes on to investigate the case in which short-range and long-range potentials coexist, i.e., when the drifts along the equi-potential contours and the quantum hopping processes coexist. The result shows that the inverse localization length always obeys the $\alpha(E) \sim |E|^s$ law, with $s \sim 2$, but introduction of a small amount of short-range scatterers switches on the quantum jumps between the orbitals on different equi-potential contours, and drastically delocalizes the states. Thus the dual role of the scattering processes manifests itself here.

The critical exponent problem has also attracted a field theoretical (nonlinear σ model) approach to the quantum Hall problem: Pruisken and coworkers[43] predicted a flow diagram that is independent of the Landau index, which is apparently confirmed by some experimental results for

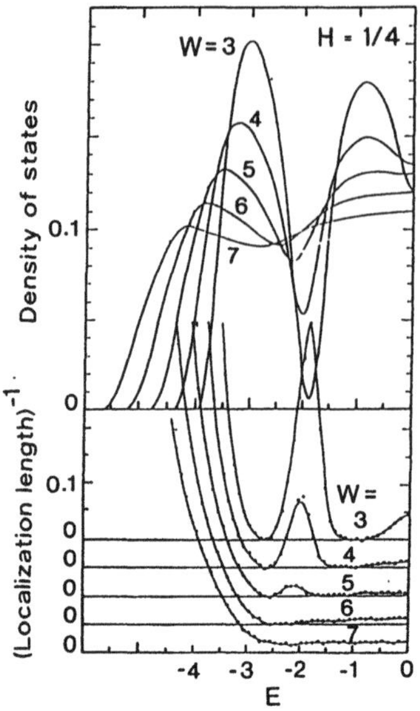

Fig. 7. Density of states and the inverse localization length in square lattice systems with H=1/4 and various degree of disorder[36].

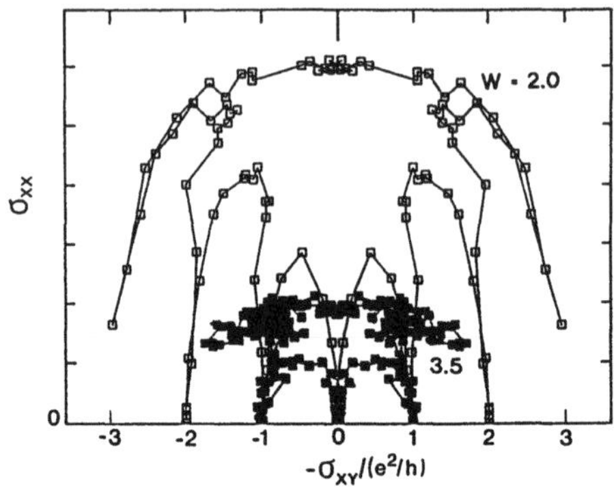

Fig. 8. Ensemble-averaged $\sigma_{xx}-\sigma_{xy}$ diagram for 24×24 square lattice systems with H=1/8, W=2.0(open squares) and W=3.5(full squares) [49].

heterostructures[44]. This is in conflict with the results obtained by Wakabayashi *et al*[45] for MOS inversion layers, which favour the finite-size scaling result[38]. To resolve the discrepancy, it seems that further numerical studies on the Landau-index dependence and the understanding of the temperature dependence of the electron-electron interaction effects are required.

A most important problem in the physics of the quantum Hall effect is the following. Since delocalized states exist for $H \to \infty$, while all the states are localized for $H=0$, we can ask what happens when H is continuously decreased from ∞ to 0, thereby increasing the mixing (overlap) of different Landau levels. Ando's Thouless-number result for lattice systems[36] (Fig.7) shows that the extended states belonging to different Landau levels shift upwards on the energy axis as the degree of the mixing (strength of the random potential divided by $\hbar\,\omega_c$) is increased, and they finally merge and disappear at a critical degree of mixing. If we turn to the quantum Hall conductivity, we can look at the $\sigma_{xx}-\sigma_{xy}$ diagram (see below) for varying degree of mixing(Fig.8) [49]. The result that the higher plateaux vanish first is consistent with Ando's result[36] that the extended states go up in energy and disappear.

4. The $\sigma_{xx} - \sigma_{xy}$ diagram

Due to the critical behaviour of localization, σ_{xy} against energy (or Landau level filling) becomes a series of step functions and accordingly the longitudinal conductivity, σ_{xx}, becomes zero except at a number of points in an infinitely large system at T=0. When T or the system size is made finite,

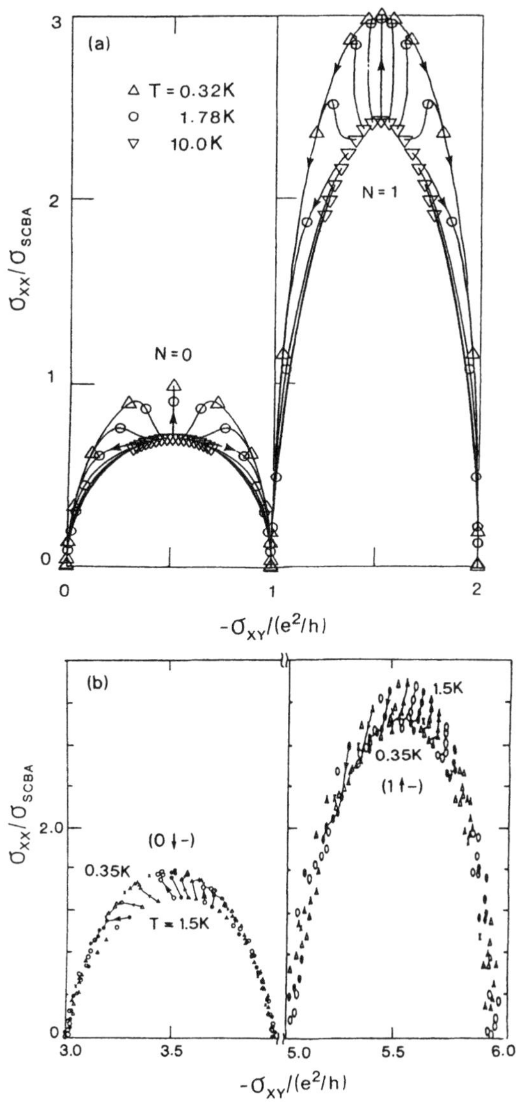

Fig. 9. Temperature-driven $\sigma_{xx} - \sigma_{xy}$ flow diagram obtained (a)theoretically[47] and (b)experimentally for a Si-MOS system (with spin-and valley-split band index indicated)[48].

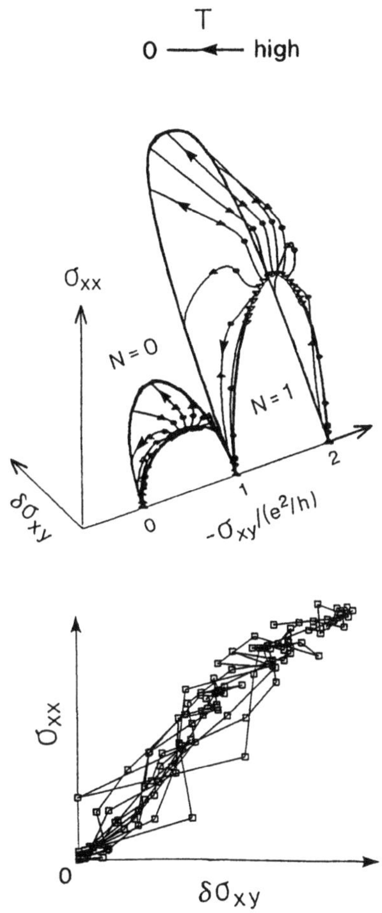

Fig. 10. Temperature-driven $\sigma_{xx} - \sigma_{xy} - \delta\sigma_{xy}$ diagram derived from numerical results[49]. Lower panel shows the projection of the numerical result for 24×24 square lattice systems with H=1/8, W=2.0 on the $\sigma_{xx} - \delta\sigma_{xy}$ plane.

these singularities will be smeared out, and we can plot a $\sigma_{xx}-\sigma_{xy}$ diagram. The nonlinear σ model predicts that the diagram be identical for every Landau level[43]. Ando's numerical result[46] unambiguously shows, in contrast, that the diagram does indeed depend on the Landau index.

For finite temperatures, the inelastic scattering length, L_ε, becomes finite, and the energy width for effectively extended states (with $L_\varepsilon <$ localization length) competes with the thermal width, k_BT. The flow diagram obtained by Aoki and Ando[47] indicates that σ_{xx} remains nonzero at the Landau level centre as $L_\varepsilon \to \infty$. It is interesting to consider why a kind of minimum metallic conductivity exists in the quantum Hall system. The calculated temperature-driven flows[47] compare favourably with experimental results obtained by Kawaji and coworkers[48] (Fig.9).

We can go one step further to include the quantum fluctuations mentioned previously to construct the $\sigma_{xx}-\sigma_{xy}-\delta\sigma_{xy}$ diagram(Fig.10) [49].

5. Fractional quantum Hall effect

There is a vast amount of numerical studies on the fractional quantum Hall effect[1,2,4,50], on which we shall only make cursory remarks. Laughlin's wavefunction has been shown to be an exact solution for short-range electron-electron interactions, and the evidence that the Laughlin's function contains the essential physics even for Coulombic interactions has been provided numerically[51].

The fractional quantum Hall problem gained a dramatic impetus when Laughlin[52] suggested that the problem is closely related to the electronic mechanism of high T_C superconductivity. The unsuspected relevance of the strong electron correlation problem in the Hubbard model to the quantum Hall problem (incompressible quantum liquid, Hofstadter butterfly, etc) has attracted a number of researchers to the area, and numerical studies are being extensively performed (see, eg, [53]).

The problem of whether the fractional charges, e^*, are observable in the fractional quantum Hall systems is related to the problem of whether the minimum metallic conductivity exists in this system, since the analysis of e^* involves the pre-exponential factor of σ_{xx}. The problem of a critical exponent of localization[54] and the conductance fluctuation[55] have also been studied experimentally for the fractional quantum Hall system.

6. Concluding remarks

Although the understanding of the quantum Hall problem has been deepened remarkably by numerical studies, there are still important open questions such as those mentioned above. For the fractional Hall systems there are theoretical attempts at a scaling theory[56] and an understanding of the role of the edge states[57]. When the randomness and the electron-electron interaction are both considered, an early numerical result[7] predicted that the ground state for sufficiently strong random potential is a Wigner glass.

One could then question a phase transition into a quantum liquid as the degree of randomness or T is varied. Numerical studies will have to cover these subjects as well.

Acknowledgements I wish to thank Prof T Ando and Prof S Kawaji for illuminating discussions and Dr D Ko for a critical reading of the manuscript.

References

[1] *Proc. 8th Int. Conf. on Electronic Properties of Two-dimensional Systems*, Grenoble, 1989, ed. by J.Y. Marzin, Y. Guldner and J.C. Maan (North-Holland, Amsterdam, 1990).
[2] G. Landwehr (ed.), *High Magnetic Fields in Semiconductor Physics II* (Springer, Berlin, 1989).
[3] H. Aoki, *Rep. Progr. Phys.* **50**(1987) 655.
[4] R.E. Prange and S.M. Girvin, *The Quantum Hall Effect*, 2nd edn (Springer, New York, 1990).
[5] H. Kamimura and H. Aoki, *The Physics of Interacting Electrons in Disordered Systems* (Oxford University Press, Oxford, 1989).
[6] H. Aoki and H. Kamimura, *Solid State Commun.* **21**(1977) 45.
[7] H. Aoki, *J. Phys. C* **10**(1977) 2583; **11**(1978) 3823; **12**(1979) 633.
[8] H. Aoki, *Solid State Commun.* **31**(1979) 999.
[9] T. Ando in *Anderson Localization* ed. by Y. Nagaoka and H. Fukuyama (Springer, Berlin, 1982) p.176; T. Ando, *J. Phys. Soc. Jpn* **52**(1983) 1740.
[10] H. Aoki in *Application of High Magnetic Fields in Semiconductor Physics* ed. by. G. Landwehr (Springer, Berlin, 1983) p.11.
[11] H.A. Fertig, *Phys. Rev. B* **38**(1988) 996.
[12] S.A. Trugman, *Phys. Rev. Lett.* **62**(1989) 579.
[13] K. von Klitzing, G. Dorda and M. Pepper, *Phys. Rev. Lett.* **45**(1980) 494.
[14] H. Aoki and T. Ando, *Solid State Commun.* **38**(1981) 1079.
[15] R.B. Laughlin, *Phys. Rev. B* **23**(1981) 5632.
[16] H. Aoki, *J. Phys. C* **15**(1982) L1227.
[17] T. Ando, *J. Phys. Soc. Jpn* **53**(1984) 3101; 3126.
[18] H. Aoki, *Phys. Rev. Lett.* **55**(1985) 1136.
[19] H. Aoki, *J. Phys. C* **16**(1983) 1893.
[20] R.J. Brown, C.G. Smith, M. Pepper, M.J. Kelly, R. Newbury, H. Ahmed, D.G. Hasko, J.E.F. Frost, D.C. Peacock, D.A. Ritchie and G.A. Jones, *J. Phys. Condens. Matter* **1**(1989) 6291.
[21] D.J. Thouless, M. Kohmoto, M.P. Nightingale and M. den Nijs, *Phys. Rev. Lett.* **49**(1982) 405. See also D.J. Thouless, *Phys. Rep.* **110**(1984) 279, D.J. Thouless in Ref.4, M. Kohmoto, *Ann. Phys.* **160**(1985) 343. Although the approach has been criticized by J. Riess, *Europhys. Lett.* **12** (1990) 253, numerical results for finite systems confirm that the flux-averaged σ_{xy} is an integer[23].
[22] Q. Niu, D.J. Thouless and Y.S. Wu, *Phys. Rev. B* **31**(1985) 3372.
[23] H. Aoki and T. Ando, *Phys. Rev. Lett.* **57**(1986) 3093.
[24] H. Aoki and T. Ando in *High Magnetic Fields in Semiconductor Physics* ed. by G. Landwehr (Springer, Berlin, 1987), p.45.
[25] H. Aoki, *J. Phys. C* **18**(1985) L67; H. Aoki, to be published.
[26] R. Rammal, *J. Physique* **46**(1985) 1345 (energy spectrum); H. Aoki, unpublished (numerical quantum Hall study).
[27] A.M. Chang, G. Timp, T.Y. Chang, J.E. Cunningham, P.M. Mankiewich, R.E. Behringer and R.E. Howard, *Solid State Commun.* **67**(1988) 769.
[28] H. Aoki, *Jpn. J. Appl. Phys.* **26**(1987), Suppl.26-3, 699.
[29] H. Aoki, to be published.
[30] T. Ohtsuki and Y. Ono, *J. Phys. Soc. Jpn* **58**(1989) 2482; **59**(1990) 637.
[31] T. Ando, *J. Phys. Soc. Jpn* **58**(1989) 3711.
[32] H. Akera and T. Ando, *Phys. Rev. B* **39**(1989) 5508; T. Ando, this volume.
[33] H. Aoki, *Phys. Rev. B* **42**(1990) 6869.
[34] E. Abrahams, P.W. Anderson, D.C. Licciardello and T.V. Ramakrishnan, *Phys. Rev. Lett.* **42**(1979) 673.
[35] H. Aoki, *Surf. Sci.* **196**(1988) 107.
[36] T. Ando, *Phys Rev. B* **40**(1989) 5325.
[37] A. MacKinnon and B. Kramer, *Phys. Rev. Lett.* **47**(1981) 1546.
[38] H. Aoki and T. Ando, *Phys. Rev. Lett.* **54**(1985) 831; T. Ando and H. Aoki, *J. Phys. Soc. Jpn* **54**(1985) 2238.
[39] H. Aoki, *Phys. Rev.* B **33**(1986) 7310. See also B. Kramer, Y. Ono and T. Ohtsuki, *J. Phys. Soc. Jpn* **58**(1989) 1705.
[40] B. Huckestein and B. Kramer, *Phys. Rev. Lett.* **64**(1990) 1437.
[41] G.V. Mil'nikov and I.M. Sokolov, *JETP Lett.* **48**(1988) 536.
[42] T. Ando, *Phys Rev. B* **40**(1989) 9965.

[43] A.M.M. Pruisken, *Phys. Rev. B* **32**(1985) 2636.
[44] H.P. Wei, D.C. Tsui, M.A. Paalanen and A.M.M. Pruisken, *Phys. Rev. Lett.* **61**(1988) 1294; A.M.M. Pruisken, *Phys. Rev. Lett.* **61**(1988) 1297; H.P. Wei, S.W. Hwang, D.C. Tsui and A.M.M. Pruisken, *Surf. Sci.* **229**(1990) 34.
[45] J. Wakabayashi, M. Yamane and S. Kawaji, *J. Phys. Soc. Jpn* **58**(1989) 1903.
[46] T. Ando, *J. Phys. Soc. Jpn* **55**(1986) 3199.
[47] H. Aoki and T. Ando, *Surf. Sci.* **170**(1986) 249.
[48] M. Yamane, J. Wakabayashi and S. Kawaji, *J. Phys. Soc. Jpn* **58**(1989) 1899.
[49] H. Aoki, to be published.
[50] T. Chakraborty and P. Pietilainen, *Fractional Quantum Hall Effect* (Springer, Berlin, 1988).
[51] F.D.M. Haldane in Ref.4.
[52] R.B. Laughlin in *Mechanisms of High Temperature Superconductivity*, ed. H. Kamimura and A. Oshiyama (Springer, Berlin, 1989) p.76.
[53] G.S. Canright, S.M. Girvin and A. Brass, *Phys. Rev. Lett.* **63**(1989) 2291; 2295.
[54] L. Engel, H.P. Wei, D.C. Tsui and M. Shayegan, *Surf. Sci.* **229**(1990) 13.
[55] J.A. Simmons, H.P. Wei, L.W. Engel, D.C. Tsui and M. Shayegan, *Phys. Rev. Lett.* **63**(1989) 1731; G. Timp, R. Behringer, J.E. Cunningham and R. E. Howard, *Phys. Rev. Lett.* **63**(1989) 2268.
[56] J.K. Jain, S.A. Kivelson and N. Trivedi, *Phys. Rev. Lett.* **64**(1990) 1297.
[57] C.W. Beenakker, *Phys. Rev. Lett.* **64**(1990) 216; A.H. MacDonald, *Phys. Rev. Lett.* **64**(1990) 220.

Scaling and the Quantum Hall Effect

A. MacKinnon

Blackett Laboratory, Imperial College of Science, Technology & Medicine,
Prince Consort Rd., London SW7 2BZ, UK

An introduction to the ideas of scaling theory is given and the description of the quantum Hall effect in terms of a 2–parameter scaling theory is considered, mainly by studying the results of numerical simulations. It is argued that the apparently contradictory results can be explained in terms of a modification of the flow diagram predicted by Pruisken *et al.* This modification is shown to be consistent with the generalised flow diagram for the fractional effect.

1. Introduction

The quantum Hall effect has now been with us for over a decade[1]. The basic result of the accurate quantisation of the Hall conductivity σ_{xy} in units of e^2/h is now well accepted and the integer QHE is now a research tool rather than a subject of research in its own right. The fractional effect however continues to yield surprises[2,3]. The effect has been the subject of several review articles and of at least 2 substantial books[4,5]. It may therefore be considered a mature subject. Nevertheless there are still a number of outstanding questions.

Very soon after the initial discovery Aoki and Ando [6] pointed out that the concept of localisation of electrons in a disordered potential could provide a mechanism by which electrons could be added to or removed from the system without changing the Hall conductivity. Soon afterwards Laughlin[7] and Halperin[8] developed gauge arguments which gave important clues to the origin of the universal nature of the effect. Later Thouless *et al*[9,10] developed a formulation of the theory based on homotopy theory which related the Hall quantum number to the Chern class (a sort of generalised winding number) of a mapping of the Brillouin zone.

At about the same time as the discovery of the QHE considerable advances were made in the theory of electronic transport in disordered solids, and, in particular, the Anderson [11] metal–insulator transition. The scaling theory of Abrahams *et al.*[12] based on earlier work of Thouless[13] and Wegner[14] applied the renormalisation group ideas which had been so successful in the treatment of conventional phase transitions.

Two approaches in particular are relevant to the topic of this paper. The application of field theoretical ideas[15,16] to the metal–insulator transition has been the subject of a considerable body of work. Unfortunately this approach,

Springer Series in Solid-State Sciences, Vol. 101
High Magnetic Fields in Semiconductor Physics III Editor: G. Landwehr

while capable of yielding significant insight, tends to be so inaccessible to the non-specialist that it is difficult to draw meaningful conclusions about the accuracy of quoted results or the significance of the various approximations.

An alternative approach is to use computer simulation combined with scaling ideas, so–called *finite size scaling*[17,18]. In this case, while the results are in principle exact, there is always some doubt about whether the system sizes used in such calculations are sufficient to represent the asymptotic behaviour of a macroscopic system.

In this paper I shall consider the results of the application of the scaling approach to the integer quantum Hall effect. Rather than present new results I shall ask a few questions which the previous work has raised. In the final section I shall consider the application to the fractional quantum Hall effect[5,19].

2. What is Scaling?

One possible approach to constructing a theory of the transport in a sample would be to start by writing down a list of all known properties of the sample. Our training as physicists should enable us to distinguish the relevant from the irrelevant properties. We know that the crystal structure, the distribution and nature of impurities, the temperature and pressure may well be required, whereas such properties as the distribution of quarks in the nuclei and the marital status of the crystal grower are probably irrelevant. Other properties may or may not be relevant; such as the type of contacts used or the address of the Lab. in which the sample was grown.

Our experience tells us how to reduce the long list of properties to a much shorter list of relevant quantities. We can go further than this. We know from statistical mechanics that we can characterise a macroscopic (theoretically infinite) system by a much smaller list of properties than we require to describe a finite system, for which geometry and fluctuations are important. This latter aspect can be made more mathematically precise.

Let us define a vector u which represents our list of properties. Then we could write an equation to represent the way in which this list changes as we increase the size of the system

$$\frac{du}{d\ln L} = f(u) \tag{1}$$

where L is the size of the system and $f(u)$ is some function of u.

The reduction in the length of our list of necessary properties will then be represented by a tendency for u to approach a subspace such as a line or a surface. This line or surface is usually called an *attractor*. We can represent the asymptotic behaviour of the system in terms of a single parameter, roughly position on a line, or perhaps 2 parameters, a surface. At this stage one of the difficulties of the approach should already be apparent. How can we be

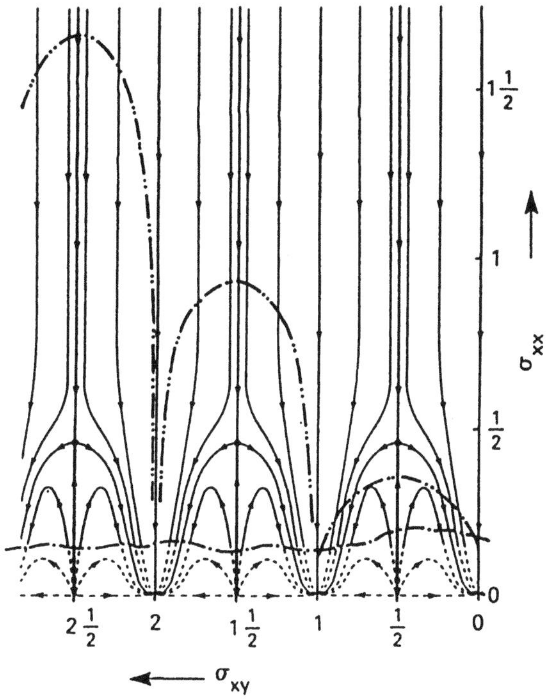

Figure 1. Theoretical renormalisation group Flow Diagram showing 2 parameter scaling for the integer QHE [20]. The – · · line shows the self–consistent Born approximation which forms the starting point for the flow and the –· line signifies the limit of the validity of (7). $(\sigma_{xx}, \sigma_{xy})$ are in units of e^2/h.

sure that a real physical system is large enough to be described by the attractor alone? This difficulty is particularly acute in the case of computer simulations.

Let us describe the attractor by a 1 or 2 dimensional vector u'. Then the behaviour of u' is described by an equation of the same form as (1). One very useful representation of (1) is in the form of a flow diagram such as figure 1 where $u' \equiv (\sigma_{xx}, \sigma_{xy})$. By replacing $\ln L$ by t we can envisage (1) as a *velocity* field. The arrows in the diagram show the change in $(\sigma_{xx}, \sigma_{xy})$ as $\ln L$ is increased.

The general features of figure 1 can be derived from quite general considerations[21]:

a) The flow must approach the limits $\sigma_{xx} = 0$ and $\sigma_{xy} = n$, in units of e^2/h.

b) σ_{xx} must be finite somewhere in each Landau level.

Of particular note in figure 1 are the so–called fixed points where the *velocity* is zero. These are located at $(\sigma_{xx}, \sigma_{xy}) = (0, n/2)$ and at $(\sigma_{xx}, \sigma_{xy}) = \left(\frac{1}{2}, n + \frac{1}{2}\right)$. By analysing the flow around the fixed points it is possible to compute the critical exponents.

Close to a fixed point u'_0 (1) can be written

$$\frac{du'}{d\ln L} = \mathbf{F}\,(u' - u'_0) \tag{2}$$

where $\mathbf{F}$ is the matrix of derivatives $\partial f_i / \partial u'_j$.

In the case of the $(\sigma_{xx}, \sigma_{xy})$ flow diagram in figure 1 the solution of (2) takes the simple form

$$\sigma_{xx} = \sigma_{xx}^* + A\left(\tau - \tau^*\right) L^{F_x} \tag{3a}$$

$$\sigma_{xy} = \sigma_{xy}^* + B\left(\tau - \tau^*\right)^2 L^{F_y} \tag{3b}$$

where $(\sigma_{xx}^*, \sigma_{xy}^*)$ are the coordinates of the fixed point, (F_x, F_y) are eigenvalues of F and τ is some parameter such as energy or disorder.

It is often useful to rewrite (3) in dimensionless form to obtain

$$\sigma_{xx} = \sigma_{xx}^* + \alpha\left(L/\xi\right)^{F_x} \tag{4a}$$

$$\sigma_{xy} = \sigma_{xy}^* + \beta\left(L/\zeta\right)^{F_y} \tag{4b}$$

form which we immediately obtain the form of the *correlation lengths* (ξ, ζ) as

$$\xi \sim \left|\tau - \tau^*\right|^{-1/F_x} \tag{5a}$$

$$\zeta \sim \left|\tau - \tau^*\right|^{-2/F_y} \tag{5b}$$

where ξ and ζ are associated with deviations in the σ_{xx} and σ_{xy} directions respectively. Hence by analysing the flow around the fixed points we can deduce the critical exponents associated with the phase transition in question: in this case the transition from one quantised state to another.

At this stage we must try to make the rather abstract discussion more concrete. In particular we should like to identify the length scales ξ and ζ and the parameter τ. To answer this we should look for a combination of system parameters which constitute the *knob(s)* which are to be turned in order to drive a finite sized sample into any position in the diagram. Clearly by varying the Fermi energy or the magnetic field it is possible to move along the σ_{xy} axis.

I know of no *knob* which will take the system towards large values of σ_{xx}. The Landau level index is only useful if the diagram is periodic in σ_{xy}, but this is the sort of information we would prefer to get out of the analysis rather than to put it in as an assumption. The disorder does nothing for decoupled Landau levels and tends to decrease σ_{xx} when the levels are coupled. There is some numerical evidence from tight–binding models that the intrinsic width of the Landau level may be useful[22] but this is zero for most models.

From this discussion we can identify ζ with the localisation length which diverges at a single energy in each Landau level. ξ, on the other hand, is an unidentified length scale, but since F_x is negative it is probably not important.

To summarise this section: the scaling analysis allows us to identify those aspects of the problem which contribute to critical behaviour and to calculate such important properties as critical exponents. The results are expected to be universal and independent of microscopic details, types of models etc.

3. Effective Field Theory

The flow diagram of figure 1 can be derived from the effective field theory originally derived by Pruisken[23,24], which is based on the effective Lagrangian

$$\begin{aligned}\mathcal{L}_{\text{eff}}[T] = &-\frac{1}{8}\sigma^0_{xx}\int d^2r\ \mathrm{Tr}\,\nabla_\mu\tilde{\mathbf{Q}}\nabla_\mu\tilde{\mathbf{Q}} + \eta\pi\rho(E)\,\mathrm{Tr}\,\tilde{\mathbf{Q}}\hat{\mathbf{s}} \\ &+\frac{1}{8}\sigma^0_{xy}\int d^2r\ \mathrm{Tr}\,\varepsilon_{\mu\nu}\tilde{\mathbf{Q}}\nabla_\mu\tilde{\mathbf{Q}}\nabla\nu\tilde{\mathbf{Q}}\end{aligned} \tag{6}$$

where $\mathbf{Q}$ are matrix fields. The derivation of (6) is discussed at some length in [24]. Here it should suffice to point out that the 2 terms are those obtained for the normal metal–insulator transition and the second term is a topological invariant which reduces to an integer (a sort of winding number).

In the so–called *dilute instanton gas* approximation analysis of (6) gives a flow diagram described by

$$\frac{d\sigma_{xx}}{d\ln L} = -\sigma_{xx}De^{-2\pi\sigma_{xx}}\cos(2\pi\sigma_{xy}) + \frac{-1}{2\pi^2\sigma_{xx}} \tag{7a}$$

$$\frac{d\sigma_{xy}}{d\ln L} = -\sigma_{xx}De^{-2\pi\sigma_{xx}}\sin(2\pi\sigma_{xy}) \tag{7b}$$

where D is a positive constant. The approximation here is expected to break down for small σ_{xx} as indicated in figure 1.

Using the fixed point analysis on (7) gives a critical exponent for the localisation length of $\nu = 2/F_y = 2$.

There is some confusion in the literature about the universality and periodicity of the flow diagram. There are numerous references to the periodicity being an essential feature. However in [24] Pruisken states that the constant D is non–universal and that the fixed point value σ^*_{xx} need not be the same for all lines $\sigma_{xy} = n + \frac{1}{2}$. It should be emphasised however that a periodic flow diagram need not imply that the physics of a real system is periodic in σ_{xx} as the starting point for the flow may be different. Indeed in practice σ_{xx} tends to increase linearly with σ_{xy}.

4. Numerical Approaches

There is a substantial body of numerical work by Ando and Aoki, summarised in a short review by Aoki[25]. Here I will only quote a couple of results of relevance to the subject of scaling. In all their work Ando and Aoki have used a model based on the solution for free electrons in a magnetic field with a potential defined using a random distribution of Gaussian potentials. Usually the coupling between Landau levels is ignored. Using the Thouless Number approach and the Kubo formula Ando [26,27] tried to calculate a flow diagram (Fig. 2). Note that the data fall on a single curve, representing a single parameter scaling rather than the 2–parameter form.

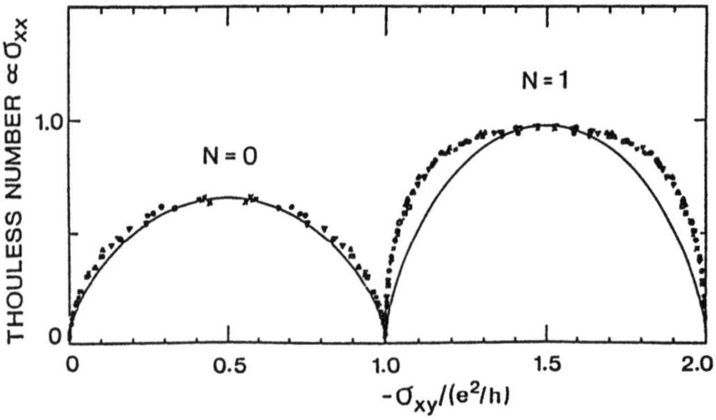

Figure 2. Numerical results from a Thouless number study [26,27]. Different symbols represent different energies and/or sample sizes. The full curve is an approximation.

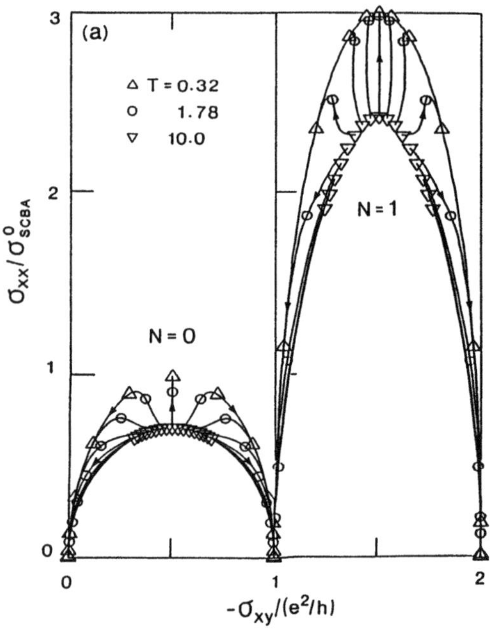

Figure 3. Theoretical Result for a temperature driven flow diagram[28]. Arrows represent flow to lower T.

In another study Aoki and Ando[28] present a temperature driven flow diagram for direct comparison with experiment (Fig. 3). In this case the flow seems to be 2–parameter but to be topologically different from fig.1. Note in particular the apparent unstable fixed point rather than the saddle point of Fig. 1.

An alternative approach has been pursued by Chalker and Coddington [28]. Using the *finite size scaling* method[17,18] on a model including the leading quantum correction to the semi–classical theory of Trugman[29]. Their results

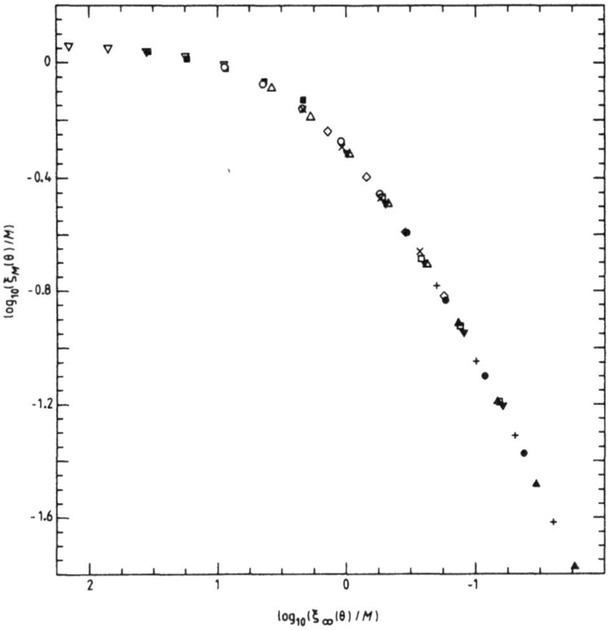

Figure 4. Finite size scaling curve of $\log_{10}\Lambda$, the renormalised localisation length (roughly $\Lambda \equiv \sigma_{xx}$) versus $\log_{10}(\zeta/L)$ for the model of Chalker and Coddington [28].

are summarised in Fig. 4. Again note that the data fall on a single common curve. The critical exponent is $\nu = 2.5 \pm 0.5$.

The present author[30] has applied the finite size scaling method to a tight–binding model and obtained results which could not be interpreted in terms of a single scaling parameter. It was possible to derive a 2–parameter flow diagram (Fig. 5) and to derive a critical exponent $\nu \approx 2$. The numbers on which this analysis was based are very quantitatively similar to those found by Ando and Aoki using a similar algorithm [31,32]. They find $\nu \leq 2$.

Another recent development which shows considerable promise is based on a new model developed by Huckestein and Kramer[33]. Again using finite size scaling they derive $\nu = 2.34 \pm 0.04$ from a one–parameter scaling curve (Fig. 6).

It may appear at first that the numerical data is inconclusive, indeed contradictory, about the validity of the 1–parameter or 2–parameter scaling. Some of this contradiction is illusory however. The models which show one parameter scaling have, in general, thrown away the data for smaller system sizes which don't fit the 1–parameter scaling hypothesis. This explains, for example, the apparent contradiction between the 2 sets of Thouless number data (Figs 2 & 3).

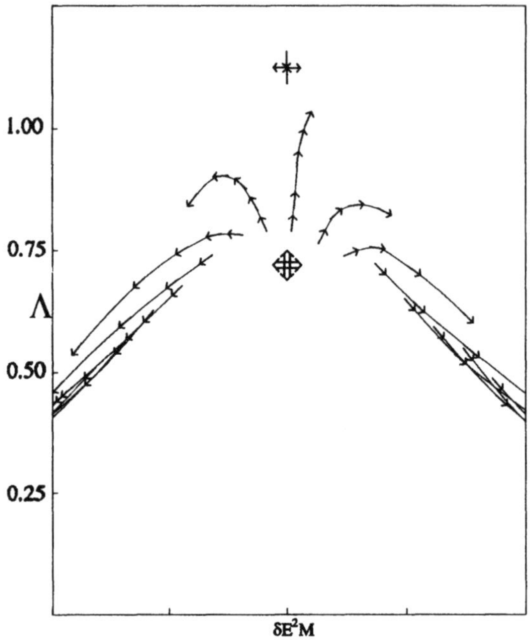

Figure 5. Renormalised localisation length Λ vs. $\delta E^2 M$, where δE is the deviation of the energy from the centre of the lowest Landau level and M is the width of the strip[30].

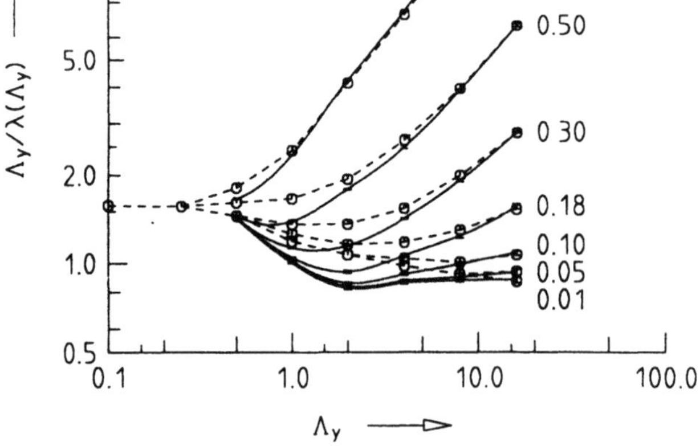

Figure 6. Model of Huckestein & Kramer[33]. Λ_y is the effective width of the system and $\lambda(\Lambda_y)$ is the localisation length.

The remarkable aspect of these data is the degree of agreement between them. Close inspection shows considerable *quantitative* agreement between results derived from a very wide range of models representing very different conditions. In fact the 2–parameter flow diagrams (Figs 2 & 5) both show flow towards a 1D attractor. Indeed this is also true of the theoretical diagram (Fig. 1) as long as the starting point is below the attractor joining the fixed points with finite σ_{xx} with those for $\sigma_{xx} = 0$.

There is however some disagreement between different authors about the meaning of the data for smaller systems. Is this data an intrinsic part of the scaling behaviour, as illustrated in Figs. 2 & 5 or simply an irrelevant deviation from the asymptotic 1–parameter behaviour as assumed by the authors of Figs. 3, 4 & 6. One justification for leaving this data out is that it seems to be weakly dependent on details such as the range of the potential fluctuations[34]. On the other hand the short range data for the different models show a similar behaviour to Figs 2 & 5, which may be represented by flow away from an unstable fixed point (or source) towards a 1D attractive line.

5. Scaling and the Fractional Effect

So far our discussion has been solely in terms of the Integer QHE. In 1985 Laughlin et al published a very speculative generalisation of Fig.1 to the fractional effect[35] (Fig. 7). The basic argument involved is that everything should behave exactly as in the Integer effect except that the positions of the fixed points along σ_{xx} should be rescaled to e^{*2}/h where e^* is the charge of the quasiparticles. However in order to fit all the features of Fig. 7 together correctly it is necessary to include an unstable fixed point between the saddle points predicted in Fig.1.

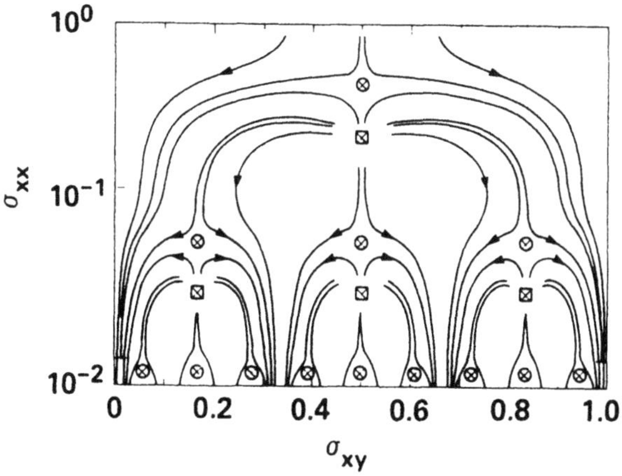

Figure 7. Generalised Flow Diagram for the Fractional QHE[35].

This diagram was largely ignored until recently when the Oxford group [36,37] succeeded in reproducing large sections of the diagram by looking at $(\sigma_{xx}, \sigma_{xy})$ as a function of temperature for a range of samples. Fig. 7 must therefore be taken seriously.

6. Conclusions and Speculation

The numerical results for the integer effect and the Fig. 7 and the experimental results for the fractional effect seem to be in agreement with each other but in disagreement with Fig. 1 on one important feature, namely the presence of a source of the flow below each of the fixed points associated with the Pruisken Theory. In the theoretical work on the fractional effect[35] this point is inserted simply as a connection between the other fixed points and is implicitly assumed to have something to do with the many body aspects. Comparison of the data for the Integer and the Fractional effects presented here suggests however that the source is an essential feature of the integer effect and that the unit which must be repeated in order to generate Fig. 7 is in fact the pair of fixed points.

Where does this leave the scaling theory? On the one hand there is nothing in the results presented here which contradicts the basic idea of 2–parameter scaling. The apparent contradiction with the Pruisken theory (Fig. 1) may well be simply due to the breakdown of the dilute instanton approximation for small σ_{xx} as indicated by the –· line in Fig. 1.

In conclusion, the basic idea of scaling theory as applied to the quantum Hall effect is alive and well. However a great deal more work is required to clear up the details and draw reliable conclusions.

References

1. von Klitzing K, Dorda G and Pepper M: Phys. Rev. Lett. **45** 494 (1980)
2. Clark R G, Haynes S R, Suckling A M, Mallett J R, Wright P A, Harris J J and Foxon C T: Phys. Rev. Lett. **62** 1536 (1987)
3. Clark R G, Mallett J R, Haynes S R, Maksym P A, Harris J J and Foxon C T: *Proc. Int. Conf. Applications of High Magnetic Fields, Würzburg* Springer Series in Solid State Sciences **87** 127 (1989)
4. Prange R E and Girvin S M: The Quantum Hall Effect (Graduate texts in Contemporary Physics) (Springer–Verlag, New York) (1987)
5. Chakraborty T and Pietiläinen P: The Fractional Quantum Hall Effect (Springer Series in Solid State Sciences **85**) (Springer–Verlag, Berlin Heidelberg) (1988)
6. Aoki H and Ando T: Solid State Commun: **38** 1079 (1981)
7. Laughlin R B: Phys. Rev. **B 23** 5632 (1981)
8. Halperin B I: Phys. Rev. **B 25** 2185 (1982)

9. Thouless D J, Kohmoto M, Nightingale M P and den Nijs M: Phys. Rev. Lett. **49** 405 (1982)
10. Thouless D J: in [4] above p101
11. Anderson P W: Phys. Rev. **109** 1492 (1958)
12. Abrahams E, Anderson P W, Licciardello D C and Ramakrishnan T V: Phys. Rev. Lett. **42** 673 (1979)
13. Thouless D J: Phys. Rep. **13** 93 (1974)
14. Wegner F J: Z. Phys. **B25** 327 (1976)
15. Wegner F: Z. Phys. **B35** 207 (1979)
16. Efetov K B: Adv. Phys. **32** 53 (1983)
17. MacKinnon A and Kramer B: Phys. Rev. Lett. **47** 1546 (1981)
18. MacKinnon A and Kramer B: Z. Phys. **53** 1 (1983)
19. Tsui D C, Störmer H L and Gossard A C: Phys. Rev. Lett. **48** 1559 (1982)
20. Wei H P, Chang A M, Tsui D C, Pruisken A M M and Razeghi M: *Proceedings of Int. Conf. on Electronic Properties of 2–dimensional Systems VI* Surf. Sci. **170** 238 (1985)
21. Khmel'nitskii D E: Piz'ma Zh. Exsp. Teor. Fiz. **82** 454 (1983) JETP Lett. **38** 552 (1983)
22. MacKinnon A: unpublished.
23. Pruisken A M M: Nuclear Physics **B 235** FS[11] 277 (1984)
24. Pruisken A M M: in ref. [4] p117 and references therein.
25. Aoki H: Rep. Prog. Phys. **50** 655 (1987) and references therein.
26. Ando T: Surf. Sci. **170** 243 (1986)
27. Ando T: J. Phys. Soc. Japan **55** 3199 (1986)
28. Chalker J T and Coddington P D: J. Phys. C: Solid State Physics **21** 2665 (1988)
29. Trugman S A: Phys. Rev. **B27** 7359 (1983)
30. MacKinnon A: J.Phys: Condensed Matter **1** 10407 (1989)
31. Aoki H and Ando T: Phys. Rev. Lett. **54** 831 (1985)
32. Ando T and Aoki H: J. Phys. Soc. Japan **54** 2238 (1985)
33. Huckestein B and Kramer B: Sol. State Commun. **71** 445 (1989)
34. Huckestein B: private communication.
35. Laughlin R B, Cohen M L, Kosterlitz J M, Levine H, Libby S B and Pruisken A M M: Phys. Rev. **B32** 1311 (1985)
36. Clark R G, Mallett J R, Usher A, Suckling A M, Nicholas R J, Haynes S R, Journaux Y, Harris J J and Foxon C T: *Proceedings of the 17th International Conference on the Electronic Properties of 2 Dimensional Systems, Santa Fe, New Mexico.* Surf. Sci. **196** 219 (1987)
37. Mallet J R, Clark R G, Harris J J and Foxon C T: *Proc. Int. Conf. Applications of High Magnetic Fields, Würzburg* Springer Series in Solid State Sciences **87** 132 (1989)

Scaling Behaviour of Doped AlGaAs/GaAs Heterostructures in the Quantum Hall Regime

S. Koch, R.J. Haug, K. von Klitzing, and K. Ploog

Max-Planck-Institut für Festkörperforschung, Heisenbergstr. 1, W-7000 Stuttgart 80, Fed. Rep. of Germany

The influence of repulsive (Be) and attractive (Si) scattering centers close to the two dimensional electron gas of AlGaAs/GaAs heterostructures on characteristic transport properties has been studied. The temperature dependent halfwidth of the Shubnikov-de Haas-peaks ΔB (T) and the maximum of the slope of the Hall resistance $\partial\rho_{xy}/\partial B$ (T) between adjacent Hall plateaus have been investigated in the temperature range from 1.1K down to 40 mK. We have studied both delta-doped and homogeneously doped samples. In contrast to results reported for InGaAs/InP-heterostructures where a scaling exponent $\varkappa = 0.42 \pm 0.04$ with $\Delta B \propto T^{\varkappa}$ and $(\partial\rho_{xy}/\partial B)^{max} \propto T^{-\varkappa}$ is found our data lead to scaling exponents which depend on the distribution and the density of the impurities. The values for $\varkappa$ are in the range of 0.36 to 0.81.

1. Introduction

Scaling of characteristic transport properties of a two-dimensional electron gas (2D EG) in the quantum Hall regime has recently gained considerable interest [1]. In the integral quantum Hall effect (IQHE) studied on InGaAs/InP - heterostructures the half width ΔB of the peaks in ρ_{xx} between adjacent integer filling factors exhibited a scaling behaviour of the form $\Delta B(T) \propto T^{\varkappa}$ with $\varkappa = 0.42 \pm 0.04$ as the exponent of the temperature T. Similarly, the maxima of the derivative of the Hall resistance ρ_{xy} with respect to the magnetic field B followed the law $(\partial\rho_{xy}/\partial B)^{max} \propto T^{-\varkappa}$ with the same exponent $\varkappa$. In the following, we will call this power law dependence on temperature "scaling behaviour". A theoretical description of this phenomenon has been given by Pruisken [2]. The value of $\varkappa = 0.42$ has been proposed to be *universal* [1].

2. Experiments

We report on results obtained on AlGaAs/GaAs- heterostructures with definitely built-in scatterers. The scatterer distribution was either homogeneous or like a delta-profile [3]. Attractive and repulsive scat-

Springer Series in Solid-State Sciences, Vol. 101
High Magnetic Fields in Semiconductor Physics III Editor: G. Landwehr

Table 1: Strength n_δ of the doping, carrier concentration n_e and mobility μ_e at T=1.5K for the samples used. Temperature exponents $\varkappa$ for Landau level N=0↓ (filling factor ν= 1- 2).

Sample	n_δ (10^{10} cm^{-2})	n_e (10^{11}cm^{-2})	μ_e (cm^2/Vs)	$\varkappa$
#1	0	2.52	302000	0.36 ± 0.04
#2	0.5	2.40	90300	0.56 ± 0.05
#3	2.0	2.10	22500	0.81 ± 0.04

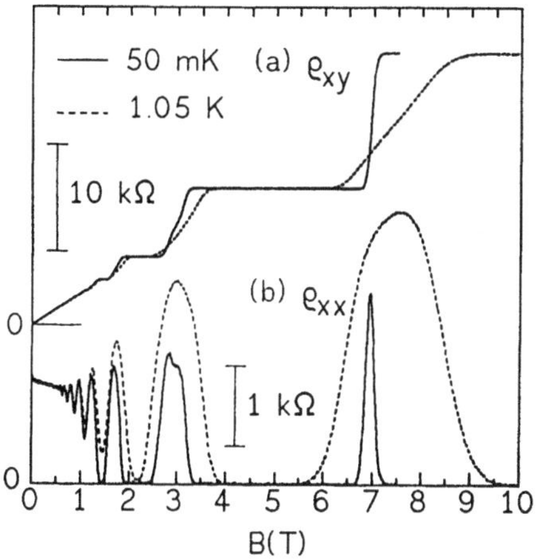

Fig. 1: Hall resistivity ρ_{xy} (a) and longitudinal resistivity ρ_{xx} (b) at 50 mK and 1.05 K for sample #3.

terers were used, realized by silicon and beryllium dopants, respectively. In this contribution, we will focus on results obtained on samples which have been δ-doped with Be, but also shortly discuss results of other structures. The samples were grown using the technique of molecular beam epitaxy (see e.g. Ref.3). After the growth process they were etched into a standard Hall bar geometry. The mobility μ_e and the electron and doping concentrations n_e and n_δ of the samples are summarized in table 1.

In Fig.1 the longitudinal resistivity ρ_{xx} and the Hall resistivity ρ_{xy} are shown for sample 3 as a function of magnetic field at temperatures of 50 mK and 1.05 K. We note that the spin splitting of the first Landau level (LL N=1) is not resolved even at the lowest temperature due to the strong doping, expressed by the correspondingly low mobility.

In Fig.2 we show the reciprocal half width $(\Delta B)^{-1}$ and the maxima of $\partial\rho_{xy}/\partial B$ in the temperature range from 40mK to 1.1K

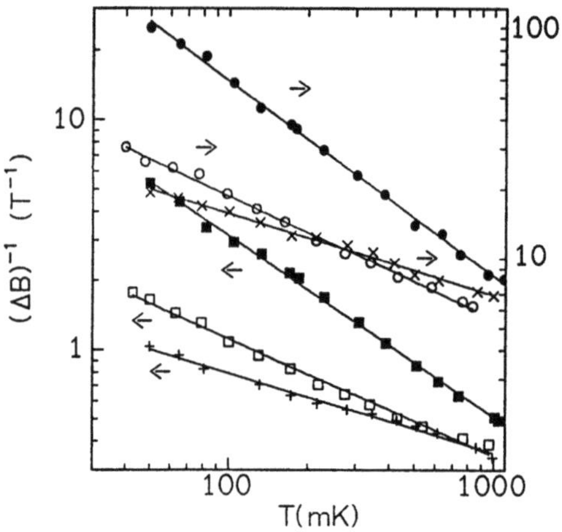

Fig. 2: Reciprocal half width $(\Delta B)^{-1}$ (left scale) and maxima of $\partial\rho_{xy}/\partial B$ (right scale) in the lowest Landau level (N= 0↓) as functions of temperature for the Be-δ-doped samples. $(\Delta B)^{-1}$: + (sample #1), □ (#2), ■ (#3). $(\partial\rho_{xy}/\partial B)^{max}$: x (#1), o (#2), ● (#3).

for LL N=0↓. The results are given for the three samples of table 1. As can be seen, we obtain a scaling behaviour in all three cases, with a value of κ ranging from 0.36 for the undoped structure to 0.81 for the strongly doped sample (cf. table 1). This increase of κ with doping is also observed in silicon δ- doped samples, although quantitatively less pronounced ($\kappa = 0.62$ for the same doping strength as in sample 3). For the LL N=1 we also see an increase of κ with doping in the homogeneously Si doped samples. For the δ-doped samples, however, in this LL κ does not change with doping.

As a summary of these experimental results we conclude that there is an increase of κ with the concentration of impurities. The effect is most pronounced in the case of the Be δ-doped samples in LL N= 0↓ where an increase of κ of a factor of more than two results. The same effect occurs in Si δ-doped samples, although the change is smaller. In the first LL we see the increase of κ with doping in the homogeneously doped samples.

3. Discussion

We discuss the phenomena described above making use of the framework given by Pruisken [2]. He uses the inelastic scattering length L_{in} (Thouless length [4]) as an effective sample size which behaves like $L_{in} \propto T^{-p/2}$ with T the temperature and an exponent p of order 1. On the other hand, the localization length ξ describes the transition from delocalized to localized states near the Fermi energy as $\xi \propto |B - B^*|^{-\nu}$ with B the magnetic field and an exponent ν of order 1. B^* is the critical field. From this the following relation for the scaling exponent $\varkappa$ is derived:

$$\varkappa = p/2\nu. \quad (1)$$

Assuming the validity of (1) higher values (than 0.42 [1]) can be explained by higher values of p or lower values of ν. Quantitative results for ν have been given by Ando and Aoki [5] and by Huckestein and Kramer [6] In Ref. 5 calculations lead to a value of $\nu \leq 2$ for the lowest LL and to $\nu \leq 4$ for the first LL. Huckestein and Kramer [6] have calculated a value of $\nu = 2.34 \pm 0.04$ for the lowest LL. The classical percolation picture yields $\nu = 4/3$ [7] while the inclusion of quantum tunneling leads to $\nu = 7/3$ [8].

To check the validity of eq. (1) one has to find reasonable values for p and ν consistent with our experimental results reported above. If we take the value $\nu = 4/3$ [7] we obtain values for p between 1.0 and 2.2, with the higher values for *higher* doping. What we expect is a value of p=1 (*dirty* metal limit) for *impure* systems and p=2 (Fermi liquid theory) for clean samples, i.e. the experiment behaves just opposite as expected. On the other hand, as can be seen above, the value of $\nu = 4/3$ is far from being generally accepted. Although it is difficult to draw definite conclusions due to the lack of experimental information about p and ν and the differing answers theorists give, we think that our experiments justify doubts whether eq. (1) can be considered as being generally valid.

In summary, systematic changes of the exponent $\varkappa$ which describes the temperature dependence of certain magnetotransport properties of the two- dimensional electron systems in specifically doped AlGaAs/GaAs- heterostructures have been found. $\varkappa$ increases with doping, depending on the Landau level and the distribution of scattering centers whereas the type of scatterers has no qualitative influence. This behaviour indicates limitations of validity of the relation $\varkappa = p/2\nu$.

We thank Dr. B. Mieck and Dr. G. Zumbach for valuable discussions.

4. References

1. H.P.Wei, D.C.Tsui, M.A.Paalanen, and A.M.M.Pruisken, Phys.Rev.Lett. 61, 1294 (1988).
2. A.M.M.Pruisken, Phys.Rev.Lett. 61, 1297 (1988).
3. K.Ploog, J.Cryst.Growth 81, 304 (1987).
4. D.J.Thouless, Phys.Rev.Lett. 39, 1167 (1977).
5. T.Ando and H.Aoki, J.Phys.Soc.Jpn. 54, 2238 (1985).
6. B.Huckestein and B.Kramer, Phys.Rev.Lett. 64, 1437 (1990).
7. S.A.Trugman, Phys.Rev.B 27, 7539 (1983).
8. G.V.Mil'nikov and I.M.Sokolov, JETP Lett. 48, 536 (1988).

Magneto-Capacitance in Two-Dimensional Electronic Systems in $Al_xGa_{1-x}As/GaAs$ Heterostructures Under the Influence of Ionized Impurities

J. Richter[1], *H. Sigg*[1;2], *K. von Klitzing*[1], *and K. Ploog*[1]

[1]Max-Planck-Institut für Festkörperforschung, Heisenbergstr. 1, W-7000 Stuttgart 80, Fed. Rep. of Germany

[2]Paul Scherrer Institut Zürich, CH-8048 Zürich, Switzerland

We have studied the magneto-capacitance of impurity-influenced two-dimensional electronic gases(2DEG's) at the interface of *AlGaAs/GaAs*-heterostructures. The 2D-carrier density as evaluated from the Shubnikov de Haas-type oscillations indicates an impurity-specific filling of the lowest spin-split Landau level. We interpret this behavior with the bound states of impurities, which have to be taken into account when considering the degeneracy of the 2DEG in the magnetic quantum limit.

The discovery of the quantum Hall effect(QHE) [1] initiated a vigorous research activity of the underlying physical properties of two-dimensional electronic gases(2DEG's).
In strong magnetic fields the otherwise constant density of states(DOS) of a 2D-electric subband at the *AlGaAs/GaAs*-interface splits up into a series of Landau levels(LL's) which alter significantly the electrical and optical properties of the 2DEG's. The width and shape of the LL's can be greatly affected by potential fluctuations which arise from either interface roughness or ionized impurities acting on the 2DEG's. According to the dominant impurity type the LL's can be asymmetrically broadened to the high or low energy side [2] and give rise to a specific shift of the Hall plateaus as observed in magneto-transport measurements performed on intentionally impurity-doped heterostructures [3] and on *Si*-MOSFETs with ionized *Na*-ions implanted in the SiO_2-barrier [4].

In the present work the investigated samples were MBE-grown *AlGaAs/GaAs*-heterostructures with a δ-layer of donors (*Si*) or acceptors (*Be*) at well defined spacings from the *AlGaAs/GaAs*-interface. The δ-doping levels ranged between $3 \cdot 10^9\ cm^{-2}$ and $4 \cdot 10^{10}\ cm^{-2}$, about an order of magnitude smaller than the investigated 2D-carrier densities. The samples were $4 \times 6mm$ in size with ohmic contacts to provide connection to the 2DEG. A 10 *nm* thick *NiCr*-gate was evaporated on top of a 100 *nm* thick Al_2O_3-layer which was included to prevent leakage currents to the 2DEG. The voltage was applied between the 2DEG and the *NiCr*-top-gate. The samples had mobilities between 3 and $10 \cdot 10^4 cm^2/V \cdot s$. We chose magneto-capacitance(MCV) as the method of investigation because it has proven to reflect quite reliably the DOS in a magnetic field [5]. Furthermore, it allows a continuous variation of the 2D-carrier concentration, and therefore a "scanning" of the Fermi level through the LL's at constant magnetic fields.

Fig. 1 shows several traces of MCV obtained by sweeping the gate voltage (i.e. the 2D-carrier density) at a constant magnetic field. The MCV-signal varies as a function of the filling factor(FF) $\nu = N_s h/eB$ of the LL's, where N_s is the 2D-carrier density and B the magnetic field. At low fields and negative applied gate voltages, $V_g \approx -2\ V$, a sharp decrease of the MCV due to the depletion of the 2DEG's is observed, the midpoint of which will be referred to as the depletion point DP. At higher magnetic fields the LL-spacing increases, the gaps between the LL's in the DOS become more pronounced and the MCV exhibits minima when the DOS between the LL's is small [5]. Noteworthy is the substantial shift of the DP with increasing magnetic field. As in the case of Shubnikov-de-Haas(SdH) oscillations in magneto-transport, the minima in MCV can be used to determine the 2D-carrier density. For various fixed magnetic fields minima in the MCV

Springer Series in Solid-State Sciences, Vol. 101
High Magnetic Fields in Semiconductor Physics III   Editor: G. Landwehr

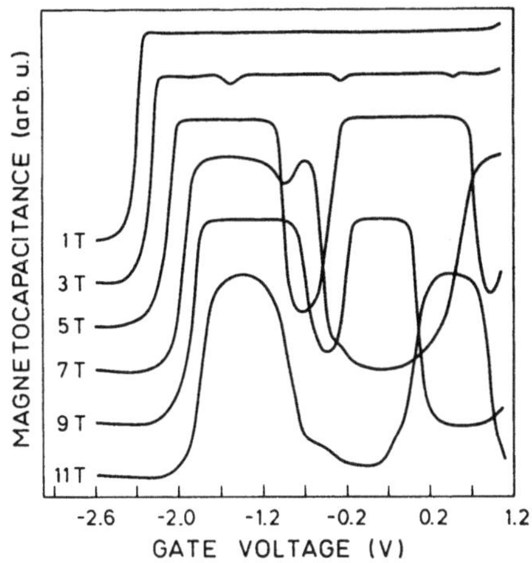

Fig. 1. MCV-Traces of a Si-δ-doped heterostructure at different magnetic fields.

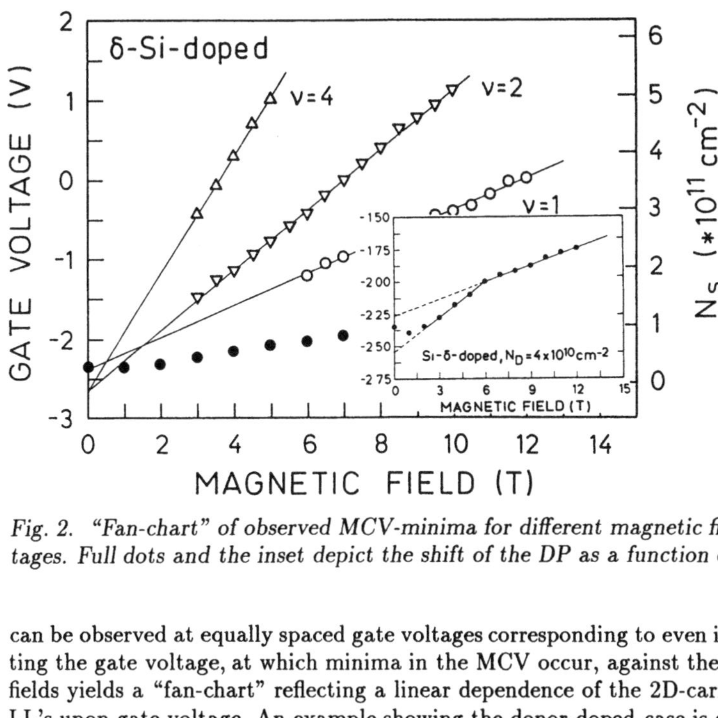

Fig. 2. "Fan-chart" of observed MCV-minima for different magnetic fields and gate voltages. Full dots and the inset depict the shift of the DP as a function of magnetic field.

can be observed at equally spaced gate voltages corresponding to even integer FF's. Plotting the gate voltage, at which minima in the MCV occur, against the applied magnetic fields yields a "fan-chart" reflecting a linear dependence of the 2D-carrier density of the LL's upon gate voltage. An example showing the donor-doped case is shown in Fig. 2.

A similar plot is obtained for the acceptor-doped case. Notice, that the point of intersection of the lines on which the minima are located is not the same for all FF's. We find that for both doping types the line obtained from the *observed* $\nu = 1$-minima is substantially different from that for $\nu \geq 2$. This is caused by the fact that for fixed magnetic fields the *observed* $\nu = 1$-minimum does not occur at half the gate voltage for $\nu = 2$, but at a different one. Conversely, at fixed gate voltages the *observed* $\nu = 1$-minimum occurs at different magnetic fields than twice the fields for $\nu = 2$.

If we use the fundamental degeneracy of a 2DEG in a magnetic field to calculate N_s on the basis of the *observed* minima in MCV, $N_s = 2.418 \cdot B \cdot \nu \cdot 10^{10} cm^{-2}$, we find a

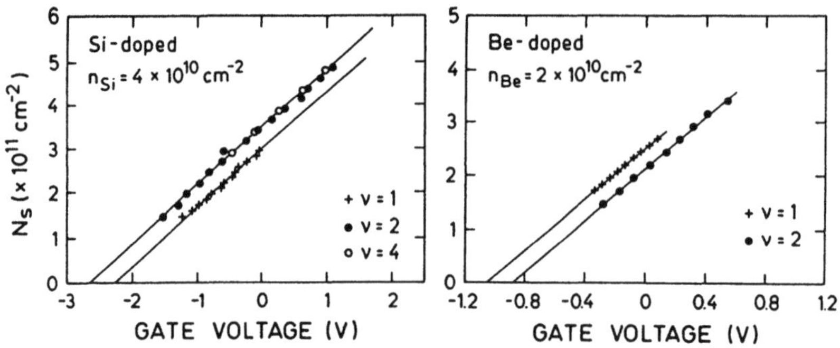

Fig. 3. 2D-carrier densities determined from MCV-minima using the degeneracy of the 2D-system for different gate voltages.

quite spectacular relationship. Fig. 3 shows the linear dependence of N_s on the applied gate voltage V_g for both doping types. However, not only one but *two* different linear relationships between N_s and V_g are obtained for both doping types. One relationship is obtained from the observation of the MCV-minima for $\nu \geq 2$ and the other one is based on the MCV-minima observed for $\nu = 1$. For donor-doping the $\nu = 1$-relationship is uniformly shifted to *lower* 2D-carrier densities (higher gate-voltage), whereas for the acceptor-doping the $\nu = 1$-relationship is shifted to *higher* 2D-carrier densities (lower gate-voltage).

The uniformly shifted $N_s(V_g)$-dependence obtained from the $\nu = 1$-minima for both dopings suggests that the observed constant carrier density difference to the $\nu \geq 2$-relationship is related to a well defined number of bound states, which is independent of the magnetic field and of the particular 2D-carrier density. In our opinion the impurity states of *both* spin levels of the lowest LL are the keystone to the understanding of the observed behavior.

In a simple picture this behavior can be explained by the formation of split-off, bound states for each spin level of the lowest LL, separated from the spin-split LL by more than half of the spin splitting. This leads to a significantly altered filling behavior of the lowest LL as compared to the higher LL's. For donor-influenced 2DEG's the filling of the lowest, spin-resolved LL is therefore not only given by the fundamental relation eB/h but, due to the additional occupation by electrons of the split-off states of the higher, spin-split LL, is raised by an amount N_D corresponding to the doping-induced, attractive donor centers. In acceptor-influenced 2DEG's the opposite will happen: at the negatively charged impurity sites the Pauli exclusion principle and the repulsive Coulomb forces prevent the population by any 2D-electrons of the lowest LL. The filling of the lowest LL is therefore decreased by an amount of N_A, the acceptor density, and the corresponding $\nu = 1$-related MCV-minimum appears at lower gate-voltages, i.e. apparently higher 2D-carrier densities. Therefore, for arbitrary magnetic fields and 2D-carrier densities, more carriers(higher voltages) are needed to fill the lowest spin-split LL for donor-doping, whereas less carriers(lower voltages) are necessary for acceptor doping. In fact, the impurity densities determined by this method, agree quite well with sheet impurity densities obtained from sample growth parameters (c.f. Fig. 3).

Another interesting observation in donor-doped heterostructures was the quite strong dependence of the depletion point DP upon magnetic field (see Fig. 1). The inset of Fig. 2 shows the "as-defined" depletion point DP as a function of magnetic field.

Note that the shift of the DP as a function of magnetic field is not continuous, but changes slope at $B \approx 6\ T$. The linear extrapolation $B < 6\ T$ yields a voltage threshold at $B = 0$, which is in good agreement with the threshold exhibited by the $\nu \geq 2$-relationship shown in Figs. 2 and 3(left) and is related to the fully depleted 2DEG. The extrapolation for $B > 6$ T yields a threshold voltage which rather corresponds to the voltage threshold exhibited by the $\nu = 1$-relationship. The two threshold values differ again by a carrier density difference corresponding to N_D, the donor dopant density.

Interestingly, the magnetic field at which the slope changes coincides with the first observation of a $\nu = 1$-minimum in MCV. Moreover, the slopes of the linear extrapolations of the two regimes also differ by a factor of two, and their values, expressed in terms of the LL-degeneracy, correspond to a FF of $\nu = 1/2$ and $\nu = 1/4$, respectively. We do not have a full explanation for this peculiar level-filling dependence of the DP. However, from a percolation model, Efros [6] has predicted a critical slope following the filling of half a LL: i.e. $\nu = 1$ at low magnetic field and $\nu = 1/2$ at high magnetic field when the spin-LL's are resolved. The discrepancy with the experiment might be explained by additional, magnetic field-induced bound states at the $AlGaAs/GaAs$-interface [7].

In conclusion we have shown that 2DEG's under the influence of ionized impurities show magnetic field localization. We further showed that magneto-capacitance can be successfully employed to determine the major impurity type in a 2DEG. The accuracy of the method allows a fair quantitative determination of the impurity density in 2DEG's.
We would like to acknowledge financial support from the Bundesministerium für Forschung und Technologie (BMFT).

References

[1] K. von Klitzing, G. Dorda, M. Pepper; *Phys. Rev. Lett.* **45**, 494 (1980).
[2] T. Ando; *J. Phys. Soc. Japan* **36**, 1521 (1974).
[3] R. J. Haug, R. R. Gerhardts, K. v. Klitzing, and K. Ploog; *Phys. Rev. Lett.* **59**, 1349 (1987).
[4] J. E. Furneaux, T. L. Reinecke; *Phys. Rev.* **B33**, 6897 (1986).
[5] D. Weiss, E. Stahl, G. Weimann, K. Ploog; K. von Klitzing; *Surf. Sci.* **170**, 285 (1986).
[6] A. L. Efros; *Solid State Comm.* **65**, 1281 (1988).
[7] J. L. Robert, A. Raymond, L. Konczewitz, C. Bousquet, W. Zawadzki, F. Alexandre, I. M. Masson, J. P. Andre, P. M. Frijlink; *Phys. Rev.* **B33**, 5935 (1986).

Localization in the Quantum Hall Effect

S. Kawaji

Department of Physics, Gakushuin University, Mejiro,
Toshima-ku, Tokyo 171, Japan

Abstract. Experimental studies carried out by the Gakushuin group during the last decade on localization in Landau levels in 2-dimensional systems of Si(001)-MOSFETs under strong magnetic fields in connection with the quantum Hall effect are reviewed and discussed in the light of recent progress.

1. Introduction

Ten years have passed since the first precision measurements of quantized Hall resistances were carried out by von Klitzing et al. [1] On this occasion it is worthwhile to recollect work we have performed in the last decade on localization in n-channel Si (001) inversion layers in connection with the quantum Hall effect.

In n-channel MOS inversion layers on Si(001) surfaces at low temperatures and in strong magnetic fields, it is possible to realize an extreme-quantum-limit condition ($k_BT < \Gamma < \hbar\omega_c$), where Γ is the broadening of a Landau level.

Experimental evidence of localization in such a system was presented first by measurements of the diagonal conductivity σ_{xx}.[2] Experiments were carried out on the effect of the source-drain fields E_{SD} on the finite gap regions in the gate voltage V_G for vanishing σ_{xx} in a Corbino disk in magnetic fields up to B=15 T at a temperature of T=1.4 K. The result showed that the sum of concentrations of localized electrons $N_{immobile}$ associated with the higher edge of the (N-1)th Landau level and the lower edge of the Nth Landau level is approximately given by $[2\pi \ell_0^2(2N+1)]^{-1}$ where N is the Landau quantum number and $\ell_0=(h/eB)^{1/2}$ is the magnetic length.

Localization in the lowest Landau level, where the σ_{xx} peak vanishes at low E_{SD}, has a different nature from localization in other Landau levels. From that standpoint, further studies were carried out on T-dependence and E_{SD}-dependence of σ_{xx} in the lowest Landau level. [3,4]

2. Measurements of the Hall conductivity in 2D systems under strong magnetic fields

Under strong magnetic fields, direct measurements of the conductivity tensor are only possible for σ_{xx} by the use of Corbino disks. There is no direct method for the measurement of the Hall conductivity σ_{xy}.

Several examples of various geometries of samples are shown in Fig. 1. The long Hall bar is a geometry for measurements of the Hall resistivity ρ_{xy} and the diagonal resistivity ρ_{xx}. The

Springer Series in Solid-State Sciences, Vol. 101
High Magnetic Fields in Semiconductor Physics III Editor: G. Landwehr

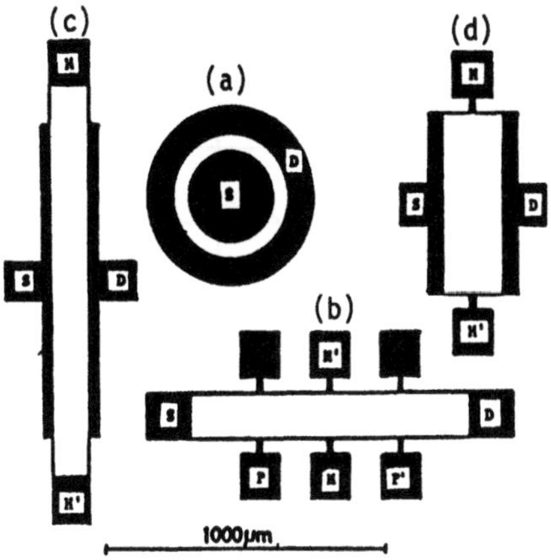

Fig. 1. Typical electrode structures for galvanomagnetic measurements for 2D systems. (a) Corbino disk, (b) Hall bar(or Hall bridge), (c) wide Hall current bar, and (d) wide Hall bridge.

tranport equations for an isotropic homogeneous 2D system lead to the following well known relations.

$$\sigma_{xx}=\rho_{xx}/(\rho_{xx}^2+\rho_{xy}^2), \quad \sigma_{xy}=-\rho_{xy}/(\rho_{xx}^2+\rho_{xy}^2) \qquad (1)$$

In 2D systems, combinations of measured quantities of ρ_{xx} and ρ_{xy} by Eq.(1) usually do not give right values of σ_{xx} and σ_{xy}. This fact can be recognized by comparing peak values of the σ_{xx} measured by using a Corbino disk with the same quantities given by Eq.(1) from measured values of ρ_{xx} and ρ_{xy}.[5]

Evaluation of the conductivity tensor components from measured sample conductance and Hall voltage in rectangular samples by solving the Hall-effect field problem also does not give the right values of σ_{xx} and σ_{xy}.[6,7]

These difficulties arise probably from effects of the sample edges on the voltage measurement.

Measurements of the short-circuited Hall current and the conductance by the use of wide Hall current bars, shown in Fig. 1, is probably the only procedure which gives the right values of conductivity tensors .[8,9]

Experimental data which will be described in the following sections have been obtained by the Hall current method with a constant source-drain voltage for samples with a peak electron mobility of about 1.5 $m^2/V \cdot s$.

3. Quantum Hall effect and localization

Preliminary experimental results of a relation between localization and quantization of the Hall conductivity were reported in ref. 8. Clear evidence of the relation between the quantum Hall effect and localization was reported in 1980 in Hakone, at the meeting on the Application of High Magnetic Fields.[10]

Figure 2(a) shows T-dependences of changes in σ_{xy} and σ_{xx} in the lowest four Landau levels of a Si-MOSFET against V_G.

Curves of σ_{xy} and σ_{xx} measured at 1.5 K in Fig. 2(a) are reproduced in Fig. 2(b).[11] In this figure, the ordinate is scaled by eB/h, the degeneracy of a Landau level, and the abscissa of σ_{xy} is scaled by e^2/h, the quantized step in the Hall conductivity. Figure 2(b) shows the quantized plateaus in

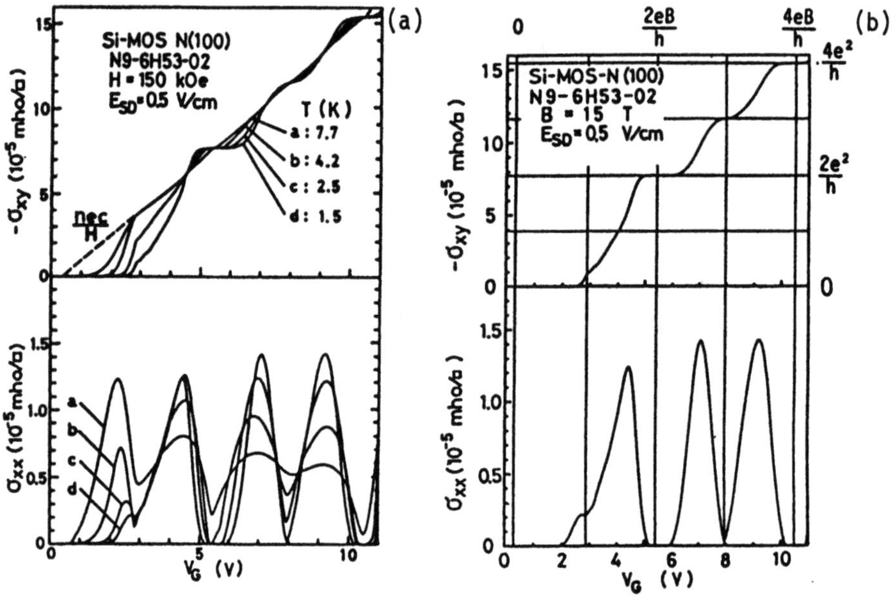

Fig. 2. Hall conductivity $-\sigma_{xy}$ and diagonal conductivity σ_{xx} vs gate voltage V_G in the lowest four N=0 Landau levels of a Si-MOSFET in a magnetic field of 15 T at different temperatures (a) [10] and at 1.5 K (b) [11].

σ_{xy} at the values of $2e^2/h$ and $4e^2/h$ and that σ_{xx} is zero in the gate voltage region corresponding to the quantized Hall plateau.

The increase in temperature from 1.5 K to 7.7 K in Fig. 2(a) gives rise to the decrease in the gate voltage region for both the quantized Hall plateau and zero σ_{xx}. At the highest temperature in Fig. 2(a), the σ_{xy} approaches a straight line which connects crossing points of the scale lines, $(ieB/h, ie^2/h)$, i being an integer, and represents a classical value of the Hall conductivity in the case of no scattering, $-N_s e/B$ where N_s is the surface electron concentration.

We note here that such a behaviour of σ_{xx} and σ_{xy} in 2D systems had been predicted theoretically by Ando, Matsumoto and Uemura in 1975 [12] as described in ref. 10. They calculated σ_{xy} in several approximations. Their result in the self-consistent Born approximation at T=0 is given by

$$\sigma_{xy} = -N_s e/B + \{\mathrm{Im}\Sigma(E_F)/\hbar\omega_c\}\sigma_{xx} \qquad (2)$$

where $\mathrm{Im}\Sigma(E_F)$ is close to Γ.

Concerning the impurity effect, Ando et al. derived important theoretical conclusions summarized as follows:

(1) When the Fermi level lies in the energy gap between the ith and (i+1)th Landau levels, the Hall conductivity is not affected by the presence of impurities and given by

$$\sigma_{xy} = -ie^2/h. \qquad (3)$$

(2) The Hall conductivity is still given by Eq.(3) even if the ith Landau level is not completely filled but σ_{xx}=0.

The experimental results in Fig. 2 clearly demonstrate the properties of the localization in Landau levels in 2D systems which Ando et al. predicted theoretically .

4. Analysis of the temperature dependence of the Hall conductivity by an effective mobility edge model

Computer simulations of localization in Landau levels were first carried out by Aoki.[13] More extensive numerical experiments performed by Ando in 1981 showed that the wave functions are exponentially localized outside the mobility edge.[14] His calculations also showed that the Hall conductivity vanishes below the lower mobility edge and increases almost linearly with energy between the lower and higher mobility edges, then takes the value $-e^2/h$.

A phenomenological analysis of the T- and N_s-dependence of σ_{xy} was carried out for the lower half part of the third lowest Landau level in Fig. 2(a) based on the following model:[15] The change in the Hall conductivity is given by

$$\Delta\sigma_{xy}(\nu,T) = -(e^2/h)[n_M(\nu,T)/N_M] \qquad (4)$$

where ν is the filling factor of the Landau level, N_M is total number of mobile or extended states and n_M is the number of electrons in the mobile states. The density of states is assumed to be constant for simplicity.

Ando extended further numerical experiments and predicted that all states are localized exponentially except those just the center of the Landau levels.[17] Ando's numerical result suggests the existence of a temperature dependence of an effective mobility edge in the experimental observations.

A new analysis of experimental Hall conductivity was carried out by investigation of T-dependence of $d\sigma_{xy}/dN_s$.[18] The derivative $d\sigma_{xy}/dN_s$ is much more sensitive to the change in temperature than σ_{xy} itself when Γ is much smaller than $\hbar\omega_c$ as expected from Eqs.(2) and (4). Figure 3(a) shows V_G-dependence of σ_{xy}, σ_{xx} and $-d\sigma_{xy}/dN_s$ in B=15 T at T=1.4 K and Fig.3 (b) shows T-dependence of $-d\sigma_{xy}/dN_s$. Assuming a function for the density of states proportional to $\exp[-2(E/\Gamma)^3]$ and using Eq.(4), the effective mobility edge E_c/Γ was determined at temperatures between 1.4 K and 4.2 K to reproduce $d\sigma_{xy}/dN_s$ in the third lowest Landau level;i.e., the $(0\downarrow+)$ Landau level.

The T-dependence of E_c/Γ was discussed in connection with the inelastic scattering time τ_e. Two possibilities were considered of the role of inelastic scattering in the determination of E_c. One is that E_c is determined by $E_c \sim \hbar/\tau_e$. If this is the case, $E_c/\Gamma \propto \tau/\tau_e$. Another possibility is that the localization is destroyed by the inelastic scattering with a cutoff length L_c given by the inelastic diffusion length $L_e=(D^*\tau_e)^{1/2}$, where D^* is the diffusion constant. In this case, E_c is determined as $L_e=1/\alpha(E_c)$ where $\alpha(E)$ is the inverse localization length of the electron wave functions. Since $D^* \sim \ell_N{}^2/\tau$ [16], we have

$$L_e \sim [(2N+1)\tau_e/\tau]^{1/2}\,\ell_0 \qquad (5)$$

If this is the case, τ/τ_e looks strongly T-dependent when we use Ando's numerical results for the localization length as a function of E/Γ. We had already known about the inelastic scattering time in the absence of strong magnetic fields by nega-

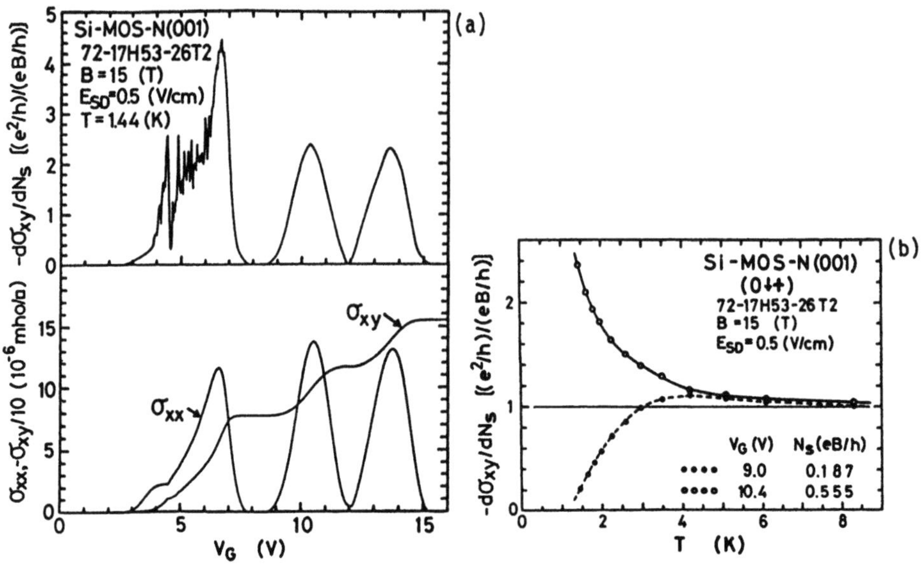

Fig. 3. Hall conductivity $-\sigma_{xy}$, diagonal conductivity σ_{xx} and $-d\sigma_{xy}/dN_s$ vs gate voltage V_G in the lowest four N=0 Landau levels of a Si-MOSFET (a) and temperature dependence of $-d\sigma_{xy}/dN_s$ in the $(0\downarrow+)$ Landau level (b). [18].

tive magnetoresistance experiments.[19] The temperature dependence of τ/τ_e is favorable for the former case but the value of τ_e is too small comparing with the weak field case. The problem remained unsolved until recently.

5. Scaling relation between the conductivity tensor components

Ando demonstrated by numerical experiments that σ_{xy} and σ_{xx} are not independent scaling variables but that they are mutually correlated.[20] This relation between σ_{xx} and σ_{xy} depends on the Landau quantum number N. Ando's result contradicts the conclusion obtained from the nonlinear σ model.[21, 22] Experimental results for the $(0\downarrow+)$ Landau level at eight temperatures between 4.2 K and 1.5 K, a part of which are shown in Fig.2(a), indicated that $(\sigma_{xx}, -\sigma_{xy})$ data at fixed gate voltages flow on a single curve at lower temperatures.[23] The experimental examination of the correlation between σ_{xx} and σ_{xy} was extended to higher N up to the $(1\downarrow-)$ level, and to lower T down to 0.35 K as shown in Fig.4.[24] It should be noted that the scale of the ordinate in Fig.4(b) is twice the size of the scale in Fig. 4(a). The flow lines in Fig.4(a) are close to those calculated by Aoki and Ando[25] based on the correlation between σ_{xx} and σ_{xy} which Ando obtained[20, 26] and the cutoff length given as

$$L_e \sim (\tau_e/\tau)^{1/2}\ell_0 . \qquad (6)$$

This cutoff length is determined by the condition that the spacing of energy levels is comparable with the broadening of each level due to inelastic scattering. Aoki and Ando[25] took

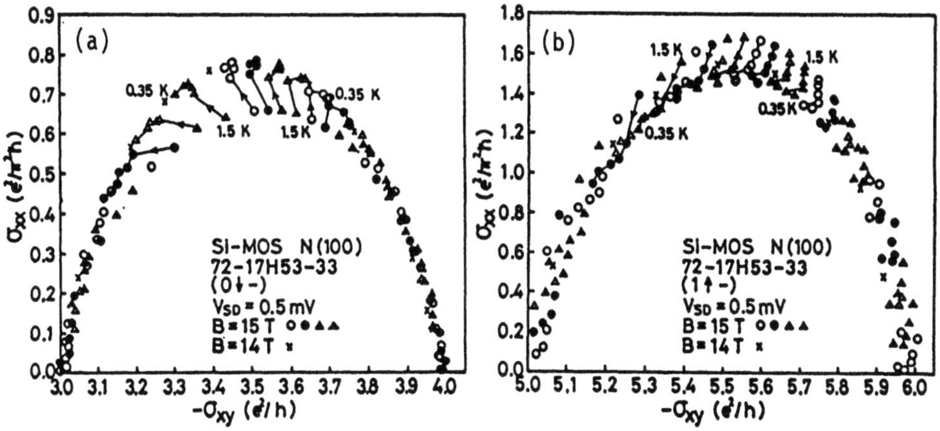

Fig. 4. Diagonal conductivity σ_{xx} vs Hall conductivity $-\sigma_{xy}$ in the (0↓-) Landau level (a) and in the (1↑-) Landau level (b) in a Si-MOSFET at temperatures of 1.5 K, 1.1 K, 0.87 K, 0.65 K, 0.50 K and 0.35 K in a magnetic field of 15 T. Crosses represent (σ_{xx}, $-\sigma_{xy}$) in 14 T at 0.35 K. [24]

Eq.(6) which gives smaller values than Eq.(5). Figure 6(b) is also close to that calculated by Aoki and Ando [25] except the behaviour near the center of the figure.

6. Critical exponent of localization studied by the temperature dependence of the Hall conductivity.

The correlation between σ_{xx} and σ_{xy} examined in ref.[24] suggested that the temperatures in refs. 15 and 18 are not low enough to analyze the T-dependence of σ_{xy} in connection with localization. Experimental data of ref. 24 were analyzed and the critical exponents of localization in Landau levels with N= 0 and 1 were derived. [27]

The Gaussian density of states given as

$$D(E) = (eB/h\Gamma)(2/\pi)^{1/2}\exp[-2(E/\Gamma)^2] \tag{7}$$

was used to calculate $-d\sigma_{xy}/dN_s$ based on Eq.(4) assuming that D(E) and E_c are symmetric with respect to the center of the Landau level and that Γ does not change with ν.

The T-dependence of the E_c/Γ values which reproduces $-d\sigma_{xy}/dN_s$ for the (0↓-) and the (1↑-) Landau levels is plotted on a log-log scale in Fig. 5. In Fig. 5, the reciprocal maximum values $[-d\sigma_{xy}/dN_s]_M^{-1}$ are added to the original plot in ref.27 since the combination of Eqs. (4) and (7) leads to the following result at the low T limit; i.e., $[(-d\sigma_{xy}/dN_s)_{MAX},\ T \rightarrow 0]^{-1} = (h/e^2)(eB/h)(8/\pi)^{1/2}(E_c/\Gamma)$. When we describe the T-dependence of E_c/Γ as $E_c/\Gamma \propto T^q$, we have q=0.25±0.04 for the (0↓-) Landau level and q=0.15±0.04 for the (1↑-) Landau level. Exponents of $[-\sigma d_{xy}/dN_s]_M^{-1}$ are q=0.18±0.05 for the (0↓-) Landau level and q = 0.09±0.05 for the (1↑-) Landau level.

The energy dependence of the localization length has been numerically studied by the Thouless number method by

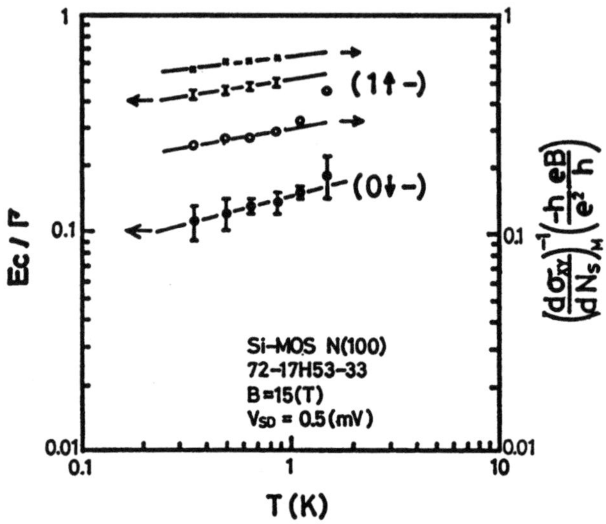

Fig. 5. Temperature dependence of the effective mobility edge E_c normalized by the broadening Γ, and of the reciprocal maximum of $-d\sigma_{xy}/dN_s$ in the (0↓-) and (1↑-) Landau level in a Si-MOSFET in 15 T. [27].

Ando [17] and by the finite size scaling method by Ando and Aoki [28] and Aoki and Ando.[29] Their results have shown a critical behaviour of the localization length as

$$\alpha(E) \propto |E|^s. \qquad (8)$$

The critical exponent is $s \gtrsim 2$ for the Landau level with N=0 and $s \gtrsim 4$ for N=1, respectively. When we take a cutoff length given by Eq.(6), the critical behaviour given by Eq.(8), and assume a relation $1/\tau_e \propto T^p$, the T-dependence of E_c/Γ is expected to be q=p/2s.

When we use s=2 and s=4 for the (0↓-) and the (1↑-) Landau level, respectively, the results in Fig. 5, i.e., q= 0.25(0.18) and q= 0.15(0.09) for the (0↓-) and the (1↑-) Landau level, respectively, give us p=1.0(0.7) for the former and p=1.2(0.7) for the latter. These results for the temperature dependence are in good agreement with the result obtained in weak magnetic fields. [19]

These results show that the critical behaviour of localization obtained by Ando and Aoki's numerical studies is consistent with our experiments.

The experiment was extended to higher magnetic fields up to 27 T and to lower temperatures down to 0.05 K [30] as shown in Fig. 6. However, similar results were obtained in the temperature dependence except that saturation appeared at temperatures lower than 0.2 K.

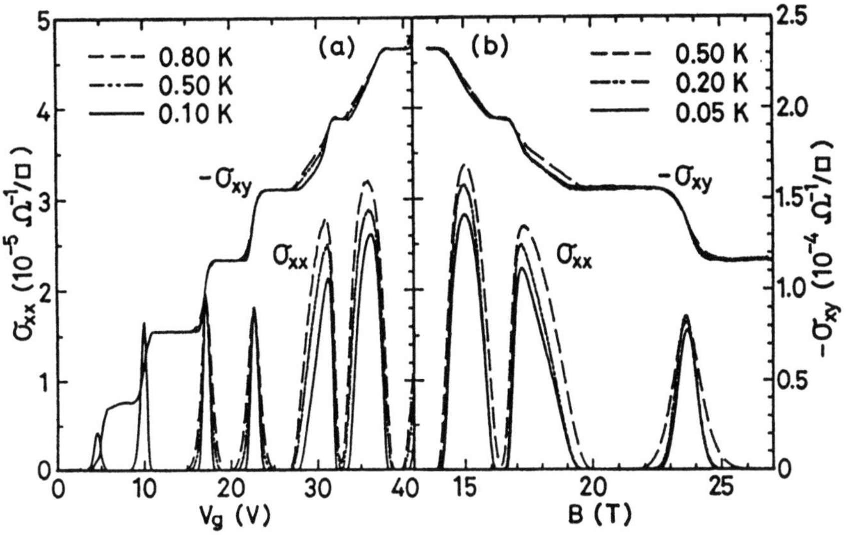

Fig. 6. Diagonal conductivity σ_{xx} and Hall conductivity σ_{xy} as a function of gate voltage V_G (a) and as a function of magnetic field B (b) at three temperatures, respectively.[30]

7. Discussion

Several problems remain unsolved of the localization in 2D systems of n-channel Si inversion layers in strong magnetic fields described in the former sections.

In the quantum Hall effect, extended states are necessary in each Landau level as well as localized states. The peak values of σ_{xx} in the $(0\downarrow+)$ and the $(0\downarrow-)$ Landau level in B=15 T below 0.5 K are about 1.6 times the SCBA value of $e^2/2\pi^2\hbar$[16] as shown in Fig. 5. The peak values in 25 T at T<0.2 K are almost the same as this value; i.e., $1.6x(e^2/2\pi^2\hbar)$ and $1.5x(e^2/2\pi^2\hbar)$ in the $(0\downarrow+)$ and the $(0\downarrow-)$ Landau level, respectively. Hikami carried out a direct summation of the perturbation series for σ_{xx} and obtained a value which is 1.4 times the SCBA value.[31] This is close to our experimental results.

The peak value of σ_{xx} in the $(1\downarrow+)$ Landau level in B=15 T at 0.35 K is close to the SCBA value for N=1. However, the peak value is still decreasing with decreasing temperature. The peak value of σ_{xx} in the same Landau level in 25 T saturates around 0.2 K and the saturation value is 0.7 times the SCBA value. The effect of the mixing of the states between Landau levels with opposite valley indices is a possible origin of the observation of low values of σ_{xx} peaks. Ando's numerical results for the level-mixing effect show that the energy of the extended states is shifted to the higher energy side with increasing the degree of the mixing[17] and finally merge and disappear at a critical degree of mixing[32]. Strongly asymmetric shape of the $(\sigma_{xx}, -\sigma_{xy})$ diagrams in the $(1\uparrow+)$ and $(1\downarrow+)$ in ref. 24 shows the presence of the level mixing. Further experiments are needed in higher magnetic fields to solve the problem. Also advanced theoretical studies as carried out by Hikami for the N=0 case are needed for the σ_{xx} peak in the N=1 Landau level.

Though the σ_{xx}(peak, N=0) increases and the σ_{xx}(peak, N=1) decreases with decreasing temperature, they have saturation values in magnetic fields up to 25T; $0.8\times(e^2/\pi^2\hbar)$ in the ($0\downarrow -$) Landau level and $1.1\times(e^2/\pi^2\hbar)$ in the ($1\uparrow -$) Landau level. And though the peak value depends on the spin direction and the valley index in the Landau levels with the same Landau quantum number, the difference between those in the N=0 and N=1 Landau level is appreciable in our experiments so far carried out.

Concerning the T-driven flow lines in (σ_{xx}, $-\sigma_{xy}$) diagrams, our results do not show many flow lines but a single curve at the low temperature limit as shown in Fig. 4 and results of experiments in higher magnetic fields [33].

The behaviour of the peak σ_{xx} values and the T-driven flow lines in (σ_{xx}, $-\sigma_{xy}$) diagrams is not in accord with the prediction of the non-linear σ model[21, 22] but in favour of Ando's numerical result [26].

Kravchenko and Pudalov [34] and Kravchenko et al.[35] carried out experiments on (σ_{xx}, $-\sigma_{xy}$) diagrams of Si-MOSFETs with high electron mobility. However, their peak values of σ_{xx} in the N=0 Landau level are much smaller than those we obtained.

Wei et al. presented an interesting result of the critical exponent for heterostructures where the exponent in T-dependence of $d\rho_{xy}/dB$ is 0.43 which does not depend on the Landau quantum number N. [36] Further, Engel et al. presented that the same exponent was obtained in the fractional quantum Hall effect.[37] These results are in conflict with our data.

It is noted here that our result is based on Hall current measurements and other groups' results are based on Hall voltage measurements. Recent measurements of potential distributions by Fontein et al.[38] are worth to be considered to understand the results of transport experiments. As a suggestive work, McEuen et al.'s recent article[39] is cited here.

In conclusion there are still many interesting problems in the transport properties of 2D systems in strong magnetic fields.

Acknowledgements The author thanks J. Wakabayashi for helpful discussions. He is also indebted to T. Ando and H. Aoki for illuminating discussions. This work is supported by Grants-in-Aid from the Ministry of Education, Science and Culture.

References

[1] K. von Klitzing, G. Dorda and M. Pepper: Phys. Rev. Lett. **45** 494(1980).
[2] S. Kawaji and J. Wakabayashi: Surf. Sci.**58** 238(1976).
[3] S. Kawaji and J. Wakabayashi: Solid State Commun. **22** 87 (1977).
[4] S. Kawaji, J. Wakabayashi, M. Namiki and K. Kusuda: Surf. Sci. **73** 121(1978)
[5] For example, see Th. Englert and K. von Klitzing: Surf. Sci. **73** 70(1978).
[6] S. Kawaji: Surf. Sci. **73** 46(1978).
[7] J. Wakabayashi and S. Kawaji: J. Phys. Soc. Jpn. **44** 1839 (1978).
[8] J. Wakabayashi and S. Kawaji: Surf. Sci.**98** 299(1980).
[9] J. Wakabayashi and S. Kawaji: J. Phys. Soc. Jpn. **48** 333 (1980).

[10] S. Kawaji and J. Wakabayashi: Physics in High Magnetic Fields (Proc. Oji Int. Seminar on the Application of High Magnetic Fields, Hakone, Sept. 1980), eds. S. Chikazumi and N. Miura(Springer Series in Solid State Sciences, Vol. 24, Springer-Verlag, 1981) p. 284.
[11] S. Kawaji: Proc. Int. Symp. Foundation of Quantum Mechanics,Tokyo, 1983, eds. F. Kamefuchi et al.(Physical Society of Japan, 1984), p.327.
[12] T. Ando, Y. Matsumoto and Y. Uemura: J. Phys. Soc. Jpn. **39** 279(1975).
[13] H. Aoki: J. Phys. **C10** 2583(1977); **C11** 3923(1978).
[14] T. Ando: Surf. Sci. **113** 182(1982.)
[15] S. Kawaji, J. Wakabayashi and J. Moriyama: J.Phys.Soc. Jpn. **50** 3839(1981).
[16] T. Ando and Y. Uemura: J. Phys. Soc. Jpn. **36** 959(1974).
[17] T. Ando: J. Phys. Soc. Jpn. **52** 1893(1983); **53** 3101, 3126 (1984)
[18] J. Moriyama and S. Kawaji: Solid State Commun. **45** 511 (1983).
[19] Y. Kawaguchi and S. Kawaji: J. Phys. Soc. Jpn. **48** 699 (1980); Surf. Sci. **113** 113(1983).
[20] T. Ando: Surf. Sci. **170** 243(1986).
[21] H. Levine, S. B. Libby and A. M. M. Pruisken: Phys. Rev. Lett.**51** 1915(1983);Nucl.Phys. **B240** [FS12] 30; 49; 71 (2984).
[22] A. M. M. Pruisken: Phys. Rev. **32B** 2636(1985).
[23] S. Kawaji and J. Wakabayashi: J. Phys. Soc. Jpn. **56** 21 (1987).
[24] M. Yamane, J. Wakabayashi and S. Kawaji: J. Phys. Soc. Jpn. **58** 1899(1989).
[25] H. Aoki and T. Ando: Surf. Sci. **170** 249(1986).
[26] T. Ando: J. Phys. Soc. Jpn. **55** 3199(1986).
[27] J. Wakabayashi, M. Yamane and S. Kawaji: J. Phys. Soc. Jpn. **58** 1903(1989).
[28] T. Ando and H. Aoki: J. Phys. Soc. Jpn. **54** 2238(1985).
[29] H. Aoki and T. Ando: Phys. Rev. Lett. **54** 831(1985).
[30] J. Wakabayashi, A. Fukano, S. Kawaji, Y. Koike and T. Fukase: Surf. Sci. **229** 60(1990).
[31] S. Hikami: Phys. Rev. **B29** 3726(1984).
[32] T. Ando: Phys. Rev. **B40** 5325(1989).
[33] J. Wakabayashi, S. Kawaji, Y. Koike and T. Fukase: unpublished.
[34] S. V. Kravchenko and V. M. Pudalov: JETP Lett. **50** No.2 (1989).
[35] S. V. Kravchenko, B. K. Medvedev, V. G. Mokerov, V. M. Pudalov, D. A. Rinberg and S. G. Semenchinsky: Surf. Sci. **229** 63(1990).
[36] H. P. Wei, D. C. Tsui, M. Paalanen and A. M. M. Pruisken: Phys. Rev. Lett. **61** 1294(1988).
[37] L. Engel, H. P. Wei, D. C. Tsui and M. Shayegan: Surf. Sci. **229** 13(1990).
[38] P. F. Fontein, P. Hendriks and J. H. Wolter: Surf. Sci. **229** 47(1990).
[39] P. L. McEuen, A. Szafer, C. A. Richter, B. W. Alphenaar, J. K. Jain, A. D. Stone, R. G. Wheeler and R. N. Sacks: Phys. Rev. Lett. **64** 2062(1990).

Full Localization of the 2D Electron Gas in Si MOSFETs at 30 mK and at High Magnetic Fields

M. D'Iorio[1], *V.M. Pudalov*[2], *and S.M. Semenchinsky*[2]

[1]Division of Physics, National Research Council of Canada, Ottawa, Canada, K1A 0R6

[2]Institute for the Metrological Service, Andreevskaya nab. 2, SU-117334 Moscow, USSR

Abstract: Scaling theory predicts that the number of delocalized states in the quantum Hall regime should be vanishingly small as the temperature approaches zero. This has been observed previously only in AlGaAs/GaAs and other types of heterojunctions and not heretofore in Si-MOSFETs. We report on measurements in low and high mobility Si-MOSFETs in the temperature range from 30 mK to 4.2 K and in magnetic fields up to 14 T. Full localization of the 2D electron gas is observed in the low temperature limit; however, the scaling behaviour of the magnetoconductivity is seen to be dependent on the Landau level number N.

1. Introduction

One of the key elements to the existence and understanding of the quantum Hall effect is the presence of delocalized or extended states n_D. Scaling theory [1] predicts that the ratio n_D/n_H of delocalized states to the Landau degeneracy where n_H = eH/hc will be vanishingly small as the temperature approaches zero. This has indeed been confirmed by many experiments on GaAs/AlGaAs and other types of heterojunctions where values of n_D/n_H as low as 0.05 or less have been reported.

However, in all experiments reported to date on Si-MOSFETs, n_D/n_H has not been observed to decrease below 0.3 - 0.4, even with an extrapolation to zero temperature. The first aim of our experimental work was to study this discrepancy.

2. Experiments

We report on measurements obtained on two samples with different peak mobility values: Si 2-1 with μ=18700 cm^2/Vs and Si 2-9b with μ = 43000 cm^2/Vs. These samples approximate two limiting cases: in the lower mobility sample, the electron interaction with the random background potential predominates; while in the other sample the electron-electron interaction plays an important if not dominant role [2]. The samples were inserted into a top loading dilution refrigerator yielding temperatures in the range from 30 mK to 4.2 K. The Hall and longitudinal resistances were measured as a function of gate voltage V_g, at constant magnetic field and temperature; an AC technique was used with channel currents in the range 10-20 nA at 13 Hz.

Springer Series in Solid-State Sciences, Vol. 101
High Magnetic Fields in Semiconductor Physics III Editor: G. Landwehr

3. Results and Discussion

3.1 Full Localization

As seen in Fig.1, at 30 mK and 14 Tesla the quantized Hall resistance plateaus are very wide and strikingly similar in appearance to those observed in GaAs/AlGaAs heterojunctions. Indeed, for the filling factor $\nu = 2$, the maximum plateau width is measured to be 0.97 n_H. Using the width of the plateaus as a measure of the number of localized states n_L, the temperature dependence of the ratio of localized states to the Landau level degeneracy n_L/n_H can be plotted for various filling factors as illustrated in Fig. 2 for sample Si 2-1. The respective values of n_D/n_H

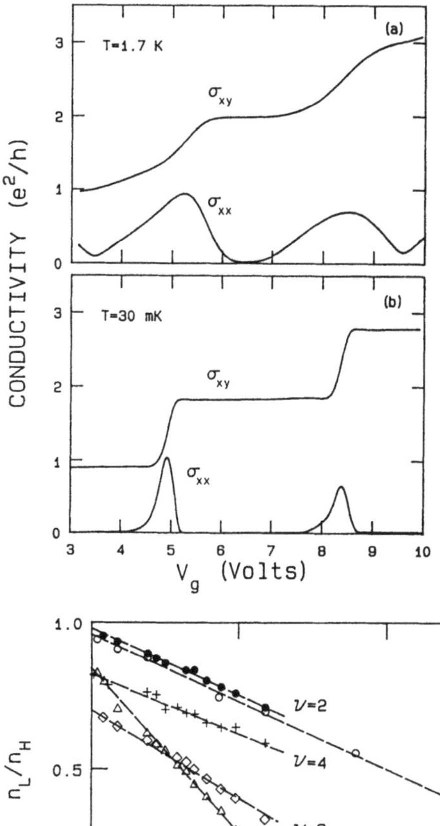

Fig. 1 Conductivities σ_{xx} and σ_{xy} versus gate voltage for sample Si 2-1 at a magnetic field of 14 Tesla, for T =1.7 K (a) and T = 30 mK (b).

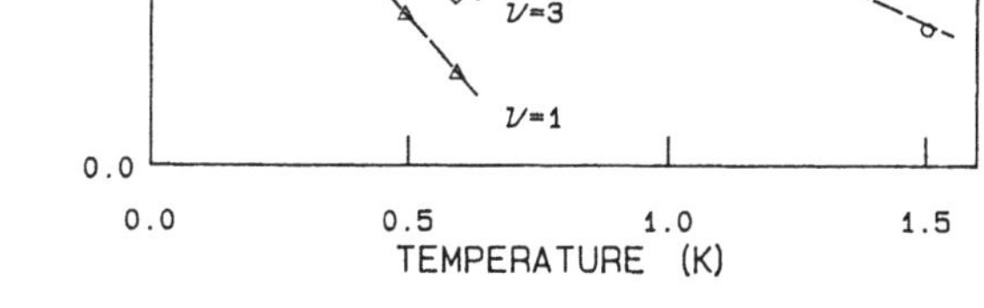

Fig. 2 Temperature dependence of the number of localized states for various filling factors. Data at two magnetic fields are shown for ν = 2 : ● for B = 14 T and ○ for B = 8 T.

extrapolated to zero temperature are 0.17 for $\nu=1$, 0.01 for $\nu=2$ and 0.16 for $\nu=4$ at 14 Tesla. These values depend only slightly on the magnetic field (see Fig.2). These new experimental results are consistent with scaling theory [1] and with previous results in GaAs/AlGaAs and InGaAs/InP heterostructures [4].

3.2 Scaling Behavior of the Magnetoconductivity

To further characterize the localization in these samples, the temperature dependence of the magnetoconductivity variation with gate voltage $\partial\sigma_{xy}/\partial V_g$ was analyzed. This quantity is expected to diverge as T^{-k} due to the divergence of the localization length $\zeta \propto (E-E_c)^{-s}$ near the centre of the Landau level E_c. In the two-parameter scaling theory [1], the exponent s is a universal critical exponent while in the numerical calculations by Aoki and Ando [3], the exponent s depends on the Landau level number N. Whereas the experimental results of Wei et al [4] for InGaAs/InP structures could be fitted to a universal exponent $k = 0.42 \pm 0.04$ independent of N, the measurements of Wakabayashi et al [5] for Si-MOSFETs yielded an exponent k which depended on the Landau level number.

Our results drawn from the two samples (Si 2-1, Si 2-9b) with different mobility are summarized in Table 1 for various filling factors ν. The fitted exponent k has been extracted from the measured temperature dependence of $\partial\sigma_{xy}/\partial V_g$ at 14 Tesla over the temperature range 0.2 - 3.2 K. We observe that (i) the exponent does not depend on the mobility of the sample and (ii) the value of k depends on the Landau level number N, as seen in Table 1.

Table 1. Fitted temperature dependence T^{-k} of $\partial\sigma_{xy}/\partial V_g$, for filling factors ν

k	ν	N
0.61 ± 0.04	1.5	1
0.60 ± 0.04	2.5	1
0.65 ± 0.04	3.5	1
0.43 ± 0.08	4.5	2
0.20 ± 0.10	5.5	2
0.34 ± 0.08	6.5	2

Our experimental results on Si MOSFETs therefore seem to contradict the scaling theory predictions [1] confirmed by experimental data on other types of heterojunctions [4]. In this respect our results are consistent with previous measurements on Si-MOSFETs [5]. However, we cannot exclude the possibility that this discrepancy could be dependent on the magnetic field strength since the electron scattering in the investigated samples is caused mostly by long-range variations of the potential for high filling factors and by short-range impurity scattering for lower values of ν. Thus the variation of k with ν observed for our samples might be expected to decrease if the magnetic field were to be increased dramatically.

4. References

1. A.M.M. Pruisken: Chap. 5 in **The Quantum Hall Effect** ed. by R.E. Prange and S.M. Girvin (Springer-Verlag, New York, 1987)
2. S.V. Kravchenko, V.M. Pudalov, D.A. Rinberg, and S.G. Semenchinsky: Phys. Rev. B (1990) in press
3. H. Aoki and T. Ando: Phys. Rev. Letters **54** , 831 (1985)
4. H.P. Wei, D.C. Tsui, M.A. Paalanen, and A.M.M. Pruisken: Phys. Rev. Letters **61**, 1294 (1988)
5. J. Wakabayashi, M. Jamane, and S. Kawaji: J. Phys. Soc. Japan **58**, 1903 (1989)

Electronic States in 2D Random Systems in High Magnetic Fields

Y. Ono[1], *T. Ohtsuki*[2], *and B. Kramer*[2]

[1]Department of Physics, Faculty of Science, Toho University, Miyama 2-2-1, Funabashi, Chiba 274, Japan
[2]Physikalisch-Technische Bundesanstalt, Bundesallee 100, W-3300 Braunschweig, Fed. Rep. of Germany

Abstract. The present status of the theoretical studies of electronic states in 2D disordered systems subject to strong magnetic fields is reviewed with stress on the fractal structures of delocalized states. The effect of confinement is discussed in connection with the behavior of edge states. The importance of the correlation length of disorder or the potential range of impurities in considering electronic structures is pointed out.

1. Introduction

Full understanding of the transport or other physical properties of quasi-2D electron systems such as MOSFETs or heterostructures requires some detailed knowledge of the wave functions in disordered potentials. As was shown by the single parameter scaling theory of the Anderson localization, all electronic states in disordered 2D systems are localized in vanishing magnetic fields, at least as far as the disorder is caused by non-magnetic impurities [1]. In disordered systems, the effect of magnetic fields has two aspects. The first is to induce a kind of localization through bending the electronic loci by the Lorentz force. The same bending effect acts to weaken or destroy the localization due to disorder. This second aspect of the magnetic field effect explained quite beautifully the negative magneto-resistance observed in many semiconductors [2,3]. Based on several theoretical investigations [4-6], however, it is now widely believed that a weak magnetic field cannot completely destroy the 2D Anderson localization. On the contrary, 2D systems in strong magnetic fields have delocalized states; there is a very delicate balance of these two aspects. The existence of delocalized states is strongly suggested in the limit of extremely strong magnetic fields, in which the electronic loci correspond to the equipotential lines of the random potential [7-13]. The theory of percolation tells us that there exists an equipotential line which extends over the whole system with infinite size [14].

The fact that, in the strong field limit where the Landau level separation is much larger than the subband width, each Landau subband has extended states only at its center has been first pointed out by one of the present authors [15] by using partial sums of diagrams for the diffusion propagator similar to Vollhard and Wölfle's treatment in vanishing magnetic field [16]. Later this was confirmed by numerical works [17-19], the perturbational series expansion method [20], the renormalization group treatment of a non-linear σ model [21], and also by experiments [22-24].

Springer Series in Solid-State Sciences, Vol. 101
High Magnetic Fields in Semiconductor Physics III   Editor: G. Landwehr

At the very tails of Landau subbands, the states are expected to be localized with a Gaussian decay [25]. The resistivity in the tail region observed experimentally shows a temperature dependence compatible with variable range hopping in Gaussian localized states [26]. In the intermediate region between the centers and the tails of Landau subbands, the states are expected to be localized exponentially as in vanishing magnetic fields [17]. As for the properties of extended states, not much is known. Recent studies, however, have revealed that they have fractal character [27,28] similar as in the critical extended states at the metal-insulator transition in vanishing fields [29,30].

In order to understand the electronic structures in disordered systems, one usually considers statistical properties of energy level spectra and wave functions. The level density and the distribution of the level spacing belong to the former. The level spacing distribution is known to reflect the essential symmetry property of the system [31,32]. The typical properties of the wave functions are the electron density distributions and the moments of the density, which are written as

$$I_n = \int d\mathbf{r} |\psi(\mathbf{r})|^{2(n+1)}, \tag{1}$$

for a normalized wave function $\psi(\mathbf{r})$. Among them, I_1 is known as the inverse participation number (IPN), which is quite often used in discussions about the localization-delocalization problems of electronic states in many papers. The system size dependences of I_n's provide us with some informations about the properties of electronic states.

2. Model

The model Hamiltonian used in many theoretical studies is written in the form,

$$H = \frac{1}{2m}(\mathbf{p} + \frac{e}{c}\mathbf{A})^2 + V(\mathbf{r}), \tag{2}$$

where $\mathbf{A}$ is the vector potential giving rise to a magnetic field perpendicular to the system, $V(\mathbf{r})$ is the random potential, and m and $-e$ are the mass and the charge of an electron. The specification of the system requires not only the Hamiltonian but also the boundary conditions of the wave functions. In the purely theoretical (or analytical) studies, the most frequently assumed boundary conditions are $\psi(x, y + L_y) = \psi(x, y)$ and $\psi(\infty, y) = \psi(-\infty, y) = 0$, and usually the period L_y is taken to be infinite in the end of calculations. If the finiteness of the system size is required, the size in the x-direction is introduced by restricting the center coordinate of the cyclotron motion X within a region $0 < X < L_x$. In numerical studies, the system size is necessarily finite except for special cases. When we are interested in the bulk properties only, the edge effects are got rid of by assuming the periodic boundary conditions in both directions;

$$\psi(x, y + L_y) = \psi(x, y); \qquad \psi(x + L_x, y) = \psi(x, y). \tag{3}$$

On the other hand, if we are interested also in the edge states, then the cylinder boundary conditions are assumed;

$$\psi(x, y + L_y) = \psi(x, y); \qquad \psi(0, y) = \psi(L_x, y) = 0. \tag{4}$$

In any of the above boundary conditions, a periodic condition is assumed in the y-direction. In this situation, it is quite convenient to choose the Landau gauge for the vector potential $\mathbf{A}$, i.e. $\mathbf{A} = (0, Bx)$ with B the field strength. It should be noted here that the second condition in eq. (3) must be modified by y-dependent phase factor because the vector potential is not a periodic function of x [17].

Another type of model is the discrete lattice model with the tight-binding Hamiltonian plus the random potential, where the magnetic field or the vector potential is introduced through the Peierls substitution of a phase factor. This model is quite convenient for numerical calculations and many useful informations e.g. about the localization length can be derived. However, we do not treat this model in the present paper in order to avoid too many discussions.

When we calculate the eigenstates of the above Hamiltonian eq. (2) numerically, the most naive way is to calculate first the Hamiltonian matrices by preparing a proper orthonormal complete set of basis functions satisfying one set of the above boundary conditions. Since the calculation of the Hamiltonian matrix elements takes a non-negligible amount of CPU time, it is more convenient, if possible, to prepare directly Hamiltonian matrices in a properly chosen space of basis functions, which reproduce essential characteristics of the matrices computed from the random Hamiltonian in real space. In fact this is possible as discussed in [37,38]. It is possible to incorporate the effect of finite correlation length of the disordered potential with this random matrix model. This model allows us to perform numerical simulations in larger system sizes.

3. Edge States

In this section, we consider the cylinder model and particularly discuss the edge states which extend almost one dimensionally along the edges. In order to study the electronic states on the cylinder surface, we have performed numerical simulations using the Hamiltonian eq.(2) with the boundary conditions eq.(4). As for the random potential $V(\mathbf{r})$, we assume a sum of Gaussian type impurity potentials for technical reasons;

$$V(\mathbf{r}) = \sum_i u_i \frac{1}{\pi a^2} \exp\left(-\frac{(\mathbf{r} - \mathbf{R}_i)^2}{a^2}\right), \tag{5}$$

where $\mathbf{R}_i$ and u_i are random variables denoting the position and the potential strength of the i-th impurity. For u_i a uniform distribution in a region $[-u/2, u/2]$ is assumed. For vanishing value of a, the potential becomes a sum of δ potentials.

In Fig. 1, examples of square amplitudes of eigenfunctions obtained for a system with $a = 0$ and $\gamma = 0.2\hbar\omega_c$, where $\omega_c = eB/mc$ is the cyclotron frequency, and

$$\gamma \equiv \sqrt{\frac{n_i u^2}{\pi \ell^2 (1 + a^2/\ell^2)}}, \qquad (n_i\text{: the impurity concentration}), \tag{6}$$

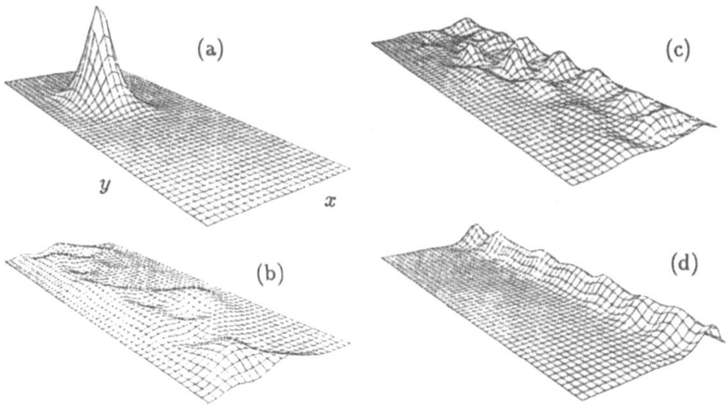

Fig. 1: Examples of electronic states in a cylinder geometry which is periodic in the y-direction and confined in the x-direction. Square amplitudes of wave functions are depicted. System parameters are $L_y = 17.1\ell$, $L_x = 7.6\ell$ and $\gamma = 0.2\hbar\omega_c$.

represents the subband width calculated within the selfconsistent Born approximation [39]. Here $\ell \equiv \sqrt{c\hbar/eB}$ is the magnetic length. The system size is specified by $L_x = 22\Delta X$ and $L_y = 52\Delta X$ with $\Delta X = 2\pi\ell^2/L_y$ the interval of the center coordinate in the Landau sense. The localized state (a) has its eigenenergy in the bottom region of the lowest bulk Landau subband and clearly shows the Gaussian form consistent with analytic calculation [25]. The state (b) is an example of delocalized ones near the center of the lowest Landau subband. The state (d) is an example of edge states which are quite similar to those in the absence of disorder except for some oscillations along the edge. Its energy lies between the lowest and the second lowest bulk Landau subbands. The state (c) has an intermediate character between the bulk extended and purely edge like states. Its energy is between the top and the center of the lowest Landau subband.

By comparing Fig. 1(d) with the edge states in the pure case, we find the edge states are rather stabel against disorder. This is because energetically degenerate states are spatially apart from each other in the case of edge states. If other states degenerate with an edge state exist near in space, quantum mechanical mixing between them will be induced by impurity potentials. This is the case when the edge states connected with the lowest Landau subband are degenerate with the bulk extended states belonging to the second lowest Landau subband, and when the two edge states belonging to different Landau subbands and existing at the same side of the system are accidentally degenerate. The latter mixing is important when one discuss the independence of different branches of edge states [40-42]. The mixing of edge states with other states is quite sensitive to the range of impurity potentials. This can be seen in Fig. 2, where the correlation between the eigenenergy ε_ν and the x-component of the center of mass $x_{\nu\nu}$ is shown for different values of a (the range of impurity potentials); $a = \ell$ on the right half and $a = 2\ell$ on the left half. In both cases, $\gamma = 0.2\hbar\omega_c$ and the system size is

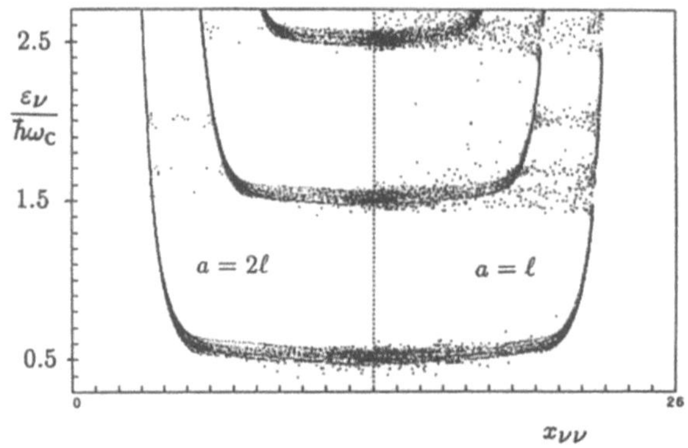

Fig. 2: Correlation between the eigenvalue ε_ν and the center of mass $x_{\nu\nu}$ (x-component) in confined systems with impurities having the range $a = \ell$ (the right half) and $a = 2\ell$ (the left half). System parameters are $L_y = 19.4\ell$, $L_x = 26\Delta X = 8.4\ell$ and $\gamma = 0.2\hbar\omega_c$.

$L_x = 26\Delta X$ and $L_y = 60\Delta X$. Details of the sample preparation are given in ref. [43]. As a matter of course, the mixing is much stronger when $a = 0$ [28]. Thus, if the effective potential range affecting the edge states is longer than the magnetic length, the independence of different branches of edge states will be satisfied in good approximation. Here, however, we should refer to the fact that a slight amount of short-ranged impurities can easily induce the mixing of the edge states in different branches. This sensitivity of the edge states to the potential range is easily understood from the consideration of the Fourier transform of the potentials. Namely, the shorter the range, the more a large momentum transfer can occur at a single scattering by a potential. This in turn makes it possible for states with different wave numbers in y-direction and therefore with different center coordinates to be mixed.

Here we should mention that edge states cannot be extended in a rigorous sense when L_y goes to infinity while L_x is fixed in the presence of disorder. This is because the system is essentially one dimensional in such a case [44].

4. Extended States in Confined Systems

Compared to the localized states, the characterization of extended states is not simple. One of the characteristics of delocalized states in disordered systems will be their fractality. For 2D systems in strong magnetic fields, Aoki [27] has first pointed out that the extended states have fractal properties like the self-similarity structures, and he estimated the fractal dimensionality to be 1.57 in the case of unconfined systems with short-ranged impurities, by employing the Hausdorff method [45]. On the other hand, percolation theories tell us that the percolating equipotential lines in 2D random systems have the dimension 1.75 [46]. The latter case is considered to correspond to the long-ranged potentials in the case of 2D disordered systems in strong magnetic fields.

We analyzed the fractal dimension D^* of electronic states in the lowest Landau subband for systems satisfying the boundary conditions eq. (4), from the system size dependences of the IPN when the ratio between L_y and L_x is fixed at 2 [28]. See the following section for the relation between the system size dependence of IPN and the fractal dimension D^*. The obtained results are $D^* = 1.39 \pm 0.04$ for a vanishing range of potentials and $D^* = 1.56 \pm 0.14$ for the range of the order of the magnetic length. From these results and the result of the percolation theory, we conjectured that D^* would increase with the potential range.

In order to understand the anisotropy of the extended states in confined systems, we have studied the behaviors of the IPN when L_x is varied while L_y is fixed and *vice versa* [47]. We obtain the effective dimension d_x^* from the former behavior and d_y^* from the latter. The energy dependences of d_x^* and d_y^* in the lowest Landau subband were studied for the case with $\gamma = 0.2\hbar\omega_c$ and $a = 0$; the system sizes were changed from $L_x = 6.5\ell$ to $L_x = 12.5\ell$ with L_y fixed at 25.1ℓ for d_x^*, and from $L_y = 15.0\ell$ to $L_y = 30.1\ell$ with L_x fixed at 7.5ℓ for d_y^*. In the edge state region, $d_x^* \approx 0$ and $d_y^* \approx 1$ as expected. d_x^* takes a maximum value ($= 0.76 \pm 0.05$) at $\varepsilon = 0.51\hbar\omega_c$ and d_y^* decreases smoothly as the energy is decreased from the edge state region. At $\varepsilon = 0.51\hbar\omega_c$, where d_x^* takes a maximum, d_y^* is 0.77 $\pm$0.09. Note that the sum of d_x^* and d_y^* at $\varepsilon = 0.51\hbar\omega_c$ is 1.53 ± 0.14 which agrees, within the statistical error, with the fractal dimension obtained in [28], i.e. $D^* = 1.39 \pm 0.04$. These results clearly show the cross over from very anisotropic (or one dimensional) extended edge states to almost isotropic extended bulk states. Furthermore it will be evident that there are intermediate states extended less anisotropically than the edge states but not so isotropically as the bulk states. Those states will typically look like Fig. 1 (c). Such states will play an important role when one considers the continuous change from bulk extended states to edge states and the relation between the edge and bulk descriptions of the quantum Hall effect [48-50].

5. Extended States in Edgeless Systems

As mentioned in the previous section, the fractal dimension of delocalized states in unconfined systems with short-ranged impurities has been studied by Aoki [27], who employed the Hausdorff dimension for the fractal dimension. In order to get deeper insight into the fractal character of bulk delocalized states, we have investigated their multi-fractality through system size dependences of the generalized IPN's I_n (eq. (1)) [38]. Edge effects are avoided by imposing the periodic boundary conditions for both directions (eq.(3)) [51]. Furthermore we have employed the random matrix model in order to carry out the simulations for larger system sizes. As for the correlation length of the random potential, which is equivalent to the potential range of a single impurity, we considered $a = 0, \ell$, and 2ℓ.

The multifractality is studied in terms of the generalized IPN's as follows. If the states are extended, I_n is expected to depend on the system size $L (= L_x = L_y)$

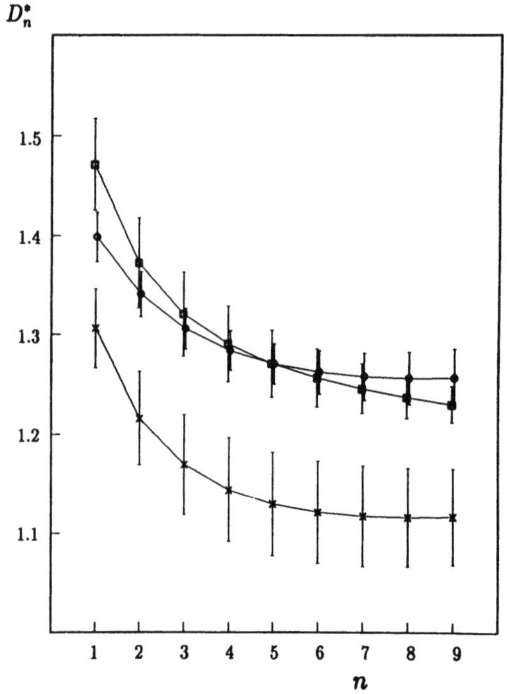

Fig. 3: Multi-fractality of most extended states in the lowest Landau subband. The fractal dimension D_n^* depends on the index n of the moment I_n. The circles, crosses and squares represent the cases with $a = 0, \ell$ and 2ℓ, respectively.

as $I_n \sim L^{\lambda(n)}$. The corresponding fractal dimension D_n^* is defined by $D_n^* = \lambda(n)/n$ and if D_n^* shows n dependence, then the states are said to be multifractal [52,53].

In Fig. 3, we show the summary of the results where the effective dimensions at the center of the lowest Landau subband, obtained by the above mentioned method, are given for three different correlation lengths [$a = 0$ (circles), ℓ (crosses), and 2ℓ (squares)] as functions of the moment index n. These data were obtained by changing the system size L from 11.2ℓ to 50.1ℓ. The subband width parameter γ was fixed to be $0.2\hbar\omega_c$. The numbers of samples were chosen so that the total number of states might exceed 3000. Because of technical reasons, we averaged I_n^{-1} instead of I_n itself over the states whose eigenenergies belong to the region $-0.05\gamma < \varepsilon < 0.05\gamma$. The error bars in Fig. 3 were estimated by the χ^2 fitting.

The multi-fractal character is clearly seen from Fig. 3. However, our previous conjecture that the fractal dimensionality may be an increasing function of the potential range could not be confirmed within the accuracy of the present calculations. In order to check whether this result is due to the artificial random matrix model, we have carried out similar calculations by the standard method, in which the random potential is first calculated by distributing impurities in real space and then the Hamiltonian matrix is calculated. The results are, however, essentially the same as in Fig. 3 [38].

6. Discussion

The present status of theoretical understanding of electronic states in 2D disordered systems in strong magnetic field was reviewed. Particular stress was put on the fractal character of delocalized states. The fractal dimensionality was discussed for two different systems; confined systems with edges and unconfined systems satisfying periodic boundary conditions in both directions. In the former case, the anisotropy of fractal characters was considered and the cross-over from the very anisotropic edge states to the almost isotropic bulk extended states was discussed. In the latter case, the random matrix model quite suitable to numerical works was employed, and the multi-fractality of the extended states was discussed. In both cases, the dependence of the fractal dimensions on the correlation length of the random potential (or equivalently the range of the impurity potentials) was considered. Unfortunately we could not confirm any systematic dependence within the present treatment.

At the moment, it is not clear enough which physical properties are affected by the fractality of extended states. As pointed out previously, the fact that the fractal dimensionality is smaller than two, the real dimensionality, means that the extended states are not space-filling [28]. Therefore, there is a possibility that localized and delocalized states can have eigenenergies quite near to each other. This is consistent with the notion that the extended states exist only at the centers of Landau subbands. In the numerical simulations of smaller system sizes, however, there exist extended states near the most extended states, and the spatial dependences of their amplitudes resemble each other as will be expected from the analogy to rather dense equipotential lines near the percolating equipotential line in 2D random potentials. This means that the phases of the corresponding wave functions are rather random, since different eigenstates of a Hamiltonian should be orthogonal to each other. The random behavior of the phases is in turn reflected presumably in the spatial fluctuations of the current distribution and then in the fluctuations of the Hall conductivity as a function of the filling factor ν in regions near $\nu = 1/2$, $3/2$, and so on. The role of detailed structures of the fractality, e.g. the multi-fractality, in determining the behavior of the physical quantities will be one of the most interesting subjects for future studies.

References

1. E. Abrahams, P.W. Anderson, C.D. Licciardello, and T.V. Ramakrishnan: Phys. Rev. Letters **42**, 673 (1979)
2. S. Hikami, A.I. Larkin, and Y. Nagaoka: Prog. Theor. Phys. **64**, 1466 (1980)
3. A. Kawabata: Solid State Commun. **34**, 432 (1980); J. Phys. Soc. Jpn. **49**, 628 (1980)
4. S. Hikami: Prog. Theor. Phys. Suppl. No. **84**, 120 (1985)
5. F. Wegner: in *Localization, Interaction, and Transport Phenomena*, ed. B. Kramer, G. Bergmann, and Y. Bruynseraede (Springer, 1985), p.99
6. Y. Ono, D. Yoshioka, and H. Fukuyama: J. Phys. Soc. Jpn. **50**,2143 (1981)
7. R. Kubo, S.J. Miyake, and N. Hashitsume: in *Solid State Physics*, ed. F. Seitz and D. Turnbull (Academic Press, New York, 1965) Vol. 17, p.269

8. M. Tsukada: J. Phys. Soc. Jpn. **41**, 1466 (1976)
9. Y. Ono: in *Anderson Localization*, ed. Y. Nagaoka and H. Fukuyama (Springer, 1982) p.207
10. S.V. Iordanskii: Solid State Commun. **43**, 1 (1982)
11. R.F. Kazarinov and S. Luryi: Phys. Rev. B**25**, 7626 (1982); S. Luryi and R.F. Kazarinov: Phys. Rev. B**27**, 1386 (1983)
12. S.A. Trugman: Phys. Rev. B**27**, 7539 (1983)
13. S.M. Apenko and Yu.E. Lozovik: J. Phys. **C18**, 1197 (1985); Sov. Phys. JETP **62**, 328 (1985)
14. D. Stauffer: *Introduction to Percolation Theory*, (Taylor and Francis, London, 1985)
15. Y. Ono: J. Phys. Soc. Jpn. **52**, 2055,3544 (1981); **52**, 2492 (1983); **53**, 2342 (1984); see also Prog. Theor. Phys. Suppl. No. **84**, 138 (1985)
16. D. Vollhard and P. Wölfle: Phys. Rev. Letters **45** 842 (1980); Phys. Rev. B**22**, 4666 (1980)
17. T. Ando: J. Phys. Soc. Jpn. **52**, 1740 (1983); **53**, 3101, 3126 (1984)
18. B. Kramer and A. MacKinnon: in *Proc. Int. Seminar " Localization in Disordered Systems ", Johnsbach/b. Dresden 1983* (Teubner, Leipzig, 1984)
19. H. Aoki and T. Ando: Phys. Rev. Letters **54**, 831 (1985); T. Ando and H. Aoki: J. Phys. Soc. Jpn. **54**, 2238 (1985)
20. S. Hikami: Phys. Rev. **B39**, 3726 (1983)
21. H. Levine, S.B. Libby, and A.M.M. Pruisken: Phys. Rev. Letters **51**, 1915 (1983); Nucl. Phys. **B240**[FS 12], 30, 49, 71 (1984)
22. N. Moriyama and S. Kawaji: Solid State Commun. **45**, 511 (1983)
23. M.A. Paalanen, D.C. Tsui, A.C. Gossard, and J.C. Hwang: Solid State Commun. **50**, 841 (1984)
24. H.P. Wei, D.C. Tsui, M.A. Paalanen, and A.M.M. Pruisken: Phys. Rev. Letters **61**, 1294 (1988)
25. Y. Ono: J. Phys. Soc. Jpn. **51**, 237 (1982)
26. G. Ebert, K. von Klitzing, C. Probst, E. Schuberth, K. Ploog, and G. Weimann: Solid State Commun. **45**, 625 (1983)
27. H. Aoki: J. Phys. **C16**, L205 (1983); Phys. Rev. **B33**, 7310 (1986)
28. Y. Ono, T. Ohtsuki, and B. Kramer: J. Phys. Soc. Jpn. 58, 1705 (1989), see also B. Kramer, Y. Ono, and T. Ohtsuki: Surf. Sci. **196**, 127 (1988)
29. C.M. Soukoulis and E.N. Economou: Phys. Rev. Letters **52**, 565 (1984)
30. M. Schreiber: Phys. Rev. **B31**, 6146 (1985)
31. M.L. Mehta: *Random Matrices and the Statistical Theory of Energy Levels* (Academic Press, New York, 1967)
32. F.J. Dyson: J. Math. Phys. **3**, 140,157, 166,1191,1199 (1962)
33. L. Schweitzer, B. Kramer and A. MacKinnon: J. Phys. C**17**, 4111 (1984)
34. H. Aoki: J. Phys. C**18**, L167 (1985)
35. D.J. Thouless, M. Kohmoto, M.P. Nightingale, and M. den Nijs: Phys. Rev. Letters **49** (1982)
36. T. Ohtsuki and Y. Ono: Solid State Commun. **65**, 403 (1988)
37. B. Huckestein and B. Kramer: Solid State Commun. **71**, 445 (1989); Phys. Rev. Letters **64**, 1437 (1990)
38. Y. Ono, T. Ohtsuki, and B. Kramer: submitted to J. Phys. Soc. Jpn.
39. T. Ando and Y. Uemura: J. Phys. Soc. Jpn. **37** 1044 (1974)
40. M. Büttiker: Phys. Rev. **B38**, 9375, 12724 (1988); Phys. Rev. Letters **62**, 229 (1989)
41. S. Komiyama and H. Hirai: Phys. Rev. **B40**, 7767 (1989)
42. H. Hirai, S. Komiyama, S. Sasa, and T. Fujii: J. Phys. Soc. Jpn. **58**, 4086 (1989)
43. T. Ohtsuki and Y. Ono: J. Phys. Soc. Jpn. **58**, 3863 (1989)
44. T. Ando: to be published in Phys. Rev. **B**

45. B.B. Mandelbrot: *The Fractal Geometry of Nature* (Freeman, San Francisco, 1982)
46. H. Saleour and B. Duplantier: Phys. Rev. Letters **58**, 2325 (1987)
47. S. Fukuda and Y. Ono: J. Phys. Soc. Jpn. **59**, 2409 (1990)
48. J. Hajdu, M. Janßen, and O. Viehweger: Z. Phys. **B66**, 433 (1987)
49. Y. Ono and B. Kramer: Z. Phys. **B67**, 341 (1987)
50. T. Ohtsuki and Y. Ono: Solid State Commun. **68**, 787 (1988); J. Phys. Soc. Jpn. **58**, 956 (1989)
51. T. Ando: Prog. Theor. Phys. Suppl. No. **84**, 69 (1985)
52. L. de Arcangelis, S. Render, and A. Coniglio: Phys. Rev. **B31**, 4725 (1985)
53. T. C. Hasley, M.H. Jensen, L.P. Kadanoff, I. Procaccia, and B.I. Shraiman: Phys. Rev. **A33**, 1141 (1986)

Localization and Scaling in the Quantum Hall Regime: Dependence on Landau Level Index and Correlation Length

B. Huckestein and B. Kramer

Physikalisch-Technische Bundesanstalt, Bundesallee 100,
W-3300 Braunschweig, Fed. Rep. of Germany

Abstract. The exponential decay length λ_M of the one-particle Greens function is calculated for a disordered two-dimensional system of finite width M under the influence of a strong magnetic field. A recursive numerical method is employed. The influence of the correlation length of the disorder potential and the Landau level index on the critical behaviour is investigated by numerical calculations and by comparing with other results available in the literature.

1 Introduction

For sufficiently high perpendicular magnetic fields the spectrum of a disordered two-dimensional system breaks into well resolved Landau bands. The localization properties of the states near the center of a Landau band are of considerable importance for the understanding of the quantum Hall effect.

A numerical scaling analysis shows that for an uncorrelated random potential the localization length $\xi(E)$ diverges in the centre of the lowest Landau band according to a power law $\xi(E) \propto |E - E_0|^{-\nu}$ with $\nu = 2.34 \pm 0.04$ [1]. Comparison with both analytical and numerical results for very large correlation lengths of the random potential [2, 3] suggested that the scaling function as well as the critical exponent ν are independent of the correlation length. In the present paper we want to compare the previous results with calculations for a random potential with a correlation length σ equal to the magnetic length $l_c = (\hbar/eB)^{1/2}$.

For higher Landau bands it has been concluded from numerical calculations that the critical exponents are different, i.e. $\nu \lesssim 4$ for the second Landau band [4]. However, in the percolation limit $\sigma \gg l_c$ it is obvious that the scaling properties cannot depend on the Landau level index [5]. We present results for a δ-correlated random potential and compare our results with previous ones.

2 The Model

A two-dimensional system of length L and width M with periodic boundary conditions in the M-direction is considered in the limit $L \to \infty$. Following the method of Ref. [1] the inverse exponential decay length

$$\lambda_M^{-1}(E) = -\lim_{k\to\infty} \frac{\pi}{k} \log |G_n(E; 0, k)| \quad (1)$$

Springer Series in Solid-State Sciences, Vol. 101
High Magnetic Fields in Semiconductor Physics III Editor: G. Landwehr

of the modulus of the propagator $G_n(E; k, k') = \langle nk|(E-H)^{-1}|nk'\rangle$ in the space of the Landau states is calculated. As in Ref. [1] the unit length was taken to be $\sqrt{2\pi}l_c$ and the unit of energy is the bandwidth Γ in the self-consistent Born approximation.

3 Finite correlation length

We have calculated $\lambda_M(E)$ for the lowest Landau band for a random potential with a correlation length $\sigma = l_c$. The system width M ranges between 1/2 and 64 and the energy between 0 and 0.5 relative to the center of the lowest Landau band. The results are shown in Fig. 1 a. For $M = 0.5$, λ_M is smaller by a factor 2 as compared to the δ-correlated potential of Ref. [1] in agreement with $\lambda_M/M \to 2/\pi(\sigma^2 + l_c^2)^{-1}$ for $M \to 0$ [4] while for the largest systems the localization length becomes smaller by a factor of about 4. For large M and $E = 0$, λ_M/M tends to the same fixed point value as in the cases $\sigma = 0$ [1] and $\sigma \to \infty$ [2]. Fig. 1 b shows the scaling function obtained by fitting the data for the δ-correlated potential from Ref. [1] for $16 \le M \le 64$ and $0.05 \le E \le 0.5$, the present data for $M = 64$ and $0.05 \le E \le 0.3$, and for the percolation limit from Ref. [2] for $32 \le M' \le 128$. The corresponding localization lengths for correlation lengths 0 and l_c are shown in Fig. 2 a. In the fit it was assumed that the critical exponents are the same in both cases and were calculated to be 2.32. A statistical test as described in [6] gives that the scaling function can be trusted while the assumption of one critical exponent is not justified even though the deviations are quite small. Increasing the number of data taken from the present calculation decreases the quality of both the fit to the scaling function and the fit of the critical exponent.

The conclusion from the observed behaviour is that system sizes of $M = 64$ are just at edge of the critical regime. Larger systems would allow for a more

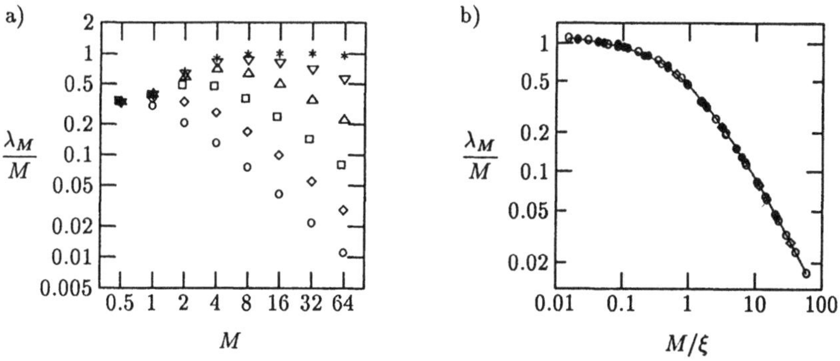

Fig. 1 a The renormalized exponential decay length λ_M/M as a function of system width M for $\sigma = l_c$ and energies $E = 0$ (∗), 0.05 (▽), 0.1 (△), 0.18 (□), 0.3 (◇) and 0.5 (○). Fig. 1 b The renormalized exponential decay length λ_M/M as a function of the scaling variable $M/\xi(E)$. Full circles (•) are data for $\sigma = 0$ [1], opencircles (○) for $\sigma \to \infty$ [2] and boxes (◇) are taken from the present calculation. The solid curve approximates the scaling function.

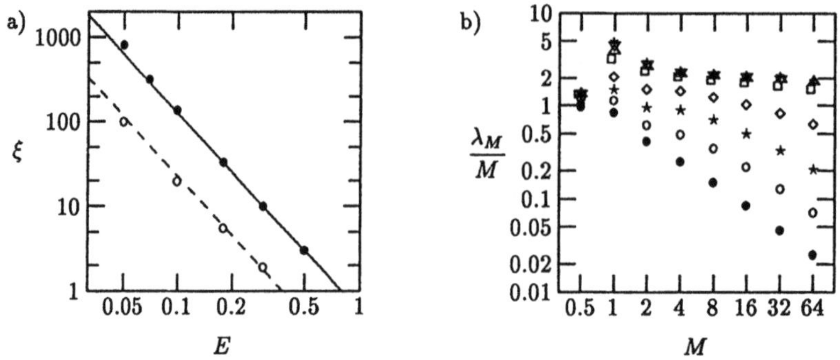

Fig. 2 a The localization length $\xi(E)$ for $\sigma = 0$ (•) and $\sigma = l_c$ (○). Fig. 2 b The renormalized exponential decay length λ_M/M as a function of system width M for $n = 1$ and energies $E = 0.01$ (∗), 0.1 (▽), 0.18 (△), 0.3 (□), 0.5 (◇), 0.65 (⋆), 0.8 (○) and 1 (•).

rigorous analysis. However, the present data do substantiate the assumption that the scaling behaviour in the lowest Landau band is independent of the correlation length of the random potential.

4 Second lowest Landau band

To get some insight into the influence of the Landau band index on the localization properties we have calculated the localization length for the $n = 1$ Landau band. From the results in Fig. 2 b it can be seen that for large system sizes M, λ_M/M shows almost no size and energy dependence for energies up to 30 % of the bandwidth. In this energy region it is not possible to obtain scaling behaviour or to perform any other reliable extrapolation to infinite system size. For larger energies there is a system size and energy dependence of λ_M/M. The energy dependence is in fact much stronger than in the lowest Landau level. From this fact it was deduced that the critical exponent is of the order of 4 in this Landau level [4, 5]. However, no scaling behaviour can be obtained from the data neither near the center of the band nor in the tails even though the present system sizes are four times larger than those previously available. Without the observation of scaling behaviour it is not meaningful to derive critical exponents. For short correlation length of the potential the present numerical analysis is inconclusive concerning the critical behaviour within higher Landau bands.

5 Conclusion

We have studied numerically in a single Landau band approximation the critical behaviour of electrons confined to the $n = 0$ and $n = 1$ Landau band, respectively. For the $n = 0$ band and a finite correlation length $\sigma = l_c$ of the random potential strictly speaking the data showed no scaling behaviour but tended towards the same scaling behaviour as the data for $\sigma = 0$ and $\sigma \to \infty$. This substantiates the assumption that the scaling behaviour within the lowest Landau level is independent of the correlation length of the potential. For the $n = 1$ Landau level and $\sigma = 0$ not even an onset of scaling behaviour was observed for the present system sizes although these are far larger than those used in previous calculations [4, 5]. It was concluded that in the absence of a scaling behaviour a meaningful calculation of the critical exponent cannot be performed and that the question of the critical behaviour in higher Landau bands still remains unsolved.

References

[1] B. Huckestein and B. Kramer, Phys. Rev. Lett. **64**, 1437 (1990), and references therein.

[2] J. T. Chalker and P. D. Coddington, J. Phys. C **21**, 2665 (1988).

[3] G. V. Mil'nikov and I. M. Sokolov, JETP Lett. **48**, 536 (1988).

[4] H. Aoki and T. Ando, Phys. Rev. Lett. **54**, 831 (1985). T. Ando and H. Aoki, J. Phys. Soc. Jpn. **54**, 2238 (1985).

[5] B. Mieck, preprint, (1990).

[6] B. Huckestein, in *Proceedings of the International workshop on Anderson transition and mesoscopic fluctuations, PTB Braunschweig, Germany, 9–12 January, 1990, Physica A (to be published).*, (1990).

Magnetic Field Induced Transitions Between Quantized Hall and Insulator States in a Dilute 2D Electron Gas

M. D'Iorio[1], *V.M. Pudalov*[2], *and S.G. Semenchinsky*[2]

[1]Division of Physics, National Research Council of Canada, Ottawa, Canada, K1A 0R6

[2]Institute for the Metrological Service, Andreevskaya nab. 2, SU-117334 Moscow, USSR

Abstract: We report on the observation of a new phenomenon: a sequence of magnetic field induced transitions between well defined Quantum Hall effect states, with a Hall resistance quantized as h/e^2 and a vanishingly small longitudinal resistance, and insulator states with longitudinal resistance exceeding 2 GΩ. This phenomenon is observed in extremely high mobility Si MOSFETs, in a range of electron concentrations corresponding to a dilute 2D electron gas in or near an activated electronic transport regime. We attribute this effect to a modulation of the metal-insulator transition by the Quantum Hall effect.

1. Introduction

The behaviour of a two-dimensional (2D) electron gas in the Quantum Hall Effect (QHE) regime has been studied extensively [1] in the range of high electron densities $n_s \geq 4\times10^{15}$ m^{-2} and high magnetic field B ~ 5 – 20 tesla. Work has also been done in the low ($\geq 2\times10^{14}$ m^{-2}) and very low ($\geq 4\times10^{13}$ m^{-2}) density regimes, in connection with the study of the fractional quantum Hall effect in GaAs/AlGaAs heterojunctions [2-5]. These experimental studies were performed in an essentially metallic range of 2D gas densities where the low scattering rate condition $\omega_c\tau >> 1$ is valid, and $\omega_c = eB/m^*$ is the cyclotron frequency and τ^{-1} is the scattering rate.

However, interesting questions remain concerning the dilute electron gas regime where $n_s \leq 1\times10^{15}$ m^{-2}. What are the limits for the existence of the QHE upon lowering the electron density and the magnetic field? Can the QHE coexist with an activated conductivity regime in the proximity of the metal-insulator transition into an insulator phase?

It can be argued that, on one hand, the resistivity of a 2D gas should increase as a function of decreasing temperature in an activated transport regime while, on the other hand, the resistivity should decrease as a function of magnetic field in the QHE regime, in those regions where the Fermi energy sweeps through energy gaps[1]. Moreover, as $\omega_c\tau$ is reduced by lowering the carrier concentration and the magnetic field, the band of extended states responsible for the dissipationless transport in the QHE regime will be pushed upwards in energy until it rises above the Fermi energy [6]. This case was analyzed theoretically by Khmel'nitskii[7], who predicted the reappearance at low magnetic fields of features in the Hall and longitudinal resistances similar to those characteristic of the QHE. The above mentioned puzzle motivated our experimental investigation of a dilute electron gas in high mobility Si MOSFETs in the low temperature-low magnetic field limit.

Springer Series in Solid-State Sciences, Vol. 101
High Magnetic Fields in Semiconductor Physics III   Editor: G. Landwehr

2.Experiments

In order to study the QHE in the dilute regime, a sample with a sufficiently high $\omega_c\tau$ value is required. A Si MOSFET sample with an extremely high peak mobility μ^{peak} = 4.2 $m^2/V\cdot s$ was chosen for this study. The carrier density could be decreased to 6×10^{14} m^{-2} with $\omega_c\tau = \mu B \sim 3$, for B = 3 tesla. The mobility curves at 30 mK and at 1.7 K are shown in Fig. 1 as a function of electron density. In this sample, the 2D electron gas is in a metallic phase at zero magnetic field in the range of concentration $n_s > 1\times10^{15}$ m^{-2} while $n_s < 1\times10^{15}$ m^{-2} corresponds to thermally activated transport. In the metallic phase, the Shubnikov-de Haas oscillations are observed over the whole range of electron densities at temperatures below 200 mK. At higher magnetic field, they are superceded by the QHE which is well defined down to $n_s \sim 1\times10^{15}$ m^{-2}. Fig. 2 shows two quantized Hall plateaus with filling factors ν = 1 and 2, and their respective R_{xx} minima in the dilute metallic region with $n_s = 1.02\times10^{15}$ m^{-2}. Even in the low concentration regime, the mobility exceeds 1 $m^2/V\cdot s$ (a value which Si MOSFETs of moderate mobility reach at higher n_s) and the value of $\omega_c\tau \sim 10$ is rather high. The features of the QHE are therefore as clearly defined as in the high-n_s / high-B regime, with the exception of a higher noise level due to the smaller current used.

An AC measurement technique at 13 Hz was used whenever the magnitude of the R_{xx} maxima was lower than 200 kΩ, while a DC technique was used to measure higher R_{xx} values. The largest amplitude of the R_{xx} maxima was found to exceed 2 GΩ over the total length of the sample L = 5 mm. All measurements were performed with very low source-drain current, typically 20-100 pA, since a notice-

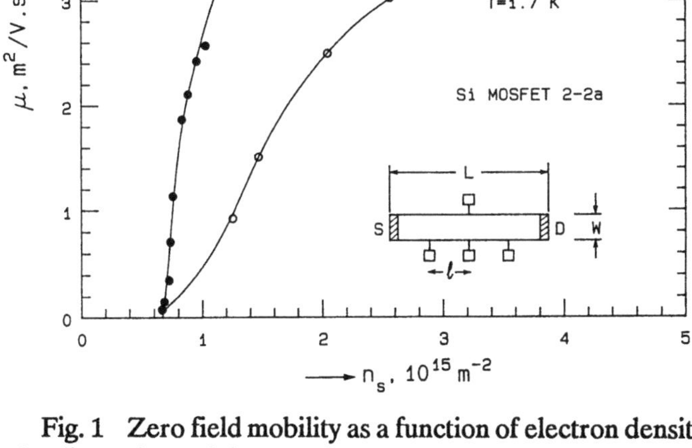

Fig. 1 Zero field mobility as a function of electron density in the dilute regime. The dimensions of the sample are L = 5000 μm, l = 1250 μm and W = 800 μm.

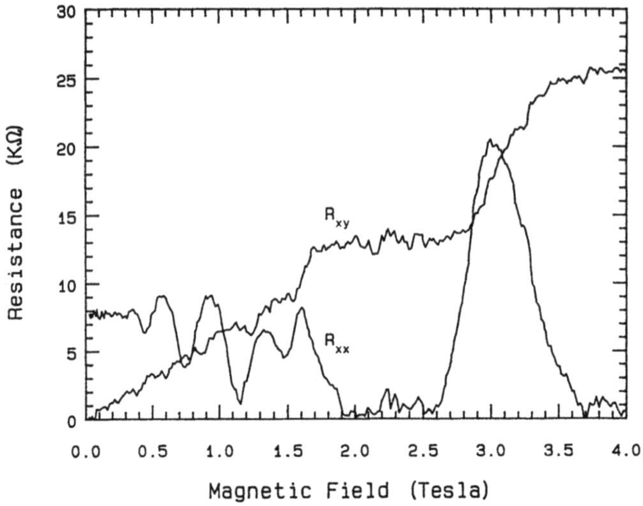

Fig. 2 Hall and longitudinal resistances as a function of magnetic field in the metallic range of density $n_s = 1.02\times10^{15}\ m^{-2}$ at 18 mK.

able current dependence of the R_{xx} maxima was observed at higher source-drain currents.

The Hall resistance was also measured using an AC technique at 13 Hz. The value of the quantized Hall resistance on a Hall plateau can be measured reliably due to the relatively small longitudinal resistance at the corresponding minimum.

3. Results

Upon lowering the carrier concentration by means of the gate voltage, we observed the unexpected growth of the R_{xx} peaks by more than five orders of magnitude for filling factors $v \sim 1.5$ and 2.5. This is illustrated in Fig. 3. These giant peaks are separated by minima, occurring at $\nu = 1$ and 2, as expected for the ordinary integer QHE. As a consequence of the growth of a maximum at $\nu \sim 1.5$, the minima at $\nu = 3$, 4 and 5 are suppressed. In the insert of Fig. 3 (B), the development of the minima at $\nu = 6$ and 8 is illustrated on an enlarged scale. The maximum at $\nu \sim 7$ replaces the unresolved minima related to the smallest valley splitting in Si.

Despite the giant R_{xx} maxima the magnitude of the Hall resistance on the plateaus is in good agreement with the quantized values $h/\nu e^2$ as seen in Fig. 3 (B). However, in the intermediate regions between the Hall plateaus, the longitudinal resistance becomes practically infinite and capacitive coupling between the 2D gas and the gate strongly affects the measured Hall voltage, changing its phase. Therefore these data are not shown in Fig. 3.

In order to convert the longitudinal resistance into resistivity values, the potential distribution was checked at six different potential contacts including the source and drain contacts; the electric potential was found to drop homogeneously and

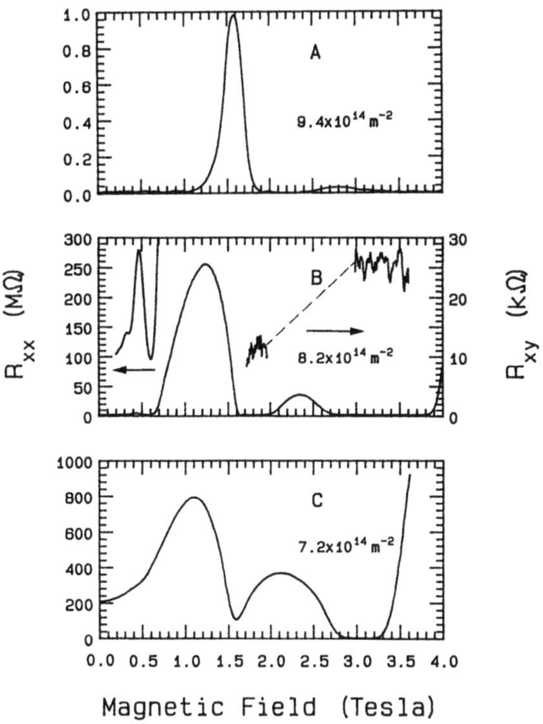

Fig. 3 Dependence of the Hall and longitudinal resistances on the magnetic field at 18 mK. The insert in (B) shows the development of novel peaks in R_{xx} around $\nu = 7$, magnified 56×. The dashed line between the $\nu = 1$ and $\nu = 2$ plateaus is meant to be a guide to the eye.

symmetrically over the entire sample as observed in the ordinary integer QHE. We are therefore convinced that: (i) the observed R_{xx} maxima are not related to a contact phenomenon and are intrinsic to the 2D gas; (ii) this effect can not be attributed to some local area within the 2D layer and must be related to the behaviour of the 2D gas over the entire length of the sample; and (iii) the measured R_{xx} oscillations are proportional to the longitudinal resistivity ρ_{xx} (see sample geometry in the insert of Fig. 1).

Upon lowering the temperature from 1 K to 18 mK, the magnitude of the R_{xx} minima decreases slowly as previously observed[1] for the integer quantum Hall effect, in contrast with the R_{xx} maxima whose values rapidly increased by 5 to 6 orders of magnitude from typical R_{xx} values (~10 kΩ) for the QHE at 1 K. The experimental results are reproducible on different cool downs and the same phenomenon has been observed in other samples of comparable quality.

4. Discussion

The infinite wall on the high field side of each curve in Fig. 3 reflects the magnetic freeze-out in the 2D electron gas in the extreme quantum limit $\nu < 1$. As seen in Fig. 3, the new effect reported here exists in a very narrow range of electron densities $n_s = 6\text{–}10 \times 10^{14}\ m^{-2}$, in a magnetic field range 0-6 tesla and at temperatures below 1 K.

The giant R_{xx} maxima ($R_{xx} > 10^9\ \Omega$) are usually characteristic of an insulator state. This state appears periodically upon sweeping the magnetic field. However, the Hall plateaus and the longitudinal resistance minima display the characteristic features of the QHE in a true metallic state. We attribute the experimental results to transitions between QHE and insulator states which are taking place homogeneously over the whole length of the sample; these transitions are stimulated by the Shubnikov-de Haas or the Quantum Hall effect modulating the background intrinsic conductivity of an initial activated transport phase.

The phase diagram for the sequence of four transitions between the QHE and insulator states at 18 mK is illustrated in Fig. 4. The straight lines are meant as a guide to the eye and show the expected positions of R_{xx} minima for the integer QHE extrapolated from the high n_s / high B region. The shaded and dotted regions

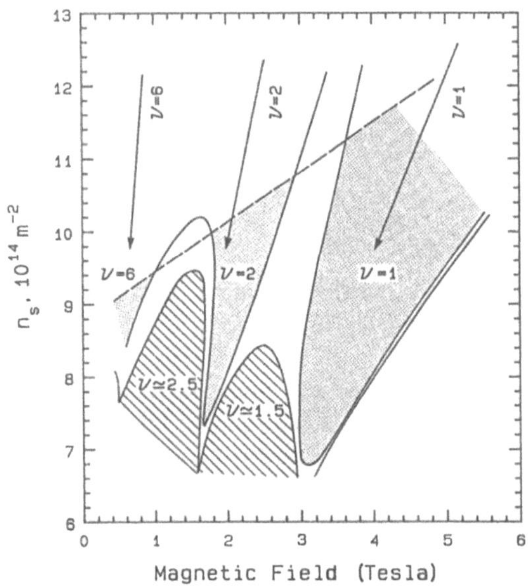

Fig. 4 Phase diagram for the coexistence of the QHE (dotted regions) and insulator states (shaded regions) at T = 18 mK. The arrows show the expected positions of the R_{xx} minima for the integer quantum Hall effect. The QHE and insulator states are confined by the contour lines R_{xx} = 10 kΩ and R_{xx} = 100 kΩ respectively. The straight dashed line corresponds to $R_{xx} = h/e^2$ and schematically separates the region of coexistence of the insulator and QHE states, above, from that of the ordinary integer quantum Hall effect, below.

correspond to the insulator and QHE states respectively. The emergence and growth of the next insulator states is observed around $\nu \sim 6$ to 8 at low fields as shown in Fig. 3 (B). The boundaries between the insulator ($\nu \sim 2.5, 1.5$) and the QHE ($\nu = 2$) states are practically vertical walls which precludes the observation of these transitions by sweeping the gate voltage.

The phenomenological explanation presented above assumes that the current distribution is homogeneous over the width W of the sample: 800 μm. With the exception of a narrow region in the mid-point of the R_{xx} minimum, the homogeneity of the current distribution in the QHE regime has been well established [1, 8-11] particularly for Si MOSFETs. However those experiments were performed in a true metallic state using rather high source-drain currents, of the order of 1 μA. An alternative explanation to the effect reported here can therefore not be rejected immediately, and relates to the presence of edge current carrying states [6]. The edge states would occupy narrow strips of width $l_B \sim 10$ nm, where l_B is the magnetic length, $l_B = \sqrt{(\hbar/eB)}$. This hypothesis could account for the large value of the R_{xx} maxima as a result of an effective width ratio $W/l_B \sim 10^5$, but it does not account for the narrow range of density and temperature over which the phenomenon exists.

It is also worth noting that the Coulomb interaction should play a significant role in accounting for this new phenomenon since it is the dominant parameter in the concentration range of interest: $e^2\sqrt{n_s}/\varepsilon \sim 70$ K where ε is the dielectric constant. The Landau level splitting energy, $\hbar\omega_c$, is approximately 10 K. Therefore, we can not reject the possibility of forming (or melting) the Wigner crystal at each boundary magnetic fields which separate the QHE and the insulator states (Fig. 3 and 4).

Further experimental investigations and theoretical calculations should shed some light on the origin of this new phenomenon in the dilute 2D electron gas.

5. References

1. The Quantum Hall Effect, edited by R. E. Prange and S. M. Girvin, (Springer-Verlag, New York, 1987).
2. E. E. Mendez, M. Heiblum, L. L. Chang and L. Esaki, Phys. Rev. B **28**, 4886 (1983).
3. V. J. Goldman, M. Shayegan and D. C. Tsui, Phys. Rev. Letters **61**, 881 (1988).
4. R. L. Willett, H. L. Stormer, D. C. Tsui, L. N. Pfeiffer, K. W. West and K. W. Baldwin, Phys. Rev. B **38**, 7881 (1988).
5. T. Sajoto, Y. W. Suen, L. W. Engel, M. B. Santos and M. Shayegan, Phys. Rev. B **41**, 8449 (1990).
6. B. I. Halperin, Phys. Rev. B **25**, 2185 (1982).
7. D. E. Khmel'nitskii, Phys. Lett. A **106**, 182 (1984).
8. T. Ando, A. B. Fowler and F. Stern, Rev. Mod. Phys. **54**, 437 (1982).
9. V. M. Pudalov and S. G. Semenchinsky, JETP Lett. **42**, 232 (1985).
10. S. G. Semenchinsky, Sov. Phys. JETP **64**, 1068 (1986).
11. O . Heinonen and P. L. Taylor, Phys. Rev. B **32**, 633 (1985).

The Frequency-Dependent Deformation of the Hall Plateaus

O. Viehweger and K.B. Efetov

Max-Planck-Institut für Festkörperforschung, Heisenbergstr. 1, W-7000 Stuttgart 80, Fed. Rep. of Germany

Abstract. We calculate the Hall conductivity in the low frequency limit for a 2d model of independent electrons in the tails of the density of states between the disorder broadened Landau levels. It is shown that there are corrections to the quantized values of σ_{yx} which are proportional to ω^2 and to the number of localized states in a given tail of the band. We relate our theoretical results to microwave experiments on GaAs-AlGaAs heterostructures.

In a recent article [1] we have calculated the frequency dependence of the Hall conductivity in the lowest localization regime of a 2d as well as a 3d noninteracting electron gas. The question arises whether the result in two dimensions may be generalized and applied to regimes where the Hall conductivity is quantized at $\omega = 0$.

Experimental data from transmission measurements of GaAs-AlGaAs heterostructures in microwave-guides [2-4] indicate that the Hall plateaus seem to persist up to frequencies above 30GHz. Nevertheless, their width decreases with increasing frequency and within the experimental accuracy it is not possible to decide whether the plateaus are really flat or whether they develop a finite slope.

We consider noninteracting electrons in two dimensions under the influence of a perpendicular magnetic field B, i.e. a system described by the one-particle Hamiltonian

$$\begin{aligned} H &= H_0 + V(r) \\ H_0 &= \frac{1}{2m}(\mathbf{p} - e\mathbf{A})^2 \quad , \quad \mathbf{A} = \frac{1}{2}B(-y, x, 0) \end{aligned} \tag{1}$$

with a white noise random potential V

$$\overline{V(\mathbf{r})} = 0 \quad , \quad \overline{V(\mathbf{r}_1)V(\mathbf{r}_2)} = \lambda\delta(\mathbf{r}_1 - \mathbf{r}_2) \tag{2}$$

The magnetic field is assumed to be strong in the sense that the disorder induced broadening of the Landau levels (LLs) is small compared to the Landau level distance. Consequently we neglect interactions between different subbands and discuss only the lowest Landau level (LLL). In the tail region the density of states reads

$$\rho(E) = \frac{2}{\sqrt{\pi}} \frac{1}{2\pi\ell^2} \frac{1}{\Gamma} \left(\frac{|\varepsilon|}{\Gamma}\right)^2 \exp\left(-\left(\frac{|\varepsilon|}{\Gamma}\right)^2\right) \tag{3}$$

with $\varepsilon = E - \hbar\omega_c/2$, $\Gamma = (2\pi\ell^2/\lambda)^{-1/2}$ and ℓ denotes the magnetic length. For later use let us recall the low frequency asymptotics of the Kubo conductivities obtained in [1] for the Fermi energy being situated in the lower tail of the LLL, i.e. $E_F \ll \hbar\omega_c/2$.

$$\sigma_{xx}(\omega) = -ie^2\omega \cdot 2\ell^2 \rho(E_F) \quad , \quad \sigma_{yx}(\omega) = \frac{e^2}{h} \frac{\hbar^2\omega^2}{\Gamma^2} \cdot 8\pi\ell^2 n \tag{4}$$

where n denotes the number of electrons per unit area. These results have been obtained in the one-instanton approximation following the method developed by Affleck [5] for the calculation of the density of states.

Springer Series in Solid-State Sciences, Vol. 101
High Magnetic Fields in Semiconductor Physics III Editor: G. Landwehr

One of the main difficulties in previous discussions of instantons between two LLs has been the occurrence of edge states. A confining potential has to be taken into account if the Kubo conductivities are calculated in coordinate representation which implies the use of the equation of motion $i\hbar v_\mu = [r_\mu, H]$. We want to present a symmetry argument which allows us to circumvent difficulties due to edge states and confining potentials on the one hand and to avoid the explicit calculation of contributions from extended states on the other hand. For that purpose we establish the relation between the complex conductivity $\sigma(\omega)$

$$\begin{aligned} \sigma(\omega) &= \beta\int_0^\infty e^{i\omega t-\eta t} < j; j(t) > dt \quad , \quad j = \frac{1}{\sqrt{2}}(j_x + i\,j_y) \\ \sigma_{xx}(\omega) &= \frac{1}{2}(\sigma(\omega)+\sigma^*(-\omega)) \quad , \quad \sigma_{yx}(\omega) = \frac{i}{2}(\sigma(\omega)-\sigma^*(-\omega)) \end{aligned} \tag{5}$$

and the force-force correlation γ_F

$$\gamma_F(\omega) = \beta\int_0^\infty e^{i\omega t-\eta t} < F; F(t) > dt \quad , \quad F = -\frac{1}{\sqrt{2}}\frac{e}{m}\left(\frac{\partial V}{\partial x} + i\frac{\partial V}{\partial y}\right) \tag{6}$$

which is given by the identity

$$\sigma(\omega) = \frac{e^2 n}{m}\frac{i}{\omega-\omega_c} + \frac{\gamma_F(\omega)}{(\omega-\omega_c)^2} \tag{7}$$

In eq.(5-7) we have used standard textbook notation (cf.[6]) and the correlations are defined in terms of the Kubo scalar product

$$< A; B >= \beta^{-1}\mathrm{Tr}\int_0^\beta d\lambda \rho(H) A(-i\hbar\lambda) B^+ \quad , \quad \beta^{-1} = k_B T \tag{8}$$

In the following it will be shown that the frequency dependent force-force correlation $\gamma_F(\omega)$ for the fully occupied LLL vanishes. Assuming for a moment the validity of this statement we are able to deduce from eq.(7) that in the right localised tail of the LLL, i.e. for $\Gamma \ll \varepsilon_F \ll \hbar\omega_c/2$, $\varepsilon_F = E_F - \hbar\omega_c/2$ the Hall conductivity is

$$\sigma_{yx}(\omega; \hbar\omega_c/2 + \varepsilon_F) = \frac{e^2}{h}\frac{1}{1-\omega^2/\omega_c^2} - \sigma_{yx}(\omega; \hbar\omega_c/2 - \varepsilon_F) \tag{9}$$

The first term in the RHS of eq.(9) is the Hall conductivity of the completely filled LLL and the second term gives the correction from localized states in leading order of the frequency expansion. In order to obtain the final result we just have to insert $\sigma_{yx}(\omega)$ from eq.(4) with n replaced by $-n + 1/2\pi\ell^2$ and for comparison with the experiments the non resolved spin degeneracy has to be taken into account.

In order to prove that $\gamma(\omega; E_F \to \infty) = 0$ in the one-band model, we write the susceptibility χ_{FF} related to the force-force correlation by

$$\gamma_F(\omega) = \frac{\chi_{FF}(\omega) - \chi_{FF}(0)}{i\omega} \tag{10}$$

in terms of Green's functions $G^\pm$

$$\chi_{FF}(\omega) = \frac{1}{2\pi i\,Ar}\int f(E)\mathrm{Tr}\Big\{\overline{FG^+_\omega F(G^+-G^-)} + \overline{F(G^+-G^-)FG^-_{-\omega}}\Big\}dE \tag{11}$$

where the notation $G^\pm_\omega = G(E + \hbar\omega \pm i\eta)$ has been used, Ar is the area and f denotes the Fermi distribution. Shifting the energies by $\pm\hbar\omega/2$ it is possible to derive from eq.(11)

$$\gamma_F(\omega) = c(E_F,\omega) - \frac{\hbar}{4\pi Ar}\int f(E)\mathrm{Tr}\Big\{\overline{FG^+_{\omega/2}FG^+_{-\omega/2}G^+} - \overline{FG^+_{\omega/2}G^+FG^+}$$
$$+\overline{FG^-_{\omega/2}G^-FG^-_{-\omega/2}} - \overline{FG^-FG^-_{-\omega/2}G^-}\Big\}dE \tag{12}$$

At $T=0$ the function c which has not been given explicitely above just depends on states with energies in the interval $]-\hbar\omega/2+E_F, E_F+\hbar\omega/2[$ around the Fermi energy and vanishes with the density of states as $E_F/\Gamma \to \infty$ in the one-band approximation. For any probability distribution of the random potential which attributes the same weight to a given configuration V and to $-V$ the corresponding sum in the integrand of eq.(12) is antisymmetric with respect to the center of the LLL before performing the average since in coordinate representation only terms proportional to

$$Im\prod_{n=1}^{N} G^+(\mathbf{r}_n,\mathbf{r}_{n+1}) \quad , \quad N \text{ even} \qquad \text{or to} \qquad Re\prod_{i=1}^{N} G^+(\mathbf{r}_i,\mathbf{r}_{i+1}) \quad , \quad N \text{ odd} \tag{13}$$

with $\mathbf{r}_{N+1} = \mathbf{r}_1$, occur as coefficients in the frequency expansion of the integrand in eq.(12). This statement can be checked explicitly noting that in the functional integral representation in terms of supervectors (cf.[1]) $\Phi = (s,\chi)$ we have

$$\prod_{n=1}^{N} G^+(\mathbf{r}_n,\mathbf{r}_{n+1}) = (-i)^N \int [d\bar{\Phi}][d\Phi]\prod_{n=1}^{N}\chi_n(\mathbf{r}_n)\bar{\chi}_n(\mathbf{r}_{n+1})\exp(-S)$$
$$S = -i\int\sum_{n=1}^{N}\bar{\Phi}_n(E-H_0-V+i\eta)\Phi_n d^2r \tag{14}$$

and H_0 can be replaced by $\hbar\omega_c/2$ in the LLL. The transformation $\varepsilon \to -\varepsilon$; $V \to -V$ maps G^+ onto $-G^-$ thereby establishing the antisymmetry stated above.

We arrive at the conclusion that the force-force correlation indeed vanishes for the completely filled LLL and conclude in agreement with eq.(9) that in the energy region $\Gamma \ll \varepsilon_F \ll \hbar\omega_c/2$ the Hall conductivity in the case of non resolved spin degeneracy reads

$$\sigma_{yx}(\omega) = \frac{e^2}{h}\left(\frac{2}{1-\omega^2/\omega_c^2} - \left(\frac{2\hbar\omega}{\Gamma}\right)^2(2-\nu)\right) \tag{15}$$

where $\nu = 2\pi\ell^2 n$ is the filling factor and $1-\nu/2$ is the fraction of unoccupied localized states in the upper tail.

In the next step we have to determine the frequency range limiting the validity of eq.(15). First we note that in the low frequency expansion presented in [1] the relative corrections form a power series in terms of Ω^2 , $\Omega = 4\hbar\omega\varepsilon_F/\Gamma^2$. Since $\varepsilon_F < \hbar\omega_c/2$, an upper bound for the Ω is given by

$$\Omega < \pi\omega\tau \simeq 1.2\cdot 10^{-12}s\,\omega\tilde{\mu} \tag{16}$$

In eq.(16) we have used the relation $\Gamma^2 = (2/\pi)\hbar\omega_c(\hbar/\tau)$ between the bandwidth, the lifetime τ and the corresponding mobility at zero magnetic field $\mu = e\tau/m$, $m = 0.067m_e$ for GaAs and $\mu = \tilde{\mu}\cdot 10^4 cm^2/Vs = \tilde{\mu}/T$. Ω^2 can be considered as a good expansion parameter provided

$$\Omega^2 \lesssim 0.1 \quad \Longrightarrow \quad \omega \lesssim \tilde{\mu}^{-1}\cdot 265\,GHz \tag{17}$$

We note that this criterion is independent of B. A strong magnetic field just favors the observability of the effect by increasing the relative LL distance $\hbar\omega_c/2\Gamma$. Thus, samples with a high electron density are most appropriate because the plateaus are shifted to higher values of the magnetic field. With the experimental parameters of [4], i.e. $\tilde{\mu} = 1.5$ and $\tilde{\mu} = 8$, we obtain the condition that eq.(15) is valid for $\omega \lesssim 175GHz$ and $\omega \lesssim 35GHz$ respectively. In this frequency range the deviation of the Hall conductivity from its plateau values can still be considered as a correction. Since it is a familiar phenomenon that for higher mobility samples the measured bandwidth is larger than the one obtained from the zero field mobility the second upper bound presented above might even be enlarged. We are of course aware of the fact that our description is based on the assumption of short range disorder although the disorder has long range in GaAs. We believe that this doesn't affect the applicability of our theory to the experiment. However, in order to obtain exact quantitative predictions it is probably not sufficient to estimate the bandwidth from the $B = 0$-mobility which in general doesn't characterize a sample of a given material uniquely.

In [4] it was assumed that only the plateau width decreases when increasing the frequency and becomes zero at some critical value of ω. Properly speaking, the notion of a critical frequency is incompatible with our result that the plateaux are deformed for any finite frequency and we think that the interpretation of the experiments of [4] given by the authors is not the only possible one. We also disagree with their conclusion that the frequency dependent deformation is the same for samples with different mobilities. It follows from eq.(15) that the effect of low mobilities is to diminish the deformation of the Hall plateaus in the GHz-range provided the overlap of adjacent LLs doesn't become too large. In high mobility samples plateaux will disappear already at lower frequencies. Since in GaAs-AlGaAs heterostructures the magnetic field has to be lowered in order to reach higher LLs and the bandwidth satisfies $\Gamma \propto \sqrt{B}$ it will be difficult to resolve the higher subbands in the low mobility samples. This observation is confirmed by the experiment of ref.4 where only the $\nu = 2$-plateau was well developed as well as in [2,3].

Finally, we want to compare the deviation of the Hall conductivity from the quantized value predicted by eq.(15) with the results of [4]. In microwave experiments the measured bolometer signal from the transmitted radiation is proportional to $\sigma_{yx}^2(\omega)$. Let b denote the distance from the center of the plateau at $\omega = 0$ in Tesla and $\omega = \tilde{\omega} \cdot 10^{2.5} GHz$. With the parameters $n = 2.1 \cdot 10^{11} cm^{-2}$, $\tilde{\mu} = 1.5$ we obtain in the vicinity of the observed $\nu = 2$-plateau

$$\sigma_{yx}^2(b;\tilde{\omega}) \simeq \frac{e^4}{h^2}\left(4 - \tilde{\omega}^2 \cdot b\right) \tag{18}$$

Inserting into eq.(18) the highest experimentally realized frequency of [4], i.e. $\omega = 62GHz$, we obtain a finite slope of about 0.04 which is already significant with respect to the experimental errors. In agreement with the arguments presented above we think that the frequency effect should be described in terms of the finite slope the former plateaus begin to develop in the GHz-range.

We conclude that our theory describes satisfactorily all characteristic features of the frequency dependent deformation of the Hall plateaus in microwave experiments. It even gives predictions concerning the influence of the mobility and the electron density which we propose to verify in experiment of higher precision.

References

[1] O.Viehweger,K.B.Efetov, submitted to J.Phys.C
[2] F.Kuchar, R.Meisels, G.Weimann, W.Schlapp, Phys.Rev. **B22**, 2965 (1986)
[3] R.Meisels, F.Kuchar, K.Y.Lim, G.Weimann, W.Schlapp, Surf.Sci. **196**, 177 (1988)
[4] L.A.Galchenkov, I.M.Grodnenskiĭ, M.V.Kostovetskiĭ,O.R.Matov, JETP Lett.**46**(11), 542 (1987)
[5] I.Affleck, J.Phys. **C17**, 2323 (1984)
[6] R.Kubo, M.Toda, N.Hashitsume, Springer Series in Solid-State Sciences **31**, Berlin,Heidelberg (1985)

Multifractal Eigenstates in the Centre of Disorder Broadened Landau Bands

B. Huckestein and L. Schweitzer

Physikalisch-Technische Bundesanstalt, Bundesallee 100,
D-3300 Braunschweig, Fed. Rep. of Germany

1 Introduction

The nature of the electronic states in the centre of disorder broadened Landau bands plays a key role for the explanation of the integer Quantum Hall Effect. The existence of states which are not exponentially localized is essential for an understanding of the transport properties. However, not much is known about the structure of these states [1–4].

In the theoretical models for two-dimensional disordered systems in strong magnetic fields almost all states are exponentially localized and the localization length $\lambda(E)$ diverges only in the centre of the Landau band [5]. Recent numerical studies find $\lambda(E) = |E - E_0|^{-\nu}$, with $\nu = 2.34$ for the lowest Landau band [6].

To get more information about the nature of these states, we have calculated different moments of the wavefunctions using a Lanczos algorithm. Applying the framework of fractal sets [7, 8] and calculating the generalized fractal dimensions D_q, a multifractal behaviour of the most extended states, which appear in the centre of the Landau bands, is established.

2 The Model

The two-dimensional disordered system of non-interacting electrons subject to a strong normal magnetic field B is described by a tight binding Hamiltonian [9, 4]. The vector potential $\mathbf{A}$ is chosen in the Landau gauge, $\mathbf{A} = (0, Bx, 0)$.

$$\begin{aligned}(H\phi)(x,y) = \varepsilon(x,y)\phi(x,y) + V(\phi(x+a,y) + \phi(x-a,y) + \\ \exp\{-i2\pi\alpha_B x/a\}\phi(x,y+a) + \exp\{i2\pi\alpha_B x/a\}\phi(x,y-a)),\end{aligned} \tag{1}$$

where $\alpha_B = a^2 eB/h$ is the number of flux quanta h/e per unit lattice cell a^2. The transfer matrix element V is taken to be the unit of energy. Periodic boundary conditions are applied in both directions and the magnetic field B is commensurate with the lattice. The $\varepsilon(x,y)$ are a set of independent random variables distributed uniformly around zero energy with width W.

Using a Lanczos algorithm for large sparse hermitian matrices [10], the eigenvalues and eigenvectors are calculated for a two-dimensional disordered system of size $M \times M$, with M up to $100\,a$, where a is the lattice constant. This

Springer Series in Solid-State Sciences, Vol. 101
High Magnetic Fields in Semiconductor Physics III Editor: G. Landwehr

means that for $\alpha_B = 1/5$ the ratio of system size M and the magnetic length $\ell = \sqrt{\hbar/eB}$ is about 112.

3 Energy Level Statistics

From the energy eigenvalues of the lowest Landau band the nearest neighbour energy spacing distribution $L(\Delta)$ has been calculated, where $\Delta = s/\bar{s}$ is the difference s of two successive eigenenergies divided by the mean energy spacing $\bar{s}$. The form of the distribution gives information about the symmetry of the Hamiltonian [11]. We find a Poisson distribution for energies situated in the wings of the Landau band, as expected for localized states, but observe a distribution, which is characteristic of an unitary Gaussian ensemble, for energies near the centre of the Landau band.
In Fig. 1 the normalized energy level distributions $L(\Delta)$ are shown together with distributions for a random, an orthogonal and an unitary Gaussian ensemble. If the energy range is restricted to an interval close to the middle of the band, $L(\Delta)$ is well approximated by an unitary Gaussian ensemble which indicates energy level repulsion. It must be noted that the energy range for which an unitary Gaussian ensemble can be observed decreases with increasing system size. This is compatible with the notion that there exist only localized states in the limit of an infinite system, except at one singular energy, but these remaining non-localized states are of measure zero.

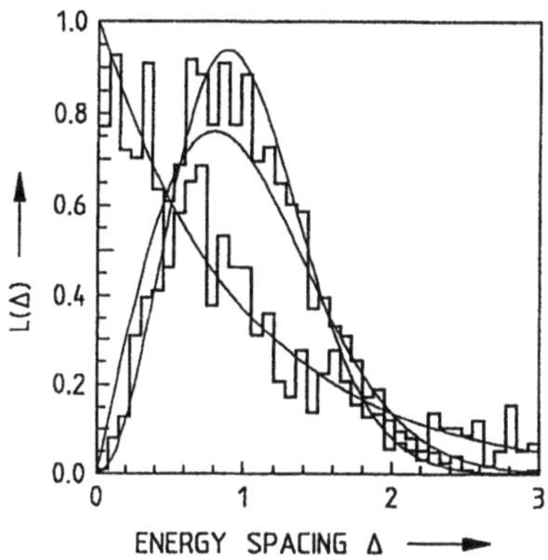

Fig. 1 Energy level distribution.

4 Fractal Behaviour of the Wavefunctions

To characterize fractal sets a probability measure is defined on the lattice [7, 8],

$$p_i(l, M, E) = \sum_{\mathbf{r} \in \Omega_i(l)} |\psi_{M,E}(\mathbf{r})|^2. \tag{2}$$

and the moments of the eigenstates can be defined by

Fig. 2 $|\psi_{M,E}(\mathbf{r})|^2$ of the most extended state.

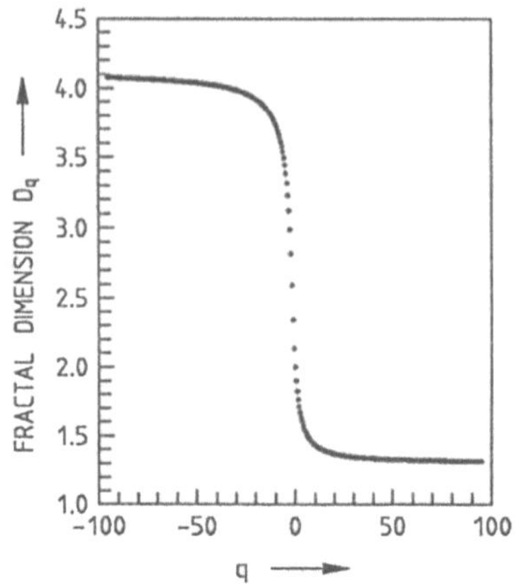

Fig. 3 Generalized fractal dimensions D_q.

$$P_q(l, M, E) = \sum_i^{N(l)} p_i^q(l, M, E), \tag{3}$$

where the $M \times M$ lattice was covered with $N(l)$ squares $\Omega_i(l)$ of length l, and with a probability measure p_i in the i-th square. For a system of size M, $\psi_{M,E}(\mathbf{r})$ is a normalized eigenvector corresponding to an eigenvalue E.

In Fig. 2, the p_i in the limit $l = a$, i. e. there is only one site in the i-th square, are plotted for a lattice of size $M = 100\,a$. The magnetic field is $\alpha_B = 1/5$ and the disorder strength is $W = 0.5V$. The area of the spots is proportional to $|\psi_{E,M}(\mathbf{r})|^2$. The generalized fractal dimensions for the most extended state are shown in Fig. 3, where the D_q are plotted versus q in the range $-95 \le q \le 95$. For $q \neq 1$, the generalized fractal dimensions D_q are defined by [7, 8]

$$P_q(l, M, E) = \left(\frac{l}{M}\right)^{(q-1)D_q} \tag{4}$$

whereas D_1 is given by

$$D_1(M, E) = \frac{\sum_i p_i(l, M, E) \log p_i(l, M, E)}{\log(l/M)}. \tag{5}$$

In the present case, equations (4) and (5) hold for the range $a \le l \le M$.

According to Halsey *et al* [8], a normalized distribution $f(\alpha(q)) = q(q-1)\, dD_q/dq + D_q$ can be defined to characterize fractal sets by considering the singularities α of these measures. The $f(\alpha)$ spectrum describes the range and the densities of the singularity strength $\alpha(q) = d[(q-1)D_q]/dq$, which is easily calculated from the generalized dimensions D_q [8]. The $f(\alpha)$ vs. $\alpha(q)$ spectrum, calculated from the D_q of Fig. 3, is shown in Fig. 4.

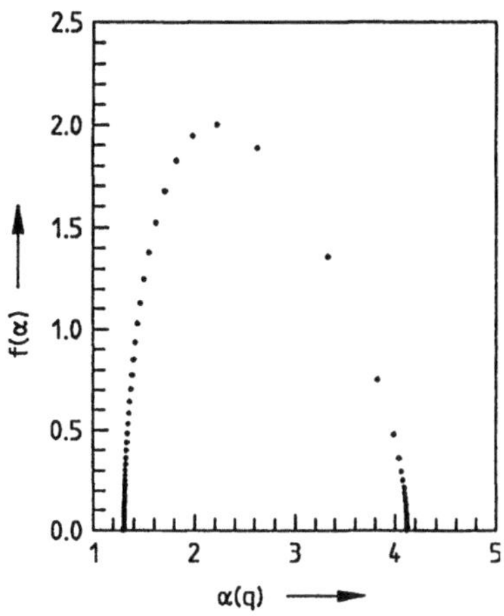

Fig. 4 $f(\alpha)$ vs. $\alpha(q)$ spectrum.

The limiting dimensions $D_{-\infty}$ and D_∞ can easily be obtained from

$$\lim_{q\to\infty} D_q = \ln(\max_{\mathbf{r}}\{|\psi_{E,M}(\mathbf{r})|^2\})/\ln(a/M), \tag{6}$$

and

$$\lim_{q\to-\infty} D_q = \ln(\min_{\mathbf{r}}\{|\psi_{E,M}(\mathbf{r})|^2\})/\ln(a/M). \tag{7}$$

We find $D_{-\infty} = 4.12$ and $D_\infty = 1.30$. The value for $D_2 = 1.71$ is the same as for smaller lattice systems [4], but is larger than those reported by Aoki [1] and Ono *et al* [3] for continuum models.
Our results show that the electronic states with localization length $\lambda(E)$ larger than the system size M are multifractal and the infinite set of the D_q or the $f(\alpha)$ vs. $\alpha(q)$ spectrum is necessary to fully characterize their structure. Since the states are not truly extended but multifractal, the influence of these states on the transport properties has to be reconsidered.

References

[1] H. Aoki, Phys. Rev. **B33**, 7310(1986)

[2] B. Kramer, Y. Ono, and T. Ohtsuki, Surf. Science **196**, 127(1988)

[3] Y. Ono, T. Ohtsuki, and B. Kramer, J. Soc. Jap. **58**, 1705(1989)

[4] R. Johnston and L. Schweitzer, J. Phys.: Condens. Matter **2**, 4137(1990)

[5] Y. Ono, J. Phys. Soc. Jpn. **51**, 2055(1982); 3544(1982)

[6] B. Huckestein and B. Kramer, Phys. Rev. Lett. **64** , 1437(1990)

[7] H. G. E. Hentschel and I. Procaccia, Physica **8D**, 435(1983)

[8] T. C. Halsey, M. H. Jensen, L. P. Kadanoff, I. Procaccia, and B. Shraiman, Phys. Rev. **A33**, 1141(1986)

[9] L. Schweitzer, B. Kramer, and A. MacKinnon, J. Phys. **C17**, 4111(1984)

[10] J. K. Cullum and R. A Willoughby, *Lanczos algorithms for large symmetric eigenvalue computations*, Birkhäuser(1985)

[11] M. L. Mehta, *Random Matrices and the Statistical Theory of Energy Levels*, Academic Press (1967)

Exact Widths and Tails for Landau Levels Broadened by a Random Potential with an Arbitrary Correlation Length

K. Broderix, N. Heldt, and H. Leschke

Institut für Theoretische Physik, Universität Erlangen-Nürnberg, Staudtstr. 7, W-8520 Erlangen, Fed. Rep. of Germany

1. Introduction and Basic Notation

This publication is concerned with the broadening of Landau levels due to disorder and some effects thereof. The underlying model can be described by the Hamiltonian

$$H := H_0 + V \quad , \quad H_0 := \frac{1}{2m}\left(\frac{\hbar\partial}{i\partial x_1}\right)^2 + \frac{1}{2m}\left(\frac{\hbar\partial}{i\partial x_2} + eB\,x_1\right)^2 \tag{1}$$

for one (spinless) electron of (effective) mass m and charge $-e$ in the infinite (x_1, x_2)-plane under the influence of a perpendicular constant magnetic field of strength B and a random potential V. The probability distribution of V is assumed to be Gaussian with

$$\overline{V(x)} = 0 \quad , \quad \overline{V(x)\,V(x')} = \sigma^2 \exp\left\{-(x-x')^2/2\lambda^2\right\} \quad , \quad x := (x_1, x_2) \quad . \tag{2}$$

Here the overbar denotes the average with respect to the probability distribution, σ is the strength and λ the correlation length of the fluctuations of the potential.

In the following E_n denotes the projection operator on the eigenspace of the unperturbed Hamiltonian H_0 belonging to the n-th eigenvalue $\varepsilon_n := (2n+1)\hbar eB/2m$, that is, to the n-th Landau level ($n = 0, 1, 2, \ldots$). The position representation of E_n is given by

$$\langle x|E_n|x'\rangle = (2\pi l^2)^{-1}\exp\left\{-[2i(x_1+x_1')(x_2-x_2') + (x-x')^2]/4l^2\right\}\mathrm{L}_n\left((x-x')^2/2l^2\right) \tag{3}$$

where $\mathrm{L}_n(\xi) := (n!)^{-1}\exp\{\xi\}\mathrm{d}^n(\xi^n\exp\{-\xi\})/\mathrm{d}\xi^n$ is the n-th Laguerre polynomial and $l := (\hbar/eB)^{1/2}$ is the magnetic length.

In the next section we present exact results on the averaged density of states

$$D_n(\varepsilon) := \langle x|\,\overline{E_n\delta(\varepsilon - E_nHE_n)E_n}\,|x\rangle \tag{4}$$

(per area) of the Hamiltonian H restricted to the n-th eigenspace of H_0. Upon integration over the real line, the density D_n is normalized to $1/2\pi l^2$. Furthermore, D_n is symmetric with respect to ε_n, that is, $D_n(\varepsilon_n + \varepsilon) = D_n(\varepsilon_n - \varepsilon)$.

2. Exact Width and Leading Tail of the Restricted Density of States

Defining the width σ_n of D_n as its standard deviation we find explicitly

$$\sigma_n^2(\lambda^2) := 2\pi l^2\int_{-\infty}^{\infty}\mathrm{d}\varepsilon\, D_n(\varepsilon)\,(\varepsilon-\varepsilon_n)^2 = \sigma^2\frac{\lambda^2}{\lambda^2+l^2}\left(\frac{\lambda^2-l^2}{\lambda^2+l^2}\right)^n \mathrm{P}_n\left(\frac{\lambda^4+l^4}{\lambda^4-l^4}\right) \tag{5}$$

where $\mathrm{P}_n(\xi) := (n!2^n)^{-1}\,\mathrm{d}^n(\xi^2-1)^n/\mathrm{d}\xi^n$ is the n-th Legendre polynomial.
Remarks:

Springer Series in Solid-State Sciences, Vol. 101
High Magnetic Fields in Semiconductor Physics III Editor: G. Landwehr

- The quantity σ_n has appeared earlier in the literature [1, 2, 3], but has not been identified as the exact width of D_n there.
- For the validity of (5) only the Gaussian correlation function (2) is needed. The Gaussian nature of the underlying probability distribution of V is not necessary.

The leading behaviour of D_n for large $|\varepsilon - \varepsilon_n|/\sigma_n$ is given by the Gaussian decay

$$\lim_{|\varepsilon|\to\infty} \frac{1}{\varepsilon^2} \ln D_n(\varepsilon + \varepsilon_n) = -\frac{1}{2\Gamma_n^2(\lambda^2)} \tag{6}$$

where the decay constant Γ_n is the solution of the variational problem

$$\Gamma_n^2(\lambda^2) := \sup_{|\psi\rangle,\, \langle\psi|E_n|\psi\rangle = 1} \overline{\langle\psi|E_n V E_n|\psi\rangle^2} = \frac{\lambda^2}{\lambda^2 + l^2}\,\sigma_n^2(\lambda^2 + l^2) \quad . \tag{7}$$

Equations (6,7) considerably generalize results known earlier [4].

The proof of (5) follows from (3). We will supply a proof of the results (6,7) in a forthcoming publication [5].

3. Implications

Estimates

$$\Gamma_n^2(\lambda^2) \le \sigma_n^2(\lambda^2) \le \sigma^2\lambda^2/(\lambda^2 + l^2) \quad , \quad \Gamma_n^2(\lambda^2) \le \sigma^2\lambda^2/(\lambda^2 + 2l^2) \tag{8}$$

Reciprocity relation

$$\sigma_n^2(\lambda^2) = (\lambda^2/l^2)\,\sigma_n^2(l^4/\lambda^2) \tag{9}$$

Curiosity

The mapping $\xi \mapsto \Gamma_n^2(\xi l^2)/\sigma_n^2(\xi l^2)$ has a stable fixed point at the golden mean $(\sqrt{5}-1)/2$.

Limit of a delta-correlated random potential (white noise)

For $\sigma^2 = \nu^2/2\pi\lambda^2$ with fixed ν^2 one gets

$$\lim_{\lambda^2\to 0} \sigma_n^2(\lambda^2) = \nu^2/2\pi l^2 \quad , \quad \lim_{\lambda^2\to 0} \Gamma_n^2(\lambda^2) = (\nu^2/4\pi l^2)\,(2n)!/(n!2^n)^2 \tag{10}$$

Absence of broadening in the high Landau-level limit

Since $\lim_{n\to\infty} \sigma_n^2(\lambda^2) = 0$ for $\lambda^2 < \infty$, the Chebychev inequality yields

$$\lim_{n\to\infty} D_n(\varepsilon + \varepsilon_n) = \delta(\varepsilon)\,/\,2\pi l^2 \quad . \tag{11}$$

A broadening only survives in the two extreme situations for the spatial extent of correlations:

- For the constantly correlated limit, $\lambda^2 = \infty$ with fixed σ^2, the density D_n is trivially given [6] as a Gaussian of width σ centered at ε_n.
- For the delta-correlated limit it is known [7] that $D_n(\varepsilon + \varepsilon_n)$ approaches a half-ellipse of width $(\nu^2/2\pi l^2)^{1/2}$ as $n \to \infty$.

4. Two Gaussian Approximations to the Restricted Density of States

Exactly knowing the width σ_n and the Gaussian decay with constant Γ_n suggests two Gaussian approximations to D_n

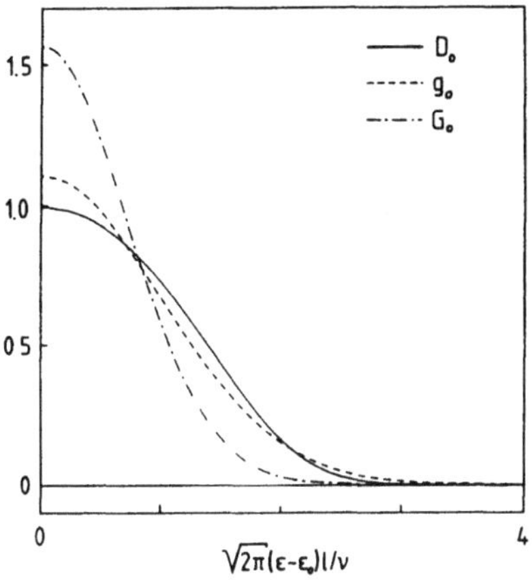

Fig. 1: The density of states D_0 and the approximations g_0 and G_0 as a function of the energy ε for white noise. The ordinate is given in units of $\sqrt{2}/\pi^2 l\nu$.

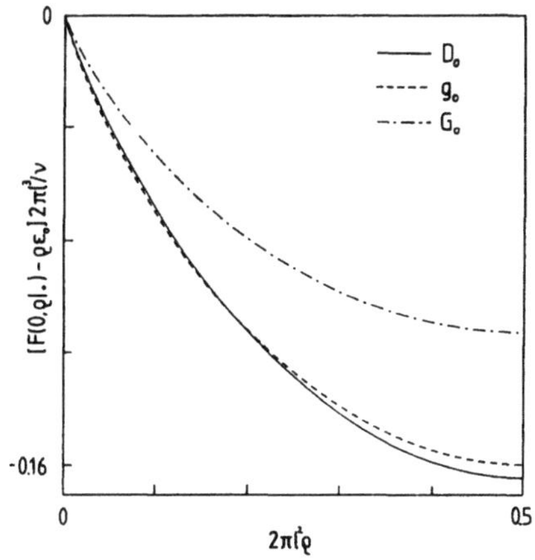

Fig. 2: The ground-state energy $F(0, \varrho|D_0)$ and the approximations $F(0, \varrho|g_0)$ and $F(0, \varrho|G_0)$ vs. the filling factor $2\pi l^2 \varrho$ for white noise. The curves are symmetric with respect to $2\pi l^2\varrho = 1/2$.

$$\begin{aligned} D_n(\varepsilon) &\approx g_n(\varepsilon) := (2\pi l^2)^{-1}[2\pi\sigma_n^2(\lambda^2)]^{-1/2} \exp\left\{-(\varepsilon-\varepsilon_n)^2/2\sigma_n^2(\lambda^2)\right\} \\ D_n(\varepsilon) &\approx G_n(\varepsilon) := (2\pi l^2)^{-1}[2\pi\Gamma_n^2(\lambda^2)]^{-1/2} \exp\left\{-(\varepsilon-\varepsilon_n)^2/2\Gamma_n^2(\lambda^2)\right\} \quad . \end{aligned} \tag{12}$$

Remarks:

- While g_n reproduces the exact width, G_n correctly reproduces the leading tail.
- Both approximations are exact for $\lambda^2 = \infty$, because $\sigma_n^2(\infty) = \Gamma_n^2(\infty) = \sigma^2$.
- In Fig. 1 we compare g_0 and G_0 to D_0 in the other limit, where D_0 is exactly known [8], namely in the delta-correlated limit.
- For any convex function K the inequality of Jensen-Peierls type

$$\int_{-\infty}^{\infty} d\varepsilon G_n(\varepsilon)\, K(\varepsilon) \leq \int_{-\infty}^{\infty} d\varepsilon D_n(\varepsilon)\, K(\varepsilon) \tag{13}$$

  holds [5]. This generalizes and sharpens an inequality in [3].
- The last inequality implies an upper bound on the averaged free energy (per area) of non-interacting electrons with one-particle Hamiltonian H restricted to the n-th eigenspace of H_0

$$F(T, \varrho|D_n) \leq F(T, \varrho|G_n) \quad . \tag{14}$$

  Here the free energy at temperature T of non-interacting fermions with particle density ϱ and one-particle density of states w is denoted by

$$F(T, \varrho|w) := \sup_{\mu}\left\{\mu\varrho - k_B T \int_{-\infty}^{\infty} d\varepsilon\; w(\varepsilon) \ln\left[1 + \exp\left\{(\mu-\varepsilon)/k_B T\right\}\right]\right\} \quad . \tag{15}$$

- For the delta-correlated limit we compare $F(0, \varrho|g_0)$ and $F(0, \varrho|G_0)$ to the exact ground-state energy $F(0, \varrho|D_0)$ in Fig. 2.

5. Two Approximations to the Unrestricted Density of States

The approximations g_n and G_n to D_n suggest the approximations

$$D(\varepsilon) \approx g(\varepsilon) := \textstyle\sum_{n=0}^{\infty} g_n(\varepsilon) \quad , \quad D(\varepsilon) \approx G(\varepsilon) := \textstyle\sum_{n=0}^{\infty} G_n(\varepsilon) \tag{16}$$

to the averaged density of states (per area)

$$D(\varepsilon) := \langle x | \overline{\delta(\varepsilon - H)} \, | x \rangle \tag{17}$$

of the unrestricted Hamiltonian H.

Remarks:

- The approximation g has already been proposed in [1] by other arguments, see also [2].
- By construction, the approximations g and G neglect, for $\lambda^2 < \infty$, the mixing of Landau levels.
- Both approximations exhibit Gaussian tails for $\varepsilon \to -\infty$ with decay constants σ_0 and Γ_0, respectively. Except for the delta-correlated limit, the leading low-energy behaviour of D is indeed known to be Gaussian, but the decay constant is σ [3].
- Inequality (13) and an argument of [3] imply for any convex function K

$$\int_{-\infty}^{\infty} d\varepsilon \, G(\varepsilon)\, K(\varepsilon) \le \int_{-\infty}^{\infty} d\varepsilon \, \textstyle\sum_{n=0}^{\infty} D_n(\varepsilon)\, K(\varepsilon) \le \int_{-\infty}^{\infty} d\varepsilon \, D(\varepsilon)\, K(\varepsilon) \quad . \tag{18}$$

- As a consequence, one gets the inequalities

$$F(T,\varrho|D) \le F(T,\varrho|\textstyle\sum_{n=0}^{\infty} D_n) \le F(T,\varrho|G) \tag{19}$$

  for the free energy $F(T,\varrho|D)$ corresponding to the unrestricted Hamiltonian H.
- In Fig. 3 we give plots of g and G. The approximation G is superior to the one given in [3] in the sense that G yields a sharper bound in (18). (For comparison the reader should note that it must read $2\lambda^2$ instead of λ^2 in Figs. 2, 3 of [3].)
- In Fig. 4 we give a plot of the approximation $M(T,\varrho,B) := -\partial F(T,\varrho|2g)/\partial B$ to the magnetization $-\partial F(T,\varrho|2D)/\partial B$, where we now have taken into account the

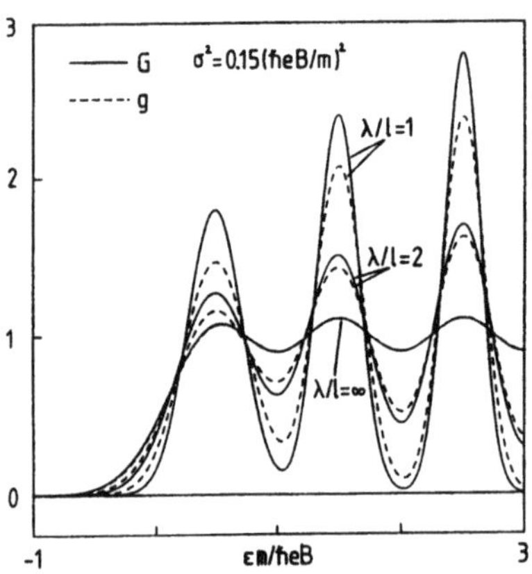

Fig. 3: The approximate densities of states g and G as a function of the energy ε for three different values of the correlation length λ. The ordinate is given in units of $m/2\pi\hbar^2$.

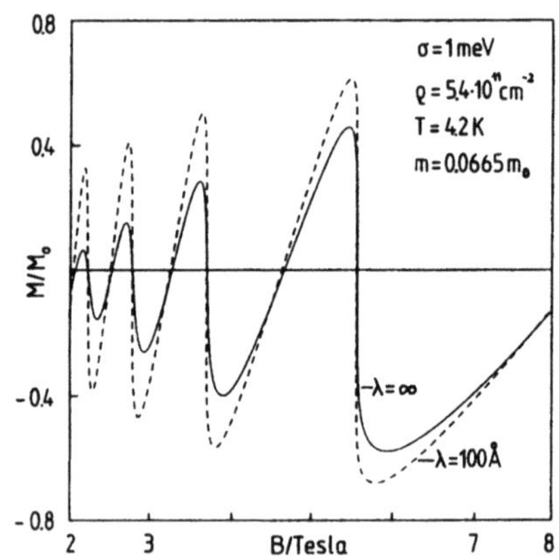

Fig. 4: The approximate magnetization $M(T,\varrho,B)$ normalized to the saturation magnetization $M_0 := \varrho\hbar e/2m$ vs. the magnetic field B for two values of the correlation length λ.

spin degeneracy by doubling the density of states. Due to above remarks M is exact for $\lambda^2 = \infty$. The latter case has previously been studied; sometimes [9], however, it has been assumed that $D(\varepsilon) = 0$ for $\varepsilon < 0$. We have adapted the values of $m = 0.0665\, m_0$ (m_0: bare electron mass), $T = 4.2$K and $\varrho = 5.4 \cdot 10^{11}/\mathrm{cm}^2$ to the experimental situation for sample 1 investigated in [10]. Furthermore, we have chosen $\sigma = 1$meV and either $\lambda = \infty$ or $\lambda = 100$Å. We have neglected the spin contribution to the Zeeman energy, because for the sample considered it is less than $\sigma/10$ for field strengths up to 8 Tesla. We have not aimed at fitting the experimental curve (a) of Fig. 1 in [10], because it is not clear to us, what is meant by the subtraction of "a small, smooth background" [10].

[1] *R. R. Gerhardts*: Z. Physik B **21**, 275 (1975); 285 (1975); Surf. Sci. **58**, 227 (1976)

[2] *K. Broderix, N. Heldt, H. Leschke*: Nuovo Cimento **11** D, 241 (1989)

[3] *K. Broderix, N. Heldt, H. Leschke*: Phys. Rev. B **40**, 7479 (1989)

[4] *K. A. Benedict*: Nucl. Phys. B **280**, 549 (1987); *W. Apel*: J. Phys. C **20**, L577 (1987)

[5] *K. Broderix, N. Heldt, H. Leschke*: In preparation.

[6] *K. A. Benedict, J. T. Chalker*: J. Phys. C **18**, 3981 (1985); *K. Broderix, N. Heldt, H. Leschke*: Z. Physik B **68**, 19 (1987)

[7] *K. A. Benedict, J. T. Chalker*: J. Phys. C **19**, 3587 (1986); *R. Salomon*: Z. Physik B **65**, 443 (1987)

[8] *F. Wegner*: Z. Physik B **51**, 279 (1983)

[9] *W. Zawadzki, R. Lassnig*: Surf. Sci. **142**, 225 (1984)

[10] *J. P. Eisenstein, H. L. Störmer, V. Narayanamurti, A. Y. Cho, A. C. Gossard, C. W. Tu*: Phys. Rev. Lett. **55**, 875 (1985)

Scattering Mechanism in the Integer Quantum Hall Effect

J. Riess

Centre National de la Recherche Scientifique, Centre de Recherches
sur les Très Basses Températures, BP 166X,
F-38042 Grenoble Cedex, France

Abstract. We consider a large class of two-dimensional systems of electrons in a static disorder potential and subject to an in-plane electric field and to a strong perpendicular magnetic field. The time evolution of the single-particle states is investigated. It is found that the macroscopic Hall current is carried by the non-adiabatic states and that quantum Hall behavior occurs, when the Fermi energy lies in a range of adiabatic levels. Linear response theory is inadequate to describe the quantum mechanical scattering process of bulk states in the quantum Hall regime. The general results are illustrated by an explicit weak disorder model, where the scattering process and the nature of dc-insulating and -conducting states can be understood in detail. The time evolution of a scattered Landau function is calculated numerically. It gives a striking illustration of the velocity increase, due to disorder, of the conduction electrons in the bulk of a quantum Hall system. This phenomenon leads to current compensation, which is crucial for the IQHE. It is caused by a special kind of nonclassical particle propagation, which results from the nonlinear components of the single-particle currents. We believe that our results improve the present microscopic understanding of the IQHE.

1. Introduction

The knowledge of the quantum mechanical scattering process is indispensable for a complete understanding of the IQHE [1]. Here we will discuss general results on the time evolution of the electronic states in a large class of systems and simultaneously illustrate these with a simple model system (which is however sufficiently realistic to be significant), where the microscopic behaviour of the scattering states can be calculated explicitely. We will see, that the quantum mechanical time evolution of the electron states shows some remarkable nonclassical features which (to the best of our knowledge) have not been considered in previous theories (see e. g. [2,3]) of the IQHE.

We consider independent electrons (charge $q < 0$, mass m), on a long strip (of width L_y) in the x-direction, subject to a potential $V(x,y)$, a magnetic field $\mathbf{B} = (0,0,B)$, and an electric field $\mathbf{E} = (0,E_y,0)$. (This strip is thought to be a subdomain of a larger, but finite, quantum mechanical system containing the leads of the measuring apparatus.) The single-particle Hamiltonian can be written in the form

$$H = (1/2\,m)\left\{\left(\frac{\hbar}{i}\frac{\partial}{\partial x}\right)^2 + \left[\frac{\hbar}{i}\frac{\partial}{\partial y} - (q/c)\,[Bx + \phi(t)/L_y]\right]^2\right\} + V(x,y), \qquad (1)$$

where $\phi(t) = -cE_yL_yt$. We impose periodic boundary conditions in y-direction (and ψ

Springer Series in Solid-State Sciences, Vol. 101
High Magnetic Fields in Semiconductor Physics III Editor: G. Landwehr

= 0 for large | x|). Such a model is justified by the results of microwave experiments [4], which seem to indicate that effects due to contacts and edge states are not essential for the occurrance of the IQHE. The potential V(x,y) does not describe a macroscopic electric field, i.e. the integrals of V(x,y) over x and over y are zero.

2. Adiabatic and Nonadiabatic States

If the field E_y is sufficiently low, all the solutions of the time-dependent Schroedinger equation are adiabatic (i. e., $\phi(t)$ in (1) can be considered as a parameter). If V(x,y) has no symmetry, e.g. in the presence of static disorder, it follows [5] from the structure of the Hamiltonian (1), that the total current of any adiabatic state is periodic in time with period $\tau = |h/(qE_yL_y)|$ and with zero average. For realistic values of E_yL_y the period τ is extremely small (e. g. $\tau = 4 \times 10^{-12}$ s for $E_yL_y = 1$ mV). Hence adiabatic states are dc-insulating. As a consequence *any dc-current must result from non-adiabatic states* (see [6] for a more detailed discussion of this point).

To illustrate this general result we will now consider an explicit model system [6] defined by a potential of the type

$$V(x,y) = V(x) + V^1(x,y), \tag{2}$$

where $V^1(x,y)$ represents some disorder potential, and V(x) is a sequence of smooth barriers and wells, whose first and second spatial derivatives (piecewise) vary slowly over a magnetic length $\lambda = (\hbar c/qB)^{1/2}$ (e.g. V(x) of Fig. 1).

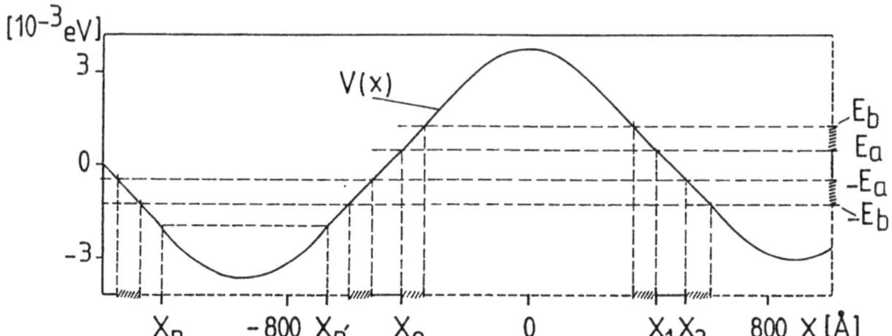

Fig. 1 Smooth substrate potential V(x) and the nature of the orbitals in the case, where in addition to V(x) a disorder potential $V^1(x,y)$, an electric field E_y and a magnetic field B in z-direction are present. The orbitals are characterized by the position of the localization centres x_p of the corresponding unperturbed ($V^1 = 0$) orbitals $\psi_p(x-x_p)$, see text. Full line regions : fully non-adiabatic (classically conducting) orbitals ; shaded regions : intermediate nonadiabatic orbitals (composed of classically and nonclassically conducting parts) ; dashed region : fully adiabatic (nonconducting) orbitals. Fig. 1 corresponds to the parameter values $(d_{pp'})^2 = 0.5 \times 10^{-6}$ (eV)^2, $L_y = 0.1$ cm, B = 6 T, $E_y = 2.37 \times 10^{-7}$ V cm^{-1}. V(x) also represents the energy E_p of the unperturbed orbitals ψ_p, see (5). For orbitals outside the dashed region unperturbed and perturbed energies coincide on the scale of the figure.

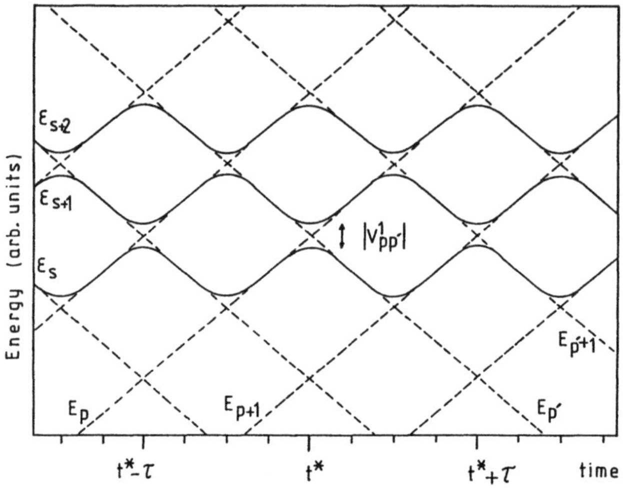

Fig. 2 Adiabatic evolution of the single-particle energies in the absence of the disorder potential $V^1(x,y)$ (dashed lines) and in the presence of $V^1(x,y)$ (full lines). Shown is the weak disorder case corresponding to Fig. 1. $\tau = |h/(qE_yL_y)|$.

In the absence of V^1 the solutions of the time-dependent Schroedinger equation have then the approximate form [6] (in the k-th Landau band):

$$\psi_{p,k}(x,y,t) = (L_y)^{-1/2}\exp(i2\pi py/L_y)u_{p,k}(x,t), \quad k, p \text{ integer,} \tag{3}$$

where $u_{p,k}(x,t)$ is the product of a Hermite polynomial and of a Gaussian $g_p(x,t)$ centered at

$$x_p(t) = chp/(qBL_y) - \phi(t)/(BL_y) + (mc^2/q^2B^2)V'(x_p). \tag{4}$$

The energies are (in good approximation)

$$E_{p,k} = \hbar\omega(k + 1/2) + V(x_p) + (m/2)\left[\frac{cV'(x_p)}{qB}\right]^2 \tag{5}$$

[the prime denotes d/dx and $\omega = |qB/(mc)|$]. In the following we consider a single band and drop the index k. Each $\psi_p(x,y,t)$ describes a particle localized at $x_p(t)$, which moves in x-direction with the constant, classical velocity $v = cE_y/B$. The corresponding energies $E_p(t)$ increase (decrease) with time if the centers $x_p(t)$ are situated on the left (right) hand side of a barrier in Fig. 1. Therefore the spectrum consists of intersecting levels (Fig. 2).

In the presence of the disorder potential V^1 the energy levels *anticross* and become individually periodic with period $\tau = h/|qE_yL_y|$ (Fig. 2). This is true quite generally for any sufficiently asymmetric potential V(x,y) in (1) [5,6]. The corresponding adiabatic states will be denoted $w_s(x,y,t)$. In the model defined by (2)

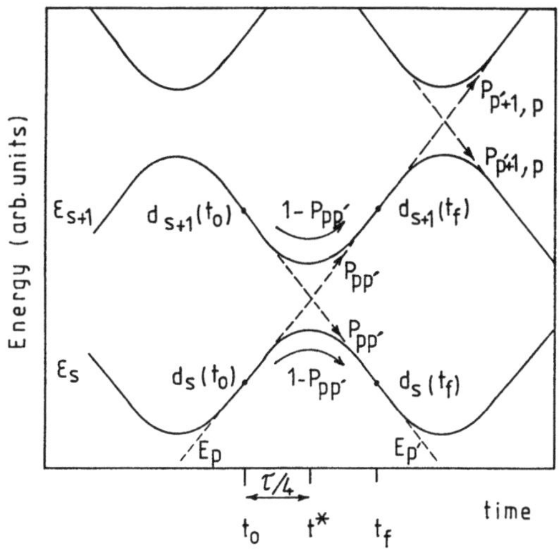

Fig. 3 Anticrossing situation around the value $t = t^*$ (mod τ). $P_{pp'}$ denotes the Zener tunnelling probability for the non-adiabatic process from t_0 to t_f. $1-P_{pp'}$ is the probability for the two corresponding adiabatic processes. $d_s(t)$ and $d_{s+1}(t)$ denote the time-dependent expansion coefficients of a weakly perturbed orbital expressed in the basis of the adiabatic orbitals $w_s(t)$ and $w_{s+1}(t)$ in the interval $t^*-\tau/4 \leq t \leq t^* + \tau/4$.

and Fig. 1 the physical parameters (in particular $V^1(x,y)$) are now chosen such that the perturbed ($V^1 \neq 0$) adiabatic levels (denoted $\varepsilon_s(t)$) in the center $[-E_b, E_b]$ of the band and the corresponding adiabatic states $w_s(x,y,t)$ can be described by a *weak disorder approximation* [6-8]. Here the states $w_s(x,y,t)$ are linear combinations $c_p(t)\psi_p(t) + c_{p'}(t)\,\psi_{p'}t)$ of only *two* unperturbed states $\psi_p(t)$, $\psi_{p'}(t)$ at a given value of t (the pair of indices (p,p') changes periodically whenever t increases by $\tau/2$), and further they are identical with a single unperturbed function $\psi_p(x,y,t)$ for times exactly half way between two successive anticrossings [6-8]. Physically such a weak disorder adiabatic state $w_s(x,y,t)$ describes a wave packet moving with the classical velocity v, but which is alternately localized at one of the two fixed sites $x_p(t^*)$ and $x_{p'}(t^*)$ situated on *different* sides of a barrier or of well. (Here t^* is the time where $E_p(t)$ and $E_{p'}(t)$ intersect.) The whole particle motion is periodic in time (with period τ), and therefore corresponds to a vanishing dc-current (see [6] for more details).

At sufficiently high values of E_y nonadiabatic transitions between the adiabatic states become possible. These may be difficult to calculate in general. But in our weak disorder case the nonadiabatic transition probabilities $P_{pp'}$ (see Fig. 3) can be obtained explicitly from the Zener formula. We get [7]

$$P_{pp'} = \exp\{-|4\pi^2(d_{pp'})^2 \exp[-2(x_p-x_{p'})^2/(2\lambda)^2]B/[V'(x_p)chE_y]|\}. \tag{6}$$

Here $d_{pp'}$ is a Fourier coefficient of V^1 defined by Eq. (21) of Ref. [6], and x_p, $x_{p'}$

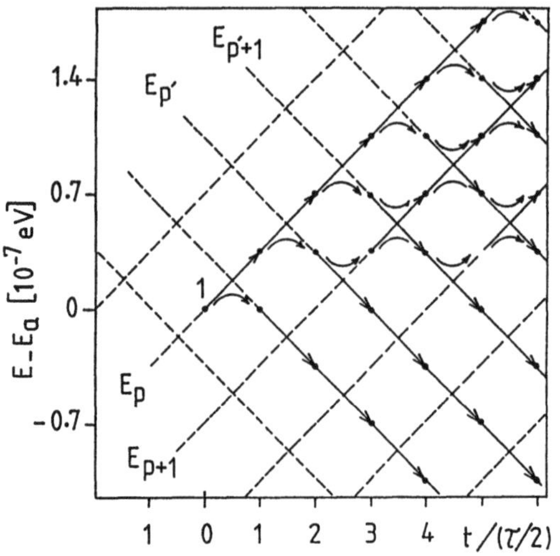

Fig. 4 Sequence of splittings which undergoes a scattering function $\psi(x,y,t) = \Sigma c_{p'}(t)\psi_{p'}(x,y,t)$ with initial condition $c_{p'}(t{=}0) = \delta_{pp'}$, $E_p(0) = E_a$. Each thick dot symbolizes the square of a (nonvanishing) expansion coefficient $|c_{p'}(t = n\tau/2)|^2$. These values are calculated according to the scheeme indicated by arrows.

are situated on opposite sides of a barrier (well) such that $E_p(t) = E_{p'}(t)$. Since $P_{pp'} = P_{p+1,p'+1}$ the probabilities only depend on the energy of intersection E of the unperturbed levels $E_p(t)$, $E_{p'}(t)$ in Fig. 2. For $E\in[-E_a,E_a]$ (center of the band) the nonadiabatic transition probabilities $P_{p,p'}$ are practically equal to one. This is called the *fully nonadiabatic zone*. For $E\in[E_a,E_b]$ or $\in[-E_a,-E_b]$ the values of $P_{pp'}$ decrease from practically one at $E = \pm E_a$ to practically zero at $E = \pm E_b$. The states in these energy zones $[-E_b,-E_a]$ and $[E_a, E_b]$ are called *intermediate states*. Here each wave function undergoes a series of splittings at each time $n\tau/2$, n integer, see Fig. 4. Due to a succession of such splitting events in the intermediate zones any initial state $\psi_p(x,y,t)$ with energy $E_p(t)$ in the *nonadiabatic region* $[-E_b,E_b]$ eventually develops into a *time-dependent* linear combination $\Sigma c_{p'}(t)\psi_{p'}(x,y,t)$ of unperturbed functions $\psi_{p'}(x,y,t)$ whose energies $E_{p'}(t)$ are contained in the whole non-adiabatic region $[-E_b,E_b]$. States outside $[-E_b,E_b]$ remain adiabatic, since their nonadiabatic transition probabilities are negligible. They do not mix with the states of the non-adiabatic region $[-E_b , E_b]$ in the course of time. In the following numerical calculations E_a and E_b have been defined such that $P_{pp'}(E_a) = 0.9999$ and $P_{pp'}(E_b) = 0.01$. For $E\in[-E_a,E_a]$ we set $P_{pp'}(E) = 1$. Outside $[-E_b,E_b]$ the nonadiabatic transitions can be neglected.

3. Time evolution of a scattered Landau Function

In our weak disorder case an intermediate wave function $\psi(x,y,t) = \Sigma c_{p'}(t)\psi_{p'}(x,y,t)$ develops according to the scheme of Fig. 4, i. e., it can be calculated at times t =

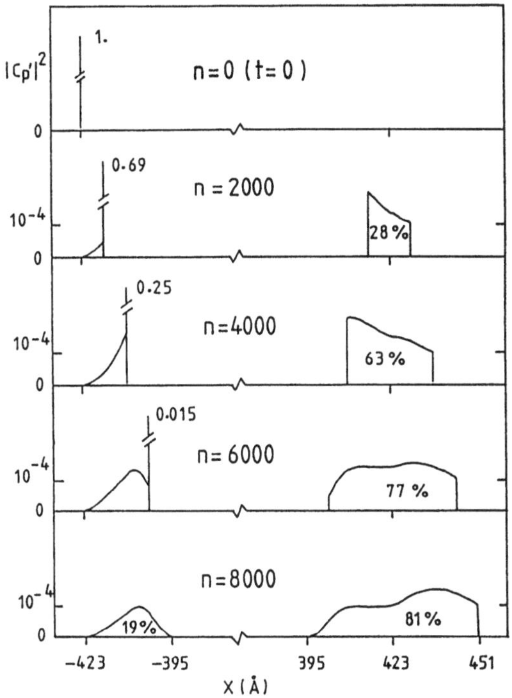

Fig. 5 Time evolution of the scattered orbital $\psi(x,y,t) = \Sigma c_{p'}(t)\psi_{p'}(x,y,t)$ with initial condition $| c_{p'}(0)|^2 = \delta_{p,p'}$. Shown are the probabilities $| c_{p'}(t)|^2$ as a function of the occupied centers $x_{p'}$ at different times $t = n\tau/2$, n integer; $\tau = 1.72 \times 10^{-7}$ s.

$n\tau/2$ by means of a matrix (see [8]) which is composed of (2x2) matrices, each of which describes a single splitting (Fig. 3). Fig. 5 shows the evolution of the single particle density $\rho(x,n\tau/2) = \int \psi\psi^*(x,y,n\tau/2)dy = \Sigma | c_{p'}(n\tau/2)|^2 u_{p'}(x)^2$ of a function $\psi(x,y,t)$, which at $t = 0$ is chosen to be a single unperturbed Landau function $\psi_p(x,y,t)$ (i. e., $c_{p'} = \delta_{pp'}$), whose center $x_p(0)$ is located at x_o on the left hand side of the potential barrier in Fig. 1, with an energy $E_p(0)$ situated at the upper edge E_a of the *fully non-adiabatic interval* $[-E_a, E_a]$. Shown are the numerical [9] values $| c_{p'}(n\tau/2)|^2$ for different times $t = n\tau/2$, $n>0$, integer. The corresponding centers $x_{p'}(n\tau/2)$, which are occupied with the probability $| c_{p'}(n\tau/2)|^2$, represent a finite set of the fixed, discrete points $x_p(0) \equiv x_p$ on the x-axis. (Their separation is 0.7×10^{-10} cm, which is not visible on the scale of the figure.)

At $t = 0$, when the energy of the original, unperturbed state $\psi_p(x,y,t)$ enters the intermediate zone $[E_a,E_b]$ at $E = E_a$, where the probabilities $P_{p,p'}$ start to become smaller than one (corresponding to $x_p(t = 0) = x_o$ on the left of the barrier) it starts to be partly scattered to the right hand side of the barrier (to x_1 and further) according to the scheme of Fig. 4. This implies that the total probability of the electron on the original side of the barrier, i. e., the sum of all the probabilities $| c_p|^2$ corresponding to occupied centers on the left hand side of the barrier, diminishes with time, whereas

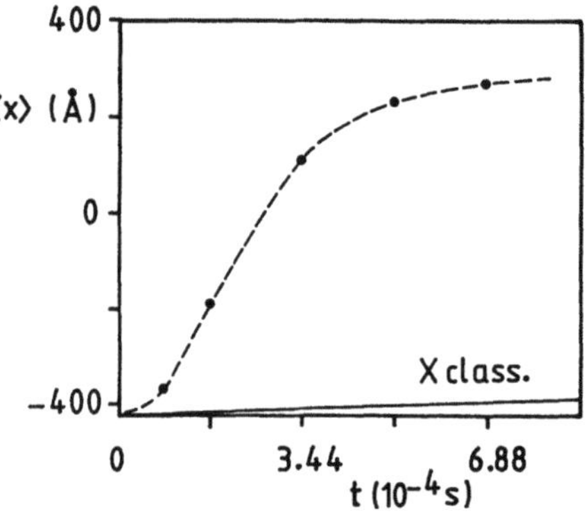

Fig. 6 Average position <x(t)> of the particle corresponding to the wave function $\psi(x,y,t)$ shown in Fig. 3. As a comparison the position $x_{class}(t)$ associated with the unperturbed, initial Landau function $\psi_p(x,y,t)$ is represented.

the total probability on the right hand side of the barrier increases. From the scattering mechanism shown in Fig. 4 one sees, that during every splitting interval of length $\tau/2$ a fraction of each probability $|c_{p'}(n\tau/2)|^2$ situated on the right hand side, and which corresponds to an unperturbed state with energy $E_{p'}(n\tau/2) \in [E_a,E_b]$, is backscattered to the original (left hand) side. This causes the tail in Fig. 5 behind the unscattered front peak on the left hand side of the barrier.

The *average position* $<x(n\tau/2)>$ of the electron at $t = n\tau/2$ is given by $\Sigma x_{p'}|c_{p'}(n\tau/2)|^2$. In the absence of disorder ($V^1 = 0$), or for fully nonadiabatic states, $<x(t)>$ is equal to $x_p(t) = x_o + (cE_y/B)t = x_{class}(t)$. This describes a motion with the constant, classical velocity $v = cE_y/B$. In the presence of the disorder V^1 the front peak on the left hand side of the barrier moves with this classical velocity, but the average position $<x(t)>$ is *ahead* of the classical position $x_{class}(t)$, i. e., the effective velocity $v_{eff} = (d/dt)<x(t)>$ of the electron is *increased* with respect to the classical value v. This can be understood from the general nature of the splitting process shown in Fig. 4. Fig. 5 and 6 illustrate this numerically: In Fig. 6 v_{eff} increases from $v = cE_y/B$ at $t = 0$ to roughly 40 times this value between $t \cong 0.5 \times 10^{-4}$s and $t \cong 3 \times 10^{-4}$s. At later times v_{eff} decreases, and it would asymptotically reach the classical value v at times, when practically all the particle density has been scattered to the right hand side of the barrier.

Such an asymptotic situation will never be reached, since the front of the occupied centers $x_{p'}$ on the *right* hand side of the barrier eventually will reach x_2, i. e., the beginning of the next intermediate zone (caused by a potential well, see Fig. 1), and the corresponding probability $|c_{p'}|^2$ will then itself be split and gradually scattered across the well, in the same way as the probability $|c_p(t{=}0)|^2 = 1$ of the incoming Landau function has been scattered across the barrier for $t > 0$. The same scenario will go on with all the other probabilities, which reach this next intermediate zone at subsequent times. As a consequence, after successive scatterings across many barriers and wells, the wave function $\psi(x,y,t)$ will be considerably spread in x-direction (and it will be located simultaneously on different sides of different barriers and wells). Further, its energy will be smeared out in the whole nonadiabatic range $[-E_b,E_b]$.

After sufficiently large times the dc-velocity v_{eff} of the electron tends to a constant value which is considerably higher than the classical value v. This increase with respect to v of the effective velocities of the dc-conducting states constitutes the so-called compensating current, which lies at the basis of the IQHE.

The weak disorder case significantly illustrates the nature of the scattering mechanism in systems described by the general Hamiltonian (1). The difference between the weak disorder and the general case lies in the fact that in the latter an adiabatic solution is a time-dependent linear combination of more than two adiabatic basis functions $w_s(x,y,t)$ (which are composed of more than two unperturbed Landau functions $\psi_p(x,y,t)$ at a given time). This leads to a more complicated time evolution matrix matrix (from t_0 to $n\tau/2$) than in the weak-disorder case. But as in the weak disorder case the nonadiabatic functions are simultaneously composed of parts, which describe propagation with the classical velocity $v = cE_y/B$ (due to nonadiabatic transitions) and of parts, which describe nonclassical propagation (originating from adiabatic components). Note that in our numerical model the states above E_b and below $-E_b$ cannot be described by the weak disorder approximation. But since they are fully adiabatic, they do not contribute to the dc-current, hence their analytic structure is not needed for the present purpose.

We remark, that screening is not considered explicitely in this article. The potential V(x,y) in the Hamiltonian (1) is supposed to be already the total, screened potential. In our explicit model system the slowly varying potential V(x) of eq. (2) could be obtained in a Hartree approximation along the lines developed in Ref. 10 (see also Ref. 6).

4. Absence of Dissipation and Quantized Hall Conductance

We briefly review some consequences of the time evolution of the states in the general case, i. e., independent of the weak disorder approximation. We assume, that the non-adiabatic states are sandwiched between adiabatic states (as in our weak disorder model, Fig. 1) and that the system is in thermodynamic equilibrium with a surrounding heat bath. We consider the case where the Fermi level E_F is contained in the range of adiabatic levels above the nonadiabatic range of the n-th broadened Landau band, and where the temperature is sufficiently low, that the non-adiabatic states below E_F (and the empty non-adiabatic states above E_F) are not affected by inelastic scattering events. From these conditions, together with the unitarity of the time evolution operator, it follows [7], that the occupation probabilities of all basis states $w_s(x,y,t)$, into which the true time-dependent states $\psi^j(x,y,t)$ can be expanded, remains unchanged in the course of time (in particular for nonadiabatic states they remain equal to one). Further, since the modulus of the adiabatic basis states $w_s(x,y,t)$ is periodic in time with period τ (which is extremely small) the total charge density in the system is periodic with period τ, i. e., it is macroscopically constant. This means that there is no charge redistribution, i. e., the presence of the field E_y does not modify the self-consistent potential V(x,y) of the Hamiltonian (1) in the course of time (in particular, no macroscopic field E_x is created, whence $\sigma_{xx} = 0$) and all adiabatic energies $\varepsilon_s(t)$ remain unchanged, periodic with period τ and occupied with unchanged probability for all times. As a consequence the system does not lose energy to the heat bath: there is no dissipation. It follows, that the macroscopic dc-current in y-direction vanishes and $\sigma_{yy} = 0$.

The total Hall current in x-direction I_x results from the sum of the current densities of all time-dependent states in the non-adiabatic zone. Since $x^j(t) = \langle\psi^j(t)| x| \psi^j(t)\rangle = \Sigma_p| c_p{}^j(t)|^2 x_p(t)$, I_x can be thought of beeing created by sums of time-dependent charges of value $q| c_p{}^j(t)|^2$ situated at the different points $x_p(t)$, which themselves move with the constant velocity $v = cE_y/B$.

In our weak disorder model which corresponds to Fig. 1 we find that (under the stated conditions) $I_x = nq/\tau = nq^2E_yL_y/h$, hence $\sigma_{xy} = nq^2/h$. This follows from the fact that there is a spatial region in the system, which is occupied by centers $x_p(t)$ of functions $\psi_p(x,y,t)$, which appear only in fully nonadiabatic states $\psi^j(x,y,t)$. Since these states are characterized by *time-independent* coefficients $| c_p{}^j(t)|^2$) and further, since $\Sigma_j | c_p{}^j(t)|^2 = 1$ and since x_p (t) moves with the velocity $v = cE_y/B$, in the considered region exactly one charge q moves from $x_p(t_o)$ to $x_{p-1}(t_o)$ in the time interval $t_o \leq t \leq t_o + \tau$. In the next time interval of length τ the orbital ψ_p is replaced by ψ_{p+1} and the scenario repeats itself in the same spatial interval, since $x_{p+1}(t+\tau) = x_p(t)$ according to (4) ($q < 0$). Therefore here (i. e., for $x_p(t_o) \leq x \leq x_{p-1}(t_o)$) a current flows in x-direction, which is equal to q/τ per occupied band (including the highest partially occupied band). Now I_x has this value at all x in the strip, since the macroscopic charge density is constant in time. (More precisely we have $\{I_x\} = P_{max}q/\tau$, where $\{..\}$ denotes the average over a time period τ and P_{max} is the highest nonadiabatic transition probability in the band. In our numerical example $P_{max} = P(E = 0) \cong 1 - 10^{-8}$.) In the general case the same situation of quantized Hall conductance occurs (under the stated conditions), if there exist at least two adjacent centers $x_p(t)$ and $x_{p+1}(t)$, such that the corresponding occupation numbers $| c_p{}^j(t)|^2$ and $| c_{p+1}{}^j(t)|^2$ are time independent (with negligeable corrections for all j) for at least a time interval of length τ.

5. Discussion

The total dc-current $I_x = nq/\tau$ of the previous section can be written as nN_Lqv, where N_L is the number of states in a Landau band. On the other hand I_x can also be expressed as the sum of all the particle currents $qv^j{}_{eff}$, where j runs over all nonadiabatic states below E_F. Since their number per band is less than N_L, the average v_{eff} must be higher than v. One therefore can say, that the nonadiabatic states carry an extra current which just compensates the loss of current from the adiabatic states and from the adiabatic components of the intermediate states. Our analysis shows that this compensating current is not created by increased *stationary* single-particle currents, as one might think on classical grounds. Instead it results from processes during which *single-particle* charge density disappears in small areas and simultaneously reappears at different locations. This nonclassical propagation results from the time-dependent scattering between localized Landau functions. This process may be very complex in the general case, but in our weak disorder model we were able to give a detailed description.

The total macroscopic Hall current $\{I_x\}$ is linear in E_y (and so are the single-particle currents associated with the fully non-adiabatic states and with the fully non-adiabatic fractions of the intermediate states). However for a correct description of the time evolution of nonadiabatic and intermediate states by an expansion with respect to E_y all orders have to be included., see also [7,8]. Now we have seen that the *full* time-dependent behaviour of each non-adiabatic state is necessary for a correct quantum mechanical description of the Hall current I_x in our systems. Therefore a linear response approximation is inadequate for a quantum mechanical theory of quantized Hall conductance described by a Hamiltonian of type (1). This result is quite remarkable, since most of such theoretical approaches presented so far have been placed into the framework of linear response theory.

Therefore our results call for a reexamination of microscopic theories of the IQHE (associated with bulk currents), which are based either on linear response theory or on the adiabatic time evolution, in particular if Hamiltonians similar to (1) (including boundary conditions) are used. Gauge invariance [11] and topological approaches [12] fall into this class. Indeed, a recent analysis [8, 13, 14] has shown, that these two theories are based on incorrect assumptions.

Quite generally the nonadiabatic transition probabilities increase with increasing E_y. Therefore the range of nonadiabatic (i. e. conducting) states increases, i. e. the width of the Hall plateaus shrinks, with increasing E_y, i. e., with increasing Hall current I_x. The IQHE breaks down, when E_y is sufficiently high, that all fully adiabatic (i. e. nonconducting) states have disappeared in the highest occupied band. If one extrapolates from the weak disorder to the general case, one is lead to the conclusion that for sufficiently high E_y each nonadiabatic state essentially describes again a linear, classical current $qv = qcE_y/B$ and that consequently, for sufficiently high Hall currents I_x, the Hall conductance essentially shows the linear, classical behaviour without plateaus (as observed in experiments).

In this review we have described the time-dependent scattering mechanism, which causes the IQHE [1]. In particular we showed that processes nonlinear with respect to the electric Hall field are essential. In the present discussion edge effects have been neglected, however such effects can be included by a suitable modification of the boundary conditions. We remark that effects caused by edge currents are successfully described by linear response theory [3]. Such an approximation can be justified within our general framework, since at the edges the electric field is usually so high, that all edge states are fully nonadiabatic. It is however important to note, that such edge states do not carry compensating currents.

References

1. K. von Klitzing, G. Dorda and M. Pepper, Phys. Rev. Lett. 45, 494 (1980)
2. For a review see e. g. *The Quantum Hall Effect*, edited by R.E. Prange and S.M. Girvin (Springer-Verlag, New York, 1987)
3. M. Buttiker in *Semiconductor and Semimetals* in the volume *Nanostructured Systems*, ed. Mark A. Reed, (Academic Press, New York, 1990).
4. F. Kuchar, in *Festkörperprobleme (Advances in Solid State Physics)*, edited by U. Rössler (Vieweg, Braunschweig 1988), Vol. 28, p. 45.
5. J. Riess, Phys. Rev. B38, 3133 (1988)
6. J. Riess, Z. Phys. B - Condensed Matter 77, 69 (1989)
7. J. Riess, Phys. Rev. B41, 5251 (1990)

8. J. Riess, J. Phys. France 51, 815 (1990)
9. J. Riess and C. Duport, unpublished.
10. V. Gudmundsson and R. R. Gerhardts, in *The Application of High Magnetic Fields in Semiconductor Physics II*, Vol 87 of *Springer Series in Solid State Sciences,* edited by G. Landwehr (Springer, Berlin, 1989), and references therein.
11. R. B. Laughlin, Phys. Rev. B23, 5632 (1981); see also R. E. Prange, section 1.7 of Ref. 2
12. For a review see D. J. Thouless, chapt. 4 of Ref. 2
13. J. Riess, Europhys. Lett. 12, 253 (1990)
14. J. Riess, Solid State Comm. 74, 1257 (1990)

Transport Theory of the Quantized Hall Effect

M. Büttiker

IBM T.J. Watson Research Center, Yorktown Heights, NY 10598, USA

Abstract: A transport theory is discussed which views the sample as a target permitting transmission and reflection of carriers. This approach allows us to include current and voltage contacts. The approach is applied to the quantized Hall effect. The approach is extended to include thermal fluctuations at equilibrium and to include fluctuations away from the average currents in the presence of transport. These fluctuations are discussed for a sample with contacts which exhibit internal reflection. It is shown that even in the presence of internal reflection there is no shot noise as long as the contact resistances are quantized.

1. Introduction

In this paper we briefly review recent progress in the formulation of a transport theory [1] of the quantized Hall effect [2]. For the first time since the discovery of the quantized Hall effect, a transport theory has come into existence which is closely related to experimental reality. We briefly review some key elements of this theory. We are concerned with standard resistance measurements [3] but also extend our considerations to a discussion of the fluctuation dissipation theorem and to a discussion of noise in the presence of current transport [4]. The theory is exemplified by discussing particularly striking experiments which show how the Hall effect can be manipulated with the help of contacts [5,6].

Much of the experimental and theoretical literature on the quantized Hall effect presents results in terms of resistivities ρ_{xx} and ρ_{xy}. Such a notation suggests that transport in high magnetic field can be characterized by only two transport coefficients, that the transport coefficients reflect a local relationship between current densities and electric field and that resistances scale linearly like $R_{xx} = \rho_{xx}L/W$ where L is the length and W the width of the conductor. The theory presented below in accordance with much experimental evidence overturns these notions.

Experiments are performed on conductors with boundaries. Contacts are used to bring current into the sample and take it out, and contacts are used to measure voltages. If inelastic scattering is not sufficiently strong for carriers to lose memory of the injection conditions, the sample geometry and the prop-

Springer Series in Solid-State Sciences, Vol. 101
High Magnetic Fields in Semiconductor Physics III Editor: G. Landwehr

erties of contacts become of importance. A theory is then needed which is able to treat the geometry and the contacts of the entire conductor [1]. It is remarkable that it is only as a consequence of our understanding of the role of contacts that the existence of very long equilibration lengths [7,8] in the quantized Hall effect has been appreciated. This in turn has led to the formulation of lower bounds for the ultimate precision of the quantized Hall effect based on the scattering properties of the contacts [9].

2. Basic Elements of a Transport Theory

The transport theory which we will use here views the conductor as a target at which incident electrons are either reflected or transmitted. Such a viewpoint is familiar from tunneling theory: That it is a valid viewpoint also for extended systems is less obvious. Landauer [10] has for many years advocated and advanced such an approach. The theory [1,3,4] discussed below is an implementation of such a viewpoint. To introduce a scattering approach we need "asymptotic regions" which permit the definition of incident and outgoing waves. It is convenient to picture these asymptotic regions as perfect leads which support a large number of quantum channels. Fig. 1 taken from Ref. 11 shows a conductor with four contacts. With each contact $\alpha = 1,2,3,..$ we must associate an asymptotic region. Fortunately, the results presented below do not depend on the detailed nature of these asymptotic regions and we can, therefore, make the discussion as simple as possible. We assume that we can characterize each quantum channel by a dispersion $E_{\alpha n}(k) = E_{\alpha n}(0) +$

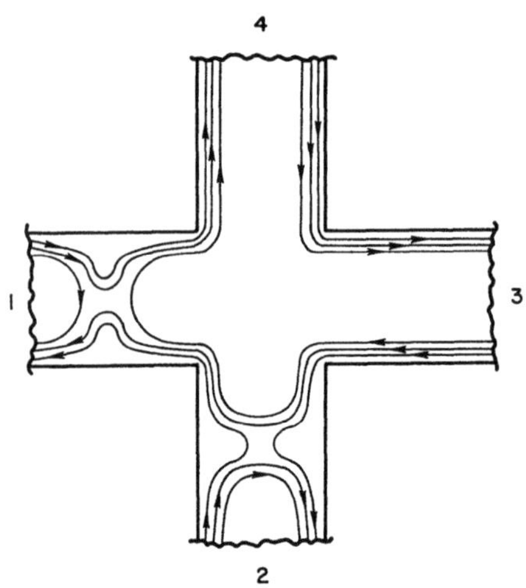

Fig. 1
Two dimensional conductor with four contacts connected to electron reservoirs at chemical potentials μ_α. Two ajacent contacts exhibit internal reflection. The faint lines depict edge states at the Fermi energy. The edge state pattern is shown for the particular case where transmission at the contacts is quantized.

$\hbar^2 k^2/2m_0$. Here $E_{\alpha n}(0)$ is the threshold for conduction of the n-th quantum channel determined by the lateral confinement. With the help of these asymptotic regions we can now define a scattering problem. The scattering matrix which gives the current amplitudes of the outgoing waves in the asymptotic region β in terms of the incident waves in the asymptotic region α is denoted by $s_{\beta\alpha}$. Physically the asymptotic regions described above play the role of electron reservoirs [10,12,3] if they support a large number M_α of quantum channels at the Fermi energy. A large number of quantum channels is needed to obtain an equilibrium distribution of carriers in a reservoir. If the carriers are incident from an asymptotic region which is at equilibrium, the carriers are distributed according to the equilibrium Fermi distribution f_α of this reservoir. (If we attempted to describe an asymptotic region with a probe which supports only a few quantum channels, we would have to provide additional considerations to specify the incident carrier distribution). It is the assumption of equilibrium reservoirs which gives us the simple results presented below. The reservoirs in Fig. 1 do not only present a carrier source or sink but also provide for dissipation and phase randomization [10,12,3].

The transfer of carriers from one reservoir to another can be described with the help of the operator,

$$\hat{\Psi} = \sum_{\alpha m} \frac{1}{\sqrt{2\pi}} \int dk \psi_{\alpha m}(k) \hat{a}_{\alpha m}(k) e^{-i(E_{\alpha m}(k) - \mu_\alpha)t/\hbar}. \tag{1}$$

Here $\psi_{\alpha m}$ is the scattering state with an incident wave of *unit amplitude* in channel m in lead α, and $\hat{a}_{\alpha m}(k)$ annihilates an electron in channel m in reservoir α with intial momentum k. On the statistical average it is only pairs of operators [13] with the same indices which have a non-zero average:

$$< \hat{a}^\dagger_\alpha \hat{a}_\beta > = \delta_\alpha \delta_\beta f_\alpha. \tag{2}$$

For convenience we take here α to denote the complete set of indices $\alpha \equiv (\alpha, m, k_{\alpha m})$. The fluctuations away from the average occupation probability given by Eq. (2) are determined by the statistical average

$$< a^\dagger_\alpha a_\beta a^\dagger_\gamma a_\delta > = \delta_{\alpha\delta}\delta_{\gamma\beta} f_\alpha (1 - f_\gamma) + \delta_{\alpha\beta}\delta_{\gamma\delta} f_\alpha f_\gamma. \tag{3}$$

To evaluate the current in lead γ we must calculate matrix elements of the current operator due to a wave incident in channel m in lead α and a wave incident in channel n in lead β

$$I(\gamma)_{\alpha\beta mn} = \frac{\hbar}{2m_0 i} \int dy_\gamma [\psi^\dagger_{\alpha m}(\gamma) \frac{d\psi_{\beta n}(\gamma)}{dx} - \frac{d\psi^\dagger_{\alpha m}(\gamma)}{dx} \psi_{\beta n}(\gamma)]. \tag{4}$$

In Eq. (3) $\int dy_\gamma$ denotes an integration over the cross section of the conductor in reservoir γ. It is convenient to introduce a normalized matrix with elements

$$\Lambda_{\alpha\beta mn}(\gamma) = (v_{\alpha m})^{-1/2}(v_{\beta n})^{-1/2} I_{\alpha\beta mn}. \tag{5}$$

A calculation shows that the matrix Λ evaluated at the Fermi energy is related to the scattering matrices by [4]

$$\Lambda_{\alpha\beta}(\gamma) = \delta_{\alpha\beta}\delta_{\alpha\gamma} 1_{\gamma} - s^{\dagger}_{\gamma\alpha} s_{\gamma\beta}. \tag{6}$$

All the results stated below can be expressed in terms of this matrix. Below we shall simply state these results without further allusions to their derivation.

3. Average Currents and Resistances

In this paragraph we relate the average net current at a probe to the chemical potentials of the reservoirs [3]. These results are then used to obtain the measured resistances. We use the follwing notation: The scattering matrices which link the incident current amplitudes at reservoir α to the outgoing current amplitudes at probe β are denoted by $t_{\beta\alpha} \equiv s_{\beta\alpha}$ and the scattering matrix which links the incident current amplitudes to the outgoing amplitudes at the same reservoir by $r_{\alpha\alpha} \equiv s_{\alpha\alpha}$. Since all incident channels at a probe are occupied up to the same chemical potential and since the incident current in a narrow energy interval is the same for all channels, it is only the traces of these matrices which enter into the final result,

$$T_{\alpha\beta} = \mathrm{Tr}(t^{\dagger}_{\alpha\beta} t_{\alpha\beta}); \quad R_{\alpha\alpha} = \mathrm{Tr}(r^{\dagger}_{\alpha\alpha} r_{\alpha\alpha}). \tag{7}$$

$T_{\alpha\beta}$ is the total transmission probability for carriers incident in probe β to reach probe α and $R_{\alpha\alpha}$ is the total reflection probability for carriers incident in lead α. The calculation sketched in Section 2 gives for the average current

$$< I_{\alpha} > = (e/h)\int dE[(M_{\alpha} - R_{\alpha\alpha})f_{\alpha} - \sum_{\beta(\neq\alpha)} T_{\alpha\beta} f_{\beta}]. \tag{8}$$

Our calculation is only valid if the chemical potentials differ only slightly from one another. We can expand the Fermi functions in Eqs. (8) away from the equilibrium Fermi function f. Eq. (8) then becomes [3],

$$< I_{\alpha} > = (e/h)\int dE(-df/dE)[(M_{\alpha} - R_{\alpha\alpha})\mu_{\alpha} - \sum_{\beta(\neq\alpha)} T_{\alpha\beta}\mu_{\beta}]. \tag{9}$$

Eqs. (9) can be used to calculate the measured resistances. If carrier flow is from reservoir α to reservoir β and the voltage is measured at lead γ and δ, the resistance is $\mathcal{R}_{\alpha\beta,\gamma\delta} \equiv (V_{\gamma} - V_{\delta})/I$. Here $I \equiv I_{\alpha} = -I_{\beta}$ is the current from carrier source to carrier sink. A voltage measurement at probe γ is characterized by the condition $I_{\gamma} = 0$ and gives $eV_{\gamma} = \mu_{\gamma}$. In the zero temperature limit, for a four-probe conductor, we find [3]

$$\mathcal{R}_{\alpha\beta,\gamma\delta} = (h/e^2)(T_{\gamma\alpha}T_{\delta\beta} - T_{\gamma\beta}T_{\delta\alpha})/D, \tag{10}$$

where D is a subdeterminant of rank 3 of the matrix of transport coefficients of Eq. (9). Finally we shall also use the two terminal resistances which can be measured at a four-terminal conductor: If the voltage measurements are made at the carrier source and sink, the resistance is

$$\mathscr{R}_{\alpha\beta,\,\alpha\beta} = (e^2/h) \tag{11}$$

$$[T_{\gamma\delta}(T_{\beta\gamma} + T_{\beta\delta} + T_{\alpha\gamma} + T_{\alpha\delta}) + (T_{\gamma\alpha} + T_{\gamma\beta})(T_{\alpha\delta} + T_{\beta\delta})]/D.$$

Note that the complicated two terminal resistance is a consequence of the additional probes. For a two-probe conductor [12] the resistance at zero temperature is $\mathscr{R} = (h/e^2)(1/T)$, where $T = Tr(t^\dagger_{21}t_{21}) = Tr(t^\dagger_{12}t_{12})$. The two-terminal resistances are symmetric under magnetic field reversal. The four-terminal resistances are not symmetric or anti-symmetric under field reversal but obey the reciprocity theorem [3], $\mathscr{R}_{\alpha\beta,\,\gamma\delta}(B) = \mathscr{R}_{\gamma\delta,\,\alpha\beta}(-B)$.

As an example, let us discuss the resistances of the conductor in Fig. 1 which exhibits two adjacent contacts with internal reflection. Conductors of this type have been investigated by van Wees et al. [5] and by van Houten et al. [5] using split gates to control the scattering at the contacts. Alternatively, gates which are isolated from the two-dimensional gas can be employed to create barriers for electron motion [14, 15] and can be used to introduce controlled scattering at contacts [6]. The faint lines in Fig. 1 depict edge states at the Fermi energy. We assume that the magnetic field is such that the Fermi energy is between the N-th and N + 1-th bulk Landau level. Therefore, there are at most N edge states which intercept the Fermi energy near the boundary of the sample. The pattern of states at the Fermi energy determines the transmission behavior of the conductor. Contacts 3 and 4 are ideal [1]: Each carrier reaching the contact from inside the conductor leaves the conductor with probability 1. Note that carriers incident from the reservoir (metallic contact) have a large probability for reflection. There are many more states at the Fermi energy in the metallic contact than inside the conductor. An ideal contact is thus characterized by the absence of *internal reflection* [1]. Contacts 1 and 2 exhibit also internal reflection. In Fig. 1 the edge state pattern is shown for the case where transmission through each of these contacts is quantized. Edge states are either totally reflected or permit transmission with probability 1. Denote the transmission probability of contact 1 by T_1 and of contact 2 by T_2. The overall transmission probability for carriers incident in contact 2 to reach contact 1 is denoted by T_{12} as in Eq. (7). Using Eq. (9) or (10) we find a Hall resistance

$$\mathscr{R}_{24,13} = (h/e^2)(\frac{T_{12}}{T_1T_2}). \tag{12}$$

Suppose that the transmission probabilities of contacts 1 and 2 are quantized and given by $T_1 = N_1$ and $T_2 = N_2$. If carriers injected into an edge state remain in this edge state for the entire distance between the contacts the overall transmission probability T_{21} is simply determined by $T_{12} = \min(T_1, T_2) = \min(N_1, N_2)$. Hence the Hall resistance [5]

$$\mathcal{R}_{24,13} = (h/e^2)(\frac{1}{\max(N_1, N_2)}) \tag{13}$$

is quantized and determined by the contact with the larger transmission probability. On the other hand the Hall resistance $\mathcal{R}_{13,24} = -(h/e^2)(1/N)$ is independent of the contact properties 1 and 2 and given by the number of bulk Landau levels below the Fermi energy. The deviation from the bulk value of the Hall resistance given by Eq. (13) is a consequence of the different filling edge of states by the current source contact and the selective detection of this non-equilibrium population [1,5] by the voltage contact 1. Eq. (13) demonstrates that the Hall effect even in high magnetic fields can be manipulated with the help of contacts. The Hall effect according to Eq. (13) is determined by the contacts only, and independent of the bulk properties of the sample!

Let us go one step further and also consider the case where the transmission probabilities at the contacts are not quantized. We follow van Wees et al. [5] and assume that it is only one edge state which is partially reflected or transmitted at a contact. Let us denote the transmission probability of contact 1 by $T_1 = N_1 + \Delta T_1$ and the transmission probability of contact 2 by $T_2 = N_2 + \Delta T_2$. Here N_1 and N_2 is the number of completely transmitted edge states and ΔT_1 and ΔT_2 is the transmission probability of the partially transmitted edge state. Note that $0 < \Delta T_i < 1$. Three differing cases have to be distinguished: (a) For $N_1 > N_2$, we find $T_{12} = N_2 + \Delta T_2$, (b) for $N_1 < N_2$, we find $T_{12} = N_1 + \Delta T_1$, and (c) for $N_1 = N_2 \equiv N_c$, we find $T_{12} = N_c + \Delta T_1 \Delta T_2$. For case (a) the Hall resistance, using Eq. (12) is

$$\mathcal{R}_{24,13} = (h/e^2)(\frac{1}{N_1 + \Delta T_1}) \tag{14}$$

For case (a) the Hall resistance is quantized if $\Delta T_1 = 0$, regardless of the value of ΔT_2. Similarly for case (b) the Hall resistance is quantized whenever the transmission probability at contact 1 is quantzied, i. e. if $\Delta T_2 = 0$. For case (c) the Hall resistance is quantized only if both $\Delta T_1 = \Delta T_2 = 0$. Note that the Hall resistance $\mathcal{R}_{13,24}$ remains quantized regardless of the contact properties of contact 1 and 2. Finally we mention that all longitudinal resistances are zero except for the resistance $\mathcal{R}_{32,41}$. For case (a) this resistance is given by

$$\mathcal{R}_{32,41} = (h/e^2)(\frac{1}{N_1 + \Delta T_1} - \frac{1}{N}). \tag{15}$$

Let us now put these results into a larger context. In the experiments of van Wees et al. [5] the contacts with internal reflection are 1.6μm apart. These experiments demonstrate, therefore, that over this distance carriers remain in the edge state into which they have been injected by the carrier source contact. Surprisingly, experiments by Komiyama et al. [7] on samples with contacts with internal reflection made by metallic diffusion let to the conclusion that carriers remain in the edge state into which they are injected even over much longer distances. Independently van Wees et al. [8] demonstrated that simply by pinching off a voltage contact the Shubnikov-De Haas peaks disappear. In the van Wees et al. [8] experiment the distance between contacts is more than $250\mu m$. As shown by Alphenaar et al. [16] and Müller et al. [6] the configuration discussed here can be used to study the transfer of carriers from one edge state to another. Understanding of the role of contacts has permitted to use them as an important tool for detailed investigations of the conduction process.

Next we mention briefly several recent advances and extensions of the theory and the experiments discussed above. This list is necessarily incomplete. Contacts at the periphery of the sample predominantly couple to one or several of the outer edge states. Faist et al. [17] demonstrate how interior contacts can be used to couple predominantly to the innermost edge state. Away from a Hall plateau we have backscattering via localized or extended states in the bulk. Each contact exhibits then internal reflection. Theories and experiment which address the transition from one plateau to another are discussed by McEuen et al. [18] and by van Son et al. [18]. We have focused on quantizing magnetic fields and the case of very small voltage differences (or small currents). The high current regime is also of interest; van Son and Klapwijk [19] have shown that we can expect injection distributions near a carrier source contact which are far from equilibrium. For further references we guide the reader to a number of recent review papers [20]. Below we discuss the thermal fluctuations at equilibrium and fluctuations in the presence of transport for the conductor of Fig. 1.

4. Current and Voltage Fluctuations at Equilibrium

The calculation outlined in Section 2 can be used to discuss fluctuations of the current away from the average value. We first discuss fluctuations at equilibrium in the absence of a net current. We refer to these fluctuations as Johnson-Nyquist noise [21,22]. These fluctuations are a consequence of thermal fluctuations of the occupation probabilities of the incident states in the reservoirs. For an instructive discussion we refer to Landauer's treatment [23] of the Johnson-Nyquist noise in a two-terminal configuration. The current fluctuations discussed below are obtained if the reservoirs are large enough to present an unlimited capacitance or if the sample is short-circuited [23]. For the multi-terminal conductors of interest here we find that the mean squared current in the frequency interval $\Delta\nu$ at probe α is given by [4]

$$< (I_\alpha)^2 > = 4\Delta\nu kT \int dE(-df/dE)(e^2/h)(\sum_{\beta(\neq\alpha)} T_{\alpha\beta}). \tag{16}$$

Eq. (16) for the case of a two-terminal conductor [23] reduces to the Johnson-Nyquist noise formula $< I^2 > = 2\Delta\nu kTG$, where we have used the two terminal expression for the conductance $G = \int dE(-df/dE)T$, with $T = Tr(t^{\dagger}t)$.

The currents at differing terminals are in general correlated. A calculation gives [4]

$$< I_{\alpha} I_{\beta} > = -2\Delta\nu(e^2/h)kT\int dE(-df/dE)(T_{\alpha\beta} + T_{\beta\alpha}). \tag{17}$$

If we compare Eqs. (16) and (17) with Eq. (9) and take into account that $\Sigma_{\beta(\neq\alpha)} T_{\alpha\beta} = M_{\alpha} - R_{\alpha\alpha}$, we see that the mean square current fluctuations are related to the symmetrized transport coefficients. Eqs. (16) and (17) are, therefore, a manifestation of the fluctuation dissipation theorem. Eq. (16) tells us that the mean square current at a probe is determined by the sum of all transmission probabilities permitting carriers to enter this probe. The current current correlations are determined by the transmission probabilities which directly connect the two probes.

Experimentally it might be simpler to investigate not the current fluctuations but the voltage fluctuations. Below we present results for the voltage fluctuations calculated in the following way: We use the relation between the averaged currents and the average chemical potentials given by Eq. (9) for the fluctuations. We consider, therefore, the fluctuations in the chemical potentials needed to counterbalance the fluctuating currents. We find that the mean square voltage difference measured between any pair of leads is given by [4]

$$< (V_{\alpha} - V_{\beta})^2 > = 4\Delta\nu kT\mathcal{R}_{\alpha\beta,\alpha\beta} \tag{18}$$

where $\mathcal{R}_{\alpha\beta,\alpha\beta}$ is a two-terminal resistance as given in Section 3. More genrelly, the correlation between voltage differences measured across two pairs of leads is given by [4]

$$< (V_{\alpha} - V_{\beta})(V_{\gamma} - V_{\delta}) > = 2\Delta\nu kT(\mathcal{R}_{\alpha\beta,\gamma\delta} + \mathcal{R}_{\gamma\delta,\alpha\beta}). \tag{19}$$

If all indices in Eq. (19) differ, the correlation is determined by a four-terminal resistance. If two indices coincide, the correlation is determined by a three-terminal resistance. For $\alpha = \gamma$ and $\beta = \delta$, Eq. (19) reduces to Eq. (18): the mean square voltage fluctuations are determined by the corresponding two terminal resistance.

Next let us discuss the thermal fluctuations for the conductor in Fig. 1. To be brief we discuss only case (a), i.e. we assume $N_1 > N_2$. The mean square current fluctuations at contact 1 and contact 2 are proportional to $N_i + \Delta T_i$, $i = 1,2$. The mean square current fluctuations at contact 3 and 4 are proportional to N. The correlations between currents at differing probes are as follows;

$$< I_1 I_4 > \propto N_1 + \Delta T_1, \tag{20a}$$

$$< I_4 I_3 > \propto 2N - N_1 - \Delta T_1, \tag{20b}$$

$$< I_3 I_2 > \propto N_2 + \Delta T_2, \tag{20c}$$

$$< I_2 I_1 > \propto N_2 + \Delta T_2, \tag{20d}$$

$$< I_1 I_3 > \propto N_1 - N_2 + \Delta T_1 - \Delta T_2, \tag{20e}$$

$$< I_2 I_4 > = 0. \tag{20f}$$

The mean square voltage fluctuations are determined by the two-terminal resistances Eq. (11). The correlations between voltage fluctuations at differing terminals are either zero or are proportional to the longitudinal resistance Eq. (15). Note that this means in particular that for a conductor with ideal contacts (i.e. if $N_1 = N_2 = N$), all correlations vanish at a Hall plateau. Noise measurements on quantum Hall conductors by Kil et al. [24] show a white noise shoulder for the mean square voltage fluctuations and do not show a white noise shoulder for the correlations.

5. Fluctuations in the Presence of Transport

Now we leave the regime of equilibrium thermal fluctuations and ask about fluctuations in the presence of current transport. We are interested in fluctuations $\Delta I_\alpha = I_\alpha - < I_\alpha >$ away from the average current. For a two terminal conductor Lesovik [25] and Yurke and Kochanski [26] have found that the excess noise is suppressed whenever the transmission matrix has eigenvalues close to one. These papers are concerned with transport in zero magnetic field. Yurke and Kochanski consider a single channel transmission and reflection problem. Lesovik considers a multi-channel problem but assumes at the outset that the scatterer does not mix channels. It is, therefore, of interest to generalize these results. Let us consider a two-terminal conductor. The elements of the scattering matrix S can be grouped together such that carriers incident from the left are characterized by a transmission matrix t_{21} and a reflection matrix r_{11} and carriers incident from the right are characterized by a transmission matrix t_{12} and a reflection matrix r_{22}. Using Eq. (6) we find in the zero temperature limit [4],

$$< (\Delta I)^2 > = 2(e^2/h)\Delta\nu \mid eV \mid \mathrm{Tr}(r^\dagger{}_{11} r_{11} t^\dagger{}_{12} t_{12}) \tag{21}$$

$$= 2(e/h^2)\Delta\nu \mid eV \mid \mathrm{Tr}(t^\dagger{}_{21} t_{21} r^\dagger{}_{22} r_{22}).$$

Eq. (21) is valid for arbitrary elastic scattering and is valid in the presence of a magnetic field. Alternatively, we could express this result with the help of the transmission matrix only. Since $r^\dagger{}_{11} r_{11} + t^\dagger{}_{12} t_{12} = 1$, it follows that the trace in Eq. (21) can also be expressed as $\mathrm{Tr}(t_{12} t^\dagger{}_{12}(1 - t_{12} t^\dagger{}_{12}))$. To make contact with the result of Lesovik we note that $t^\dagger{}_{12} t_{12}$ is a hermitian matrix and can, therefore, be brought into a diagonal form. Denote the eigenvalues of the matrix $t^\dagger{}_{12} t_{12}$ by $T_n(B)$ and the eigenvalues of the matrix $r_{11}{}^\dagger r_{11}$ by $R_n(B)$. Since $R_n(B) + T_n(B) = 1$ and $R_n(B) = R_n(-B)$, the eigenvalues $T_n(B)$ are also symmetric functions of the magnetic field. Eq. (21) for the excess noise takes the form,

$$< (\Delta I)^2 > = 2(e^2/h)\Delta\nu \mid eV \mid \sum T_n(1 - T_n). \tag{22}$$

This is the result given by Lesovik [25]. If all the eigenvalues of the transmission matrix are small compared to one, Eq. (22) reduces to the standard expression for shot noise

$$< (\Delta I)^2 > = 2e\Delta\nu I. \tag{23}$$

Therefore, whenever the transmission matrix has eigenvalues equal or comparable to 1, Eqs. (21) and (22) tell us that the shot noise will be smaller than expected from the standard result. Interestingly, experiments on double barrier resonant tunneling structures [27] and experiments on constrictions by Li et al. [28] indeed do show a reduction of the shot noise below the value given by Eq. (23).

From the calculation scheme outlined in Section 2 we can also derive an expression for the current fluctuations in the presence of transport for multiprobe conductors. We find in the zero temperature limit

$$< \Delta I_\alpha \Delta I_\beta > = 2\Delta\nu(e^2/h)\int dE \sum_{\gamma,\,\delta(\gamma\neq\delta)} Tr(s^\dagger_{\alpha\gamma} s_{\alpha\delta} s^\dagger_{\beta\delta} s_{\beta\gamma}) f_\alpha(1 - f_\beta). \tag{24}$$

The Fermi factors in Eq. (24) serve only to multiply the coefficients in Eq. (24) with the appropriate chemical potential difference. The Fermi function f_α at zero temperature is equal to 1 for energies below the chemical potential μ_α and is zero for energies E above the chemical potential. For the mean square fluctuations it is seen that the coefficients in Eq. (24) are real since each s matrix occurs together with its hermitian conjugate. For the correlations of current fluctuations at differing probes, the coefficients in Eq. (24) are in general not real. That the entire result is real and negative can be seen by invoking the unitary relations $\Sigma_\delta s_{\alpha\delta} s^\dagger_{\beta\delta} = 0$ if $\alpha \neq \beta$. Again we calculate the voltage fluctuations by using Eq. (9) for the fluctuating currents and fluctuating voltages instead of the averaged quantities.

Before returning to our example consider a single point contact in a high magnetic field [29]. Suppose that for this two-terminal configuration the transmission probability is given by $T = N + \Delta T$ where N corresponds to the number of perfectly transmitted edge channels and ΔT is the partial transmission due to, at most, one edge state. In this case, the shot noise is determined solely by the partially transmitted edge channel, $< (\Delta I)^2 > = 2(e^2/h)\Delta\nu\Delta T\Delta R \mid eV \mid$. The voltage fluctuations are $< (\Delta V)^2 > = 2(h/e^2)\Delta\nu(\Delta T\Delta R/(N + \Delta T)^2) \mid eV \mid$. Therefore, the excess voltage fluctuations increase as the the number of channels decreases.

We now return to our example, the conductor in Fig. 1. We first consider the situation when carrier flow is form probe 2 to probe 4. Using Eq. (24) we find

$$< (\Delta I_1)^2 > = 2(e^2/h)\Delta\nu[\Delta T_2\Delta R_2(\mu_2 - \mu_3) + \Delta T_1\Delta R_1(\mu_1 - \mu_3)], \tag{25a}$$

$$< (\Delta I_2)^2 > = 2(e^2/h)\Delta\nu\Delta T_2\Delta R_2(\mu_2 - \mu_3), \tag{25b}$$

$$< (\Delta I_4)^2 > = 2(e^2/h)\Delta\nu\Delta T_1\Delta R_1(\mu_1 - \mu_3), \tag{25c}$$

and $< (\Delta I_3)^2 > = 0$. We re-emphasize that Eq. (25) is a consequence of a net carrier trasnport from contact 2 to contact 4. Thus the current is driven by the chemical potential difference $\mu_2 - \mu_4$. Note that for the conductor of Fig. 1, we have $\mu_3 = \mu_4$ and $(N_1 + \Delta T_1)(\mu_1 - \mu_4) = (N_2 + \Delta T_2)(\mu_2 - \mu_4)$. We could, therefore, express Eq. (25) in terms of the chemical potential difference $(\mu_2 - \mu_4)$ between carrier source and sink or as a function of the net current driven through the sample. There is no backscattering into probe 3, and consequently the mean square current fluctuations at this probe are zero. Partial reflection at contact 2 give rise to fluctuations both at contact 2 and at contact 1. Partial refelection at contact 1 gives rise to fluctuations at contact 1 and at contact 4. The current current correlations which are non-vanishing are

$$< \Delta I_1 \Delta I_2 > = - 2(e^2/h)\Delta\nu\Delta T_2\Delta R_2(\mu_2 - \mu_3), \tag{26a}$$

$$< \Delta I_1 \Delta I_4 > = - 2(e^2/h)\Delta\nu\Delta T_1\Delta R_1(\mu_1 - \mu_3). \tag{26b}$$

As in the case of a single point contact, excess fluctuations in the multiterminal conductor of Fig. 1 occur only if there is partial reflection and transmission of an edge state. If transmission at all the contacts is quantized, i.e. if $\Delta T_1 = \Delta T_2 = 0$ there is no excess noise. This is in particular the case if all contacts are ideal, i.e. if $N_1 = N_2 = N$. Remarkably there is no excess noise even if the contacts allow for internal reflection as long as internal reflection is quantized.

Consider next an experiment which analyzes the voltage fluctuations between probes 1 and 3 in the presence of a carrier flow from probe 2 to probe 4. Using the current correlations as given above and using Eq. (9) for the fluctuating voltages and currents instead of the averaged quantities, we find

$$< (\Delta V_1 - \Delta V_3)^2 > = 2\Delta\nu(h/e^2)\frac{\Delta R_1\Delta T_1}{(N_1 + \Delta T_1)^2}(\mu_1 - \mu_3). \tag{27}$$

The mean square voltage fluctuations vanish if the Hall resistance $\mathcal{R}_{24,13}$ given by Eq. (14) is quantized, i.e. if $\Delta T_1 = 0$ independent of ΔT_2. The mean squared voltage fluctuations measured between leads 1 and 3 are only sensitive to the shot noise produced at contact 1 if $N_1 > N_2$.

In this paper we have presented a transport theory for multiprobe conductors and have applied it to a conductor in a high magnetic field with two adjacent contacts with internal reflection. We found that this conductor exhibits no excess noise if the internal reflection is quantized. Conductors with a simpler geometry have been analyzed in Ref. 4. The absence of noise proportional to the current driven through the sample is a very remarkable property of the quantum Hall effect. To our knowledge an experimental demonstration of the lack of shot noise in the quantized Hall regime has not been given.

We also point out that it would be of interest to analyze the shot noise in the fractional quantum Hall effect. Transport theories for the fractional quantum Hall effect, extending the point of view taken in this paper, have been provided by Beenakker [30], MacDonald [30], and Chang [30]. Since the suppression of shot noise depends sensitively on the eigenvalue structure of the transmission matrix, an analysis of the shot noise of the fractional quantized Hall effect could be very revealing.

References

1. M. Büttiker, Phys. Rev. **B38**, 9375 (1988).
2. K. von Klitzing, Rev. Mod. Phys. **58**, 9375 (1985).
3. M. Büttiker, Phys. Rev. Lett. **57**, 1761 (1986); IBM J. Res. Develop. **32**, 317 (1988); H. U. Baranger and A. D. Stone, Phys. Rev. **B40**, 8169 (1989); O. Viehweger, Z. Phys. **B77**, 135 (1989); P. N. Butcher, J. Phys. **C2**, 4869 (1990).
4. M. Büttiker, (unpublished); in the proceedings of the "20-th International Conference on Physics in Semiconductors", edited by J. Joannopoulos (World Scientific Publishing Co., Singapore).
5. B. J. van Wees, E. M. M. Willems, C. J. P. M. Harmans, C. W. J. Beenakker, H. van Houten, J. G. Williams, C. T. Foxon and J. J. Harris, Phys. Rev. Lett. **62**, 1181 (1989); H. van Houten, C. W. J. Beenakker, J. G. Williams, M. E. I. Brockaart, P. H. M. van Loosdrecht, B. J. van Wees, J. E. Moij, C. T. Foxon, and J. J. Harris, Phys. Rev. **B39**, 8556 (1989).
6. G. Müller et al., D. Weiss, S. Koch, K. von Klitzing, H. Nickel, W. Schlap, and R. Lösch, (unpublished).
7. S. Komiyama, H. Hirai, S. Sasa, and F. Fujii, Solid State Commun. **73**, 91, (1990); S. Komiyama, H. Hirai, S. Sasa, and S. Hiamizu, Phys. Rev. **B40**, 12566 (1989); S. Komiyama and H. Hirai, Phys. Rev. **B40**, 7767 (1989).

8. B. J. van Wees, E. M. M. Willems, L. P. Kouwenhoven, C. J. P. M. Harmans, H. G. Williamson, C. T. Foxon, and J. J. Harris, Phys. Rev. **B39**, 8066 (1989); R. J. Haug and K. von Klitzing, Europhysics Lett. **10**, 489 (1989).

9. H. Hirai and S. Komiyama, J. Appl. Phys. **68**, 655 (1990).

10. R. Landauer, Z. Phys. **B21**, 247 (1975); Z. Phys. **B68**, 217 (1987).

11. M. Büttiker, in "Nanostructure Physics and Fabrication", M. A. Reed and W. P. Kirk, eds. (Academic Press, Boston, 1989). p. 319.

12. Y. Imry, in "Directions in Condensed Matter Physics", G. Grinstein and G. Mazenko, eds. (World Scientific, Singapore, 1986), p. 101.

13. L. D. Landau and E. M. Lifshitz, "Statistical Physics", Part 1, (Pergamon Press, Oxford, 1980). p. 354

14. S. Washburn, A. B. Fowler, H. Schmid, and D. Kern, Phys. Rev. Lett. **61**, 2801 (1988).

15. R. J. Haug, A. H. MacDonald, P. Streda and K. von Klitzing, Phys. Rev. Lett. **61**, 2797 (1988).

16. B. W. Alphenaar, P. L. McEuen, R. G. Wheeler, and R. N. Sacks, Phys. Rev. Lett. **64**, 677 (1990); T. Martin and S. Feng, Phys. Rev. Lett. **64**, 1971 (1990).

17. J. Faist, H. P. Meier and P. Gueret, (unpublished).

18. P. L. McEuen, A. Szafer, C. A. Richter, B. W. Alphenaar, J. K. Jain, A. D. Stone, R. G. Wheeler, R. N. Sacks, Phys. Rev. Lett. **64**, 2062 (1990); P. C. von Son and T. M. Klapwijk, (unpublished).

19. P. C. van Son and T. M. Klapwijk, Europhys. Lett. **12**, 429 (1990); P. C. van Son, G. H. Kruithof, and T. M. Klapwijk, (unpublished).

20. M. Büttiker, in "Nanostructured Systems" in Semiconductor and Semimetals, edited by M. Reed, (Academic Press, Orlando, Florida). (unpublished); C. W. J. Beenakker and H. van Houten, in "Nanostructured Systems" in Semiconductor and Semimetals, edited by M. Reed, (Academic Press, Orlando, Florida). (unpublished); A. M. Chang, "The Hall Effect in Quantum Wires", (unpublished);

21. J. B. Johnson, Phys. Rev. **29**, 367 (1927).

22. H. Nyquist, Phys. Rev. **32**, 229 (1928).

23. R. Landauer, Physica **D38**, 226 (1989).

24. A. J. Kil, R. J. J. Zijlstra, M. F. H. Schuurmans, J. P. Andre', Phys. Rev. **B41**, 5169 (1990); A. J. Kil, Thesis.

25. G. B. Lesovik, JETP Lett. **49**, 594 (1989).

26. B. Yurke and G. P. Kochanski, Phys. Rev. **B41**, 8184 (1990).

27. Y. P. Li, A. Zaslavsky, D. C. Tsui, M. Santos, and M. Shayegan, Phys. Rev. **B41**, 8388 (1990).

28. Y. P. Li, D. C. Tsui, J. J. Hermans, J. A. Simmons, G. W. Weiman, (unpublished).

29. M. Büttiker, Phys. Rev. **B41**, 7906 (1990) and references therin; B. J. van Wees, L. P. Kouwenhoven, H. van Houten, C. W. J. Beenakker, J. E. Mooij, C. T. Foxon, and J. J. Harris, Phys. Rev. **B38**, 3625 (1988).

30. C. W. J. Beenakker, Phys. Rev. Lett. **B39**, 216 (1990); A. H. MacDonald, Phys. Rev. Lett. **64**, 220 (1990); A. M. Chang, Solid State Commun. **74**, 871 (1990); Experiments probing regions of differing filling factors are reported by A. M. Chang and J. E. Cunningham, Solid State Commun. **72**, 651 (1989) and by L. P. Kouwenhoven et al. Phys. Rev. Lett. **64**, 685 (1990).

Edge Channel Transport Under Quantum Hall Conditions

G. Müller[1], D. Weiss[1], S. Koch[1], K. von Klitzing[1], H. Nickel[2], W. Schlapp[2], and R. Lösch[2]

[1]Max-Planck-Institut für Festkörperforschung, Heisenbergstr. 1, W-7000 Stuttgart 80, Fed. Rep. of Germany

[2]Forschingsinstitut der Deutschen Bundespost Telekom, W-6100 Darmstadt, Fed. Rep. of Germany

By selectively populating current carrying edge states using negatively biased Schottky gates we investigate adiabatic transport in the Quantum Hall regime. By the use of two additional Schottky gates we have the possibility to electrically connect or disconnect two additional Hall probes to the region where first adiabatic transport is observed. The now experimentally observed change from adiabatic to equilibrated transport demonstrates the importance of ohmic contacts as energy and phase randomizing reservoirs. The versatile sample geometry also allows the experimental investigation of the bend resistance and its dependence on current level. The experiments show strong evidence for current carrying edge states.

The Landauer-Büttiker-formalism [1,2]has been successfully applied to the Quantum Hall effect explaining the quantized resistance values as long as the Hall voltage is small compared to $\hbar\omega_C/e$. Within this picture the transport is governed by the different electrochemical potentials of the contacts which are connected by one-dimensional states at the edges of the sample. The number of the edge channels involved is given by the filling factor b in the bulk of the two-dimensional electron gas (2DEG).

An ideal contact within this picture populates all emitted edge channels up to its electrochemical potential μ_i equally. A non ideal contact, on the other hand, populates the outgoing channels unequally and reflects part of the incoming channels. Such nonideal contacts together with the long equilibration length observed for selectively populated edge channels influence the accuracy of the Quantum Hall plateaus.

We systematically investigate a situation where we selectively populate the edge channels using Schottky gates. Schottky gates are much easier to prepare compared to point contacts [3] used in previous experiments and of course better controllable than disordered contacts [4] and therefore allow the performance of the experiments in a clean and simple way. We demonstrate for the first time the ability of Schottky gates to achieve adiabatic transport between two Schottky gates over macroscopic distances. Here adiabatic means transport with a nonequal distribution of the net current among the different available edge states in contrast to equilibrated transport.

The layout of our devices is sketched in Fig.1. We have evaporated two gates (NiCr/Au) across the Hall bar. The magnetic field B is perpendicular to the 2DEG. Using contact 1 and 4 as current source and sink and contacts 2 and 3 as voltage probes, respectively, one can calculate this resistance within the Landauer - Büttiker - formalism: $R^{ad}_{14,23} = (h/e^2)(g^{-1} - b^{-1})$. This is valid for perfect adiabatic transport. To achieve equilibrated transport one experimental possibility is to use a large enough gate finger spacing d in Fig.1a. Then, due to interchannel scattering there is a coupling between the transmitted and the circulating edge states. Another possibility is to incorporate additional contacts to the region between the two gate fingers. Because of the strong scattering within the contacts the incident edge states equilibrate. The equilibrated value for the four point resistance is given by $R^{eq}_{14,23} = (2h/e^2)(g^{-1} - b^{-1})$, which is twice as much as the "adiabatic resistance" $R^{ad}_{14,23}$. Thus by connecting or disconnecting the reservoirs to the Hall bar this resistance should change by a factor of two.

Springer Series in Solid-State Sciences, Vol. 101
High Magnetic Fields in Semiconductor Physics III   Editor: G. Landwehr

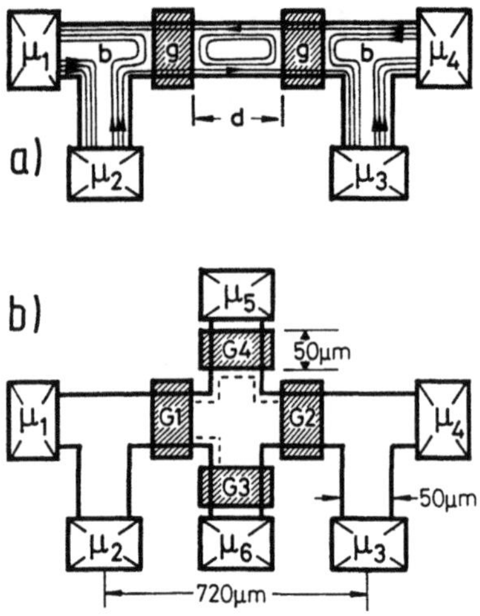

Fig.1. Schematic layout of the investigated samples. In structure (a) the edge channels for a filling factor $g = 1$ under both gates and $b = 3$ in the bulk are sketched. The four gates (b) G1, G2, G3, G4 with their corresponding filling factors g_1, g_2, g_3, g_4 underneath. Relevant lengths (dashed line): G1 $\rightarrow$ G2 $= 110\mu m$, G1 $\rightarrow$ G3 $= 45\mu m$.

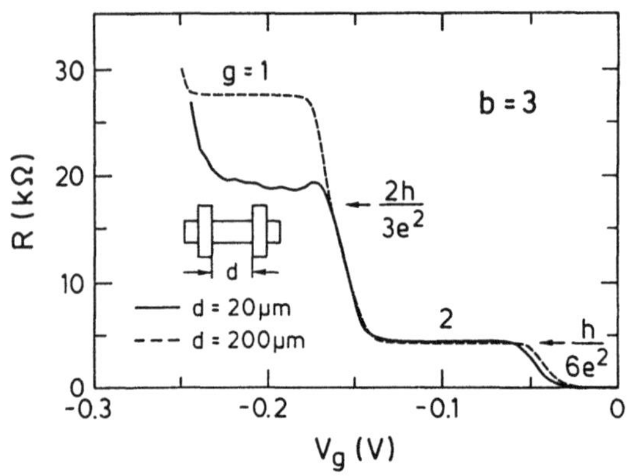

Fig.2. $R_{14,23}$ as function of gate voltage. The expected plateau values for complete adiabatic transport between the two gates are marked by arrows.

The carrier densities N_s in our GaAs/AlGaAs heterojunction devices range between $1.7 \times 10^{11} cm^{-2}$ and $2.0 \times 10^{11} cm^{-2}$, the mobilities between $1.1 \times 10^6 cm^2/Vs$ and $1.4 \times 10^6 cm^2/Vs$ at liquid helium temperature. The alloyed AuGe/Ni contacts have contact resistances of typically 400Ω. The experiments are carried out in a $^3He/^4He$ dilution refrigerator at about $30mK$. The resistances are measured by applying an ac-current ($10nA, 10Hz$) and measuring the corresponding voltage drops (accuracy 3%) by Lock-in techniques. In Fig.2 we demonstrate the ability of Schottky gates (using the geometry of Fig.1a) to selectively populate edge channels between the two gates for the filling factor $b = 3$ in the bulk. The resistance $R_{14,23}$ as a function of the gate voltage V_g applied to the two Schottky gates shows a series of plateaus corresponding to an integer filling factor g under the gates. The spinsplit edge channels are resolved. The resistance plateau corresponding to $g = 2$ has the full adiabatic value not only for a gate finger spacing of $20\mu m$ but also for $200\mu m$. At $g = 1$ and for $d = 20\mu m$ one observes nearly full adiabatic transport. However, for $g = 1$ and $d = 200\mu m$ the plateau moves towards the equilibrated value since interchannel scattering pushes the edge channel population

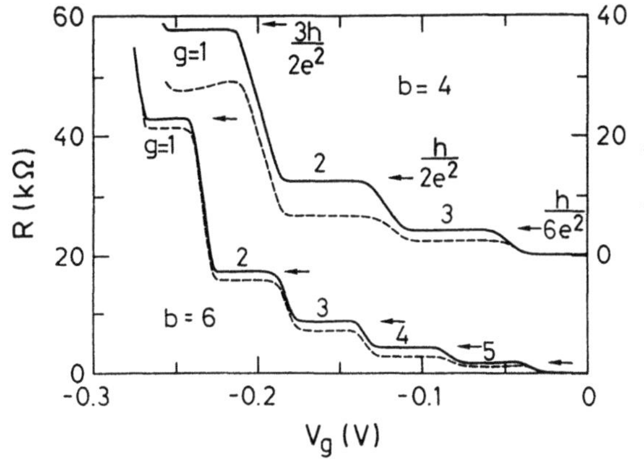

Fig.3. $R_{14,23}$ for connected (solid line) and disconnected (dashed line) Hallprobes (contacts 5 and 6 in Fig.1b) as a function of the gate voltage V_g applied to G1 and G2 ($g_1 = g_2 = g$) for filling factors $b = 4$ (upper curves) and $b = 6$ (lower curves) in the bulk. The arrows indicate the expected plateau values for equilibrated transport.

towards equilibrium. Interchannel scattering depends on the spatial separation of the edge channels [5] and is exponentially suppressed with increasing edge channel separation. The edge channel spacing is largest between the two highest occupied Landau levels (Spin neglected) and explains the long mean free path observed for $g = 2$ in agreement with previous experiments [6].

The role of contacts in the Quantum Hall regime can be demonstrated by connecting or disconnecting metallic reservoirs (again using Schottky gates) to the region between the two gate fingers as is sketched in Fig.1b. The results with the Hall probes connected ($g_1 = g_2 < b = g_3 = g_4$) are displayed in Fig.3 (solid lines) where $R_{14,23}$ is plotted as a function of the negative gate voltage applied to $G1$ and $G2$ for $b = 4$ (upper curves) and $b = 6$ (lower curves). With the Hall probes connected $R_{14,23}$ has the equilibrated value indicated by arrows thus demonstrating convincingly the energy and phase randomizing of the edge channels in the contacts. For $b = 4$ we are able to switch to nearly perfect adiabatic transport by electrically disconnecting the contacts 5 and 6 (see for example $g = 2$, dashed line). For $b = 6$ we do not observe the full adiabatic plateau value for disconnected Hall contacts (dashed line) indicating that interedge channel scattering is not negligible.

Another example nicely demonstrating the Landauer-Büttiker edge channel picture is the bend resistance $R_{12,34}$ (inset Fig.4). If edge channels starting from contact 1 are deflected and move adiabatically along the negatively biased gates towards 4 the quantized resistance $R^{ad}_{12,34} = (h/e^2)(g^{-1} - b^{-1})$ is expected. For equilibrated transport $R^{eq}_{12,34} \equiv 0$. The bend resistance is displayed in Fig.4. For $b = 3$ and $g = 2$ $R_{12,34} = h/6e^2(\pm 3\%)$ is measured. Increasing the current level increases the coupling of the edge channels with electrochemical potential μ_1 (from contact 1) with those coming from contact 2 and the bend resistance moves towards 0.For a reversed magnetic field the bend resistance (now edge channels moving clockwise in Fig.1.a) $R_{12,34} \equiv 0$.

In summary we have demonstrated that Schottky gates provide an excellent tool to selectively populate edge channels in versatile geometries. We have demonstrated the phase and energy randomizing character of ohmic contacts and first presented results of adiabatic transport in spin polarized edge channels.

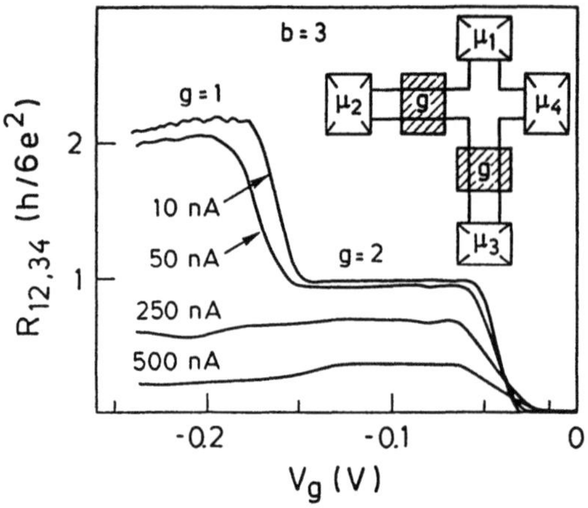

Fig.4. Bend resistance as fuction of gate voltage applied to both gates (shaded areas). The different traces correspond to different applied currents.

References

[1] R. Landauer, IBM J.Res.Dev. **1**, 223 (1957).

[2] M. Büttiker,Phys. Rev. Lett. **57**, 1761 (1986).

[3] B. J. van Wees, E. M. M. Willems, C. J. P. M. Harmans, C. W. J. Beenakker, H. van Houten, J. G. Williamson, C. T. Foxon, and J. J. Harris, Phys. Rev. Lett. **62**, 1181 (1989).

[4] S. Komiyama and H. Hirai, Phys. Rev. **B40**, 7767 (1989), and S. Komiyama, H. Hirai, S. Sasa, and T. Fujii, Solid Sate Comm. **73**, 91 (1990).

[5] T. Martin and S. Feng, Phys. Rev. Lett. **64**, 1971 (1990) and J. J. Palacios and C. Tejedor, to be published.

[6] B. W. Alphenaar, P. L. McEuen, R. G. Wheeler, and R. N. Sacks, Phys. Rev. Lett. **64**, 677 (1990).

Effect of Disorder and Gate Barrier on Edge States

T. Ohtsuki[1] *and Y. Ono*[2]

[1]Physikalisch-Technische Bundesanstalt, Bundesallee 100, W-3300 Braunschweig, Fed. Rep. of Germany
[2]Department of Physics, Faculty of Science, Toho University, Miyama 2-2-1, Funabashi, Chiba 274, Japan

Abstract. The behavior of edge states in the quantum Hall regime is studied using numerical methods. The mixing between edge states and bulk extended states, as well as the mixing among edge states in different Landau levels, are shown to depend strongly on correlation lengths of random potentials. The properties of edge states in the presence of a gate barrier are also presented. When the guiding center energy of the edge states in the n-th Landau level, $\Delta E_\nu \equiv \epsilon_\nu - (n+1/2)\hbar\omega_c$, satisfies $\Delta E_\nu << U_g$, where U_g is the strength of the barrier potential, the edge states are completely reflected, while they are transparent when $\Delta E_\nu > U_g$. In the intermediate region ($\Delta E_\nu \lesssim U_g$), the edge states are shown to have significant amplitude in the gate barrier region due to quantum tunneling.

1. Introduction

Recent microfabrication technology has made possible to prepare very narrow 2-dimensional samples, and many interesting experiments have been done in the presence of high perpendicular magnetic fields [1,2]. The results of these experiments are interpreted by the recently proposed transmission approach to the quantum Hall effect [3]. In this approach, the role of contacts is explicitly taken into account, and the edge states play an important role. The quantum mechanical properties of edge states, however, are not well understood. We present the results of a numerical calculation of the edge states in the presence of impurities or a gate barrier.

2. Model

We consider the following Hamiltonian,

$$H = \frac{1}{2m}\left(\mathbf{p} + \frac{e}{c}\mathbf{A}\right)^2 + U_c(x) + V(\mathbf{r}) + V_g(y), \tag{1}$$

with $U_c(x)$ a hard wall confinement potential which is 0 within the system $0 < x < L_x$, and ∞ otherwise. We assume the Gaussian form for the random potential,

$$V(\mathbf{r}) = \sum_i \frac{u_i}{\pi a^2} \exp\left(-\frac{(\mathbf{r}-\mathbf{r}_i)^2}{a^2}\right), \tag{2}$$

where u_i, x_i and y_i are random variables distributed uniformly within the region $[-u/2, u/2]$, $[0, L_x]$ and $[0, L_y]$, respectively. When the range of the potential a is much larger than the magnetic length $l (\equiv \sqrt{c\hbar/eB})$, the eigenstates become similar to the equipotential lines, while when $a << l$, the quantum jump of cyclotron orbits becomes considerably large. The Landau band width Γ is estimated from the self-consistent Born approximation [4] as $\sqrt{n_i u^2/\pi(l^2+a^2)}$, with n_i the density of impurities.

Springer Series in Solid-State Sciences, Vol. 101
High Magnetic Fields in Semiconductor Physics III Editor: G. Landwehr

We also construct a gate barrier potential $V_g(y)$ in the middle of the system,

$$V_g(y)\begin{cases} = U_g, & \text{if } L_y/2 - W < y < L_y/2 + W, \\ = 0, & \text{otherwise} \end{cases}, \tag{3}$$

where $2W$ is the length of the gate barrier potential.

Due to the hard wall confinement potential $U_c(x)$, the eigenfunction Ψ_ν should satisfy the boundary condition in the x-direction $\Psi_\nu(0,y) = \Psi_\nu(L_x,y) = 0$. We further impose the generalized periodic boundary condition in the y-direction, $\Psi_\nu(x,y) = \exp(i\theta)\Psi_\nu(x,y+L_y)$. By changing the value of θ, the energy distance of edge states in different Landau levels can be set to be much closer than that obtained by the simulation where the value of θ is fixed.

3. Results

First, we discuss properties of edge states in the presence of disorder but in the absence of the gate barrier, i.e., $U_g = 0$. The value of Γ is set to be $0.2\hbar\omega_c$, and the impurity density n_i is chosen so that the normalized density $C_i \equiv 2\pi l^2 n_i$ should be 5. In Fig. 1, we plot the correlation between the eigenenergy ϵ_ν and the center of mass of the eigenfunction $x_{\nu\nu}(\equiv \int dxdy|\Psi_\nu|^2 x)$. In the left half of the figure we set $a = 2l$, while in the right half, a small number (20%) of impurities are replaced by short ranged ($a = 0$) ones. We see that the edge states are not influenced by disorder as long as the system includes only long ranged impurities. This is due to the fact that the matrix elements $H_{k,k'}$ between the edge states with the center coordinate X_k and $X_{k'}$ are proportional to the Gaussian factor [5],

$$H_{k,k'} \propto \exp(-(X_k - X_{k'})^2 a^2/4l^4). \tag{4}$$

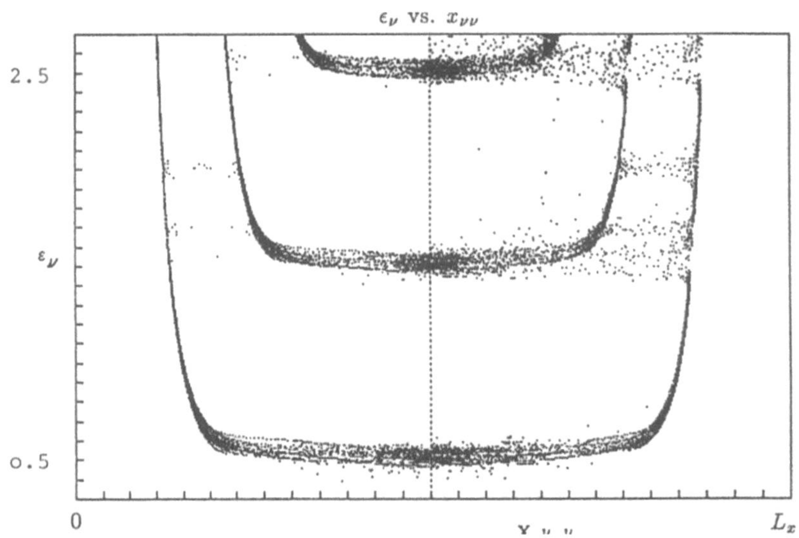

Figure 1. Correlation between the eigenenergy ϵ_ν and the center of mass of the eigenfunction $x_{\nu\nu}$. The energy is scaled by $\hbar\omega_c$, while $x_{\nu\nu}$ is scaled by $\Delta X(\equiv 2\pi l^2/L_y)$. The width L_x is $26\Delta X$, while the length in the y-direction L_y is $60\Delta X$. The magnetic length l is $3.09\Delta X$. In the left half of the figure, i.e., for $x_{\nu\nu} < L_x/2$, we set $a = 2l$, while in the right half, 20% scatterers are replaced by the short ranged ones. The value of θ is changed from 0 to 2π with 20 intervals.

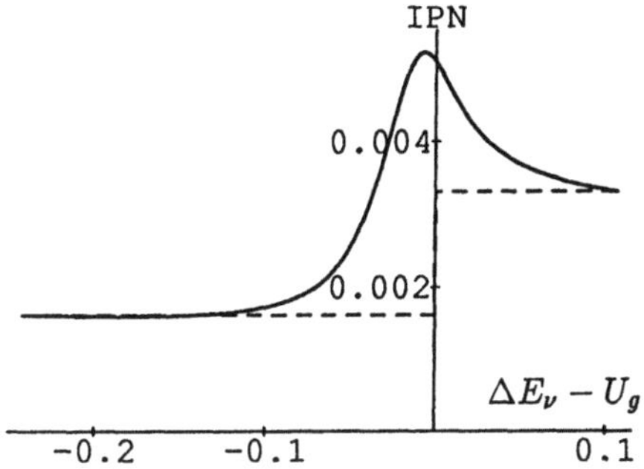

Figure 2. IPN vs. $\Delta E_\nu - U_g$. The abscissa is scaled by $\hbar\omega_c$. The broken line indicates the classical behavior. The structure around $\Delta E_\nu \approx U_g$ shows the existence of the quantum tunneling. The larger value of IPN in this region means that there exists a strong local current (vortices) in the gate barrier region.

This matrix elements also contain a factor which decreases Gaussianly with the spatial separation of the edge states [6]. However, since the magnetic length l is rather large ($\approx 100\text{Å}$) and the cyclotron energy $\hbar\omega_c$ is small ($<<$eV), the spatial separation of the edge states in different Landau levels will not be much larger than the magnetic length. On the other hand, the guiding center of the edge states in a certain Landau subband and that of the states in another Landau subband are separated by about $2l$ even if the confinement potential is a hard wall type one, and the factor in (4) becomes small enough to prevent the mixing. Note that the stability of edge states is destroyed by the inclusion of the short range impurities. This suggests that most of impurity potentials in the experiments are of long ranged nature.

We then investigate the properties of the edge states in the presence of a gate barrier. To concentrate on the effect of the gate barrier potential, we treat the case without impurities, i.e., $\Gamma = 0$. The values of U_g and W are set to be $0.5\hbar\omega_c$ and $2l$, respectively. We define the guiding center energy ΔE_ν,

$$\Delta E_\nu \equiv \epsilon_\nu - \left(N + \frac{1}{2}\right)\hbar\omega_c. \tag{5}$$

When ΔE_ν is larger than U_g, the edge states are classically expected to be transparent to the gate barrier, while they are expected to be reflected when $\Delta E_\nu < U_g$. Quantum mechanically, however, when ΔE_ν is smaller but close to U_g, the states are expected to penetrate into the barrier region due to tunneling and show structure in this energy region. To discuss the guiding center energy dependence of the edge states, we calculate the inverse participation number (IPN, $\equiv 2\pi l^2 \int dxdy|\Psi_\nu|^4$). The more the electronic states are localized, the lager the IPN. Therefore, the IPN becomes the smaller value when the edge states are reflected by the gate barrier and extended along both of the system edges, while it shows a larger value when they are transparent and extended only along one of the edges.

In Fig. 2, we plot IPN vs. $\Delta E_\nu - U_g$. The broken line corresponds to the behavior expected from the classical picture. We see that the IPN shows a maximum in the region $\Delta E_\nu \approx U_g$. This structure comes from the fact that there exists a strong localized current (vortices) in the gate barrier region.

4. Summary and Discussion

In this paper, we have shown that the edge states are stable against long range type impurities, but unstable against short range type ones [5]. From the experimentally obtained result that the edge states are quite stable [2], it is speculated that only the long range type impurities exist in the experimental samples, since a small number of short range impurities is enough to destroy the stability of edge states and cause mixing among edge states as well as mixing between the edge states and bulk extended states. In our simulation, $L_y = 20l$ and the value of θ is changed from 0 to 2π with 20 intervals, and the energy level distance of the edge states in different Landau levels are 1/20 times smaller than the case where the value of θ is fixed to be 0. Since the energy level intervals of edge states are proportional to $1/L_y$, the effective system size is $400l$, which is about $4\mu m$. That is, the elastic mean free path of edge states can be larger than μm for long range type impurities.

We have also calculated the electronic states in the presence of a gate barrier. It is shown that the IPN of the edge states shows a maximum before the states become transparent to the gate barrier. This is due to the quantum tunneling which leads to the formation of vortices in the gate barrier region.

When we apply a uniform electric field and calculate the non-equilibrium Hall current density from the linear response theory [7], we can show that the current does not flow in the gate barrier region before these quantum tunneling states are occupied. As these states begin to be occupied, the current density shows vortices in the gate barrier region, and very strong spatial fluctuations of the current density are seen in the gate barrier region. When all of these quantum tunneling states are occupied and the Fermi level cuts the edge states which are completely transparent to the gate barrier, the above fluctuations disappear and the current flows uniformly in the bulk region. That is, the current is locally compensated [7,8] due to these quantum tunneling states. The investigation of detailed effects of these states on the transport properties is one of the problems left for the future.

References

1. R.J. Haug, A.H. Macdonald, P. Streda and K. von Klitzing: Phys. Rev. Lett. **61** (1988) 2797.
2. B.J. van Wees, E.M.M. Willems, C.J.P.M. Harmans, C.W.J. Beenakker, H. van Houten, J.G. Williamson, C.T. Foxon and J.J. Harris: Phys. Rev. Lett. **62** (1989) 1181, S. Komiyama, H. Hirai, S. Sasa and S. Hiyamizu: Phys. Rev. **B40** (1989) 12566, B.W. Alphenaar, P.L. McEuen, R.G. Wheeler and R.N. Sacks: Phys. Rev. Lett. **64** (1990) 677.
3. P. Streda, J. Kucera and A.H. MacDonald: Phys. Rev. Lett. **59** (1987) 1973, J.K. Jain and S.A. Kivelson: Phys. Rev. Lett. **60** (1988) 1542, M. Büttiker: Phys. Rev. **B38** (1988) 9375,12724, Phys. Rev. Lett. **62**(1989) 229, Proceedings of the Symposium on Nanostructure Physics and Fabrication, (Texas, March 1989), ed. W.P. Kirk, Academic Press.
4. T. Ando and Y. Uemura: J. Phys. Soc. Jpn. **54** (1974) 959.
5. Y. Ono, T. Ohtsuki and B. Kramer: J. Phys. Soc. Jpn. **58** (1989) 1705, T. Ohtsuki and Y. Ono: J. Phys. Soc. Jpn. **58** (1989) 3863.
6. T. Martin and S. Feng: Phys. Rev. Lett. **64** (1990) 1971.
7. T. Ohtsuki and Y. Ono: J. Phys. Soc. Jpn. **58** (1989) 2482, **59** (1990) 637.
8. T. Ando: J. Phys. Soc. Jpn. **58** (1989) 3711.

Effect of Nondissipative Edge Currents on the Magnetoresistance of a Two-Dimensional Electron Gas at High Magnetic Fields

S.I. Dorozhkin[1,2], *S. Koch*[1], *K. von Klitzing*[1], *and G. Dorda*[3]

[1]Max-Planck-Institut für Festkörperforschung, Heisenbergstr. 1, W-7000 Stuttgart 80, Fed. Rep. of Germany

[2]Institute of Solid State Physics, Academy of Sciences of the USSR, Chernogolovka, Moscow District, SU-142432 Moscow, USSR

[3]Siemens Forschungslaboratorien, W-8000 München, Fed. Rep. of Germany

Abstract. It is experimentally shown that the magnetoresistance $R_{xx}(\upsilon)$ at noninteger filling factors υ in Si MOSFETs and GaAs/AlGaAs heterostructures with large aspect ratio is equal to the difference of the Hall resistances: $\delta R_{xy}=h/e^2\upsilon_1-R_{xy}(\upsilon)$, where υ_1 is an integer corresponding to the nearest well developed quantum Hall state. Our model of nondissipative current flow near the sample edges, which dominates at low temperature and low currents, explains our results and predicts the Hall resistance difference to be an upper limit for R_{xx} independent of the sample size.

Since the discovery of the quantum Hall effect [1] a number of experimental results appeared which cannot be explained in terms of a bulk phenomenon independent of the boundary conditions. In this paper we are concerned only with an effect of the sample size on the average diagonal resistivity ρ_{xx}. To explain qualitatively the first experimental observations [2,3] of this effect, ideas of nondissipative edge currents were used [3,4]. These currents were considered either as macroscopic [3] or as quantum edge states [5] similar to quasiclassical skipping orbits. In addition to previous observations [2,3] we report experimental results which demonstrate that in samples with a large aspect ratio L/W not the resistivity ρ_{xx} but the magnetoresistance R_{xx} is the important characteristic which obeys the relation $R_{xx}\cong\delta R_{xy}$. Then we propose a model based on the idea of macroscopic edge currents [3] which explains our result and gives a number of predictions for the upper limit for the R_{xx} value and on current distributions in the quantum Hall regime. This model specifies the conditions under which the current through the bulk of the sample dominates. Interpreted in terms of this model our results particularly indicate that vanishing magnetoresistance at integer filling factors is a result of nondissipative current flow in the bulk of the sample.

A similar experimental observation ($R_{xx}\cong\delta R_{xy}$) has been reported very recently [6] for GaAs/AlGaAs heterostructures but only for the transition region between the Hall steps with filling factors 2 and 3. The authors [6] explained the effect quantitatively in terms of edge states [5] introducing a number of assumptions for a transmission matrix. In this paper we would like to demonstrate the possibility of another approach to this problem. A comparison of our results with the edge state model [5] will be presented elsewhere.

The measurements of R_{xy} and R_{xx} were carried out on two-dimensional electron channels of Si MOSFETs and of GaAs/AlGaAs heterostructures. The peak mobility in the Si MOSFETs was about 3×10^4 cm^2/Vs (T=1.5K). The mobility in the heterostructure was about 5×10^5 cm^2/Vs at an electron concentration $n_s=1.4\times10^{11}$ cm^{-2}. In the case of the Si MOSFETs we could compare the results for the devices with the same width W=40 μm but different distances between

Springer Series in Solid-State Sciences, Vol. 101
High Magnetic Fields in Semiconductor Physics III   Editor: G. Landwehr

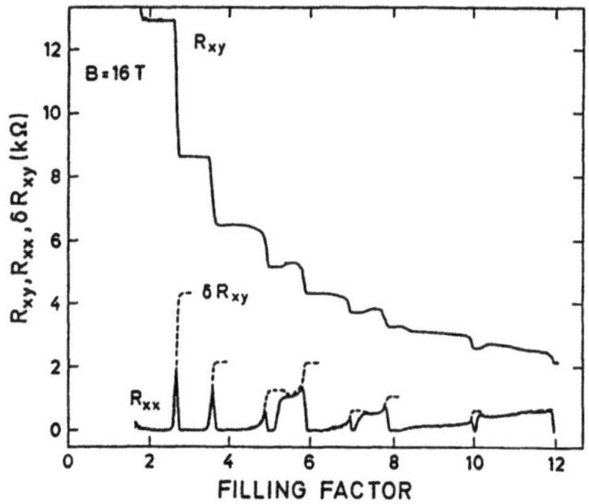

FIG. 1. Experimental dependencies of the Hall resistance R_{xy} and the resistance R_{xx} on the filling factor at a magnetic field B=16 T for a long Si MOSFET. T=30 mK, current I=1 nA. Comparison of R_{xx} with δR_{xy} (dashed line).

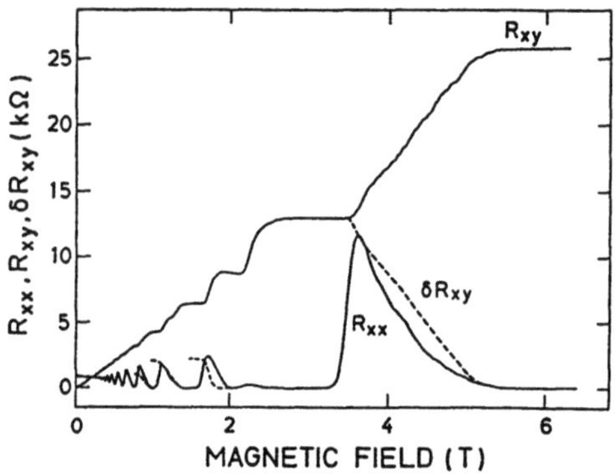

FIG. 2. Experimental dependencies of R_{xy} and R_{xx} on the magnetic field for the AlGaAs/GaAs heterostructure. T=0.56 K, current I=10 nA. Comparison of R_{xx} with δR_{xy} (dashed line).

potential probes: L_1=80 μm and L_2=2880 μm. For the GaAs/AlGaAs heterostructure W=240 μm and L=2400 μm. Typical experimental traces for the long MOSFET are presented in Fig.1 as a function of the filling factor which was varied with the gate voltage. For the following discussion we divide the υ range into the intervals (υ_l, υ_u) between filling factors 2 and 3, 3 and 4, 4 and 6, 6 and 8, 8 and 12. Then we compare the $R_{xx}(\upsilon)$ dependence within the intervals (υ_l, υ_u) with the value $\delta R_{xy} \equiv h/e^2\upsilon_l - R_{xy}(\upsilon)$. This comparison clearly demonstrates that these values are approximately equal at all υ not close to integer values different from υ_l. An increase of temperature to about 1 K results in the disappearance of the R_{xx} minima at the filling factors υ=5,7,10 which improves the agreement between R_{xx} and δR_{xy}.

The results for the GaAs/AlGaAs heterostructure (Fig.2) also demonstrate the existence of a similar effect. But in this case a decrease of temperature down to 50 mK results in a strong narrowing of the transition regions between the Hall steps and a considerable decrease in the R_{xx} peak values. This decrease already exists at 0.56 K for the resistance at filling factors in between 2 and 3 (B≅2.2T).

Comparing the results for the short and the long MOSFETs we do not observe any essential difference for the R_{xy} values. They coincide within about 5%. The ratio of the sample resistances drastically changes with magnetic field from 36, as expected from the geometry, at B=0 to a typical value of about 2 at high fields. This means that the long sample resistivity ρ_{xx} at high magnetic fields is more than a factor of 10 smaller than that of the short one.

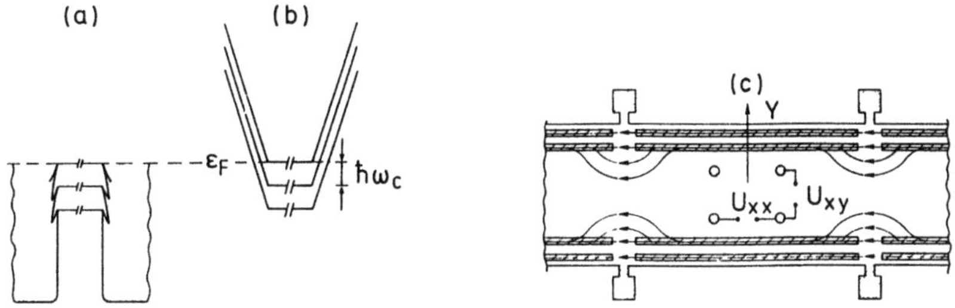

FIG. 3.(a,b) Schematic representation of the magnetic levels near the sample edges in the contact region of the sample (a) and far from it (b). Only the three lowest levels with energies in the bulk lower than the Fermi energy ε_F are shown. (c) Schematic of the current distribution in a sample with nondissipative edge regions (shaded) and half-integer filling factor $\nu=2.5$ in the bulk of the sample. The small inner contacts are discussed in the text.

To explain our experimental results we propose a model which is based on the following assumptions.

1. Following the ideas of Ref. [3] we assume the existence of depleted regions with filling factors smaller than ν, near the sample edges due to the confining potential (Fig. 3b). We consider these regions as macroscopic and characterize them by the resistivities ρ_{xx} and ρ_{xy} depending on ν.
2. We believe that the quantum Hall effect can exist in all the regions with integer filling factors i due to the position of the Fermi energy in the energy gaps between magnetic levels, i.e. in these regions $\rho_{xyi}=h/e^2i$ and ρ_{xxi} goes to zero.
3. The potential probes break at least nondissipative current paths along the edge regions with integer filling factors (Fig. 3a, 3c).
4. In a sample consisting of three parts with different filling factors ν,ν_0,ν ($\nu_0<\nu$), connected in series, the resistance introduced by the boundaries is given by the relation: $R=\rho_{xy}(\nu_0)-\rho_{xy}(\nu)$ [7] (under the condition $\rho_{xy}\gg\rho_{xx}$).
5. We can estimate the resistance corresponding to the current flow through the edge region with an integer filling factor i as $R_i=\rho_{xxi}L/W_i+[\rho_{xyi}-\rho_{xy}(\nu)]$, where W_i is the effective width of this region and $\rho_{xy}(\nu)$ is the Hall resistivity in the bulk of the sample. The first term in this relation includes the possible finite resistivity of the edge region. The second one describes the "contact" resistance corresponding to the current flow from the bulk of the sample into the edge region and back (see Fig. 3c). To get an idea about the current distribution in such a sample and about the sample resistance we should compare the resistances corresponding to the current flow through the different edge regions and the bulk of the sample. The bulk current dominates when $\rho_{xx}(\nu)L/W < \min\{R_i\}$ for $i<\nu$. This should be the case in the samples with a small value of L/W. In the samples with large aspect ratio the edge region with the minimal resistance R_l ($i=\nu_l$) short-circuits the main part of the sample length (see Fig. 3c). We may expect for this region $\rho_{xxl}L/W_l\cong 0$. As a result we obtain the sample resistance

$$R_{xx}(\nu)\cong\rho_{xy}(\nu_l)-\rho_{xy}(\nu)\cong h/e^2\nu_l-R_{xy}(\nu).$$

In the case of vanishing resistances $\rho_{xxl}L/W_l$ the R_{xx} should be equal to the difference between the nearest Hall step and the $R_{xy}(\nu)$ value. Our experimental results for the MOSFETs demonstrate that some of the edge regions (i=5,7,10) are not effective in shunting. This might be due to the finite resistances $\rho_{xxl}L/W_l$ of these regions originating from the smaller energy gaps. Then it seems to be natural to relate the zero resistance values at $\nu=i=5,7,10$ to the current flow in the bulk of the sample with vanishing resistance $R_{xx}(\nu=i)=\rho_{xx}(\nu=i)L/W\cong0$ due to the much larger width $W \gg W_1$ of the current carrying region.

On the basis of the proposed model we can make a number of predictions which can be experimentally tested.

1. The four-terminal resistance measured between neighboring voltage probes at noninteger ν cannot exceed a value of the order of the Hall resistance difference between adjacent Hall steps, approaching this value with increasing aspect ratio. If a sample is divided by voltage probes into a number of parts its resistance is the sum of the resistances of these parts.
2. For samples with a large aspect ratio L/W we predict that at noninteger filling factors the current will flow through the bulk of the sample only near the potential probes (see Fig.3c.). Far from them it will concentrate near the samples edges, so that all potential differences between the inner contacts shown in Fig. 3c. will be zero. At integer filling factors our results provide an indication of the current flow in the bulk of the sample. In this case we should expect a nonzero Hall voltage U_{xy} between the inner potential probes placed in the appropriate geometry (Fig. 3c), as was observed in the experiments [8,9].
3. A reduction of the carrier density around the voltage probes by application of the gate voltage, as in Ref. [10], restores the continuity of the edge regions with the largest filling factors which should result in disappearence of the "contact" resistance and in the vanishing R_{xx} value.
4. Using theoretical values [11] for conductivities σ_{xx} and σ_{xy} at half-integer filling factors ν, in the presence of short-range scatterers only, it is easy to get the following estimation for the aspect ratio at which the effect of the edge regions becomes important under these conditions:

$$L/W > \frac{2}{\pi}\,\frac{(N+1/2)^2+\pi^2\nu^2/4}{\nu_1 (N+1/2)} - \frac{\pi}{2}\,\frac{\nu}{N+1/2}$$

Here N is the Landau level number. For example, for the typical case in the Si MOSFETs N=1 and $\nu=4.5$, the right-hand side of this relation is about 0.8. This estimation shows that the sample size may influence the measured resistivity even in rather short samples.

One of us (S.I.D.) gratefully acknowledges a grant from the Alexander von Humboldt Foundation.

References.

1. K. von Klitzing, G. Dorda, and M. Pepper, Phys. Rev. Lett. **44**, 479 (1980).
2. K. von Klitzing, G. Ebert, N. Kleinmichel, H. Obloh, G. Dorda, and G. Weimann, Proc. of 17th Int. Conf. Physics of Semiconductors, San Francisco (1984), Ed. J.D. Chadi, W.A. Harrison, Springer Verlag, New York, p. 271.
3. B.E. Kane, D.C. Tsui, and G. Weimann, Phys. Rev. Lett. **59**, 1353 (1987).
4. R.J. Haug and K. von Klitzing, Europhys. Lett. **10**, 489 (1989).
5. M. Büttiker, Phys. Rev. **B 38**, 9375 (1988).
6. P.L. McEuen, A. Szafer, C.A. Richter, B.W. Alphenaar, J.K. Jain, A.D. Stone, R.G. Wheeler, and R.N. Sacks, Phys. Rev. Lett. **64**, 2062 (1990).
7. D.A. Syphers, and P.J. Stiles, Phys. Rev. **B 32**, 6620 (1985).
8. G. Ebert, K. von Klitzing, and G. Weimann, J. Phys. C **18**, L257 (1985).
9. H.Z. Zheng, D.C. Tsui, and Albert M. Chang, Phys. Rev. **B 32**, 5506 (1985).
10. B.J. van Wees, E.M.H. Willems, L.P. Kouwenhoven, C.J.P.M. Harmans, J.G. Williamson, C.T. Foxon, and J.J. Harris, Phys. Rev. **B 39**, 8066 (1989).
11. T. Ando, A.B. Fowler, and F. Stern, Rev. Mod. Phys. **54**, 437 (1982).

Nuclear Spin-Lattice Relaxation Under the QHE Conditions in the Edge States

I.D. Vagner[1], *T. Maniv*[2], *and T. Salditt*[1]

[1]Max-Planck-Institut für Festkörperforschung, Hochfeld-Magnetlabor, BP 166X, F-38042 Grenoble Cedex, France
[2]Department of Chemistry and the Solid State Institute, Technion, Haifa 32000, Israel

The nuclear spin-lattice relaxation process in semiconductors and metals is usually caused by the hyperfine interaction [1] between the nuclear spins and the conduction electron spins (Fig.1).

The energy needed to reverse the spin of an electron in the external magnetic field H_0, $\Delta E_{el} = 2g\mu_B H_0$ is larger than the energy needed to reverse the nuclear spin $\Delta E_{nucl} = 2g_{nucl}\mu_n H_0$ by a factor $M_n/m_e >> 1$. M_n and m_e being the nuclear and free electron mass, μ_B and μ_n the Bohr and the nuclear magneton respectively.

Thus, the simultaneous electron-nuclear spin flip-flop can take place only if there exists an energy reservoir external to the spin system. In an isotropic 3D metal this is the kinetic energy of the electron (Korringa relaxation [1]). In a strong magnetic field, $\hbar\omega > k_B T$, the kinetic energy perpendicular to the field is quantized. Therefore, the kinetic energy parallel to the field should change in order to ensure the energy conservation of the process. This is, by definition, impossible for the ideal 2D system (the vertical process in Fig.2).

In Ref.[2] a theory of the nuclear spin-lattice relaxation under the quantum Hall effect conditions was presented, based on the assumption that the Korringa relaxation process may be responsible for the nuclear spin-lattice relaxation in heterojunctions with a two-dimensional electron gas and in superlattices. It was shown there, that the magnetic field dependence of T_1^{-1} is similar to that of ρ_{xx}: T_1^{-1} is maximal when the chemical potential is in the Landau levels (Fig.2 in Ref.[2]) and T_1^{-1} is minimal when the chemical potential lies within the localized states [3]. It was concluded there, that the magnetic field dependence of $T_1^{-1}(B)$ can be used to study the electronic properties of a 2D electron gas.

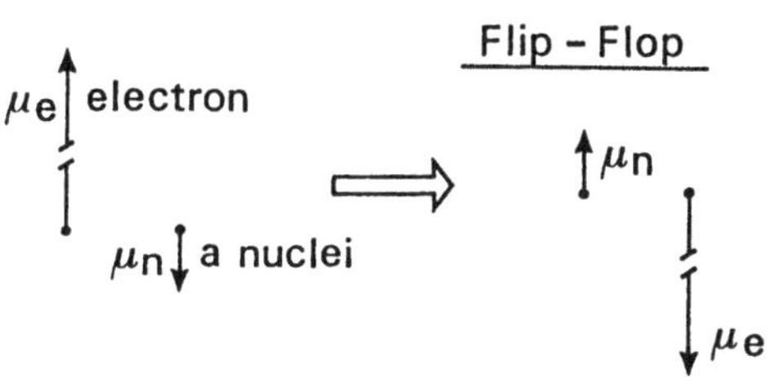

Fig.1: The nuclear spin-lattice relaxation process caused by the hyperfine interaction

Springer Series in Solid-State Sciences, Vol. 101
High Magnetic Fields in Semiconductor Physics III Editor: G. Landwehr

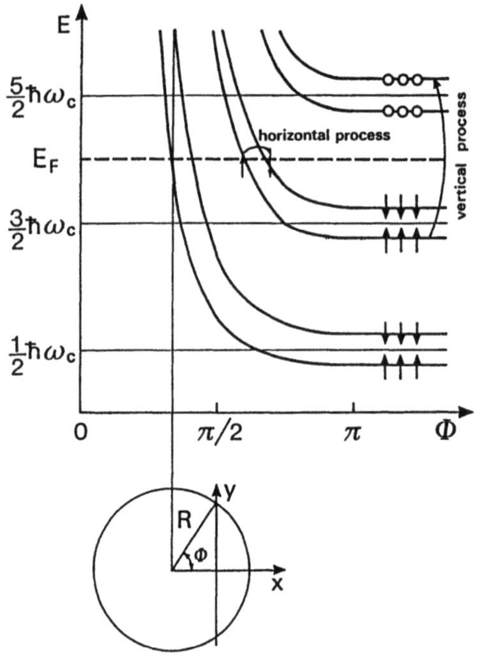

Fig.2 Energy spectrum of edge states: The energy dispersion of the first 3 Landau levels is shown for the spin-up and the spin-down levels. Only transitions that correspond to horizontal processes are energetically possible. Vertical processes are forbidden due to the energy gap.

Very recently the magnetic field dependence of T_1^{-1} under QHE conditions was successfully measured by Berg *et al.*[4]. It was observed there that T_1^{-1} is, indeed, due to the Korringa relaxation. These measurements show clearly a close similarity between the magnetic field dependence of $\rho_{xx}(B)$ and T_1^{-1} (Fig.2 in Ref.[4]).

An important question in the integer and fractional quantum Hall effect is the role of the edge states [5] : electronic orbits magnetically confined to the sample boundary. While the number of edge states may be small compared to the "bulk" states, their contribution to the nuclear spin relaxation can be very important, since they possess a homogeneous energy spectrum. In order to estimate the contribution of the edge states in the Korringa process: $T_1^{-1} \propto k_B T\, g^2(E_F)$, where $g(E_F)$ is the density of the electronic states in the vicinity of the Fermi energy, we choose a simple, quasiclassical model for the edge states.

Let us consider an edge state characterized by the angle ϕ in a long strip with sides L_x, L_y, $L_x \ll L_y$ (see Fig.2). The time interval between two successive collisions with the boundary is

$$T = \frac{2\phi}{\omega_c} \tag{1}$$

A Bohr-Sommerfeld quantization of such an orbit yields the following energy dispersion

$$E_n(\phi) = \frac{\left(\frac{1}{2}+n\right) \hbar\omega_c \, \pi}{\phi - \sin(\phi)\cos(\phi)} \tag{2}$$

The center of orbit is at a distance x_o away from the edge

$$x_o = -\sqrt{2n+1} \; l_H \; \frac{\cos(\phi)\sqrt{\pi}}{\sqrt{\phi - \cos(\phi)\sin(\phi)}} \tag{3}$$

where $l_H = \sqrt{\hbar/m\omega_c}$ is the magnetic length.

The density of states is

$$g_{edge}(E) = \frac{1}{L_x L_y} \frac{dN}{dE} = \rho_0 \, \hbar\omega_c \, \frac{1}{L_x} \sum_{n=0}^{n_F} \frac{dx}{d\phi_n} \frac{d\phi_n}{dE_n} \tag{4}$$

where $\rho_0 = \frac{m}{2\pi\hbar^2}$ is the 2D density of states without magnetic field.

Thus

$$g(E_F) = \rho_0 \frac{l_H}{L_x} \sum_{n=0}^{n_F} \frac{\phi_n}{\sin(\phi_n)} \sqrt{\frac{\phi_n - \cos(\phi_n)\sin(\phi_n)}{(2n+1)\pi}} \tag{5}$$

where ϕ_n is determined by the intesection of $E_n(\phi)$ and the fermi energy E_F.

It follows that the spin-lattice relaxation rate is proportional to square of the ratio between the magnetic length and the width of the strip.

$$T_1^{-1} \propto l_H^2 \propto H^{-1} \tag{6}$$

Let us now compare different contributions to T_1^{-1} in a 2D electron system under strong magnetic fields. The main mechanism for the nuclear spin relaxation, in the interplateau region, is the usual Korringa relaxation via the "bulk" electronic states. In the "plateau" region, i.e. when the chemical potential is in the localized states [3], the nuclear relaxation time is long, however finite, and the mechanisms contributing to T_1 are not yet clear. In Ref [2] a Lorentzian form was assumed for the energetical δ-function smeared by collisions. The Gaussian form of the Landau level broadening was used and the magnetic field dependence of the electronic g-factor was considered in Ref.[4].

Here we have analyzed the contribution of the edge states to $T_1^{-1}(B)$, when the Fermi energy lies within the localized states. We note, that the influence of the edge states on T_1 can be of primary importance in the mesoscopic systems, where the Larmor radii are comparable to the system's size. This can be studied by varying the size of the system.

Acknowledgement

We are indebted to Dr.S.Meshkov for illuminating discussions, and to Prof.P.Wyder and Prof.K.v.Klitzing for their interest in this work. This research was supported by a grant from the G.I.F, the German-Israeli Foundation for Scientific Research and Development, Grant no. *G*-112-279.7/88.

References

[1] J.Winter, Magnetic Resonance in Metals (Clarendon, Oxford, 1971)
[2] I.D.Vagner and T.Maniv, Phys.Rev.Lett., **61**, 1400 (1988)
[3] Note, that the definition of the "localized" states may be different for the transport phenomena, as conductivity, and for the nuclear spin-lattice relaxation.
[4] A.Berg, M.Dobers, R.R.Gerhardts and K.v.Klitzing, Phys.Rev.Lett.,**64**, 2563 (1990)
[5] M.Büttiker, Phys.Rev.B**38**, 9375(1988) and references therein.

Low-Frequency Response and Dissipationless Edge Currents in the Integral QHE

I.M. Grodnensky[1], *D. Heitmann*[1], *K. von Klitzing*[1], *and A. Yu. Kamaev*[2]

[1]Max-Planck-Institut für Festkörperforschung, Heisenbergstr. 1, W-7000 Stuttgart 80, Fed. Rep. of Germany

[2]Institute of Radioengineering and Electronics, Academy of Sciences of the USSR, Marx Avenue 18, SU-103907 Moscow, GSP-3, USSR

A novel dynamical response of a two-dimensional electron gas (2DEG) in a strong magnetic field B to an in-plane ac electric field has been observed at frequencies f which are smaller than the known characteristic frequencies. This response manifests itself by an increase of the electric field penetration in the sample and appears above a new characteristic frequency f_E which depends on the conductivity σ_{xx} and the sample size L. We shown that this effect arises from an induced edge potential in a 2DEG which spatial distribution is characterized by a length $l_E(\sigma_{xx}, f)$. We have studied two different cases: (i) l_E is of macroscopic scale and comparable with the sample size, (ii) l_E is of microscopic scale. The first case occurs when $f \sim f_E$. In this case we find experimentally a proportionality which allows us to determine σ_{xx} by measuring f_E. The second case corresponds to high $f \gg f_E$. In this case the response exhibits an unusual behaviour which we interpret as a manifestation of dissipationless edge currents in macroscopic samples.

GaAs-AlGaAs heterostructures of square and rectangular form of the side lengths 1-6 mm have been studied in a nonresonant radiofrequency measurement cell at temperatures T between 4.2 and 1.5 K. [1]. The sample is positioned on a plastic base between two electrodes. One of them is connected to the generator, the other one to the receiver. At different f the B dependencies of the normalized amplitude $U(f = const, B) = E(f = const, B)/E(f = const, B = 0)$ and phase $\Delta\vartheta(f = const, B) = \vartheta(f = const, B) - \vartheta(f = const, B = 0)$ on the field $E \cdot exp[i(\omega t - \Delta\vartheta)]$ near the sample have been measured. This field results from the vector sum of $\vec{E}_G$ and $\vec{E}_S$, i.e., from the fields of the generator electrode and the induced charge distribution in the 2D system, respectively.

In the original experiment we have swept B at constant f. In Fig. 1 we have reconstructed the normalized amplitude $U(B = const)$ vs. f. One can observe that at a rather sharply defined frequency f_E the normalized amplitude decreases significantly. These f_E positions are indicated by the arrows for different ν. We associate this decrease of the signal with an increasing electric field penetration in the sample. The effect is more pronounced at the lowest filling factor ν and is sensitive to the temperature and the sample quality. The strength increases with increasing f and depends on L where L is the side length between the electrodes. It is important to note that the effect depends on the B direction when the electrodes are placed asymmetrically with respect to the sample. The frequency f_E, which characterizes this novel response is much smaller than known characteristic frequencies in 2DEG.

An interesting finding is, as demonstrated in Fig. 1a, that there is a direct proportionality between the experimental values of $f_E(B)$ (points) and $\sigma_{xx}(B)$ (solid line). The latter was measured at the same frequency on a Corbino disk sample. The proportionality constant has the dimension of a length which is, as we will show later, related to the sample length. However, it is important to note that f_E does not depend on the Hall conductivity σ_{xy}.

As can be seen from Fig. 1b the experimental dependence of $f_E(L)$ (points) decreases with increasing L. These results can be explained by the existence of a length $l_E(\sigma_{xx}, f)$

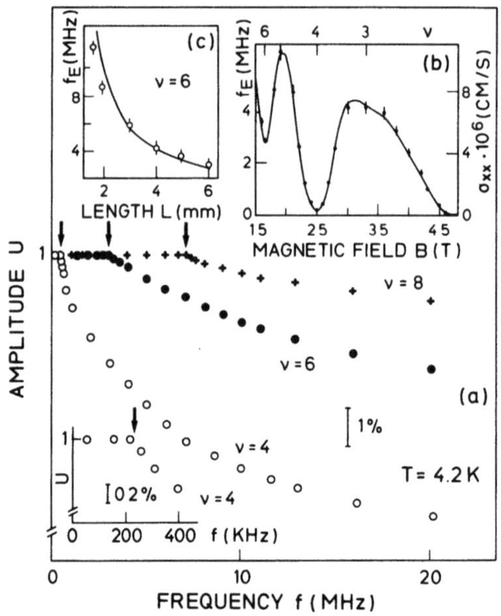

Fig. 1 (a) Normalized experimental amplitude vs frequency f. A sharply defined onset for a decreased response defines a characteristic frequeny f_E. (b) The experimental f_E vs. B dependence indicates a direct proportionality between f_E and the experimental conductivity σ_{xx}. (c) Comparison of the experimental f_E vs. sample length L with the calculation in the text.

which characterizes the spatial distribution of the induced edge potential in a 2D system. To demonstrate this let us consider the response of a 2D electron gas in a disk of radius R which is placed in the center of the xy-plane of an external electric field with potential

$$\varphi_{ext} = E\, x\, e^{-i\omega t} \tag{1}$$

Metal plates are placed at $z = d$ and $z = -d$. The spaces $0 < z < d$ and $-d < z < 0$ are filled by a dielectric medium of permittivity ε. Then we have the following system of equations:

$$\rho = (2\varepsilon\varepsilon_0/d)\varphi_{ind}\,, \tag{2}$$

$$\frac{\partial\rho}{\partial t} + div\vec{j} = 0\,, \tag{3}$$

$$j_\alpha = -\sigma_{\alpha\beta}\nabla_\beta(\varphi_{ext} + \varphi_{ind})\,, \qquad (\alpha,\beta) = (x,y) \tag{4}$$

with the boundary condition $j_r(r = R) = 0$. ε_0 is the permittivity of the vacuum. We assume $\varphi_{ind} = V\, e^{-i\omega t}$ (analog expressions hold for ρ and $\vec{j}$) and then obtain an equation for $V(r,\psi)$ in a polar system of coordinates

$$\frac{1}{r}\frac{\partial}{\partial r}[r\frac{\partial V}{\partial r}] + \frac{1}{r^2}\frac{\partial^2 V}{\partial\psi^2} + [\frac{1+i}{l_E}]^2\, V = 0 \tag{5}$$

with the boundary condition

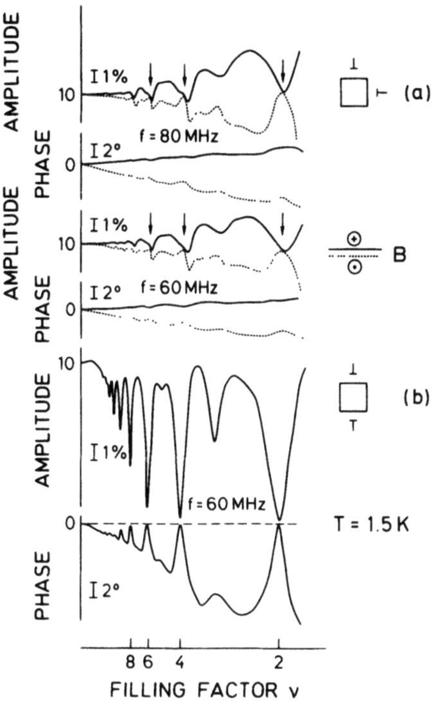

Fig. 2 Amplitude and phase in different probe electrode positions.

$$R\sigma_{xx}[E\ cos\psi + \frac{\partial V}{\partial r}\ (r = R)]\ +\ \sigma_{xy}[-E\ Rsin\psi\ +\ \frac{\partial V}{\partial \psi}\ (r = R)]\ =\ 0\ , \tag{6}$$

where

$$l_E^2\ =\ d\sigma_{xx}\ /\ 2\pi\varepsilon_0\varepsilon f\ . \tag{7}$$

We look for a solution of (5) in the form $V(r,\psi) = g(r) \cdot f(\psi)$ and obtain for strong magnetic fields, $\sigma_{xy} \gg \sigma_{xx}$,

$$V(r,\psi)\ =\ -E\ R\ \frac{J_1[(1+i)r/l_E]}{J_1[(1+i)R/l_E]}\ cos\psi\ , \tag{8}$$

where J_1 is the Bessel function.

The important result is the exponential dependence of (8) on r which leads to a sharp onset in the dependence $U(f)$ near f_E. The theoretical dependence $f_E(L)$ in Fig. 1c (solid line) agrees well with the experimental results. We have used this effect to determine σ_{xx} in the integral QHE regime. We measured f_E and used the equality $l_E(f_E) = L/2$ which follows from the analysis of (8).

At high frequencies, $f \gg f_E$, in high quality samples l_E can be of the order of l_B. This follows from (7) if $\sigma_{xx}(f) = \sigma_{xx}(f = 0)$. In this case we observe an unusual behaviour of the measured signal (Fig. 2) which occurs near integer filling factors indicated by the arrows. For probing electrodes in the position (a) the amplitudes are equal for both B directions. For the arrangement (b) the phase shift is practically zero. In both cases the signals do not depend on f in a wide f-region. From these results we can conclude that the induced edge current is in phase with the external field. We interpret this fact as a manifestation of a dissipationless dynamical edge current in the integral QHE in macroscopic samples.

We thank D.E. Khmelnitzkii for illuminating discussions, and G. Weimann for providing us with high-quality samples.

References

[1] V.A. Volkov, D.V. Galchenkov, L.A. Galchenkov, I.M. Grodnensky, O.R. Matov, and S.A. Mikhailov, *JETP Lett.* **44** (1986) 655. L.A. Galchenkov, I.M. Grodnensky, and A.Yu. Kamaev, *Sov. Phys. Semicond.* **21** (1987) 1330. I.M. Grodnensky, and A.Yu. Kamaev, Surf. Science **229** (1990) 522.

Effects of High Frequency Delocalization on the Quantum Hall Effect

R. Meisels[1], *K.Y. Lim*[1], *F. Kuchar*[1], *G. Weimann*[2], *W. Schlapp*[3], *V. Rampton*[4], *P. Beton*[4], *L. Eaves*[4], *J.J. Harris*[5], *and C.T. Foxon*[5]

[1]Institut für Festkörperphysik, Universität, A-1090 Vienna, Austria and L. Boltzmann Institut für Festkörperphysik, A-1060 Vienna, Austria
[2]Walter Schottky Institut, TU München, Am Coulombwall, W-8046 Garching, Fed. Rep. of Germany
[3]Forschungsinstitut der Deutschen Bundespost Telekom, W-6100 Darmstadt, Fed. Rep. of Germany
[4]Department of Physics, University of Nottingham, UK
[5]Phillips Research Laboratories, Redhill, Surrey RH1 5HA, UK

Abstract. The width of the QHE plateaus ΔB is investigated in GaAs/-AlGaAs using DC and microwave (MW) techniques. ΔB is narrower at MW frequencies than at DC and saturates below 2.4K. No fractional plateaus were observed at MW frequencies where they appeared at DC.

In this contribution we present DC and high frequency experiments on GaAs/AlGaAs heterostructures investigating the width of the Hall plateaus – the characteristic features of the Quantum Hall Effect – in terms of the localization model. In a Hall plateau the Fermi level lies within localized states in the tails of a Landau level. Previous experiments on the maximum slope in ρ_{xy} and the half width in ρ_{xx} between the plateaus were also interpreted on the same basis, but only considered the situation where the Fermi level lies in the range of delocalized states near the center of a Landau level [1].

The DC and high-frequency experiments regarding the width of the Hall plateaus were performed on MBE grown single heterostructures of AlGaAs/GaAs in the temperature range 0.3-4.2 K. The sample properties are listed in the table below.

Table 1. Dc parameters of the AlGaAs/GaAs samples at T=4.2 K.

Sample	1	2	3	4
$n_s(10^{11}cm^{-2})$	5.2	2.9	2.5	1.4
$\mu(10^5cm^2/Vs)$	1.4	1.2	1.7	6.5

In the high-frequency case a contactless transmission technique was applied where the Hall conductivity σ_{xy} is measured.[2] The essential part of the experimental set-up consists of crossed rectangular wavegui-

Springer Series in Solid-State Sciences, Vol. 101
High Magnetic Fields in Semiconductor Physics III Editor: G. Landwehr

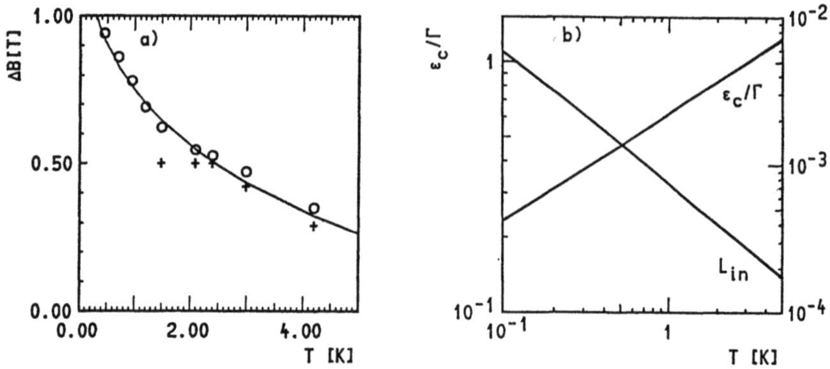

Fig.1: a: i=4 Hall plateau width ΔB at DC(o) and at 35Ghz(+) and fit to the DC data (solid line) vs. temperature. b: Values for mobility edge ε_c (in units of Landau level width Γ) and inelastic scattering length L_{in} from the DC fit.

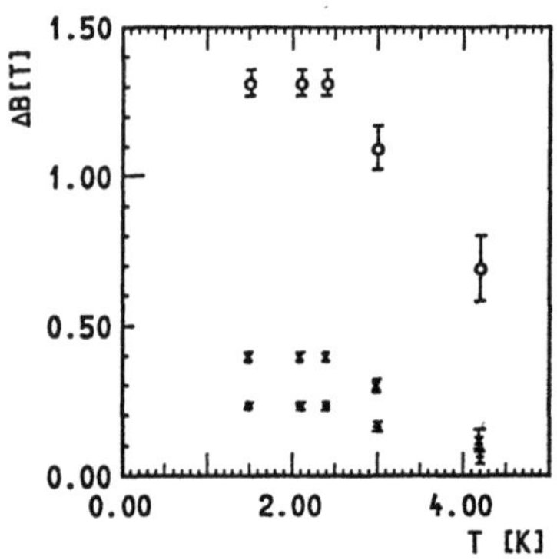

Fig.2: Widths of the i=2 plateaus of samples 2 (o), 3 (×), 4 (+) at 35 GHz.

des which act as polarizer and analyser. The MW power transmitted through the sample is proportional to ${\sigma_{xy}}^2$.

Fig.1a shows results of the temperature dependence of the i=4 plateau width of sample 1 for DC and f=35 GHz. Due to the noise level in the MW experiment the plateau width is determined by a 0.5% deviation from the average value (0.05% in the DC case). With 0.5% in the DC case ΔB is wider by 0.15T at $T > 2.4$K. Above about 2.4 K the temperature dependences are very similar. Below 2.4 K the MW plateau width remains constant. This is a general behavior also observed in samples 2-4 with the i=2 plateau (Fig.2). Fig.1a also shows a fit to the DC data using the localization model of Huckestein et al.[3] outlined below.

The Landau levels are broadened due to disorder, their width being Γ. Around their centers E_N, N is the level index, the electronic states are delocalized. They are localized in the tails. Only the delocalized states contribute to transport and to the QHE.

The boundary between localized and delocalized states defines the mobility edge ε_c; states are localized if their energy E obeys $|E - E_N| > \varepsilon_c$. If the Fermi level E_F lies there, the Hall conductivity is quantized and forms a plateau. At $T = 0$ states are localized as long as their localization length $\xi(E) < L_s$, L_s is the sample length. Only very close to E_N, where $\xi(E)$ diverges like $|E - E_N|^{-\nu}$, there are delocalized states. For $T > 0$, L_s is replaced by $L_{in}(T) = \sqrt{D\tau_{in}(T)}$. L_{in} and τ_{in} are the inelastic scattering length and time, resp., D is the diffusion constant. τ_{in} is assumed $\propto T^{-p}$. Using the equality $\xi(E_N \pm \varepsilon_c(T)) = L_{in}(T)$, $\varepsilon_c(T)$ is calculated[3] and, by integrating over the density-of-states, the width of the Hall plateau can be determined. The best fit over the whole temperature range (Fig.1a) is obtained with $p = 2$ using $\nu = 2.34$. The value of $p = 2$ is supported by the theory of Fukuyama et al.[4] which yields, apart from a logarithmic prefactor, the same exponent.

In Fig.1b the temperature dependence of the mobility edge relative to the center of the Landau level (actually an average for the two involved) is plotted for the DC case. Near zero temperature the mobility edge approaches the center of the Landau level. Also shown is $L_{in}(T)$ where the theoretical result by Fukuyama et. al. for $\tau_{in}(T)$ has been used. D was set equal to $D(B = 0)$. At $T = 2.4$K, L_{in} is 3.3μm. When using $D(B)$ of Ref.[5], L_{in} is smaller by a factor of 50. In our high-frequency experiments the electron states are probed by their interaction with photons. Therefore they are effectively delocalized if they are within $\hbar\omega$ of the center of the Landau level[6]. This effect competes with the temperature dependent inelastic scattering ($\tau_{in} \propto T^{-p}$) and explains the constant Hall plateau widths at lower temperatures where the frequency effect dominates. If $\hbar\omega \geq \Gamma$ the Hall plateaus disappear[6].

For high frequency experiments on the FQHE samples from wafer G641 showing FQHE plateaus in DC experiments[7] were used. Without illumination carrier concentrations of n_s = 2.5-4.5 $\times$ 10^{10}cm^{-2} and mobilities μ of 2-4 $\times$ 10^6Vs/cm^2 were measured on Hall bars from the same wafer. The sample properties varied due to inhomogeneities. By illumination with red light from a LED a persistent increase of n_s and μ can be achieved. Under saturating illumination n_s = 6-9 $\times$ 10^{10}cm^{-2} and μ = 3-7 $\times$ 10^6Vs/cm^2). Fig.3 shows the results of the high-frequency (32.7 GHz) experiments without illumination.

Without illumination no plateaus, IQHE or FQHE, can be seen. Under illumination, the i=1 plateau appears and a weak structure at i=2, but no fractional plateaus were found. This contrasts with the prominent d.c IQHE and FQHE plateaus[7].

Similar considerations like those applied to the samples of lower mobility are used again. In the high-mobility samples there is less disorder reducing the Landau level width Γ. Also, due to the smaller n_s the Hall plateaus occur at lower fields reducing $\Gamma \propto \sqrt{B}$[5] further. Only the highest i=1 plateau can be seen and only if n_s is increased sufficiently

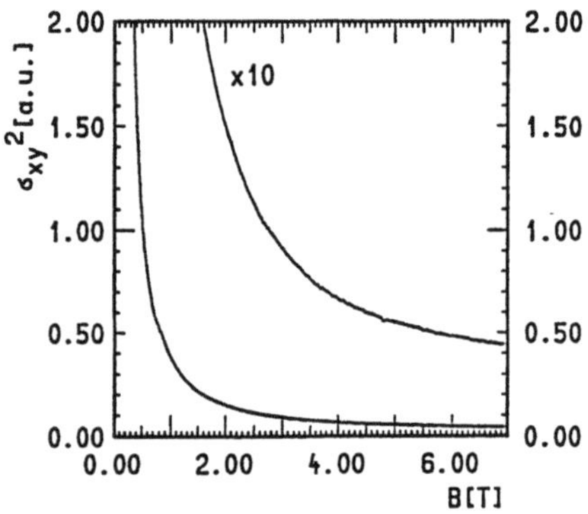

Fig.3: Transmitted MW (32.7GHz) intensity $\propto \sigma_{xy}^2$ vs. magnetic field B for sample G641[7] (no illumination) at 0.45K.

by illumination. The absence of the FQHE plateaus at microwave frequencies can have two different reasons: (a) The electrons can be exited above the fractional gap by absorption of microwave photons. (b) Some disorder is necessary to cause localized states and to pin the Fermi level within the fractional gap for the formation of a plateau. The high frequency absorption again causes delocalization and a breakdown of the FQHE. Regarding (a) it is not clear whether a excitation across the fractional gap by photons isn't forbidden in the first place. Process (b) is more probable as Engel et al. have shown that the same arguments using localization applied to the IQHE[1] are also successful in the case of the FQHE[8].

Acknowledgement: This work was partly supported by the "Fonds zur Förderung der wissenschaftlichen Forschung", Austria, project 6437.

References:

[1] H.P.Wei et al., Phys.Rev.Lett. **61**,1294(1988).
[2] F.Kuchar,R.Meisels, Phys.Rev.B **33**,2965(1986).
[3] B.Huckestein et al., Springer Series in Sol.St.Sci. (eds. F.Kuchar, H.Heinrich, G.Bauer), in print.
[4] H.Fukuyama, E.Abrahams, Phy.Rev.B **27**,5976(1983).
[5] T.Ando, et al., Rev.Mod.Phys. **54**,439(1982).
[6] F.Kuchar et al., Proc. 20th ICPS, Thessaloniki, 1990, to be published.
[7] R.G.Clark et al., Surf.Sci. **229**,25(1990).
[8] L.Engel et al., Surf.Sci. **229**,13(1990).

Screening and Transport Properties in a Triangular Well

H. Nielsen

Physics Laboratory, University of Copenhagen,
Universitetsparken 5, DK-2100 Copenhagen Ø, Denmark

1. Introduction

Since dissipative transport properties always depend on matrix elements containing the screened or selfconsistent potential, there is a close connection between the form of the screened potential and the transport coefficients. In the quantum Hall regime the current may be dissipationless. However, the potential still plays a role because of the possibility of a metal insulator transition of the Mott type /1/. The condition for such a transition is $eV^{sc} > kT$ where V^{sc} is the screened potential and kT may be interpreted as the thermal energy belonging to the guiding center of an electron. If the filling factor in the insulator regions have exact integer or fractional values then the quantum Hall effect (QHE) can be explained and its accuracy depends on the nature of the transition region between the insulator and the metallic domains. This problem is not understood.

Near the boundaries where the confinement potential dominates the condition $eV^{sc} > kT$ will always be fulfilled and the Mott transition thus legitimates the use of exact filling factors for the edge currents. A sufficient narrow channel connected with broader regions which act as reservoirs may be an insulator domain and therefore have a charge density which is different from the density in the bulk of the broad regions.

In an earlier work this idea was used to investigate the breakdown of the QHE in the high current regime, especially the observed linear relation between the critical current and $B-B_i$ where B_i corresponds to i filled Landau levels in the broad regions /2/. In this work we shall mainly investigate the low current regions, where periodic fluctuations in the magnetoresistance and asymmetric Shubnikov de Haas effect has been observed both in the integer and the fractional regime /3/. Finally we show that for a heterostructure without magnetic field the density dependent screening parameter may be essential in order to explain the wellknown experimental fact that the mobility increases with the density n as a power law /4/.

2. The quantum Hall regime

A simple expression for V^{sc} can be found in the inter regime if a Hartree treatment and the Thomas–Fermi approximation for a two dimensional electron gas (2 DEG) is applied. Then the screening parameter is $Q_{TF} = \frac{e^2}{2\kappa\epsilon_0} D_T$ where κ is the relative dielectric constant ($\kappa \cong 12$ for GaAs) and $D_T = (\frac{\partial n}{\partial \mu})_T$ is the thermodynamic density of states. μ is the chemical potential and n is the density (in m^{-2}). For a 2 DEG in a triangular well subjected to a quantizing magnetic field the energy levels are of the form $\epsilon = (i - 1/2)\hbar\omega_c + \gamma_s n^{2/3}$ where i is a positive integer. The first term is the quantized kinetic energy in the x–y plane and the last term where γ_s is a constant dependent of the subband index s represent the z–dependent part of the Hamiltonian. $(\frac{\partial n}{\partial \mu})_T$ can now be calculated by implicit differentation of

$$n = \frac{2}{2\pi l_B^2} f\left(\frac{(i - 1/2)\hbar\omega_c + \gamma_0 n^{2/3} - \mu}{kT}\right) \tag{1}$$

Springer Series in Solid-State Sciences, Vol. 101
High Magnetic Fields in Semiconductor Physics III Editor: G. Landwehr

where f is the Fermi function, $l_B = \sqrt{\frac{\hbar}{eB}}$ is the magnetic length and only the lowest subband s = 0 is considered for simplicity.

In the low temperature limit we find

$$Q_{TF} = \frac{e^2}{2\kappa\epsilon_0} \cdot \frac{2}{3} \cdot n^{1/3} \tag{2}$$

which disappears as $n \to 0$ as expected since no electrons means no screening. For a single Fourier component V_q^{ext} we have $V_q^{sc} = \frac{V_q^{ext}}{\kappa(1 + Q_{TF}/q}$ and the transition condition $eV_q^{sc} > kT$ gives

$$eV_q^{ext} > kT\,(\kappa + \frac{3}{4}\frac{e^2 n^{2/3}}{\epsilon_0 \gamma_0 q}) \tag{3}$$

which means that the Mott transition always takes place at sufficiently low temperatures. Using (3) numerical calculation shows that for T = 1K a typical dimension of an insulating or nonmetallic region is of the order $1\mu = 10^{-6}$m. Such a region with a well defined filling factor may be defined as a mesoscopic system in a quantizing magnetic field.

In the fractional filling regime it is not possible to find a simple expression for the screened potential since the charged excitations can only be treated like fermions if a long range statistical correlation is introduced. However, even in this case the z–dependent part of the Hamiltonian gives rise to a special n–dependent term i the free energy. Therefore it may be possible that even in this case it can be shown that the screened potential is independent of T in the low temperature limit, so that regions with a well defined fractional filling factor exists. In fact we shall show that the periodic magnetoresistance fluctuations observed in /3/ near $\nu = 1/3$ can be understood by using this idea. In /3/ the fluctuations in the magnetoresistance were explained by resonance tunnelling through quasi bound states with dimensions of the order of the electrical width of the channel which was 0.8 μ i.e of the order $10^2\, l_B$. Since the dimensions of a closed equipotential orbit is large compared to l_B the energy levels E_n can be obtained semiclassically from the Bohr Sommerfeld condition and the classical expression for the energy

$$J = m^* \frac{E_r(r_n)}{B} \cdot 2\pi r_n + e^* B \pi r_n^2 = nh \tag{4}$$

$$E_n = \frac{1}{2} m^* \left(\frac{E_r(r_n)}{B}\right)^2 + e^* V(r_n) \tag{5}$$

where for simplicity circular orbits are considered. r_n is the radius for the n'th state, $E_r(r_n) = -\frac{\partial V}{\partial r}\Big|_{r_n}$ is the radial field necessary to compensate the Lorentz force and e* and m* is the effective charge and mass of the "particle" in the quasi bound state. $\frac{E_r}{B}$ is the velocity of its guiding center. From (4) we get

$$r_{n+1} - r_n = \frac{\hbar}{(e^* B r_n + m^*(\frac{E_r}{B} + \frac{r_n}{B}\frac{dE_r}{dr})} \tag{6}$$

where an estimate based on the Poisson equation for E_r allows us to neglect $\frac{E_r}{B}$ compared to the two other terms which are of equal order of magnitude.

If the magnetic field is changed to B + ΔB and the resonnance tunnelling occurs at a definite energy belonging to the movement of the guiding center ΔB has to be obtained from $E_{n+1}(B + \Delta B) = E_n(B)$ or $\Delta B \frac{\partial E_{n+1}}{\partial B} = \frac{\partial E_n}{\partial r_n} \cdot (r_{n+1} - r_n)$. Using (5) and (6) we get

$$\Delta B = \left\{\frac{\frac{e^* B}{m^*} + \frac{1}{B}\frac{dE_r}{dr}}{\frac{e^* B}{m^*} + \frac{1}{B}\frac{dE_r}{dr}}\right\} \frac{\hbar B}{(\frac{\partial E_r}{\partial B} - \frac{E_r}{B}) \cdot m^* r_n} \tag{7}$$

which shows that the terms containing the effective charge cancel so that the final expression becomes

$$\Delta B = \frac{\hbar}{m^* r_n} \cdot \frac{B}{\left(\frac{\partial E_r}{\partial B} - \frac{E_r}{B}\right)} \tag{8}$$

According to /3/ ΔB is found to be three times larger for $\nu \approx 1/3$ than for integer filling.

Formula (8) shows that this is not related to the charge e^* but depends on the form of the function $E_r(B)$. In /2/ it was assumed that when the filling factor in the broad regions connected to the channel is close to an integer then the channel contains an extra charge density $\rho_1 = -e\left(\frac{eB}{h} - \frac{eB_i}{h}\right)$ since the narrow channel is a nonmetallic region with integer filling factor. The strong influence of the confining potential in the narrow region means that the transition condition $eV^{sc} > kT$ is fulfilled in the whole width of the channel. A corresponding assumption for the fractional regime could bring agreement between (8) and the experiment. The tunnelling particle could then simply be an ordinary electron and the transition condition could in a natural way explain the extra charge which creates the radial field in the channel.

A further experimental test of the assumption could be the observation of a linear relation between the critical current for breakdown and $B - B_\nu$ in the regime $\nu \approx \frac{1}{3}$. However we expect that such an experiment would be very difficult to perform.

3. Transport without magnetic field

Here the thermodynamic density of state is calculated from $n = D \cdot (E_F - \gamma_0 n^{2/3})$ where $D = \frac{m^*}{\pi\hbar}$ is the two dimensional density of states. This gives

$$\left(\frac{\partial n}{\partial \mu}\right)_T = \frac{D}{1 + \frac{2}{3} D \gamma_0 n^{-2/3}} \tag{9}$$

which again gives for the screening parameter $Q_{TF} \sim n^{1/3}$ since the last term in the denominator dominates for $n < 10^{16}\ m^{-2}$ /2/.

According to /5/ the potential from a remote impurity varies as l_{sc}^2/r^3, where the screening length is $l_{sc} = Q_{TF}^{-1}$.

Therefore the squared absolute value of the matrixelement for scattering $|M_{kk'}|^2 = |\langle k | V^{sc} | k' \rangle|^2$ varies as Q_{TF}^{-4}. For the scattering rate we then get

$$\frac{1}{\tau} = \frac{2\pi}{\hbar} |M_{kk'}|^2 \cdot D \sim Q_{TF}^{-4} \sim n^{-4/3} \tag{10}$$

which finally gives for the mobility $\mu = \frac{e\tau}{m^*}$ a variation with density of the form $\mu \sim n^{4/3}$.

The best experiments give $\mu \sim n^{1.6}$ /4/. In a more careful theoretical treatment also the n–dependence of the form factors has to be taken into account. Such an investigation is in progress. However, the conclusion seems to be that the major part of the density dependence of the mobility comes from the screening parameter.

For completeness we also write down D_T for the case where the second subband starts to be occupied:

$$\left(\frac{\partial n}{\partial \mu}\right)_T = \frac{2D}{1 + \frac{2}{3}(\gamma_1 + \gamma_0) D n^{-1/3}} \tag{11}$$

from which $|M_{kk'}|^2$ can be estimated.

References

1. A.L. Efros, Solid State Comm. 70, 253 (1989)
2. H. Nielsen, Semic.Sci.Techn. 5, 604 (1990)
3. J.A. Simmons, et.al. Phys.Rev.Lett. 63, 1731 (1989)
4. M.A. Paalanen et.al. Phys.Rev. B29, 6003 (1984)
5. F. Stern, W.F. Howard Phys.Rev. 163, 816 (1967)

The Quantum Hall Effect in the Non-Isolated Quantum Well Approximation

A. Raymond and H. Sibari

Groupe d'Etude des Semiconducteurs, URA 357, UM2, Place E. Bataillon, F-34095 Montpellier Cedex 2, France

We assume that in modulation doped heterostructures (MDH) the two dimensional electron gas (2 DEG) is a non-isolated system. As a consequence the Fermi level remains constant but both the 2DEG density N and the electrical subbands vary when the magnetic field varies. We performed the calculations for GaAs-GaAlAs heterojunctions (GaAs-H), only one occupied subband, in the triangular well approximation (TWA) taking into account the g factor enhancement and assuming that localized electrons do not participate in the conduction processes. We calculated the components of the conductivity tensor in quantizing magnetic fields and deduced ρ_{xx} and ρ_{xy}. The theoretical results have been compared to a large number of experimental ones.For helium or pumped helium temperature experiments (T > 1K), we obtained a good agreement for both ρ_{xx} and ρ_{xy}, without any fitting parameter, for a large number of samples with mobility varying between 4×10^4 and 10^6 cm^2/Vs.

Figure 1 shows the magnetic field dependence of N obtained under a pressure of 8.8. kbar. N has been determined. assuming that, like in 3D systems, the localized electrons do not participate in the Hall conductivity.

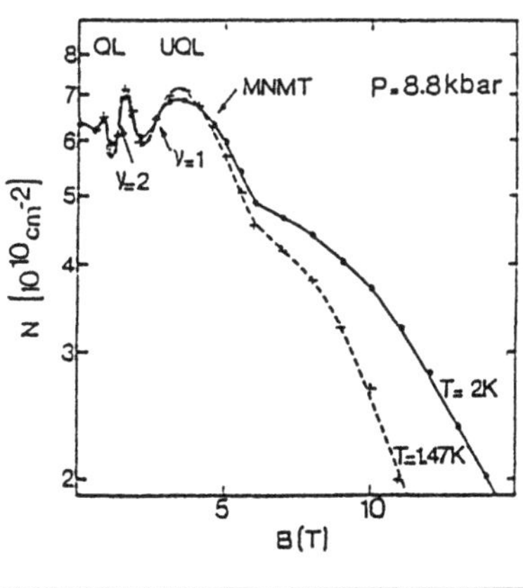

Fig. 1. Experimental B dependence of N for P = 8.8 kbar. $N=(\rho_{xy}B)/e(\rho_{xx}^2+\rho_{xy}^2)$

In the quantum limit (QL) N clearly exhibits oscillations with a period exactly equal to the one of the Shubnikov-de-Haas effect and QHE. To explain these oscillations with a 10% increase of N in the QHE regime, we have assume that the 2DEG is a non isolated system.
This assumption was first made by Baraff and Tsui[1] after the original paper on the QHE of v. Klitzing et al[2].
Nevertheless the commonly accepted model[3] assumes, in order to account for the quantization of ρ_{xy} in plateaus, that the 2DEG is isolated : the total density of the 2D electrons N is magnetic field independent and the finite width of the plateaus is explained by the existence of localized states..
In this paper we develop the model proposed by Raymond and Karray[4]. We assume that a transfer of electrons is possible between the 2DEG and one or several reservoirs and as a consequence that the Fermi level of the whole system keeps a constant value when B varies. The existence of gaps in the energy spectrum of the 2DEG must induce oscillations of the density of conducting electrons when B varies. We will show that these variations of N give rise to plateaus of ρ_{xy} with the exact quantization h/ie^2, without invoking the commonly accepted form of the DOS with a background density of localized states and a very small percentage of extended states.

At helium or pumped helium temperatures (T > 1K) the plateaus of ρ_{xy} are quite narrow and the transition between two adjacent plateaus is smooth (Fig.2). We will study the transport coefficients in this temperature range. As a consequence of the variation of the charge density N when B varies, the electric field F is magnetic field dependent.
This means that the equilibrium of the heterojunction must be recalculated for each value of B and that, although the Fermi level is in first approximation independent of B, E_F-E_o varies because the lowest electrical subband energy E_o varies (see equation 1).
The calculations have been made in the TWA assuming that E_o is the only populated subband :

$$E_o=(\hbar^2/2m^*)^{1/3}\ (3/2\ \pi\ e\ F\ x\ 0.7587)^{2/3} \qquad (1)$$

where F is the mean electric field.
We have considered non interacting electrons in a spherical parabolic energy band and used a DOS in the form of Gaussian peaks (see equation 2).

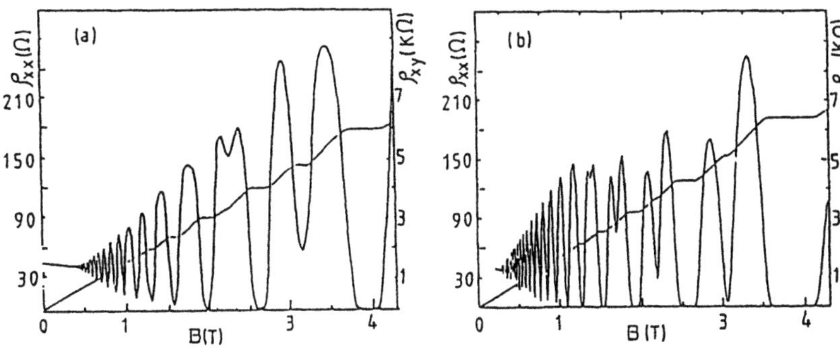

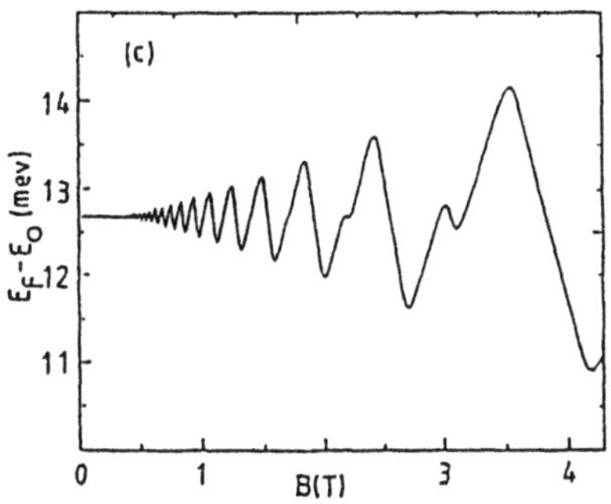

Fig.2 (a) Experimental B dependence of ρ_{xx} and ρ_{xy} at T=1.4K ($N=3.610^{11}cm^{-2}$, $\mu=4.2410^{5}cm^{2}/Vs$) (b) ρ_{xx} and ρ_{xy} obtained from the proposed model (m^*/m_o = 0.07, $\varepsilon/\varepsilon_o$=12.91, $N_A-N_D=2.10^{14}$ cm^{-3} and Φ_d = 1.425 eV) (c) Magnetic field dependence of $E_F - E_o$ from the model.

We made the hypothesis that all the 2D electron-states are non localized states. N is then given by :

$$N = \sum_{s} \sum_{n=0}^{\infty} \frac{eB}{h} \left(\frac{2}{\pi}\right)^{1/2} \frac{1}{\Gamma} \int_{-\infty}^{+\infty} \frac{\exp -2\left(\frac{E-E_{ns}}{\Gamma}\right)^{2}}{1+\exp\left(\frac{E-E_F}{kT}\right)} dE \quad (2)$$

with $E_{ns} = E_o+(n+1/2 + \theta s/2)\hbar\omega_c$, $S = \pm 1$ and $\theta = (g^*m^*/(2m_o)$
Our study showed that, with no ambiguity, the best agreement for ρ_{xx} was obtained for the broadening parameter Γ independent of B ($\Gamma \sim 1/\mu$).The g factor is calculated taking into account the spin splitting enhancement by the exchange interaction of electrons[5].
We have two unknown quantities N and E_o and two independent equations, equations 1 and 2. We have performed iterative numerical calculations of N and E_o for each value of B.
We have calculated σ_{xy} in the relaxation time approximation by using the Drude model : $\sigma_{xy} = -(Ne)/B$ for quantizing magnetic fields ($\omega_c \tau >> 1$).

To calculate the longitudinal conductivity σ_{xx} we use[6] :

$$\sigma_{xx} = \frac{e^2}{\pi^2 \hbar} \int_{-\infty}^{+\infty} \left(- \frac{df}{dE} \right) \sum_{n,s} (n + \frac{1}{2}) \exp\left[-\left(\frac{E - E_{ns}}{\Gamma} \right)^2 \right] dE \qquad (3)$$

ρ_{xx} and ρ_{xy} can be then calculated by using the well known relations for an isotropic system. For most of the investigated samples we have obtained a very good agreement between the experimental and theoretical magnetic field dependence of both ρ_{xx} and ρ_{xy} in a large magnetic field range and a large temperature range without any fitting parameter.

As an example, Fig. 2 corresponds to the case of a MBE sample at 1.4K.The magnetic field dependence of $E_F - E_o$ shown in Fig. 2c is very similar to the one observed for the chemical potential by measuring the contact potential difference[7]. We can notice that by using this technique Nizhankowski et al. have pointed out an oscillatory dependence of N versus B which allowed then to explain the QHE without invoking the localization concept.

REFERENCES

1. G.A. Baraff and D.C. Tsui, Phys. Rev. B24, 2274, (1981).
2. K.von Klitzing et al. Phys. Rev. Lett. 45, 494, (1980).
3. "The Quantum Hall Effect". Graduate Texts in Contemporary Physics.Ed.R.E. Prange and S.M.Girvin, Springer Verlag, (1987).
4. A.Raymond and K. Karraï,J.de Phys.C5,n°11,T.48, 491,(1987).
5. T. Ando and Y. Uemura, J. Phys. Soc. Jpn. 37, 1044 (1974).
6. T. Ando and Y. Uemura, J. Phys. Soc. Jpn. 36, 959, (1974).
7. V.I. Nizhankowski et al, Sov. Phys. JETP, 63, 776, (1986).

QUILLS in a Corbino Geometry

L. Bliek

Physikalisch-Technische Bundesanstalt, Bundesallee 100,
W-3300 Braunschweig, Fed. Rep. of Germany

In analogy to a calculation by Eaves and Sheard for QUILLS processes in rectangular samples, corresponding formulae for Corbino rings are derived and are compared to experimental results.

1. Introduction

In quantum Hall effect samples with narrow constrictions, effects have been observed [1,2] that Eaves and Sheard have attributed to quasielastic interlandaulevel scattering (QUILLS) processes [3,4] . These effects included unusually high critical curents for the breakdown of the dissipationless quantum Hall effect and step-like structures in the magnetoresistance of the samples. Haug, von Klitzing and Ploog [5] have published very similar experimental results for ring-shaped samples, i.e. for a pseudo Corbino geometry. Therefore, the equivalent of the Eaves and Sheard model for a ring geometry is treated below and the results are compared to the experimental ones.

2. The 2-Dimensional Schrödinger Equation in Crossed Electric and Magnetic Fields for Rotational Symmetry

Instead of the asymmetric gauge used by Eaves and Sheard, the symmetric gauge is more suited to our purpose. The 2-D Schrödinger equation in polar coordinates (r, φ) one then obtains has been treated by Dingle [6] , but for the present purpose a slightly different approach is followed here.

Putting $\Psi(r,\varphi) = e^{il\varphi}\, Y(r)$ and introducing an additional radial electric potential $V(r)$, one has

$$\left\{\frac{\hbar^2}{2m}\left[-\frac{\partial}{\partial r^2} - \frac{1}{r}\frac{\partial}{\partial r} + \frac{l^2}{r^2} + \frac{e^2B^2}{4\hbar^2}r^2 - \frac{eBl}{\hbar}\right] + eV(r)\right\} Y(r) = \varepsilon\, Y(r). \tag{1}$$

As for the rectangular geometry treated by Eaves and Sheard there is, apart from a possible confining potential $\sim r^2$ which is not useful for our purpose, only one shape for the electric potential that one can readily deal with namely

$$V(r) = a_{-2}\, r^{-2} \tag{2}$$

Any other dependence on r poses serious mathematical difficulties. Also as for the rectangular case, it is hard to conceive a 2-dimensional charge distribution that, in a 3-D world, will produce this electric potential.

Springer Series in Solid-State Sciences, Vol. 101
High Magnetic Fields in Semiconductor Physics III Editor: G. Landwehr

The equation is solved by

$$\Psi_{N,l} = e^{il\varphi}\, r^{\sqrt{l^2 + \frac{2a_{-2}em}{\hbar^2}}}\, e^{\frac{-eBr^2}{4\hbar}} \left[\sum_{n=0}^{N} C_n\, r^{2n} \right] \tag{3}$$

and

$$\varepsilon_{N,l} = \hbar\omega \left[N + \frac{1}{2} - \frac{1}{2} l + \frac{1}{2} \sqrt{l^2 + \frac{2a_{-2}em}{\hbar^2}} \right] \tag{4}$$

with $\omega = \frac{eB}{m}$. For large l one has

$$\varepsilon_{N,l} \approx \hbar\omega \left[N + \frac{1}{2} + \frac{a_{-2}\, em}{2\hbar^2 l} \right] = \hbar\omega \left[N + \frac{1}{2} \right] + \frac{a_{-2} e^2 B}{2\ \hbar l}, \tag{5}$$

$\Psi_{N,l}$ has its maximum amplitude at r_o, with

$$r_o{}^2 = \frac{2\hbar}{eB} \sqrt{l^2 + \frac{2a_{-2}\, em}{\hbar^2}}\ , \tag{6}$$

As for the rectangular geometry, the electric potential acts on the position of the wave function which is related to its (angular) momentum. The spread of the wave function is

$$\Delta r = \sqrt{(2N+1)\, \frac{\hbar}{eB}} \tag{7}$$

which differs from Dingle's result [6], but is identical to the spread used by Eaves and Sheard [3,4].

3. QUILLS

For a QUILLS process to occur, states for which n differs by one unit have to be degenerate and the wave functions have to have sufficient overlap. To express this we use the condition given by Eaves and Sheard [3,4] and eq.(7). This leads to

$$\Delta V_c = a_{-2} \left(\frac{1}{r_i{}^2} - \frac{1}{r_a{}^2}\right) = \frac{\hbar B}{2m}\, r^3\, \frac{1/r_i{}^2 - 1/r_a{}^2}{\sqrt{(2N+3)\frac{\hbar}{eB}} + \sqrt{(2N+1)\frac{\hbar}{eB}}}\ . \tag{8}$$

If $r_i - r_a \ll r_i, r_a$ this result becomes identical to the one derived by Eaves and Sheard [3,4]. For the Corbino rings used by Haug et al.[5] , however, this is not the case.

With the assumption that QUILLS transitions will take place where the electric field is highest, i.e. near the inner side of the ring, a comparison of eq. (8) with the corresponding result for the rectangular geometry leads to an effective width W* which may be considerably smaller than the width $W = r_a - r_i$ of the ring. For the rings used by Haug et al. [5, 7] the ratio of W* to W ranges from 0.95 to 0.26.

The condition that l has to change by an integer number n_l leads to:

$$n_l = l_1 - l_2 \approx \left(\sqrt{2N+1} + \sqrt{2N+3}\right) \cdot \frac{r_1 + r_2}{2} \cdot \sqrt{\frac{eB}{\hbar}} \ . \qquad (9)$$

This is the result obtained by Eaves and Sheard for the rectangular geometry, but with the length of the rectangular bar replaced by $\pi(r_1+r_2)$ which becomes $2\pi r_i$ if the QUILLS transitions are oncemore assumed to take place near the inner side of the ring.

The QUILLS transitions will provide a radial current I_R since the change in l corresponds to a change in position. One finds

$$I_R = \frac{n_T n_l}{n_o^2} \, i^2 \, \frac{e^2}{h} \, \Delta V \ . \qquad (10)$$

Combining eqs. (9) and (10) leads one to expect periodic structures in the product of the conductivity and $\sqrt{B}$, instead of the periodic structures in the product of the resistance and $\sqrt{B}$ that one observes in the rectangular case.

4. Comparison with experimental results

From the data published by Haug et al. [5] on critical velocities one can derive critical currents for their samples. For the rectancular samples the critical currents decrease with increasing sample width from ~3,5 % at 10 μm to ~2,5 % at 80 μm of the critical currents predicted by the QUILLS model. For the Corbino samples, there is an increase from ~2,5 % for the 10 μm sample with r_i = 0,3 mm to ~3 % for the 100 μm sample with r_i = 0,80 mm.

From Fig. 2 of their paper one can obtain the positions of the additional steps on the B scale and the resistance values they correspond to and compare them to eqs. 9 and 10. One finds with reasonable accuracy

$$\sqrt{B} = n_l \cdot 3 \cdot 10^{-3} \ T^{1/2} \qquad (11)$$

and n_l values around 1200, whereas the QUILLS model according to eq. 9 predicts

$$\sqrt{B} = n_l \cdot 3{,}1 \cdot 10^{-5} \ T^{1/2} \ . \qquad (12)$$

The numerical constant should be propertional to $1/r_i$ which turns out to have an effective value about 100 times larger than expected.

For the quantity $\sqrt{B}/R$ at the small steps one finds

$$\frac{\sqrt{B}}{R} = n_R \cdot 5{,}2 \cdot 10^{-7} \ T^{1/2} \ \Omega^{-1} \qquad (13)$$

with n_R between 1 and 44. The QUILLS model predicts

$$\frac{\sqrt{B}}{R} = \frac{n_T}{n_o} \cdot 3{,}8 \cdot 10^{-9}\ T^{1/2}\ \Omega^{-1} \tag{14}$$

which follows from eqs. 9 and 10. The numerical constant is proportional to $r_i\ /(r_a^{\ 2} - r_i^{\ 2})$ and again the effective value is about 100 times the expected one.

The general conclusion one may draw is that the inhomogeneous current flow that one finds on calculating for a Corbino geometry the equivalent of the Eaves and Sheard model for QUILLS in a rectangular geometry, reduces the effective sample width and thereby makes the discrepancy between theory and experiment smaller. Yet, this inhomogeneity is not sufficient to explain the experimental observations which qualitatively agree with the QUILLS model but for which there is a quantitative difference of a factor around 100. As for wide rectangular samples one therefore has to look for other reasons for inhomogeneous current flow.

References

[1] L.Bliek, G.Hein, D.Jucknischke, V.Kose, J.Niemeyer, G.Weimann, W.Schlapp: Surface Science **196**, 156-164 (1988)

[2] L.Bliek, G.Hein, V.Kose, J.Niemeyer, G.Weimann, W.Schlapp: The Applic. High Magnetic Fields in Semicon. Phys., Proc. of the Int. Conf. Würzburg 1986, G.Landwehr ed., Springer Series in Sol. State Sciences **71**, Springer-Verlag Berlin, Heidelberg 1987, 113-117

[3] L.Eaves, F.W.Sheard: The Applic. High Magnetic Fields in Semicon. Phys., Proc. of the Int. Conf. Würzburg 1986, G.Landwehr ed., Springer Series in Sol. State Sciences **71**, Springer-Verlag Berlin, Heidelberg 1987, 118-121

[4] L.Eaves, F.W.Sheard: Semicond. Science and Technol. **1**, 346-349 (1986)

[5] R.J.Haug, K.v.Klitzing, K.Ploog: The Applic. High Magnetic Fields in Semicon. Phys. II, Proc. of the Int. Conf. Würzburg 1988, G.Landwehr ed., Springer Series in Sol. State Sciences **78**, Springer-Verlag Berlin, Heidelberg 1989, 185-189

[6] R.B.Dingle: Proc. Roy. Soc. **A 211**, 500-516 (1952)

[7] R.J.Haug, priv. comm.

Coulomb-Driven Destruction of Quantum Hall States in a Double Quantum Well

G.S. Boebinger

AT&T Bell Laboratories, Murray Hill, NJ 07974, USA

Abstract. Magnetotransport studies of double quantum well structures with interwell tunneling resolve the energy gap between the symmetric and antisymmetric electron states. The new quantum Hall states which originate from the symmetric/antisymmetric (SAS) energy gap are found to be missing in high perpendicular magnetic fields. Increasing barrier thickness between the two wells reduces the threshold magnetic field for the missing states. Recent calculations find that coulomb interactions drive the collapse of the SAS energy gap at high magnetic fields, in semi-quantitative agreement with experiment.

1. Introduction

The integral [1] and fractional [1,2] quantum Hall effects were both discovered in high mobility two-dimensional electron systems. Considerable recent interest focusses on structures which allow controlled introduction of degrees of freedom associated with the third dimension. The double quantum well (DQW) is the simplest of these structures and preserves both high electron mobility and external gating of the electron density in each layer. Interlayer coulomb interactions in a multi-layer structure are predicted to lead to even denominator fractional quantum Hall states [3-5] and increased stability of a new collective state, perhaps the Wigner crystal. [6-9] Experimentally, both the integral [10] and fractional [11] quantum Hall effects have been observed in multi-layer systems with substantial interlayer tunneling. In this paper, we review recent experiments [12] and calculations [8,9] on double quantum wells which demonstrate the coulomb-driven destruction of integral quantum Hall states in high magnetic fields.

2. Sample Description

Of the DQW samples studied, the three discussed herein consist of nominally identical quantum wells separated by barriers of different widths. The samples are grown by molecular beam epitaxy and consist of two GaAs wells of width d_W=139Å, separated by a $Al_{0.3}Ga_{0.7}As$ barrier of thickness d_B=28Å, 40Å, or 51Å (see inset of Figure 1). The electrons are provided by remote delta-doped donor layers ($N_{Si}=6\times10^{11}cm^{-2}$) set back from each side of the DQW by ~600Å-thick $Al_{0.3}Ga_{0.7}As$ spacer layers.

Springer Series in Solid-State Sciences, Vol. 101
High Magnetic Fields in Semiconductor Physics III Editor: G. Landwehr

Table 1: Sample parameters

d_B(Å)	n_{2D}(cm^{-2})	μ(cm^2/V-s)	Δ_{SAS}^{calc}(K)	Δ_{SAS}^{meas}(K)	missing states
28	4.2×10^{11}	740,000	17.3	17.5±0.5	—
40	3.8	630,000	8.1	7.7±0.5	ν = 1
51	3.9	350,000	3.9	—	ν = 1,3

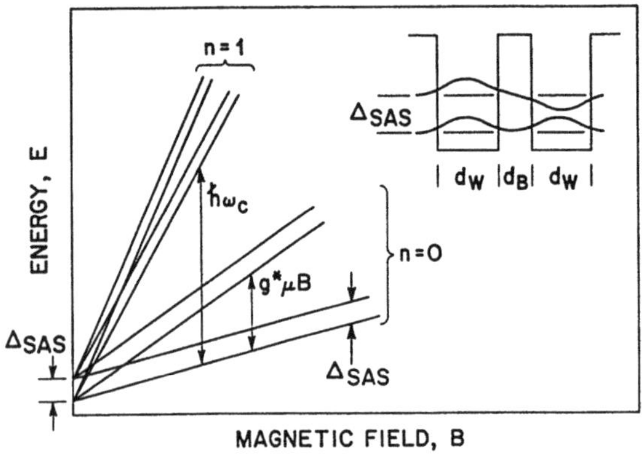

Figure 1: Single electron energy diagram for a double quantum well. The cyclotron ($\hbar\omega_c$), Zeeman ($g^*\mu_B B_{total}$), and symmetric-antisymmetric (Δ_{SAS}) energies are indicated. The symmetric and antisymmetric states of the double quantum well are shown in the inset. [Ref. 12]

When the electron density is balanced between the two quantum wells (hereafter referred to as "balanced"), tunneling mixes the single-well electron states to form symmetric and antisymmetric DQW states separated by an energy gap, $\Delta_{SAS}=E_{AS}-E_S$, where E_S and E_{AS} are the energies of symmetric and antisymmetric states. In a magnetic field perpendicular to the 2D planes, $B_\perp$, both the symmetric and antisymmetric states give rise to a fan of spin-split Landau levels, shown in Fig. 1 for the n=0 and n=1 Landau levels. The energy level filling factor is defined as the number of occupied energy levels, $\nu=n_{2D}h/eB_\perp$, where n_{2D} is the total electron density (i.e., $n_{2D}/2$ per well if the densities are balanced). In the single electron picture of Fig. 1, a quantum Hall effect will occur when the Fermi energy lies in any of the energy gaps resulting from the three relevant energies labelled in Fig. 1: the cyclotron energy, $\hbar\omega_c=\hbar eB_\perp/m^*$; the Zeeman energy, $g^*\mu_B B_{total}$; and the symmetric-antisymmetric energy, Δ_{SAS}. Table 1 contains the total electron densities, n_{2D}, and mobilities, μ, for the three samples at T=0.3K along with the calculated Δ_{SAS}^{calc}, resulting from self-consistent solution of the one-dimensional DQW Schrodinger equation and Poisson's equation.

3. Experimental Results

3.1 Observation of the SAS Gap

Figure 2 contains the Shubnikov-deHaas (SdH) oscillations in the longitudinal magnetoresistance, ρ_{xx}, for the d_B=51Å sample. The beating observed in the SdH oscillations evidences two electron states occupied with similar electron densities. A positive backside gate bias, V_{BSG}, shifts the beating node at B~1T to lower magnetic fields, indicating that (i) there are two spatially separated layers of highly mobile electrons in our samples and (ii) the two electron densities are becoming more balanced. At V_{BSG}=+175V in the d_B=51Å sample, the electron densities are balanced between the two quantum wells. Higher V_{BSG} re-introduces the SdH beating.

The electron density in each state is measured from two distinct peaks in the Fourier transform of ρ_{xx} versus inverse magnetic field. Figure 3 contains the measured densities versus applied backside gate bias for the d_B=40Å sample. In the absence of interwell tunneling, the two densities would vary linearly with

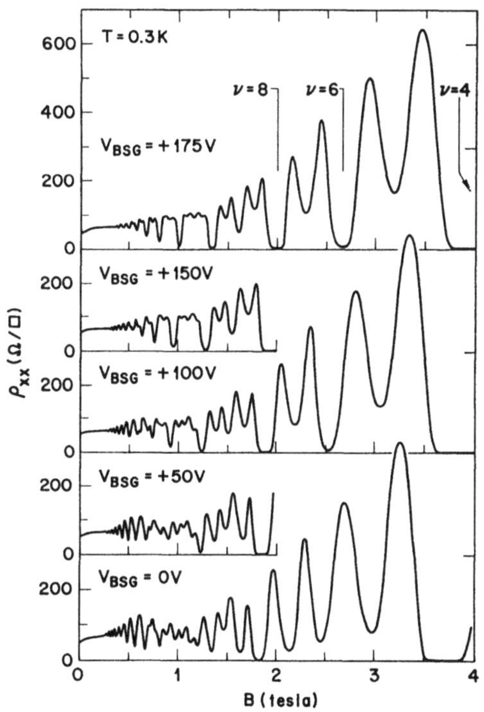

Figure 2: Shubnikov-deHaas oscillations versus backside gate bias, V_{BSG} for the d_B=51Å sample. The beating of the oscillations indicates different densities of two occupied electron states. For V_{BSG}=+175V, the electron densities of the two quantum wells are balanced.

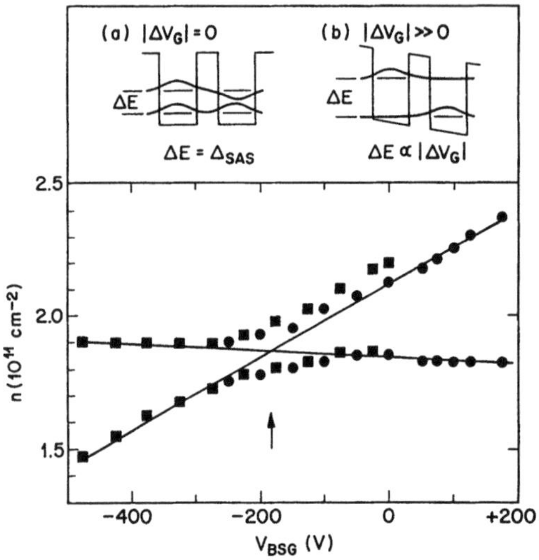

Figure 3: Electron densities versus backside gate bias, V_{BSG}, for the d_B=40Å sample. The lines depict the dependence expected in the absence of interwell tunneling. Circles and squares indicate different cooling runs in which the zero-bias interwell electric field is different. At V_{BSG}=−190V (arrow), the double quantum well is balanced, shown in inset (a). Inset (b) depicts the out-of-balance condition which suppresses mixing of the single well states.

applied voltage (intersecting lines in Fig.3). This behavior is seen when the wells are greatly out of balance (inset 3b), since the energy gap between the decoupled single-well states is proportional to the electric field between the wells. However, when the double quantum well is balanced (inset 3a), there remains a finite, minimum density difference (arrow in Fig.3) resulting from the higher electron density in the symmetric state, n_S, than the antisymmetric state, n_{AS}. In addition to demonstrating the experimental resolution of the Δ_{SAS} energy gap, this minimum density difference provides a measurement of Δ_{SAS}: $\Delta_{SAS}^{meas}/(E_F-E_S)=(n_S-n_{AS})/n_S$. In two of our samples, Δ_{SAS}^{meas} was measurable and found to be in excellent agreement with the calculated values (see Table 1).

3.2 Assignment of the SAS gap to the ν=odd states

As in a single quantum well, whenever E_F lies in an energy gap, there is a minimum in ρ_{xx} accompanied by a plateau in the Hall resistivity, ρ_{xy}. The assignment of each observed quantum Hall state to one of the energy gaps of Fig. 1 is made unambiguous by tilting the samples in a magnetic field. When an in-plane magnetic field is applied to a balanced double quantum well (see Figure 4), the quantum Hall states at ν=4n+2 (n=0,1,2...) are found to strengthen (deeper ρ_{xx} minima and broader ρ_{xy} plateaus). Since an in-plane magnetic field increases

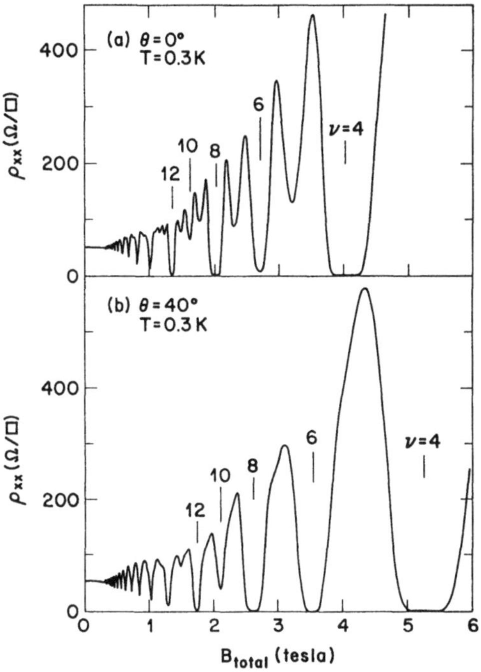

Figure 4: Shubnikov-deHaas oscillations of the balanced d_B=51Å sample in the presence of an in-plane magnetic field. θ=0° corresponds to magnetic field normal to the 2D planes. The minima at ν=6, 10 and 14 (corresponding to the Zeeman gap) become deeper, while all minima at ν=odd (corresponding to the SAS energy gap) are destroyed. [Ref. 12]

the Zeeman energy (proportional to B_{total}) at a given filling factor (fixed $B_{\perp}$), we ascribe the states at ν=4n+2 to the Zeeman energy gap. By contrast, each quantum Hall state at ν=odd in this sample is *destroyed* by a ~1.5T in-plane magnetic field. We therefore identify these states with the Δ_{SAS} energy gap, which the data suggest would collapse due to decreased tunneling between quantum wells in the presence of an in-plane magnetic field. [13]

3.3 Destruction of Δ_{SAS} gap in perpendicular magnetic fields.

The experimental results discussed to this point demonstrate that the DQW samples behave in accordance with the single-electron energy diagram of Fig.1 at low magnetic fields. Figure 5 contains the most striking result of our experiments: the low temperature ρ_{xx} and ρ_{xy} data from the balanced d_B=51Å sample in a perpendicular magnetic field to B=23T. Note that there are well-formed plateaus at all integers, 4<ν<16 (see the 3× enlarged data in the inset). Strong fractional quantum Hall states at ν=2/3, 4/3, and 8/3 also exist. Despite

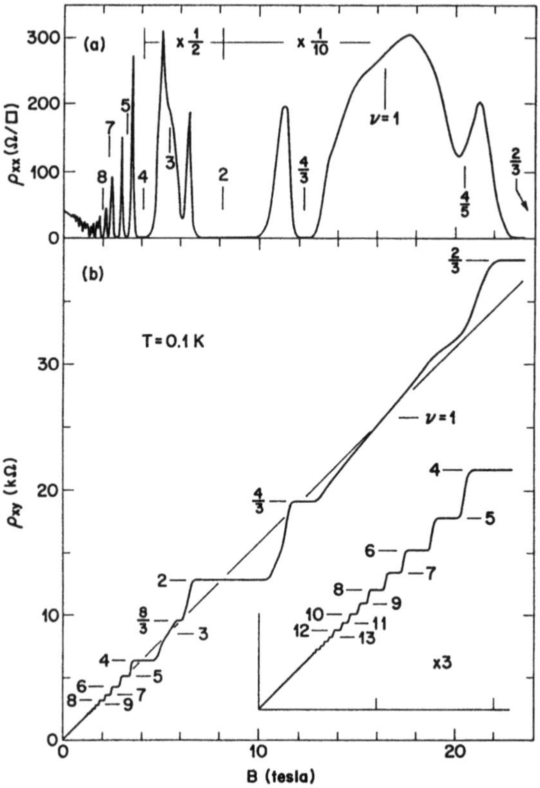

Figure 5: Longitudinal resistivity, ρ_{xx}, and Hall resistivity, ρ_{xy}, for the balanced d_B=51Å sample in a perpendicular magnetic field. Note that there is no evidence of quantum Hall states at ν=1 and ν=3. The inset is an expansion of the low field ρ_{xy} data. For clarity, the ρ_{xx} data at 4<B<8T has been reduced by a factor of 2, at B>8T by a factor of 10. [Ref. 12]

the fact that this high-mobility sample resolves all three single electron energies ($\hbar\omega_c$, $g^*\mu B$, and Δ_{SAS}) and, additionally, the collective excitation gap of the fractional quantum Hall effect, *there is no evidence of a quantum Hall state at either* ν*=3 or* ν*=1*. Particularly striking is the well-developed ν=8/3 fractionally quantized state and the missing ν=3 quantized state at nearly identical magnetic fields.

The other two balanced samples show different behavior at ν=1 and ν=3. In the d_B=40Å sample, the ν=3 state is visible; however, there is no evidence of the ν=1 state. Finally, in the d_B=28Å sample, all the quantum Hall states at integer ν<16 are well developed. The missing quantum Hall states are summarized in Table 1. The data demonstrate that high magnetic fields perpendicular to the quantum wells can collapse the Δ_{SAS} energy gap and the associated ν=odd quantum Hall states. Furthermore, the mechanism for the destruction of the quantum Hall states depends strongly on the thickness of the interwell barrier.

The fractional quantum Hall states in Fig. 5 offer further evidence for this interpretation. Normally, if the SAS energy gap were intact, the observation of the ν= 2/3, 4/3 and 8/3 fractional states would imply the ν=5/3 and 7/3 states should also exist. Instead, there is no evidence of fractional quantum Hall states at ν=5/3 and 7/3 in Fig. 5. If, however, the Δ_{SAS} gap is collapsed in high magnetic fields, the DQW becomes energetically identical (in a single electron picture) to a system of two independent single quantum wells with no interwell tunneling. In this case, the filling factor ν, as defined in this letter, would represent filling factor $\nu/2$ in each single quantum well. It follows that the well-developed quantized states at ν=2/3, 4/3 and 8/3 would correspond to the well known 1/3, 2/3, and 4/3 states existing independently in each well. Similarly, the lack of fractional quantum Hall states at ν=5/3 and 7/3 in the DQW results from the absence of fractional states at filling factors 5/6 and 7/6 in the individual wells.

Before discussing the effects of coulomb interactions in a DQW system, we emphasize that there is no evidence of the missing states of Table 1 from detailed ($\Delta\theta$~2°) angular studies near θ=0°. The accidental in-plane magnetic fields are an order of magnitude smaller than those required to destroy the ν=odd states at lower magnetic fields (see Fig. 4) for both the d_B=40Å and d_B=51Å samples. We also point out that the missing states of Table 1 are not due to the Fermi energy lying at a degeneracy of two energy levels. This effect has been observed in a single quantum well with two occupied subbands [14] and, in fact, underlies the beating nodes observed in the SdH oscillations. It cannot, however, explain the missing states in the DQW samples: (i) For example, in the d_B=51Å sample, for the $\Delta_{SAS}=g^*\mu B$ degeneracy to occur at B=16.5T (and thus obliterate the ν=1 state), the enhanced g^* factor at ν=1 must be $\lesssim$0.34, *smaller* than the bulk GaAs value. Similarly, setting $\Delta_{SAS}=\hbar\omega_c$ at B=5.5T (to remove the ν=3 state) implies m^* more than 13 times its GaAs value. (ii) The in-plane magnetic field data of Fig. 4 identifies the ν=4n+2 states in our samples with the Zeeman gap and the ν=odd states with the SAS energy gap. As Figure 1 illustrates, the observed ν=odd states correspond to the SAS energy gap only if the $\Delta_{SAS}=g^*\mu B$ degeneracy occurs at a magnetic field smaller than the observed ν=15 state at B~1T.

4. Calculated collapse of the Δ_{SAS} energy gap

Recent calculations have been performed which suggest that coulomb interactions could create a high magnetic field regime in which the Δ_{SAS} energy gap has collapsed. [8,9] In the single-electron model of Fig. 1, each electron is shared equally between the two quantum wells. At ν=odd, the lowest energy excitation consists of exciting an electron from a symmetric state to an antisymmetric state. Mixing antisymmetric states into the many-body wavefunction allows local charge density fluctuations to be established between the two quantum wells. [An extreme case would be the complete admixture of symmetric and antisymmetric wave functions to confine an electron entirely within a single quantum well.] As a result, electrons in the same well can become more strongly correlated than

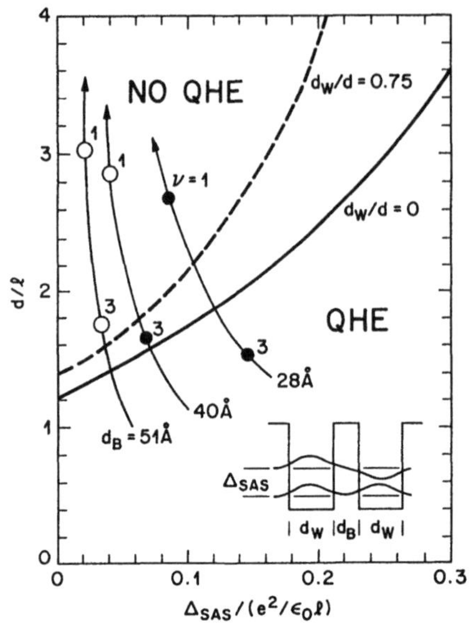

Figure 6: Phase diagram for the quantum Hall effect in a double quantum well (see inset). Δ_{SAS} is the symmetric/antisymmetric energy gap, $d=d_W+d_B$ is the inter-well spacing and $\ell=(\hbar c/eB)^{1/2}$ is the magnetic length. The heavy solid line is the calculated phase boundary of the n=0 Landau level for collapse of the energy gap above the ground state. The dashed line includes a simple finite well thickness correction. The experimental data for ν=1 and 3 from the three samples of Table 1 are represented by open/filled circles which denote the missing/observed quantized states. The arrows indicate increasing magnetic field. [Ref. 8]

electrons in different wells. If the coulomb energy reduction per antisymmetric state exceeds Δ_{SAS}, the symmetric to antisymmetric excitation gap will vanish, thereby destroying the quantum Hall effect. If Δ_{SAS} is held fixed while coulomb interactions are increased, at the point where the energy gap vanishes there is a *transition* to a new ground state with qualitatively different correlations. In the data of Fig. 5, the missing quantum Hall plateaus at strong magnetic field result when coulombic effects, which scale as $e^2/\varepsilon_0\ell \propto B^{1/2}$, drive the DQW through this transition to the new ground state. A calculation done in the single mode approximation [8] on a model two-layer system, results in a phase diagram for the destruction of the quantum Hall effect in a DQW (Figure 6) which is in semi-quantitative agreement with the experimental results of Table 1. A Hartree-Fock calculation [9], using the same DQW parameters as in Table 1, finds that a transition to charge density wave or Wigner crystal state will occur at d_B^c~23Å, compared to the experimental result: 28Å<d_B^c<40Å.

5. Conclusion

In conclusion, we reveal a high magnetic field regime in double quantum well structures which is evidenced by missing quantum Hall states at energy level filling factors ν=odd. For samples of increasing barrier thickness between the two wells, first the ν=1 state is destroyed, then both the ν=1 and ν=3 states are destroyed. This destruction is attributed to coulomb interactions which collapse the SAS energy gap in strong perpendicular magnetic fields, giving rise to a new correlated electron ground state. Calculations in both the single mode and Hartree-Fock approximations find that coulomb interactions can collapse the Δ_{SAS} energy gap by mixing antisymmetric electron states into the many electron ground state at ν=odd. These coulomb interactions give rise to a new correlated state which depends critically on the additional electron degree of freedom in the third dimension which is provided by the double quantum well.

6. Acknowledgements

The experiments described herein were collaborations with H.W. Jiang, L.N. Pfeiffer, K.W. West, and A. Passner. The single mode approximation calculation was a collaboration with A.H. MacDonald and P.M. Platzman. I also thank J.P. Eisenstein, H.L. Stormer, D.C. Tsui, and J.S. Moodera for loaning equipment and J.P. Eisenstein, H.A. Fertig, E.S. Hellman, H.L. Stormer, and R.L. Willett for valuable discussions. Some experiments were performed at the Francis Bitter National Magnet Laboratory.

7. References

1. For review, "The Quantum Hall Effect", R.E. Prange and S.M. Girvin, eds., Graduate Texts in Contemporary Physics, Springer-Verlag (New York, 1987).
2. For review, "The Fractional Quantum Hall Effect", by Tapash Chakraborty and P. Pietilainen, Springer Series in Solid State Sciences 85 Springer-Verlag, (Heidelberg, 1988) and G.S. Boebinger, Physica B155 (1989), 347.
3. T. Chakraborty and P. Pietilainen, Phys. Rev. Lett. 59, (1987) 2784. Herein, ν refers to density of electrons in each layer and, thus, is one-half of our ν.
4. D. Yoshioka, A.H. MacDonald and S.M. Girvin, Phys. Rev. B39, (1989) 1932.
5. A.H. MacDonald, Proc. of the 8th Int'l. Conf. on the Electronic Properties of Two-dimensional Electron Systems, Grenoble, 1989, to be published.
6. H.C.A. Oji, A.H. MacDonald and S.M. Girvin, Phys. Rev. Lett 58, (1987) 824.
7. H.A. Fertig, Phys.Rev. B40 (1989), 1087.

8. A.H. MacDonald, P.M. Platzman, and G.S. Boebinger, Phys. Rev. Lett. 65, (1990), accepted for publication.

9. Luis Brey, preprint.

10. H.L. Stormer, J.P. Eisenstein, A.C. Gossard, W. Wiegmann, and K. Baldwin, Phys. Rev. Lett. 56 (1986), 85.

11. H.L. Stormer, G.S. Boebinger, D.C. Tsui, and C.W. Tu, unpublished.

12. G.S. Boebinger, H.W. Jiang, L.N. Pfeiffer, and K.W. West, Phys. Rev. Lett. 64, (1990), 1793.

13. J. Smoliner, W. Demmerle, G. Berthold, E. Gornik, G. Weimann, and W. Schlapp, Phys. Rev. Lett. 63 (1989), 2116.

14. Y. Guldner, J.P. Vieren, M. Voos, F. Delahaye, D. Dominguez, J.P. Hirtz, and M. Razeghi, Phys. Rev. B33 (1986), 3990.

Metrological Aspects of the Quantum Hall Effect

E. Braun

Physikalisch-Technische Bundesanstalt, Bundesallee 100,
W-3300 Braunschweig, Fed. Rep. of Germany

Abstract: Ten years after the discovery of the quantum Hall effect, a short review of the application of the effect in metrology, including recent experimental results, is given. Experimental problems and newly developed methods concerned with precision measurement are briefly described. Finally, the CIPM's recommendation for the representation of the ohm, which has been in effect since January 1st 1990, is explained.

1. Introduction

One fascinating aspect of the quantum Hall effect (QHE) - see Fig. 1 - at the time of its discovery in 1980 was the low uncertainty with which it was possible to reproduce the quantized Hall resistances (QHR) /1/. This encouraged metrologists to make further improvements and there has been rapid progress in the decade following the discovery /2,3/. While von Klitzing started with a relative uncertainty in the order of 10^{-5}, the lowest uncertainty values today are several times 10^{-9} /4,5/.

The precision measurement of a QHR involves two main steps:

- the comparison of the QHR with a resistance standard and
- the calibration of this standard in terms of the SI ohm.

The improvement in decreasing the uncertainty mentioned above is mainly due to the first of these steps. The uncertainty of the calibration of the standard resistance depends on the realization of the SI ohm. The lowest uncertainty for this realization is obtained by deriving the ohm from the capacitance of a calculable capacitor as shown by Thompson and Lampard /6/. The relative uncertainty obtained with this method at the time of von Klitzing's discovery was in the order of 10^{-7} and since then it has not improved by more than about a factor of two. Results for precision measurements of QHRs in terms of the SI ohm, including a realization of the ohm with a total relative uncertainty of less than 10^{-7} have been published only by three institutes /7,8,9/. At least twice as many other institutes have published results in terms of their as-maintained unit of resistance with a similar or even smaller uncertainty /10/.

In the following the problems involved in the precision measurement of a quantized Hall resistance and the methods which have been developed will be briefly summarized. Following this the reproducibility of a QHR and its application in monitoring the drift rate of resistances, finding new ways of realizing the ohm and determining the finestructure constant will be described. Finally, the conclusions arrived at by the comittees of the Meter Convention considering that a QHR can be reproduced with a lower uncertainty than that by which its value is given in terms of the SI unit are explained. In the formulation of a recommendation /11/ the von Klitzing constant R_K was defined as "the quotient of the Hall potential difference V_H divided by current corresponding to the plateau i=1 in the quantum Hall effect"

$$R_K = V_H(1)/I. \tag{1}$$

Springer Series in Solid-State Sciences, Vol. 101
High Magnetic Fields in Semiconductor Physics III   Editor: G. Landwehr

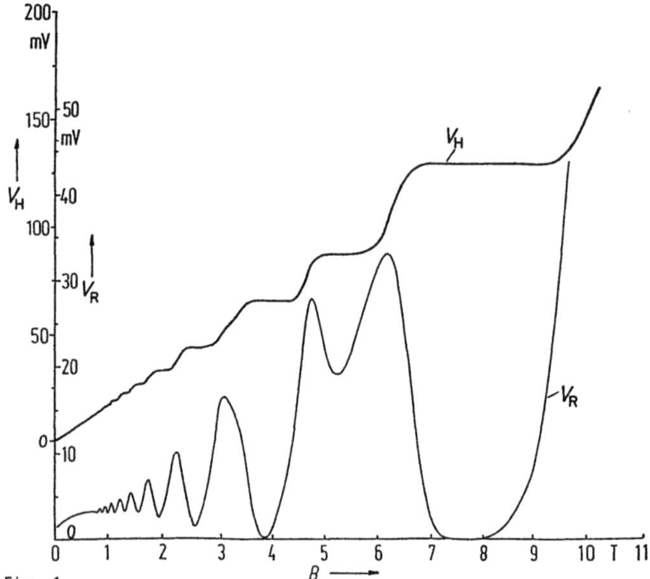

Fig. 1

Typical measurement result for the QHE of a GaAs-$Ga_xAl_{1-x}As$ heterostructure. The voltage drop V_R in the direction of the current and the Hall voltage V_H are plotted as a function of the magnetic flux density B. For those magnetic fields for which V_H reaches its quantized values, V_R shows pronounced minima and may even vanish below the experimental detection limit.

2. Experimental aspects

Before the discovery of the QHE, measurements on the highest precision level were only required for the comparison of decimal values or approximately decimal values of resistances. New methods have meanwhile been developed for the QHRs occurring at the steps i=2 and i=4, $R_H(2)$ and $R_H(4)$, near 12 906.4 Ω and 6453.2 Ω respectively. In order to link the odd value of $R_H(2)$ or $R_H(4)$ with a decimal value, Hamon resistor networks, cryogenic current comparators, and Josephson potentiometers are mainly used. Hamon resistor networks are well-established devices used in classic resistance measuring techniques /12/. They are now constructed for these special ratios. The two other methods are based on effects occurring at low temperatures such as superconductivity and the Josephson effect. The QHE has provided the impulse for their development to very high precision methods. They are shown schematically in Fig. 2 and Fig. 3.

In order to obtain a relative uncertainty of 10^{-8} or less in the comparison of resistances in the kΩ range, the insulation resistance of the total circuit must be higher than 10^{12} Ω. One major problem which arises in the measurement is the dependence of wire-wound standard resistors on temperature, pressure, dissipated power and time. For precision measurements of QHRs GaAs-GaAlAs heterostructures and Si MOSFETs have mainly been used. However, each sample must be carefully checked to injure that the requirements for a precision measurement are fulfilled, e.g. the longitudinal resistivity ρ_{xx} must be below a certain value (Fig. 1). In the application of the QHE as described in chapter 6, the "technical guidelines for reliable measurements of the quantized Hall resistance", as published in /16/, must be observed.

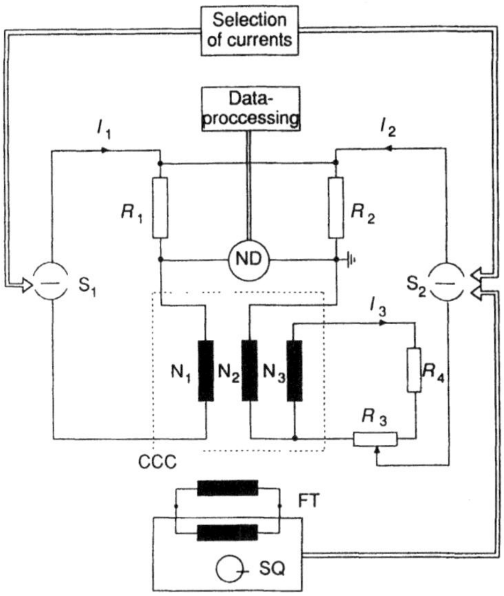

Fig. 2

Scheme of a measuring circuit for the ratio of two resistances R_1 and R_2 based on a cryogenic current comparator (CCC) as developed by Delahaye et al. /13/. S_1 and S_2 are current sources. The currents I_1 and I_2 are electronically nically controlled with their ratio close to the desired one, (N_1/N_2). N_1 and N_2 are the windings of the CCC. The ratio of the currents is finely adjusted using the output signal of the SQUID (SQ) in combination with a flux transformer (FT). This signal is due to the resulting flux produced by the two windings. The measurement is done by feeding a small current I_3 through the winding N_3 and adjusting the resistors R_3 and R_4 until the null instrument (ND) indicates the balance of the voltage drops at R_1 and R_2.

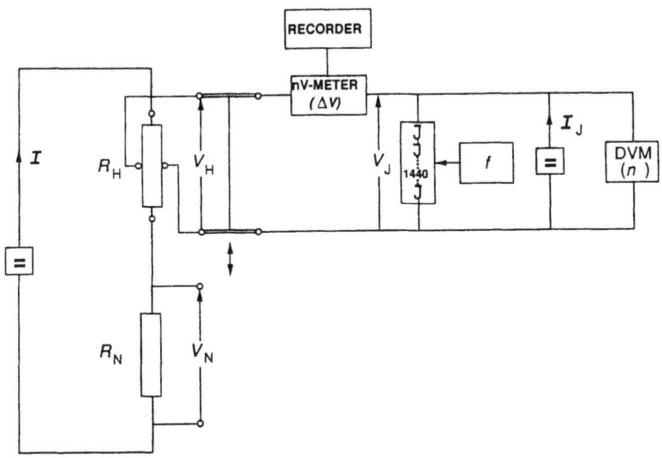

Fig. 3

Scheme of a Josephson potentiometer /14/ based on a series array of 1440 Josephson junctions /15/. The current I_J is fed through the array and the step n at which it should operate is monitored by a digital volt meter (DVM). The array is irradiated with a microwave of frequency f. The Josephson voltage V_J almost balances the Hall voltage V_H or the voltage drop V_R at a standard resistance R_N. The small remaining voltage ΔV is measured by a nV meter.

3. Reproducibility of a QHR

As an example of the low uncertainty in the reproducibility of a QHR, a result obtained by Delahaye et al. at the BIPM is shown in Fig. 4 /17/. It shows the ratio of R_K/R_0, where R_K is the von Klitzing constant and R_0 is a reference resistance, as a function of time obtained at different step numbers, different temperatures, and with different samples. The slope is due to the drift of R_0 which is found to be a relative value of $2.5 \cdot 10^{-9}$ per day.

The standard deviation for the reproducibility of the eleven values is $3 \cdot 10^{-9}$, independent of the different parameters.

Before the discovery of the QHE the units of resistance at the national institutes - maintained by groups of standard resistors - were assumed to be independent of time. However, Cage et al. /18/ showed that the US legal ohm, Ω_{NBS}, for example, decreased by 52.9 nΩ per year (Fig. 5).

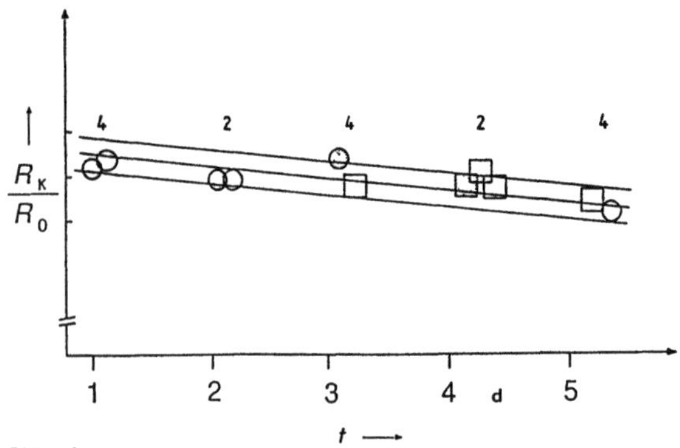

Fig. 4

Comparison of several measurements of the ratio of the von Klitzing constant R_K to a reference resistance R_0 of 100 Ω using two different GaAs samples (circle and square), different temperatures (blank sign: 0.5 K, dotted sign: 1.5 K) and various steps. The measurements on one day were made using either step i = 4 or i = 2. The lines indicate the relative standard deviation of $3 \cdot 10^{-9}$ with respect to the adjusted straight line which is due to the drift of R_0.

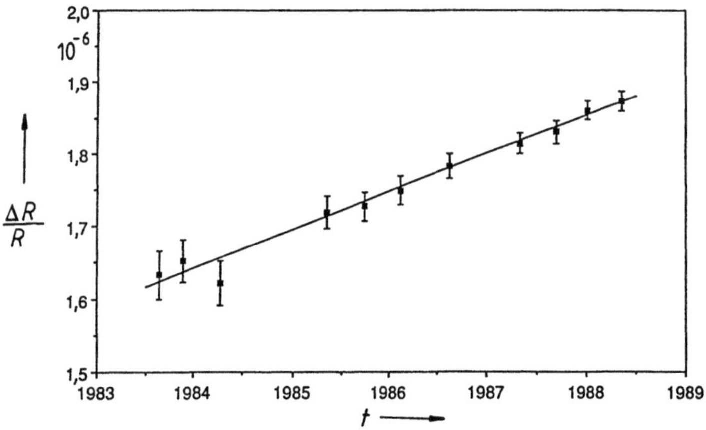

Fig. 5

Measurements of the von Klitzing constant in terms of Ω_{NBS} as a function of time. ΔR/R is the relative deviation of the measured value from a reference value of 25 812,8 Ω_{NBS}. From the data it follows that Ω_{NBS} decreases by (52.9 ± 4) nΩ per year.

4. Determination of the fine-structure constant

On the assumption that the von Klitzing constant R_K is identical to h/e^2, the QHE makes it possible to determine the fine-structure constant α in two ways. Since the inverse fine-structure constant is then given as:

$$\alpha^{-1} = 2 \cdot R_K/(\mu_o c) \qquad (2)$$

where $\mu_o = (4\pi \cdot 10^{-7} \pm 0)\ \mathrm{VsA^{-1}m^{-1}}$ according to the definition of the ampere and $c = (299\ 792\ 458 \pm 0)\ \mathrm{ms^{-1}}$ according to the definition of the meter, the fine-structure constant is obtained from the von Klitzing constant - with the same relative uncertainty - by a simple calculation. As R_K must be determined in SI units, a realization of the ohm is necessary in order to determine α. Besides the direct method via a calculable capacitor, indirect methods based on the precision measurement of fundamental constants have also been developed.

4.1 Determination of α using the QHE and the realization of the ohm via the calculable capacitor

Fig. 6 shows the result of the determination of the von Klitzing constant at eleven national institutes and at the Bureau International des Poids et Mesures (BIPM) /10/. These results were obtained in terms of the as-maintained units of resistance at the various institutes and were then converted into values which are now given in terms of Ω_{69-BI}(1987-10-20). This was the as-maintained unit of resistance at the BIPM on the 20th of October 1987, when the most recent international comparison of resistance standards took place.

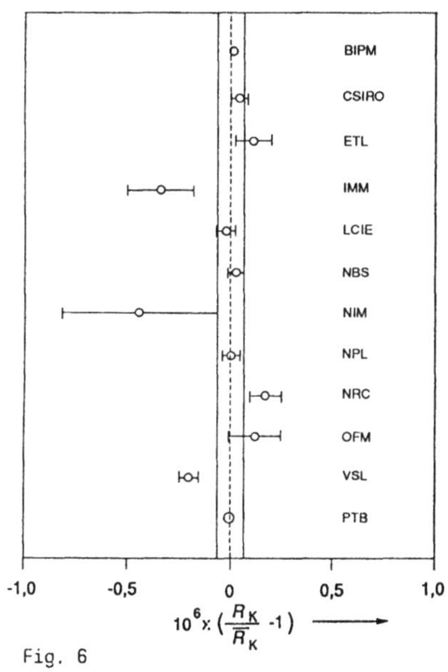

Fig. 6

Comparison of determinations of the von Klitzing constant at national institutes and at the BIPM /10,19,20/. All experimental results have been converted into terms of Ω_{69-BI} on the basis of the international intercomparison of 1-Ω resistances on October 20th, 1987. The dashed line is the weighted mean value $\bar{R}_K$. The two adjacent vertical lines indicate the standard deviation of $7 \cdot 10^{-8}$.

List of abbreviations of the names of laboratories:

BIPM	Bureau International des Poids et Mesures, Paris, France
CSIRO	Commonwealth Scientific and Industrial Research Organization, Division of Applied Physics, Lindfield, Australia
ETL	Electrotechnical Laboratory, Ibaraki, Japan
IMM	Institute de Métrologie D. I. Mendéléev, Leningrad, UDSSR
LCIE	Laboratoire Central des Industrie Electriques, Fontenay-aux-Roses, France
NBS	National Bureau of Standards (now NIST, National Institute of Standards and Technology), Gaithersburg, USA
NIM	National Institute of Metrology, Beijing, China
NPL	National Physical Laboratory, Teddington, Great Britain
NRC	National Research Council of Canada, Ottawa, Canada
OFM	Office Fédéral de Métrologie, Switzerland
VSL	Van Swinden Laboratory, Netherlands.

Ω_{69-BI}(1987-10-20) in terms of Ω is well known from intercomparisons of resistance standards between the BIPM and the CSIRO and from the realization of the ohm at the CSIRO /10/. From the weighted mean of these measurements one obtains

$$R_K = 25\,812.807\cdot(1+(57\pm67)\cdot10^{-9})\ \Omega = (25\,812.8085\pm0.0017)\ \Omega \qquad (3)$$

leading to a value for the fine-structure constant of

$$\alpha^{-1} = 137.036\,0047 \cdot (1 \pm 67\cdot10^{-9}). \qquad (4)$$

4.2 Determination of α by precision measurement of QHE and other fundamental constants in terms of laboratory units

While the uncertainty obtainable with the method described above is limited by the uncertainty of the realization of the SI ohm via a calculable capacitor, the following method does not include this realization.

The simultaneous measurement in terms of laboratory units (e.g.: A_L, V_L, Ω_L) of the von Klitzing constant R_K, the Josephson constant K_J, and of the gyromagnetic ratio of the proton, γ_p', allows α to be determined via the following relation /21/

$$\alpha^{-3} = \{\mu_p'/\mu_B\}\cdot\{K_J\}_L\cdot\{R_K\}_L\Big/\left(2\cdot\{\mu_o\}\cdot\{R_\infty\}\cdot\{\gamma_p\}_L\right) \qquad (5)$$

(μ_p'/μ_B proton magnetic moment in Bohr magnetons, R_∞ Rydberg constant. $\{x\}$ and $\{x\}_L$ stands for the numerical value of x, measured in terms of the SI-unit or a laboratory unit, respectively.)

The most recent value for α^{-1} determined with this method was obtained at the National Institute of Standards and Technology (NIST) /22/:

$$\alpha^{-1} = 137.035\,9840\cdot(1 \pm 37\cdot10^{-9}). \qquad (6)$$

The values for α given in (4) and (6) should be compared with those obtained without making use of the QHE.

4.3 Comparison with values for α otherwise obtained

The following values for α are based on properties of the electron, the proton and the neutron:

Derivation of α from the anomalous magnetic moment of the electron: From the experimental result obtained by van Dyck, Schwinberg and Dehmelt for the anomalous magnetic moment of the electron, a_e, /23/ and a series expansion up to the fourth order, Kinoshita /24/ obtained for α

$$\alpha^{-1} = 137.035\,9914\cdot(1 \pm 8.1\cdot10^{-9}). \qquad (7)$$

So far this value for α has the lowest relative uncertainty.

Derivation of α from the gyromagnetic ratio of the proton, the Josephson constant and an realization of the Ω: The relation

$$\alpha^{-2} = \{c\}\cdot\{\mu_p'/\mu_B\}\cdot\{K_J\}_L\Big/\left(4\cdot\{R_\infty\}\cdot\{\gamma_p'\}_L\cdot\{\Omega_L\}\right) \qquad (8)$$

was used to obtain the following value for the fine-structure constant at NIST /22/

$$\alpha^{-1} = 137.035\,9770\cdot(1 \pm 56\cdot10^{-9}). \qquad (9)$$

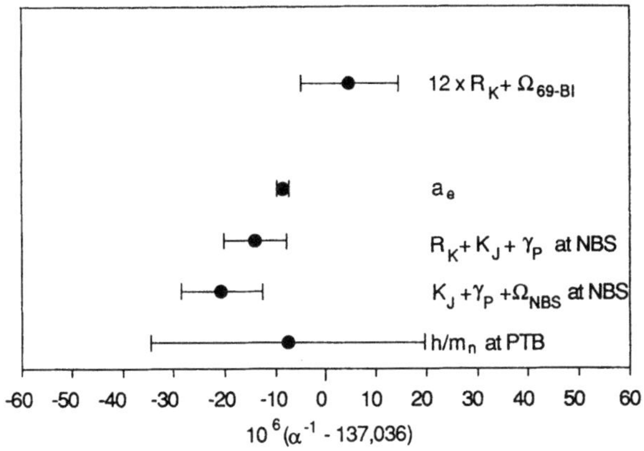

Fig. 7
Comparison of values for the inverse fine-structure constant α^{-1} determined with either the QHE and Ω_{69-BI} or via the precision measurement of fundamental constants. For details see text.

Derivation of α from the quotient of the Planck constant and the neutron mass: A third way to obtain α^{-1} stems from an experiment at PTB, where the quotient of the Planck constant and the neutron mass m_n has been determined /25/. As shown in /26/ α^{-1} can be obtained by means of the following equation

$$\alpha^{-2} = c\cdot(\mu_p'/\mu_B)\Big/\Big(2\cdot R_\infty\cdot(\mu_n/\mu_p)\cdot(\mu_p'/\mu_N)\cdot(h/m_n)\Big) \tag{10}$$

(μ_N neutron magnetic moment)

With the result obtained for h/m_n /25/ we have

$$\alpha^{-1} = 137.035\ 993\cdot(1 \pm 2\cdot 10^{-7}). \tag{11}$$

However, the relative uncertainty of this route for determining α is at present much higher than the other ones. All these values are plotted in Fig. 7. As can be seen, there is not a total overlap of all the values. This figure in particular could be interpreted as showing that the relation $R_K = h/e^2$ is not valid on a level of relative uncertainty of $1\cdot 10^{-7}$. However, the indirect determinations are only based on measurements in one laboratory. More measurements in other laboratories are necessary to clarify the situation.

5. Realization of the SI ohm by means of the QHE

As shown in 4.2 the simultaneous measurement of R_K, K_J and γ_p in terms of laboratory units allows the fine-structure constant to be determined. The same experimental results can be used to calculate the numerical value of the laboratory resistance unit Ω_L in terms of the SI unit Ω /21/

$$\{\Omega_L\}^3 = \{c\}^3\cdot\{\mu_o\}^2\cdot\{\mu_p'/\mu_B\}\cdot\{K_J\}_L\Big/\Big(16\cdot\{R_\infty\}\cdot\{\gamma_p\}_L\{R_K\}_L{}^2\Big) \tag{12}$$

This method is in competition with the one based on the calculable capacitor. Ω_L can be obtained in this way with a lower uncertainty if the relative uncertainty of the γ_p determination is less than three times the relative uncertainty of the ohm realization with the calculable capacitor.

If we take the experimental results obtained at NIST /22/ as an example, we have for Ω_{NBS} on the date of 1988-04-03,

$$\Omega_{NBS} = (1 - (1.689 \pm 0.037)\cdot 10^{-6})\ \Omega.$$

The value obtained with the calculable capacitor is

$$\Omega_{NBS} = (1 - (1.600 \pm 0.022)\cdot 10^{-6})\ \Omega.$$

6. The CIPM's recommendation for the representation of the ohm by means of the QHE

It has been pointed out that the limitation in the uncertainty of the measurement of a QHR is given by the uncertainty of the realization of the ohm which is at present based on the calculable capacitor. It has been shown that a QHR can be reproduced with a relative uncertainty of about one order of magnitude lower than the relative uncertainty of the realization of the ohm. This situation has led to a discussion of whether there should be a change in the definition of the ohm.

However, the SI system is a system of units where base units are defined and the other units are derived uniquely according to physical laws. The ampere, the unit of electric current, is a bace unit, while the ohm is a derived unit. It is not possible to keep the ampere as a base unit and to change the definition of the ohm.

When the Josephson effect was discovered a similar discussion arose with respect to the volt. But the result was not a change in the SI system. A recommendation was passed to the national laboratories by the Comité Consultatif d'Electricité (CCE), which is a part of the Meter Convention, for the common use of the same value for - what is now called - the Josephson constant. Unfortunately, not all countries followed the recommendation, and four different values were in use. More over the value recommended by the CCE was, as we now know, too high by a relative value of about $7\cdot 10^{-6}$.

The Comité International des Poids et Mesures (CIPM) therefore followed a proposal made by the CCE at its meeting in October 1988, adopting two recommendations concerning the use of the QHE and the Josephson effect for the representation of the ohm and the volt.

The recommendation for the representation of the ohm is shown in Fig. 8 /11/. From a review /27/ of all the determinations of the von Klitzing constant, including those based on the assumption that R_K is identical with h/e^2, the value 25 812.807 Ω was chosen and recommended to be used for the representation of the ohm. It was designated as R_{K-90}, where the index 90 indicates that this value has been in use since January 1st, 1990.

It is assumed that this value agrees within a one-standard-deviation uncertainty of $2\cdot 10^{-7}$ with the von Klitzing constant, R_K, if the measurement follows the previously mentioned technical guidelines. It has also been decided that the recommended value will not be changed in the foreseeable future, even if it turns out that a value closer to the true value could be chosen.

Meanwhile this recommendation has been adopted by all members of the Meter Convention. Those institutes which are not able to derive their as-maintained unit of resistance from a QHR are linked up to these fixed points of resistance through a comparison of resistance standards with an institute which has the equipment, for example, the BIPM.

The main advantages of this recommendation are:

- **International uniformity.** Calibrations in all countries have the same numerical value.

Representation of the ohm by means of the quantum Hall effect

Recommendation 2 (CI-1988)

The Comité International des Poids et Mesures,

acting in accordance with instructions given in Resolution 6 of the 18th Conférence Générale des Poids et Mesures concerning the forthcoming adjustment of the representations of the volt and the ohm,

considering

– that most existing laboratory reference standards of resistance change significantly with time,

– that a laboratory reference standard of resistance based on the quantum Hall effect would be stable and reproducible,

– that a detailed study of the results of the most recent determinations leads to a value of 25 812.807 Ω for the von Klitzing constant, R_K, that is to say, for the quotient of the Hall potential difference divided by current corresponding to the plateau $i = 1$ in the quantum Hall effect,

– that the quantum Hall effect, together with this value of R_K, can be used to establish a reference standard of resistance having a one-standard-deviation uncertainty with respect to the ohm estimated to be 2 parts in 10^7, and a reproducibility which is significantly better,

recommends

– that 25 812.807 Ω exactly be adopted as a conventional value, denoted by R_{K-90}, for the von Klitzing constant, R_K,

– that this value be used from 1st January 1990, and not before, by all laboratories which base their measurements of resistance on the quantum Hall effect,

– that from this same date all other laboratories adjust the value of their laboratory reference standards to agree with R_{K-90},

– that in the use of the quantum Hall effect to establish a laboratory reference standard of resistance, laboratories follow the most recent edition of the technical guidelines for reliable measurements of the quantized Hall resistance drawn up by the Comité Consultatif d'Electricité and published by the Bureau International des Poids et Mesures, and

is of the opinion

– that no change in this recommended value of the von Klitzing constant will be necessary in the foreseeable future.

Fig. 8

Recommendation 2 (CI-1988) of the Comité International des Poids et Mesures for the representation of the ohm by means of the QHE.

- **Independence of time.** Calibrations at all times have the same numerical values.
- The result of a calibration is in the **best possible agreement with the SI unit.**

It should be pointed out that the recommendation does not answer the question whether the von Klitzing constant is equal to h/e^2. This remains to be clarified in future experiments with lower uncertainties.

References

/1/ K.v. Klitzing, G. Dorda, and M. Pepper: Phys. Rev. Lett. **45**, 494 (1980).

/2/ E. Braun, E. Staben, K.v. Klitzing: PTB-Mitteilungen **90**, 350 (1980).

/3/ L. Bliek, E. Braun, H.-J. Engelmann, H. Leontiew, F. Melchert, W. Schlapp, B. Stahl, P. Warnecke, G. Weimann: PTB-Mitteilungen **93**, 21 (1983).

/4/ All uncertainties stated in this paper are on a basis of 1 σ, which means the level of confidence is about 68 %.

/5/ F. Delahaye, A. Satrapinsky, T. Witt: IEEE **IM-38**, 256 (1989). See also proceedings of CPEM '90, to be published in IEEE **IM-40** (1991).

/6/ A.M. Thompson and D.G. Lampard: Nature **177**, 888 (1956).

/7/ G.J. Slogett, W.K. Clothier, and B.W. Ricketts: Phys. Rev. Lett. **57**, 3237 (1986).

/8/ A. Hartland, R.G. Jones, and D.J. Legg: CCE Document **88-9** (1988).

/9/ J.Q. Shields, R.F. Dziuba, and H.P. Layer: IEEE **IM-38**, 249 (1989).

/10/ T.J. Witt, F. Delahaye, D. Bournaud: IEEE **IM-38**, 279 (1989).
/11/ T.J. Quinn: Metrologia **26**, 69 (1989).
/12/ B.V. Hamon: J. Sci. Instrum. **31**, 450 (1954).
/13/ F. Delahaye, D. Reymann: IEEE **IM-34**, 316 (1985).
/14/ P. Warnecke; J. Niemeyer, F.W. Dünschede, L. Grimm, G. Weimann, W. Schlapp: IEEE **IM-36**, 249 (1987).
/15/ J. Niemeyer, L. Grimm, W. Meier, J.H. Hinken, E. Vollmer: Appl. Phys. Lett. **47**, 1222 (1985).
/16/ F. Delahaye: Metrologia **26**, 63 (1989).
/17/ Private comm. See also /5/.
/18/ M.E. Cage, R.F. Dziuba, C.T. van Degrift, D. Yu: IEEE **IM-38**, 263 (1989).
/19/ Zh. Zhang: private comm. Talk given at the "Meeting of Chinese Electrical Units", Chendu, Sichuan Province of China, September 15 - 18, (1989).
/20/ E. Braun, P. Gutmann, P. Klinzmann, H. Leontiew, W. Schlapp, P. Warnecke, G. Weimann: PTB-Mitteilungen **99**, 119 (1989).
/21/ V. Kose: PTB-Mitteilungen **92**, 249 (1982). E. Braun, V. Kose, F. Melchert, and P. Warnecke: Application of High Magnetic Fields in Semiconductor Physics, Proceedings, Grenoble, France 1982 ed. by G. Landwehr. Springer-Verlag Berlin, Heidelberg, New York, Tokyo (1983).
/22/ M.E. Cage, R.F. Dziuba, R.E. Elmquist, B.F. Field, G.R. Jones, P.T. Olsen, W.D. Phillips, J.Q. Shields, R.L. Steiner, B.N. Taylor, and E.R. Williams: IEEE **IM-38**, 284 (1989).
/23/ R.S. van Dyck, Jr., P.B. Schwinberg, and H.G. Dehmelt: Phys. Rev. Lett. **59**, 26 (1987).
/24/ T. Kinoshita: IEEE **IM-38**, 172 (1989).
/25/ E. Krüger, W. Nistler, W. Weirauch: PTB-Mitteilungen **99**, 318 (1989).
/26/ W. Wöger: PTB-Bericht **E-12**, 149 (1979).
/27/ B.N. Taylor, and T.J. Witt: Metrologia **26**, 47 (1989).

The Use of an Exciton Probe of a Two Dimensional Electron Gas in the Quantum Hall Regime

W. Chen, M. Fritze, and A.V. Nurmikko

Division of Engineering and Department of Physics, Brown University, Providence, RI 029112, USA

The photoluminesence spectra in (In,GaAs) modulation doped single quantum wells are studied under conditions where the Fermi level is near the n=2 conduction subband. A many-body exciton is observed from this subband which is strongly influenced by a 2-dimensional electron gas. Strong variations occur in external magnetic fields in both the exciton spectrum and its amplitude of which the latter correlates with the longitudinal Hall resistance in the integer quantum Hall regime.

As shown in this conference series as well as other literature, magneto-optical studies of a quasi-two dimensional (2D) electron or hole gas in semiconductor heterostructures have been widely applied to extract information about both single electron and many-body interactions [1], [2], [3]. We have studied a 2D electron gas in an (In,Ga)As quantum well structure under particular circumstances which strongly suggests a direct link between specific features in photoluminescence (PL) spectra and the longitudinal Hall resistance in the QHE regime. In particular, we employ recombination transition which originates from a normally unoccupied n=2 conduction subband level in proximity to the Fermi level (E_F) of the 2D gas in the n=1 subband.

Our structures were one-sided modulation doped n-type $Ga_{0.70}Al_{0.30}As/In_{0.15}Ga_{0.85}As/GaAs$ single quantum wells (MDQW), with (In,Ga)As quantum well thickness $L_w=150$ A. **Figure 1** shows low temperature, low excitation (< 10 mW/cm^{-2}) PL traces of two samples with electron densities $n_s=8.0\times10^{11}$ cm^{-2} and $n_s=1.1\times10^{12}$ cm^{-2} (obtained from Hall measurements) [4]. The spectrum of the former for the $n_c=1$ conduction band to $n_v=1$ valence band transition shows a well defined cut-off at an energy indicative of the Fermi level in the $n_c=1$ conduction subband. For the latter, the contribution from the $n_c=1$ subband merges into intense, strongly excitation power dependent, emission from the $n_c=2$ conduction subband. The MDQW structures are asymmetric both from their compositional structure and the one-sided doping; hence the parity selection rule ($\Delta n=0$) is weakened and the PL spectra at low temperatures involve thermalized holes in the $n_v=1$ valence subband (see inset of Fig. 1). Furthermore, the lattice mismatch strain in these (pseudomorphic) samples splits the valence band extremum by approximately 60 meV so that only the 'light-hole' band (with in-plane 'heavy-hole' dispersion) is of relevance here.

For the sample with higher sheet density in Fig. 1, the electron density is such that E_F is near the $n_c=2$ subband edge but **slightly below it** (by 4 ± 1 meV). Then the $n_c=2$-->$n_v=1$ emission, involving **nonequilibrium** electron-hole pairs, is exciton-like while very strongly influenced by (many-body) Coulomb effects of the energetically proximate 2D equilibrium electron gas [5]. In fact this Coulomb interaction of photoholes with electrons is quite analogous to the the 'Fermi-edge singularity' studied earlier in single subband occupation conditions [6]. Here, the n=2 exciton is analogous to an **edge singularity occurring at k ~ 0** and the need for hole localization is relaxed. Its temperature dependence is consistent with that of a weakly bound many-body excitons [6].

Springer Series in Solid-State Sciences, Vol. 101
High Magnetic Fields in Semiconductor Physics III Editor: G. Landwehr

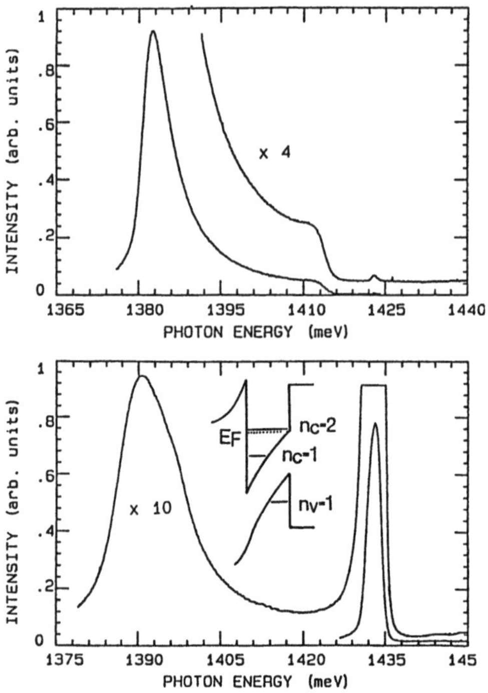

Figure 1: Photoluminescence spectra of two (In,Ga)As MDQW's at T= 1.8 K of different electron density (two magnifications): n_s=8.0x10^{11} cm^{-2} (upper trace) and n_s=1.1x10^{12} cm^{-2} (lower trace). The inset shows a schematic of the quantum well band profile and relevant energy levels at B=0.

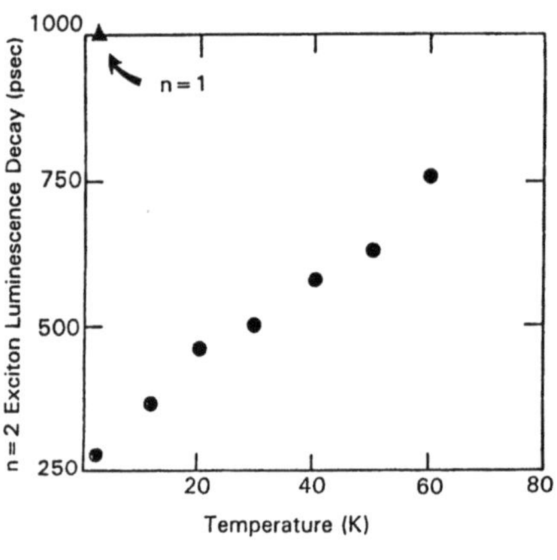

Figure 2: Temperature dependence of the photoluminescence decay time for sample of n_s=1.1x10^{12} cm^{-2} in the vicinity of the n_c=2 --> n_v=1 transition (low temperature datapoint for the n_c=1 transition is also included).

The large oscillator strength of this exciton at low temperatures is also reflected in the (radiative) lifetimes. **Figure 2** shows the lifetime of the n_c=2-->n_v=1 emission as a function of temperature (note also the data point for the n_c=2 emission).

In a perpendicular magnetic field the n_c=1 subband derived transition is readily Landau quantized. The PL originating from the n_c=2 subband in the sample with n_s=1.1x10^{12} cm^{-2}, on the other hand, shows that it represents a Coulombically bound state being only subject to diamagnetic shifts while retaining its narrow linewidth. The key effect of the magnetic field, however, pertains to the **amplitude** of the n=2 exciton-like transition, producing pronounced amplitude variations strictly periodic in 1/B. The width of the peaks (in 1/B) is approximately one third of the field equivalent width of the n=1 Landau levels. We have established a link between the amplitude variations of the magnetoexciton and

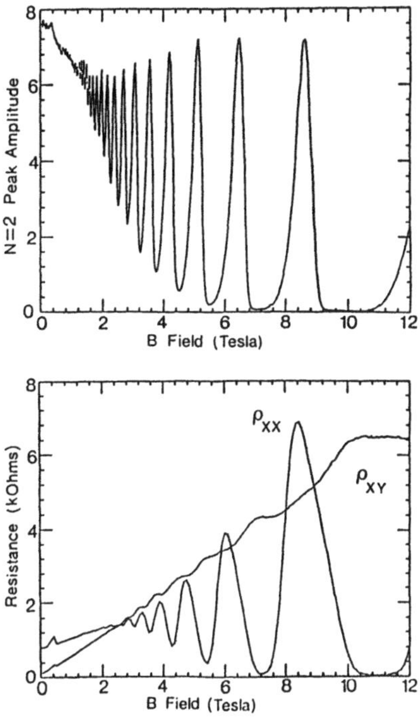

Figure 3: Variations of the $n_c=2$ magneto-exciton amplitude as a function of magnetic field at T= 1.8 K (upper trace) and compared with Hall measurements on the same sample (lower trace). A scale factor of x2.5 is applied to ρ_{xx}.

transport phenomena in simultaneous Hall measurements as shown in **Figure 3** [4]. Note the the correlation between the longitudinal Hall resistance ρ_{xx} in the QHE regime and the 'optical Shubnikov-de Haas' oscillations (OSdH) in the exciton amplitude. The (electrical) SdH effect gives a check on the electron density of $1.1\text{x}10^{12}$ cm^{-2}, in full agreement with the OSdH. The OSdH oscillations remain very well defined with excellent signal-to-noise ratio to low magnetic fields (B ~ 1T) while yielding large amplitude ratios of maxima-to-minima (better than 1000:1 for the highest oscillation in Fig. 4).

The strong field induced modulation of the magnetoexciton requires that a Coulomb coupling between the 2D electron gas and the nonequilibrium electron-hole pair exists; that is, zero field conditions where E_F is very near the $n_c=2$ subband edge but where negligible equilibrium electron density resides in the $n_c=2$ subband. Apart from the Coulomb or many-body exciton enhancement at $k \sim 0$, the interband optical transition probability for a $n_c=2$-->$n_v=1$ transition is also enlarged by the strong quantum well asymmetry. The asymmetry in our one-sided modulation doped SQW samples can be approximated as a potential consisting of a asymmetric triangular part (containing the $n_c=1$ and $n_v=1$ subbands) and a square well part (for the $n_c=2$ subband); a calculation shows up to an order of magnitude larger one-electron interband matrix element for the $n_c=2$ -->$n_v=1$ transition.

In our interpretation, the periodic build-up and quenching of the 'magneto-exciton' amplitude is related to two sequential factors as follows. First, the crossing of a $n_c=1$ Landau level with the n=2 exciton provides a means for hybridization of the two conduction band states. The hybridization provides a means of a Coulomb channel for coupling the magneto-exciton to the 2D electron gas. At the same time, the pronounced changes in the magnetoexciton amplitude occur at field values which are **higher** than those for the actual level 'crossing'. The importance of the coincidence with the maxima in ρ_{xx} shows intuitively that the existence of the magneto-exciton (in the spirit of the edge singularity) is most effectively realized when the Fermi level is within the **delocalized** states of the uppermost occupied Landau level (excess electron-hole concentration is much lower than the 2D electron

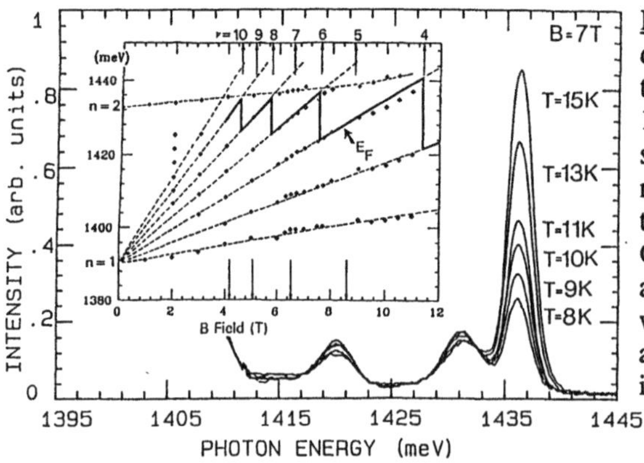

Figure 4: Illustration of temperature dependence of the magneto-exciton transition over the interval of T= 8-15 K in a field of 7 Tesla. The inset shows the fanplot of the n=1 and n=2 transition energies (dashed lines to guide the eye), together with the OSdH maxima (arrows at bottom) and filling factors (arrows at top), as well as the calculated position of E_F at T=0 K over the field range of interest (bold line).

density). That is, the maxima in OSdH oscillations coincide with **odd** (delocalized) integer Landau level filling factors in the $n_c=1$ conduction subband within 2-3% accuracy (at field values where the spin degeneracy is not yet lifted). These key observations are summarized in the inset of **Figure 4** which shows the fanplot of the n=1 and n=2 transition energies, together with the OSdH maxima and filling factors, as well as the calculated position of E_F at T=0 K for the equilibrium electrons over the field range of interest.

Since we use radiative recombination as a means to derive the signal from the exciton (which is turn is a probe for the 2D electron system) a question arises as to the particular way that the luminescence amplitude varies, i.e. where does the photoexcitation go in the amplitude minima of the OSdH oscillations. We have performed initial time resolved PL experiments in magnetic fields which suggest that **formation** of the many-body exciton is the particular key kinetic step which yields the phenomena as in Fig. 3.

The role of the localized and delocalized states on the interaction of the $n_c=2$ magneto-exciton and the 2D electron gas can be further seen from the temperature dependence of the OSdH oscillations. For example, by fixing the magnetic field at an amplitude minimum, thermally activated behavior is observed. As an example, **Figure 4** shows how this occurs at B= 7 Tesla where a finite overlap between a $n_c=1$ Landau level and the $n_c=2$ exciton already exists at T=2K. Note the strong increase with temperature of the exciton amplitude (shown here for T=8-15 K) while Landau level spectra are very little influenced. The temperature dependence of the exciton amplitude fits an exponential over nearly two orders of magnitude yielding an activation energy for the case of Fig. 5 of approximately 1.5 meV. Note that we are not simply observing the thermal excitation of electrons into the $n_c=2$ subband. Depending on the B-field 'bias' value we have measured such energies in the range of 1-5 meV. This is in the range of values obtained from T-dependence of ρ_{xx} in (In,Ga)As MDSQW's [7].

We are grateful to D. Ackley, C. Colvard, and H. Lee, who, while at Siemens Princeton Laboratories produced the HEMT structures. We thank B.B. Goldberg at Boston University for expert advice on transport measurements. We also appreciate the interest of S. Schmitt-Rink of AT&T Bell Laboratories and J. Müller of Rutgers University. This work was supported by ONR, NSF, and an IBM Materials Research Award.

References:

[1] M.S. Skolnick, K.J. Nash, S.J. Bass, P.E. Simmonds, and M.J. Kane, Solid State Comm. **67**, 637 (1988), I. V. Kukushkin and V.B. Timofeev, Surf. Sci. **196**, 196 (1988); B. Goldberg, D. Heiman, M.J. Graf, D.A. Broido, A. Pinczuk, C.W. Tu, J.H. English, and A.C. Gossard, Phys. Rev. **B38**, 10131 (1988)

[2] D. Heiman, B.B. Goldberg, A. Pinczuk, C. Tu, A. Gossard, and J. English, Phys. Rev. Lett. **61**, 605 (1988); A. Pinczuk, J.P. Valladares, D. Heiman, A. Gossard, J. English, C. Tu, L. Pfeiffer, and K. West, Phys. Rev. Lett. **61**, 2701 (1988)

[3] F. Meseguer, J.C. Maan, and K. Ploog, Phys. Rev. **B35**, 2505 (1987); C.H. Perry, J.M. Worlock, M.C. Smith, and A. Petrou, in High Magnetic Fields in Semiconductor Physics, ed. G. Landwehr (Springer-Verlag, Berlin 1987) p. 202; I.V. Kukushkin, K. v.Klitzing, and K. Ploog, Phys. Rev. **B37**, 8509 (1988); R. Cingolani, W. Stolz, Y.H. Zhang, and K. Ploog, Int. Conf. on Modulated Semiconductor Structures, Ann Arbor MI, July 1989; R. Stepniewski, W. Knap, A. Raymomd, G. Matrinez, T. Rotger, J.C. Maan, and J.P. Andre, in High Magnetic Fields in Semiconductor Physics II, ed. G. Landwehr (Springer-Verlag, Berlin 1989)

[4] W. Chen, M. Fritze, A.V. Nurmikko, D. Ackley, C. Colvard, and H. Lee, Phys. Rev. Lett. **64**, 2434 (1990)

[5] J.F. Müller, A.E. Ruckenstein, and S. Schmitt-Rink, Phys. Rev. Lett. **65**, xxx (1990)

[6] M.S. Skolnick, J.M. Rorison, K.J. Nash, D.J. Mowbray, P.R. Tapster, S.J. Bass, and A.D. Pitt, Phys. Rev. Lett., **58**, 2130 (1987); J.S. Lee, Y. Iwasa, and N. Miura, Semic. Sci. and Techn. **3**, 675 (1988); G. Livescu, D.A.B. Miller, D.S. Chemla, M. Ramaswamy, T.Y. Chang, N. Sauer, A.C. Gossard, and J.H. English, IEEE J. Quantum Electr. **QE-24**, 1677 (1988)

[7] H. Wei, A.M. Chang, D. Tsui, and M. Razeghi, Phys. Rev. **B32**, 7016 (1985)

Quantum Wells in Tilted Magnetic Fields

G. Marx, K. Lier, and R. Kümmel
Physikalisches Institut der Universität Würzburg,
W-8700 Würzburg, Fed. Rep. of Germany

Abstract: The energy spectrum and the density of states are computed for an isolated quantum well (QW) in a tilted magnetic field. Furthermore, energy spectra and charge density distributions are calculated for quantum wells and superlattices (SLs) in parallel magnetic fields.

1 Introduction

Recently, a new method has been developed to compute the electronic structure of QWs of any potential shape in tilted magnetic fields. It is applied here to a square well potential. For a parabolic potential well analytical solutions have been obtained previously by Maan [1]. In QWs and SLs with magnetic fields parallel to the potential walls [1, 2, 3] localized interface states are to be expected [4]. They are found indeed in the charge density distribution. It remains to be seen how they influence the optical properties of band gap modulated systems.

We choose the z-direction to be perpendicular to the walls of the QW. In the general case the tilted magnetic field is $\mathbf{B} = (B\cos\vartheta, 0, B\sin\vartheta)$. For the vector potential we choose a gauge such that $\mathbf{A} = (0, xB\sin\vartheta - zB\cos\vartheta, 0)$. For parallel fields we have $\vartheta = 0$. The scalar potential $U(z)$ is 0 for $|z| < a/2$; in the case of an isolated QW $U(z) = V$ for $a/2 < |z| < L_z$ and in $|z| \geq L_z$ we set $U(z) = \infty$ where L_z is so large that its value does not influence the spectrum of the bound states with energies $E < V$; in the case of a SL $U(z) = V$ for $a/2 < |z| < 3a/2$, and the periodicity is $2a$.

2 Isolated QW in Tilted Magnetic Fields

For the wave function we make the ansatz $\Psi(\vec{r}) = 1/\sqrt{2\pi}\exp\{ik_y y\}\psi(x,z)$. After substituting x by $x - \hbar k_y/(eB\sin\vartheta)$ we obtain the Schrödinger equation

$$(H_x + H_{xz} + H_z)\psi(x,z) = E\psi(x,z) \tag{1}$$

where

Springer Series in Solid-State Sciences, Vol. 101
High Magnetic Fields in Semiconductor Physics III Editor: G. Landwehr

$$H_x = (\hbar^2/2m)(-D_x^2 + (eB/\hbar)^2 x^2 \sin^2\vartheta),$$
$$H_z = (\hbar^2/2m)(-D_z^2 + (eB/\hbar)^2 z^2 \cos^2\vartheta) + U(z).$$

$H_{xz} = -2(\hbar^2/2m)(eB/\hbar)^2 z\cos\vartheta x \sin\vartheta$ couples the hamiltonian H_x of a free harmonic oscillator with that of a QW in a parallel magnetic field (for $k_y = 0$) i.e. H_z. $D_x \equiv \partial/\partial_x$ etc. Since k_y does not appear explicitly in Eq. (1), we get the same eigenvalues for different k_y, i.e. each eigenvalue has a degeneracy (per unit area) of $eB\sin\vartheta/(2\pi\hbar)$.

The calculation of the eigenfunctions and eigenstates is done in two steps:

a) First we assume H_{xz} to be zero. Then we can separate Eq. (1) and solve $H_x\eta(x) = E^x\eta(x)$ by the harmonic oscillator eigenfunctions $\eta_\ell(x)$ with the eigenvalues $E_\ell^x = (\ell + \frac{1}{2})\hbar(eB/m)\sin\vartheta$.

In order to solve $H_z\varphi(z) = E^z\varphi(z)$ we develop its eigenfunctions $\varphi_n(z)$ in terms of the eigenfunctions of an infinite potential well of width $2L_z$. Thus, we obtain a representation of H_z by matrix elements formed with the trigonometric functions. This matrix is being transformed into a tridiagonal matrix by the Lanczos method [5] and finally diagonalized numerically.[1]

b) In a second step we take the functions $\eta_\ell(x)$ and $\varphi_n(z)$ as a complete basis for the expansion of the solution $\psi(x,z)$ of Eq. (1). The advantage of this procedure is that H_x and H_z are already diagonal, whereas H_{xz} can be arranged into a band matrix and diagonalized easily.

We consider a $GaAs - Ga_{1-x}Al_xAs$ QW of width a = 15nm, 60% band–offset and $x = 0.20$ so that the potential height is V = 147meV. An effective mass $m = 0.0665m_0$ is assumed. For these parameters and for B = 15T we get the spectrum and the density of states shown in Figs. 1 and 2.

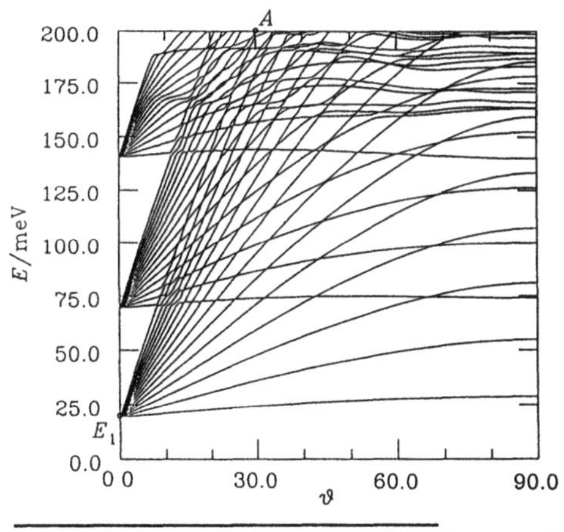

Figure 1: Energy spectrum of a QW vs. the tilt angle ϑ.

[1]All numerical calculations were done on the VAX 8810 and VAX 6000 of the Rechenzentrum der Universität Würzburg.

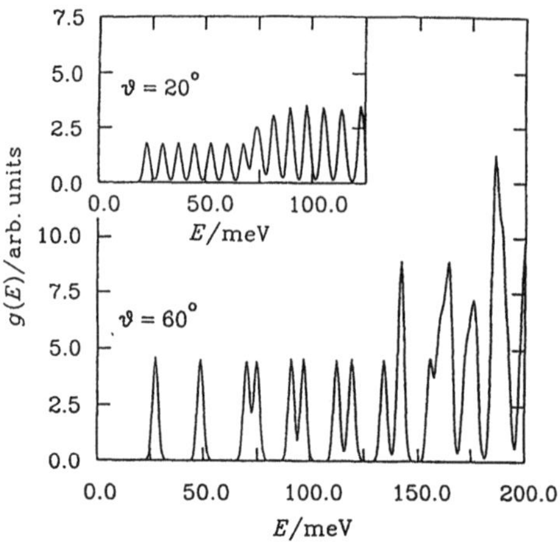

Figure 2: Density of states for $\vartheta = 20^o$ and $\vartheta = 60^o$ (with the delta functions being replaced by Gaussians of line width 2meV).

For the sake of clarity Fig. 1 only shows the tilt angle dependence of those energy levels which evolve from the 15 lowest states of the quasi continuum in each of the first three subbands at tilt angle $\vartheta = 0$. Therefore, a complete spectrum exists only to the right of the straight line which connects the points E_0 and A in Fig. 1. The density of states $g(E)$ at tilt angle $\vartheta = 20^o$ is already close to that of the quasi two dimensional electron gas at $\vartheta = 0$, whereas for $\vartheta = 60^o$ $g(E)$ indicates the approach to the Landau level structure as $\vartheta \rightarrow 90^o$.

3 QW and SL in Parallel Magnetic Fields

In the case of a parallel magnetic field, i.e. $\vartheta = 0$, we make the ansatz $\Psi(x, y, z) = 1/(2\pi) \exp\{ik_x x\} \exp\{ik_y y\} \varphi(z)$ and the Schrödinger equation for $\varphi(z)$ becomes

$$\left(\hbar^2/(2m)\left(k_x^2 - \mathrm{D}_z^2 + (eB/\hbar)^2 (z - z_0)^2\right) + U(z)\right) \varphi(z) = E\varphi(z), \quad (2)$$

where $z_0 := \hbar k_y/(eB)$ is the center of orbit coordinate.

3.1 Quantum Well

In order to solve the Schrödinger Eq. (2) we proceed as in the second part of step a) in section 2. Using the same set of QW parameters as in section 2 we obtain the subband structures shown in Fig. 3 for $B = 5$T and $B = 15$T. The flattening out of the third ($B = 15$T) subband in the vicinity of $z_0/a = 1.0$ indicates a localized interface state [4] as can be seen explicitly in the charge density distribution of Fig. 4b. Such a localized state also shows up at $B = 5$T for the fourth energy level of Fig. 4a. The existence of these localized states might have been guessed from the (shallow) depressions of the total potential formed by the superposition of the square well and the harmonic oscillator potential. Furthermore, in Fig. 4a at the seventh energy level there is another localized state formed by interferences.

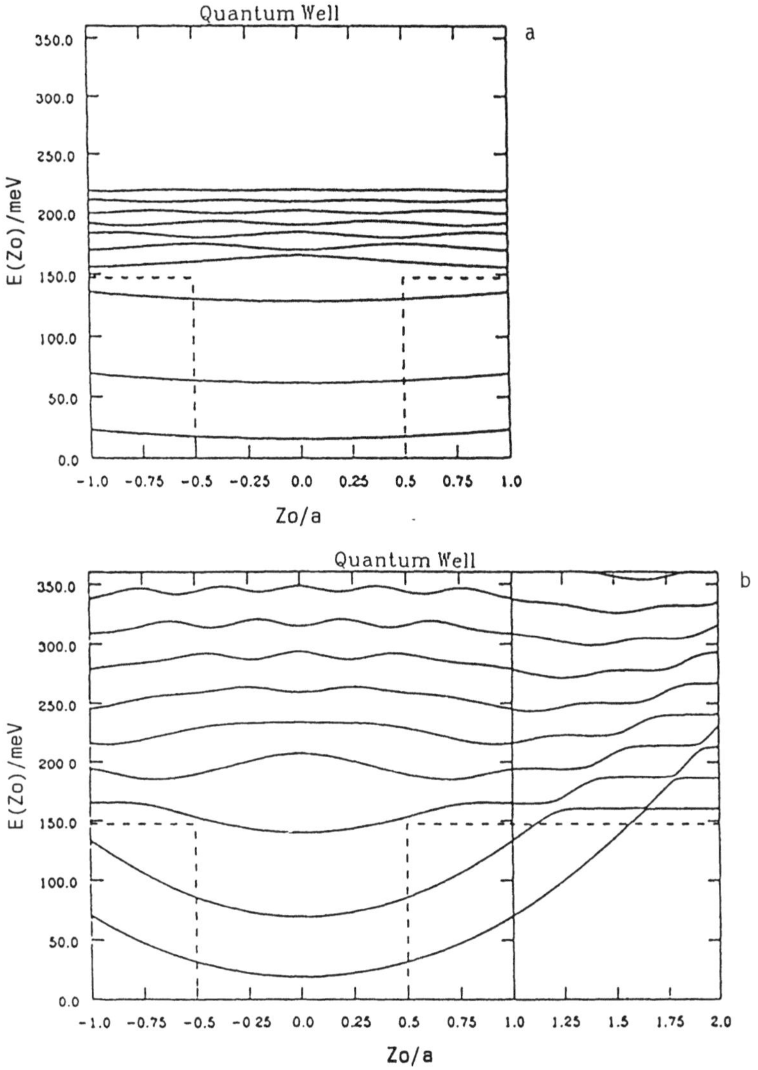

Figure 3: Subband structure of an isolated QW for a) $B = 5\text{T}$ and b) $B = 15\text{T}$; $z_0 = \hbar k_y/(eB)$. Here and in Fig. 5 the QW potential is indicated by dashed lines.

3.2 Superlattice

Finally let us see how band structure and charge density distribution change, if one takes into account the coupling of neighboring QWs in SLs, where the QW parameters are the same as in section 2. Since the SL potential is periodic we have to expand $\varphi(z)$ in terms of harmonic oscillator functions [7] which are defined in the whole range of the SL for $-\infty < z < \infty$. Fig.

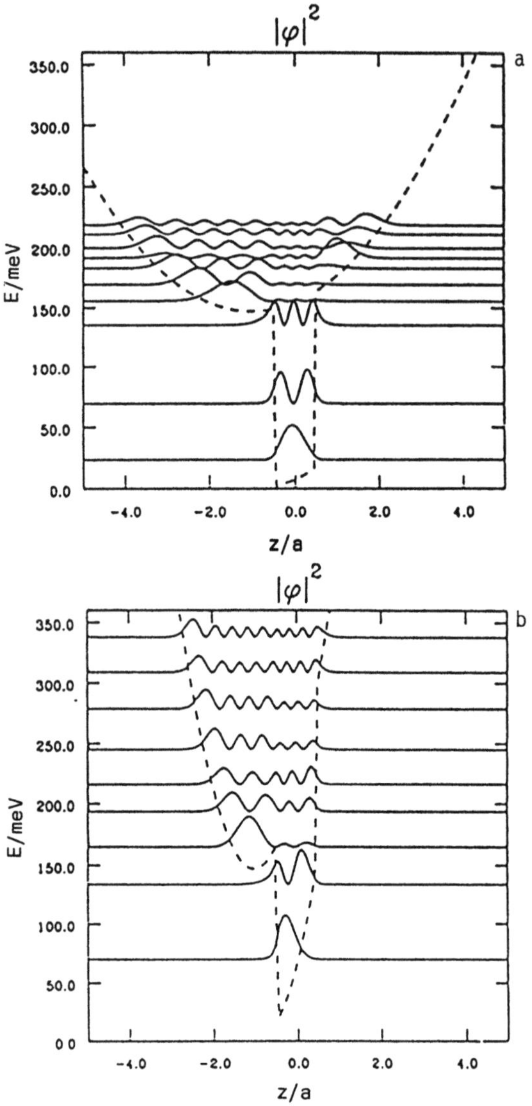

Figure 4: QW–electron density $|\varphi(z)|^2$ plotted in the total potential at the corresponding energy levels $E(z_0/a = 1.0)$ for a) $B = 5\text{T}$ and b) $B = 15\text{T}$.

5 shows the energy spectrum in the first Brillouin zone. A similar spectrum would be obtained if in Fig. 3b one reflects the part of the spectrum for $z_0/a > 1.0$ at $z_0/a = 1.0$. The main difference is the opening up of minigaps in Fig. 5 because of tunneling between the QWs of the SL.

The charge density distributions plotted in Fig. 6 show a localized barrier state (Fig. 6a) as in an isolated QW and states localized in neighboring QWs (Fig. 6b).

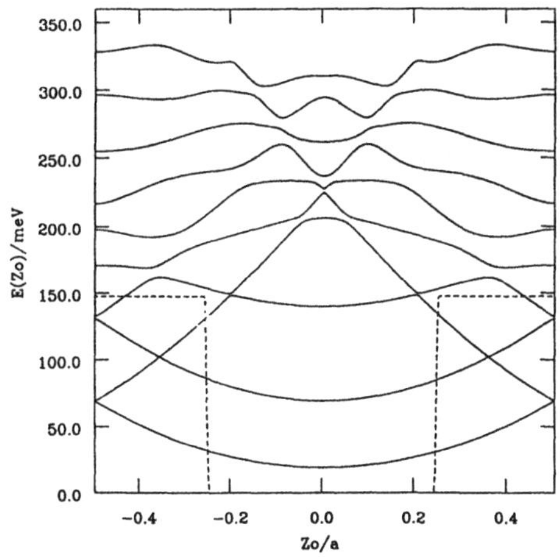

Figure 5: SL–subband structure for $B = 15$T.

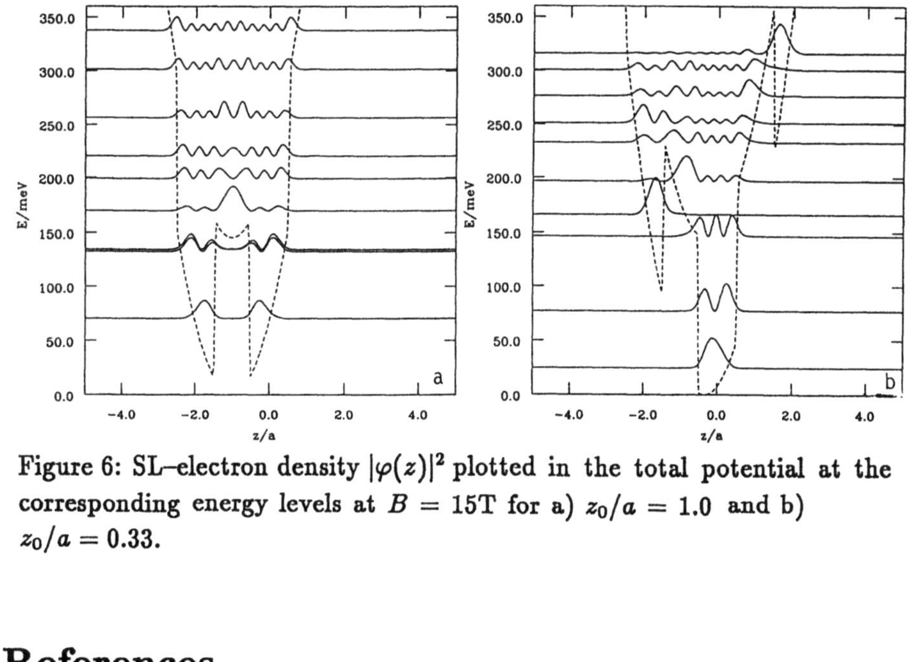

Figure 6: SL–electron density $|\varphi(z)|^2$ plotted in the total potential at the corresponding energy levels at $B = 15$T for a) $z_0/a = 1.0$ and b) $z_0/a = 0.33$.

References

[1] J. C. Maan, Springer Series in Solid State Sci. **53** (Springer, Berlin 1984), p. 183,

[2] I. Yoshino, H. Sakaki, T. Hotta, Surf. Sci. **142**, 326 (1984),

[3] B. Huckestein, R. Kümmel, Z. Phys. B **66**, 475 (1987),

[4] E. A. Johnson, A. MacKinnon, C. J. Goebel, J. Phys. C: Solid State Phys. **20**, L521 (1987),

[5] R. Haydock, V. Heine, M. J. Kelly, J. Phys. C: Solid State Phys. **5**, 2845 (1972),

[6] J. K. Maan, Festkörperprobleme/Adv. in Solid State Phys. **27** (Vieweg, Braunschweig 1987), p. 137,

[7] T. Ando, J. Phys. Soc. Jpn. **50**, 2978 (1981).

Magnetothermal Properties of a 2DEG in the Quantum Hall Effect Regime

W. Zawadzki and M. Kubisa

Institute of Physics, Polish Academy of Sciences,
PL-02-668 Warsaw, Poland

1. Introduction

This paper reviews thermodynamical properties of a 2DEG in the presence of a quantizing magnetic field, which can be described using only the knowledge of the density of states. In 1983, Zawadzki and Lassnig [1] presented a theory describing behavior of the Fermi energy, the magnetization, the specific heat and the thermoelectric power in the QHE regime. Hope was expressed that the investigation of these effects, which do not depend on electron scattering, might contribute to a better understanding of the density of states and the mobility edges. All the above mentioned properties have been observed since and the above hope turned out to be justified. This progress is the subject of the present review. The theory is based mainly on ref. [1]. The experimental data are shown as examples without the pretence of completness. In the second part we consider the so called reservoir hypothesis.

We consider noninteracting electrons in a parabolic, spherical energy band at a finite temperature T. Spin degeneracy is included, but it is assumed that $g^*= 0$ (incorporation of the spin splitting into the theory is straightforward). We assume further that only one electric subband is populated. The density of states is taken in the form of Gaussian peaks (without background),

$$\rho(\varepsilon) = \frac{2}{2\pi L^2} \sum_n \left(\frac{2}{\pi}\right)^{1/2} \frac{1}{\Gamma} \exp\left[-2\left(\frac{\varepsilon-\lambda_n}{\Gamma}\right)^2\right] \qquad (1)$$

where $L^2=c\hbar/eH$; $\lambda_n=\hbar\omega_c(n+1/2)$ and Γ is the broadening parameter (assumed to be constant).

2. Electron Density. Fermi Energy.

The electron number in 1 cm^2 is

$$N = A\sum_n \left(\frac{2}{\pi}\right)^{1/2} \frac{1}{\gamma}\int_0^{\infty} \frac{1}{1+e^{z-\eta}}\, e^{-2y_n^2}\, dz \qquad (2)$$

where $A=eH/c\hbar\pi$, $y_n=(z-\theta_n)/\gamma$, and $z=\varepsilon/kT$, $\eta=\zeta/kT$, $\theta_n=\lambda_n/kT$, $\gamma=\Gamma/kT$. The filling factor is defined as $\nu=N/A$, denoting the number of occupied Landau levels (LLs). The condition of a constant electron density in the sample leads to the integral equation (2) for the Fermi energy $\zeta(H)$. Fig. 1 shows this dependence calculated for $m^*=0.065m_0$, $N=8\times10^{11}cm^{-2}$, $\Gamma=0.5$meV and T=6K.

The Fermi level jumps almost vertically between LLs, where the density of states vanishes. Fig. 2 shows the measured $\zeta(B)$ in a $Ga_{1-x}Al_xAs$ heterojunction [2], while Fig. 3 shows the same as measured in an Si inversion layer [3]. Other measurements show similar behavior, cf. [4]. The not abrupt variation of $\zeta(B)$ between LLs seen in Fig. 2, can be interpreted as an evidence

Springer Series in Solid-State Sciences, Vol. 101
High Magnetic Fields in Semiconductor Physics III Editor: G. Landwehr

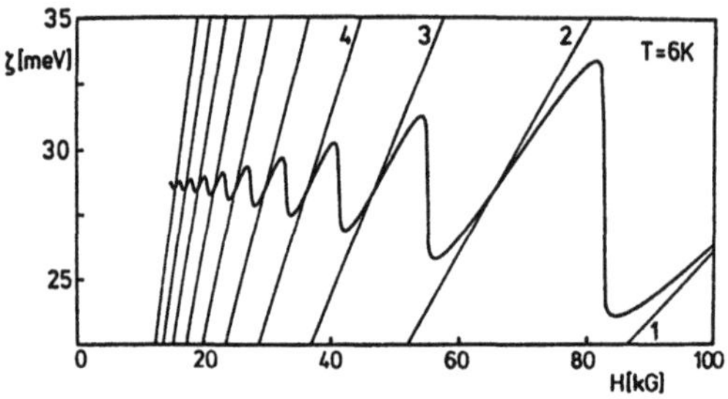

Fig. 1. The Fermi energy versus magnetic field, calculated for 2DEG in GaAs at a constant electron density and T = 6K. The Landau levels are also indicated. After ref. [1].

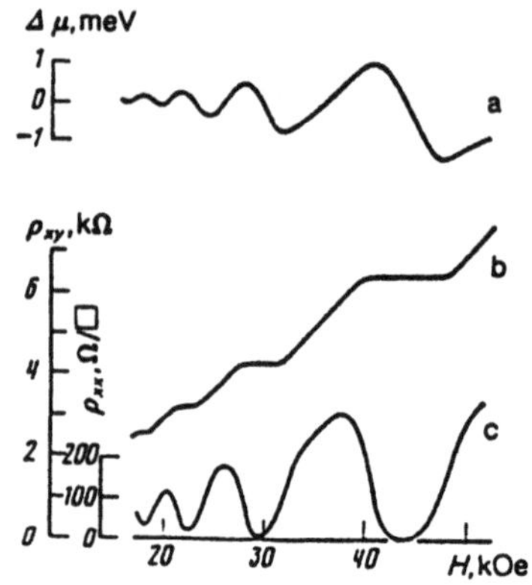

Fig. 2. The chemical potential (a), the Hall effect (b), and the magnetoresistance (c) versus magnetic field, measured on GaAs-GaAlAs heterojunction. After Nizhankovskii et al. [2].

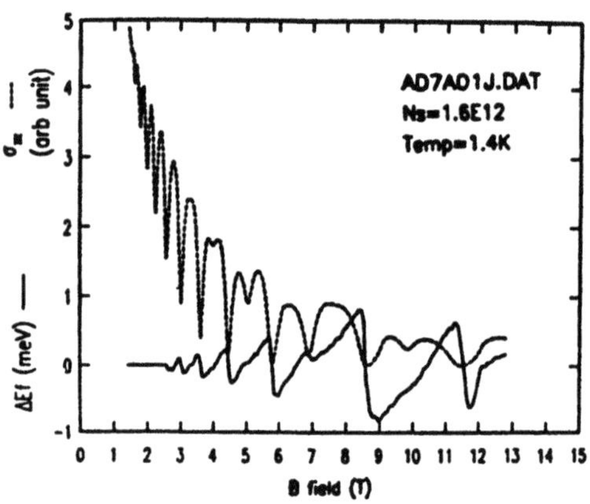

Fig. 3. The chemical potential and σ_{xx} versus magnetic field, measured on Si inversion layer. After Fang et al. [3].

for a nonvanishing density of states between LLs, or for a change of concentration of 2DEG in the well [2].

3. Magnetization

The free energy of the system is

$$F = N\zeta - kT\int \rho(\varepsilon)\ \ln\left[1 + \exp\left(-\frac{\varepsilon-\zeta}{kT}\right)\right] d\varepsilon. \qquad (3)$$

The magnetization is M=-dF/dH. After some manipulations one obtains

$$M = akT\sum_n \left(\frac{2}{\pi}\right)^{1/2}\frac{1}{\gamma}\int_0^\infty \ln(1+e^{\eta-z})\, e^{-2y_n^2}\left(1+4\frac{\theta_n}{\gamma}y_n\right)dz \qquad (4)$$

where $a=e/\hbar c\pi$. Figure 4 shows the magnetization calculated using Eq. (4) for the above conditions. As follows from Figs. 1 and 4, the magnetization oscillations follow quite closely those of the Fermi level. On the other

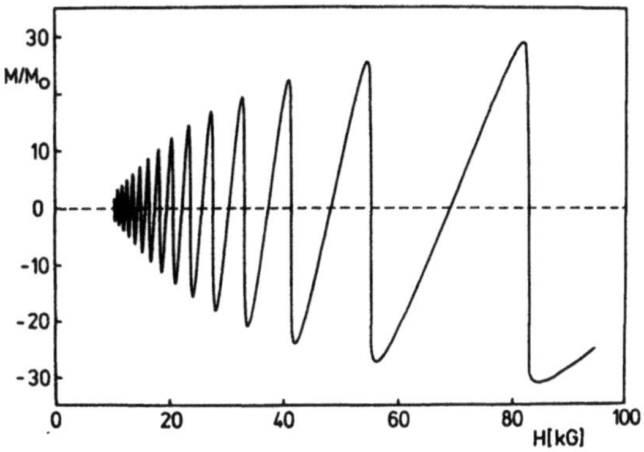

Fig. 4. Magnetization of 2DEG versus magnetic field, calculated for the same conditions as in Fig. 1 and T = 4.2K. M_0=kT(e/πħc). After ref. [1].

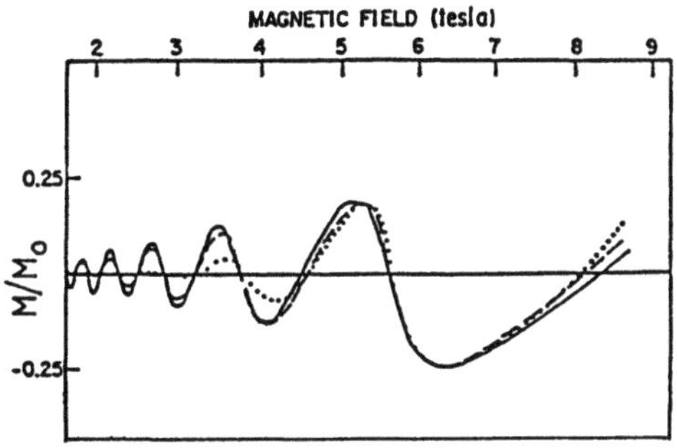

Fig. 5. Normalized magnetization versus magnetic field, measured on a multilayer sample of GaAs-GaAlAs. Dotted and dashed lines are fits assuming different level broadening. After Eisenstein et al. [6].

hand the Shubnikov-de Haas oscillations of magnetoresistance have maxima when LLs cross the Fermi energy. Thus, there should be a phase shift between the magnetization and the ShdH oscillations. This is in fact observed, cf. [5]. Figure 5 shows the experimental magnetization, together with theoretical fits. No background density of states between LLs has been assumed in the theory.

4. Specific Heat

The specific heat is given in general as

$$C_v = \int_0^{\infty} \frac{df}{dT} \; (\varepsilon-\zeta)\rho(\varepsilon)d\varepsilon \tag{5}$$

in which f is the Fermi function. We have

$$\frac{df}{dT} = - \frac{\partial f}{\partial \varepsilon} \left(\frac{\varepsilon-\zeta}{T} + \frac{d\zeta}{dT}\right) \tag{6}$$

The dependence $\partial\zeta/\partial T$ at a constant concentration is determined by differentiating Eq. (2) with respect to T and using Eq. (6). This leads to $\partial\zeta/\partial T=-L_1/L_0$, and

$$C_v = kA\left(L_2 - \frac{L_1^2}{L_0}\right) \tag{7}$$

where

$$L_r = \sum_n \left(\frac{2}{\pi}\right)^{1/2} \frac{1}{\gamma} \int_0^{\infty} \frac{e^x}{(1+e^x)^2} \, x^r e^{-2y_n^2} dz \tag{8}$$

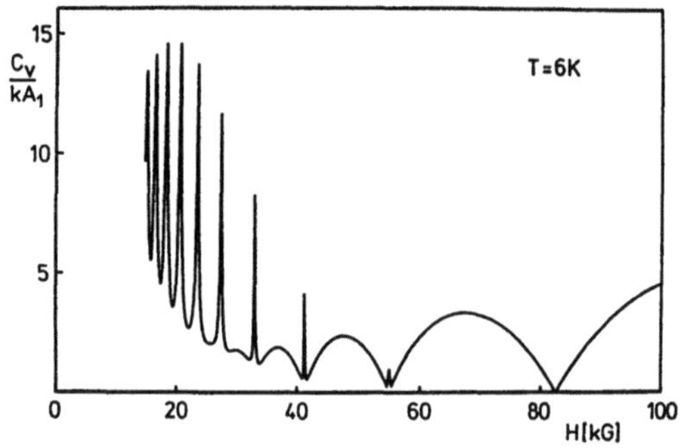

Fig. 6. Specific heat of 2DEG versus magnetic field, calculated for the same conditions as for Fig. 1. $A_1 = kH/\pi\hbar c$ for H = 1kG.

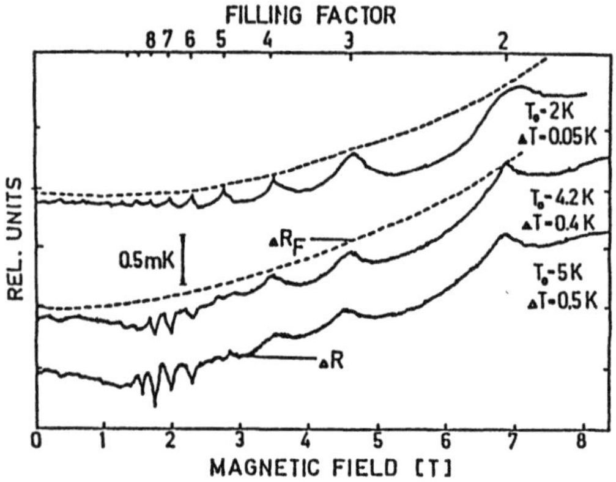

Fig. 7. Specific heat of a 2DEG versus magnetic field, measured on a multilayer sample of GaAs-GaAlAs. Intralevel and interlevel contributions to C_v are seen. After Gornik et al.

in which $x=z-\eta$ and other quantities have been defined above. Fig. 6 shows the specific heat computed for the above parameters at T=6K. The specific heat of a 2DEG is seen to consist of two contributions. At higher magnetic fields, for $\hbar\omega_c >> kT$, only the intralevel thermal excitations contribute to C_v. At weaker magnetic fields the interlevel excitations begin to come into play if the temperature is not too low. They are of importance when the Fermi energy is between two LLs and have the form of spikes since, as seen in Fig. 1, the Fermi energy jumps between LLs within a narrow range of the field. At low temperatures the interlevel spikes disappear. If the above effects are to be observable, C of a 2DEG should be comparable to that of the lattice, cf. [1]. Fig. 7 shows the experimental specific heat of a 2DEG, measured by the heat pulse technique. In agreement with the theoretical predictions, at low T only intralevel contributions to C_v are observed, while at somewhat higher T the interlevel spikes are seen as well. The best fit to the data has been obtained by assuming a constant flat background in the density of states in addition to the Gaussian peaks. However, a resonable fit could be obtained also without the background. The specific heat of a 2DEG has been measured also by a calorimetric method [8]. The best fit to the data has been obtained by assuming an oscillatory width of the Gaussian peaks without background. It has also been verified that at low temperatures the amplitude of C_v goes to zero, in agreement with the theory [1].

5. Thermoelectric Power

A quantum theory of thermo-magnetic transport phenomena offers some difficulties, since in the presence of a temperature gradient the system is not homogeneous. As a consequence, the automatic application of the Kubo method led in the past to results which did not satisfy the Onsager symmetry relation, violated the third law of thermodynamics, ect. These paradoxes and puzzles were resolved by Obraztsov [9], who showed that, in order to obtain a correct description of the off-diagonal components of thermomagnetic tensors, one should explicitly include in the theory a contribution of the magnetization. This is related to the fact that the microscopic surface currents, which determine the Landau magnetization of conduction electrons, make a significant contribution to the macroscopic current density when a temperature gradient is present. At high magnetic fields, i.e. for $\omega_c\tau\gg1$, the diagonal components of the transport tensor may be neglected with respect to the off-diagonal ones. The latter do not depend on electron scattering in the high-field limit. Taking into account the contribution of magnetization M one obtains for the off-diagonal component of the macroscopic thermoelectric tensor

$$\beta_{xy} = \beta^0_{xy} + c\frac{dM}{dT} = \frac{c}{H} S \tag{9}$$

where β^0_{xy} determines the microscopic current density. When β^0_{xy} is calculated using the standard methods of the density matrix, the equality (9) is obtained, in which S is the entropy of the electron gas. The thermoelectric power becomes

$$\alpha(H) \simeq \frac{\beta_{xy}}{\rho_{xy}} = -\frac{S}{eN} \tag{10}$$

The entropy: $S=-(\partial F/\partial T)_v$, can be calculated from Eq.(3) to give

$$S = kA \sum_n \left(\frac{2}{\pi}\right)^{1/2} \frac{1}{\gamma} \int_0^\infty \left[\ln(1+e^{-x}) + \frac{x}{1+e^x}\right] e^{-2y_n^2}\, dz \tag{11}$$

where, as before, $x=z-\eta$. The completely filled levels (for which $x\ll0$) give a vanishing contribution to the entropy. N is given in Eq.(2), so that $\alpha(H)$ can be computed in the no-scattering limit. Measuring α at a constant electron concentration one determines directly the entropy of the electron gas. Fig.8 shows $\alpha(H)$ of a 2DEG calculated for the above parameters. The predictions agree quite well with experimental observations, cf. Fig.9. The

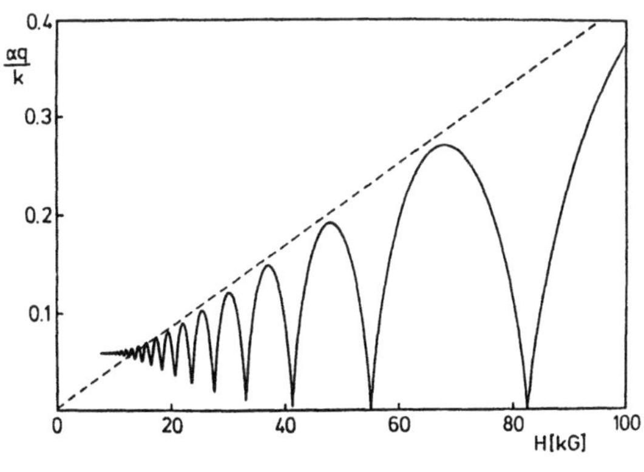

Fig.8. Thermoelectric power of a 2DEG versus magnetic field, calculated for the same conditions as for Fig.1. The dashed line indicates maxima values of $(-e/k)(\ln2)/\nu$. After ref. [1].

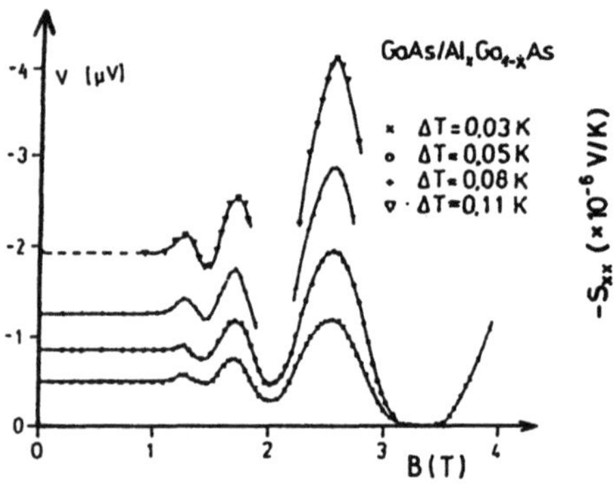

Fig. 9. Thermal voltage versus magnetic field at different temperature gradients ΔT, measured on a GaAs-GaAlAs heterostructure. The bath temperature T = 4.2K. After Obloh et al. [10]

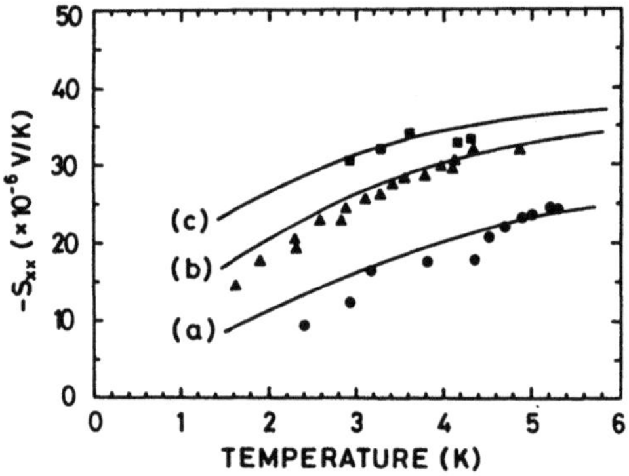

Fig. 10. Peak value of the thermoelectric power α_{xx} at n=1 versus temperature for three different GaAs-GaAlAs heterostructures. The solid curves are calculated. After Obloh et al. [11].

main difference is that the experimental curve has a plateau of the vanishing values. This is usually interpreted as an effect of localization.

It has been also shown in ref. [1] that the amplitude of $\alpha(H)$ for a given LL is a universal function of kT/Γ where as before, Γ is the level broadening. This property can be used to determine Γ if the temperature dependence of α is known. Such a procedure is shown in Fig. 10. However, the theoretically predicted temperature dependence of α is not always confirmed experimentally, cf. ref. [10]. This is probably due to the restrictive theoretical assumptions (e.g. that Γ is independent of H and T), and the phonon-drag effect.

6. The Reservoir Hypothesis

In the previous sections we quoted a theoretical description of magnetothermal properties based on the assumption that the concentration N of the 2DEG remains constant as magnetic field changes. In this picture the plateaus of the quantum Hall effect, the zeros of the Shubnikov-de Haas effect as well as those of the thermoelectric power (cf. Fig. 9) are attributed to the localization regions in the density of states. When the Fermi level traverses the localized region the electron transport coefficient vanish, while ρ_{xy} has very well defined plateaus.

In 1981 Baraff and Tsui [13] proposed an explanation of the plateaus in QHE based on a transfer of electrons into and out the potential well containing the 2DEG. The argument is simple. If σ_{xy}=Nec/H at all fields, i.e. if the Hall effect measures the electron concentration at all fields, N must be changing in order to keep σ_{xy}=const. as H changes. This is known as the reservoir hypothesis, since in this picture there is a need for a reservoir to provide or take the electrons. As the number of electrons in the well changes, the well changes shape (in a triangular well the electric field $E \sim N^{2/3}$). The modification of the well may influence the electron transfer, so that the problem has to be solved self consistently. In particular, the modification of the well changes the position of the electric subband in which the 2DEG is located on the energy scale. This description requires a specific model for the heterostructure. A number of

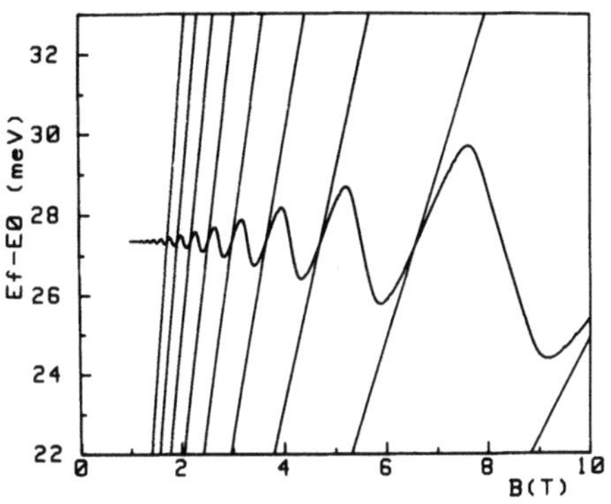

Fig.11. Difference between the Fermi energy and the bottom of the lowest electric subband versus magnetic field, calculated for constant Fermi energy and the parameters of a 2DEG in GaAs at T = 6K.

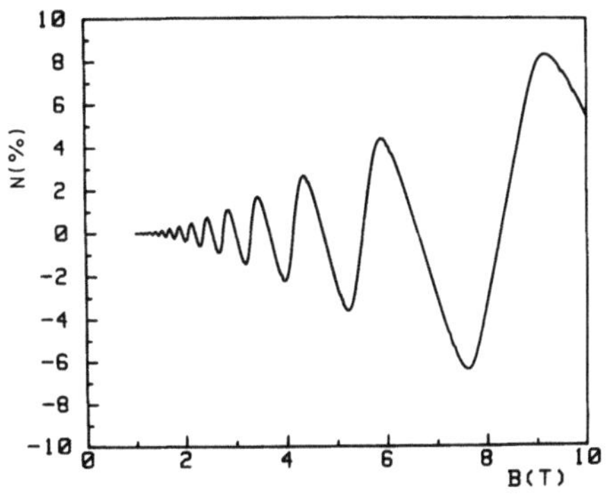

Fig.12. Change of density of the 2DEG versus magnetic field, consistent with the results of Fig.11.

papers, both theoretical and experimental followed the reservoir hypothesis, e.g. [2,14,15,16].

Below we present theoretical results on magneto-thermal properties of a 2DEG in a triangular potential well, calculated under the assumption of a constant Fermi level. Since at a constant concentration N the Fermi level oscillates as the magnetic field is swept (cf. Fig.1), it is clear that for ζ=const. N must oscillate. This in turn causes oscillations of the subband energy E_0, so that the energy interval $\zeta - E_0$ oscillates. We do not consider here the nature of a possible reservoir. Also, the details of the calculation are not given [17]. The computations have been performed for the same parameters as those quoted above, in order to facilitate a comparison between the two models.

It can be seen from Fig.11 that the Fermi level does not jump between LLs as abruptly as in case of N=const. (cf. Fig.1). This is due to the compensating effect of the concentration transfer, as shown in Fig.12. Under the assumption of N=const. the not-so-steep changes of ζ are interpreted as due to an additional density of states between LLs. The reservoir interpretation explains quite well the smooth oscillations of the magnetization (cf. Fig.13 and Fig.5), as well as the plateau of vanishing values of the thermoelectric power (cf. Fig.14 and Fig.9). Finally, it gives the plateaus of the quantum Hall effect (cf. Fig.15) assuming that the Hall effect measures the electron concentration at all magnetic fields, i.e. using the formula ρ_{xy} = B/Ne. In this interpretation one does not need to invoke the localization concept to obtain exactly the quantized values of the QHE.

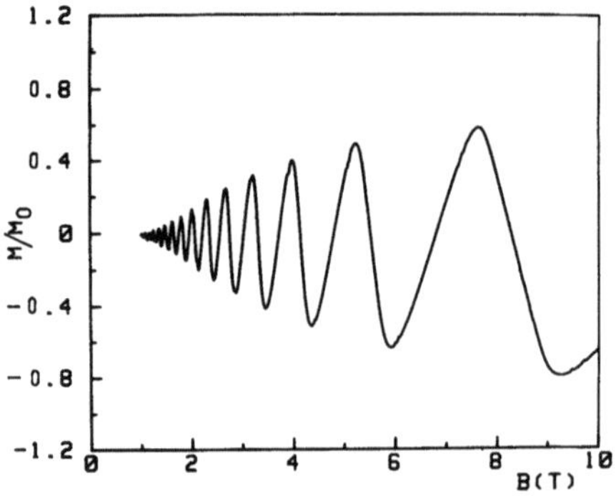

Fig. 13. Magnetization of the 2DEG versus magnetic field, calculated for the conditions of Figs. 11 and 12.

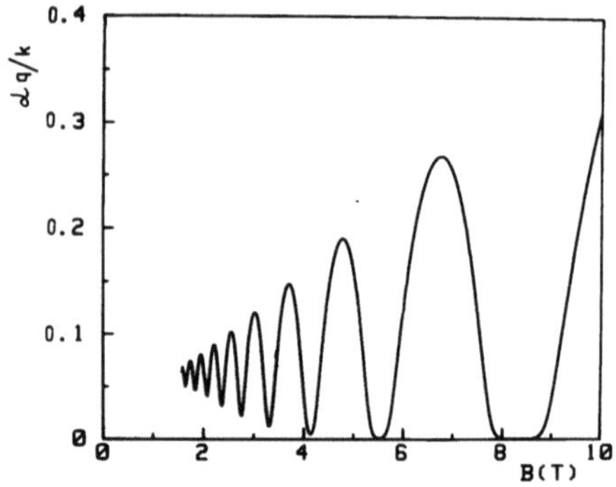

Fig. 14. The thermoelectric power of the 2DEG versus magnetic field calculated for the conditions of Figs. 11 and 12.

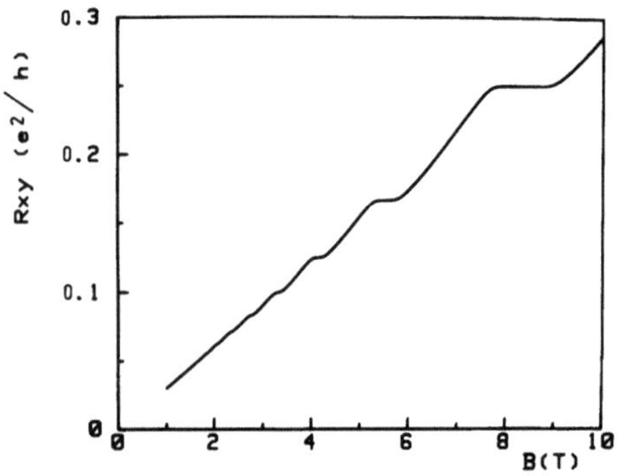

Fig. 15. The Hall effect of the 2DEG versus magnetic field, calculated for the conditions of Figs. 11 and 12 using the relation: $\rho_{xy} = B/Ne$.

7. How to Observe Mobility Edges? - A Proposal

It was suggested in ref. [1] that one should combine investigations of electron transport phenomena (related to delocalized electron states in the energy spectrum) with those discussed above (related to the total density of states, both localized and delocalized) in order to gain information on mobility edges in the localization hypothesis. Although the magneto-thermal effects contributed considerably to the knowledge of the density of states, the combination of transport and nontransport effects has not been really exploited. Below we propose an idea, which could lead to a direct observation of the mobility edges, based on such a combination.

Fig. 16 shows a standard schematic picture of the density of states with shaded areas indicating the localized parts of the spectrum. The behavior of the Fermi energy as a function of magnetic field is governed by the total density of states. As H increases, ζ rises gradually when traversing the peak of the density of states and then drops rapidly between points a and a', cf. Fig. 1. (If one assumes an additional density of states between LLs or an electron transfer, this drop is less abrupt, cf. Fig. 11, but the qualitative picture is the same.) On the other hand, the plateau of the QHE or the vanishing value of the magnetoresistance occur when the Fermi energy is between the points b and b', i.e. in a wider interval of energy (and a corresponding magnetic field). In other words, if, as it is commonly assumed, the extended electron states occupy only a

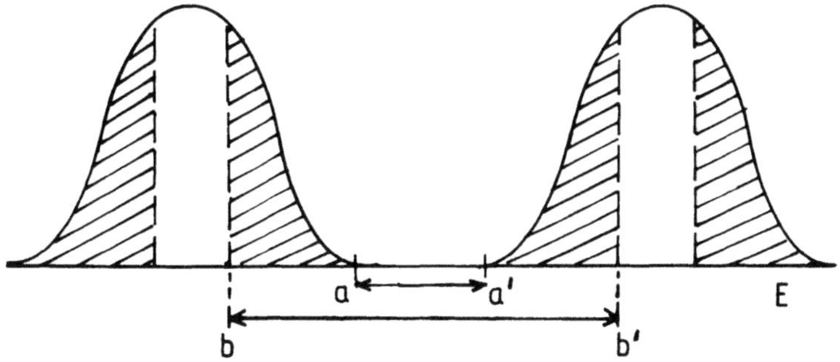

Fig. 16. Density of states of 2D electrons in a magnetic field. The shaded areas indicate localized regions. The Fermi energy drops in the a-a' region, while no transport occurs in the b-b' region.

small part of the energy spectrum, the plateau of QHE or magnetoresistance should be distinctly wider than the region in which the Fermi energy drops. Thus we propose to measure at low temperatures $\zeta(B)$ and $\rho_{xy}(B)$ [or $\rho_{xx}(B)$] and to compare the two. In principle, Figs. 2 and 3 show just that for a GaAs- GaAlAs heterojunction and an Si inversion layer. However, the effect of the mobility edges is not seen. A precise experiment of this type could discriminate between the localization and the reservoir hypotheses.

Acknowledgements: It is our pleasure to thank Professor André Raymond for numerous elucidating discussions.

7. References

1. W. Zawadzki and R. Lassnig, Surface Sci. 142, 225 (1984); W. Zawadzki, Solid State Commun. 47, 317 (1983); W. Zawadzki and R. Lassnig, Solid State Commun. 56, 537 (1984).
2. V.I. Nizhankovskii, V.G. Mokerov, B.K. Medvedev, and Yu.V. Shaldin, Sov. Phys. J.E.T.P. 63, 776 (1986).
3. F.F. Fang, J.J. Nocera, J. Luo, and T.P. Smith, in Proceed 19th Inter. Conf. Phys. Semicond. (Ed. W. Zawadzki), Institute of Physics, Warsaw 1988, p.169.
4. T.P. Smith, B.B. Goldberg, M. Heiblum, S.L. Wright, and P.J. Stiles, Phys. Rev. B33, 1529 (1986); V.M. Pudalov, S.G. Semenchinskii, and V.S. Edelman, J.E.T.P. Lett. 39, 474 (1974).
5. T. Haavasoja, H.L. Störmer, D.J. Bishop, V. Narayanamurti, A.C. Gossard, and W. Wiegmann, Surface Sci. 142, 294 (1984).
6. J.P. Eisenstein, H.L. Störmer, V. Narayanamurti, A.Y. Cho, and A.C. Gossard, Phys. Rev. Lett. 55, 875 (1985).
7. E. Gornik, R. Lassnig, G. Strasser, H.L. Störmer, and A.C. Gossard, Surface Sci. 170, 277 (1986).
8. J.K. Wang, J.H. Campbell, D.c. Tsui, and A.Y. Cho, Proceed. 19th Inter. Conf. Phys. Semicond. (Ed. W. Zawadzki), Institute of Physics, Warsaw 1988, p.173.
9. Yu.N. Obraztsov, Sov. Phys. Sol. State 6, 331 (1964); ibid. 7, 455 (1965)
10. H. Obloh, K. von Klitzing, and K. Ploog, Surface Sci. 142, 236 (1984).
11. H. Obloh, K. von Klitzing, K. Ploog, and G. Weimann, Surface Sci. 170, 292 (1986).
12. T.H.H. Vuong, R.J. Nicholas, M.A. Brummell, J.C. Portal, M. Razeghi, K.Y. Cheng, and A.Y. Cho, Surface Sci. 170, 298 (1986).
13. G.A. Baraff and D.C. Tsui, Phys. Rev. B24, 2274 (1981).
14. J. Bok and M. Combescot, Solid State Commun. 47, 611 (1983).
15. O.V. Konstantinov, O.A. Mezrin, and A.Ya. Shik, Sov. Phys. Semicond. 17, 675 (1983).
16. A. Raymond and K. Karrai, Journ. de Physique, 48, Suppl.C5, 491 (1987).
17. W. Zawadzki and M. Kubisa, to be published.

Part II

Fractional Quantum Hall Effect

Tilted-Field Effect, Optical Transitions and Spin Configurations of the Fractional Quantum Hall States

T. Chakraborty[1] *and P. Pietiläinen*[2]

[1]Max-Planck-Institut für Festkörperforschung, Heisenbergstr. 1, W-7000 Stuttgart 80, Fed. Rep. of Germany

[2]Department of Theoretical Physics, University of Oulu, SF-90570 Oulu 57, Finland

We discuss an important effect of the magnetic field *tilted* from the direction perpendicular to the electron plane viz., the subband-Landau-level coupling, in the fractional quantum Hall effect regime. We also present new theoretical results on recombination radiation in two-dimensional electron systems which might be an interesting route to investigate the spin states of various filling fractions.

1. Introduction

The standard theoretical explanation of the fractional quantum Hall effect (FQHE) is that, at certain filling fractions of spin and Landau levels the two-dimensional electron system condenses into an incompressible electron fluid where the elementary excitations are fractionally charged quasiparticles and quasiholes [1]. One interesting property of this state i.e., the spin reversed ground state and excitations predicted theoretically [1,2] have received strong support from recent experimental work [3] in tilted magnetic fields. In the experiment of Eisenstein et al., a sharp change in the tilt-angle dependence of the activation energy was observed for the filling fraction $\nu = \frac{8}{5}$ (electron-hole symmetric to $\frac{2}{5}$). This is described as a transition from a spin-unpolarized ground state at small angles to a polarized state at larger angles. The linear behavior of the activation energy at the two different ground states were identified with the Zeeman energy, and they appear because of the presence of spin-reversed quasiparticles and quasiholes [1,2].

At low magnetic fields, the ground state spin configurations for a few filling fractions, as predicted theoretically, are given below:

$$\uparrow\uparrow \frac{1}{3}\left(\frac{5}{3}\right) \qquad \left(\frac{4}{3}\right)\frac{2}{3}\uparrow\downarrow$$
$$\uparrow\downarrow \frac{2}{5}\left(\frac{8}{5}\right) \qquad \left(\frac{7}{5}\right)\frac{3}{5}\updownarrow$$

where $\uparrow\uparrow, \uparrow\downarrow$ and $\updownarrow$ denotes the fully spin-polarized ($S = \frac{1}{2}N$), unpolarized ($S = 0$) and partially-polarized ($S \neq 0, \frac{1}{2}N$) respectively. S is the total spin of the system and N is the total number of electrons. The spin assignments for $2 > \nu > 1$ can be obtained from the electron-hole symmetry which, in the absence of Landau level mixing is between ν and $2 - \nu$ (between ν and $1 - \nu$ if spin degrees of freedom is ignored). The absence of electron-hole symmetry between $\frac{1}{3}$ and $\frac{2}{3}$ was first noticed in the tilted-field experiments by Haug et al. [4]. Although spin degrees of freedom can explain the breaking of electron and hole symmetry between $\frac{1}{3}$ and $\frac{2}{3}$ filling factor, these two fractions appear at rather high magnetic fields. Therefore the breaking of electron-hole symmetry in this case might have some other origin.

Springer Series in Solid-State Sciences, Vol. 101
High Magnetic Fields in Semiconductor Physics III Editor: G. Landwehr

2. Tilted-Field effect

In this section, we describe one of the major effects of the tilted magnetic field–the subband-Landau-level coupling and its influence on the FQHE. In a two-dimensional electron gas (2DEG), confinement of electron motion in the direction perpendicular to the electron plane results in the formation of discrete subbands. Upon application of a perpendicular magnetic field, each subband is further quantized into discrete Landau levels. When the magnetic field is tilted from the direction normal to the plane, the perpendicular and parallel motions of electrons are coupled and one can not treat the subbands and Landau levels separately. In the following, we present a theory for the tilted-field effect on FQHE, where our starting point is the single-electron theory due to Maan [5].

We consider a situation where the electrons are confined to the $x-y$ plane by a parabolic potential well $V(z) = \mathcal{A}z^2$ and the magnetic field is applied in the $x-z$ plane. We choose the gauge such that the vector potential is $\mathbf{A} = (0, xB_z - zB_x, 0)$, where $B_x = B\sin\theta, B_z = B\cos\theta$, and θ is the tilt angle. The single-particle Schrödinger equation is then solved by an appropriate rotation of the coordinates. The energy eigenvalues are then that of two harmonic oscillators with frequencies

$$\omega_{1,2} = \left[\tfrac{1}{2}\left(\omega_c^2 + \omega_0^2\right) \pm \tfrac{1}{2}\left\{\omega_c^4 + \omega_0^4 + 2\omega_0^2\left(\omega_x^2 - \omega_z^2\right)\right\}^{\frac{1}{2}}\right]^{\frac{1}{2}} \tag{1}$$

where $\omega_c = eB/m^*, \omega_0 = (2\mathcal{A}/m^*)^{\frac{1}{2}}, \omega_x = \omega_c \sin\theta, \omega_z = \omega_c\cos\theta$, and m^* is the effective mass. It should be mentioned that the Landau-level degeneracy and the filling factor depend only on the component of the magnetic field normal to the confinement plane. The magnetic field dependence of $\hbar\omega_{1,2}$ is shown in Fig. 1 for different values of the tilt angle. For $\theta = 0°$, i.e., the magnetic field being perpendicular, every subband splits into Landau levels as discussed above. For the magnetic field at any other angle, subband separation depends on the magnetic field and also the Landau level separation on the tilt angle.

Our next task is to evaluate the Coulomb interaction energy of the electron system. Toward this end we use the exact diagonalization scheme for a finite number of electrons in a periodic rectangular geometry [1,6]. The single-electron states appropriate for the present case is [7]

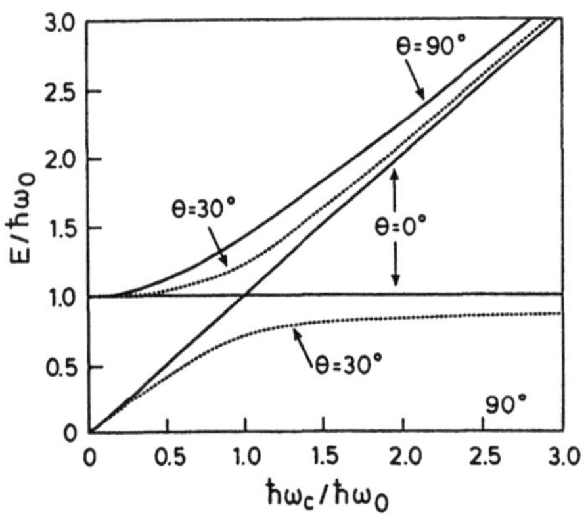

Fig. 1
The eigenfrequencies of the Landau levels for an electron in a parabolic potential well for different values of magnetic field and tilt angle θ

$$\Phi_j(\mathbf{r}) = \left[\frac{1}{bl_1 l_2 \pi}\right]^{\frac{1}{2}} \sum_{k=-\infty}^{\infty} \exp\Big\{ i\,(X_j + ka)\, y/\ell_0^2 - [(X_j + ka - x)\cos\phi + z\sin\phi]^2/2l_1^2 - [(X_j + ka - x)\sin\phi - z\cos\phi]^2/2l_2^2 \Big\} \tag{2}$$

where a and b are the two sides of the rectangular cell. The magnetic length is $\ell_0 = (\hbar/m^*\omega_z)^{\frac{1}{2}}$, and $l_1 = (\hbar/m^*\omega_1)^{\frac{1}{2}}$, $l_2 = (\hbar/m^*\omega_2)^{\frac{1}{2}}$. The center coordinate of the cyclotron motion is, as usual, $X_j = (2\pi j/b)\ell_0^2$. Also the angle of rotation of the coordinates to separate the variables in the single-particle Schrödinger equation [5] is

$$\phi = \tfrac{1}{2}\arctan\left[\frac{\sin(2\theta)}{\cos(2\theta) - \mathcal{A}\frac{2m^*}{e^2}}\right].$$

Index j which describes the linear momentum in y direction, can have values $1 < j \leq m$, where m is the Landau-level degeneracy. The two-electron part of the Hamiltonian which include the Coulomb interaction is now written [7] as

$$\begin{aligned}
A_{j_1,j_2,j_3,j_4} &= \tfrac{1}{2}\int d\mathbf{r}_1 \int d\mathbf{r}_2 \Phi_{j_1}^*(\mathbf{r}_1)\Phi_{j_2}^*(\mathbf{r}_2)\, v(\mathbf{r}_1 - \mathbf{r}_2)\Phi_{j_3}(\mathbf{r}_2)\Phi_{j_4}(\mathbf{r}_1) \\
&= \frac{1}{2ab}\sum_{\mathbf{q}}\sum_{s}\sum_{t} \delta_{q_x,2\pi s/a}\delta_{q_y,2\pi t/b}\delta'_{j_1-j_4,t}\frac{2\pi e^2}{\epsilon q} \\
&\times \exp\left[2\pi i s(j_1 - j_3)/m - \pi(s^2 + \lambda^2\Omega_1^2 t^2)/(m\lambda\Omega_1)\right] \\
&\times \frac{2}{\sqrt{\pi}} I(s,t)\delta'_{j_1+j_2,j_3+j_4}
\end{aligned} \tag{3}$$

where $v(r)$ is the Coulomb interaction in the periodic rectangular geometry, the Kronecker δ with prime means that the equality is defined modulo m, the summation over q excludes $q_x = q_y = 0$, and ϵ is the background dielectric constant. The last term in (3) is written explicitly as follows

$$\begin{aligned}
I(s,t) = \int_0^\infty \exp\left[-z^2 - 2\left\{\frac{\pi}{m\lambda}\frac{s^2 + \lambda^2 t^2}{\Omega_3 - \Omega_2^2/\Omega_1}\right\}^{\frac{1}{2}} z\right] \\
\times \cos\left[2\left\{\frac{\pi}{m\lambda}\frac{1}{\Omega_3 - \Omega_2^2/\Omega_1}\right\}^{\frac{1}{2}}\frac{\Omega_2}{\Omega_1} s z\right] dz
\end{aligned} \tag{4}$$

where

$$\begin{aligned}
\Omega_1 &= \frac{\omega_1}{\omega_z}\cos^2\phi + \frac{\omega_2}{\omega_z}\sin^2\phi \\
\Omega_2 &= \left[\frac{\omega_2}{\omega_z} - \frac{\omega_1}{\omega_z}\right]\sin\phi\cos\phi \\
\Omega_3 &= \frac{\omega_1}{\omega_z}\sin^2\phi + \frac{\omega_2}{\omega_z}\cos^2\phi.
\end{aligned}$$

The aspect ratio a/b is denoted by λ. As a check, it is easy to verify that when the tilt angle $\theta = 0$ and the strength of the potential $V(z)$ goes to infinity, i.e., $\omega_0 \to \infty$, we obtain $\phi = 0, \Omega_1 \to 1, \Omega_2 \to 0$, and $\Omega_3 \to \infty$.

In the case of the magnetic field perpendicular to the electron plane there is electron-hole symmetry in the lowest Landau level. In that case, the Hamiltonian does not have any explicit dependence on the magnetic field and in units of potential energy, the properties of $\frac{1}{3}$ and $\frac{2}{3}$ filling factors are always the same. In the present case such an *ideal* situation, however, does not exist, and in order to study the angular

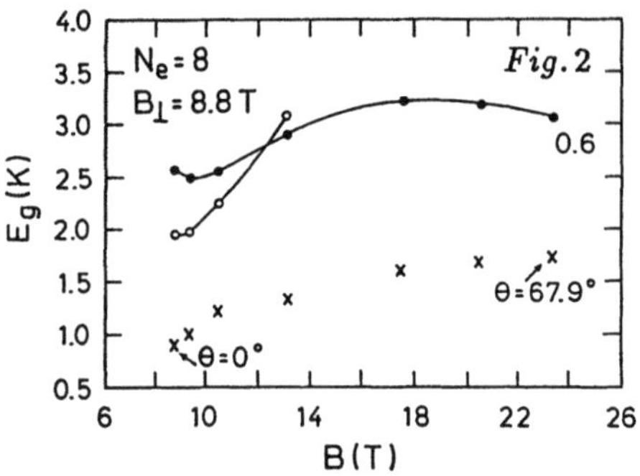

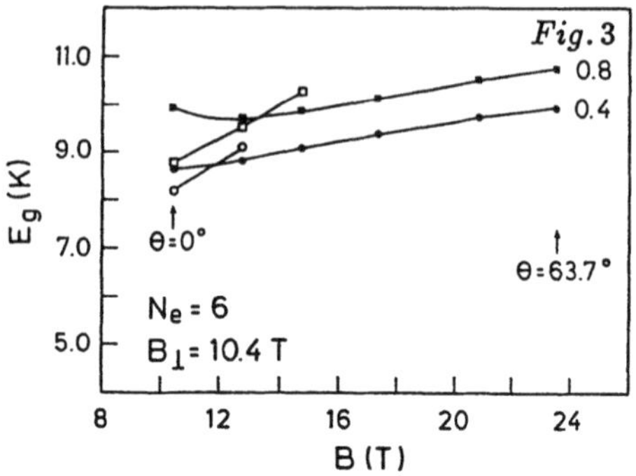

Fig. 2 Energy gap E_g (in K) for $\frac{2}{3}$ filling factor as a function of total magnetic field B (in units of tesla) for $\omega_0/\omega_z = 0.6$. The experimental results (denoted as $\times$) are from Ref. 4. The open points correspond to spin-reversed excitations

Fig. 3 Same as in Fig. 2 but for $\frac{1}{3}$ filling of the lowest Landau level

dependence of the energy gap for these two filling fractions, we need to consider the different frequencies appearing in (1) for the two filling fractions. This is obviously a direct consequence of the tilted magnetic field.

We consider the electrons to be in the lowest Landau level and also in the lowest subband. The parameters of the parabolic confinement potential are adjusted such that they correspond to the subband energy of a triangular potential with the Fang-Howard choice of trial wave function [1]. We have considered a six-electron system for the filling factor $\nu = \frac{1}{3}$ and an eight-electron system for $\nu = \frac{2}{3}$. The electrons are considered to be fully spin polarized. For $\frac{1}{3}$ filling, spin reversal in the ground state is not important. But for $\frac{2}{3}$ filling at low magnetic fields, the ground state is known from theory and experiments to be spin unpolarized ($S = 0$). Therefore our present study is valid only at high magnetic fields.

The method of calculating the quasiparticle–quasihole energy gap (energy required to create a pair of quasiparticles of opposite charge well separated from each other) is described in detail earlier [1,2]. The many-electron Hamiltonian is diagonalized numerically to obtain the ground state energy per particle at a certain filling factor. The energy gap is then obtained from the discontinuity of the chemical potential at that filling fraction [1,2].

The results for the energy gap (in Kelvin) at $\frac{2}{3}$ filling of the lowest Landau level are shown in Fig. 2 as a function of the *total* magnetic field. The experimental results of Haug et al. [4] are also shown for comparison. The angular dependence of the energy gap for large values of the tilt angle is qualitatively the same as found in the experiment. For small tilt angles (i.e., for small values of the total magnetic field), the energy gap shows an unexpected upward bend. This behavior presumably does not have any physical significance since here the fully spin-polarized excitations do not have the lowest energy. The excitation energy in this region can in fact be reduced by introducing the spin-reversed quasiparticles [1,2]. These are also plotted in Fig. 2, where the Zeeman energy ($g = 0.5$) is already included. The angular dependence of the resulting gap is now in good agreement with the experimentally observed behavior. Experimental results for $\nu = \frac{2}{3}$ by Furneaux et al. [4] are also consistent with our theoretical results.

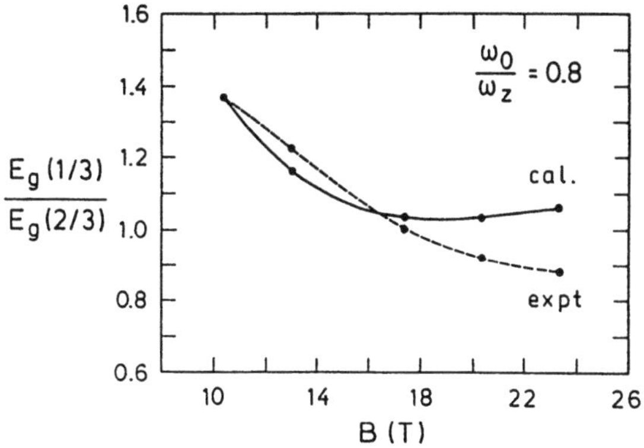

Fig. 4
The ratio of the energy gap at $\frac{1}{3}$ and $\frac{2}{3}$ filling fractions. The calculated energy gaps are for $\omega_0/\omega_z = 0.8$

The results for $\frac{1}{3}$ filling are presented in Fig. 3. Although the gap does not decrease with increasing tilt angle as observed in the experiment, the energy gap as a function of magnetic field is rather flat and its angular dependence is *different* from that of $\frac{2}{3}$ filling. In Fig. 4 we have plotted the energy gap ratio for $\frac{1}{3}$ and $\frac{2}{3}$ filling fractions from the experimental data of Ref. 4 and compared with our present theoretical results. In order to see the magnetic field dependence of this ratio, we have fitted the calculated value with the experimental result at the lowest magnetic field considered ($\sim 10T$). Except in the high magnetic field region the calculated magnetic field dependence is found to agree qualitatively with the experimentally observed behavior.

We should mention that even in the untilted-field case the magnetic field dependence of the gap is still not clear. With very high quality samples, a better quantitative agreement between theoretical and experimental data has been achieved recently [8]. The experimental data for $\frac{1}{3}$ in Ref. 4 are not from samples of quality similar to those of Ref. [8]. Moreover, there are only two data points available for $\frac{1}{3}$. Therefore, unlike the $\frac{2}{3}$ filling, here the trend is not clear. The discrepancy in the high-magnetic-field region of Fig. 4 may be in part due to this experimental uncertainty. More experiments with better quality samples are undoubtedly needed for this filling fraction.

In closing this section, we have studied the effect of tilted field on the FQHE, where the subband-Landau-level coupling is fully taken into account. The importance of spin-reversed excitations at low magnetic fields is also clearly demonstrated.

3. Optical Transitions

Magneto-optical transitions in the 2DEG might be an interesting route to investigate the spin polarization of the FQHE states just described. Recent optical experiments [9–11] have been very successful in providing new and interesting information in the FQHE regime. However, a proper theoretical study of the optical transition in a strongly interacting electron system is still not available. To this end, Bychkov and Rashba have recently reported a model calculation for a three electron cluster and short-range interactions [12]. In their work, the recombination process involves the capture of one of the electrons by a local center which is charge neutral in the initial state and acquires a charge in the final state. The two other electrons then move in the field of that charge. The prime intention of that work was to study the relation among the intensities of the optical transitions to the ground state of the system, in particular, the differences in the emission from various spin polarization of the initial state. They noticed that the intensity distribution depends on the initial spin of the cluster. For example, if the initial state of the cluster is spin-partially-polarized ($S = \frac{1}{2}$), transitions to the singlet final state will be the most intense.

In the present paper, we report on our work which goes beyond the model calculation described above. We consider a finite-electron system in a periodic rectangular geometry as the initial state. In the final state, one electron is missing in the system and the other electrons are moving in the field of a charged impurity. The spatial extent of the impurity center is assumed to be small compared to the magnetic length ℓ_0. The ground state and excitations for a system of four electrons with a charged impurity in the present geometry was studied earlier by Zhang et al. [13]. It should be mentioned that, in Ref. 13 as well as in the present work, the impurity is assumed to be in the electron plane. In the actual systems, the impurities are located away from the electron plane. In order to incorporate this fact, Zhang et al. assumed $Z \sim 0.1$ where Ze is the impurity charge. The resulting potential of the defect is thereby weakened. In the present study we also consider this value of the impurity charge.

The transition probability W is proportional to the square of the overlap of the initial and final states. This is evaluated as follows: In the initial state (in occupation representation)

$$|\psi^{\rm i}\rangle = \sum_{\{j_i\}} c^{\rm i}_{\{j_i\}} |j_1, j_2, \cdots, j_N\rangle. \tag{5}$$

Let us assume that the impurity is in a cell at position $\mathbf{R}$. For convenience, the position of the impurity is fixed in the middle of the cell. The initial state with one electron at position $\mathbf{R}$ is

$$|\psi^{\rm i}\rangle = \sum_{\{j_i\}} c^{\rm i}_{\{j_i\}} \sum_j \phi_j(\mathbf{R})\, a_j |j_1, j_2, \cdots, j_N\rangle \tag{6}$$

where $\phi_j(\mathbf{r})$ is the single-particle wave function and a_j destroys one particle from state j. Stated differently, we can introduce an *overlap operator* of the type, $\mathcal{O} = \sum_j \phi_j(\mathbf{R}) a_j$. The final state is simply

$$|\psi^{\rm f}\rangle = \sum_{\{j_i\}} c^{\rm f}_{\{j_i\}} |j_1, j_2, \cdots, j_{N-1}\rangle. \tag{7}$$

The overlap is then taken over the final and initial states.

We consider a four-electron system in the initial state and at the filling fraction $\nu = \frac{2}{3}$. In the periodic rectangular geometry the ground state is multiply degenerate which arises due to the center of mass degeneracy [14]. In the final state this degeneracy is lifted because the impurity interaction mixes the momentum eigenstates. In Fig. 5, we plot the intensity where we sum over all the degenerate initial states:

$$W \sim \delta(E - E^{\rm f} + E^{\rm i}) \sum_m \left| \langle \psi^{\rm f} | \psi^{\rm i}_m \rangle \right|^2, \tag{8}$$

for various values of the energy difference (in units of $e^2/\epsilon\ell_0$) between the initial and final states. We have considered all different spin polarizations ($S = 0, 1$ and 2) in the initial state. As we have seen in section 1, at low magnetic fields the ground state is spin unpolarized ($S = 0$). The intensities in this case (solid lines) are well separated in the energy range from other spin polarizations of the initial state. While the transition from the fully spin polarized ($S = 2$) initial state is the most intense, transitions from

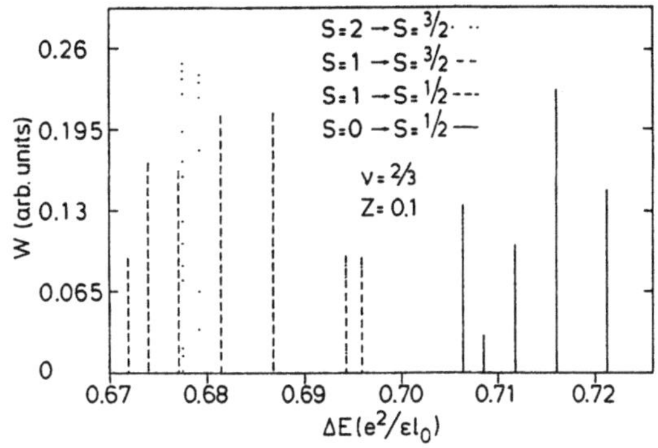

Fig. 5 The intensity (in arbitary units) of transitions from the initial state at $\nu = \frac{2}{3}$ for various spin polarizations as a function the energy difference between the initial and final states

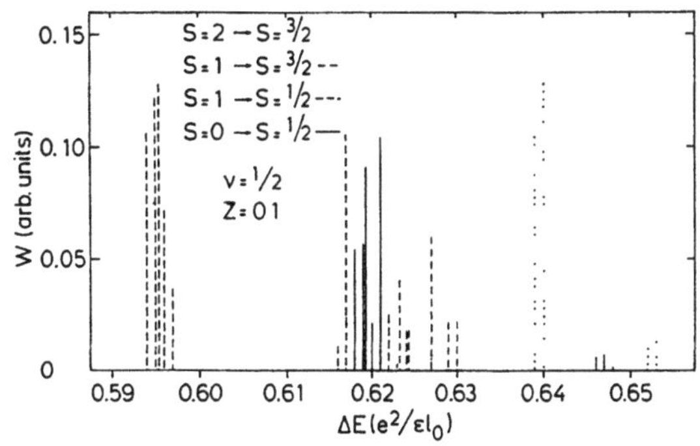

Fig. 6
Same as in Fig. 5 but for $\nu = \frac{1}{2}$ in the initial state

the unpolarized state is also quite comparable. Except in the spin polarized state, transition to an excited state is more intense than that in the ground state.

In Fig. 6, we present the intensity distribution for $\nu = \frac{1}{2}$. This filling factor is particularly interesting because here no FQHE has been observed so far, and the precise nature of the state is also not clear [15]. The intensity distribution is very different from that of Fig. 5. The major difference to note is that, unlike the $\frac{2}{3}$ filling factor, here the intensities for transitions from various spin states are distributed all over the energy range considered. It is not clear if the segregation of the intensity lines for the ground state from the other spin states as seen for the $\frac{2}{3}$ filling, and its absence at $\nu = \frac{1}{2}$, could be taken as an indication for the incompressible states. Further work would be needed for larger systems and for other filling factors to settle this point.

To summarize, from our preliminary results on the optical transitions, we can make the following tentative statements: the emission spectrum is very much filling factor dependent. It also depends on the spin configuration of the initial state. It is expected that the emission spectrum might provide an interesting means to study the ground state and excitations of the highly correlated two-dimensional electron fluid. Results for larger systems and other filling factors will be published elsewhere.

Acknowledgements

The work on the tilted-field effect was done in collaboration with V. Halonen (Oulu).

References

1. Tapash Chakraborty and P. Pietiläinen: *The Fractional Quantum Hall Effect*, Springer, Heidelberg, 1988
2. Tapash Chakraborty: Surf. Sci. **229**, 16 (1990); Tapash Chakraborty, P. Pietiläinen, F. C. Zhang: Phys. Rev. Lett. **57**, 130 (1986)
3. J. P. Eisenstein, H. L. Störmer, L. Pfeiffer, K. W. West: Phys. Rev. Lett. **62**, 1540 (1989); J. P. Eisenstein, H. L. Störmer, L. N. Pfeiffer, K. W. West: Phys. Rev. B**41**, 7910 (1990); see also, R. G. Clark, S. R. Haynes, A. M. Suckling, J. J. Harris, C. T. Foxon: Phys. Rev. Lett. **62**, 1536 (1989); R. G. Clark, S. R. Haynes, J. V. Branch, A. M. Suckling, P. A. Wright, P. M. W. Oswald, J. J. Harris, C. T. Foxon: Surf. Sci. **229**, 25 (1990)
4. R. J. Haug, K. von Klitzing, R. J. Nicholas, J. C. Maan, G. Weimann: Phys. Rev. B**36**, 4528 (1987); see also, J. E. Furneaux, D. A.Syphers, A. G. Swanson: Phys. Rev. Lett. **63**, 1098 (1989)
5. J. C. Maan, in *Two Dimensional Systems, Heterostructures and Superlattices*, edited by G. Bauer, F. Kuchar, H. Heinrich (Springer, Heidelberg 1984), p. 183
6. D. Yoshioka, B. I. Halperin, P. A. Lee: Phys. Rev. Lett. **50**, 1219 (1983); F. C. Zhang, Tapash Chakraborty: Phys. Rev. B**30**, 7320 (1984)
7. V. Halonen, P. Pietiläinen, Tapash Chakraborty: Phys. Rev. B**41**, 10202 (1990)
8. R. L. Willett, H. L. Störmer, D. C. Tsui, A. C. Gossard, J. H. English: Phys. Rev. B**37**, 8476 (1988); G. Boebinger, A. M. Chang, H. L. Störmer, D. C. Tsui: Phys. Rev. Lett. **55**, 1606 (1985)
9. D. Heiman, B. B. Goldberg, A. Pinczuk, C. W. Tu, A. C. Gossard, J. H. English: Phys. Rev. Lett. **61**, 605 (1988)
10. I. V. Kukushkin, V. Timofeev, K. von Klitzing, K. Ploog: in *Advances in Solid State Physics*, vol. 28, Ed. U. Rössler (Vieweg, Braunschweig, 1988) p. 21
11. I. V. Kukushkin, A. S. Plaut, K. von Klitzing, K. Ploog: Surf. Sci. **229**, 447 (1990); and to be published
12. Yu. A. Bychkov, E. I. Rashba: Sov. Phys. JETP **69**, 430 (1989)
13. F. C. Zhang, V. Z. Vulovic, Y. Guo, S. Das Sarma: Phys. Rev. B**32**, 6920 (1985)
14. F. D. M. Haldane: Phys. Rev. Lett. **55**, 2095 (1985)
15. H. W. Jiang, H. L. Störmer, D. C. Tsui, L. N. Pfeiffer, K. W. West: Phys. Rev. B**40**, 12013 (1989)

Model Calculations for the Fractional Quantum Hall Effect

R.H. Morf

Paul Scherrer Institut Zürich, CH-8048 Zürich, Switzerland

Abstract. The hierarchical theory of the fractional quantum Hall effect has been successful in predicting the quantum numbers of higher order fractional states. Here, we show that it can also be used to obtain accurate energies and gaps. Indeed, the large gaps experimentally observed at $\nu = 2/5$, $3/7$, $4/9$, etc, follow from the detailed form of the quasiparticle interaction, which can be determined accurately. The absence of a polarized state at $\nu = 4/11$ also appears to be explained in this way.

Introduction

It is by now widely accepted that the basic physical principles behind the formation of the steps of the Hall conductivity σ_{xy} at rational multiples of e^2/h are well captured by Laughlin's wavefunctions for the ground state at filling fraction $\nu = 1/m$ and its generalizations for excited states which exhibit charged excitations with charge e/m [1], [2],[3].

The theoretical picture of higher order fractional quantum Hall states, e.g at $\nu = \frac{2}{5}, \frac{2}{7}, \frac{3}{7}$ or $\frac{4}{9}$ is still somewhat controversial. In particular, some people doubt the correctness and even more the usefulness of the hierarchical construction of higher order states, first introduced by Haldane [4] and by Halperin [5]. In this picture, the $\nu = 2/5$ state is assumed to result from the condensation of the quasiparticles with charge $e/3$ into an incompressible fluid state similar to the Laughlin fluid of electrons in the parent $\nu = 1/3$ state. Invoking this principle iteratively, further daughter states at $\nu = 3/7, 4/9$ are interpreted as condensates of quasiparticle systems with charge $e/5$ and $e/7$ in the respective parent state at $\nu = 2/5$ and $3/7$. In a very strict interpretation of this theory, one would expect that the appearance of a Hall plateau at some higher order hierarchical level would require that all its parent states with their associated Hall plateaux would be present as well [6]. So far, to my knowledge, no exception to this rule has been observed.

Exact calculations for small systems have provided a great deal of information about the structure of ground and excited states [7]. The hierarchical scheme does make precise predictions for the quantum numbers of the ground and excited states in finite systems. For example in the spherical geometry [4], for a system of N_e electrons, a quantum Hall state at filling fraction ν is predicted to occur for 'magic numbers' $N_\Phi = N_e/\nu + C(\nu)$ of the magnetic flux

Springer Series in Solid-State Sciences, Vol. 101
High Magnetic Fields in Semiconductor Physics III   Editor: G. Landwehr

Φ, which is measured in units of the flux quantum $\Phi_0 = hc/e$. The nontrivial prediction consists, apart from the filling fractio ν, in the size independent term $C(\nu)$. For example daughter states of the $\nu = 1/3$ state at filling fractions $\nu = 2/7$ and $2/5$, formed from quasiholes and quasiparticles, respectively, are predicted at $C(\frac{2}{7}) = -2$ and $C(\frac{2}{5}) = -4$.

These and further predictions of the hierarchy have been verified in finite system studies by Haldane and Rezayi [8] and by d'Ambrumenil and myself [9]: For a sequence of states at filling fraction ν, the energy depends smoothly on the system size, provided N_Φ and N_e are varied according to the predictions of the hierarchy. A change in the number of flux quanta N_Φ by modifying $C(\nu)$ leads to ground states that are, typically, *not* translation invariant and which are identified as excited states of nearby fractional quantum Hall states. Their size dependent energy can be used to obtain the energy of excitations.

It is of great interest to find out if, beyond this, the hierarchy is really fit to provide accurate quantitative predictions for energies and gaps, or if an alternative theory might be needed (cf. e.g. [10]).

In this review, I shall discuss some recent work I have carried out together with P. Béran on the hierarchical theory of higher order fractional quantum Hall states [11]. It is based on the phenomenological wave functions, introduced by Halperin [5], which describe systems of fractionally charged particles in a magnetic field. These obey fractional statistics. I will show that their interaction differs significantly from that of point charges. This leads to substantial modifications in the prediction of ground state energies and gaps, as compared to the original estimates by Halperin [5] and by Zhang [12]. In particular, the large gaps in the polarized states at 2/5, 3/7, 4/9 etc., observed experimentally, are traced back to the fact that pairs of quasiparticles are found to exhibit a very large energy in a state with minimum relative angular momentum $L = L_{min}$ and quite small energy in the state $L = L_{min} + 2$ [4].

Quantitative description of higher order fractional states

Following Halperin, at the first level of hierarchy, the energy of the daughter states of the $\nu_0 = 1/m$ state at filling factor $\nu_1 = 1/(m-1/p_1)$ can be expressed as

$$\mathcal{E}(\nu_1) = (1 + 1/2mp_1)\mathcal{E}(\nu_0) + \tilde{\epsilon}_1^{+} /p_1 + \tilde{\mathcal{U}}_{p_1} /p_1. \qquad (1)$$

Symmetry conditions require m to be odd and $p_1, p_2, ..$ even. Here and in the following, the subscripts 0 and 1 refer to the level of hierarchy. The quantity $\mathcal{E}(\nu)$ is the energy per electron and $\tilde{\epsilon}_1^{+}$ is the proper energy of a quasiparticle [13]. They are both known accurately from the study of finite systems [14,9,15,16]. The last quantity $\tilde{\mathcal{U}}_{p_1}$, the interaction energy per quasiparticle of the quasiparticle Laughlin-condensate is less well known and many people have expressed doubts if it can be computed reliably [10].

The interaction energy $\widetilde{U}_{p_1}$ has been first computed by Halperin [5]. Treating the quasiparticles as *point* charges and using *pseudo-wavefunctions* with fractional statistics to describe their condensate states, he obtained the intriguing result that the energy (per flux unit) is a continuous function of filling fraction, for $0 < \nu \leq 1$, yet with discontiuous slope at all rational fractions with odd denominator [5]. The pseudo-wavefunctions were derived, in this work, from generalizations of Laughlin's wave function, in which some of the electrons are grouped in pairs [17,13], giving rise to quasiparticles, and by integrating out the coordinates of unpaired electrons. Their lowest energy states would again be Laughlin states of quasiparticles, which in the disk geometry, have the form [5],

$$\Psi_{m,p_1}^{(N_1)}(Z_1,...,Z_{N_1}) = \prod_{1\leq i<j\leq N_1} (Z_i - Z_j)^{p_1-\frac{1}{m}} \prod_{k=1}^{N_1} e^{-\frac{|Z_k|^2}{4m}} \tag{2}$$

where $Z_k = Y_k + iY_k$ is the position of quasiparticle k in complex notation, and measured in units of the magnetic length l_0 of the electrons. Throughout this paper, microscopic wave functions will be denoted by small, pseudo wavefunctions by capital letters.

In the case, that the interaction between the quasiparticles can be represented by a pair interaction, the energy of the quasiparticle system in state 2 can be expressed in terms of an integral over the pair correlation function, which can easily be calculated by Monte Carlo [13]. Using the Coulomb interaction of point charges with charge e/m, the energy of the $\nu = 2/5$ state becomes $\mathcal{E}(2/5) = -0.424$, where as unit of energy we use $e^2/\epsilon l_0$, with l_0 the magnetic length of the electrons and ϵ the dielectric constant. On an absolute scale, this estimate may seem to be in reasonable agreement with the extrapolation of diagonalizations for small systems [18] $\mathcal{E}(2/5) = -0.433$. However, the relevant energy scale (cf. [5] characterizing quantum effects, is given by the deviation of these results from the classical plasma value $\mathcal{E}_{plasma}(2/5) \approx -0.4429$ [19]. Thus, quantum effects have been overestimated by about a factor of two.

What is the source of this difference? Is the description in terms of a parent state condensate at $\nu = 1/3$ plus a quasiparticle condensate 1 unjustified? Are the pseudo-wavefunctions 2 to be blamed? Or, is it simply the interaction of the quasiparticles, that is poorly represented by the Coulomb interaction of point charges?

That the use of the fractional statistics (anyon) representation of quasiparticle systems is natural, we wish to illustrate with the instructive example of a pair of quasiparticles, attached to electrons with reversed spin [17,13,20]. For this case, a microscopic wave function can be written easily,

$$\begin{aligned}\psi_L^{(N_1)}(z_1,...,z_N) &= (z_1 - z_2)^L \prod_{i\neq 1}(z_i - z_1)^{-1} \prod_{j\neq 2}(z_j - z_2)^{-1} \\ &\times\Big[\prod_{1\leq i<j\leq N} (z_i - z_j)^m \prod_{k=1}^{N} e^{-\frac{|z_k|^2}{4}}\Big]. \end{aligned}\tag{3}$$

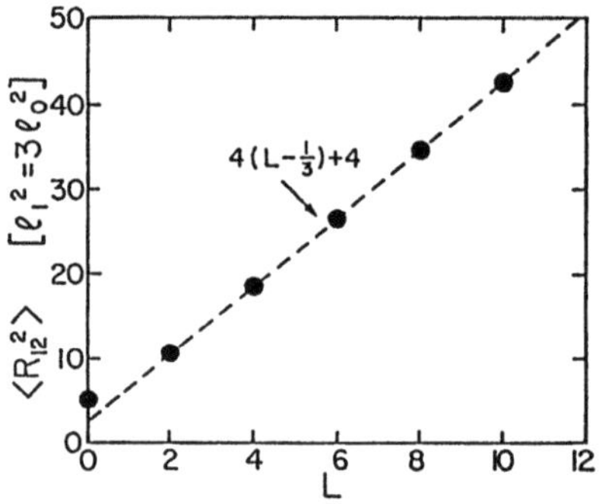

Figure 1. **The mean-square separation of quasiparticles in state 3 is plotted as a function of L. Dots are from Monte Carlo calculations with up to 64 electrons on the surface of a sphere. The dashed line corresponds to the behaviour for pseudo-wavefunction 2.**

The term in square brackets is the Laughlin wave function for the state at $\nu = 1/m$ and L determines the relative angular momentum of the two quasiparticles, which, owing to the first two products, are created at the position of down-spin electrons 1 and 2. This wavefunction is the ground state for electrons interacting via the hardcore interaction. It can be easily evaluated by Monte Carlo. The condition of antisymmetry with respect to exchanges of electrons 1 and 2 demands L to be even, in the same way as p_n must be even in eqs. (1,2). In Figure 1, we illustrate the mean square quasiparticle separation $< R_{12}^2 >$, which is just the separation of down-spin electrons 1 and 2, in this state for $m = 3$ and various values of relative angular momentum L. With the exception of the case $L = 0$, the separation of the quasiparticles in state 3 scales as $< R_{12}^2 > \sim (4L^* + 4)l_1^2$, where $l_1 = \sqrt{3}l_0$ is the magnetic length of quasiparticles with charge $e/3$ and L^* is an effective *fractional* angular momentum, $L^* = L - \frac{1}{3}$. The observed L–dependence of the quasiparticle separation in state 3 is thus consistent with Halperin's pseudo wavefunction 2,

$$\Psi_L(Z_1, Z_2) = (Z_1 - Z_2)^{L-\frac{1}{3}} e^{-\frac{|z_1|^2+z_2|^2}{4m}}, \tag{4}$$

up to an arbitrary phase factor, which consistent with the Berry phase argument by Arovas et al. [21], is taken to be unity [22].

Let us now turn to the problem of the quasiparticle interaction. According to previous investigations [16,13] the charge density of the quasiparticle in the polarized state at $\nu = 1/3$, is not at all point-like, but actually ring shaped, revealing an internal structure which ought to affect the interaction of quasiparticles at short distance. We thus have to devise a method for treating the quasiparticle interaction more accurately.

In the next section, I will discuss a scheme developed by Béran and myself [11] to determine the short distance behaviour of the quasiparticle interaction. Although we shall limit ourselves to two-body interactions, our results for energies and gaps are consistent with exact diagonalizations for small systems.

Determination of the Quasiparticle Interaction

The interaction between quasiparticles differs in two ways from the Coulomb repulsion of point charges: (i) Multipolar corrections originate from the finite size of quasiparticles. For large quasiparticle separation, these corrections can be calculated approximately from the charge ditribution of isolated quasiparticles. (ii) At short distances (of order of the magnetic length l_0), complicated quantum mechanical polarization effects occur. Fortunately, exact calculations using systems with only few electrons allow to investigate this situation In collaboration with P. Béran, I have carried out detailed calculations with up to nine electrons confined to the surface of a sphere [14,11].

Following Halperin [5], the quantum Hall state at filling factor $\nu = 1/(m - 1/p_1)$ consists of a $\frac{1}{m}$ state of N_0 electrons plus a system of $N_1 = N_0/p_1$ quasiparticles in an incompressible fluid state described by equation (2). We discuss both type of quasiparticles, those in the fully polarized system as well as those associated with an electron with reversed spin. The latter actually becomes the lowest energy quasiparticle excitation of the $\frac{1}{3}$ state at low magnetic field [20]. Using this kind of excitation in the hierarchical construction leads to a spin unpolarized state at $\frac{2}{5}$ for $p_1 = 2$ or a partially polarized state at $\frac{4}{11}$ for $p_1 = 4$.

We assume that the interaction of quasiparticles can be approximated by a pair potential $\tilde{V}(|Z_i - Z_j|)$. This interaction $\tilde{V}(R)$ is defined in such a way that the energy $\tilde{V}_L$ of two quasiparticles in a state with given relative angular momentum L described by *pseudo*-wavefunction Ψ_L

$$\tilde{V}_L = \langle\Psi_L|\,\tilde{V}\,|\Psi_L\rangle/\langle\Psi_L|\Psi_L\rangle \tag{5}$$

coincides with the interaction energy of the corresponding *microscopic* two quasiparticle state, which can be obtained from numerical diagonalizations. This is most easily illustrated in the spherical geometry: For a given number of electrons $N_e = N_0$, the Laughlin state at filling fraction $\nu = 1/m$ occurs for a number of flux units of $N_\Phi = m(N_0 - 1)$, and has an angular momentum $L_{tot} = 0$. Decreasing the number of flux unis by n introduces n quasiparticles with charge e/m. A one quasiparticle state has angular momentum $L_{tot} = N_e/2$, while two-quasiparticle states have a total angular momentum $0 \leq Ltot \leq N_e$, corresponding to a relative angular momentum L between the quasiparticles

$$L = N_e - L_{tot} \tag{6}$$

In the following, we will use superscripts to denote the number of quasiparticles in a given state. For example, $E_L^{(2)}(N_0)$ denotes the lowest energy of an N_0 electron system with $L_{tot} = N_0 - L$ with two flux units below the corresponding N_0 electron Laughlin state, i.e. with 2 quasiparticles with relative angular momentum L. This energy is composed of the energy of the underlying pure Laughlin state $E^{(0)}(N_0)$, twice the proper energy $\tilde{\epsilon}_1^+$ of an isolated quasiparticle and finally the actual energy of interaction of the two quasiparticles. We use this to define as 'exact microscopic' interaction energy $\tilde{V}_L^{(mic)}$ by,

$$\tilde{V}_L^{(mic)} = \lim_{N_0 \to \infty} (E_L^{(2)}(N_0) - E^{(0)}(N_0)) - 2\tilde{\epsilon}_1^{+}. \quad (7)$$

Problems associated with the extrapolation to the bulk limit $N_0 \to \infty$ are discussed in references [14,18,9].

Similarly, the proper energy $\tilde{\epsilon}_1^{+}$ of a single quasiparticle (cf. equation 1) is defined by

$$\tilde{\epsilon}_1^{+} = \lim_{N_0 \to \infty} (E^{(1)}(N_0) - E^{(0)}(N_0)). \quad (8)$$

For the quasiparticle of the polarized system and of the one involving the spin reversal of an electron, we obtain respectively $\tilde{\epsilon}_1^{+} = 0.077 \pm 0.001$ and $\tilde{\epsilon}_1^{+} = 0.050 \pm 0.001$ (cf. also reference [20]).

Exact calculations for the matrix elements $\tilde{V}_L$ with large L are not possible. However, since they involve quasiparticles at large separation, we can approximate their interaction by the Coulomb repulsion of *rigid*, although extended, quasiparticle charges. For this purpose we use Monte Carlo results for the charge density $\rho(r)$ of a single quasiparticle obtained previously by the author both for the polarized and the spin-reversed case (cf. [16,13]). From these, a multipole expansion of the interaction between these extended charges is defined [23]. Expectation values $\tilde{V}_L^{(\lambda)}$ of equation (5) are listed in table 1 for small L. Here, λ denotes the order at which the multipole expansion of $\tilde{V}(R)$ is truncated. For $L \geq 4$, the hexadecapole moment leads to negligible corrections. In the case of quasiparticles of the polarized system, the agreement of the estimates $\tilde{V}_6^{(mic)}$ and $\tilde{V}_6^{(5)}$ confirms the validity of the multipole expansion of $\tilde{V}$ in the evaluation of matrix elements with $L \geq 6$. For quasiparticles involving electrons with reversed spin, a reasonable agreement is found already for $L = 2$.

Table 1: Interaction energy of two quasiparticles with relative angular momentum L. $\tilde{V}_L^{(\lambda)}$ is the expectation value (5) using multipole expansion of order $r^{-\lambda}$. $\tilde{V}_L^{(mic)}$ are extrapolations of microscopic results cf. eq. (7). Numbers in parentheses denote the extrapolation uncertainty in units of the last digit.

		Polarized		Reversed spin	
L	$\tilde{V}_L^{(1)}$	$\tilde{V}_L^{(5)}$	$\tilde{V}_L^{(mic)}$	$\tilde{V}_L^{(5)}$	$\tilde{V}_L^{(mic)}$
0	0.1318		0.094(1)		0.089(1)
2	0.0231	0.0186	0.0114(3)*	0.0211	0.019(1)
4	0.0162	0.0146	0.0206(9)*	0.0157	0.013(2)
6	0.0132	0.0123	0.0118(16)	0.0129	

By contrast,the 'microscopic' results for the polarized state show dramatic effects, marked by an asterisk in Table 1: While the energy of the $L = 2$-state is reduced from its point particle value by a factor of 2, for the $L = 4$-state, it is actually increased by about 25 percent, leading to an $L = 4$ value which is *substantially larger* than that for $L = 2$.

Results

We first discuss our results for the polarized and unpolarized $\frac{2}{5}$ states. In table 2 are listed the hierarchical estimates of the groundstate and of the gap assuming point-like quasiparticles (first line) and using a more accurate representation of the interaction at short distance (second line). These last estimates are in good agreement with the extrapolation of results of diagonalization, which are given in the third line. For the polarized state, the accurate treatment of the interaction of quasiparticles greatly improves the agreement for the groundstate energy. Actually, this result agrees even better with the enrgy $\mathcal{E}(2/5) = -0.4310$ of the trial-state of reference [18]. This may be accidental, although we might also conclude, that the trial state might is actually a microscopic realization of the hierarchical construction [18]. The values for the gap in table 2 are obtained from trial wavefunctions for particle- and hole-like excitations at level $s = 2$ within the quasiparticle system at level $s = 1$ [24]. Our results for the energies of the polarized and unpolarized $\frac{2}{5}$ states can also be used to estimate the value B_c of the magnetic field at which, due to the Zeeman term, the polarized state becomes the groundstate. In the situation of the experiment of reference [25], for an electron density $n_e = 2.3 \times 10^{11} cm^{-2}$ and using the values $\epsilon = 12.8$ and $g = 0.4$, we obtain $B_c = 9.1 \pm 1.6$ Tesla, whereas extrapolations of diagonalizations lead to a smaller value $B_c = 5.4 \pm 1$ Tesla [26]. The experimental value is $B_c = 7$ Tesla.

Table 2: Groundstate energy per electron and gap of the polarized ($\uparrow\uparrow$) and unpolarized ($\uparrow\downarrow$) $\frac{2}{5}$ states. Point-charge approximation is denoted by $\tilde{V}_L^{(1)}$. $\tilde{V}_L^{(mic)}$ denotes hierarchical results based on $\tilde{V}_L^{(mic)}$ for $L \leq 4$ and $\tilde{V}_L^{(5)}$ for $L > 4$. The last row lists results of exact diagonalizations. These energies should be compared to that of a classical plasma $\mathcal{E}_{Plasma} = -0.4429$, cf. [19].

	2/5 $\uparrow\uparrow$		2/5 $\uparrow\downarrow$	
	$\mathcal{E}_{GS}$	ϵ_{gap}	$\mathcal{E}_{GS}$	ϵ_{gap}
$\tilde{V}_L^{(1)}$	-0.4220(2)	0.082(3)	-0.4355(3)	0.082(3)
$\tilde{V}_L^{(mic)}$	-0.4313(8)	0.073(4)	-0.4411(15)	0.059(4)
Exact	-0.4335(9)	0.061(15)	-0.4393(6)	

We also investigated the state at filling factor $\nu = 4/11$, which results for the choice $p_1 = 4$, for both types of quasiparticles. Indeed, adding quasiparticles involving electrons with reversed spin leads to a partially polarized state with total spin $S = N_0/4$. Our estimates for the energies of these states are consistent with results of diagonalizations (see table 3). While the partially polarized $\frac{4}{11}$ state may have a small gap (see table 3), no gap is found in the polarized $\frac{4}{11}$ state, consistent with results of diagonalizations for 4 and 8 electrons where the groundstates occur at $L_{tot} = 2$ and a large number of excited states are nearby [11,27].

Table 3: Groundstate energy per electron and gap of the polarized ($\uparrow\uparrow$) and partially polarized ($\updownarrow$) $\frac{4}{11}$ states given as in table 2. The exact results for the polarized state are derived from systems with $N_0 = 4, 8$ those for the partially polarized state are the result for $N_0 = 6$. The classical plasma value is $\mathcal{E}_{Plasma} = -0.4256$ in this case.

	4/11 $\uparrow\uparrow$		4/11 $\updownarrow$	
	$\mathcal{E}_{GS}$	ϵ_{gap}	$\mathcal{E}_{GS}$	ϵ_{gap}
$\tilde{V}_L^{(1)}$	-0.4139(4)	0.0041(2)	-0.4207(4)	0.0041(2)
$\tilde{V}_L^{(mic)}$	-0.4134(3)	~ 0	-0.4211(4)	0.003(2)
Exact	-0.4158	~ 0	-0.4217	

To summarize our results, in the spin polarized case, the interaction of quasiparticles in the $\frac{1}{3}$ state differs strongly from the Coulomb interaction at short separation. It is characterized by a particularly small pair-state energy $\tilde{V}_L$ for relative angular momentum $L = 2$ while those for $L = 0$ and $L = 4$ are very large. As a result, the stability of the quasiparticle Laughlin state (2) with $p_1 = 2$ is enhanced, while those with $p_1 \geq 4$ are destabilized. On the other hand, the interaction of quasiparticles involving electrons with reversed spin is not very different from the repulsion of point charges and allows the existence of both unpolarized $\frac{2}{5}$ and partially polarized $\frac{4}{11}$ states.

In view of our results, the experimental observation of particularly large gaps for the polarized systems at $\nu = \frac{2}{5}, \frac{3}{7}, \frac{4}{9}, ..., etc.$ appears to have a simple explanation within the hierarchical picture. Indeed, this will result if the interaction between quasiparticles at level $s + 1$ in the state with $p_s = 2$ is again characterized by strong $\tilde{V}_0^{(s+1)}$ and weak $\tilde{V}_2^{(s+1)}$ pseudopotentials. The ring shape of quasiparticles may be an indication for this to happen. Such a ring shape has been observed in numerical calculations [9,24] also at $\nu = \frac{2}{5}$ and $\nu = \frac{3}{7}$ and is probably a general feature for the sequence $\frac{s+1}{2s+3}$. This point can be studied by generalizing our trial wavefunctions of one- and two-quasiparticle states to higher hierarchical levels. While analytical evaluation of these becomes quite involved due to the fractional statistics term, Monte-Carlo calculation poses no problems.

I have benefitted from numerous helpful discussions with B.I. Halperin, P. Béran and N. d'Ambrumenil.

References

[1] R.B. Laughlin, Phys. Rev. Lett. **50** (1983) 1395.

[2] The Quantum Hall Effect, eds. R.E. Prange and S.M. Girvin, Springer 1987.

[3] R.B. Laughlin, in [2]

[4] F.D.M. Haldane, Phys. Rev. Lett. **51** (1983) 605.

[5] B.I. Halperin, Phys. Rev. Lett. **52** (1984) 1583; **52** 2390 (E) (1984).

[6] F.D.M. Haldane in [2] p.303.

[7] Indeed, in his discovery of his trial wave function, Laughlin was helped by his exact study of three electrons in a large magnetic field, cf. R.B. Laughlin, Phys. Rev. **B27**(1983) 3383

[8] F.D.M. Haldane and E.H. Rezayi, Phys. Rev. Lett. **54** (1985) 237.

[9] N. d'Ambrumenil and R.Morf, Phys. Rev. **B40** (1989) 6108.

[10] J.K. Jain, Phys. Rev. Lett. **63** (1989) 199; Phys. Rev. **B41** (1990) 7653.

[11] P. Béran and R. Morf, Phys. Rev. **B**, to be published

[12] F.C. Zhang, Phys. Rev. **B34** (1986) 5598.

[13] R.Morf and B.I.Halperin, Phys. Rev. **B33** (1986) 2221.

[14] F.D.M. Haldane and E.H. Rezayi, Phys. Rev. Lett. **54** (1985) 237.

[15] G. Fano, F. Ortolani and E. Colombo, Phys. Rev. **B34** (1986) 2670.

[16] R.Morf and B.I.Halperin, Z. Phys. **B68** (1987) 391.

[17] B.I.Halperin, Helv. Phys. Acta **56** (1983), 75.

[18] R.Morf, N.d'Ambrumenil and B.I.Halperin, Phys. Rev. **B34** (1986) 3037.

[19] This value is calculated with the interpolation formula of D. Levesque, J.J. Weiss and A.H. MacDonald, Phys. Rev. **B30** (1984) 1056.

[20] T. Chakraborty, P. Pietiläinen, F.C. Zhang, Phys. Rev. Lett. **57** (1986) 130; E.H. Rezayi, Phys. Rev. **B36** (1987) 5454.

[21] D. Arovas, J.R. Schrieffer and F. Wilczek, Phys. Rev. Lett. **53** (1984) 722.

[22] The somewhat larger value at $L = 0$ shows that there are corrections at short separation. A smooth cut-off at short separation can be introduced in the definition of expectation values of 4, which allows to obtain the correct value for $< R_{12}^2 >$ for $L = 0$.

[23] The charge density, as computed e.g. by Monte Carlo, contains the effects of cyclotron motion. A deconvolution with the kernel $e^{-r'^2/2m}$ is necessary to define "intrinsic" multipole moments of the quasiparticle. For the quasiparticle of the polarized system, we obtain $q_2^{(e)} = -3 \pm .006$ and $q_4^{(e)} = -30 \pm 3$. For the reversed spin case, we get $q_2^{(e)} = -.68 \pm .01$, $q_4^{(e)} = 4.32 \pm 3$. $q_0^{(e)} = 1/3$ in both cases. The unit of length is l_0.

[24] Again, the particle-like excitation at level $s = 2$ is characterized by a ring shaped quasiparticle density analogous to the electronic density of the quasiparticle at level $s = 1$. For the quasihole, pseudo trial wavefunctions are used which are obtained by multiplying wavefunction (2) by $\prod_{i=1}^{N_1} Z_i$. For the quasiparticle, a generalization of the pair-wavefunction of equation (2.18) of reference [13] to the case of fractional statistics (cf. 2) is employed. Details will be published.

[25] J.P. Eisenstein, H.L. Störmer, L. Pfeiffer, W. West, Phys. Rev. Lett. **62** (1989) 1540.

[26] Exact results have been quoted for $N_0 \leq 6$ by X.C. Xie, Yin Guo and F.C. Zhang, Phys. Rev. **B40** (1989) 3487. In addition, we use our result $\mathcal{E}^{\uparrow\downarrow}(2/5) = -0.44022$ for $N_0 = 8$, cf. E.H. Rezayi in [20].

[27] C. Gros and A.H. MacDonald, Phys. Rev. **B42** (1990) 9514.

Magneto-Optical Evidence for Fractional Quantum Hall States Down to Filling Factor 1/9 and for Wigner Crystallization

I.V. Kukushkin[1,2], *H. Buhmann*[3], *W. Joss*[3], *K. von Klitzing*[1], *G. Martinez*[4], *A.S. Plaut*[1], *K. Ploog*[1], and *V.B. Timofeev*[2]

[1]Max-Planck-Institut für Festkörperforschung, Heisenbergstr. 1, W-7000 Stuttgart 80, Fed. Rep. of Germany
[2]Institute of Solid State Physics, Academy of Sciences of the USSR, Chernogolovka, Moscow District, SU-142432 Moscow, USSR
[3]Max-Planck-Institut für Festkörperforschung, Hochfeld-Magnetlabor, BP 166X, F-38042 Grenoble Cedex, France
[4]Service National des Champs Intenses, Centre National de la Recherche Scientifique, BP 166X, F-38042 Grenoble Cedex, France

Abstract. Discontinuities have been observed in the energy dependence on magnetic field of the luminescence line corresponding to two-dimensional (2D) electron-hole recombination in GaAs-AlGaAs heterojunctions at filling factors ν=2/3, 1/3, 4/5, 3/5, 2/5, 1/5, 1/7 and 1/9. We associate these energy position shifts with discontinuous behaviour of the chemical potential due to condensation of the 2D electrons into an incompressible Fermi liquid state and thereby evaluate the energy gaps at fractional filling factors down to 1/9. Below a critical filling factor (ν_{cr}=0.28) and a critical temperature (T_{cr}=1.4K at 26T) an additional line has been observed in the spectrum which grows in intensity with decreasing ν and dominates the luminescence spectrum at ν<1/11. Its appearance is accompanied by a strong reduction in integrated intensity, while its relative intensity decreases sharply at ν=1/5, 1/7 and 1/9. The lack of correlation between ν_{cr} and the disorder related properties of the system indicates the intrinsic nature of the line which we propose signals the formation of a pinned Wigner solid.

1. Introduction

One of the most important outstanding questions in the physics of two-dimensional (2D-) electronic system is the nature of the ground state of the interacting electrons in a perpendicular magnetic field (H). In the extreme quantum limit, at fractional filling factors $\nu=hn_s/eH$ (n_s is the electron concentration) the ground state is known to be an incompressible Fermi liquid[1-3]. However, on further reducing ν, a phase transition is expected to occur at $\nu=\nu_{cr}$ to the energetically favoured Wigner solid. Theoretically predicted values of ν_{cr} have ranged from 1/3 to 1/10 [4-7] and were sensitive to the approximations used in the calculations. So far, observation of this changeover in the experiment has not been unambiguously identified. In this paper we describe magneto-optical experiments performed in this extrem quantum limit.

2. Fractional Quantum Hall Effect

The Fractional Quantum Hall Effect (FQHE) is now understood to occur as a result of condensation of two-dimensional electrons into an incompressible quantum fluid at filling factors $\nu=p/q$ (p,q integers and q odd) with quasi-particle excitations separated from the ground state by energy gaps Δ_G[1,2,3]. Experimental measurements of these energy gaps provide a rigorous test of the theories, which also predict that at sufficiently small ν (at $\nu \simeq 1/10$ [2]) a liquid to solid transition should occur when the ground state energy of the Wigner crystal becomes lower than that of the quantum fluid[2,3]. So far, observation of this changeover has proved elusive and fractional states down to $\nu=1/7$ have been detected[8]. The usual method for experimental determination of the FQHE gaps is activated magnetotransport [9-12]. However, in the region of small ν these measurements become extremely difficult due to strong localization at low temperatures and high magnetic fields. In this work we measure the FQHE gaps by means of magneto-optics which is much less sensitive to localization.

Optical measurements in the FQHE regime were first undertaken on Si-MOSFETs[13] and then in GaAs-AlGaAs quantum wells[14]. Although in the Si-MOSFETs the measurements allowed determination of the quasi-particle gaps this has so far not been achieved in a GaAs structure. Here we report the first optical measurements of the FQHE gaps at $\nu=2/3$, 1/3, 4/5, 3/5, 2/5, 1/5, 1/7, 1/9 in GaAs-AlGaAs single heterojunctions (SH). These have been determined from the discontinuous dependence on the magnetic field of the energy position of a luminescence line.

The radiative transition studied was that due to a 2D electron recombining with a hole bound to an acceptor (B line[15]) from a Be monolayer located in the GaAs (buffer width 50-100 nm) at a distance of 25nm from the interface of a series of GaAs-$Al_xGa_{1-x}As$ (x=0.28-0.32) SH (mobility $\mu \simeq 10^6$ cm^2/V s under continuous illumination [16]). It was possible to change the concentration n_S by the excitation power[16] and for samples 1 and 2 the variation was $1.9\text{-}2.5\times10^{11}$ cm^{-2} and $0.54\text{-}1.2\times10^{11}$ cm^{-2}, respectively. The high quality of our samples was confirmed by the observation of FQHE states at $\nu=p/q$ (with q=3,5) in magnetotransport measured both in the dark and under continuous illumination. The samples were mounted in a He^3 cryostat with optical access via a fibre-optic system. A sample temperature (measured by a coplanar RuO_2 resistance thermometer) of 340 mK was attained in the dark (400mK-600mK under continuous illumination) in magnetic fields up to 28T. Photoluminescence was excited using an Ar^+ laser (10^{-4}-10^{-3}W) and detected by a cooled GaAs photomultiplier and a triple spectrometer with 0.06meV resolution.

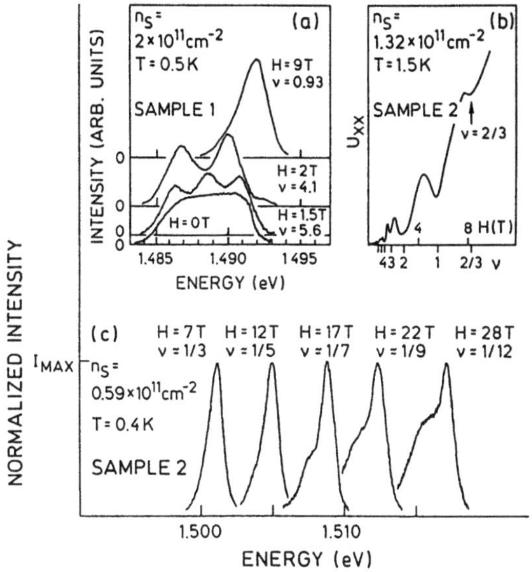

Figure 1: The luminescence spectra measured (a) at 0.5K in sample 1 and (c) at 0.4K in sample 2 at various magnetic fields. (b) The transport Shubnikov-de Haas oscillations, measured at 1.5K in sample 2 under continuous illumination.

In Fig.1 we show luminescence spectra from two samples at various magnetic fields. The whole 2D energy spectrum below the Fermi energy, as observed at zero field, splits in transverse low magnetic fields into Landau levels (Fig.1a). From the relative intensities of the different Landau levels and their dependence on magnetic field we were able to evaluate n_S to an accuracy better than 2%. In a higher quality sample (sample 2), the luminescence linewidth at low fields is approximately two times smaller than that of sample 1; its high field behaviour is shown in Fig.1c. We will discuss the shape of the line measured at high field from sample 2 further below. Shubnikov-de Haas oscillations measured in sample 2 with a Hall-bar geometry are also shown in Fig.1b. They were taken under continuous illumination hence the background of strong positive magneto-resistance due to parallel bulk photoconductivity. The high quality of this sample is here confirmed by the observation of a minimum at $\nu=2/3$ at a rather high temperature (1.5K) and the rather low magnetic field of 8T (compare with Ref.17).

The plot of the peak energies as a function of magnetic field (Fig.2a) appears linear above $\nu=2$. However below, distinct steps are apparent in the magnetic field dependence of the luminescence line position. The largest occurs between $\nu=2$ and $\nu=1$, which is the result of enhancement of the electronic spin splitting (the so-called g-factor enhancement[18]). It can be seen from this figure that as the magnetic field is further increased additional abrupt changes in the spectral position of the line are observed at low temperatures in the vicinity of $\nu=4/5$, 2/3, 3/5, 2/5 and 1/3. The size of these steps are small compared to the cyclotron energy (which mainly determines the dependence of the spectral position on H). Therefore in Fig.2b

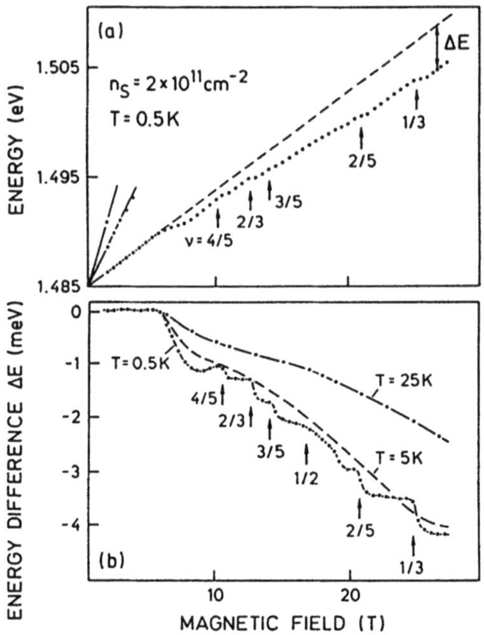

Figure 2: (a) The dependence of the luminescence line peak position on magnetic field measured at 0.5K for sample 1. (b) ΔE(H) measured for different temperatures.

we have replotted the same data but this time as the energy shift (ΔE) from the line drawn through the low-field data. Here the discontinuities at ν=4/5, 2/3, 3/5, 2/5 and 1/3 can be seen much more clearly. Note that in the vicinity of ν=p/q we also observe a small (≈10%) broadening of the luminescence line. In addition, there appears to be a broad feature in ΔE(H) around ν=1/2. In this sample on raising the temperature to 5K all the discontinuities associated with odd denominator ν disappear except that due to ν=1/3 which is severely weakened at 5K and disappears completely at higher temperatures. Also to be noted is the persistence of the feature around ν=1/2 at 5K and even at 25K. The low sensitivity to temperature of this feature, which (contrary to those at odd ν) spreads over a wide region of ν around 1/2, is analogous to that reported on transport measurements[17]. The general dependence of ΔE on H in Fig.2b which becomes less negative with rising temperature is due to magnetic field induced localization and can be used as a measure of the amount of disorder in a sample (instead of μ).

From a series of samples and a range of concentrations we have detected anomalies in the energy position at ν=2/3, 1/3, 4/5, 3/5, 2/5, 1/5, 1/7 and 1/9 as depicted in Fig.3a. In Fig.3b we compare the dependence of ΔE on H measured at three different temperatures in the sample 2. At 0.4K we observe discontinuities at all ν=1/q down to 1/9 but practically no change in position at ν=1/11. By 1.2K the steps in ΔE(H) at ν=1/7 and 1/9 have disappeared while those associated with fractions with larger gaps remain. It is important to note that the actual position of the line corresponding to the data points for 1.2K (broken line in Fig.3b) cuts through the center

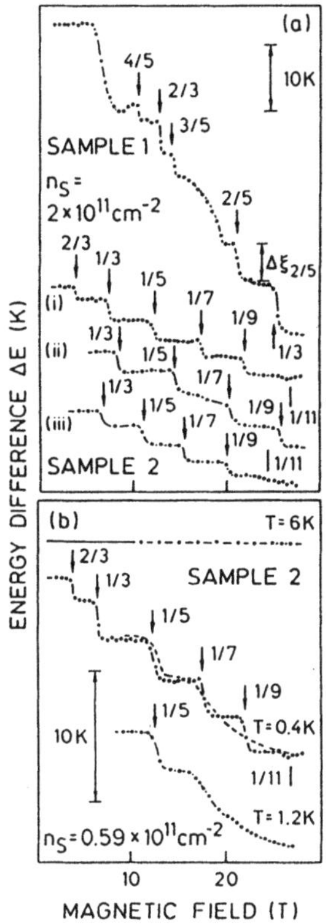

Figure 3: (a) ΔE(H) measured for samples 1 and 2 at different concentrations:
(i) $0.59\cdot10^{11}cm^{-2}$,
(ii) $0.70\cdot10^{11}cm^{-2}$,
(iii) $0.54\cdot10^{11}cm^{-2}$.
(b) ΔE(H) measured for sample 2 at different temperatures. The broken line depicts the actual position of the data measured at 1.2K relative to that at 0.4K.

of the steps observed at 0.4K. At 6K the slope of the data at high field does not differ from that at low field. This indicates that the degree of localization in sample 2 is much less than in sample 1.

According to the microscopic theory proposed by Laughlin[2] the ground state energy of the 2D electron system E(N) (N is the total number of electrons) at T=0 demonstrates cusps at $N=N_F$ corresponding to $\nu=p/q$. This results in a discontinuous change in the chemical potential $\xi=\Delta E/\Delta N$ at this point such that $\Delta\xi_{p/q}=(\Delta E/\Delta N)_{N_F}=q\ \Delta_G$, where Δ_G is the FQHE gap corresponding to $\nu=p/q$. Magneto-optics allows determination of the values $\Delta\xi_{p/q}$ since the spectral position of the luminescence line is directly influenced by changes of the chemical potential[13]. Taking the size of the step as $\Delta\xi_{p/q}$, as depicted in Fig.3a, we have plotted in Fig.4 the so-determined Δ_G. These gaps depend strongly on sample quality [9,10] and we therefore show those for sample 1 and sample 2 in figures 4a and 4b separately. The gap values and their dependence on H obtained from sample 1 are comparable to that measured[9] by activated Shubnikov-de Haas on samples with mobility, $\mu=0.4\text{-}1\times10^6 cm^2/V\ s$. On the other hand, the gaps measured in sample 2 correspond to the magnetotransport values obtained from extremely high

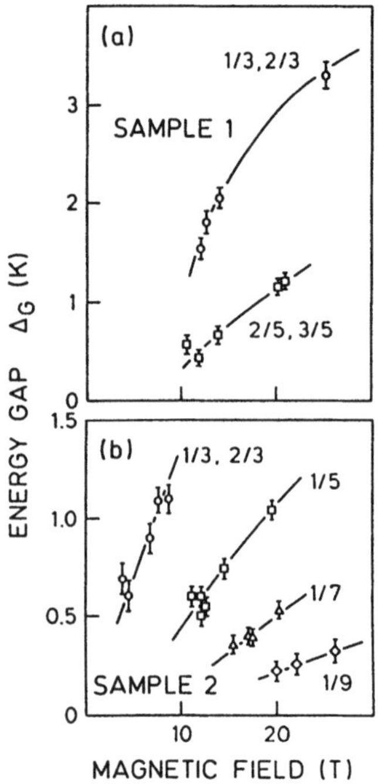

Figure 4: The FQHE gaps vs. magnetic field measured for different fractional values of ν for (a) sample 1 and (b) sample 2.

quality samples, μ=0.1-1$\times10^7$ cm^2/V s [10-12]. Note the existence of a finite threshold magnetic field due to residual disorder which strongly depends on the sample quality and is very small for sample 2.

We have studied by magneto-optics the FQHE in the low density, high magnetic field regimes. We observe discontinuous behaviour in the spectral position of the luminescence line at ν=2/3, 1/3, 4/5, 3/5, 2/5, 1/5, 1/7, 1/9 from which we have determined the energy gap values for the FQHE states ν=p/q with q=3,5 and for the first time for ν=1/7 and 1/9. The gaps obtained for q=3,5 are in agreement with those published from activated transport measurements[5-8] for similar quality samples.

3. Wigner crystallization

Up to now magneto-transport and radio-frequency spectroscopic techniques[19] have been used to investigate the phase transition from a liquid to the Wigner solid state. From transport experiments, it is now well established that at ν=1/3 and 1/5 the ground state of the system is still an incompressible Fermi liquid, giving rise to a quantized Hall resistance and the corresponding minimum in the diagonal resistivity[1,10]. And even at 1/7 evidence of a fractional state has been reported[8]. The radio-frequency absorption measurements[19], on the other hand, suggest that a value of the critical filling factor for Wigner crystallization is ν_{cr}=0.23. Here we report the observation of

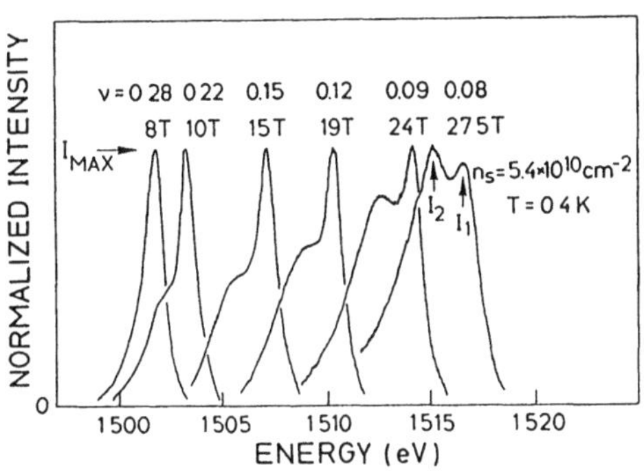

Figure 5: The luminescence spectra measured at 0.6K in sample 2 at various magnetic fields.

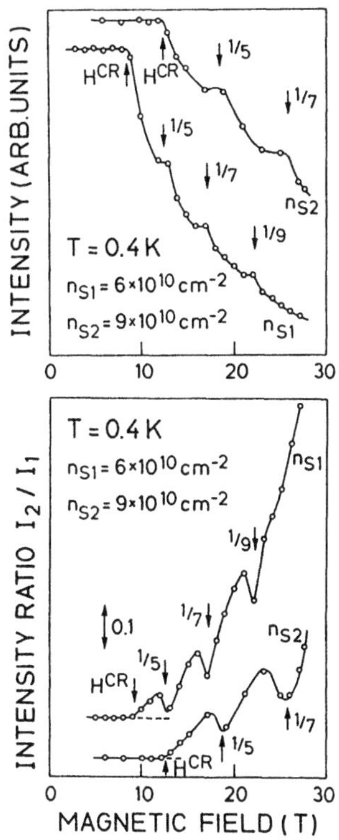

Figure 6: (a) The dependence of intensity ratio I_2/I_1 on magnetic field measured for two different concentrations. (b) The dependence of the luminescence integrated intensity on magnetic field measured for two different concentrations.

a new luminescence line appearing at $\nu<\nu_{cr}$ and below a critical temperature (T_{cr} depending on ν) which weakens in relative intensity at ν=1/5, 1/7 and 1/9 and which we associate with the formation of a Wigner solid.

In Fig.5 we show luminescence spectra recorded at $n_S=5.4\cdot10^{10}cm^{-2}$ at various magnetic fields. Analogous spectra measured at $n_S=6.0\cdot10^{10}cm^{-2}$ are shown in Fig.1c. It can be seen that in both cases above a certain magnetic field (H^{CR}) an additional line (I_2) appears, shifted by ≃1.4 meV to lower energy, which grows as the field is increased until at $\nu<1/11$ it dominates the spectrum. Its appearance is accompanied by an abrupt decrease of the integrated luminescence signal. In Fig.6 both the intensity ratio I_2/I_1 (Fig.6a) and the total integrated intensity (Fig.6b) are plotted as a function of magnetic field for two different concentrations. It can be seen from this figure that for a given concentration the value of H^{cr}, obtained from both plots is practically identical. It is our belief that at exactly $\nu=1/5$, 1/7 and 1/9, when the 2D electrons condense into incompressible Fermi liquid, the intensity of the new line actually drops to zero. That an increase in the integrated intensity is observed at this point is due to a simultaneous enhancement in intensity of I_1 line.

The most important property of the new line is that it can be characterized by two critical parameters - ν_{cr} and T_{cr}. The first statement follows from the experimentally obtained fact that H^{CR} linearly depends on concentration of 2D-electrons. From the slope of the dependence of H^{CR} on n_S, we obtain $\nu_{cr}=0.28\mp0.02$. We also found that the dependence of the ratio I_2/I_1 on temperature demonstrates a threshold behaviour at $T=T_{cr}$. For example, at H=26T ($\nu=0.09$) the line I_2 disappears from the luminescence spectrum abruptly at $T_{cr}=1.4K$.

We associate the appearance of the additional line in the emission spectra and the accompanying abrupt decrease in the integrated intensity with the formation of a (most probably polycrystalline) pinned Wigner solid. In accordance with this interpretation, the two lines I_1 and I_2 correspond to radiative recombination of 2D electrons from the liquid (at $\nu=1/5$, 1/7 and 1/9 the liquid becomes incompressible) and solid phase, respectively. That I_2 lies lower in energy than I_1 agrees with the expected lower energy of the solid state compared to that of the liquid one. We conclude from the tendency of the new line to vanish from the luminescence spectra at $\nu=1/5$, 1/7 and 1/9 and from the simultaneous enhancement in I_1 intensity that at these fractional values of ν the ground state of the system is still an incompressible Fermi liquid. That I_2 does not completely disappear from the luminescence spectrum at these values of ν may be due to a small amount of concentration inhomogeneity which is unavoidable in a real system. In such case, local values of ν in different parts of the sample will not be exactly identical.

The extent of the wavefunction of localized electron in 2D plane is defined by the magnetic length, $l_H=(\hbar/eH)^{1/2}$. In our case the holes participating in recombination processes are bound to acceptors and are thereby localized in all the three dimensions. Thus as the magnetic field is increased and the electrons become more localized in the plane parallel to the interface, the probability of an electron finding itself in the vicinity of a hole decreases, causing a reduction in the luminescence intensity. In less high quality samples (in which we did not observe the above described features that we associate with Wigner crystallization) in which the Landau level width, determined from the luminescence linewidth, was larger than 3meV, the luminescence intensity started decreasing already just below $\nu=1$ and the reduction itself was gentler. At higher temperatures above T_{cr} where the electrons become mobile no reduction in luminescence intensity was observed at all. We therefore believe that strong magnetically induced localization is responsible for sudden drop in integrated luminescence intensity below $\nu=\nu_{cr}$. It is most important however to demonstrate that this localization has an intrinsic origin rather than that of disorder. With this aim in mind we will show that ν_{cr} is totally independent on the parameters that give a measure of the disorder in the system. Quantitatively a disorder can be characterized by the Landau level width Γ (for example at $\nu=1$) and also by the parameter ΔE (measured at fixed H) obtained from the dependence on magnetic field of the line spectral position (see section 2 and Fig.2). The dependence of ν_{cr} on ΔE and Γ measured for different n_S shows no influence of these disorder related parameters on critical value of ν. This thus confirms the intrinsic nature of the observed localization-like phenomena, thereby ruling out localization on random potential as the cause.

4. Acknowledgements

We gratefully acknowledge the support of P.Wyder and of G.Arnand, J.Dumas, A.Fischer, M.Hauser, P.Sala, J.-C.Vallier and C.Warth. One of us (I.V.K.) would like to thank Alexander von Humboldt Stiftung for financial support.

5. References:

1. D.C.Tsui, H.L.Störmer, and A.C.Gossard, *Phys.Rev.Lett.* **48**, 1559 (1982)
2. R.B.Laughlin, *Phys.Rev.Lett.* **50**, 1395 (1983)
3. For a recent review see: T.Chakraborty and P.Pietiläinen, *The Fractional Quantum Hall Effect* (Springer-Verlag, New York, 1988)
4. P.K.Lam and S.M.Girvin, *Rhys. Rev.* **B30**, 473 (1984)
5. D.Levesque, J.J.Weiss and A.H.MacDonald, *Phys. Rev.* **B30**, 1056 (1984)
6. S.T.Chui, T.M.Hakim and K.B.Ma, *Phys. Rev.* **B33**, 7110 (1986)
7. S.Kivelson, C.Kallin, D.P.Arovas and J.P.Schrieffer, *Phys. Rev.* **B36**, 1620 (1987)
8. V.J.Goldman, M.Shayegan and D.C.Tsui, *Phys.Rev.Lett.* **61**, 881 (1988)
9. G.S.Boebinger, H.L.Störmer, D.C.Tsui, A.M.Chang, J.C.Hwang, A.Y.Cho, C.W.Tu, and G.Weimann, *Phys.Rev.* **B36**, 7919 (1987)
10. R.L.Willett, H.L.Störmer, D.C.Tsui, A.C.Gossard and J.H.English, *Phys. Rev.* **B37**, 8476 (1988)
11. J.R.Mallett, R.G.Clark, R.J.Nicholas, R.Willett, J.J.Harris and C.T.Foxon, *Phys.Rev.* **B38**, 2200 (1988); R.G.Clark, J.R.Mallett, S.R.Haynes, J.J.Harris and C.T.Foxon *Phys. Rev. Lett.* **60**, 1747 (1988)
12. T.Sajoto, Y.W.Suen, L.W.Engel, M.B.Santos and M.Shayegan, *Phys. Rev.* **B41**, (15th April 1990)
13. I.V.Kukushkin, V.B.Timofeev, *Pis'ma Zh. Eksp. Teor. Fiz.* **44**, 179 (1986), [*JETP Lett.* **44**, 228 (1986)]
14. D.Heiman, B.B.Goldberg, A.Pinczuk, C.W.Tu, A.C.Gossard and J.H.English, *Phys.Rev.Lett.* **61**, 605 (1988)
15. I.V.Kukushkin, K.von Klitzing, K.Ploog and V.B.Timofeev, *Phys.Rev.* **B40**, 7788 (1989)
16. I.V.Kukushkin, K.von Klitzing, K.Ploog, V.E.Kirpichev and B.N.Shepel, *Phys. Rev.* **B40**, 4179 (1989)

17. H.W.Jiang, H.L.Störmer, D.C.Tsui, L.N.Pfeiffer and K.W.West, *Phys Rev.* **B40**, 12013 (1989)
18. T.Ando, A.B.Fowler and F.Stern, *Rev. Mod. Phys.* **54**, 437 (1982)
19. E.Y.Andrei, G.Deville, D.C.Glattli, F.I.B.Williams, E.Paris and B.Etienne *Phys. Rev. Lett.* **60**, 2765 (1988)

On Some Peculiarities of Light Absorption in 2-d Wigner Crystals in High Magnetic Fields

S.V. Iordanskii and B.A. Muzykantskii

European Branch of L.D. Landau Institute for Theoretical Physics
at ISI Foundation, Villa Gualino, Viale S. Severo 65,
I-10133 Torino, Italy

Abstract. It is shown that light absorption in a 2D Wigner crystal is possible not only with creation of magnetoexcitons with zero momentum but also with finite momentum in the reciprocal Wigner lattice. The magnitude of such an absorption is small due to Kohn's theorem and is estimated using nonparabolic terms in the electron dispersion, and small terms describing the interaction with phonons and impurities. The same process was estimated for inelastic light scattering, where the effect is large.

Light absorption by 2D electrons in strong magnetic fields has been investigated rather thoroughly. The most complete and clear physical picture was developed for fully filled Landau levels (Lls) by using the concept of the magnetoexciton (ME), with conserved momentum $\vec{p}$ and energy $\epsilon(\vec{p})$ [1-3]. In this work we consider the problem of light absorption as well as light scattering due to formation of MEs in 2D Wigner crystals (Wcs) at large lattice spacing. This problem may be of interest because there is some experimental evidence for the existence of such a crystal at small Ll filling [4].

Contrary to the work of Dyckman [5] we consider not only the vicinity of $\vec{p} = 0$ but the whole dispersion curve of the ME, which splits into seperate bands due to the periodicity of the Wigner crystal.

To be specific we consider a hole-Wc with a small density of holes in almost filled Lls. Assuming a cyclotron frequency much larger than the average Coulomb interaction of electrons it is easy to find the equation of motion for the ME creation operator

$$A^+(\vec{p}) = \frac{1}{\sqrt{N_e}} \sum_k e^{-ip_x k} a^+_{k+\frac{p_y}{2},n+1} a_{k-\frac{p_y}{2},n} : \tag{1}$$
$$i\frac{\partial A^+(\vec{p})}{\partial t} = \epsilon(\vec{p}) A^+(\vec{p}) - \frac{1}{2} \int \frac{d^2q}{(2\pi)^2} V_{eff}(\vec{p},\vec{q}) A^+(\vec{p}-\vec{q}) \tilde{\rho}(\vec{q}) .$$

Springer Series in Solid-State Sciences, Vol. 101
High Magnetic Fields in Semiconductor Physics III Editor: G. Landwehr

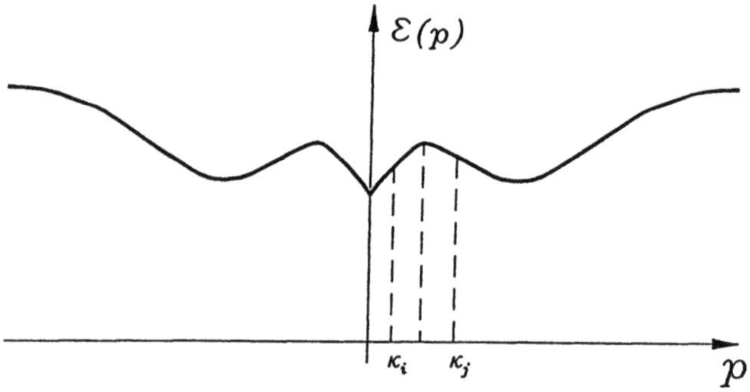

Figure 1: Magnetoexciton dispersion curve for $n = 0$ filled Ll

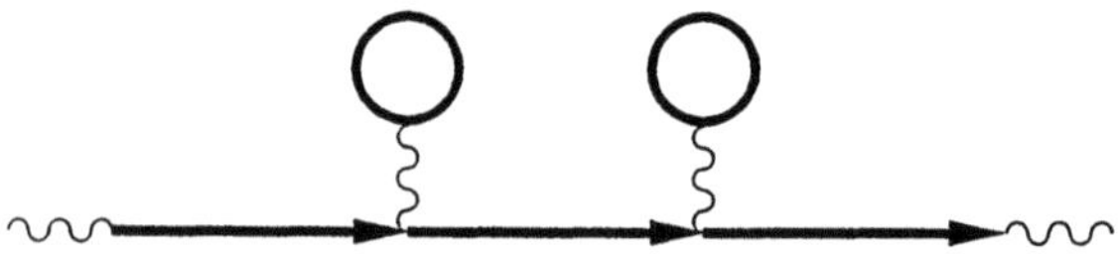

Figure 2: Diagram for light absoption on satellites ⟹-magnetoexciton propagator, ○-hole density

Here $a^{+}_{k,n}, a_{k,n}$ are electronic operators for states on Ll n in the Landau gauge, N_e is the number of 2D electrons, $\tilde{\rho}$ is the operator with the hole density, $V_{eff}(\vec{p}, \vec{q})$ is a rather cumbersome expression for the effective interaction of the ME with hole density defined by Fourier transformation of the e - e interaction and wave functions of Ll states. For an ideal Wc the average hole density has δ-function singularities at $\vec{K}_l$ on the reciprocal lattice $\langle\tilde{\rho}(\vec{q})\rangle = \sum_l \delta(\vec{q}-\vec{K}_l)\,\rho(\vec{K}_l)$. Because of the small density of holes we can use the simple mean-field approximation to find corrections to the ME dispersion curve from equation (1). The main effect consists of splitting the ME spectrum into seperate bands. Essentially we must draw the original ME dispersion curve in the reduced momentum space (first Brillouin zone) with a small gap on the boundary as shown in Fig. 1.

The process of light absorption can be described as the creation of a virtual ME with zero momentum and its scattering by the inhomogeneous periodical density of the Wc with the creation of a ME with momentum $\vec{K}_l$ on the reciprocal Wigner lattice and energy $\epsilon(\vec{K}_l)$. The corresponding diagram is shown in Fig. 2.

We see that in a Wc it is possible to absorb light at frequencies $\epsilon(\vec{K}_l)$ with $\vec{K} \neq 0$, contrary to the case of fully filled Lls, because there is a possibility of transferring momentum to the Wc. The main difficulty here is connected with Kohn's theorem, which states that all interaction effects cancel for zero momentum MEs. Kohn's theorem is valid for a parabolic electron dispersion and absence of any external fields acting on the electrons. Indeed it is easy to show in that case that $V_{eff}(p = 0, q) = 0$.

We considered the ac conductivity near the satellites $\hbar\omega_c = \epsilon(\vec{k}_i)$ due to a) nonparabolic corrections, b) external crystalline phonons, c) impurities. The imaginary part of the ME Green function entering as a factor into ac conductivity (see e.g.[2]) can be estimated as

$$\delta G(\omega, p=0) = \left(\frac{\Delta\epsilon}{\omega_c}\right)^2 \sum \left(\frac{e^2}{\chi} p\right)^2 \left|\frac{\rho(\vec{k}_i)}{\omega-\omega_c}\right|^2 ImG_0(\omega, \vec{k}_i ..)$$

where $G_0(\omega, \vec{k}) = [\omega - \epsilon(\vec{k}) + i/\tau]^{-1}$ is the ME Green function, τ the lifetime of the ME, χ the dielectric constant, and l_h the magnetic length.

In the case a) $\Delta\epsilon = \Delta\omega_c$ is the shift of cyclotron frequency due to nonparabolic corrections, in the case b) $\Delta\epsilon = \Delta\omega_c(\omega_c/\omega_0)$, where ω_0 is the optical phonon frequency, and $\Delta\omega_c$ is the shift of the cyclotron frequency due to interactions with phonons. The estimate for process c) will be given elsewhere.

All these effects are rather small and the observation of the satellite lines in the absorption will be difficult. But for inelastic light scattering when the light frequency is near the band gap there is no restriction due to Kohn's theorem. For an interband ME (IME) consisting of a hole in some Ll for the valence band and an electron in the same Ll for the conduction band it is easy to find an equation of motion analogous to (1), describing interaction with the Wc lattice and creation of a ME for almost filled Lls in the conduction band. The corresponding process of inelastic light scattering is shown in Fig.3. The inelastic light scattering rate will be proportional to

$$T_{\omega\to\omega'} \sim \int \frac{S(\vec{k}, \omega)\, ImG_0(\omega-\omega'-\Omega, -\vec{k})}{|(\omega-E(0))(\omega-E(\vec{k})-\Omega)(\omega'-E(0))|^2} |V_{eff}(\vec{k})|^4 \frac{d\vec{k}}{(2\pi)^2}\frac{d\Omega}{2\pi}, \quad (2)$$

where $V_{eff}(\vec{k})$ is the effective interaction of the order of the lattice Coulomb interaction, $E(\vec{k})$ is the IME dispersion, $S(\vec{k}, \omega)$ is the dynamic form factor of the 2D electron system. In a Wc we have $S(\vec{k}, \omega) = S(\vec{k})\,\delta(\Omega)$ so there

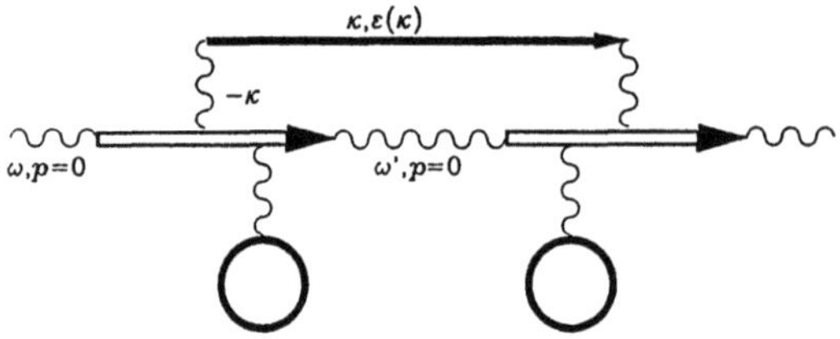

Figure 3: Diagram for inelastic light scattering ⟹-interband magnetoexciton propagator

is a momentum transfer to the Wc without energy transfer and the estimate for the scattering rate will be according to (2):

$$T_{\omega\text{-}\omega'} = \sum_{\vec{K}_1} |\tilde{\rho}(\vec{K}_1)|^2 \left|\frac{e^2}{\chi_0 l_h (\omega - E(\vec{k}))}\right|^2 |M_0|^2 ImG_0(\omega - \omega'.\vec{k})$$

where M_0 is a matrix element for a transition with a light frequency change $\omega - \omega' = \epsilon(0) = \omega_c$. We see that there is no small factor preventing observation of satellites in this case, except a small hole density.

The dynamic form factor $S(\vec{k}, \omega)$ entering (2) is not known in general. Experimental estimates give a small value of the gap in fractional quantum Hall states compared to $e^2/\chi_0 l_h$, that is small compared to the energy shift due to the formation of MEs with nonzero momentum. Therefore in the transition probability $T_{\omega\to\omega'}$, the frequency shift will probably deviate much from $\epsilon(\vec{k})$. The scattering will be concentrated near the "roton" gap shift even in liquid states. These considerations can explain probably the results of experiments [6] of light scattering in very pure GaAs heterojunctions.

References

[1] Yu.A. Bychkov, S.V. Iordanskii, G.M. Eliashberg: Pis'ma Zh. Eksp. Teor. Fis. **33**, 143 (1981); JETP Lett. **33**, 143 (1981)

[2] C. Kallin, B.I. Halperin: Phys. Rev. B**30**, 5655 (1984)

[3] C. Kallin, B.I. Halperin: Phys. Rev. B**31**, 3635 (1985)

[4] E.Y. Andrei, G. Deville, D.G. Glattli, F.I.B. Williams, E. Paris, B. Etienne: Phys. Rev. Lett. **60**, 2765 (1988)

[5] M. Dyckmann: J. Phys. C. **15**, 7379 (1982)

[6] A. Pinczuk, J.P. Valladares, D. Heiman, A.C. Gossard, J.H. English, C.W. Tu, L. Pfeiffer, K. West: Phys. Rev. Lett. **61**, 2701 (1988)

Optical Detection of Integer and Fractional QHE in GaAs: Extension to the Electron Solid

R.G. Clark[1], *R.A. Ford*[1], *S.R. Haynes*[1], *J.F. Ryan*[1], *A.J. Turberfield*[1], *P.A. Wright*[1], *C.T. Foxon*[2], *and J.J. Harris*[2]

[1]Clarendon Laboratory, University of Oxford, Parks Road, Oxford OX1 3PU, UK

[2]Philips Research Laboratories, Redhill RH1 5HA, UK

1. Introduction

Ten years on from the discovery of the integer QHE (IQHE) [1], and the subsequent discovery of condensation of electrons into an incompressible liquid - the fractional QHE (FQHE) [2], experiments are still largely restricted to electrical transport studies. Whilst there is a report of an optical measurement of the FQHE in a Si structure at 1.5K [3], optical experiments that probe the QHE in GaAs have remained a major challenge [4,5]. An important advance was made by the observation of anomalies in the energy of photoluminescence near $\nu = 2/3$ and $\nu = 1$ from a GaAs multiple quantum well at 0.4K [6]. In this paper we report a definitive detection of both the integer and fractional QHE in GaAs, using intrinsic bandgap photoluminescence, by a study of integer states from $\nu = 1$ to 10 and of the $\nu = 2/3$ hierarchy out to the 5/9 daughter state, in an ultra-high mobility single heterojunction at 120mK [7]. At higher fields the photoluminescence spectra are mapped out at 80mK through FQHE states of the $\nu = 1/3$ and 1/5 hierarchies. Of particular interest is a new photoluminescence peak that becomes clearly resolved in the region $\nu < 1/5$. The appearance of this peak at $\nu \simeq 1/5$ correlates with the rapid onset of a substantial non-linear out-of-phase conduction that has recently been shown to arise from threshold behaviour that may be associated with crystallisation [8, 9]. It also correlates in filling factor with the onset of a resonant radio-frequency absorption in this sample which maps out a phase boundary in the B-T plane [8].

2. Experiment

The GaAs/$Ga_{.68}Al_{.32}As$ heterojunction (1600Å spacer) is shown schematically in the Fig. 1(a) inset. The GaAs layer (5000Å) is very weakly n-type (~ 10^{14} cm^{-3}) producing flat bands that give sharp luminescence lines. The sample was mounted in the dilute phase of a dilution refrigerator at 25mK. Optical fibres delivered laser light at 740nm (i.e. below the GaAlAs bandgap) with an intensity < 10^{-4} Wcm^{-2}, and transmitted luminescence to a spectrograph/CCD with a spectral resolution of 0.05meV. In the high-field optical experiments a laser operating at 785nm was used. A sample temperature of ~ 40mK was achieved with the laser off; this increased to ~ 120mK under continuous illumination in the low-field ($\nu > 1/2$) experiment,

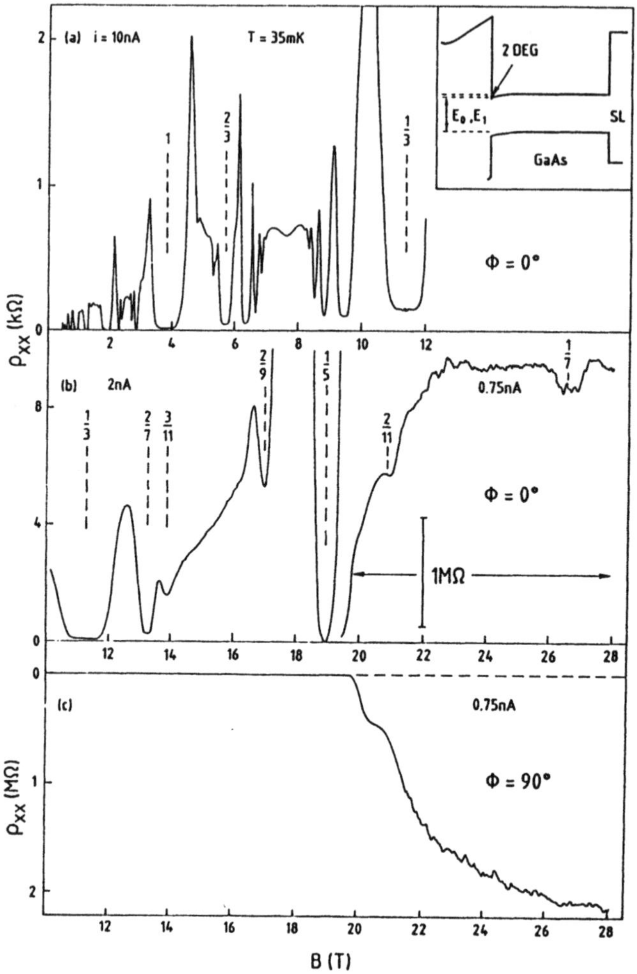

Fig 1 Transport characterisation of the modulation-doped GaAs/GaAlAs heterojunction shown in the inset: $n_s = 9.2 \times 10^{10} cm^{-2}, \mu = 9 \times 10^6 cm^2 V^{-1} s^{-1}$. A superlattice buffer (SL) has been included to prevent carbon acceptors reaching the 2DEG. $\Phi = 0°$ and 90° refer to the in-phase and out-of-phase signals respectively. T = 35mK. The data were measured in the Grenoble hybrid magnet, in collaboration with CEN, Saclay [9]

and to ~ 80mK in the high-field ($\nu < 1/2$) experiment, as determined by comparison with transport measurements at elevated refrigerator temperatures with the laser off.

The transport characterisation of a sample from the same wafer, at 35mK without laser illumination, is shown in Fig. 1. Sharp FQHE ρ_{xx} structure associated with the $\nu = 2/3$, 1/3 hierarchies (out to 5/9, 5/11 respectively) is resolved in Fig. 1(a). In Fig. 1(b), a strong $\nu = 1/5$ state is observed, together with the 2/9 daughter state. Significantly, there is a rapid onset of an out-of-phase component (Fig. 1(c)) and a substantial increase in resistance close to the high-field side of the $\nu = 1/5$ minimum. A weak ρ_{xx} minimum forms at $\nu = 2/11$ (daughter state of the 1/5 parent) and a broader, weak minimum is also resolved at $\nu = 1/7$. The ρ_{xx} structure at 2/11 and 1/7 is not robust to even small temperature increase and is absent at T > 50mK.

3. Integer Quantum Hall Effect

Luminescence spectra obtained at 120mK show two lines, E_0 and E_1, which we assign to recombination of electrons in n = 0 and n = 1 subbands of the confining potential (see Fig. 1 inset). The latter is more intense due to the greater electron wavefunction penetration into the GaAs layer where there is overlap with the photoexcited hole wavefunctions. On increasing the magnetic field, both lines shift to higher energy as expected for $l_e = 0 \rightarrow l_h = 0$ inter-Landau level transitions; a non-linear variation of energy with field at low fields suggests that the states have excitonic character. More dramatic, however, is a remarkable field-induced modulation of the E_0 and E_1 intensities.

Peak intensities of E_0 and E_1 are shown in Fig. 2(a), together with ρ_{xx} (Fig. 2(b)) measured simultaneously. The ρ_{xx} data show well-developed minima at integer filling $\nu = 1$ to 10, and at fractional states $\nu = 4/3$ and 5/3. The luminescence data reveal that the E_1 line has sharp intensity maxima at integer ν, especially at $\nu = 1$ where it undergoes a dramatic enhancement by ~ 35. In contrast, the E_0 intensity is weak and largely field-independent up to $\nu = 2$. For $\nu < 2$ the E_0 intensity first increases, shows a broad maximum, with a small but distinct dip at $\nu = 4/3$, and then decreases to a small value at $\nu = 1$.

The results of measurements made at higher temperatures are shown in Fig. 3. At 1.5K the sharp structure in E_1 has given way to oscillations, that are similar to magneto-exciton optical Shubnikov-de-Haas oscillations [10], and which become even more distinct at 4.2K. The E_0 signal shows broader oscillations, with minima at $\nu = 4$, 2 and 1, that are out of phase with E_1. At 4.2K the strong enhancement of E_1 and suppression of E_0 at $\nu = 1$ have disappeared. At 77K there is no evidence of the QHE.

The most important factor that determines the intensity of the E_0 line is probably the screening response of the electrons. An electron can scatter between states in a partly filled Landau level in the n = 0 subband to screen the potential of a photoexcited hole, increasing the electron-hole wavefunction overlap; for a filled level screening is suppressed. Since at low temperatures the $l_e = 0$ level is full for $\nu \geq 2$, the E_0 signal is expected to be weak and approximately field independent in this range, as observed. For $1 < \nu < 2$ the $l_e = 0$ level is partly filled and screening becomes active, consistent with the observed increase in E_0 intensity. At $\nu = 1$ screening is strongly suppressed and the E_0 intensity falls very close to zero. For $\nu < 1$ the E_0 intensity is expected to increase again, as screening increases; this is confirmed by the measurements described below, but there are important anomalies at $\nu = p/q$ fractional filling.

The enhanced radiative recombination intensity of E_1 at integer ν can arise both from increased occupancy of the E_1 subband and from enhanced optical transition rates. The former depends on the density of vacant n = 0

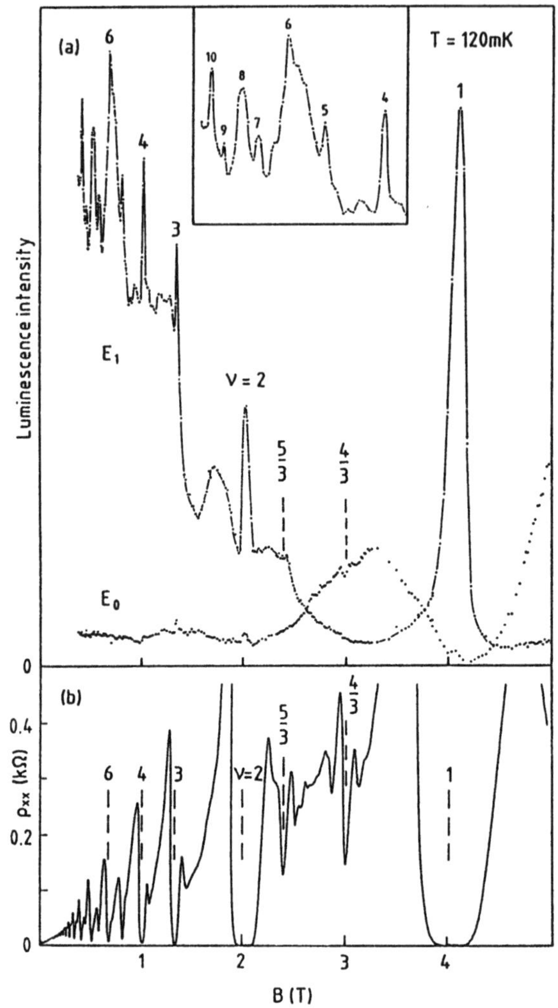

Fig. 2 (a) Peak intensities of the E_0 and E_1 luminescence lines at 120mK as a function of magnetic field for $\nu \lesssim 1$; the inset shows the region $\nu = 10$ to 4 on an expanded scale; 2(b) Simultaneous ρ_{xx} data.

states, and on the relaxation rate from E_1 to E_0, which can be reduced dramatically when thc Fermi level approaches the $l_e = 0$ level of E_1 due to the sharp cut-off of the electron-phonon interaction at low energy (≤1meV) [11]. The electron-hole recombination rate also depends on screening; in this case we expect the excitonic character of the E_1 state to be weakened due to screening by n = 0 electrons, so that the E_1 intensity will be enhanced at integer and fractional filling where screening is suppressed. The disappearance of E_1 intensity spikes at higher temperatures is consistent with this interpretation.

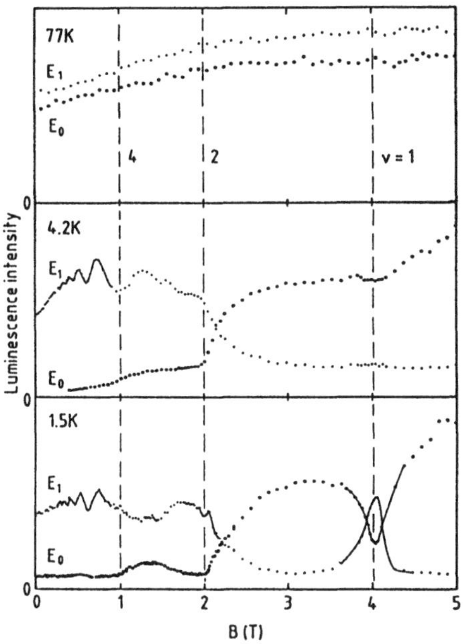

Fig. 3 Magnetic field dependence of the peak intensities of the E_0 and E_1 lines at 1.5K, 4.2K, and 77K.

4. Fractional Quantum Hall Effect

In Fig. 4 we compare E_0 and E_1 intensities with ρ_{xx} data measured simultaneously for $\nu < 1$. Also shown are ρ_{xy} measurements of a Hall bar sample from the same wafer in the absence of laser illumination. The ρ_{xx} data show well-defined FQHE states in the $\nu = 2/3$ hierarchy: the resolution and structure out to $\nu = 5/9$ are mirrored in remarkable detail in the E_0 intensity. These results provide *definitive* evidence for optical detection of the FQHE. Whereas E_0 shows intensity minima at these fractions, E_1 develops intensity maxima (as for the IQHE). The strengths of the E_0 minima and E_1 maxima decrease as the filling factor is swept through higher-order fractions in the $\nu = 2/3$ hierarchy, similar to ρ_{xx}. The optical and transport structure are not in precise field alignment (see dashed lines in Fig. 4): E_1 maxima occur at fields above $\nu = p/q$ filling, whereas E_0 minima occur at lower fields. This is suggestive of localisation effects. In fact, the maxima and minima correlate well with the high- and low-field extents of the ρ_{xy} plateaux.

The energy of E_0 also changes dramatically as the field is varied. Fig. 5 shows this behaviour for the $\nu = 2/3$ and $\nu = 1$ regions. The effect around $\nu = 2/3$, shown in the lower section of the figure, is striking. At B = 6.25T ($\nu = 0.64$) a doublet is resolved, shown in the lower inset (position b) with a splitting of 0.16meV (1.8K). The lower component of the doublet lies on a line extrapolated from lower fields (arrows mark energies corresponding to this line). With decreasing field the higher-energy component becomes the more intense, and the relative strength of the lower component rapidly diminishes (inset spectra a - e). The peak of the luminescence remains above the extrapolated line for a range of fields spanning $\nu = 2/3$, before shifting down to

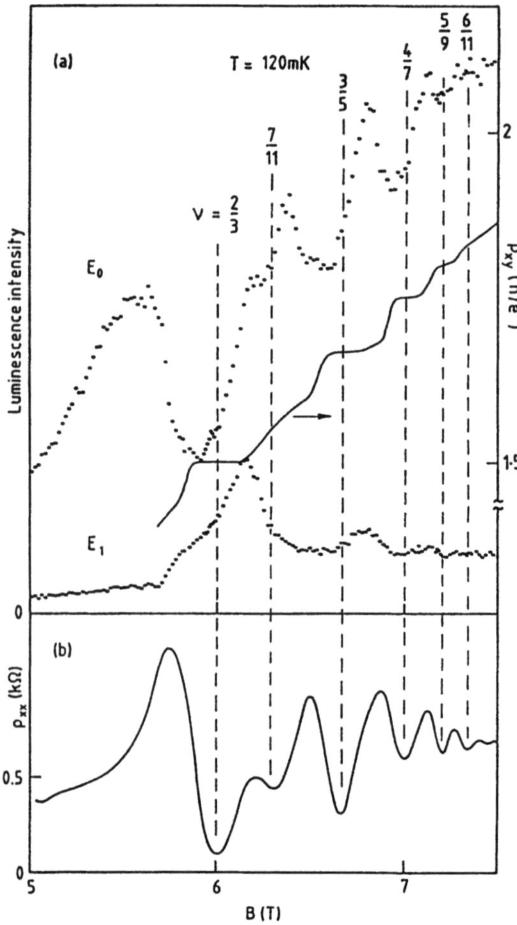

Fig. 4(a) Peak intensities of E_0 and E_1 at 120mK for $\nu < 1$; (b) Simultaneous ρ_{xx} data. Also shown in (a) is a ρ_{xy} trace from a Hall bar sample at a similar temperature in the absence of laser illumination.

the line at B = 5.75T (ν = 0.7), although a weaker upper component is still discernible at this (inset spectrum g) and lower fields. At $\nu = 1$ the luminescence peak is shifted upwards in energy by 0.2meV (2.4K) and an upper component becomes resolved again (inset spectra h-j), a further 0.2meV higher in energy. The doublet is not well resolved at lower fields, but the shift to higher energy is maintained. In contrast, E_1 is unshifted from an essentially linear B-dependent behaviour throughout this entire range of magnetic field.

In Fig. 6 we compare the E_0 energy and the E_0, E_1 intensities for fields around ν = 2/3 at three temperatures: 120mK, 430mK and 1.3K. Vertical lines mark the extreme limits of the ρ_{xx} ν = 2/3 minimum. This comparison provides convincing evidence that the optical anomalies are associated with the ν = 2/3 ground state: the intensity of E_0 decreases sharply at the low-field limit, whereas the intensity of E_1 decreases sharply at the high-field limit. At

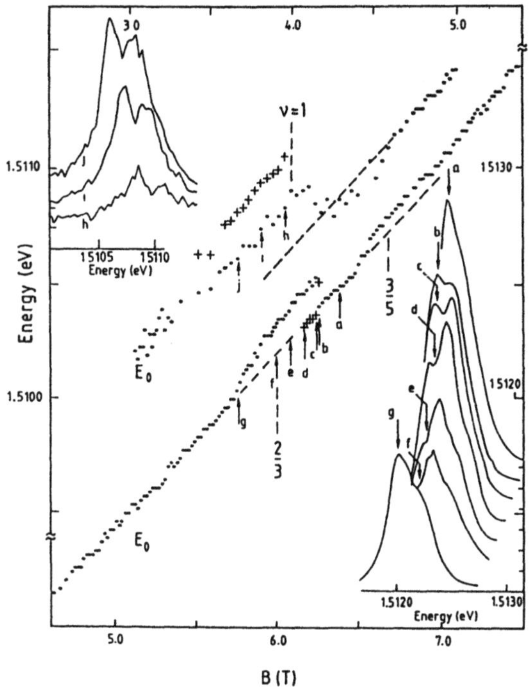

Fig. 5 Magnetic field dependence of the energy of the E_0 luminescence line. Bottom and right axes refer to lower section (ν = 2/3 hierarchy) top and left axes refer to its low-field continuation (to the ν = 1 region). Inset spectra show E_0 doublet structures resolved at ν = 2/3 and 1. Solid circles denote the more intense component.

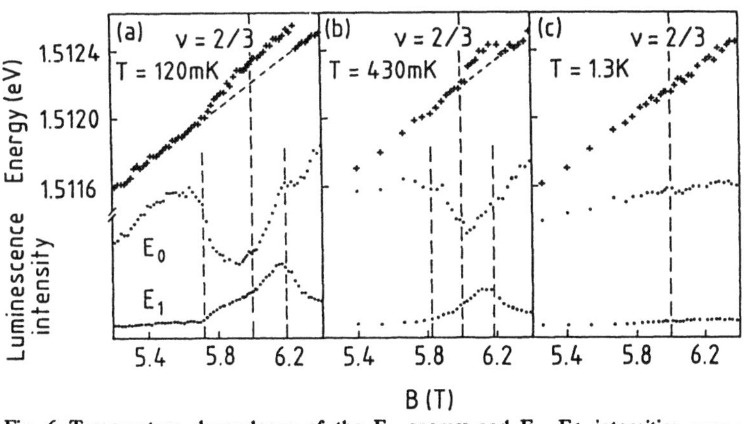

Fig. 6 Temperature dependence of the E_0 energy and E_0, E_1 intensities near ν = 2/3 at (a) 120mK, (b) 430mK, (c) 1.3K. The vertical broken lines adjacent to ν = 2/3 mark the extreme extent of the ν = 2/3 ρ_{xx} minimum.

120mK the correlation of the transport data, luminescence intensities and the E_0 splitting is excellent. At 430mK the field region in question has narrowed. At 1.3K all indications of the $\nu = 2/3$ state have vanished from both the optical and the transport data. It is significant, however, that the intensity and energy anomalies that occur at $\nu = 1$ are clearly resolved at T = 1.5K, and the splitting is still observed at 4.2K.

The E_0 line is a probe of the electronic ground state and so it may in principle give information about quasiparticle energies. Kukushkin and Timofeev [3] observed optical shifts of +0.34meV (4K) and -0.26meV (3K) above and below $\nu = 7/3$ (in filling factor) in a Si MOSFET; these shifts were interpreted to be $3\epsilon_e$ and $3\epsilon_h$ respectively, where ϵ_e, ϵ_h are the quasielectron, quasihole creation energies, which agreed quantitatively with measurements of the $\Delta_{7/3}$ transport gap. On the other hand, Heiman et al [6] observed a shift of 0.7meV (8.1K) in the luminescence spectrum of a GaAs quantum well at $\nu \approx 2/3$ which could not be identified as a quasiparticle energy. In the present case, our observed doublet separation of 0.16meV (1.8K) at $\nu = 2/3$ differs in magnitude from the GaAs quantum well measurement [6], and is smaller than the transport gap $\Delta_{2/3}$ = 4.6K in our sample. It is clear that the splitting observed at $\nu = 2/3$ is not simply related to the quasiparticle energies.

5. Extreme quantum limit

Under laser illumination (785nm) at high field, a field-dependent electron density depletion occurs, although well-resolved FQHE structure persists which is indicative of high mobility. The effect is shown in Fig. 7: both in- and out-of-phase ρ_{xx} data are shown in Fig. 7(a), and in-phase ρ_{xy} data are shown in Fig. 7(b). The FQHE ρ_{xx} and ρ_{xy} structure can be used to independently determine the field-dependent density, which is also plotted in Fig. 7(b). With the laser on, the density remains relatively constant up to 9T, above which there is an approximately linear decrease with field ~ 0.85×10^{10} cm^{-2} T^{-1}. The mechanism for this density depletion is not understood. Significantly, we again find a rapid onset of an out-of-phase ρ_{xx} component that occurs close to the high field side of the $\nu = 1/5$ minimum, as in Fig. 1.

Peak luminescence intensities in the region $\nu < 1/2$ are shown in Fig. 8, and energies in Fig. 9. Selected spectra are shown in Fig. 10. We first discuss the region $1/5 < \nu < 1/2$. Doublet structure in the E_0 line now occurs over a substantial range of fields, whereas the E_1 line is of low intensity, and disappears near $\nu = 1/3$. The weaker B-dependence of the E_1 energy is consistent with a decreasing $E_0 - E_1$ separation caused by reduction of the confining potential as the density decreases. There are several points to note:

(i) There is no obvious correlation of E_0 intensity minima in Fig. 8 with fractions of the $\nu = 1/3$ hierarchy, in contrast to the $\nu = 2/3$ hierarchy, and there are no E_1 intensity maxima. However, the E_0 lineshape is complex: two components are observed in the range $1/5 < \nu < 1/2$, but lack of resolution prevents detailed lineshape analysis. It is possible that there is underlying behaviour consistent with intensity minima in one component ocurring at fractional states in the hierarchy.

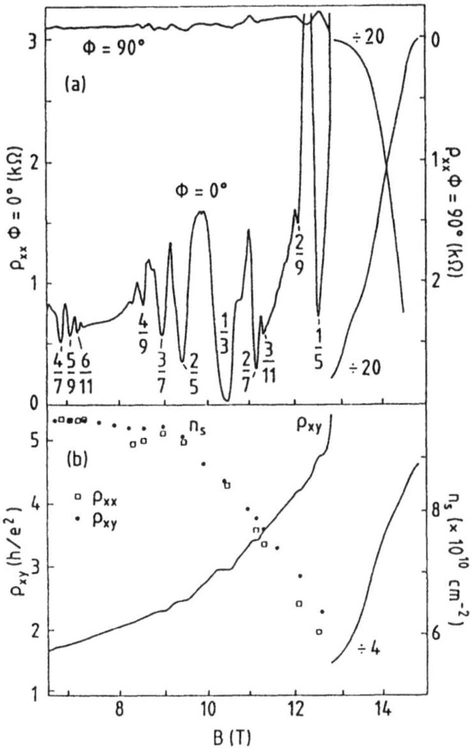

Fig. 7 High-field transport characterisation of the sample at T = 80mK, which shows density depletion in the presence of laser illumination. Φ = 0° and 90° refer to the in-phase and out-of-phase signals respectively. (a) ρ_{xx} data showing rapid onset of out-of-phase component at ν <1/5. (b) ρ_{xy} plateaux, and sheet density inferred from ρ_{xx} and ρ_{xy}.

(ii) The E_0 peak intensity is mainly associated with the upper component of the doublet for ν < 3/5 (Figs. 5 and 9) until in the region close to ν = 1/3 it switches to the lower component. This switch is seen clearly in the spectra shown in Fig. 10, and is opposite to the behaviour observed in the ν = 2/3 region. For ν < 2/7, the peak E_0 luminescence reverts to the lower component.
(iii) From Figs. 9 and 10 there is an indication that one component of the doublet dominates in the region of strong fractional states (absence of + data).
(iv) Due to the density depletion that occurs for B > 9T (Fig. 7(b)), we might expect small, gradual changes in the energies of the E_0, E_1 lines at high field; the E_0 energy, however, appears to depend linearly on field up to the highest field.

At $\nu \simeq$ 1/5 (B = 12.3T) a high-energy shoulder appears on the E_0 line, which develops into a separate, distinct peak at higher fields, as shown in Figs. 8 - 10 (see arrows in Fig. 10). Further structure appears beyond B = 14T (also indicated by arrows in Fig. 10). There are again several points to note:
(i) The appearance of this high-energy peak at $\nu \simeq$ 1/5 coincides with the onset of non-linear, out-of-phase conduction in the current source transport measurement (see Fig. 7), an effect which has been shown to arise from a threshold voltage in the I-V characteristics [8, 9]. (See also ref. 12 for other work on threshold phenomena). This coincidence was also observed on repeating the entire experiment in this sample at unsaturated density ($7 \times 10^{10} cm^{-2}$). The onset of out-of-phase conduction has been shown to

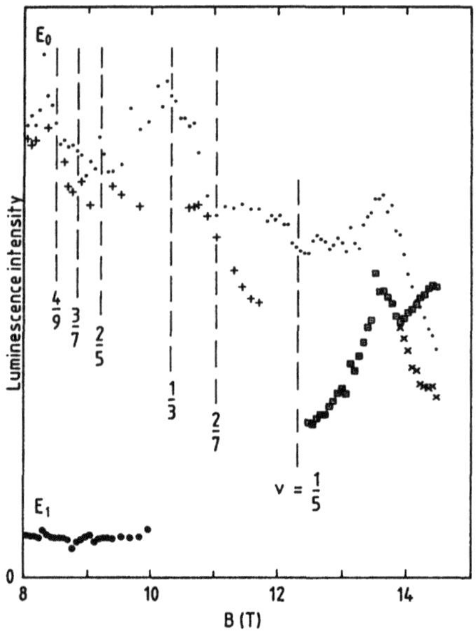

Fig. 8 Peak luminescence intensities as a function of field at T = 80mK in the extreme quantum limit. The stronger component of the E_0 doublet is marked • and the weaker + . At $\nu \lesssim 1/5$ a high- energy line appears (⊡); x marks a weaker intermediate peak which is also resolved.

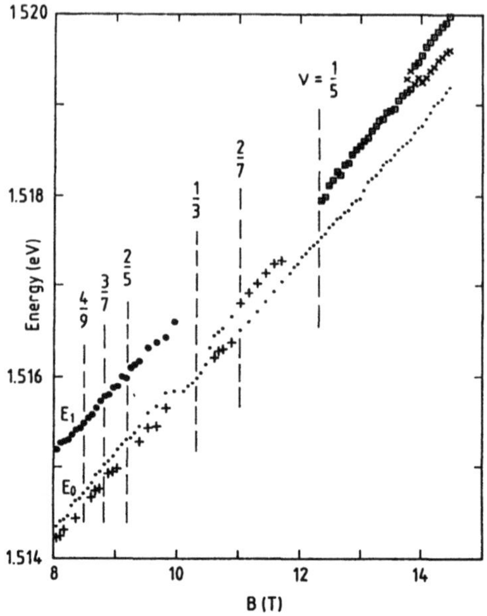

Fig. 9 Energies of the photoluminescence peaks at T = 80mK in the extreme quantum limit. The label convention is the same as that used in Fig. 8.

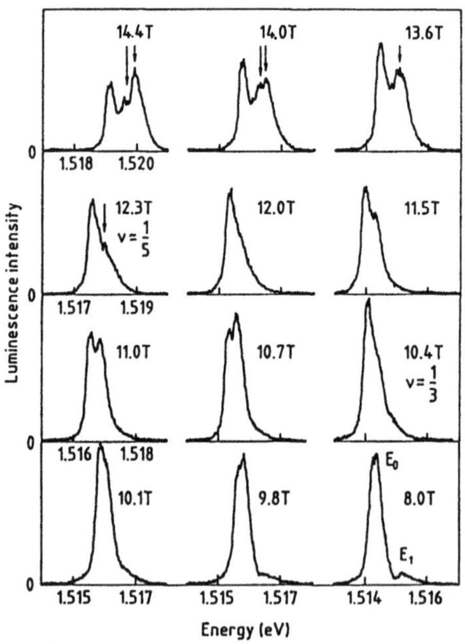

Fig. 10 Photoluminescence spectra obtained at T = 80mK in the extreme quantum limit. Arrows mark structure that develops at filling factors $\nu \lesssim 1/5$.

coincide with the appearance in this sample of resonant radio-frequency absorption [8]. Magnetically-induced Wigner solid formation in a weak random potential can account for these two phenomena [8], which suggests that the new luminescence peak may be associated with the same physics.

(ii) In Fig. 9 it is interesting to extrapolate the E_0 high-energy component and the E_1 line, to the region $\nu < 1/5$. This suggests that the "new" peak may arise from a re-emergence of the E_1 line; the weaker intermediate peak could arise from spin splitting of either E_0 or E_1.

(iii) Within an interpretation that the highest energy luminescence peak at high field *is* associated with E_1 re-emergence and, more straightforwardly, the lowest energy peak with E_0, the steady collapse of the E_0 intensity and rise in E_1 intensity for $\nu < 1/5$, would by analogy with the definitive FQHE data (E_0 minima, E_1 maxima at fractional filling) be consistent with the onset of crystallisation.

6. Conclusions

Our experiment demonstrates optical detection of the integer and fractional QHE. Whilst data associated with the $\nu = 2/3$ FQHE hierarchy are definitive, spectroscopy in the region of the $\nu = 1/3$ hierarchy is complex. In the extreme quantum limit ($\nu < 1/5$) photoluminescence structure develops that may be associated with the onset of crystallisation. A theory that extends to the conditions of our experiment is required to understand quantitatively the precise origin of the photoluminescence structure observed.

References

1. K. von Klitzing, G. Dorda and M. Pepper, Phys. Rev. Lett. 45, 494 (1980)
2. D.C. Tsui, H.L. Störmer and A.C. Gossard, Phys. Rev. Lett. 48, 1559 (1982)
3. I.V. Kukushkin and V.B. Timofeev, Pis'ma Zh. Eksp. Teor. Fiz 44, 179 (1986) [JETP Lett. 44, 228 (1986)]
4. C.H. Perry, J.M. Worlock, M.C. Smith and A. Petrou, in: "High Magnetic Fields in Semiconductor Physics", G. Landwehr, ed., Springer-Verlag, New York (1987), p202
5. R. Stepniewski, W. Knap, A. Raymond, G. Martinez, T. Rötger, J.C. Maan and J.P. André, in: "High Magnetic Fields in Semiconductor Physics", G. Landwehr, ed., Springer-Verlag, New York (1989), p62
6. D. Heiman, B.B. Goldberg, A. Pinczuk, C.W. Tu, A.C. Gossard and J.H. English, Phys. Rev. Lett. 61, 605 (1988)
7. A.J. Turberfield, S.R. Haynes, P.A. Wright, R.A. Ford, R.G. Clark, J.F. Ryan, J.J. Harris and C.T. Foxon, Phys. Rev. Lett. 65, 637 (1990)
8. F.I.B. Williams, E.Y. Andrei, R.G. Clark, G. Deville, B. Etienne, C.T. Foxon, D.C. Glattli, J.J. Harris, E. Paris and P.A. Wright, in: "Localisation and Confinement of Electrons in Semiconductors", G. Bauer, F. Kuchar and H. Heinrich, eds., Springer-Verlag, New York (1990), to be published

9. F.I.B. Williams, G. Deville, E.Y. Andrei, D.C. Glattli, J.J. Harris, C.T. Foxon, P.A. Wright, R.G. Clark, B. Etienne, O. Probst and C. Dorin, to be published
10. W. Chen, M. Fritze, A.V. Nurmikko, D. Ackley, C. Colvard and H. Lee, Phys. Rev. Lett. 64, 2434 (1990)
11. P.A. Maksym, private communication
12. R.L. Willett, H.L. Störmer, D.C. Tsui, L.N. Pfeiffer, K.W. West, M. Shayegan, M. Santos and T. Sajoto, Phys. Rev. 40, 6432 (1989)

Energy Shifts, Intensity Minima, and Line Splitting in the Optical Recombination of Electrons in the Integer and Fractional Quantum Hall Regimes

B.B. Goldberg[1,†], *D. Heiman*[2], *A. Pinczuk*[3], *L. Pfeiffer*[3], *and K. West*[3]

[1]Physics Department, Boston University, Boston, MA 02215, USA
[2]MIT Francis Bitter National Magnet Laboratory, Cambridge, MA 02139, USA
[3]AT&T Bell Laboratories, 600 Mountain Avenue, Murray Hill, NJ 07974, USA

Energy shifts in the electron-hole recombination energy and minima in the peak intensity at integer and fractional filling factors occur in the luminescence from ultra high mobility GaAs single quantum wells and heterojunctions. At Landau and spin gaps the magnetic field regions of the energy shifts and intensity minima broaden as the temperature is reduced, in consort with the transport Hall resistance. This relates the optical anomalies directly to the position of the Fermi energy in localized transport states. In the fractional quantum Hall regime a sharp intensity minimum and peak shift is observed at $\nu = 2/3$, while higher-field fractions are characterized by a splitting in the luminescence, with the higher-energy component dominant at higher fields. The response of the 2D electron gas to the perturbation of the hole is an important consideration, and is studied by varying the quantum well width, whence it is found that correlation effects are reduced relative to vertex corrections as the well width is increased.

1. Introduction

Both the integral and fractional quantum Hall effects are characterized by an energy gap in the extended state spectrum at the Fermi energy. In the former, the gap is caused by the magnetic Landau or spin quantization of the energy levels, while in the latter the gap is caused by a condensation of the electrons into an incompressible liquid. [1] When the Fermi energy lies in a gap, the longitudinal resistance is zero due to the absence of low-energy excitations. In both optics and transport, the response of the electrons to perturbations is strongly modified. For example, the electrons are ineffective in screening the potential fluctuations due to remote ionized impurities, causing a broadening in the density of states. [2] In optical processes, the reduced polarizability is apparent in the alteration of the local potential of a photoexcited hole. [3] Examination of the emission energy and oscillator strength of an electron recombining with a screened hole is a powerful tool to study the integer and fractional Hall effects, since the event probes the surrounding electrons.

Luminescence in the fractional quantum Hall effect (FQHE) regime has been observed in Si [4] and GaAs [3] and is of considerable current interest. [5]

---

† Visiting Scientist at the Francis Bitter National Magnet Laboratory.

At issue are the changes in the electron-hole recombination energies and intensities caused by the condensed state, and whether the quasi-particle gap or its effect on the electronic response is primarily responsible for the observations. The role of the photoexcited hole, whether free or bound, as a perturbation to the electrons in the initial and final state of the recombination process is an important consideration. In this work, we report on 7 samples in which electron-hole recombination was studied with the coexistence of flats in ρ_{xy} and deep minima in ρ_{xx} at the filling factors $\nu = 2$, 1, 2/3, 3/5, 4/7, 3/7, 2/5, 1/3, and 2/7. The luminescence displays energy shifts accompanied by intensity minima when E_F is in an integral or fractional gap. The energy shifts at fractional filling are distinguished by the return to the kinetic energy tuning curve at both higher and lower fields. We study the response of the 2d electrons to the hole by varying the quantum well (QW) width from 250Å to 500Å to single heterojunctions (SHJ) and find that for increasing electron-hole separations, the energy shift at $\nu = 1$ changes sign, indicating the dominance of different contributions to the many-body recombination energy.

2. Results

2.1 Experimental

These experiments consist of simultaneous measurement of the transport resistivity components and photoluminescence spectra from one-side doped GaAs/AlGaAs single QWs (250Å well width; $\mu = 2-5 \times 10^6 \mathrm{cm}^2/\mathrm{V}$-s, $n_s = 1.0-2.9 \times 10^{11} \mathrm{cm}^{-2}$) and SHJs ($\mu = 10 \times 10^6 \mathrm{cm}^2/\mathrm{V}$-s, $n_s = 1.7 \times 10^{11} \mathrm{cm}^{-2}$). The temperature-dependent minima in resistivity (ρ_{xx}) was used to monitor the electron temperature, and heating was prevented by keeping the excitation intensity below $I < 5 \times 10^{-4} W/cm^2$. Since the changes in intensity of the emission at fractional and integer gaps are an important measure, care was taken to avoid B field-dependent intensity variations created by changes in absorption (due to higher Landau levels tuning through the excitation energy). It was found that some laser sources introduce sharp intensity variations, so filtered sources of medium band-width (10 – 40nm) were employed. Fiber-optic sample holders allowed measurements to be made in B-fields to 30T and temperatures to 110mK.

2.2 Integer quantum Hall regime

Plateaus in the recombination energy and minima in the intensity at the filling factors $\nu = 1$, 2 are revealed in the series of spectra plotted as a function of B in Fig. 1a. Two peaks are readily apparent above 4T, and we will confine our discussion to the lower-energy transition. Polarization analysis of emission spectra show that the lower-energy transition is the $+\frac{1}{2} \rightarrow +\frac{3}{2}$ ($0^+ \rightarrow 0^+$) while the higher energy transition is the $-\frac{1}{2} \rightarrow -\frac{3}{2}$ ($0^- \rightarrow 0^-$). [6] This assignment is confirmed by noting that the higher energy transition weakens beyond $\nu = 1$, where the $-\frac{1}{2}$ (0^-) electron level de-populates, and is significantly reduced for

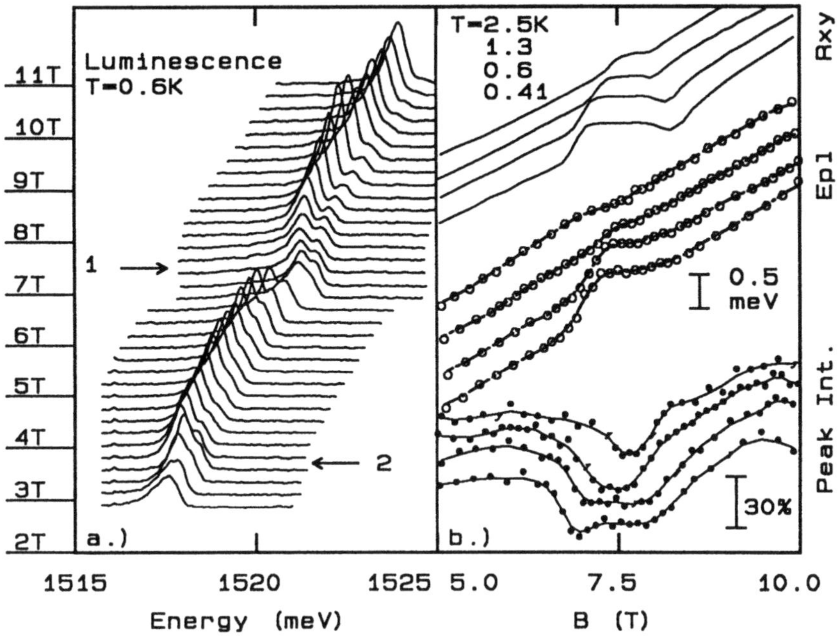

Figure 1: **a:** Emission spectra of sample (A) with $\mu = 3.2 \times 10^6 \text{cm}^2/\text{V-sec}$ and $n_s = 1.83 \times 10^{11} \text{cm}^{-2}$ at T= 0.56K excited by a 1580meV laser source ($p = 1.2 \times 10^{-4} \text{W/cm}^2$). The B-field is shown by following the spectral baseline to the scale at the left. **b:** Transition energy E_{PL} (circled dots), peak intensity (solid dots), and Hall resistance R_{xy} (solid) at $\nu = 1$ for (A) excited with $p = 8 \times 10^{-5} \text{W/cm}^2$, 10nm bandwidth source centered at 780nm at T= 2.5, 1.3, 0.6, and 0.41K. The E_{PL} curves are offset by +0.5 meV each, and the peak intensity and R_{xy} curves by arbitrary constant values.

decreasing temperature, as expected for thermal population of the the upper spin-state.

At temperatures above (T$\gtrsim$ 2.8K) we observe near linear behavior of the emission energy versus field, showing no transition to the predicted magneto-exciton insulator, [7] and indicating that luminescence plateaus observed in other studies may arise from the poorer mobility of the samples (10 to 100 times smaller). [8] As the temperature is lowered, the energy shifts and intensity minima develop, increasing in width in B in the same manner as the width of the quantized Hall flat in ρ_{xy} and zero resistance state ρ_{xx}. This directly correlates the position of the Fermi level with respect to the localized and extended transport states and the optical anomalies in recombination. [3,8,9]

Fig. 1b presents a detailed study at the spin-gap ($\nu = 1$). The intensity minima are quite robust; they exist independent of excitation source and sample carrier density, appearing whenever E_F is in localized states. At each temperature from 2.5K to 410mK, the approximate B field widths of the optical

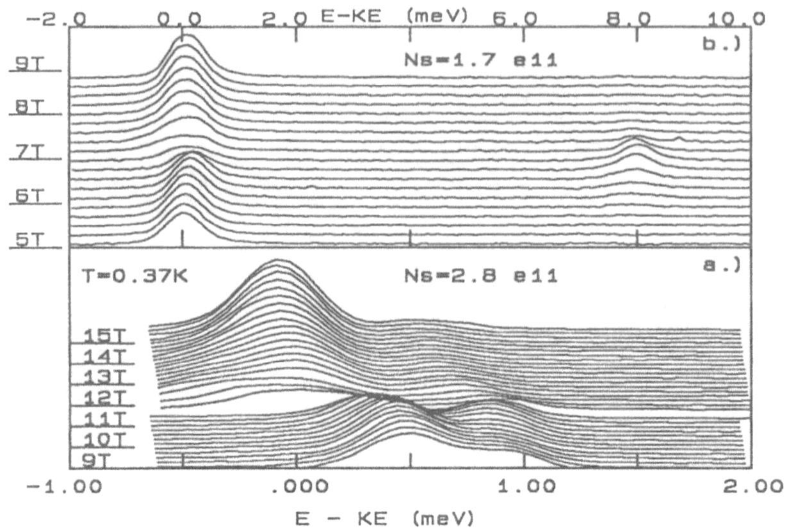

Figure 2: Spectra from wide 400$\AA$ QWs showing the red shift at $\nu = 1$. The kinetic energy term, a linear fit to the high-field peak position is subtracted to show the peak shifts. **a:** Sample (B) with $n_s = 2.8 \times 10^{11}\text{cm}^{-2}$, (the exciton peak is off the plot to the right) and **b:** sample (C) with $n_s = 1.7 \times 10^{11}\text{cm}^{-2}$ showing the increase in exciton emission when the ground state is quenched.

plateaus and intensity minima are within 30% of the corresponding Hall flat (Fig. 1b). The extent of quantized ρ_{xy} delineates the region in B over which E_F lies in a region of localized states. As the temperature is reduced, fewer extended states are within k_BT of E_F, causing the plateau width of ρ_{xy} to increase. The correspondence between the transport Hall flat, intensity minima and recombination energy plateau widths observed upon cooling the 2D electrons associates the optical and transport phenomena with a common origin – the precise position of the Fermi energy in localized transport states.

Previously observed blue shifts in luminescence [3] and absorption [10] at $\nu = 1$ can only now be recognized as due to localized transport states. Well-to-well density fluctuations and the preponderance of localized states in low mobility samples could manifest as a splitting and emergence of new, blue shifted lines.

As a function of increasing well width, and likely electron-hole separation, the energy shift at $\nu = 1$ goes from blue to red. Fig. 2 displays the luminescence from 400$\AA$ QWs at $\nu = 1$, where the kinetic energy term has been subtracted. The red shift at $\nu = 1$ at 11.7T in Fig. 2a is ~ 0.5meV, while the lower density sample in Fig. 2b has a shift of ~ 0.3meV.

As a result of the strong reduction in the ground-state emission when E_F is in a gap, wide wells of 400$\AA$, 500$\AA$, and SHJs always exhibit a the sharp increase in the excited state luminescence (Fig. 2b and Fig. 3). [5,9] The increase

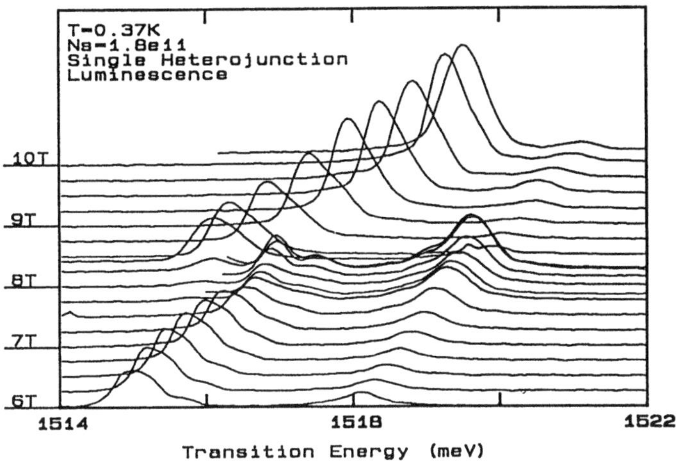

Figure 3: Spectra from a single heterojunction of $n_s = 1.9 \times 10^{11}\text{cm}^{-2}$ and $\mu = 8 \times 10^6\text{cm}^2/\text{V-sec}$ (sample D) which exhibits both a strong red shift at $\nu = 1$ (8.4T) and an increase in exciton emission.

in exciton emission is due to the quenching of the ground-state coupled with a constant photoexcited hole density and the lack of non-radiative processes. Furthermore, it shows that the hole is a critical consideration in both QWs and SHJs, though its role changes as a function of electron-hole separation as we discuss below. In wider samples (500Å) and the single heterojunction in Fig. 3, the exciton existed for a broad range in magnetic field. Note also the large red shift in the single heterojunction (Fig. 3), which approaches ~ 2.0meV.

Theoretical calculations of the influence of screening on electron-hole recombination do not account for the intensity minima, energy shifts, or high-field splitting. [11] The main many-body contributions to the single particle electron and hole energies are the screened-exchange and the Coulomb-hole self-energy terms. For electrons, the exchange interaction is reduced by screening while the Coulomb-hole term is enlarged, and the two terms nearly cancel as a function of filling factor. For holes, so few exist in our experiment ($< 10^{-5} \times n_s$) that the exchange term is irrelevant, leaving only the Coulomb-hole term. This correlation-hole for the hole describes the additional electron density in the vicinity of the hole. The screening oscillates as E_F passes from the level centers to the gaps, causing the hole self-energy to oscillate and thus E_{PL} to exhibit upward cusps at even-integral filling factors. Effectively, the screening length diverges since low-energy excitations at the Fermi surface are not available with E_F in a gap, creating a large change in the hole correlation energy. The calculations use a phenomenological model of the density of states but do not include the effects of carrier localization, and the integer and fractional Hall effects are beyond their scope. [11]

The present data shows that the passage of E_F into localized states at the edge of the ρ_{xy} flat creates an abrupt change in the response of the electrons to the presence of the hole. In the absence of correlation effects, the oscillator strength should actually increase when the polarizability is reduced, since weaker screening causes a larger vertex correction (excitonic binding), yielding both greater electron-hole overlap and a red shift in E_{PL}. This is complicated at integral and fractional gaps, since the dearth of states at E_F promotes the importance of *inter-* over *intra*-level excitations, increasing the coupling to higher Landau levels. Changes in screening introduce antithetical changes in the vertex and correlation hole terms – the polarization clouds of a closely spaced electron-hole pair cancel. [12]

Our results indicate that in quantum wells of width 250Å or less, the energy shift at $\nu = 1$ and 2 is dominated by correlation effects. Near $\nu = 1$ the upper electron spin-state has a much lower occupancy, which increases the vertex correction and reduces the electron self-energy exchange term. Yet the emission from both spin-states exhibit the same blue shift at $\nu = 1$ (Fig. 1a), indicating that the correlation-hole of the hole is the primary mechanism. While for QWs of 400Å and 500Å well widths, as well as SHJs, the energy shift is likely dominated by vertex corrections. This is shown not only in the red shift, but also in the intensity response of the upper spin-state. In the wide wells and heterojunctions, both the upper spin-state as well as the exciton increase in intensity at $\nu = 1$, as expected from vertex corrections and population effects and the fact that the ground state becomes quenched. In addition, the temperature dependence makes clear that a complete picture can only emerge with the proper inclusion of localization into the current theories.

2.3 Fractional quantum Hall regime: $\nu = 2/3$

Fig. 4b shows simultaneous development of the minima in peak intensity and ρ_{xx} at $\nu = 2/3$ for temperatures T$\leq$ 1.3K and Fig. 4c shows the correlation of the minima in intensity and transport for a lower density sample, even in the event that the density changes as a function of magnetic field. The reduction in luminescence peak intensity at fractional filling factors strengthens as the temperature is lowered, but ceases further action for temperatures well below the activation energy ($\sim$ 0.8K for $\nu = 2/3$ and about 3K for $\nu = 1/3$).

Fig. 4a and Fig. 5 detail the sudden peak shift which always accompanies the intensity minima in the region of $\nu = 2/3$. The peak position shifts abruptly to the blue just at $\nu = 2/3$ by $\sim$ 0.1meV, and is accompanied by a $\sim$ 20% increase in width. The line is too broad to distinguish between a simple shift and the appearance of a higher-energy component. Significantly, the peak narrows and returns to the kinetic energy line on either side of the 2/3 minima. The shift weakens to $\Delta E^{\frac{2}{3}} \sim$ 0.04meV at 0.77K and ceases entirely above 1.4K, concurrent with the disappearance of the ρ_{xx} minima.

An interesting feature of the luminescence at $\nu = 2/3$ is the strong reduction in the intensity of the upper spin-state which may be related to direct coupling of excited state levels in the FQHE ground state as predicted by Jain, *et al.* [13] The quenching of the upper spin-state emission at $\nu = 2/3$ reach values as high as 50%. This is shown in the spectra of Fig. 5a and Fig. 5c,

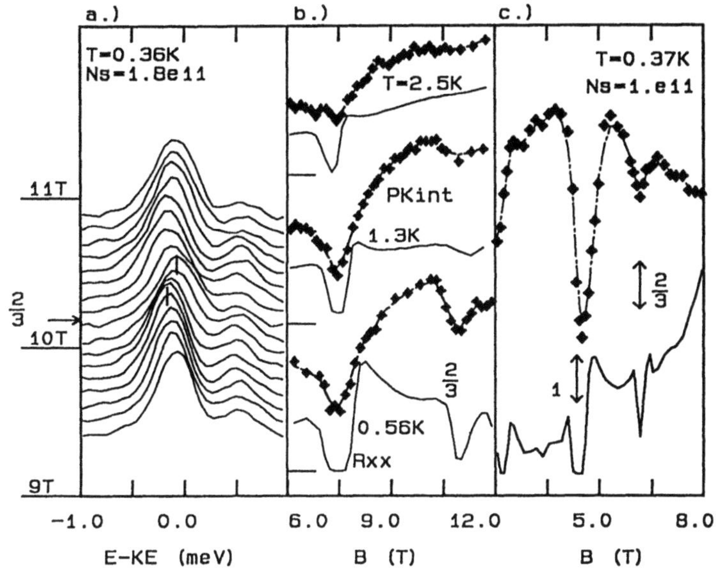

Figure 4: a: Luminescence about $\nu = 2/3$ in sample (E), 250Å with $\mu = 3.3 \times 10^6 \text{cm}^2/\text{V-sec}$ and $n_s = 1.80 \times 10^{11} \text{cm}^{-2}$ at T= 0.35K. Note the peak shift which occurs *just* at 2/3. b: Peak intensity and R_{xx} vs B as a function of T. c: Peak intensity and R_{xx} for a low density sample (F).

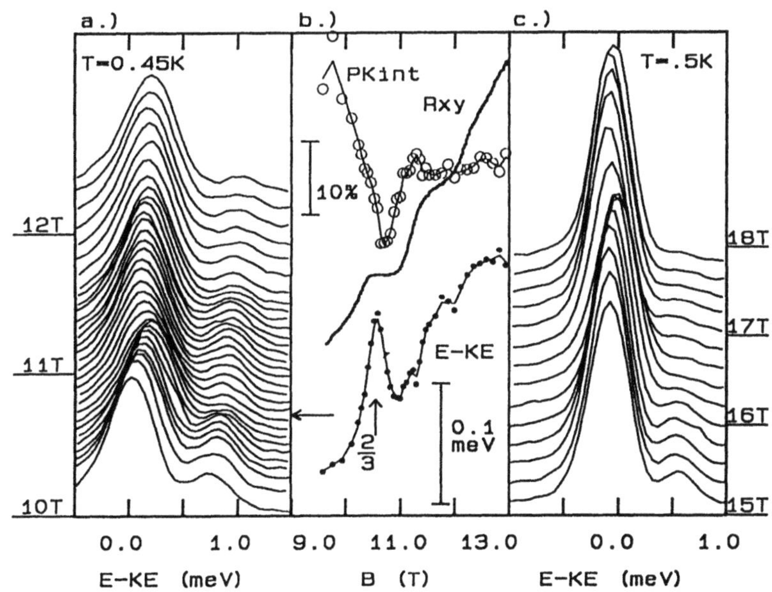

Figure 5: The peak shift at $\nu = 2/3$ from the emission in a: are plotted vs B along with R_{xy} and the intensity in b for sample (G) with $n_s = 1.8 \times 10^{11} \text{cm}^{-2}$. The 400Å high density sample (B) (Fig. 2a) shows the same shift at 2/3 in c.

where the latter shows the peak shift and quenching of both the ground state and upper spin-state emission occurring at $\nu = 2/3$ at 16.3T. Not surprisingly, an increase in the exciton emission and changes in the recombination lifetime have been observed at $\nu = 2/3$ and $\nu = 1$. [14]

In general, the minima in intensity and peak shift at $nu = 2/3$ is always seen in samples which retain strong minima in the simultaneous transport. Even those samples which show a reduction in carrier density due to illumination (Fig. 4c), but have strong minima in ρ_{xx} at $\nu = 2/3$, exhibit these effects. It is interesting to note, however, that in samples which do not exhibit good transport at $\nu = 2/3$, due either to parallel conduction or optically activated trap states, the emission at $\nu = 2/3$ does not show a shift or intensity minima. Deep, stable minima in ρ_{xx} measured simultaneously appears to be a prerequisite to identifying optical shifts with fractional Hall states. Finally, we note that the activation energy measured from transport under illumination is $E_a^{\frac{2}{3}} = 1.1 \pm 0.1$K, corresponding closely with the numerical value for the blue shift.

Significant differences exist between the well width dependence of the optics in the integer and fractional regimes. At $\nu = 1$ and 2, both the energy shift in ground-state recombination and upper spin-state emission intensity depend on quantum well width consistently explained by the change of correlation and vertex corrections. At $\nu = 2/3$ however, the emission blue shift and the reduction in upper-spin state emission appear independent of well width, indicating fundamental differences in the response of the electrons to perturbations in the fractional Hall regime.

2.4 Fractional quantum Hall regime: $\nu = 2/5,\ 1/3$

In the high-field region about the FQHE states $\nu = 2/5$ and $1/3$ a clear splitting ($\Delta E^{\frac{1}{3}} \approx 0.3$meV) of the luminescence is seen (Fig. 6). At T= 1.3K (Fig. 6, left), the splitting is observed only about $\nu = 1/3$, concurrent with a weak minima in ρ_{xx} *solely at that fraction.* The splitting exists only at temperatures low enough that fractional states are seen, and consist of components that exchange oscillator strength as a function of B. As the temperature is reduced, the higher-energy peak contains a greater percentage of the total oscillator strength, and extends over a larger range in B, similar to our earlier work on low mobility materials. [3] Also appearing at T$\leq$ 0.5K are additional splittings of weaker character in the region of $\nu = 2/5$ and $3/7$ (Fig. 6, right).

The optical signature of the FQHE in the high-field region is characterized by a line splitting with an energy spacing of $\sim$ 0.3meV and an emergence of the high-energy component which was seen in many samples. The transport activation energy is $E_a^{\frac{1}{3}} = 3.1 \pm 0.2$K (0.27meV), yielding a quasi-particle gap of $\sim$ 6.2K. Note that both the splitting and $E_a^{\frac{1}{3}}$ are about a factor of 3 greater than the shift and $E_a^{\frac{2}{3}}$ observed at $\nu = 2/3$. In some samples the splitting was confined to a very narrow field region (ΔB= 0.5T) with the high-energy component dominant at higher fields.

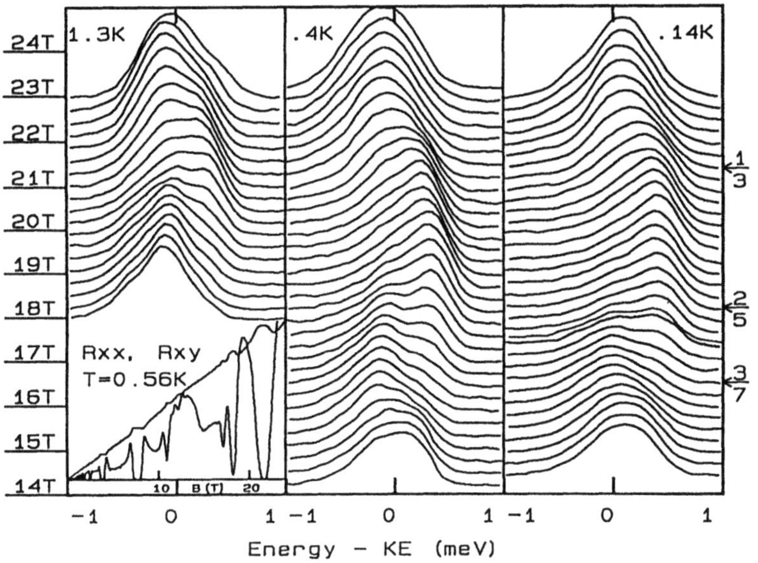

Figure 6: Spectra of (A) at $\nu = 2/5$ and $1/3$ with $\frac{1}{2}\hbar\omega_c$ subtracted at T=1.3K, 0.4K, and 140mK. At T= 1.3K the splitting is observed only about a narrow range at $\nu = 1/3$ concurrent with a minima in ρ_{xx} solely at 1/3. As the temperature is reduced, the high-energy component is both more dominant and extends over a larger field range. Inset: R_{xx} vs. B at T= 0.56K taken simultaneously with spectra.

The optical data in the integer regime show that the screening of the hole is important in determining the emission energy, and we are therefore reluctant, in the absence of a complete theory, to assign the peak shifts at $\nu = 2/3$ and the splitting at $\nu = 1/3$ directly to any quasi-particle gap. In addition, the splitting in the high-field region is unchanged over a significant range in field, whereas one would expect the quasi-particle energy gap ($^3\Delta$) to reach a maximum at a specific filling factor ν_c and diminish on either side. It is likely that the shift at $\nu = 2/3$ exists over a narrow field range because there are no other FQHE states of similar strength nearby, while both the 2/5 and 2/7 states in the neighborhood of $\nu = 1/3$ have measured activation energies exceeding 1K, leading to a much broader range in B over which the high-energy component is observed. Since the luminescence probes the states of the 2D electrons which respond to the hole over a distance of a few screening lengths, [15] small changes in the local environment of the hole between different FQHE ground states may have a negligible effect on the emission in the high-field region.

2.5 Electron Solid?

In samples with densities less than $1 \times 10^{11}\text{cm}^{-2}$, a novel break-up of the emission into discrete lines is observed. Fig. 7 shows this as a function of decreasing temperature. We are without a full explanation at this time, but

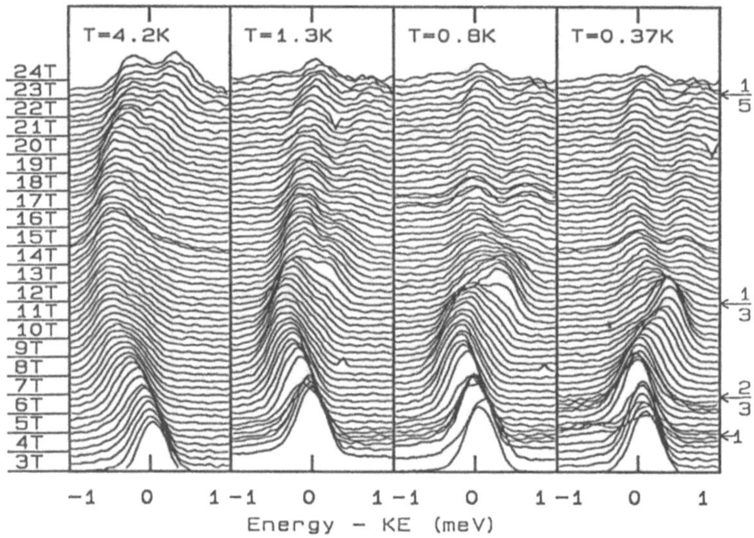

Figure 7: Emission vs B and T for (F). A blue-shift at $\nu = 1/3$ is readily apparent for T$\lesssim$ 1.0K, as well as a quenching at $\nu = 1$ (Fig. 4c). Note that the break-up of the emission into discrete peaks occurs at lower fields (beyond 1/3) for decreasing temperature, and is correlated to the large increase in ρ_{xx} seen at the onset of the electron solid.

would like to note that the onset of the break-up occurs at the B-field position corresponding to a rapid rise in ρ_{xx}. [16]

3. Conclusions

These measurements show a direct correlation between the position of the Fermi energy in IQHE and FQHE gaps and the optical recombination energies and intensities. The energy shifts and intensity minima at integral filling factors indicate physical phenomena beyond the abrupt change in screening of the hole potential by the 2D electrons and make evident that localization effects play a fundamental role. At $\nu = 2/3$ there is a minimum in the recombination intensity and a shift in the peak energy, and we surmise that the quasi-particles are involved in screening the hole potential over some suitable distance. In the regime of the fractions $\nu = 2/5$ and 1/3, evidence of a splitting coupled to the fractional quantum Hall state is presented. The energy shifts are distinguished by their return to the high temperature tuning curve away from fractional filling. The data show that the hole plays a substantial role in recombination in QWs and SHJs, and indicate that new physics can be learned from the response of the correlated 2D electron system to external perturbations.

We thank L. Rubin and B. Brandt of the Francis Bitter National Magnet Laboratory. We acknowledge valuable conversations with S. Schmitt-Rink. Work supported in part by National Science Foundation contracts DMR-8807682 and DMR-8813164.

References

1. D.C. Tsui, H.L. Störmer, and A.C. Gossard, Phys. Rev. Lett. **48**, 1559 (1982); R.B. Laughlin, Phys. Rev. Lett. **50**, 1395 (1983).
2. T. Ando and Y. Murayama, J. Phys. Soc. Jpn. **54**, 1519 (1985); S. Das Sarma and X.C. Xie, Phys. Rev. Lett. **61**, 738 (1988).
3. B.B. Goldberg, D. Heiman, A. Pinczuk, C.W. Tu, A.C. Gossard and J.H. English, Surf. Sci. **196**, 209 (1988); D. Heiman, B.B. Goldberg, A. Pinczuk, C.W. Tu, A.C. Gossard and J.H. English, Phys. Rev. Lett. **61**, 605, (1988).
4. I.V. Kukushkin and V.B. Timofeev, Pis'ma Zh. Eksp. Teor. Fiz. **44**, 179 (1986) [JETP Lett. **44**, 228 (1986)]; I.V. Kukushkin and V.B. Timofeev, Surf. Sci. **196**, 196 (1988).
5. B. B. Goldberg, D. Heiman, A. Pinczuk, L. Pfeiffer, and K. West, to appear in Phys. Rev. Lett.; A. J. Turberfield, S. R. Haynes, P. A. Wright, R. A. Ford, R. G. Clark, J. F. Ryan, J. J. Harris, and C. T. Foxon, to appear in Phys. Rev. Lett.; I. V. Kukushkin, A. S. Plaut, K. v. Klitzing, K. Ploog, H. Buhmann, W. Joss, G. Martinez, and V. B. Timofeev, 20th *ICPS.*
6. B.B. Goldberg, D. Heiman, M.J. Graf, D.A. Broido, A. Pinczuk, C.W. Tu, J.H. English, and A.C. Gossard, Phys. Rev. B **38**, 10131 (1988).
7. G.E.W. Bauer, Phys. Rev. Lett. **64**, 60 (1990).
8. C.H. Perry, J.M. Worlock, M.C. Smith, and A. Petrou, in *High Magnetic Fields in Semiconductor Physics*, ed. by G. Landwehr (Springer-Verlag, Berlin, 1987), p. 202; M.S. Skolnick, K.J. Nash, S.J. Bass, P.E. Simmonds, and M.J. Kane, Sol. State Comm. **67**, 637 (1988); H. Yoshimura and H. Sakaki, Phys. Rev. B **39**, 13024 (1989).
9. Excited-state recombination in pseudomorphic InGaAs also appears to show a correlation with magneto-transport; W. Chen, M. Fritze, A. V. Nurmikko, D. Ackley, C. Colvard, and N. Nouri, Phys. Rev. Lett. **62**, 1000 (1990).
10. B. B. Goldberg, D. Heiman, A. Pinczuk, Phys. Rev. Lett. **10**, 1102 (1989).
11. T. Uenoyama and L.J. Sham, Phys. Rev. B **39**, 11044 (1989); S. Katayama and T. Ando, Sol. State Comm. **70**, 97 (1989).
12. S. Schmitt-Rink, D.S. Chemla and D.A.B Miller, Adv. in Phys. **38**, 89 (1989).
13. J. K. Jain, Phys. Rev. Lett. **63** 199, (1989).
14. M. Dahl, *et al.* to be published.
15. Yu.A. Bychkov and É.I. Rashba, Zh. Eksp. Teor. Fiz. **96**, 757 1989; [Sov. Phys. JETP **69**, 430 1989].
16. R. L. Willett, H. Störmer, D. Tsui, L. Pfeiffer, K. West, and K. Baldwin, Phys. Rev. B 38 7881 (1989).

Magnetic Field Dependent Lifetime of Photoexcited Electrons at a Heterojunction

P.A. Maksym

Department of Physics, University of Leicester,
Leicester LE1 7RH, UK

Abstract. A one electron calculation is used to show that the lifetime of photoexcited electrons at a heterojunction oscillates with magnetic field and the influence of the electron-electron interaction is discussed.

1. Introduction

Recent experiments [1] on the optical detection of the Quantum Hall Effect involve luminescence emission from electrons in the lowest subband (lsb) and the next upper subband (usb). The usb electrons can also relax to the lsb (by emitting acoustic phonons) so it is important to understand the physics of this competing non-radiative process. This is the purpose of the present work.

2. Energy dependence of acoustic phonon emission

The first step in calculating the lifetime is to calculate the rate, $R_n(\epsilon)$, at which an electron in the N=0 Landau level of the usb makes transitions to the nth Landau level of the lsb, with energy loss ϵ. This is found in the usual way from the golden rule. The lsb envelope function is taken to have the Fang-Howard form and a model usb function is used which is orthogonal to the lsb function and has 1 node. This leads to the function $(3b/2)^{1/2}bz(1-bz/3)exp(-bz/2)$ which is believed to incorporate the relevant physics; b is taken to be $0.02\mathring{A}^{-1}$. Apart from this the calculation is similar to the one described by Toombs et al [2] .

Results for $R_n(\epsilon)$ for the deformation potential interaction in GaAs are shown in figure 1. There is one very sharp peak: moving $\sim$ 0.5 meV to the low energy side or $\sim$ 1.5 meV to the high energy side is sufficient to change the rate by *two* orders of magnitude. Physically, this reflects the ability of the electrons to transfer momentum to the phonons. Indeed the position of the peak can be estimated by considering typical momentum values. These depend on the dimensions of the electron wave function. For perpendicular emission the relevant length is the width, d, of the envelope function, while for parallel emission the length scale is set by the

Springer Series in Solid-State Sciences, Vol. 101
High Magnetic Fields in Semiconductor Physics III Editor: G. Landwehr

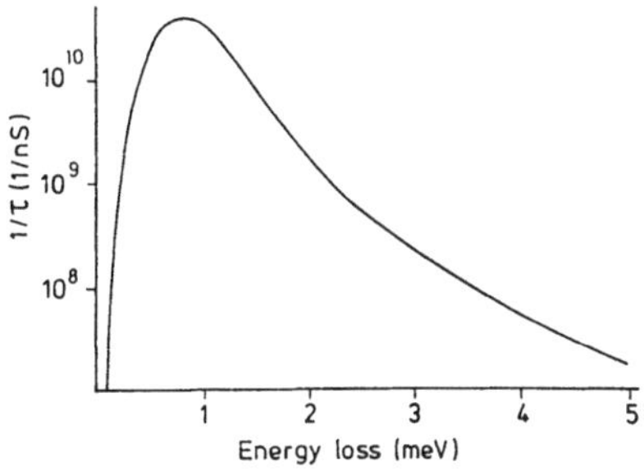

Figure 1: $R_n(\epsilon)$ for $n = 0$ and $B = 1T$

cyclotron length, l. Typically, $l \sim 100\AA$ and $d \sim 100 - 200\AA$ so that $2\pi/d$ or $2\pi/l$ corresponds to a phonon energy of $\sim 1 - 2$ meV.

3. Lifetime of non-interacting electrons

The total rate at which electrons leave the usb is obtained by summing the product of $R_n(\epsilon)$ and the density of empty states in the lsb. The Landau level broadening is assumed to be gaussian with a width of $0.1\sqrt{B}$ meV. Results for the magnetic field dependence of τ are given in figure 2. To give a measure of the lifetime of an arbitrary electron, the results for spin up and spin down are combined in the form $1/\tau = 1/\tau_\uparrow + 1/\tau_\downarrow$. The density of electrons is taken to be such that $\nu = 1$ filling occurs at B = 4T. τ clearly oscillates but the position of the peaks is sensitive to the subband energy difference. The origin of the oscillations is the oscillation of the Fermi level, but the energy dependence of R_n plays a critical role. In the 4 meV case the peaks occur close to even integer filling in the lsb. Consider the situation when the filling factor is just less than an even integer, for example just less than 4. In this case the N=2 Landau level in the lsb is empty and lies below the N=0 level in the usb. However, it is only about 0.5 meV below the usb N=0 level so the energy loss and the transition rate are both small, leading to a large lifetime. With increasing magnetic

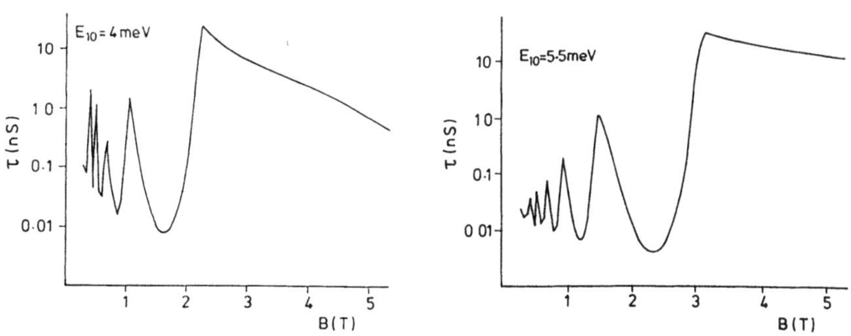

Figure 2: τ v. B for subband energy differences as indicated

field the N=2 level of the lsb crosses the N=0 level of the usb and empty states in the N=1 level of the lsb become available. The energy difference is then more favourable so electrons can easily make the transition. The lifetime therefore decreases but increases again as the N=1 level of the lsb approaches the N=0 level of the usb. If the subband energy difference is greater the peaks in τ occur further away from even integer filling because the crossing points occur at correspondingly higher fields.

4. Effects of interaction

A key question is whether the electron-electron interaction modifies the energy spectrum to such an extent that the lifetime of an electron in the usb is affected. The case of odd filling is particularly interesting because the one electron theory then predicts a very short lifetime. This case is examined here with the aid of finite size calculations for the case of $\nu = 1$.

Results for 11 interacting electrons in a rectangular cell with aspect ratio 0.8 are shown in figure 3. The subband energy difference is 3.5 meV and the Fang-Howard parameter is $0.03\mathring{A}^{-1}$. The figure shows dispersion relations for spins 11/2 and 9/2. S=11/2 is relevant to relaxation of a spin up electron in the usb. In this case there is only one state when all the electrons are in the lsb (the point at k=0), while if one electron is in the usb the dispersion relation is flat and *6 meV* higher. The S=9/2 case is relevant to relaxation of a spin down electron. The lower branch of the spectrum corresponds to the spin wave excitation of a full Landau level, while the upper branch is the dispersion obtained when one electron is in the usb. The energy difference is small at large k and this is thought to be the limit relevant to relaxation. The reason is that $k_y l^2$ for the spin wave gives the separation of the spin down electron and the hole in the spin up Landau level [3]. Large separations are expected for a dilute population of photoexcited electrons.

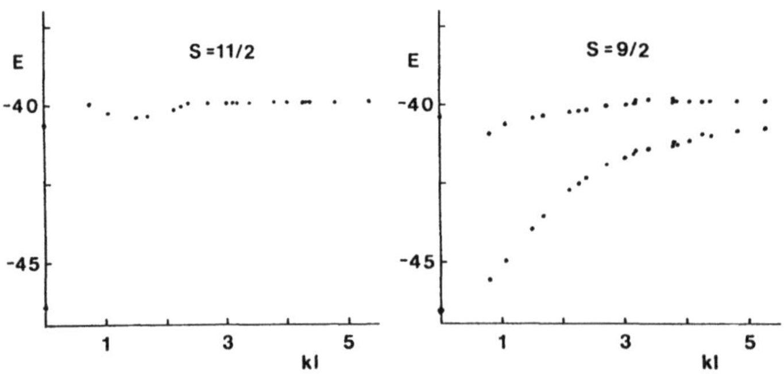

Figure 3: Energy spectrum (meV) of 11 interacting electrons at $\nu = 1$

The physics that emerges from the finite size calculation is that of an increase of the effective energy difference for spin up electrons and a decrease for spin down electrons. In either case this could correspond to unfavourable values of $R_n(\epsilon)$ with a consequent increase in lifetime. Ideally, this should be checked by direct calculation but this has not proved possible for sufficiently large systems.

5. Conclusion

The calculated non-radiative lifetime of photoexcited electrons in the usb oscillates with magnetic field. This raises the question of whether lifetime effects are important in the experiments of the type reported by Turberfield et al. In addition, the physics of inter-subband relaxation is interesting in its own right and might yield information on the electron-phonon interaction if it could be probed experimentally.

References

[1] A J Turberfield, S R Haynes, P A Wright, R A Ford, R G Clark, J F Ryan, J J Harris and C T Foxon, submitted to Phys. Rev. Lett.

[2] G A Toombs, F W Sheard, D Neilson and L J Challis, Solid State Commun. **64** 578 (1987)

[3] C Kallin and B I Halperin, Phys. Rev. B**30**, 5655 (1984)

Dependence of the Fractional Quantum Hall Effect Energy Gap on Electron Layer Thickness

J. Jo[1], *Y.W. Suen*[1], *M. Santos*[1], *M. Shayegan*,[1], *and V.J. Goldman*[2]

[1]Princeton University, Princeton, NJ 08544, USA
[2]State University of New York, Stony Brook, NY 11794, USA

A fundamental characteristic of the fractional quantum Hall effect (FQHE) is the existence of an energy gap (Δ) separating the ground state from its elementary quasiparticle and quasi-hole excitations [1]. In this paper we report an experimental study of the dependence of Δ for the FQHE states at Landau-level filling factors ν=1/3 and 2/3 on the thickness of the electron layer. We have determined Δ from the temperature dependence of the magnetoresistance minima at ν=1/3 and 2/3 in a variable-width electron system realized in a selectively-doped, parabolic AlGaAs quantum well. We observe a dramatic decrease in the measured Δ for the FQHE states at these filling factors with increasing electron layer thickness. This dramatic decrease of Δ may manifest the collapse of FQHE electron correlations expected in the theory of the FQHE as either the electron layer thickness is increased or the subband separation is decreased.

The structure used in this study was grown on an undoped (100) GaAs substrate by molecular beam epitaxy and mainly consists of a 3000Å -wide, undoped, parabolically graded $Al_xGa_{1-x}As$ well bounded on both sides by undoped (spacer) and doped layers of $Al_yGa_{1-y}As$ (y > x). The Al composition (x) is zero at the well center and is quadratically varied near the well center to produce a parabolic band profile. Owing to the self-consistent electrostatic potential, the electrons in the parabolic well screen the quadratic potential, and a system of nearly uniformly distributed electrons in a flat potential is obtained [2]. In this system, the electron layer thickness increases with increasing areal density (n_s) in the well so that the effective three-dimensional density and the Fermi energy remain essentially constant [2-4]. This is in contrast to the case of the two-dimensional electron system (2DES) at heterointerfaces where the width of the electron layer usually has a weak dependence on n_s.

In Fig. 1 we show the results of our self-consistent calculations for the charge distribution in our structure [3,4]. The calculations were performed by solving the Poisson and Schrödinger equations (for each n_s), and taking into account the exchange-correlation via local-density-functional approximation. As expected, the width w_e (which we define as the full-width at half-maximum) of the electron layer increases with n_s [5]. Note also that with increasing n_s and w_e, the separation between electric subbands decreases and more subbands become occupied. Quantitative evidence for the realization of the designed electron system is in fact provided by comparing the calculated densities of these subbands with those experimentally determined from a Fourier analysis of the Shubnikov-de Haas oscillations in the magnetoresistance at low magnetic fields [4]. We have done such an analysis and have found the experimental results in excellent agreement with the calculations, verifying the realization of the electron systems indicated in Fig. 1.

In Fig. 2a an example of the longitudinal magnetoresistance (ρ_{xx}) measured for this structure is shown. The high quality of the electron system is evidenced by the observed higher order FQHE states in addition to the fundamental 1/3 and 2/3

Springer Series in Solid-State Sciences, Vol. 101
High Magnetic Fields in Semiconductor Physics III Editor: G. Landwehr

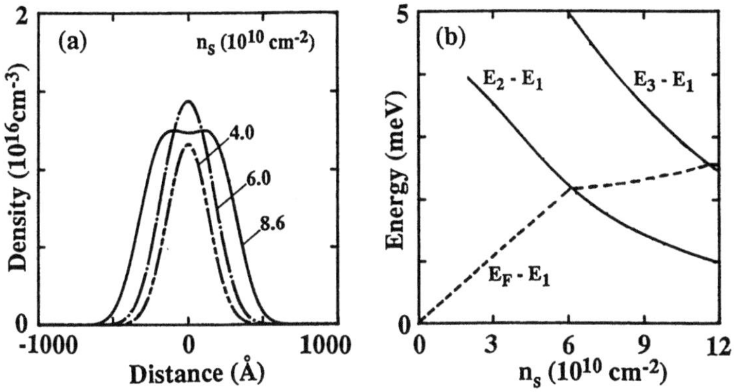

Fig. 1 The self-consistently calculated charge distribution for several n_s is shown in (a). In (b) the calculated electric subband energies and the Fermi energy are shown as a function of n_s.

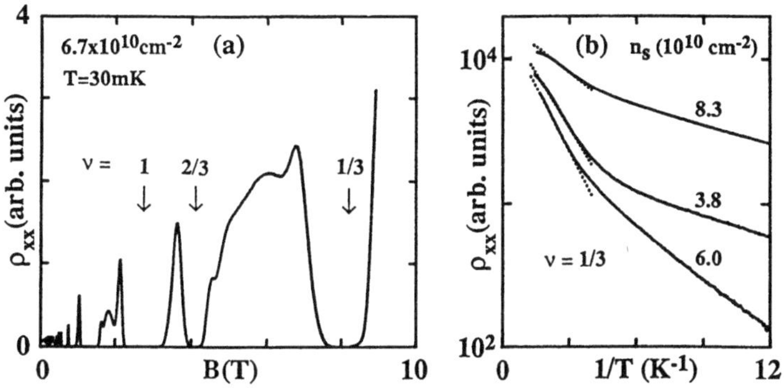

Fig. 2 The ρ_{xx} data for n_s=6.7x10^{10}cm^{-2} is shown in (a). In (b) the temperature dependence of ρ_{xx} at ν=1/3 is shown for three different n_s. The slopes of the dotted lines are used to determine the energy gap (Δ).

states. With increasing n_s, however, the ν=1/3 and 2/3 states become weaker and the higher order states disappear. In Fig. 2b we show the temperature dependence of ρ_{xx} minimum at ν=1/3 for several densities. In our experiments we varied n_s in the well by a combination of careful cooling to cryogenic temperatures and illumination by a red light-emitting diode. Similar to most of the previously reported data for the FQHE in 2DES at GaAs/AlGaAs heterojunctions, the data show an activated behavior, $\rho_{xx} \sim \exp(-\Delta/2T)$, only in a limited temperature range [6]. We used the slopes of the dotted lines drawn through the intermediate temperature range in Fig. 2b to determine Δ. The results are shown in Fig. 3. We also analyzed our data following the work of Boebinger et al. [6] by fitting the data to an expression containing an activated term and a term arising from hopping conduction which dominates at low temperatures. The values of Δ obtained from the fittings are somewhat larger than those determined directly from the slopes of the straight lines. The qualitative behavior of Δ and especially its dependence on n_s, however, remain the same for both types of analyses.

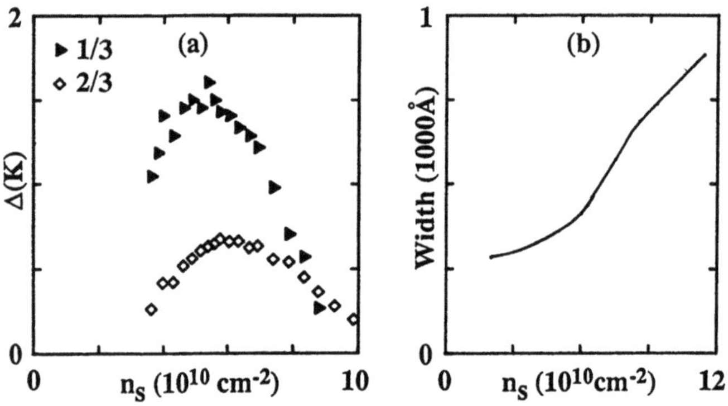

Fig. 3 The dependence of the measured energy gap (Δ) on electron density (n_s) is shown in (a). In (b) the calculated electron layer width is shown as a function of n_s.

To interpret the data, we first summarize what is known about Δ in FQHE. For an *ideal* 2DES, the theoretically expected behavior for Δ at $\nu=1/3$ and 2/3 is $\Delta \simeq 0.1 e^2/\epsilon l_o \propto B^{1/2}$ where ϵ is the dielectric constant and $l_o=(\hbar/eB)^{1/2}$ is the magnetic length [1]. Experimentally measured Δ for 2DES in GaAs/AlGaAs heterostructures, on the other hand, are substantially smaller than $0.1e^2/\epsilon l_o$ [6]. The discrepancy between the experiment and theory has been attributed to the effective reduction of Δ by disorder, finite electron layer thickness, and Landau-level mixing in the non-ideal 2DES. It is not clear how to account for the effect of disorder quantitatively [7]; however, it is an experimental fact that samples with less disorder (higher mobility) have larger Δ [6,8]. The reduction of Δ with increasing layer thickness is attributed to the weakening of the Coulomb interaction at small distances between electrons when the layer thickness is comparable to or larger than the interelectron spacing. Calculations of the effective electron-electron interaction for a 2DES with finite layer thickness have indicated a substantial reduction in both the ground state energy and Δ for the FQHE state at $\nu=1/3$ [9,10]. Calculations have also shown that the inclusion of the higher Landau levels further reduces Δ [11].

Although the experimentally measured Δ in 2DES at the GaAs/AlGaAs interface are significantly smaller than $0.1e^2/\epsilon l_o$ and do not follow a $n_s^{1/2}$ (or $B^{1/2}$) dependence, experiments have shown that, for samples of roughly similar quality, Δ increases with n_s [6,8]. This is in sharp contrast to our data summarized in Fig. 3. We believe that the drastic reduction of Δ with increasing n_s observed in our structure is related to the increase of w_e and, in part, to the decrease in the subband spacing (Fig. 1b) [12]. The lack of any theoretical calculations for Δ in a system like ours, with a very wide electron layer distribution and small subband separation, precludes us from a more quantitative comparison of our data with theory. It is worth mentioning, however, that the reduction in Δ that we observe for a relatively small increase in n_s or w_e, or a slight decrease in subband energy spacing, is more substantial than a simple extrapolation of the existing calculations [9-11] would predict.

In summary, we have measured Δ for the FQHE in an electron system with a variable electron layer thickness. We observe a significant decrease in Δ as the thickness of the electron layer is increased and the subband energy spacing is reduced. We emphasize that the electron distribution in our structures is controllable and yet relatively simple (Fig. 1). We hope our results will stimulate future calculations of Δ for these structures so that a more quantitative comparison between experiment and theory can be made.

We thank D.C. Tsui for advice and encouragement. Support of this work by NSF grant Nos. ECS-8553110, DMR-8705002, and DMR-8958453, and the ARO grant No. DAAL03-89-K-0036, is acknowledged. M. Shayegan and V.J. Goldman also acknowledge support by the Alfred P. Sloan Foundation.

References

1. For example, see: T. Chakraborty and P. Pietlainen, *The Fractional Quantum Hall Effect* (Springer, Berlin, 1988).
2. M. Shayegan, T. Sajoto, M. Santos, and C. Silvestre, Appl. Phys. Lett. **53**, 791 (1988); M. Sundaram, A.C. Gossard, J.H. English, and R.M. Westervelt, Superlattices Microstruct. **4**, 683 (1988).
3. T. Sajoto, J. Jo, H.P. Wei, M. Santos, and M. Shayegan, J. Vac. Sci. Technol. **B7**, 311 (1989).
4. T. Sajoto, J. Jo, M. Santos, and M. Shayegan, Appl. Phys. Lett. **55**, 1430 (1989).
5. Throughout this paper, w_e refers to the calculated *zero-field* electron layer thickness. In a single-electron picture, at fractional filling factors, the electrons occupy only the lowest subband. Assuming that only one subband is occupied, our self-consistent calculations show that the high-field layer thickness is slightly ($<10\%$) reduced with respect to w_e.
6. G.S. Boebinger, H.L. Störmer, D.C. Tsui, A.M. Chang, J.C.M. Hwang, A.Y. Cho, C.W. Tu, and G. Weimann, Phys. Rev. **B36**, 7919 (1987).
7. The effect of disorder has been considered only in a phenomenological manner [see, e.g., A.H. MacDonald, K.L. Liu, S.M. Girvin, and P.M. Platzman, Phys. Rev. **B33**, 4014 (1985); A. Gold, Europhys. Lett. **1**, 241 (1986); **1**, 479 (1986); P.M. Platzman, Phys. Rev. **B39**, 7985 (1989)].
8. R.L. Willett, H.L. Störmer, D.C. Tsui, A.C. Gossard, and J.H. English, Phys. Rev. **B37**, 8476 (1988).
9. A.H. MacDonald and G.C. Aers, Phys. Rev. **B29**, 5976 (1984).
10. F.C. Zhang and S. Das Sarma, Phys. Rev. **B33**, 2903 (1986).
11. D. Yoshioka, J. Phys. Soc. Jpn. **55**, 885 (1986).
12. The measured mobility in our structure in the density range $5\text{x}10^{10}<n_s<10^{11}\text{cm}^{-2}$ is nearly constant ($\simeq 3.6\text{x}10^5\text{cm}^2/\text{Vs}$). Therefore, we do not believe that disorder is responsible for the significant decrease in Δ that we observe in the same density range.

Surface Acoustic Wave Studies of the Fractional Quantum Hall Regime

R.L. Willett, M.A. Paalanen, L.N. Pfeiffer, K.W. West, R.R. Ruel, and D.J. Bishop

AT&T Bell Laboratories, 600 Mountain Avenue, Murray Hill, NJ 07974, USA

Surface acoustic wave propagation on high quality AlGaAs/GaAs heterostructures is examined in the fractional quantum Hall regime. The response to the piezoelectric field of the sound wave by the electron system is found to be similar in the fractional quantum Hall states and the integral quantum Hall states. However, deviation from this behavior is observed at Landau level filling $\nu = 1/2$, where sound propagation is markedly different from that observed in the neighboring filling factor range and in fractional quantum Hall states. In addition, anomalous sound propagation is also observed in the small filling factor range $\nu \lesssim 1/5$, where electron solid formation is suspected. The properties of these two distinct SAW anomalies are described and their possible origins discussed.

The purpose of our experiments is to study the dynamic response of electrons in a 2DES to a time and spatially varying electric field. This response is potentially most interesting in the extreme quantum limit where the fractional quantum Hall effect[1] is present and where the electron solid[2] has been proposed. The technique we employ is similar to that used by Wixforth et. al.[3] in which a surface acoustic wave is propagated on a GaAs/AlGaAs heterostructure. The 2DES interacts with the piezoelectric field of the SAW, affecting the sound amplitude and velocity and allowing inference of the frequency and wave-vector dependent conductivity from the SAW properties. In our experiments we have used SAW tranducers producing several harmonics so that a range of ω and k can be examined. In addition, our extremely high mobility samples are able to support electron correlation effects at mK temperatures.

At FQHE states we observe similar ultrasound results to those seen in IQHE states, where the electron response can be modeled using the d.c. sheet conductivity. This modeling breaks down at two distinctive filling factors regimes: at $\nu = 1/2$, the center of the lower spin split Landau level, and at $\nu \lesssim 1/5$ where electron solid formation has been suggested. These findings of anomalous sound propagation will be the focus of our study.

We have studied a total of 6 high quality GaA/AlGaAs single interface heterostructures with areal electron densities $<2 \times 10^{11}$ cm^{-2} and with high mobilities ranging from $\sim 1-4 \times 10^{6}$ cm^{2}/V $-$ sec. Each sample was grown with typically 5000 Å between the top of the cap layer and the 2DES. This distance was considerably less than our shortest wavelength of 2.5μm: the SAW penetrates only about one wavelength from the surface. The cap layer, dopant layer, spacer and 2D gas region were etched on either end of the samples leaving 2DES free zones where SAW transducers were patterned. The rectangular mesa remaining between the transducers provided a 2DES path of $\sim$ 4mm. Six indium contacts were diffused into the periphery of the 2DES mesa to allow transport measurement using standard lock-in techniques. The interdigital SAW transducers were patterned using electron-beam lithography and evaporated directly onto the samples, with wavelengths ranging from 8 to 32 μm. Typically the fundamental, third, and fifth harmonics up to 1.4GHz were observed, although one transducer set allowed observation of harmonics up to the eleventh. At kHz repetition rates, SAW pulses of ~ 1 μsec duration were launched across the 2DES mesa, with amplitude and frequency measured using standard boxcar integration and homodyne detection.[4] Sample temperatures down to ~ 50 mK and magnetic fields up to 12T were achieved in a dilution refrigerator/magnet system.

Typical longitudinal conductivity, SAW amplitude and velocity traces are shown in Figure 1 as a function of magnetic field. The more dominant features of the sound properties are associated with the FQHE features, in particular with the fractional series 1/3, 2/5, 3/7, 4/9 observed in the amplitude.

Springer Series in Solid-State Sciences, Vol. 101
High Magnetic Fields in Semiconductor Physics III Editor: G. Landwehr

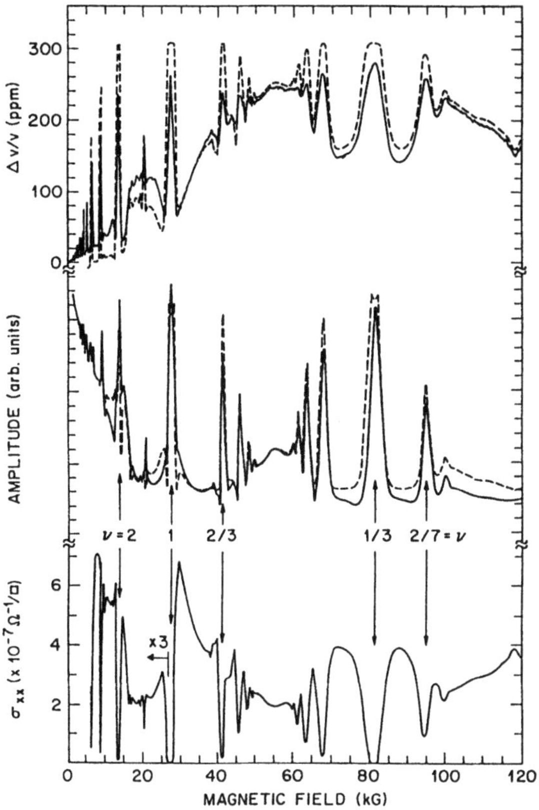

Figure 1. Conductivity, surface acoustic wave amplitude and velocity shift versus magnetic field at 160 mK and 235 MHz in sample #1. Solid lines are measured values, dashed lines are results of the conductivity model using the measured σ_{xx} (B) and $\sigma_m = 4 \times 10^{-7}\Omega^{-1}$.

As shown by the previous work in the IQHE regime,[3] the SAW interaction with the 2D electron layer may be modeled using the d.c. conductivity, $\sigma_{xx}(\omega = 0)$. In this picture,[5–7] the piezoelectric field produced by the SAW interacts with the 2D electron layer and the electron system responds in a manner characterized by a relaxation time represented in the sheet conductivity, σ_{xx} (B). According to this model, the attenuation coefficient Γ [amplitude$\sim$ exp(-Γx)] and the sound velocity v can be calculated from the conductivity using respectively

$$\Gamma = \frac{k(\alpha^2/2)(\sigma_{xx}/\sigma_m)}{1 + (\sigma_{xx}/\sigma_m)^2}$$

and

$$\frac{\Delta v}{v} = \frac{v(\sigma_{xx}) - v_o}{v_o} = \frac{\alpha^2/2}{1 + (\sigma_{xx}/\sigma_m)^2}$$

where α is the effective piezoelectric coupling coefficient[8], $\alpha^2/2 = 3.2\mathrm{x}10^{-4}$, $\sigma_m = v\times(\epsilon_o + \epsilon_s)$ and ϵ_o, ϵ_s are the dielectric constants of the vacuum and semiconductor. Plotted in Figure 1 are our best fits of the above formulas to the data using the measured d.c. conductivity σ_{xx} and adjusting the parameter $\sigma_m = 4 \times 10^{-7}(\Omega/\square)^{-1}$ which is reasonably close to the theoretical value. The agreement between the measured sound features and the model results is generally good. This agreement holds throughout the range of frequencies tested, from 90 MHz to 1.4 GHz, with a slight increase in σ_m necessary for good fit at higher frequencies. However, two

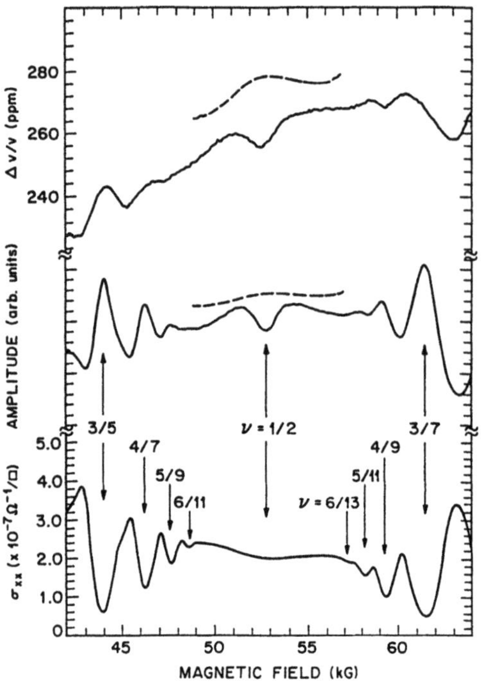

Figure 2. Conductivity, SAW amplitude and velocity shift at 700 MHz and 50 mK over a magnetic field range centered around filling factor $\nu = 1/2$. The solid lines are measured values and the dashed segments are conductivity model results as in Figure 1 using $\sigma_m = 5.5 \times 10^{-7}\Omega^{-1}$ and slightly offset from the measured trace for clarity. Sample # 1.

striking exceptions to agreement with the conductivity model occur at ν=1/2 and at $\nu \lesssim 1/5$. These anomalies are the focus of our studies and are discussed below.

I. ν=1/2

The first focus of our study is the pronounced feature at $\nu = 1/2$ in the sound measurements. Figure 2 shows transport and ultrasound data in a magnetic field range centered around filling factor one-half, with FQHE states marked. As expected, when the conductivity drops in the FQHE states, the sound velocity and amplitude increase. Correspondingly, over the broad minimum in σ_{xx} around $\nu = 1/2$, the sound properties increase except immediately near $\nu = 1/2$. Here, the sound velocity and amplitude show a distinct feature with a sharp minimum, in contrast to the broad *maximum* predicted by the conductivity as marked in the Figure. Reguardless of the d.c. conductivity model, given the sound results at integer and fractional ν it is surprising that such a striking feature should appear at filling factor 1/2. The magnitude of this effect is clearly as large as the sound response to several of the higher order fractions (4/9, etc.). The width of the feature is about 2 kG and appears in both frequency and amplitude roughly as an inverted Lorenztian.

The temperature dependence of the strength of the minimum at $\nu = 1/2$ in sound velocity, as defined in the inset, is shown in Figure 3 for the three observable SAW harmonics in one sample. From the figure it is seen that the higher frequency SAW measurements demonstrate stronger anomalies at $\nu = 1/2$ which persist to higher temperatures.

The electron areal density influences the strength of the effect at ν=1/2. In Figure 4 the magnitude of the 1/2 anomaly is plotted versus temperature for two densities in the same sample. The higher density state displays a stronger effect over the entire range of temperatures that allow observation of the anomaly.

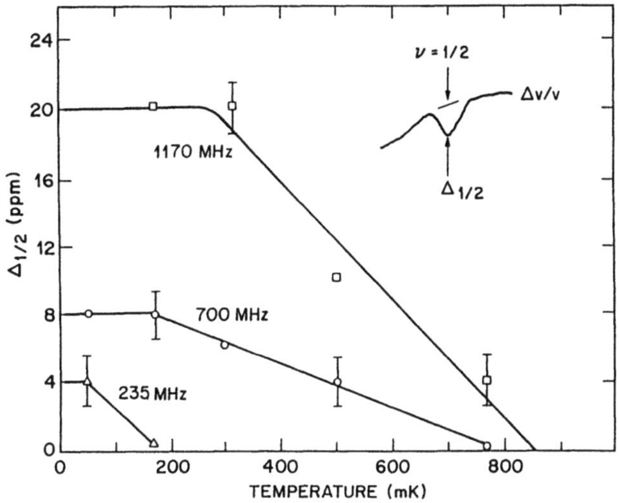

Figure 3. Magnitude of 1/2 anomaly versus temperature in sample #1. The inset defines the anomaly magnitude. The lines are a guide for the eye.

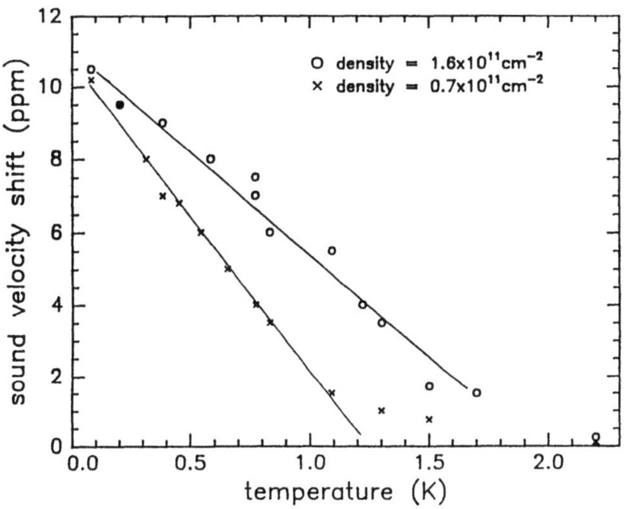

Figure 4. Magnitude of anomaly at ν=1/2 versus temperature for two densities of sample #2 using SAW frequency of 1350 MHz. The lines through the data are a guide for the eye.

All six samples tested displayed the sound anomaly at $\nu = 1/2$. Given this finding, we closely examined the sound propagation in several low density samples at $\nu = 9/2, 7/2, 5/2, 3/2$ and $\nu = 1/4$. No feature is discernible except at $\nu = 3/2$, where a similar but weaker minimum was found in both the amplitude and the sound velocity.

The sound propagation anomaly at $\nu = 1/2$ represents an additional, higher frequency conductivity $\sigma_{xx}(k,\omega) > \sigma_{xx}(\omega = 0)$ which is present over a narrow filling factor range. Accordingly the feature at $\nu = 1/2$ is clearly distinct from the FQHE in that the sound response is opposite to that observed in FQHE states. In addition, no quantization is manifested in ρ_{xy} at $\nu = 1/2$.

The cause of this anomaly at $\nu = 1/2$ is presently not known. To explore the sources of the sound anomaly one must consider several alternative explanations. First, recent localization studies[9] indicate that the magnetic field range of the extended states at $\nu = 1/2$ obeys $\Delta B \sim T^{\kappa}$ where κ=0.42. Experimentally it is observed that the magnetic field extent of the anomaly does not change appreciably with over an order of magnitude increase in the temperature. Therefore localization effects are probably not responsible for the $\nu = 1/2$ anomaly.

Another possibility at $\nu = 1/2$ is the presence of a distinct collective electron state, possessing an excited mode to which the sound couples over our range of ω and k. In this possibility, the energy scale of the condensate is established to be $\sim$1K by the temperature range over which we observe the 1/2 anomaly in Fig. 3. The simplest excitations of a noncondensed 2DEG are plasma oscillations with no applied B field and cyclotron resonances with an applied B field. The SAW frequencies used here are well below both the cyclotron frequency ($\sim 10^{13}$Hz) at ν=1/2 and the short wavelength ($\lambda \sim 1\mu$m) plasma frequencies ($\sim 10^{12}$Hz) of the 2DES. Therefore, any proposed condensed electron state must have both a condensation energy consistent with the 1K scale of the data and a relatively soft excitation mode in the GHz range to explain the coupling we observe.

Such a collective electron state might be a charge density wave (CDW) state or a Wigner lattice. In our system a collective mode to which the SAW couples may be a shear wave mode induced in the CDW. For this to be consistent with our data, very broad shear modes, due to inhomogeneities, would necessarily exist, with the width of the mode presumably ω and k dependent. In this case the linear sound dispersion of the SAW parallels the $\omega \sim k^{3/2}/B$ behavior of the electron lattice,[10]. In this picture the SAW couples to the shear mode more effectively at higher frequencies but does not cross the shear dispersion with $\omega_{SAW} > \omega_{CDW}$ for our range of ω: crossing of the SAW and shear mode dispersions would appear as a double minimum structure in amplitude rather than a single minimum. Without explicit evidence of shear mode crossing by the SAW, experimental conclusion of coupling to a Wigner lattice is highly speculative. Turning to the theoretical studies of Wigner lattice formation at ν=1/2, trial wavefunctions[11] at ν=2/7 to 5/11 show marked energy reduction in the liquid state versus the CDW state in the HF approximation approaching ν=1/2. At ν=5/11 this difference is $\sim$2K for our system. It is difficult to imagine such an increase in the energy per particle sufficient to describe the Wigner crystal as the ground state and yet not have this manifested as a gross feature in simple transport, where the conductivity is relatively calm over this ν range near 1/2. Given that both Wigner lattice formation may be energetically unfavorable at 1/2 and that we do not observe distinct mode crossing, it is strongly suggested that the anomaly at 1/2 is not due to electron lattice formation.

A new type of condensed state which is compressible should be considered. Our finding that increased electron density allows observation of the feature to higher temperatures (Fig. 4) may imply that the Coulomb interaction is important. Both ultrasound absorption and decreased velocity indicate compressibility. Any speculation upon the nature of such a state must crucially examine the precise filling factor dependence.

Our favored description of the state at $\nu = 1/2$ giving us these anomalous sound properties is a picture based in the quasiparticle excitation calculations of Halperin[12]. In an iterative formula using pairwise Coulomb interactions of the quasiparticles, quasiparticle-hole potential energies were generated over the filling factor range $0<\nu<1$. In this energy versus ν approximation a marked upward pointing cusp at $\nu = 1/2$ implies instability (negative compressibility), remedied by break up of the 2D electron system into small regions of larger and smaller density. This spontaneous breaking of the translational symmetry lowers the 2D electron system free energy as the resultant electron patches are separately more stable. The energy scale of this picture is roughly correct according to our data; the temperature below which the anomaly is observed in Figures 3 and 4 is $\sim$1K. In Halperin's calculation, to descend from the 1/2 peak to near the ν extent of our 1/2 feature ($\sim$2%), a difference in energy of $\sim$1K is traversed. This lowering of the free energy is balanced by an increase in the Coulomb energy producing clusters or producing a 2DES density fluctuation (CDW with λ>interelectron separation). If the cluster size or CDW wavelength is L, one can very crudely model the Coulomb energy of the system as $E_{COUL} \sim e^2(\Delta\nu)^2 nL$ while assessing the free energy gain as $E_{COND} \sim E_0(\Delta\nu)$ where E_0 is the CDW energy/particle[11] at 1/2. The sum of these energy terms can be minimized with respect to a density variation ($\Delta\nu$), as occurs in the phase separation process. This gives us a rough estimate of the cluster size L as $\sim 1\mu$m (excluding consideration of the unknown surface energy). This estimate is consistent with the length scales of our system.

The question remains within this picture of spontaneously broken translational symmetry, to what specifically is the SAW coupling in order that we see the $\nu=1/2$ anomaly. The electron clusters should nucleate at impurity sites, such as background inpurities in the GaAs or near fluctuations in the ionized donor layer. These imperfections will act as pinning sites for the electron clusters. Consequently, the center of mass motion of the cluster about the pinning site provides a low frequency excitation mode to which the SAW couples. This pinning should have a broad frequency distribution as it reflects the spatial distribution of the defects in the system as well as the distribution of cluster sizes.

In conclusion, we have observed anomalous SAW propagation at $\nu = 1/2$. While FQHE and IQHE filling factors displayed sound amplitude and velocity consistent with d.c. conductivity measurements of the 2DES, at $\nu = 1/2$ a distinctly different response to the SAW field occurs. This feature is most pronounced at low temperatures (<100 mK) and high SAW frequencies (>500 MHz). The cause of this effect is yet undetermined, however, it is strongly suggested that spontaneously broken translational symmetry may be the cause.

II. $\nu \lesssim 1/5$

The second focus of our studies is the small filling factor range above and below $\nu=1/5$. It is proposed that in this system of quenched kinetic energy, at sufficiently high B (low ν) the 2DES will condense into an electron solid[2]. The filling factor onset of solidification in the infinite field limit has been predicted[13] to be near $\nu=1/6$. However, in the low density samples we employ this number is expected to be somewhat lower. A large increase in d.c. resistance is expected as solidification occurs, and this effect has been extensively characterized[14]. This property is only a sensitive and not a specific signal of Wigner crystallization; magnetic localization may likewise present as an insulating state in a similar ν range. A specific sign of Wigner lattice formation is propagation of a shear mode with the magneto-phonon dispersion $\omega \sim q^{3/2}/B$. Our sound measurement scheme is designed to detect such a mode. As the magnetic field is varied in the proposed Wigner lattice regime, a crossing of the sound dispersion and magneto-phonon dispersion should appear as an additional absorption channel in the system. The following data represent our preliminary investigations into the Wigner lattice filling factor range using the SAW technique.

A prominent discrepancy exists between the ultrasound measurements and the expected sound properties from the d.c. conductivity; see Figure 5. Both below and above 1/5 the longitudinal resistivity grows extremely large, and this has been presumed to signal the possible presence of the electron lattice. At these magnetic field points the sound measurement gives much lower sound velocity shift values than expected from the d.c. transport. This suggests that over this range the high frequency conductivity is much larger than the the d.c. conductivity. The raw SAW amplitude results are consistent with this picture; both immediately below and above 1/5 the amplitude is large and nearly saturated, as expected for large high frequency conductivity.

The B field onset of this discrepancy between the measured SAW velocity and the sound velocity calculated from the d.c. transport moves to higher magnetic fields as the temperature increases. Over the limited temperature range displayed in Figure 6, the measured sound velocity remains low where the d.c. resistivity was large in Figure 5. As the temperature increases the point in B field where the conductivity model deviates from the measured sound properties moves to higher fields. At higher temperatures than shown, the velocity shift increases throughout the small ν range. These results indicate that the presence of the additional high frequency conductivity is a function of both temperature and B field.

The frequency dependence of the SAW velocity shift is shown in Figure 7. At high B fields, the sound velocity shift is larger for lower frequencies. According to the conductivity model, one can infer that the conductivity, $\sigma_{xx}(\omega)$, decreases with decreased frequency in the small ν range. Extending this trend to even lower frequencies suggests that in the d.c. limit the sound velocity shift should be large at small ν. This is consistant with the conductivity model result (Fig. 5).

From these preliminary results we do not see evidence for a distinct mode crossing in the Wigner lattice regime. Rather we observe sound properties that suggest substantial high frequency conductivity ($\sigma_{xx}(\omega>0)>\sigma_{xx}(\omega=0)$) in the small filling factor range. This small filling factor range is roughly coincident with the theoretical predictions of where Wigner lattice formation should occur. Specifically our data show a complicated picture around 1/5 in which

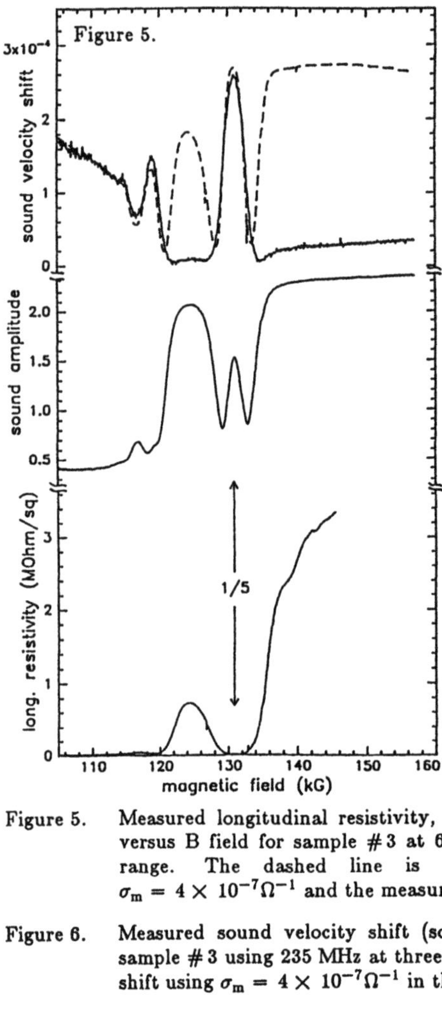

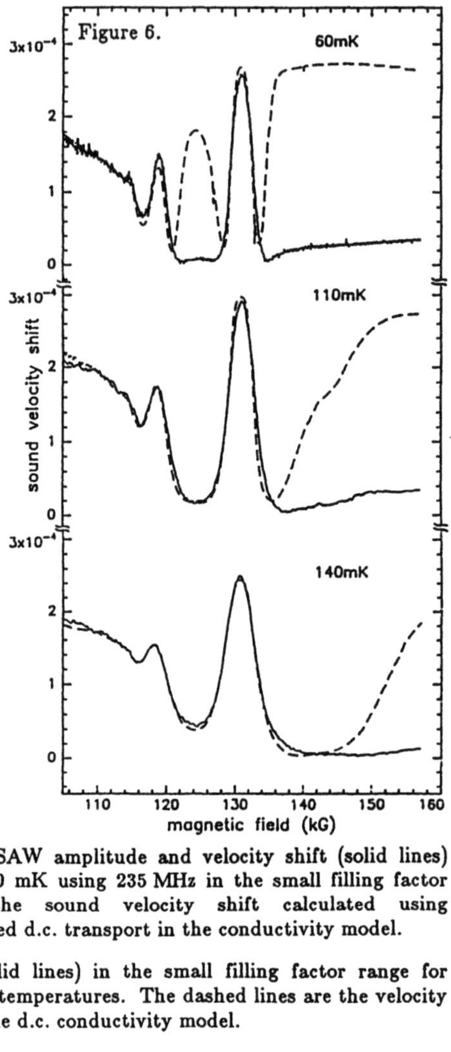

Figure 5. Measured longitudinal resistivity, SAW amplitude and velocity shift (solid lines) versus B field for sample # 3 at 60 mK using 235 MHz in the small filling factor range. The dashed line is the sound velocity shift calculated using $\sigma_m = 4 \times 10^{-7}\Omega^{-1}$ and the measured d.c. transport in the conductivity model.

Figure 6. Measured sound velocity shift (solid lines) in the small filling factor range for sample # 3 using 235 MHz at three temperatures. The dashed lines are the velocity shift using $\sigma_m = 4 \times 10^{-7}\Omega^{-1}$ in the d.c. conductivity model.

additional high frequency conductivity is seen both below and above 1/5, but the FQHE dominates at this principle fraction. Also, the region of the discrepancy between $\sigma_{xx}(\omega=0)$ and $\sigma_{xx}(\omega>0)$ is dependent upon B and T as expected in Wigner lattice formation: as T increases the onset in B should increase.

The presence of additional high frequency conductivity may well be an indication of electron solid formation. In this picture the lower hybrid or shear mode is significantly broadened and the dispersion relation is contorted by the presence of impurities[15]. This results in a broad conductivity peak centered at finite frequencies throughout the wavevector range. Our data may represent a gradual ascent of this peak as we increase frequency, but without crossing of the peak. Observation of this mode crossing is crucial for explicit evidence of electron solidification. However, the data thus far strongly suggest that if the Wigner lattice is present, disorder is playing a significant role and the system may be better described as a glassy state.

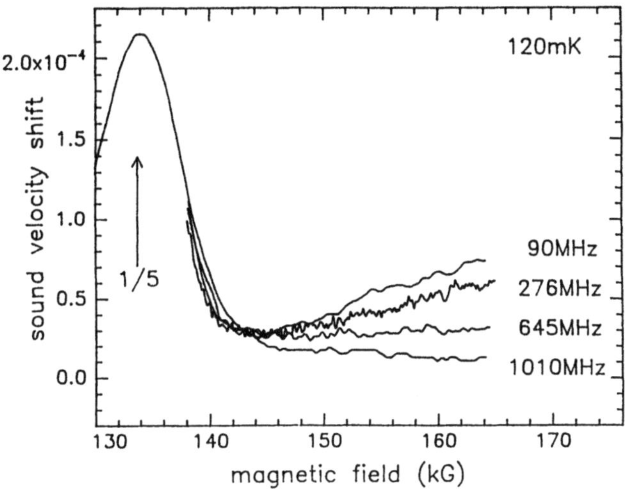

Figure 7. SAW velocity in the small filling factor range for sample #4 at 120mK; fundamental, third, seventh, and eleventh harmonics are shown.

References

1. D. C. Tsui, H. L. Stormer, and A. C. Gossard, Phys. Rev. Lett. *48*, 1559 (1982).
2. Y. Lozovik and V. Yudson, Pis'ma Zh. Eksp. Teor. 22, 26(1975) [JETP Lett. 22, 11(1975)].
3. A. Wixforth, J. P. Kotthaus, G. Weimann, Phys. Rev. Lett. *56*, 2104, (1986).
4. Sec. e.g. J. Heil, J. Kouroudis, B. Lüthi, and P. Thalmaier, J. Phys. C. *17*, 2433 (1984).
5. A. R. Hutson, D. L. White, J. Appl. Phys. *33* 40 (1962).
6. P. Beirbaum, Appl. Phys. Lett. *21*, 595 (1972).
7. K. A. Ingebrightsen, J. Appl. Phys. *41*, 454 (1970).
8. T. W. Grudkowski, M. Gilden, Appl. Phys. Lett. *38*, 412 (1981).
9. H. P. Wei, D. C. Tsui, M. A. Paalanen, A. M. M. Pruisken, Phys. Rev. Lett. *59*, 1776 (1987).
10. H. Fukuyama, Solid State Commun.17,1323(1975).
11. A.H. MacDonald, G. C. Aers, and M. W. C. Dharmawardana, Phys. Rev. B31, 5529(1985).
12. B. I. Halperin, Phys. Rev. Lett, *52*, 1583 (1984).
13. P. K. Lam and S. M. Girvin, Phys. Rev. B30, 473 (1984).
14. R. L. Willett, H. L. Stormer, D. C. Tsui, L. N. Pfeiffer, K. W. West, K. W. West, K. W. Baldwin, Phys. Rev. B38, 7881(1988).
15. H. Fukuyama and P. A. Lee, Phys. Rev. B18, 6245(1978).

Analogy Between Fractional Quantum Hall Effect and Commensurate Flux Phase States on a Lattice

F.V. Kusmartsev

Institut für Theoretische Physik der Universität zu Köln,
W-5000 Köln 41, Fed.Rep.of Germany, and
L.D. Landau Institute for Theoretical Physics, Moscow,
Kosygina 2, V-334, USSR

We study the analogy between fractional Quantum Hall Effect and commensurate flux phases on a lattice. We propose a variational approach for the study of these chiral flux phase states. We classify antiferromagnetic chiral flux phase states and evaluate analytically the corresponding expectation values of the t-J Hamiltonian at any value of filling. There are two types of chiral flux phase states: commensurate and fractional. We have found that state with the minimum energy is a commensurate flux phase state for which the value of the flux of the orbital magnetic field equals the filling and the total spin of this system equals zero.

The discovery of the Quantum Hall Effect and High-Tc Superconductivity opens a new stage in the develop of physics. An understanding of the mechanism of high-temperature superconductivity may perhaps come from improved studies of the ground state properties of the t-J model[1] and of its excitations. The ground state of this model may be generated by fermions which condense on a two-dimensional lattice and carry spontaneous orbital currents [2,3]. In Refs.[4,5] the Kalmeyer-Laughlin state [2] has been generalized to include the case of the square lattice. It has been also shown that the fractional quantum Hall state is equivalent to the RVB state [6]. This picture may be suitably described by introducing the so-called flux phase states, which break the discrete translational symmetry and parity of the lattice [7-9]. In order to describe flux phase states, in Ref.[10] an explicit singlet wave function has been proposed, which is a Gutzwiller projection of the Slater determinant of the noninteracting wave function, describing fermions in a uniform magnetic flux (so-called commensurate flux phase CFP). Some additional interesting ideas about the construction of the wave function describing flux phase states in a Hubbard Hamiltonian with large U have been presented in Ref. [5] where the Gutzwiller projection is carried out on appropriate Jastrow functions, rather than on determinants as in Ref.[10].

Springer Series in Solid-State Sciences, Vol. 101
High Magnetic Fields in Semiconductor Physics III Editor: G. Landwehr

In ref.[11] it has been shown within a renormalized mean field theory[12] that the exchange energy is minimized in CFP states, when the fictitious flux exactly equals the filling. As we will show below in the framework of our approach, this result can be reproduced by direct variational estimation of the expectation value of the exchange energy. In Ref.[13] a CFP state on a finite 4*4 cluster with two holes has been investigated and a qualitive agreement with renormalized mean field theory was obtained. In particular, the magnetic energy exhibits an absolute minimum when the flux exactly equals the filling. However, these results are not completely consistent with the ones obtained in Ref.[14] by means of Monte Carlo calculations. In Ref.[14] it has been shown that the flux phase has lower energies than the energy in the usual Gutzwiller wave function for certain ranges of electron density near half filling and coupling $J/t \sim 1$.

Our analysis is based on the assumption that the ground state correlations and excitations of the model are due to the presence of a spontaneous gauge field which connects our approach with Laughlin's treatment of the fractional quantum Hall effect [15] and Kalmeyer-Laughlin state at half filling [2,4,5]. Following Ref. [5] we choose a many-body wave function as a Gutzwiller projection of the generalized Laughlin function matched, for simplicity, to a two-dimensional square $L \times L$ lattice. Following Ref. [16] we impose periodic boundary conditions on the phase of the single particle density matrix and in this way we obtain the flux quantization that gives us a classification of chiral flux phase states. The flux per plaquette, consisting of the flux of the orbital and of the fictitious magnetic fields, is quantized and may carry either integer multiples of an elementary quantum Φ_e, i.e., a flux $\Phi_n = n\Phi_e$ with $n = 0, 1, 2, ...$, or fractional portions of size $\Phi_n = (n/L)\Phi_e$ or $(n/[L+2])\Phi_e$ with $n = 1, 2, ...$ except such n for which the ratios (n/L), $(n/[L+2])$ become integers. We call the states with fractional quantum numbers as a fractional chiral flux states. These states, indeed, break the lattice symmetries, parity P, translation T, charge conjugation C and the composite symmetries except the CPT symmetry which is still conserved. The existence of such states in antiferromagnets has been predicted in Ref.[9].

We concentrate on the study of the standard two-dimensional t-J model

$$H = -t \sum_{<i,j>\sigma} a^{+}_{i\sigma} a_{j\sigma} + J \sum_{<i,j>} S_i S_j \quad , \tag{1}$$

with the electron hopping-integral t, and Heisenberg constant J. This strength factor measures the exchange interaction between electrons on neighboring lattice sites. The Gutzwiller projected operator $a_{i\sigma}^{+} = c_{i\sigma}^{+}(1 - n_{i,-\sigma})$ where $c_{i\sigma}^{+}(c_{i\sigma})$ creates (destroys) a fermion with spin projection σ ($\sigma = +$ or $-$) at a lattice site i , $n_{i\sigma}$ is the occupation number operator $a_{i\sigma}^{+}a_{i\sigma}$, and the operator S_i is the spin operator. The summations in eq.(1) extend over the lattice sites i and — as indicated by $< i, j >$ — over the associated next nearest sites j.

The microscopic treatment of the fractional quantum Hall effect by Laughlin [15] provides a fruitful prescription for tailoring a set of correlated wave functions which may approximately describe the ground state and excited states of the t-J model (1). We assume the occurrence of a spontaneous gauge field which generates the correlations presented in the eigenstate of the t-J model and which accounts for the orbital currents on the lattice. Exploiting this concept, we are led to the ansatz

$$\Psi_{lmq}(z_{1+}, z_{2+}, ..., z_{N+}; z_{1-}, z_{2-}, ..., z_{N-}) = N_{lmq} \prod_{1=i<j}^{N_+} (z_{i+} - z_{j+})^l$$
$$\prod_{1=i<j}^{N_-} (z_{i-} - z_{j-})^m \prod_{i,j=1}^{N_+,N_-} (z_{i+} - z_{j-})^q P_G(i) \Psi(z_{i+}) \Psi(z_{j-}) \tag{2}$$

for an approximate representation of the ground state (and excited states) of the t-J model. The functions Ψ_{lmq} depend on the coordinates $z_j = x_j + iy_j$ of lattice site j and specify the location of a fermion with a given spin projection. The operator $P_G(i)$ is the usual Gutzwiller projector: $P_G(i) = 1 - n_{i+}n_{i-}$. The function $\Psi(z)$ is the single particle wave function:

$$\Psi(z) = \exp(-\frac{|z|^2}{4l_0^2} - i\frac{\phi_B}{4}) \ , \tag{3}$$

wherein the magnetic length l_0 characterizes the effect of a spontaneous gauge field [11] inducing orbital currents on the lattice. The additional phase $\phi_B/4$ is associated with a fictitious magnetic field B given by $\phi_B = 2\int A ds$, where the gauge $A = (-y, x, 0)B$ is used. Expression (2) generalizes the familiar Laughlin form of the wave function. The factor N_{lmq} normalizes the function (2) to unity. For ensuring the complete antisymmetry of ansatz (3), the exponents l and m must be odd integers, and the parameters q, l_0, and B are trial parameters. The Hamiltonian (1) may describe the dynamics of N_+ electrons with spin projection $\sigma = +$ and N_- electrons with $\sigma = -$. For simplicity, we limit ourselves to a study of the important case where $N_- = N_+ \equiv N$.

Using different values for l and m, one can describe the total hierarchy of anion states proposed in Refs.[17-19]. For simplicity we choose $l = m = 1$ and $q = 0$. We omit all calculations (for details, see [20]). The introduction of the fictitious magnetic field B as a trial parameter gives the minimal energy which corresponds to the marginal extremum of the expectation value. For the ground state we obtain the following formula:

$$E_{gr} = -4t\frac{\rho(1-\rho)}{(1-\rho/2)}L^2\exp(-\rho k/2) - J\frac{\rho^2}{(1-\rho/2)^2}L^2\exp(-\rho k) \quad . \tag{4}$$

At the half-filling ($\rho = 1$), we have for the Heisenberg Hamiltonian the ground state energy which is higher than Neel's state energy:

$$E_{gr} = -4JL^2 exp(-\pi/2). \tag{5}$$

Here $< S_iS_{i+1} >= -0.313$, but for Neel's state $< S_iS_{i+1} >$ =-0.333 [14]. At any value of the filling ρ one can calculate the spin-spin correlator:

$$< S_iS_{i+n} >= -\frac{3\rho^2}{8(1-\rho/2)^2}\exp(-\rho k n^2) \quad . \tag{6}$$

where $n = 1, 2, 3, \ldots$. Thus our flux phase state is a spin-spin antiferromagnetic short-range correlated state. On the other hand due to the periodical behavior of the flux phase of the one-body density matrix for the states constructed here, there is an off-diagonal long-range order [16], i.e. there is strong long-range correlation in the flux phase. Such state does not have a minimal energy at half-filling but nevertheless it can play a role in the hole doping when $\rho < 1$. The states with non-integer quantum numbers, identified also as fractional flux phase states, are characterized by an energy E_{fr} higher than the ground state energy (16),

$$E_{fr} = -J\frac{\rho^2}{(1-\frac{\rho}{2})^2}L^2\exp(-\rho k) \quad . \tag{7}$$

The states, described in the present work, generate spontaneous currents and diamagnetic moments which, effectively, screen the electron-electron spin correlations. The magnetic length of the associated magnetic field plays the role of a Debye radius $l_0^2 \sim 1/\rho$. The excitations of the states obtained here will have this special "plasma" character, that is, they will have a gap which is proportional to the filling ρ and long-range "special Langmuir wave condensate". It would be very interesting to determine the connection between this "magnetoplasmon" and spinons and holons.

The fractional flux phase states presented here break the discrete symmetries such as the parity P, translation T, charge conjugation C and their double products. The PCT symmetry is, however, conserved. We have classified and estimated the ground state energy for the different flux phase states which occur in the t-J model. One can conclude that such fractional flux phase states have anion statistics, breaking T,P and C symmetries. In the limit $t \to 0$ the antiferromagnetic flux phase states breaking such symmetries have been predicted by Wiegmann [9].

The present study gives the structure of flux phase states and the relationship between the properties of the t-J model and the mechanism of the generation of orbital current on the lattice. We are able to explore the structure of the ground state of the t-J model by employing the concept of a spontaneous magnetic field which generates appropriate correlated many-body wave functions of the generalized Laughlin form in conjunction with the fictitious magnetic field added to Gutzwiller projections.

Our results are consistent with those obtained on the basis of the boson-vortex-Skyrmion duality [17-19]. According to such a duality transformation approach, the fermion splits into a boson and a vortex. In the case considered in our paper ($l = m = 1$ and $q = 0$), we have for the state with minimal energy that the each pair of hard core bosons has one-vortex of the statistical flux (the flux of the orbital magnetic field) and an one-vortex of the fictitious magnetic field B . Both vortexes have an one-quantum of flux. On average the statistical flux-density cancels the fictitious magnetic flux-density and the pairs of bosons condense. Such condensation acquires long-range phase coherence. According to Yang [16], it is due to periodical boundary conditions and, as a consequence, due to the flux quantization. For the fractional flux phase states, the average statistical flux-density does not cancel the fictitious magnetic flux-density and, as a result, we have a fractional value of the total flux per elementary plaquette of the lattice. Such a fractional flux corresponds to uncompensative vortexes. It means that these states describe excitations of the vortex-Skyrmion type. But at half-filling $\rho = 1$ the energy of all states coincides. Among these fractional flux phase states there is one with half-quantum of the total flux per particle which has been proposed by Mele [5]. Using the different values for l, m and q, one can also obtain the total hierarchy of possible anion states proposed in [17-19]. This hierarchy is equavalent to Fractional Quantum Hall Hierarchy. The details of such equivalence will be published elsewhere.

In conclusion, we would like to point out that the field B ($B \neq 0$) makes a multiplicate structure on the proposed many-body wave function. This reflects the fact that in many particle system the motion of one electron fill the other ones. It is similar to the creation of some mean field in the system. Recently Fetter and Hanna [22] have shown in Hartree approximation that in anyon gas such kind of field should appear. The value of magnetic field B relates to density of vorteces. This conclusion coinsides with the present treatment. Besides the possible excitations of plasmon type in this two-dimensional antiferromagnetic system, which we attribute to considered flux phase states, there may also exist an acousto-plasmon branch of excitations related to the multicomponent system (pairs of hard-core bosons and vorteces). The sound velocity is probably proportional to the filling factor ρ. This sound is relevant to excitations of holons in the system. This problem calls for a further investigation.

I am very grateful to F.D.M.Haldane,A.I.Larkin, Dung-Hai Lee, V.L.Pokrovskii and M.L.Ristig for very useful discussions. This work has been supported by the A.v.Humboldt foundation.

REFERENCES

[1] T.M.Rice, Physica Scripta T29,72 (1989)
[2] V. Kalmeyer, R. B. Laughlin, Phys. Rev. Lett. 59, 2095 (1987)
[3] J. Affleck, J. B. Marston, Phys. Rev. B37, 3774 (1988)
[4] Z.Zou, B.Doucot, B.S.Shastry, Phys. Rev.B39, 11424, (1989)
[5] E.J.Mele, Phys. Rev. B40, 2670, 1989
[6] P.W.Anderson, Science 235, 1196 (1987)
[7] X. G. Wen, F. Wilczek, A. Zee, Phys. Rev. B39, 11413 (1989)
[8] P. W. Anderson, Physica Scripta T27, 60 (1989)
[9] P. Wiegmann, Physica Scripta T27, 160 (1989)
[10] P.W.Anderson, B.S.Shastry, D.Hristopoulos, Phys. Rev. B40, 8939 (1989)
[11] P.Lederer, D.Poilblanc, T.M.Rice, Phys. Rev. Lett. 63, 1519 (1989)
[12] F.C.Zang, C.Gros,T.M.Rice, H. Shiba, Supercond. Sci. Tech. 1,36 (1988)
[13] D.Poilblanc, Y.Hasegawa, T.M.Rice, Phys. Rev. B41, 1949 (1990)
[14] S.Liang, N.Trivedi, Phys.Rev.Lett. 64, 232 (1990)

[15] T. Chakraborty, P. Pietilainen, The Fractional Quantum Hall Effect, Springer Ser. Solid State Sci., Vol. 85, Berlin, 1988 and references cited therein.
[16] C.N.Yang, Rev. Mod. Phys. 34, 694 (1962)
[17] M.P.A.Fisher, D.H.Lee, Phys. Rev. B39, 2756 (1989)
[18] D.H.Lee, M.P.A. Fisher, Phys.Rev.Lett. 63, 903 (1989)
[19] D.H.Lee, C.L.Kane, Phys. Rev. Lett. 64, 1313 (1990)
[20] F.V.Kusmartsev, Phys.Rev. B43, 6132, 1991
[21] F.V.Kusmartsev, M.L.Ristig, Phys. Rev B44 (1991) 5351
[22] A.Fetter and C.B. Hanna, Conservation lows and anyons: Hartree approximation, preprint, Stanford, (1991)

Fractional Quantum Hall Effect Experiments in Pulsed Magnetic Fields

J.R. Mallett[1,†], *P.M.W. Oswald*[1], *R.G. Clark*[1], *M. van der Burgt*[2], *F. Herlach*[2], *J.J. Harris*[3], *and C.T. Foxon*[3]

[1]Clarendon Laboratory, University of Oxford, Parks Road, Oxford OX1 3PU, UK

[2]Laboratorium voor Lage Temperaturen en Hoge-Veldenfysika, KU Leuven, Celestijnenlaan 200 D, B-3030 Leuven, Belgium

[3]Philips Research Laboratories, Cross Oak Lane, Redhill, Surrey RH1 5HA, UK

[†]Present address: KULeuven, Celestijnenlaan 200D, B-3030Leuven, Belgium

Abstract. Pulsed magnetic fields up to 40 T have been used for transport studies of the fractional quantum Hall effect at temperatures to below 400 mK (10–40 ms pulse duration). For filling factors $\nu < 1$, FQHE structure is resolved in ρ_{xx} at $\nu = 1/3, 2/3, 1/5, 2/5, 3/5, 2/7, 3/7, 4/7$ and 4/9 that is comparable to measurements at steady fields. Activation energy studies of the 1/3, 2/3 (and 2/5) states in pulsed fields to 20 T establish a quantitative basis for extension to the extreme quantum limit, in intense fields $\gg$ 30 T.

1 Introduction

The combination of high magnetic fields and low temperatures is required for many experiments in solid state physics. For magnetic fields beyond the range of current hybrid magnets (31 T), pulsed fields must be used. We combine temperatures below 400 mK, obtained using a specially-designed ^{3}He cryostat, with pulsed fields up to 40 T. This unique facility is used to study the fractional quantum Hall effect (FQHE) in GaAs/GaAlAs heterojunctions in the extreme quantum limit. To date, FQHE experiments in pulsed fields have been limited to pumped ^{4}He temperatures [1]. In our low temperature experiments, activation studies of the $\nu = 1/3$, 2/3 (and 2/5) FQHE states are carried out to establish a quantitative basis for transport measurements in this extreme environment. In shots to 40 T, a well-resolved $\nu = 1/5$ state is observed with an indication of weak structure at $\nu = 1/7$. Overall, the quality of the data indicate that quantitative pulsed field experiments at dilution refrigerator temperatures can be contemplated, to study the FQHE under conditions of intense magnetic fields and the Wigner crystal regime in a hitherto unattainable region of the (B, T) plane. Specific problems that arise are identified.

2 Apparatus

The magnetic field is generated by discharge of a 0.5 MJ, 5 kV capacitor bank (2–40 mF, adjustable) into a coil of reinforced copper wire. A simplified di-

Springer Series in Solid-State Sciences, Vol. 101
High Magnetic Fields in Semiconductor Physics III Editor: G. Landwehr

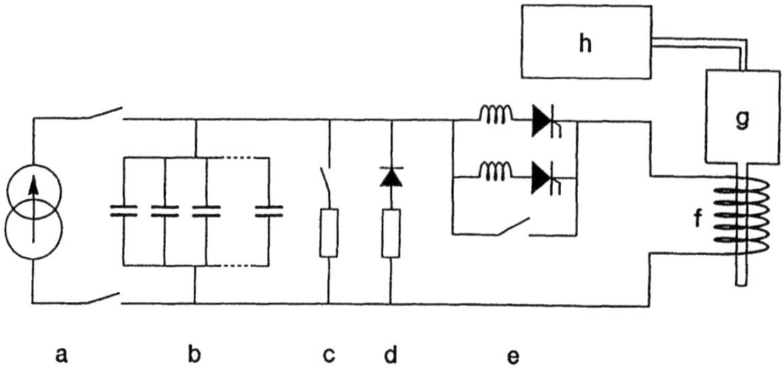

Figure 1: Schematic diagram of the pulsed field facility: (a) 38 kW, 10 kV power supply, (b) 40 mF, 5 kV capacitor bank, (c) discharge resistor, (d) crowbar circuit, (e) mechanical and thyristor switches, (f) pulsed field coil, (g) ^{3}He cryostat, (h) measurement system

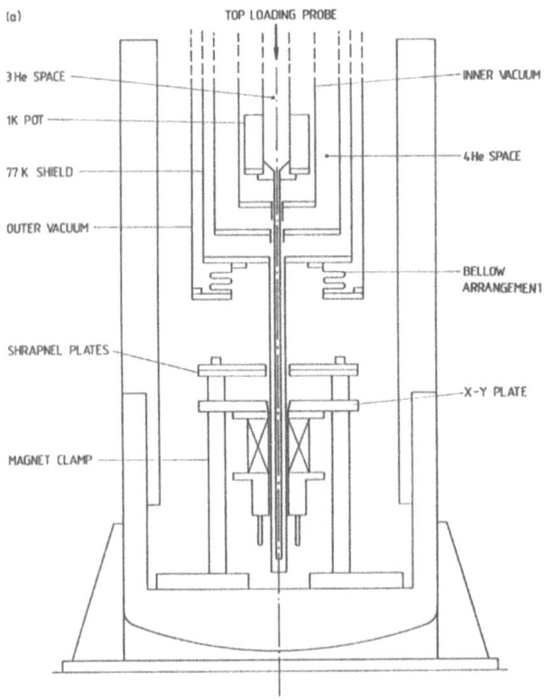

Figure 2: (a) Cryostat assembly showing the glass tails in the magnet, (b) photograph showing cryostat, inner tail and magnet

agram of the installation at KULeuven [2] is shown in Fig. 1. The crowbar circuit (d) is used for extending the pulse duration, depending on the value chosen for the crowbar resistor.

For the experiments described below, a liquid nitrogen-cooled 40 T magnet (f) is used [3]. This has a 20 mm bore, 63 mm outer diameter and is 57 mm long. Recent 20 mm bore coils of Oxford and Leuven have given 55 T for a CuNb coil and 58 T in a glass-fibre reinforced Cu coil respectively, and they will be used to extend the range of our experiments.

The unique design features of the Clarendon ^{3}He system (g in Fig. 1) are in the tail section as shown in Fig. 2a. To reduce eddy current heating, four concentric glass tails (40–50 cm long with 1 mm wall thicknesses and separations) are used, which are spray-painted black. The sample space diameter is 5 mm. Since the cryostat tail is immersed in liquid nitrogen, the room temperature outer vacuum can is terminated above the tail section by a set of bellows, which allows for thermal contraction.

The use of replaceable tails, safety valves and shrapnel plates minimises damage to the main part of the cryostat in the event of a coil explosion. A photograph of the cryostat and its innermost tail is shown in Fig. 2b. The liquid

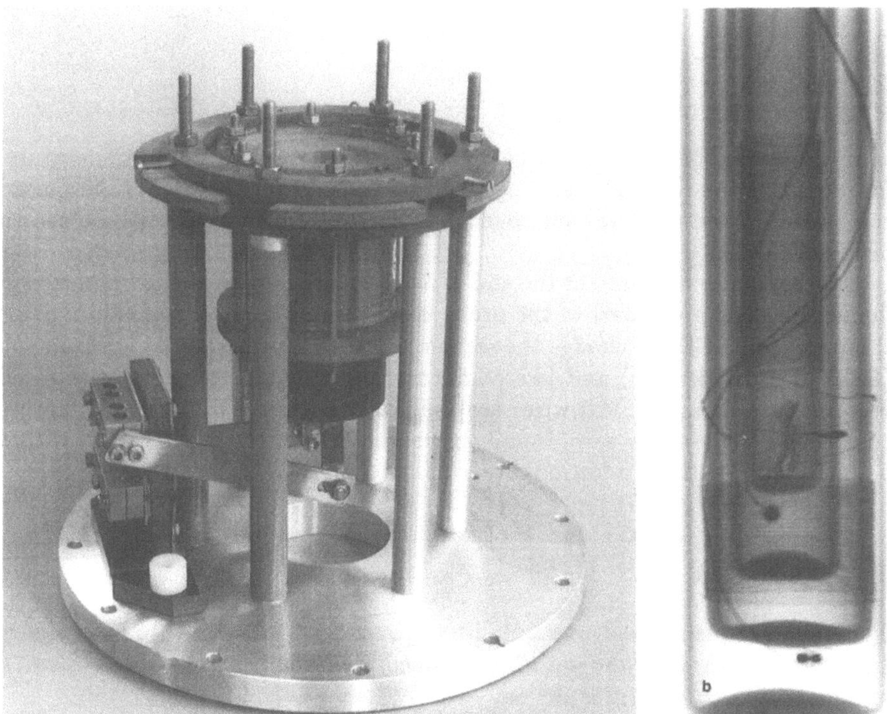

Figure 3: (a) Mounted pulsed field coil,(b) X-ray photograph of concentric glass tails

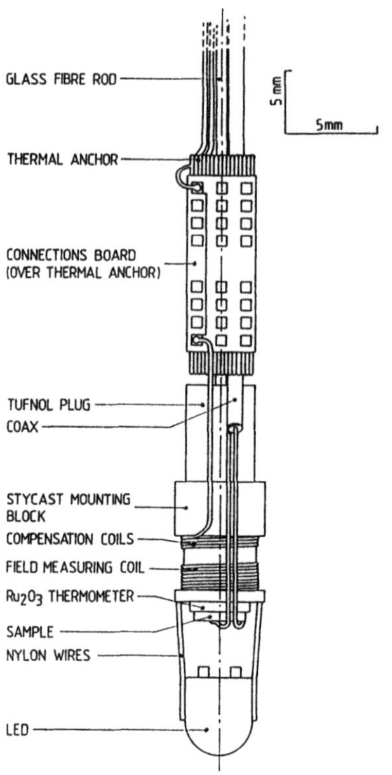

Figure 4: Sample mounting arrangement

nitrogen dewar for the coil can be split into two halves at room temperature, for mounting the coil and inserting the glass tails. A photograph of the pulsed coil assembly and an X-ray photograph of the assembled glass tails are shown in Fig. 3.

The sample is mounted at the end of a top-loading probe and is immersed in liquid ^{3}He. The lower part of the probe is shown in Fig. 4. A ruthenium oxide thermometer, LED (to change the sample electron concentration), pick-up coil (for field measurement) and two compensation coils (to compensate induced voltages) are mounted. All wires are thermally anchored at 1.2 K. Miniature coaxial cable is used for sample and thermometer. For the other wires, twisted pairs are used with an additional thermal anchor.

3 Experiment

The main sample studied was a Hall-bar geometry GaAs/GaAlAs heterojunction, G635, with dark and saturated concentration $n = 0.8$–1.45×10^{11} cm^{-2} and mobility $\mu = 4.0$–7.2×10^6 cm^2/Vs.

The measurement electronics are shown in Fig. 5. A 100 nA DC current is used and both longitudinal (V_{xx}) and transverse (V_{xy}) voltages are recorded. The raw data contains a large induced voltage from the sample and its wiring. This can be subtracted from V_{xx} and V_{xy} prior to amplification using a voltage

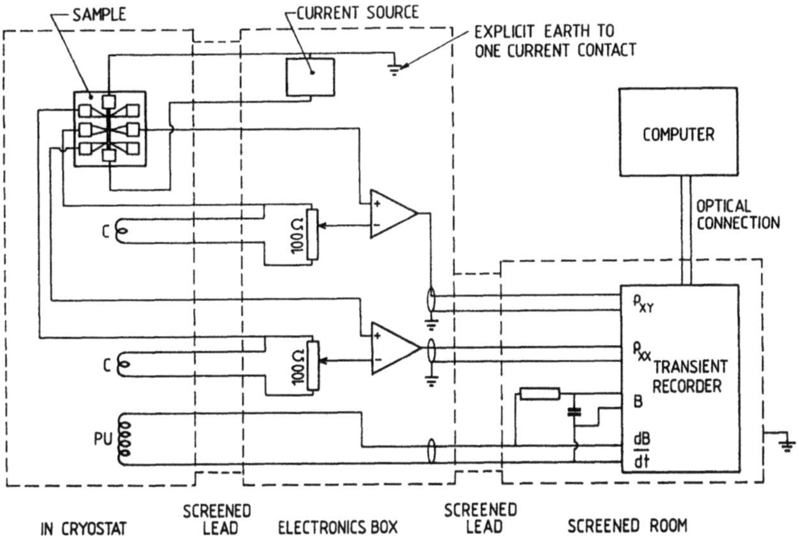

Figure 5: Circuit diagram of the measurement electronics

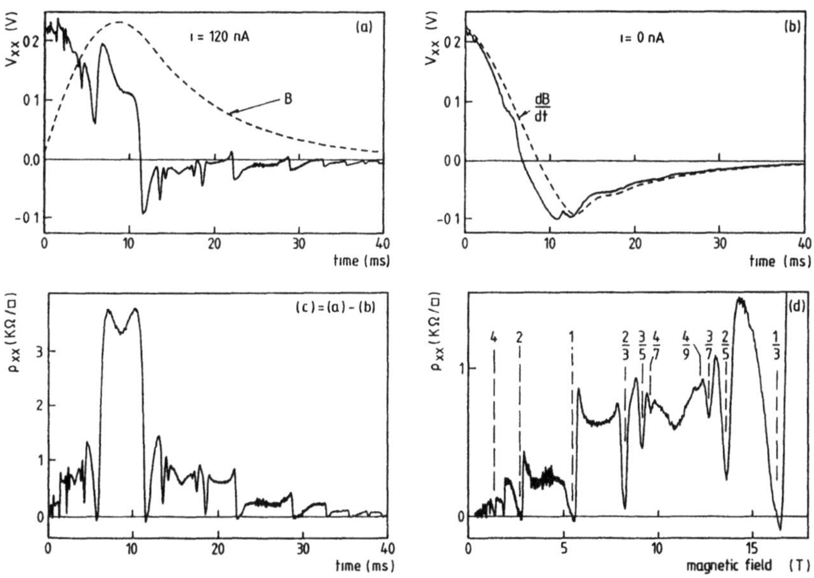

Figure 6: (a) Raw data and field profile, (b) data without applied sample current compared with dB/dt, (c) ρ_{xx} from the subtraction (a)-(b) versus time, (d) ρ_{xx} down sweep versus magnetic field

proportional to dB/dt derived from two compensation coils close to field centre. A four channel transient recorder with a fibre optic link to the computer is used for data recording.

The data acquisition and analysis sequence is summarised in Fig. 6. The raw V_{xx} data at 385 mK are shown in Fig. 6a. Since no compensation is used, both integer and fractional QHE structure are seen superimposed on the dB/dt background. Analogue compensation was found to contribute extra noise and capacitive effects to the signal, and it does not totally eliminate the background. Numerically subtracting a proportion of the pick-up voltage leaves the same remaining background. Instead the probe was carefully wired, and the background eliminated by subtracting from the raw data a second identical shot *without* applied sample current. This is shown in Fig. 6b, where it can be seen that the 'zero current' shot is largely dB/dt, but with an additional component showing induced QHE structure. Fig. 6c and 6d show the results of the Fig. 6a-b subtraction plotted against time and field respectively. The magnetic field is derived by either analogue or numerical integration of the voltage from the pick-up coil (PU). Subtraction of the induced QHE structure is important for quantitative measurements. In Fig. 6d the $\nu = 2/3$ hierarchy out to 4/7 and the $\nu = 1/3$ hierarchy out to the 4/9 daughter state are resolved, and there is a broad V-shaped minimum at $\nu = 1/2$.

4 Activation Studies of the Fractional QHE in Pulsed Fields

To carry out quantitative activation energy measurements of the FQHE, the sample temperature and the height of ρ_{xx} minima need to be accurately known. This provides a stringent test on mK pulsed field experiments of low dimensional semiconductors. Although the results in Fig. 6d qualitatively resemble those from DC experiments, a detailed study of the pulsed field data identifies problems characteristic of the fast sweep rates. These are: (i) the 'zero' baseline for ρ_{xx} measurements is less well defined, (ii) FQHE ρ_{xx} minima have different heights (above baseline) for the up and down sections of the pulse, (iii) Hall traces (ρ_{xy}) no longer have flat quantized plateaux, and (iv) a shift in field is observed between up- and down-going QHE structure (Fig. 7a and b).

The dips and peaks of the Hall trace (Fig. 7a) show a change in polarity between up/down traces characteristic of induced current effects. The fact that integer and strong fractional ρ_{xx} minima do not always correspond to the nominal $\rho_{xx} = 0$ baseline at base temperature (this effect is only small in Fig. 7b but can be pronounced) is due to errors in the cancellation of induced current effects between shots with and without applied current.

The systematic field shift observed in Fig. 7b between up/down traces is more pronounced when fast sweep rates are used and also when the sample resistance becomes high (in shots to high fields). This discrepancy largely arises from an RC phase shift of the signal, where R is the two-terminal sample resistance and C is the overall capacitance in the measurement. It is represented by $V_{in} = V_{out} + (dV_{out}/dt)RC$, where V_{in} and V_{out} are the input and output voltages of an RC-integrator. This can be modelled by computer using C as a fitting parameter. The Fig. 7b data corrected for this RC effect are shown in Fig. 7c. Note that all ρ_{xx} minima are brought into field coincidence; this is achieved using the same C value, of order of that measured on the probe (700 pF). In addition, the FQHE ρ_{xx} minima for up and down traces have

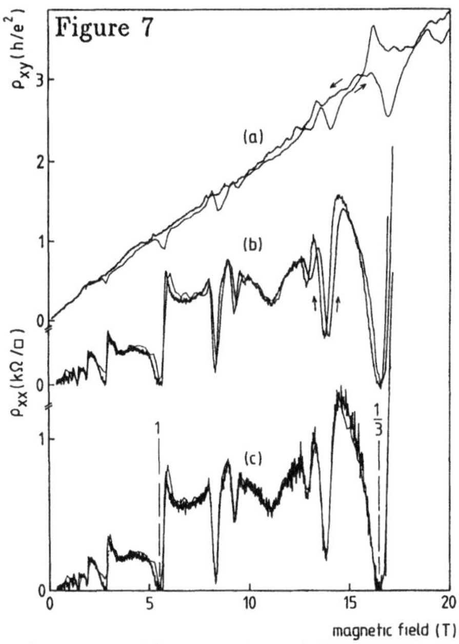

Figure 7: (a) ρ_{xy} plot, (b) ρ_{xx} trace showing shifts in the field positions of the minima between up and down sweeps, (c) RC corrected ρ_{xx} trace

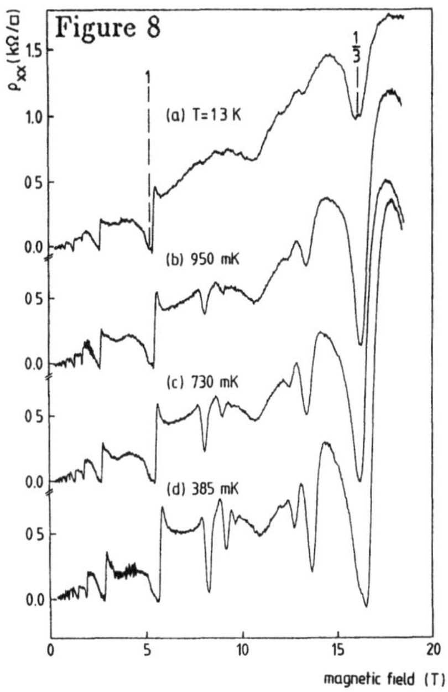

Figure 8: Temperature dependence of ρ_{xx}, using pulse durations of ~40 ms

the same depth following this analysis, which suggests that ρ_{xx} anomalies are largely accounted for by these capacitive effects and more importantly that there is little heating of the 2D electron system in pulses of ~ 10 ms to ~ 20 T.

For large dB/dt's, the anomalies in the data increase and are difficult to separate from effects that would be caused by sample heating; the best data is taken with the longest feasible pulses. This is attained using an assymetric pulse shape (Fig. 6a), making the down sweep a slow exponential decay by shorting the resistance in series with the crowbar diode.

In our activation studies, ρ_{xx} is measured from a baseline that is extrapolated from the ρ_{xx} = '0', $\nu = 1$ and 2 minima. For the lowest temperatures this extrapolates reproducibly to the $\nu = 1/3$ ρ_{xx} minimum. A calibrated ruthenium oxide thermometer was used to measure the temperature of our sample immediately prior to the pulse. Although the magnetoresistance of this thermometer is small, its resistance changed significantly during the pulse and was therefore impractical for in-field thermometry; warming occurred during the first millisecond of the sweep, with the return to the bath temperature taking at least 10–15 ms. In contrast, the depths of well-resolved temperature-sensitive ρ_{xx} minima are equal for up and down sweeps (after RC correction). This indicates that any heating of the 2D electron system is small under the conditions of the activation study. The pretrigger thermometry is therefore used as a reasonable estimate of the electron temperature in field. It is clear

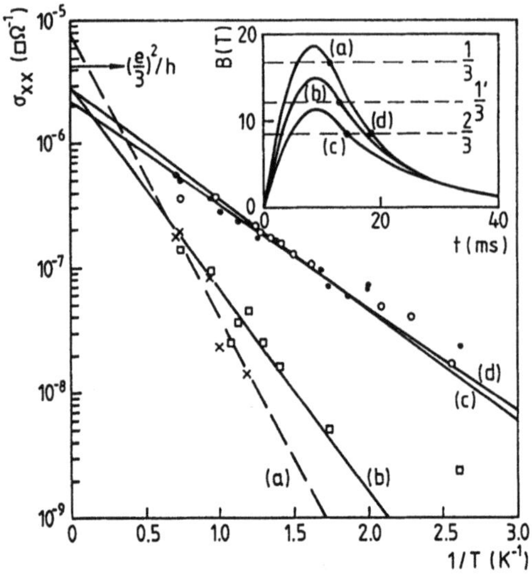

Figure 9: (a) $\nu = 1/3$ at $n = 1.3 \times 10^{11}$ cm^{-2} (b) $\nu = 1/3$ at $n = 1.0 \times 10^{11}$ cm^{-2} (c) and (d) $\nu = 2/3$ at $n = 1.3 \times 10^{11}$ cm^{-2}. The inset shows the position of the data on the field profiles

that 'self-thermometry' in the sample and overlap experiments in DC fields are required in pulsed field studies.

The temperature dependence of FQHE ground states in G635 at saturated concentration is shown in Fig. 8. Corresponding Arrhenius plots for the strong 1/3 and 2/3 fractions are shown in Fig. 9, where 1/3 data taken at dark concentration are also included. Fig. 9a and b correspond to 1/3 data at $n = 1.3$ and 1.0×10^{11} cm^{-2} respectively, where the notation $1/3'$ is used to indicate the lower concentration. For the $n = 1.0 \times 10^{11}$ cm^{-2} the $2/3'$ minimum is weak and not activated in our temperature regime. Fig. 9c and d both represent $\nu = 2/3$ at $n = 1.3 \times 10^{11}$ cm^{-2} using sweeps to different fields, as shown in the inset.

Extrapolated intercepts $\sigma_{xx}^c = \sigma_{xx}(1/T = 0)$ obtained from DC field activation data have been shown to satisfy the relation $\sigma_{xx}^c \simeq (e/q)^2/h$ at $\nu = p/q$ in both the fractional (q = 3, 5, ...) [4,5] and integer (q = 1) [6] regimes, providing an indirect measurement of the fundamental charge associated with activated conduction. It is therefore of interest to convert the pulsed field ρ_{xx} data to σ_{xx} using the relation $\sigma_{xx} = \rho_{xx}/({\rho_{xx}}^2 + {\rho_{xy}}^2)$. Since the ρ_{xy} measurements in pulsed fields are problematic, the ideal values $\rho_{xy} = (q/p)h/e^2$ at p/q filling are used with the measured ρ_{xx} data for Fig. 9. The slope of the Arrhenius plot is a measure of half the FQHE quasielectron-quasihole pair energy gap Δ. Results for $\Delta_{1/3,1/3',2/3,2/5}$ and corresponding σ_{xx}^c values are summarised in Table 1. The 1/3 fit in Fig. 9 (dashed line) is subject to the greatest error, since the activated region of this data was in a difficult temperature regime for the ^{3}He system, due to the strength of the 1/3 state at

Table 1. Intercept and activation energy values for G635.

ν	σ_{xx}^c (e^2/h)	$\Delta/2$ (K)
1/3	1.7/9	5.2
1/3′	0.67/9	3.8
2/3	0.67/9	2.1
2/3	0.51/9	1.8
2/5	0.71/25	1.1

high fields. For this 1/3 state ρ_{xx} structure at base temperature did not define an unambiguous minimum positon and fell below the '0' baseline extrapolated from the integer ρ_{xx} minima. The 2/3 data in Fig. 9d are subject to greater errors than the Fig. 9c data, due to the larger dB/dt. Overall the Fig. 9b, 9c and 2/5 data are in reasonable agreement with FQHE results on similar samples measured in DC fields, both with respect to the magnitude of Δ and the value of σ_{xx}^c. There is more scatter in the pulsed field measurements in Fig. 9, which should be reduced by the use of ultra-fast AC lock-in techniques currently being developed. Our interest is to extend this basis to new physics in the high field region $B > 30$ T.

5 High Field Studies (> 30 T)

High field sweeps to 31 and 37 T in G635 at 390 mK are shown in Fig. 10. For the trace to 31 T, a 1/5 minimum is resolved, which is absent in the 37 T trace. The 1/3 and 2/7 structure are also weaker in the 37 T shot.

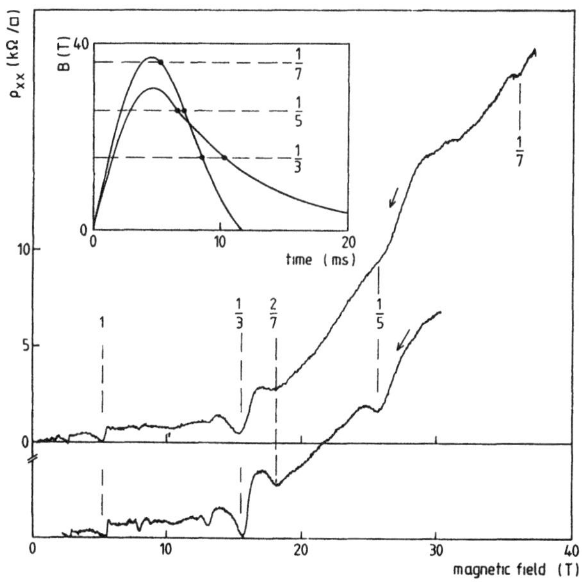

Figure 10: ρ_{xx} up to $\nu = 1/5$ and $\nu = 1/7$ positions

The data up to 37 T possibly shows some weak ρ_{xx} structure at $\nu = 1/7$. This structure cannot be unequivocally attributed to a $\nu = 1/7$ FQHE state for which there is evidence in other experiments [7], since the results at high fields are less reliable than those at lower fields. For higher fields the pulse duration must be shortened (see Fig. 10 inset) to avoid overheating the coil. Therefore the dB/dt background is greatly increased and the ρ_{xx} structure due to induced currents can be as large as that due to the applied current. Additional noise is often seen in the data and small differences in the peak field of the shot with and without current can be important.

6 Conclusions

These experiments demonstrate that pulsed magnetic fields can be successfully combined with the required mK temperatures for quantitative transport studies of low dimensional semiconductors; the fractional QHE is a particularly stringent test. Extension of this work to fields up to 60 T, dilution refrigerator temperatures and optical measurements is planned. Quantitative investigations of the region where Wigner crystallization is anticipated is of particular interest, but some experimental problems remain.

7 Acknowledgements

We would like to acknowledge the expert help of H. Jones, who designed and built the coil, L. Van Bockstal and G. Heremans. This work is financially supported by NFSR (*Nationaal Fonds voor Wetenschappelijk Onderzoek*), Belgium and SERC, UK. JRM has a fellowship from the *Onderzoeksraad* of KULeuven.

References

[1] R.E. Horstman, E.J. van den Broek, J. Wolter, A.P.J. van Deursen and J.P. André: In *Proceedings of the 17th International Conference on the Physics of Semiconductors, San Fransico, California, USA, August 6-10, 1984*, ed. by J.D.Chadi and W.A.Harrison (Springer-Verlag, New York 1985) p.295

[2] F. Herlach, L. Van Bockstal, M. van der Burgt and G. Heremans: Physica B **155**, 61 (1989)

[3] H. Jones, F. Herlach, J.A. Lee, H.M. Whitworth, D.J. Jeffrey, D. Dew-Hughes and G. Sherratt: IEEE Trans. Magn. **24**, 1005 (1988)

[4] R.G. Clark, J.R. Mallett, S.R. Haynes, J.J. Harris and C.T. Foxon: Phys. Rev. Lett. **60**, 1747 (1988)

[5] A. Sachrajda, R. Boulet, Z. Wasilewski and P. Coleridge: to be published

[6] J.V. Branch, R.G. Clark, S.J. Collocott, C. Andrikidis, G. Griffiths, J.J. Harris, C.T. Foxon and J. Chalker: submitted to Phys. Rev. Lett.;
R.G. Clark, S.R. Haynes, J.V. Branch, A.M. Suckling, P.A. Wright, P.M.W. Oswald, J.J. Harris and C.T. Foxon: Surf. Science **229**, 25 (1990)

[7] V.J. Goldman, M. Shayegan and D.C. Tsui: Phys. Rev. Lett. **61**, 851 (1988);
R.G. Clark, R.A. Ford, S.R. Haynes, J.F. Ryan, A.J. Turberfield, P.A. Wright, C.T. Foxon and J.J. Harris: these proceedings

Part III

Quantum Wires and Quantum Dots

Quantum Wires in Magnetic Fields

T. Ando and H. Akera

Institute for Solid State Physics, University of Tokyo,
7-22-1 Roppongi, Minato-ku, Tokyo 106, Japan

A review is given of recent theoretical studies in transport properties of quantum wires in magnetic fields. The topics cover the localization of edge states, its crossover from one to two dimensions, effects of boundary roughness scatterings on the Hall effect and the magnetoresistance, and the ballistic conduction at quantum point contacts.

1. Introduction

Recent developments in the technology of ultrafine lithography and crystal growth, such as molecular-beam epitaxy, have enabled fabrication of ultra-narrow wires at semiconductor heterostructures. The width of the wires is much smaller than the electron mean free path and comparable to the Fermi wavelength, giving only a few occupied subbands. Such quantum wires have revealed quite new features of transport, such as the Aharonov-Bohm effect [1,2], conductance fluctuations [3,4], and anomalies in the low-field Hall effect [5-13]. In addition, the quantum wires provide various fundamental problems concerning their transport properties. The purpose of the present paper is to give a brief review of some of the recent theoretical studies on the transport properties of quantum wires in magnetic fields.

One of the key features in the transport in high magnetic fields is edge states which are localized in the vicinity of the wire boundary. Section 2 deals with the localization of the edge states due to mixing with bulk Landau levels and its dependence on the wire width. Effects of inherent boundary roughness scatterings are discussed in Sec. 3. The scattering gives rise to an interesting current distribution among different subbands, which manifests itself as a suppression of the weak-field Hall effect and also a positive magnetoresistance. In Sec. 4, results of quantum-mechanical calculations of the ballistic transport at quantum point contacts are presented. It will be shown that the subband wave function can be directly observed through the magnetic-field dependence of the conductance at a series of two point contacts.

2. Edge States – Mixings and Localization

It is well known that states are all localized in one dimension however small the randomness may be. The situation can change drastically when a strong magnetic field H is applied perpendicular to the wires. In strong magnetic fields, ie., when $L \gg l$ or $\hbar\omega_c \gg E_L$ with $E_L=(\hbar^2/2m)(\pi/L)^2$ and $\omega_c=eH/mc$, the subbands are well separated into edge states and bulk Landau levels [14-22], where L is the wire width and l is the magnetic length given by $l=(c\hbar/eH)^{1/2}$. The edge states correspond to classical skipping orbits and are well confined in the vicinity of the wire edges. They become more and more extended with increasing magnetic field as long as their energy is away from those of bulk Landau levels, because the overlapping of two edge states moving in opposite directions becomes smaller and probabilities of the backscattering are reduced

Springer Series in Solid-State Sciences, Vol. 101
High Magnetic Fields in Semiconductor Physics III Editor: G. Landwehr

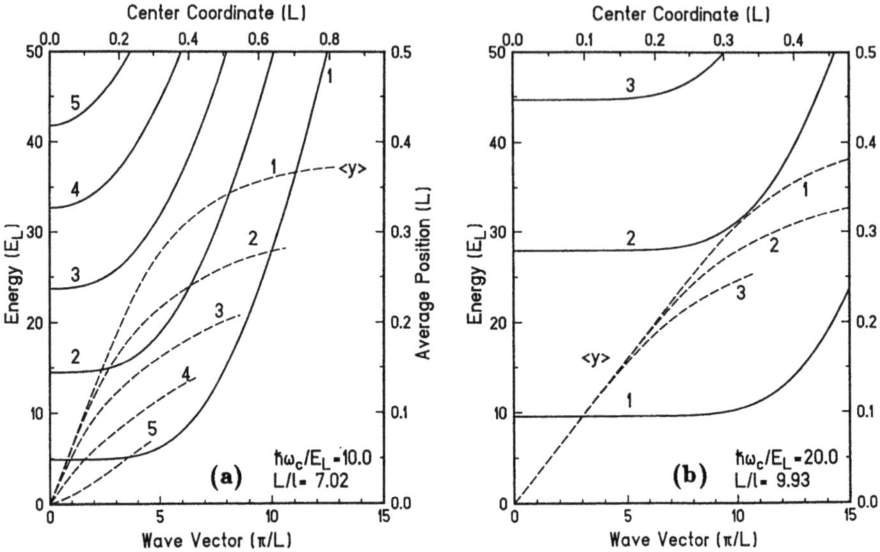

Fig. 1 Examples of the energy dispersion calculated in a lattice model. (a) $\hbar\omega_c/E_L=10$ and (b) 20. The solid lines represent the subband energy and the dashed lines the average value of y.

considerably. When their energy lies in the region of broadened bulk Landau levels, they are mixed strongly with bulk states and localized.

The extended edge states are one of the key features in gedanken experiments used for an explanation of the quantum Hall effect. As a matter of fact, the quantization of the Hall resistance into h/je^2 ($j=1,2,\ldots$) can be explained easily under the assumption that the edge states are not influenced by scatterings (see reviews on the quantum Hall effect [23-25] and references cited therein).

Quite recently, a numerical study of the localization of the edge states has been performed in high magnetic fields [26]. Use is made of a lattice model which simulates a continuum system and the randomness is introduced through fluctuation of site-diagonal energies. Figure 1 gives examples of the calculated subband-energy dispersion for $\hbar\omega_c/E_L=10$ ($L/l=7.02$) and for $\hbar\omega_c/E_L=20$ ($L/l=9.93$). The energy of the bottom of the low-lying subbands is slightly smaller than $(N+1/2)\hbar\omega_c$ ($N=0,1,\ldots$), due to the presence of a slight nonparabolicity in the lattice model.

Examples of calculated density of states and the inverse localization length $\alpha(E)$ in strong magnetic fields are given in Fig. 2. Effects of scatterings are characterized by $\Gamma_L/\hbar$ representing the probability of backscatterings for electrons with wave vector $k=\pi/L$ in the lowest subband or the corresponding mean free path Λ_L. With the increase of the magnetic field, the density of states becomes closer to that in two dimension at the energies of the bulk Landau levels, while it is given well by that of the unperturbed edge states for energies between the Landau levels. The broadening of the bulk Landau levels has been calculated in the self-consistent Born approximation [27-29] as $\Gamma_H=[(2/\pi)\hbar\omega_c\Gamma]^{1/2}$, where Γ is the broadening in two dimensions in the absence of a magnetic field and related to Γ_L through $\Gamma=(2\pi/3)\Gamma_L$.

We should note that the energy dependence of $\alpha(E)$ is completely different from that in the cylinder geometry for which there are no edge states. In the cylinder geometry, the inverse localization length takes a local minimum near the center of each

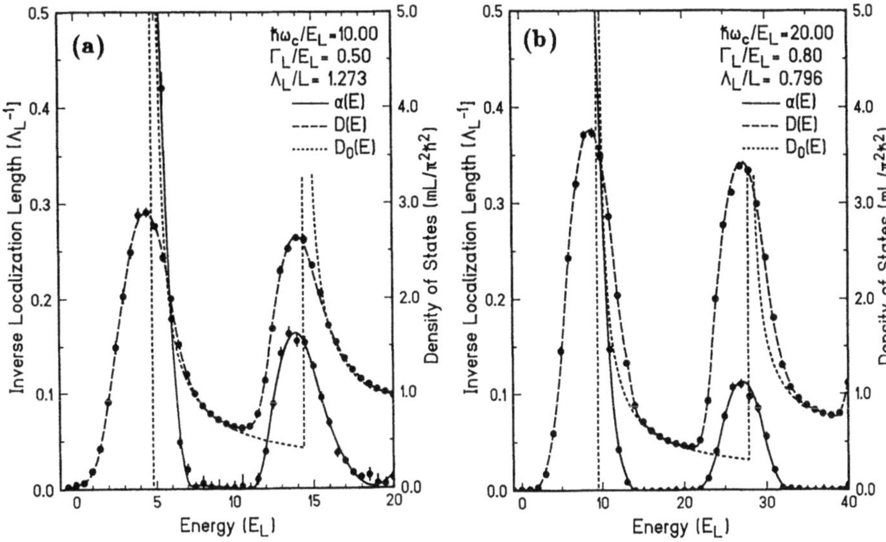

Fig. 2 Calculated inverse localization length (solid line) in units of Λ_L^{-1} and density of states (dashed line). The dotted line represents the density of states in the absence of scattering. (a) $\hbar\omega_c/E_L = 10$ and $\Gamma_L/E_L = 0.5$. (b) $\hbar\omega_c/E_L = 20$ and $\Gamma_L/E_L = 0.8$.

Landau level [$\alpha \sim 0.4/L$ for the first excited Landau level, for example] and becomes large at the spectral gaps between adjacent Landau levels ($\alpha \sim l^{-1}$) [30,31]. In the quantum wire, on the other hand, the inverse localization length becomes extremely small when the energy is away from the density of states of the bulk Landau levels. There is a remarkable difference in the behavior between the lowest and first excited Landau levels, which is the reflection of the obvious fact that the edge states are present only above the lowest Landau level. In the energy region corresponding to the first excited Landau level, $\alpha(E)$ becomes large due to strong mixings with bulk states.

Figure 3 gives $\alpha(E)$ in the energy range corresponding to the first excited Landau level for varying width L/l for a fixed value of Γ_H ($\Gamma_H/\hbar\omega_c = 0.2$). The energy is measured from the bottom of the first excited Landau level in the absence of randomness and normalized by Γ_H. The energy corresponding to the density-of-states peak is shifted to the lower energy side due to the quantum-mechanical level-repulsion effect and is close to $-0.25\Gamma_H$. It is clear that $\alpha(E)$ decreases with increasing width, showing that the edge states become more and more extended for wider wires.

The dependence of $\alpha(E)$ on L reflects how the amplitude of the edge-state wave function decays inside the wire away from the boundary. Figure 4 shows the logarithm of $\alpha(E)$ as a function of the wire width for energies within the broadened bulk Landau states. The results can be fitted to $\alpha(E) \propto \exp(-\beta L)$ except those for $L/l = 5$ which exhibits asymmetry due to the presence of a small probability of the backscattering (especially through the more extended edge states associated with the first excited Landau level). This suggests that the wave function decays exponentially away from the boundary. Obtained $\beta(E)$'s are listed in the figure. The extent of the edge states in the vicinity of the center of the first excited Landau level is roughly $\beta^{-1} \sim 15l$. We should note that $\beta(E)$ does not exhibit strong dependence on energy. This fact suggests that it is not determined by the localization length of the Landau levels in two dimension.

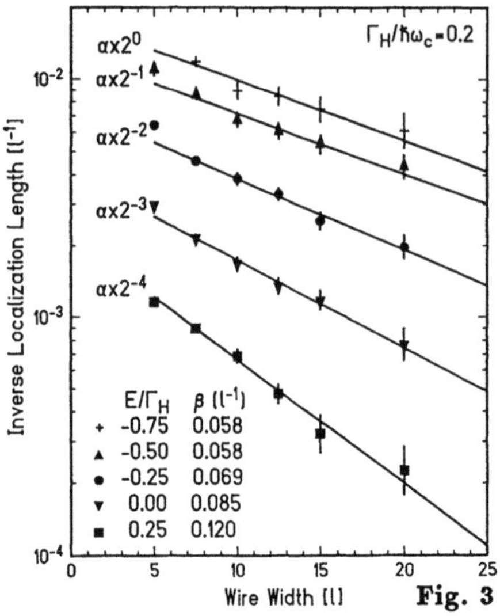

Fig. 3

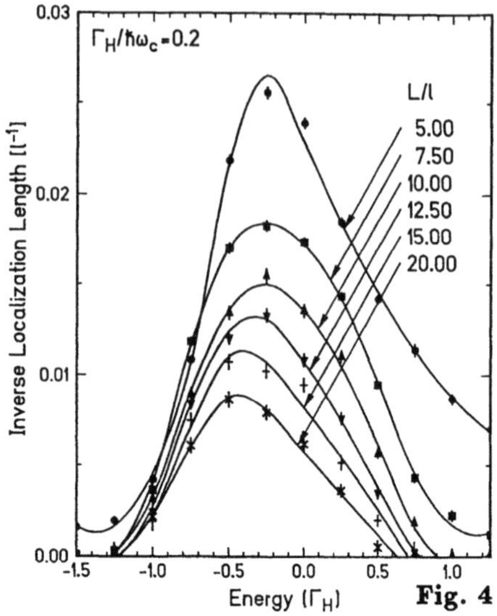

Fig. 4

Fig. 3 Calculated width dependence of $\alpha(E)$ for $\Gamma_H/\hbar\omega_c = 0.2$. The energy is measured from that of the first excited Landau level in two dimension and normalized by Γ_H.

Fig. 4 Logarithm of $\alpha(E)$ as a function of the wire width. The decay rate $\beta(E)$ obtained by fitting to $\alpha \propto \exp(-\beta L)$ is given in units of l^{-1}.

The edge states are always more extended than the bulk states. This does not mean that the Hall current is carried by the edge states in the quantum Hall effect in the two-dimensional limit. In the two-dimensional system at nonzero temperatures, the diagonal element of the conductivity tensor σ_{xx} is always nonzero and the requirement for the least dissipation leads to a homogeneous Hall field. In the presence of the Hall field, the current is carried by the bulk extended states and the contribution from the edge states is negligible. Note that the Fermi level lies always in the range of bulk Landau levels for wide wires in high magnetic fields. This is because the density of states corresponding to the edge states is negligibly small and the electron concentration remains fixed instead of the chemical potential when the magnetic field is varied.

3. Boundary Roughness Scattering

Boundary roughness is considered to be inherent to the artificially confined quantum wires and can constitute a major cause of scatterings. Classically, effects of boundary roughness are treated in terms of a single parameter p representing the probability of a specular scattering [32]. It is clear that such a treatment is inadequate and a quantum mechanical calculation of roughness effects is necessary in quantum wires. Let $\eta_{jk}(y)$ be the normalized wave function for the motion in the y direction perpendicular to the wire direction. Then, the boundary roughness $\Delta_\pm(x)$ gives rise to the perturbation whose matrix element is given by [33]

$$(jk|\mathcal{H}'_\pm|j'k') = \frac{\hbar^2}{2m}\frac{\partial\eta^*_{jk}}{\partial y}\bigg|_{\pm L/2}\frac{\partial\eta_{j'k'}}{\partial y}\bigg|_{\pm L/2}\int dx\,\Delta_\pm(x)\exp[-i(k-k')x]. \tag{3.1}$$

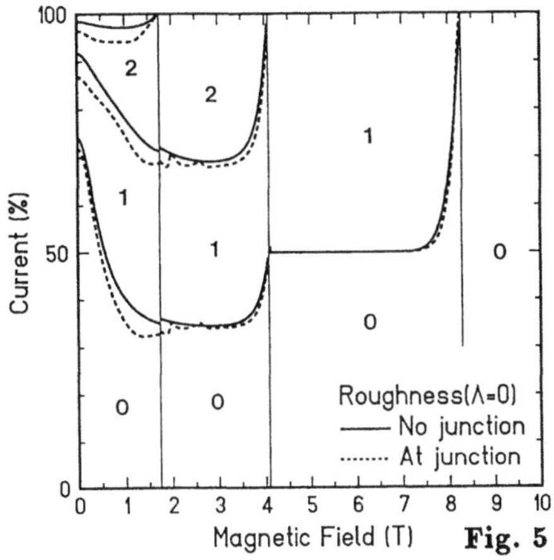

Fig. 5

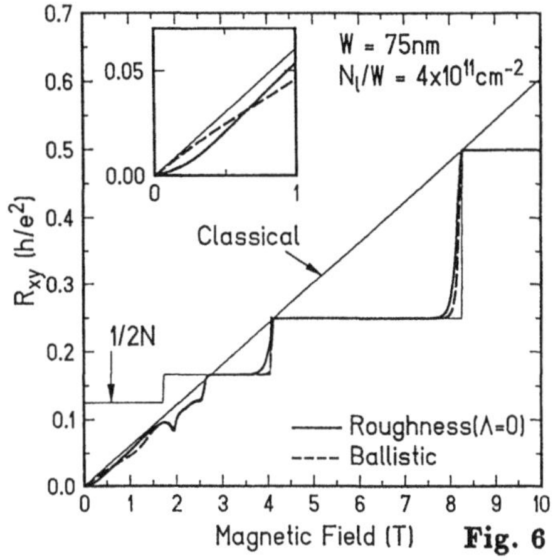

Fig. 6

Fig. 5 Current distribution among subbands for scatterings from boundary roughness. The solid lines represent the distribution in a single wire and the dashed line that in the crossed-wire geometry. $L=750$ Å.

Fig. 6 Calculated Hall resistance for scatterings from boundary roughness. $L=750$ Å. Inset: blowup of the region of low magnetic fields.

The roughness is conventionally characterized by its root mean square deviation Δ and correlation length d, i.e.,

$$\begin{aligned} &\langle \Delta_+(x)\rangle = \langle \Delta_-(x)\rangle = \langle \Delta_+(x)\Delta_-(x')\rangle = 0, \\ &\langle \Delta_+(x)\Delta_+(x')\rangle = \langle \Delta_-(x)\Delta_-(x')\rangle = \Delta^2\exp[-(x-x')^2/d^2]. \end{aligned} \tag{3.2}$$

It is straightforward to calculate the conductivity if we use the Boltzmann transport equation. The results show that the boundary roughness causes a very peculiar current distribution among subbands in the absence of a magnetic field [34]. Figure 5 presents an example of the calculated current distribution in the limit $d \ll \lambda_F$ with $d\Delta^2$ fixed for a wire with width 750 Å. In the low field region below $\sim$0.5T, the current is mainly carried by the ground subband. This is because the derivatives of the wave functions at the boundaries are close to $(\eta_n')^2 \propto n^2$ and larger for higher subbands. It is closely related to the fact that the resistivity vanishes in the classical case [35-38]. In high magnetic fields, the current is carried by different subbands almost equally except at subband onsets. This can again easily be understood if we consider the fact that the forward scatterings among subbands become dominant for edge states.

This peculiar current distribution among subbands causes various interesting phenomena. Figure 6 shows an example of the calculated Hall resistance in a crossed wire geometry in the presence of strong boundary roughness [34]. The Hall resistance is strongly suppressed from the classical linear field dependence in the weak-field region. The reason is quite clear: The lowest subband corresponds classically to the state moving predominantly in the direction of the wire and has a smaller probability of being transmitted into the wire corresponding to the Hall probes, leading to the reduction of the Hall voltage. Note that the rounding of corners at the crossing of the two wires is not taken into account in the calculation. Therefore, the present roughness effect is completely different from the so-called nozzle effect [39] or the scrambling effect [40] in which the large rounding of the corners is essential.

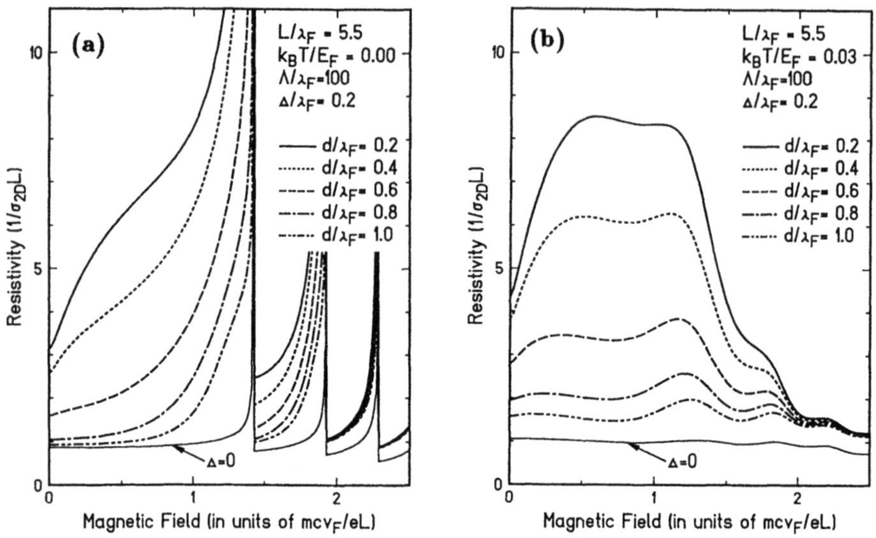

Fig. 7 Calculated magnetoresistance as a function of the magnetic field for a wire with width $L/\lambda_F = 5.5$ for different correlation lengths d. (a) $k_BT/E_F = 0$. (b) 0.03. The root-mean-square deviation is fixed at $\Delta = 0.2\lambda_F$ and the mean free path in two dimension is $\Lambda = 100\lambda_F$. Note that the same results can be obtained for different values of Δ and Λ as long as $\Delta^2\Lambda/\lambda_F^3$ remains the same.

Another interesting phenomena is a positive magnetoresistance. This occurs due to the fact that the lowest subbands, carrying most of the current in the absence of a magnetic field, become equally affected by the boundary roughness in magnetic fields. As a matter of fact, Fig. 5 demonstrates that the current distribution becomes rapidly equalized among subbands with increasing magnetic field. Figure 7 gives an example of the resistivity calculated using the Boltzmann equation in the presence of strong boundary roughness scatterings. The resistivity increases first and then decreases with magnetic field, although a strong quantum oscillation appears when the Fermi level crosses subband bottoms at vanishing temperature. This positive magnetoresistance becomes much clearer at nonzero temperatures where the oscillation is smeared out. The positive magnetoresistance due to effects of boundary roughness has been observed in wires formed on a GaAs/AlGaAs heterointerface [41]. There have been reported various classical calculations in the case of the completely diffuse scattering, $p=0$, [35-38], which have successfully explained a similar positive magnetoresistance observed in metallic thin films [42].

4. Quantum Point Contacts

The quantum point contact is an ideal system where the quantum effect on the transport can be directly studied. The typical example is the quantization of the conductance into the integer multiple of $e^2/\pi\hbar$ [43,44]. Following the interesing experimental finding, there have appeared various calculations of the conductance across narrow constrictions [45-56]. The conductance of various structures consisting of several point contacts has been measured [57,58] and analyzed in terms of the collimation effect [59].

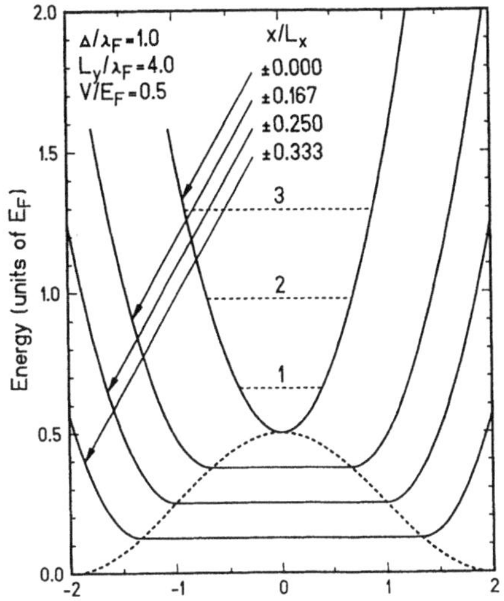

Fig. 8

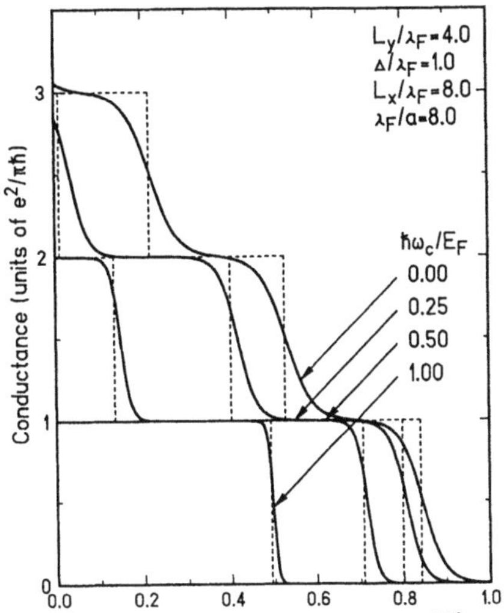

Fig. 9

Fig. 8 The potential profile of a model single quantum point contact. The bottom potential is raised gradually starting at $x = -L_x/2$, takes a maximum V at the middle $x = 0$, and decreases. The width of the channel, L_y at $x = \pm L_y/2$, becomes narrower in the central region and at $x = 0$ the confinement potential is parabolic, i.e., $E_F(y/\Delta)^2 + V$. The horizontal dotted lines represent the subband bottoms corresponding to the narrowest region.

Fig. 9 An example of the calculated conductance at a quantum point contact as a function of the energy of the point-contact bottom V. The dotted lines represent the quantized conductance in the adiabatic limit, obtained by the number of the allowed channels.

Quite recently, a numerical method has been developed to perform quantum mechanical calculations of the conductance at the point contacts. The method is based on the conventional Green's function technique combined with Landauer's conductance formula [60] and allows us to study magnetic-field effects as well as scattering effects. Figure 8 shows the potential profile of the model point contact used for calculations. Figure 9 gives an example of the conductance as a function of the energy of the point-contact bottom for different magnetic fields. It is clear that the conductance quantization becomes better in higher magnetic fields as has been observed in experiments [61] and discussed theoretically [62,63]. This is closely related to the appearance of edge channels discussed in the previous sections, which reduces backscattering probabilities at the point contact and the electron motion becomes more adiabatic.

Combinations of several point contacts are particularly interesting, because they provide means to measure the subband wavefunctions directly [64]. Figure 10 schematically shows the configuration of a series of two point contacts. Infinitely long wires are attached at both top and bottom sides. Figure 11 gives the calculated conductance as a function of the magnetic field or the distance of the two point contacts divided by the classical cyclotron radius. The channel number at the collector contact is fixed at 3 and that of the emitter is varied. The conductance depends essentially on the channel number only except for the presence of oscillations due to quantum interference. The interference pattern is almost chaotic in the sense that it depends sensitively on the detailed form of the confinement potential as well as the distance between the emitter

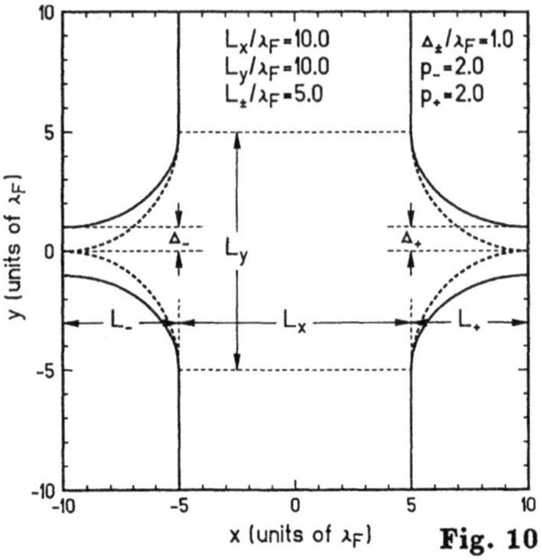

Fig. 10

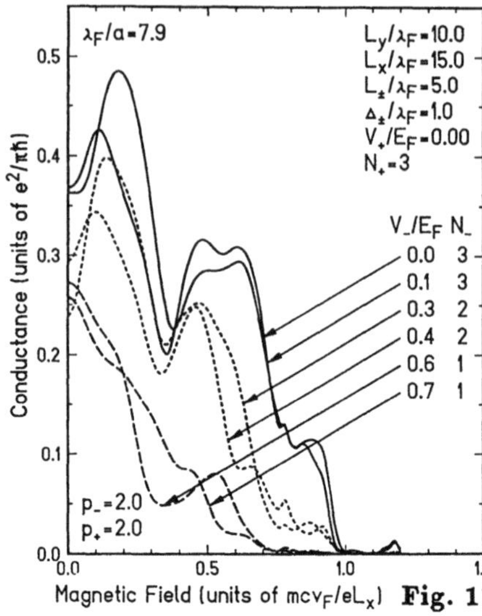

Fig. 11

Fig. 10 A schematic illustration of a series of two point contacts. Electrons are injected from the left hand side and collected at the the right hand side. Only out-going electron waves are allowed at the top and bottom. At the left (-) and right (+) point contact, the potential is parabolic and given by $E_F(y/\Delta_\pm)^2+V_\pm$. The boundary changes according to $x^p+y^p=a^p$ $(a>0)$ with $p=p_\pm$ in the vicinity of the contacts.

Fig. 11 An example of the calculated conductance across the two point contacts as a function of a magnetic field for different numbers of the channels at the emitter point contact. The channel number at the collector contact is fixed as 3.

and collector contacts. Despite such interference oscillations we can clearly identify the form of the subband wavefunctions at the emitter quantum point contact in agreement with experiments [64].

Acknowledgments

The work is supported in part by the Industry-University Joint Research Program "Mesoscopic Electronics" and by the Grant-in-Aid for Scientific Research on Priority Area "Electron Wave Interference Effects in Mesoscopic Structures" from Ministry of Education, Science and Culture, Japan.

References

[1] K. Ishibashi, K. Nagata, K. Gamo, S. Namba, S. Ishida, K. Murase, M. Kawabe, and Y. Aoyagi, Solid State Commun. **61**, 385 (1987).

[2] M. Mizuno, K. Ishibashi, S.-K. Noh, Y. Ochiai, Y. Aoyagi, K. Gamo, M. Kawabe, and S. Namba, Jpn. J. Appl. Phys. **28**, L1025 (1989).

[3] K. Ishibashi, Y. Takagaki, K. Gamo, S. Namba, S. Ishida, K. Murase, Y. Aoyagi, M. Kawabe, Solid State Commun. **64**, 573 (1987).

[4] G. Timp, P.M. Mankiewich, P. deVegvar, R. Behringer, J.E. Cunningham, R.E. Howard, H.U. Baranger, J.K. Jain, Phys. Rev. B **39**, 6227 (1989).

[5] G. Timp, A.M. Chang, P.M. Mankiewich, R.E. Behringer, J.E. Cunningham, T.Y. Chang, and R.E. Howard, Phys. Rev. Lett. **59**, 732 (1987).
[6] J.A. Simmons, D.C. Tsui, and G. Weimann, Surf. Sci. **196**, 81 (1988).
[7] A.M. Chang, G. Timp, T.Y. Chang, J.E. Cunningham, P.M. Mankiewich, R.E. Behringer, and R.E. Howard, Solid State Commun. **67**, 769 (1988).
[8] Y. Takagaki, K. Gamo, S. Namba, S. Takaoka, K. Murase, S. Ishida, K. Ishibashi and Y. Aoyagi, Solid State Commun. **69**, 811 (1989).
[9] M.L. Roukes, A. Scherer, S.J. Allen, Jr., H.G. Craighead, R.M. Ruthen, E.D. Beebe, and J.P. Harbison, Phys. Rev. Lett. **59**, 3011 (1987).
[10] G. Timp, H.U. Baranger, P. de Vegvar, J.E. Cunningham, R.E. Howard, R.E. Behringer, and P.M. Mankiewich, Phys. Rev. Lett. **60**, 2081 (1988).
[11] C.J.B. Ford, T.J. Thornton, R. Newbury, M. Pepper, H. Ahmed, D.C. Peacock, D.A. Ritchie, J.E.F. Frost, and G.A.C. Jones, Phys. Rev. B **38**, 8518 (1988).
[12] C.J.B. Ford, S. Washburn, M. Büttiker, C.M. Knoedler and J.M. Hong, Phys. Rev. Lett. **62**, 2724 (1989).
[13] A.B. Chang, T.Y. Chang and H.U. Baranger, Phys. Rev. Lett. **63**, 996 (1989).
[14] M. Heuser and J. Hajdu, Z. Phys. **270**, 289 (1974).
[15] L. Schweitzer, B. Kramer and A. MacKinnon, J. Phys. C **17**, 4111 (1984); A. MacKinnon, L. Schweitzer and B. Kramer, Surf. Sci. **142**, 189 (1984).
[16] M. Jonson, Phys. Scr. **32**, 435 (1985).
[17] J. Hajdu, M. Janssen, and O. Viehweger, Z. Phys. B **66**, 433 (1987).
[18] R. Johnston and L. Schweitzer, Z. Phys. B **70**, 25 (1988).
[19] Y. Ono and T. Ohtsuki, Z. Phys. B **68**, 445 (1987).
[20] Y. Ono and B. Kramer, Z. Phys. B **67**, 341 (1987).
[21] Y. Ohtsuki and Y. Ono, Solid State Commun. **65**, 403 (1988); **68**, 787 (1988); J. Phys. Soc. Jpn. **58**, 956 (1989).
[22] B. Kramer, Y. Ono, and T. Ohtsuki, Surf. Sci. **196**, 127 (1988); Y. Ono, T. Ohtsuki, and B. Kramer, J. Phys. Soc. Jpn. **58**, 1705 (1989).
[23] T. Ando, Prog. Theor. Phys. Suppl. **84**, 69 (1985).
[24] *The Quantum Hall Effect*, edited by R.E. Prange and S.M. Girvin (Springer, New York, 1987).
[25] H. Aoki, Rep. Prog. Phys. **50**, 655 (1987).
[26] T. Ando, Phys. Rev. B **42**, 5626 (1990).
[27] T. Ando and Y. Uemura, J. Phys. Soc. Jpn. **36**, 959 (1974); T. Ando, *ibid.* **36**, 1521 (1974); **37**, 622 (1974); **37**, 1233 (1974).
[28] T. Ando, J. Phys. Soc. Jpn. **38**, 989 (1975).
[29] T. Ando, Y. Matsumoto, and Y. Uemura, J. Phys. Soc. Jpn. **39**, 279 (1975).
[30] H. Aoki and T. Ando, Phys. Rev. Lett. **54**, 831 (1985).
[31] T. Ando and H. Aoki, J. Phys. Soc. Jpn. **54**, 2238 (1985).
[32] K. Fuchs, Proc. Camb. Philos. Soc. **34**, 100 (1938).
[33] For the treatment of boundary roughness, see, for example, T. Ando, A.B. Fowler, and F. Stern, Rev. Mod. Phys. **54**, 437 (1982).
[34] H. Akera and T. Ando, Phys. Rev. B **41**, 11967 (1990).
[35] D. MacDonald and K. Sarginson, Proc. Roy. Soc. London A **203**, 223 (1950).
[36] M. Ya. Azbel, Sov. Phys. JETP **17**, 851 (1963).
[37] E. Ditlefsen and J. Lothe, Philos. Mag. **14**, 759 (1966).
[38] See also, A.B. Pippard, in *Magnetoresistance in Metals* (Cambridge Univ. Press, United Kingdom, 1989) and references cited therein.
[39] H.U. Baranger and A.D. Stone, Phys. Rev. Lett. **63**, 414 (1989).
[40] C.W.J. Beenakker and H. van Houten, Phys. Rev. Lett. **63**, 1857 (1989).
[41] T.J. Thornton, M.L. Roukes, A. Scherer, and B.P. Van de Gaag, Phys. Rev. Lett. **63**, 2128 (1989).
[42] K. Forsvoll and I. Holwech, Philos. Mag. **9**, 435 (1964).
[43] B.J. van Wees, H. van Houten, C.W.J. Beenakker, J.G. Williamson, L.P. Kouwenhoven, D. van der Marel, and C.T. Foxon, Phys. Rev. Lett. **60**, 848 (1988).

[44] D.A. Wharam, T.J. Thornton, R. Newbury, M. Pepper, H. Ahmed, J.E.F. Frost, D.G. Hasko, D.C. Peacock, D.A. Ritchie, and G.A.C. Jones, JPC **21**, L209 (1988).
[45] L. Escapa and N. Garcia, J. Phys. Condens. Matter **1**, 2125 (1989); N. Garcia and L. Escapa, Appl. Phys. Lett. **54**, 1418 (1989).
[46] E.G. Haanappel and D. van der Marel, Phys. Rev. B **39**, 5484 (1989); D. van der Marel and E.G. Haanappel, *ibid.* **39**, 7811 (1989).
[47] G. Kirczenow, Solid State Commun. **68**, 715 (1988); J. Phys. Condens. Matt. **1**, 305 (1989); E. Castano and G. Kirczenow, Solid State Commun. **70**, 801 (1989).
[48] A. Szafer and A.D. Stone, Phys. Rev. Lett. **62**, 300 (1989).
[49] E. Tekman and S. Ciraci, Phys. Rev. B **39**, 8772 (1989); **40**, 8559 (1989).
[50] Song He and S. Das Sarma, Phys. Rev. B **40**, 3379 (1989).
[51] Y. Avishai and Y.B. Band, Phys. Rev. B **40**, 12535 (1989).
[52] A. Kawabata, J. Phys. Soc. Jpn. **58**, 372 (1989).
[53] I.B. Levinson, JETP Lett. **48**, 301 (1988).
[54] A. Matulis and D. Segzeda, J. Phys. Condens. Matt. **1**, 2289 (1989).
[55] L.I. Glazman, G.B. Lesovik, D.E. Khmelnitskii, R.I. Shekhter, JETP Lett. **48**, 238 (1988).
[56] M.C. Payne, J. Phys. Condens. Matt. **1**, 4931 (1989); **1**, 4939 (1989).
[57] D.A. Wharam, M. Pepper, H. Ahmed, J.E.F. Frost, D.G. Hasko, D.C. Peacock, D.A. Ritchie, and G.A.C. Jones, J. Phys. C **21**, L887 (1988).
[58] L.W. Molenkamp, A.A.M. Staring, C.W.J. Beenakker, R. Eppenga, C.E. Timmering, J.G. Willamson, C.J.P.M. Harmans, and C.T. Foxon, Phys. Rev. B **41**, 1274 (1990).
[59] C.W.J. Beenakker and H. van Houten, Phys. Rev. B **39**, 10445 (1989).
[60] R. Landauer, IBM J. Res. Dev. **1**, 223 (1957); Philos. Mag. **21**, 863 (1970).
[61] B.J. van Wees, L.P. Kouwenhoven, H. van Houte, C.W.J. Beenakker, J.E. Mooij, C.T. Foxon, and J.J. Harris, Phys. Rev. B **38**, 3625 (1988).
[62] L.I. Glazman and A.V. Khaetskii J. Phys. Condens. Matt. **1**, 5005 (1989).
[63] K.B. Efetov, J. Phys. Condens. Matt. **1**, 5535 (1989).
[64] M. Okada, M. Saito, K. Kosemura, T. Nagata, H. Ishiwari, and N. Yokoyama, to be published in *Proc. 5th Int. Conf. Superlattices and Microstructures, Berlin, 1990.*

Coulomb-Regulated Conductance Oscillations in a Disordered Quantum Wire

A.A.M. Staring[1,*], *H. van Houten*[1], *C.W.J. Beenakker*[1], *and C.T. Foxon*[2]

[1]Philips Research Laboratories, 5600 JA Eindhoven, The Netherlands
[2]Philips Research Laboratories, Redhill, Surrey RH1 5HA, UK
*Also at: Eindhoven University of Technology, NL-5600 MB Eindhoven, The Netherlands

Abstract Disordered quantum wires have been defined by means of a split-gate lateral depletion technique in the two-dimensional electron gas in GaAs–AlGaAs heterostructures, the disorder being due to the incorporation of a layer of beryllium acceptors in the 2DEG. In contrast to the usual aperiodic conductance fluctuations due to quantum interference, *periodic* conductance oscillations are observed experimentally as a function of gate voltage (or density). No oscillations are seen in the magnetoconductance, although a strong magnetic field dramatically enhances the amplitude of the oscillations periodic in the gate voltage. The fundamentally different roles of gate voltage and magnetic field are elucidated by a theoretical study of a quantum dot separated by tunneling barriers from the leads. A formula for the periodicity of the conductance oscillations is derived which describes the regulation by the Coulomb interaction of resonant tunneling through zero-dimensional states, and which explains the suppression of the magnetoconductance oscillations observed experimentally.

1. Introduction

Aperiodic conductance fluctuations due to quantum interference are commonly observed in disordered conductors small compared to the phase coherence length [1]. One characteristic aspect of these universal conductance fluctuations is the fundamental similarity between a conductance trace as a function of gate voltage (or density), and that as a function of magnetic field. Both traces represent a "fingerprint" of the sample-specific impurity potential. The origin of the duality between density and magnetic field is that both variables affect the *phase* of the conduction electrons, which for a particular closed trajectory depends on the Fermi wavelength (determined by the density, and thus by the gate voltage) and on the enclosed flux. This density–magnetic-field duality is quite general. For example, it also applies to the conductance quantization of a quantum point contact [2] and to the quantum Hall effect [3]. The experimental and theoretical results presented in this paper pertain to a new transport regime where gate voltage and magnetic field play an entirely different role, due to the effects of the charging energy associated with the addition

Springer Series in Solid-State Sciences, Vol. 101
High Magnetic Fields in Semiconductor Physics III   Editor: G. Landwehr

of a single electron to a conductance-limiting segment of a disordered quantum wire.

Experimentally, we investigate a phenomenon first observed by Scott-Thomas et al. [4] in ultra-narrow channels defined in the electron inversion layer in silicon. They reported remarkable conductance oscillations periodic in the gate voltage (or the electron gas density), in the absence of a magnetic field. It was concluded that the periodicity of the oscillations corresponded to the addition of a single electron to a conductance-limiting segment of the narrow channel, with a length determined by the distance between two strong scattering centers. The effect was tentatively attributed to the formation of a charge density wave. A similar effect was seen subsequently in narrow channels in inverted GaAs-AlGaAs heterostructures [5], and was given the same interpretation. As an alternative explanation, it was proposed by two of us [6] that the characteristic features of the experiment might be due to the Coulomb blockade of tunneling [7] — a single electron effect studied extensively in metals where quantum interference effects are negligible. More recently, Wingreen and Lee [8] studied the interplay of the Coulomb blockade and resonant tunneling by a self-consistent solution of the Schrödinger and Poisson equation in a narrow channel geometry.

In the present paper we explore the relative importance of single-electron charging effects and of resonant tunneling by focusing on the different roles of gate voltage and magnetic field. As a novel experimental system for these investigations we use a conventional GaAs–AlGaAs heterostructure in which a layer of compensating impurities is incorporated in the 2DEG during growth. Such impurities were chosen because they are likely to form strongly repulsive scattering centers, which might act as tunnel barriers. We note that a certain degree of compensation was also present in the inversion layers of Ref. [4] and in the channels defined by lateral p-n junctions of Ref. [5]. In our system a narrow channel is defined electrostatically in the two-dimensional electron gas by means of a split gate on top of the heterostructure.

Theoretically, we extend previous work [9,10,11] by considering the combined effects of Coulomb interactions, gate voltage variations, and of a magnetic field on resonant tunneling through a quantum dot. This is relevant to our experiments (and to related experiments [4,5,12,13]) to the extent that one channel segment, delimited by two strong scattering centers, effectively limits the channel conductance. In addition, it is a model for experiments on the Aharonov-Bohm effect in individual quantum dots [14,15,16].

2. Experiments

The quasi one-dimensional electron gas used for the experiments described in this work is obtained by electrostatic confinement of the two-dimensional electron gas (2DEG) in a GaAs–AlGaAs heterostructure using a split-gate technique [17]. On top of the heterostructure, which is mesa-etched in the form of a Hall bar, a pattern of gold gates is defined using electron-beam lithography. The insets of Figs. 1 and 3 show a top view of the two geometries studied. At

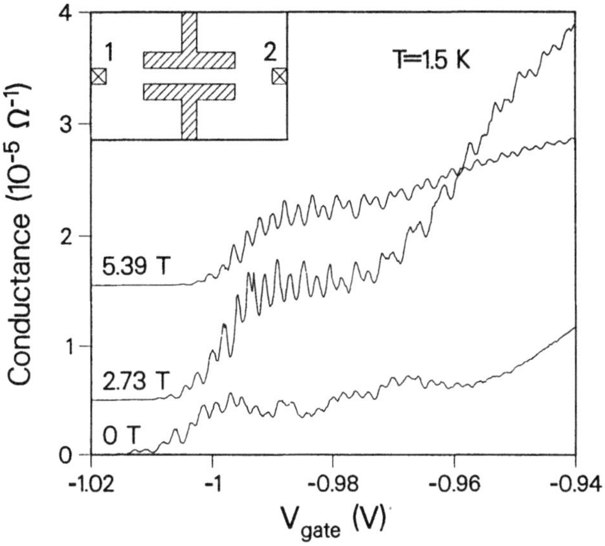

Figure 1: *Two-terminal conductance versus gate voltage at 1.5 K of a 3 μm long split-gate quantum wire (inset, the shaded parts represent the gates while the contacts are labeled 1 and 2). The curves for different magnetic fields are offset vertically for clarity (zero conductance is reached at −1.02 V gate voltage).*

the depletion threshold of the 2DEG (-0.3 V), the quantum wires thus defined are nominally 0.5 μm wide, while their lengths vary from 1 μm to 16 μm; the side probes (if present) have a nominal width of 0.5 μm. Both the width and electron concentration of the wire decrease with gate voltage V_g. Pinch-off (as evidenced by the conductance) is typically reached at $V_g \approx -1$ V. One wire of 1 μm nominal width was also studied, having a pinch-off gate voltage on the order of −2 V. The results obtained with this wire were similar to those obtained with the 0.5 μm wires.

The heterostructure is of a conventional type and consists of the following layers, which are subsequently grown on top of a semi-insulating substrate by molecular beam epitaxy: A 1 μm thick GaAs buffer layer, a 20 nm undoped AlGaAs spacer layer, a 40 nm AlGaAs layer doped to 1.33×10^{18} cm^{-3} with Si, and an undoped 20 nm GaAs capping layer. The Al fraction in the AlGaAs layers is 33%. Disorder was introduced deliberately into the 2DEG by incorporating in the GaAs a planar doping layer of beryllium at 25 Å from the heterointerface, with a sheet concentration of 2×10^{10} cm^{-2}. The electron sheet concentration n_s of the wide 2DEG is 2.7×10^{11} cm^{-2}, with a mobility of about 8×10^4 cm^2/Vs (at 4.2 K). Contact to the 2DEG is made by alloyed AuGeNi ohmic contacts, located along the edges of the 1 mm × 0.3 mm Hall bar.

The measurements were performed with the samples in the mixing chamber of a dilution refrigerator at temperatures between 50 mK and 1.5 K. A conventional double ac lock-in technique, with an excitation voltage kept below kT/e in order to avoid electron heating, was used to determine the conductance of the quantum wires as a function of gate voltage and magnetic field. The field was oriented perpendicular to the 2DEG and had a maximum strength of 7.5 T. The gate voltage was swept at a rate of 10^{-4} V/s or less.

We now give an overview of the main results of our experiments, concentrating on the phenomenology, and defer a discussion of a mechanism which can account for these results to the next section. Fig. 1 shows the two-terminal conductance of a 6 μm long quantum wire at a temperature of 1.5 K for three different magnetic fields. Periodic oscillations as a function of the gate voltage can be seen in these traces. Calculations of Laux et al. [18] for a similar geometry indicate that the 1D electron density (per unit length) depends approximately linearly on the gate voltage. We thus conclude that the oscillations are periodic in the 1D electron density. The fact that it is still possible to observe the oscillations at the relatively high temperature of 1.5 K, in combination with their number (there are about 30 oscillations with a period of 2.2 mV resolved), will prove to be an important clue to their origin, as will be detailed in the next section. The *period* is insensitive to a magnetic field. Nevertheless, a magnetic field is seen to have a variety of effects. The amplitude of the oscillations in strong fields is enhanced above the zero-field case, as is the average conductance. The pinch-off gate voltage is shifted towards zero. The conductance peaks, in this particular sample, have a tendency to regroup in a doublet-like structure consisting of a stronger and a weaker peak.

On lowering the temperature to 50 mK the oscillations are better resolved, as is shown in Fig. 2. The insets show the Fourier transforms of the corresponding conductance traces, and clearly demonstrate that the dominant oscillation has a B–independent frequency of 450 V^{-1} (the trace at 7.47 T has a slightly increased frequency of 500 V^{-1}). Additionally, a second peak in the Fourier transform emerges at about half the dominant frequency as the field is increased. This second peak corresponds to the amplitude modulation of the peaks, which is most clearly seen in the trace at 5.62 T where high and low peaks alternate in a doublet-like structure.

Fig. 3 displays the dependence of the conductance oscillations on the magnetic field for the middle section of a device of the geometry shown in the inset. This particular sample does not exhibit periodic oscillations in the absence of a magnetic field, but only for $B \gtrsim 1$ T. Remarkably, very pronounced oscillations are seen at 5 T, in sharp contrast to the weak random conductance fluctuations in zero field. Between 2 T and 3 T short-period (0.5 mV) oscillations are observed in this sample, in addition to the slower dominant oscillations with a period of 2.2 mV which persist over the entire magnetic field range from 1 T up to 7.5 T. At high magnetic fields, traces of these short-period oscillations return.

The period of the oscillations does not correlate with the length of the quantum wire. We conclude this from measurements on a number of wires with lengths varying from 1 μm up to 16 μm. Sometimes the oscillations were not quite periodic, even in a magnetic field. An example of this behavior is

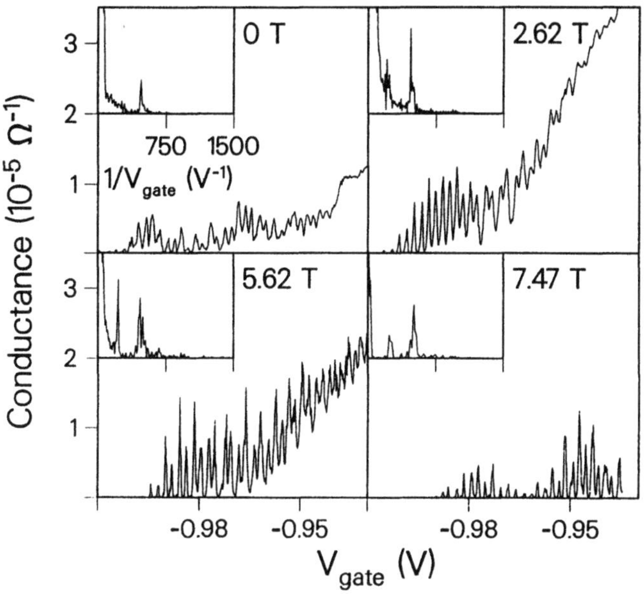

Figure 2: *Conductance versus gate voltage at 50 mK of the same device as in Fig. 1. Insets: Fourier transforms of the data, with the vertical axes of the 0 T and 7.47 T curves magnified by* 2.5×, *relative to the 2.62 T and 5.62 T traces.*

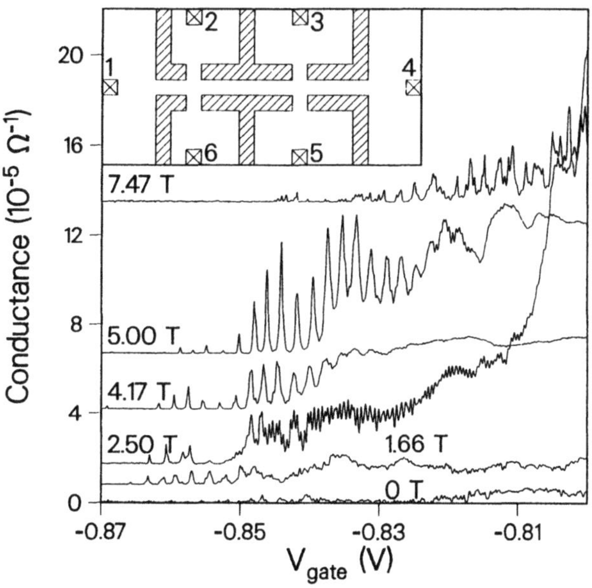

Figure 3: *Development of the conductance oscillations with magnetic field at 50 mK, for a device of the type shown in the inset. The current was passed through contacts 1 and 4, while the voltage was measured between contacts 2 and 3. The curves are offset vertically for clarity.*

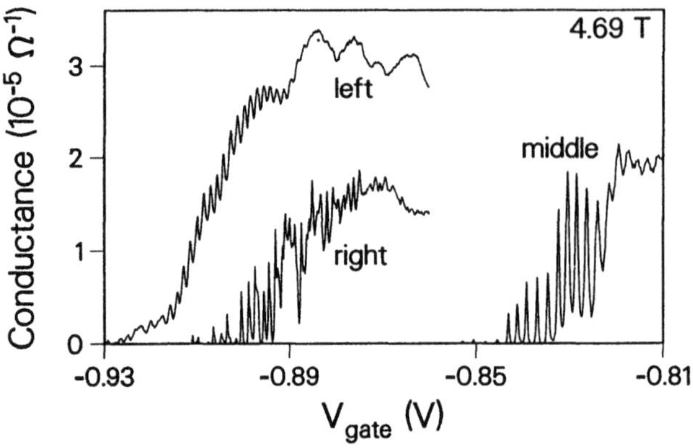

Figure 4: *Conductance at 4.69 T of the three sections of the device shown in Fig. 3 with lengths of 2* μm *(left), 6* μm *(middle) and 4* μm *(right). The current and voltage contacts used were, respectively, (1,2) and (1,6) (left), (2,3) and (6,5) (middle), and (3,4) and (5,4) (right).*

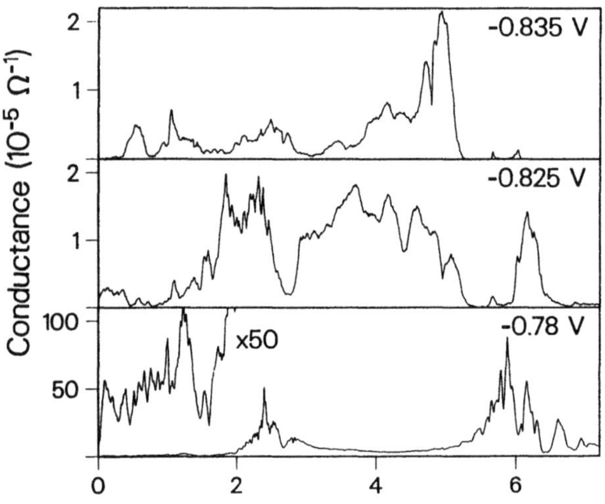

Figure 5: *Magnetoconductance of the device shown in Fig. 3, again using contacts 1 and 4 as current source and drain, and 2 and 3 as voltage probes.*

shown in Fig. 4 (right). It is also clear from this Figure that the middle section of this device determines the total two-terminal conductance (G_{14}).

Whereas the conductance as a function of gate voltage at a fixed magnetic field shows periodic oscillations, no such behavior is observed in the opposite case where the magnetic field is varied and the gate voltage is fixed. As is shown in Fig. 5 the magnetoconductance shows essentially random fluctuations, in contrast to the periodic oscillations seen in Figs. 1–4. Note the extreme sensitivity of these magnetoconductance fluctuations to a small shift in the gate voltage.

3. Theory and Discussion

A theory able to account quantitatively for all of the experimental observations is likely to require a full treatment of the electron–electron interactions. The charge density wave phenomenon [4,5,12] may play a role in such a theory, which however does not yet exist. Our present goal, in the spirit of Ref. [6], is to investigate to what extent the remarkable periodicity of the oscillations as a function of gate voltage, and the absence of regular oscillations in the magnetoconductance, may be explained in terms of single-electron tunneling.

Since quantum effects are known to be important in semiconductor nanostructures [19], it is natural to first consider whether resonant tunneling through zero-dimensional states in a "quantum dot", defined by a conductance-limiting segment of the channel (see Fig. 6), might by itself be able to account for the gate-voltage periodic oscillations. Field et al. [12] argued against such a mechanism, because of the absence of the expected spin-splitting of the peaks in a strong magnetic field, and also because the peaks would most likely not be periodic in V_g. We arrive at the same conclusion, and put forward an additional compelling argument. At a temperature as high as 1.5 K we still find clear oscillations (see Fig. 1), although some thermal smearing is evident in the data (compare with Fig. 2). The width of the thermal smearing function at this temperature is $4kT \approx 0.5$ meV, so that the energy level separation in the case of resonant tunneling would have to be somewhat larger, say around 2 meV. Since each conductance peak would correspond to the depopulation of a single discrete level, the Fermi energy $E_F = 10$ meV at channel definition would then imply a maximum number of about 5 peaks in the full gate-voltage range from definition to complete pinch-off. Clearly, a much larger number of peaks is observed in our experiments, thereby demonstrating that resonant tunneling can not by itself account for the conductance oscillations.

We now discuss, following Ref. [20], how the charging energy associated with the transfer of single electrons modifies the mechanism of sequential res-

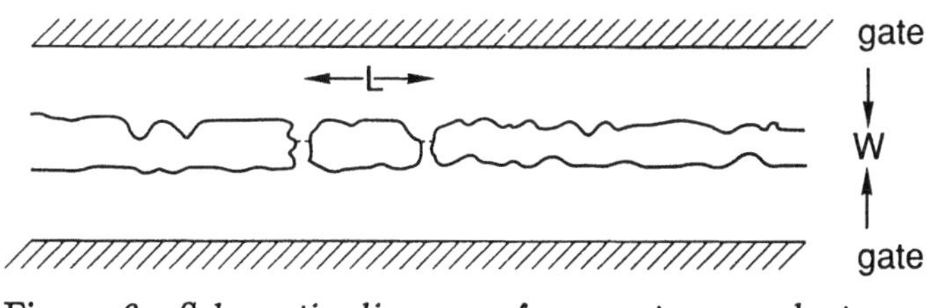

Figure 6: *Schematic diagram of a quantum conductance-limiting segment of quantum wire, of width W and length L, separated from the remainder of the wire by tunneling barriers (dotted lines). The short segment can be regarded as a quantum dot, with discrete energy levels.*

onant tunneling through zero-dimensional states. As shown schematically in Fig. 6, we model the conductance limiting segment by a "quantum dot", separated by tunneling barriers from the leads. The single-electron levels in this dot are denoted by E_p ($p = 1, 2, \ldots$), measured relative to the local conduction band bottom. These levels, which can each contain only one electron of given spin, depend on V_g and B, but are assumed to be independent of the number of electrons N in the dot [21]. The ground state energy of the dot contains a contribution from the occupied single-electron levels, and from the electrostatic energy $\int_0^{-Ne} \phi(Q)dQ$. Here $\phi = Q/C + \phi_{\text{ext}}$ is the potential difference between the dot and the leads due to a charge Q on the dot and due to an external potential ϕ_{ext} from the gate electrode and from the ionized donors in the heterostructure. The capacitance C of the dot to the leads is in our geometry dominated by the dot–gate capacitance. The ground state energy becomes:

$$U(N) = \sum_{p=1}^{N} E_p + \frac{(Ne)^2}{2C} - Ne\phi_{\text{ext}}. \tag{1}$$

Tunneling through the dot requires the transfer of a single electron with Fermi energy E_F from one of the leads into the dot. In the absence of electron-electron interactions, the resulting change in energy of the dot is simply the energy of the lowest unoccupied energy level, E_{N+1}. On resonance $E_{N+1} = E_F$, and tunneling can proceed without increasing the ground state energy of the system (leads plus dot). This picture changes, however, because of the effects of the charging energy. The condition for resonant tunneling now becomes [20]

$$U(N+1) - U(N) = E_F, \tag{2}$$

which is the general condition for equality of the electro-chemical potential $\Delta U/\Delta N$ in dot and leads. Combining Eqs. (1) and (2), we find (replacing N by $N-1$)

$$E_N^* \equiv E_N + \frac{e^2}{C}(N - \frac{1}{2}) = E_F + e\phi_{\text{ext}}. \tag{3}$$

The left hand side of Eq. (3) defines a renormalized energy level E_N^*. The renormalized level spacing relevant for transport $\Delta E^* = \Delta E + e^2/C$ is enhanced above the bare level spacing by the charging energy e^2/C. A comparison between the bare energy levels and the renormalized energy levels is shown in Fig. 7, from which it is clear that the latter are much more regularly spaced than the former.

Experimentally, the conductance peaks are spaced by $\delta V_g \approx 2$ mV. This is interpreted as the gate voltage change needed to induce a charge of one electron in the dot. The dot–gate capacitance is thus $e/\delta V_g \approx 10^{-16}$ F, which we assume to be approximately the same as the dot–lead capacitance C. Consequently, the renormalized level spacing $\Delta E^* \gtrsim e^2/C \approx 2$ meV, an energy which is consistent with the temperature dependence of the conductance, discussed in the previous section. The length L of the quantum dot may be estimated from the gate-voltage range between channel definition and pinch-

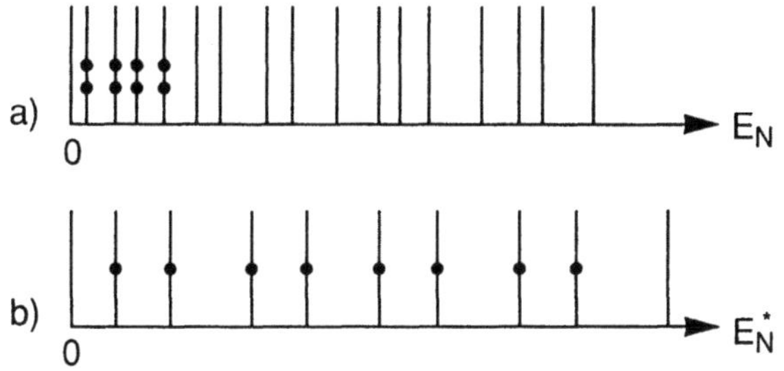

Figure 7: *Diagram of the bare energy levels (a) and the renormalized energy levels (b) for the case* $e^2/C \approx 2\langle\Delta E\rangle$. *The renormalized level spacing is much more regular than the bare level spacing. Note that the spin degeneracy of the bare levels is lifted by the charging energy.*

off: $\Delta V_g = en_s W_0 L/C \approx 1$ V, where W_0 and n_s are the width and electron concentration in the channel at definition. From the above estimate for C we find $L \approx 500$ nm. The width of the dot is estimated to be about $W \approx 40$ nm in the gate voltage range of interest. The bare level spacing for a dot of this area is $\Delta E \approx (mLW/\pi\hbar^2)^{-1}$, with $m = 0.065\, m_e$. Consequently, $\Delta E \approx 0.2$ meV, a full order of magnitude smaller than the elementary charging energy e^2/C, and two orders smaller than E_F [22]. This difference between the bare and renormalized level spacing explains how a large number of peaks in a trace of conductance as a function of gate voltage can be reconciled with the weak temperature dependence noted in the previous section. In addition, it accounts for the regularity of the conductance oscillations: since $e^2/C \gg \Delta E$, the renormalized level spacing ΔE^* is constant. Gate-voltage periodic peaks result from Eq. (3), provided that the 1D electron density varies linearly with V_g. The absence of peak splitting in a strong magnetic field is explained similarly: $\Delta E_{spin} = g\mu_B B \ll e^2/C$, so that the spin degeneracy is removed at $B = 0$ by the charging energy, see Fig. 7.

One would expect to observe Aharonov-Bohm magnetoconductance oscillations for a singly-connected quantum dot in a strong magnetic field. The reason is that such a dot is effectively doubly connected if the magnetic length l_m is much smaller than the dot radius R, due to the presence of circulating edge states. The Aharonov-Bohm (AB) effect in such a dot may be interpreted as resonant tunneling through zero-dimensional states [14,23]. In the absence of Coulomb interaction, the period ΔB of the AB oscillations for a hard-wall dot of area LW is $\Delta B = h/eLW$ (it may be larger for a soft-wall confining potential [14]). Such oscillations have indeed been observed in large quantum dots [14,15,16], but in our experiment, at high magnetic fields, no clear oscillations with the estimated $\Delta B \approx 0.2$ T are found. While random quantum-interference effects in the remainder of the wire and the effect of the magnetic field on the tunneling rates may be of importance, we here want to discuss the role of the electrostatic charging energy, which is dominant in small quantum dots. As pointed out in Ref. [20], each AB oscillation corresponds to an increase of the number of electrons in the dot by one. One can show from

Eq. (3) that the period of the magnetoconductance oscillations is enhanced due to charging effects, according to [20]

$$\Delta B^* = \Delta B \left(1 + \frac{e^2}{C\Delta E}\right), \tag{4}$$

where ΔE represents the energy level spacing of the circulating edge states. Sivan and Imry [23] estimate $\Delta E \approx \hbar\omega_c l_m/2R$ for a hard-wall dot. Under the conditions of our experiment, taking $2R = \sqrt{LW}$ and $B = 3$ T, we estimate $\Delta E \approx 0.5$ meV, so that $\Delta B^* \approx 5\Delta B \approx 1$ T. This will be further enhanced by the softness of the confining potential. The rapid AB oscillations in the magnetoresistance are therefore suppressed, notwithstanding the fact that oscillations can still be observed easily in a conductance trace as a function of gate voltage. The insensitivity of the period of the latter oscillations to a strong magnetic field is explained by the fact that the renormalized level spacing $\Delta E^* \approx e^2/C$ is approximately B-independent.

4. Conclusions

One major conclusion of our study is that Coulomb effects *regulate* resonant tunneling through a single conductance-limiting segment in a disordered quantum wire. The occurrence of periodic conductance oscillations as a function of gate voltage is thus explained. In particular, it is clarified how a large number of oscillations can be reconciled with a weak temperature dependence. The absence of regular magnetoconductance oscillations is interpreted as a signature of a more general phenomenon: the violation of the duality between density and magnetic field due to Coulomb interaction. It remains to clarify the rich variety of effects of the magnetic field on the amplitude of the oscillations, which the present study has revealed, as well as the curious doublet structure induced in one of the samples by a magnetic field. We surmise that these may be related to the influence of the magnetic field on the tunneling rates through the barriers forming the conductance-limiting segment. Also, it is necessary to consider the role of spin in this context in more detail.

Acknowledgements We acknowledge the efforts of C.E. Timmering towards sample fabrication, and wish to thank R. Eppenga, L.W. Molenkamp, and J.G. Williamson for stimulating discussions. Furthermore, we thank M.F.H. Schuurmans and J.H. Wolter for continuous support and encouragement.

References

[1] B.L. Al'tshuler, Pis'ma Zh. Eksp. Teor. Fiz. **41**, 530 (1985) [JETP Lett. **41**, 648 (1985)]; P.A. Lee and A.D. Stone, Phys. Rev. Lett. **55**, 1622 (1985).

[2] B.J. van Wees, H. van Houten, C.W.J. Beenakker, J.G. Williamson, L.P. Kouwenhoven, D. van der Marel, and C.T. Foxon, Phys. Rev. Lett. **60**, 848 (1988); B.J. van Wees, L.P. Kouwenhoven, H. van Houten, C.W.J. Beenakker, J.E. Mooij, C.T. Foxon, and J.J. Harris, PRB38, 3625 (1988); D.A. Wharam, T.J. Thornton, R. Newbury, M. Pepper, H. Ahmed, J.E.F. Frost, D.G. Hasko, D.C. Peacock, D.A. Ritchie, and G.A.C. Jones, J. Phys. C **21**, L209 (1988).

[3] K. von Klitzing, G. Dorda, and M. Pepper, Phys. Rev. Lett. **45**, 494 (1980).

[4] J.H.F. Scott-Thomas, S.B. Field, M.A. Kastner, H.I. Smith, and D.A. Antoniadis, Phys. Rev. Lett. **62**, 583 (1989).

[5] U. Meirav, M.A. Kastner, M. Heiblum, and S.J. Wind, Phys. Rev. B **40**, 5871 (1989).

[6] H. van Houten and C.W.J. Beenakker, Phys. Rev. Lett. **63**, 1893 (1989).

[7] K.K. Likharev, IBM J. Res. Dev. **32**, 144 (1988), and references therein.

[8] N.S. Wingreen and P.A. Lee, presented at the *NATO Adv. Study Inst. on Quantum Coherence in Mesoscopic systems* (Les Arcs, 1990).

[9] L.I. Glazman and R.I. Shekhter, J. Phys. Condens. Matter **1**, 5811 (1989).

[10] D.V. Averin and A.N. Korotkov, Zh. Eksp. Teor. Fiz. [Sov. Phys. JETP] (to be published); A.N. Korotkov, D.V. Averin, and K.K. Likharev, in *Proc. 19th Int. Conf. on Low Temperature Physics* (Physica B, to be published).

[11] M. Amman, K. Mullen, and E. Ben-Jacob, J. Appl. Phys. **65**, 339 (1989).

[12] S.B. Field, M.A. Kastner, U. Meirav, J.H.F. Scott-Thomas, D.A. Antoniadis, H.I. Smith, and S.J. Wind, preprint.

[13] U. Meirav, M.A. Kastner, and S.J. Wind, preprint.

[14] B.J. van Wees, L.P. Kouwenhoven, C.J.P.M. Harmans, J.G. Williamson, C.E. Timmering, M.E.I. Broekaart, C.T. Foxon, and J.J. Harris, Phys. Rev. Lett. **62**, 2523 (1989).

[15] R.J. Brown, C.G. Smith, M. Pepper, M.J. Kelly, R. Newbury, H. Ahmed, D.G. Hasko, J.E.F. Frost, D.C. Peacock, D.A. Ritchie, and G.A.C. Jones, J. Phys. Condens. Matter **1**, 6291 (1989).

[16] D.A. Wharam, M. Pepper, R. Newbury, H. Ahmed, D.G. Hasko, D.C. Peacock, J.E.F. Frost, D.A. Ritchie, and G.A.C. Jones, J. Phys. Condens. Matter **1**, 3369 (1989).

[17] T.J. Thornton, M. Pepper, H. Ahmed, D. Andrews, and G.J. Davies, Phys. Rev. Lett. **56**, 1198 (1986); H.Z. Zheng, H.P. Wei, D.C. Tsui, and G. Weimann, Phys. Rev. B **34**, 5635 (1986).

[18] S.E. Laux, D.J. Frank , and F. Stern, Surf. Sci. **196**, 101 (1988).

[19] C.W.J. Beenakker and H. van Houten, *Quantum Transport in Semiconductor Nanostructures*, in *Solid State Physics*, H. Ehrenreich and D. Turnbull, eds., (Academic Press, New York), to be published.

[20] C.W.J. Beenakker, H. van Houten, and A.A.M. Staring, submitted to Phys. Rev. Lett.

[21] A. Kumar, S.E. Laux, and F. Stern, preprint.

[22] For an elongated dot it is more appropriate to assume that only one transverse mode is present, in which case $\Delta E = (h/4L)(E_F/2m)^{\frac{1}{2}}$, assuming hard-wall boundary conditions. This leads to the same estimate for ΔE, however.

[23] U. Sivan and Y. Imry, Phys. Rev. Lett. **61**, 1001 (1988).

Magnetoconductance of Si MOSFET Quantum Wires: Weak Localization and Magnetic Depopulation of 1D Subbands

J.R. Gao[1], *C. de Graaf*[1], *J. Caro*[1], *S. Radelaar*[1], *M. Offenberg*[2], *V. Lauer*[2], *J. Singleton*[3], *T.J.B.M. Janssen*[3], *and J.A.A.J. Perenboom*[3]

[1]Delft Institute for Microelectronics and Submicron Technology, Delft University of Technology, Lorentzweg 1, 2628 CJ Delft, The Netherlands

[2]Institute of Semiconductor Electronics, Aachen Technical University, W-5100 Aachen, Fed. Rep. of Germany

[3]High Field Magnet Laboratory, University of Nijmegen, Nijmegen, The Netherlands

The magneto-electric transport in narrow (≈70 nm), parallel multiple quantum wires in Si MOSFET's has been studied at different gate voltages and in magnetic fields up to 20 T. For high gate voltages the low field magnetoconductance due to weak localization can be described well by existing theory, that we have utilized to extract the channel width and the inelastic diffusion length. From quantum oscillations in the conductance as a function of magnetic field we find that the plot of subband index n vs inverse magnetic field is nonlinear due to magnetic depopulation of 1D subbands.

1. Introduction

Currently, the electric and magneto-electric transport in quasi one-dimensional electron systems (1DES) or quantum wires is a topic of intense study [1-10]. In such systems the electron transport properties are modified compared to the two-dimensional (2D) case due to the formation of 1D subbands (quantum confinement). In a perpendicular magnetic field the subbands of a 1DES develop into hybrid magneto-electric subbands. The spacing of these subbands increases with magnetic field, giving rise to magnetic depopulation. This phenomenon has been observed in narrow channels in III-V heterostructures [1-3], but so far not in Si MOSFET's. The weak localization (WL) correction to the Boltzmann conductance in quantum wires also attracted much attention[5-9]. In theoretical papers [8,9], depending on the applied model, different refinements are derived for the WL correction to the conductance of quantum wires. Hu and O'Connell [9] state that their theory describes the experimental results of Hiramoto et al. [7] better than a theory that neglects 1D subband effects. The present status is that the (magneto)conductance due to WL in quantum wires remains to be explored further.

In this paper we report magnetoconductance measurements on Si MOSFET multiple quantum wires operating in a regime where several subbands are occupied. In low fields the magnetoconductance shows the suppression of WL. In high fields quantum oscillations are observed that show clear characteristics due to magnetic depopulation of 1D subbands.

2. Devices

The devices used are dual-gate Si MOSFET's with 240 parallel inversion channels to average out universal conductance fluctuations. A cross section of a device is shown in the inset of Fig. 1. The inversion channel length is 24 μm. Details of the device structure and its fabrication can be found elsewhere [10]. By applying a positive voltage V_{GU} to the upper Al gate, while simultaneously applying a negative voltage V_{GL} to the lower polysilicon grating gate, conducting channels are formed under the gaps between the grating lines.

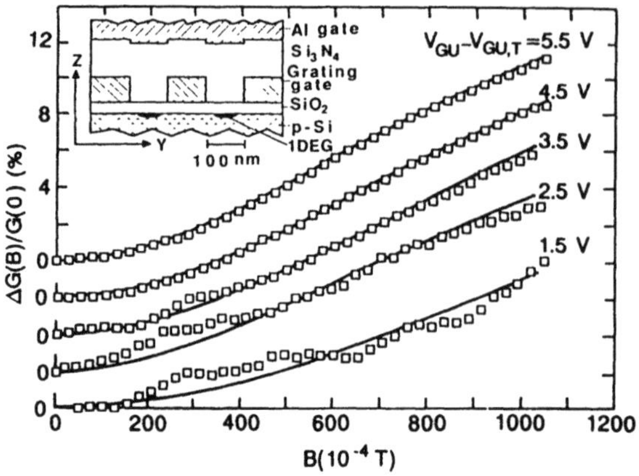

Fig. 1. Relative conductance increase as a function of magnetic field at five values of V_{GU} - $V_{GU,T}$. V_{GL}=-0.1 V, T=0.6 K. The curves are fits of Eq. (1) to the data. The inset shows a magnified cross section of a device.

3. Results and discussion

Figure 1 shows low field magnetoconductance (LFMC) data of device U4 measured at 0.6 K and for five values of $\Delta V_{GU}=V_{GU}-V_{GU,T}$. Here the threshold voltage $V_{GU,T}$ is the value of V_{GU} where the first subband starts filling. In Fig. 1 $\Delta G(B)=\delta G(B)-\delta G(0)$ is the conductance increase due to magnetic suppression of WL, normalized to the zero field conductance value. The function $\delta G(B)$ is the field dependent WL correction to the conductance. For ΔV_{GU}=5.5 V six to seven 1D subbands are occupied. This was estimated from the high field experiment described below.

Assuming that the narrow channel device operates in the regime of 1D WL (i.e. the inelastic diffusion length L_{in} exceeds the channel width W) and lacking an expression for the quantum wire case, we analysed the LFMC data using the expression of Al'tshuler and Aronov [11]:

$$\Delta G(B) = (N e^2 / \pi \hbar L) \{ L_{in} - [1/L_{in}^2 + W^2 / 12 l_B^4]^{-1/2} \} \qquad (1)$$

Here N is the number of channels, L the channel length and $l_B = (\hbar/2eB)^{1/2}$ the magnetic length. Equation (1) does not include boundary scattering ($l_e < W$, where l_e is the elastic mean free path). The curves in Fig. 1 are fits of Eq. (1) to the data points. Fitted values of L_{in} and W are compiled in Table I, together with other parameters.

For ΔV_{GU}=5.5 V and 4.5 V the fitted curves describe the experimental data excellently and the values of the fitted parameters are also very reasonable. The widths of 69 nm and 73 nm are substantially narrower than 100 nm (gap width between grating lines), which is due to fringing fields contributing to the confining potential. The values of L_{in} for ΔV_{GU}=5.5 V and 4.5 V justify the use of a 1D expression for the LFMC and agree with values extrapolated from LFMC data of single channel Si MOSFET's [12]. The estimated value of l_e in this voltage regime is 40-50 nm, so Eq. (1) is applicable here. For ΔV_{GU}=3.5 V and lower, fitted curves start to deviate from the data points. The fitted parameters also become unphysical (e.g. W > 100 nm). This may partly be attributed to a lower signal to noise ratio (fluctuations in the data point "curves" are noise). The tendency of producing increasingly unphysical fit parameters with decreasing ΔV_{GU}, however, is a real effect. This was checked by fitting Eq. (1) to other LFMC data at low ΔV_{GU} values with different noise patterns. A possible explanation of this finding is that Eq. (1) is only valid for $k_F l_e >> 1$. For ΔV_{GU}=3.5 V and lower, this condition starts to be violated. Another possibility is that for lower subband occupancies (see Table I) 1D subband effects in the LFMC show up.

In high magnetic fields we measured $\partial G/\partial V_{GU}$ of several multiple channel devices at 0.2 K and as a function of field. Figure 2 shows traces for device D1 taken at four different V_{GU}, the field ranging up to 20 T. Each of the curves exhibits three or four quantum oscillations. Starting at the highest field we counted the number of maxima in each curve using the index n. A plot of the index n vs 1/B is shown in Fig. 3. The dashed lines in the figure are straight lines through 1/B=0 fitted to the lowest two data points in 1/B. For each V_{GU} a line

TABLE I. Parameters characterizing the experimental data and fitted curves of Fig. 1: inelastic diffusion length L_{in}, channel width W, zero field conductance G(0), the product $k_F l_e$ and the number of occupied subbands $n_{max}+1$ (with uncertainty ± 1). $k_F l_e$ (k_F is the Fermi wavevector) was derived from a 2D expression.

ΔV_{GU} (V)	L_{in} (nm)	W (nm)	G(0) (S)	$k_F l_e$	$n_{max}+1$
1.5	42	167	6.9×10^{-5}	1.3	2
2.5	50	266	1.5×10^{-4}	2.7	3
3.5	104	108	2.3×10^{-4}	4.3	4
4.5	165	73	3.9×10^{-4}	7.3	5
5.5	195	69	4.9×10^{-4}	9.2	7

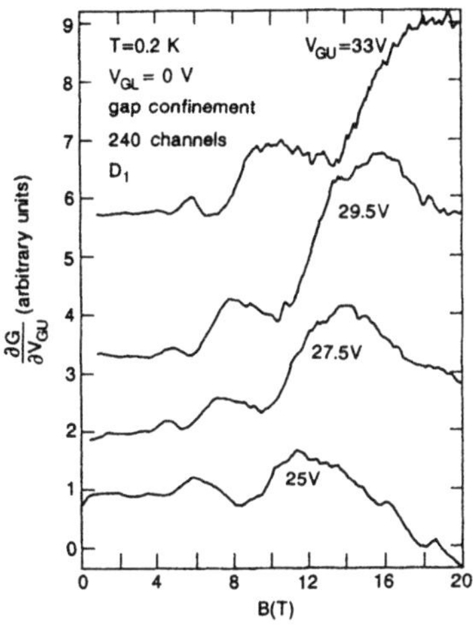

Fig. 2. Measured derivative $\partial G/\partial V_{GU}$ of device D1 as a function of magnetic field applied perpendicular to the plane of the quantum wires at four values of V_{GU}. Curves have a vertical offset.The gate voltage modulation is 0.35 V.

connecting the lowest two data points (these are in the regime where the channel width exceeds the cyclotron radius, i.e. in the 2DES-like regime) would already extrapolate well to 1/B=0, as expected for the Shubnikov-de Haas effect in a 2DES. At higher 1/B values the dependence of n on 1/B shows the nonlinear behaviour characteristic of magnetic depopulation of 1D subbands. This is the first observation of this effect in Si MOSFET quantum wires.

In the analysis of the oscillations we assumed a confinement potential in the y-direction consisting of a flat bottom of width t and parabolic walls, $V(y)=m^*\omega_0^2(|y|-t/2)^2/2$, for $|y|>t/2$ and zero otherwise. We fitted the parameters $\hbar\omega_0$, E_F and t in the usual way [10], using the expression for the magneto-electric subband levels $E_n(B)$ derived by the semi-classical WKB method [4]. In this procedure we also used the two n=4 data points that were previously neglected [10] because of the relatively large uncertainty in the position of the corresponding maxima. We found that optimum fitting is obtained for zero bottom width, so a pure parabola gives the better description of the confinement. The fitting quality is indicated by the position of the crosses in Fig. 3. The fit parameters E_F and subband spacing $\hbar\omega_0$ are summarized in Table II, together with the effective channel widths W_{eff} and electron densities N_{1D}. W_{eff} was obtained from the potential width at E_F that equals $W_{eff}=(8E_F/m^*\omega_0^2)^{1/2}$ and N_{1D} equals $W_{eff}N_{2D}$ ($N_{2D}=2m^*E_F/\pi\hbar^2$).

In Table II we notice a non-monotonous change of the parameters in the last three columns. This can be understood from the uncertainty in the n=4 data points. Further, there is a trend that $\hbar\omega_0$ increases as E_F increases, accompanied with a trend of decreasing W_{eff}. Although this is contra-intuitive to us, we can not yet judge this effect, since the model

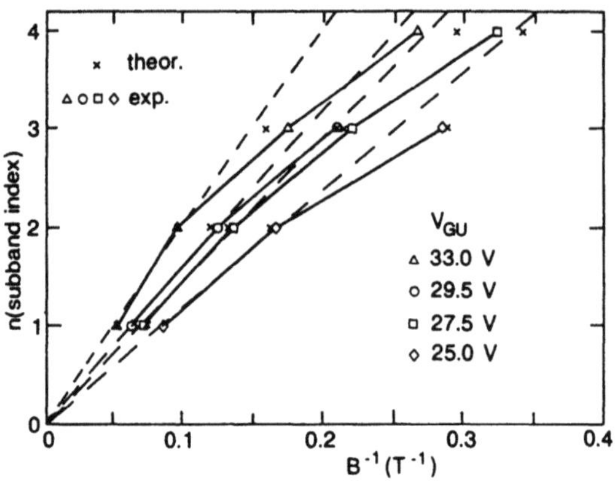

Fig. 3. Plot of subband index n vs 1/B deduced from the oscillations in Fig. 2. A pronounced deviation from a 2D behaviour is indicated by the dashed lines. The theoretical results are indicated by crosses "x".

TABLE II. Listed quantities are derived Fermi energy E_F, subband spacing $\hbar\omega_0$, effective channel width W_{eff} and electron density N_{1D}, as a function of V_{GU}.

V_{GU} (V)	E_F (meV)	$\hbar\omega_0$ (meV)	W_{eff} (nm)	N_{1D} ($10^7 cm^{-1}$)
25.0	11.0	2.3	82	1.4
27.5	12.8	2.2	92	1.8
29.5	15.1	3.2	69	1.7
33.0	17.9	3.4	71	2.0

potential is only an approximation of the real potential, of which the shape and its dependence on V_{GU} and V_{GL} is unknown. Nevertheless, the effective channel widths in Table II are consistent with a width of 70 nm derived from the LFMC experiments. The values of subband spacings are very reasonable. This can be concluded by considering a 70 nm wide infinite square well potential, that has subband spacings 0.40 (2n-1) meV ($n \geq 2$).

In conclusion we have measured the low field magnetoconductance of Si MOSFET multiple quantum wires. For the high V_{GU} (large $k_F l_e$) the data can be described very well by the theory of Al'tshuler and Aronov and the fitted channel width is 70 nm. In high fields oscillations in $\partial G/\partial V_{GU}$ show clear characteristics of magnetic depopulation of 1D subbands.

We thank the Solid State Physics Group of the Faculty of Applied Physics for the use of their dilution refrigerator. This work is part of the research program of the "Stichting voor Fundamenteel Onderzoek der Materie (FOM)", which is financially supported by the "Nederlandse Organisatie voor Wetenschappelijk Onderzoek (N.W.O)".

References

1. K.-F. Berggren, T.J. Thornton, D.J. Newson and M. Pepper, Phys. Rev. Lett. **57**, 1769 (1986).
2. H. van Houten, B. J. van Wees, J.E. Mooij, G. Roos and K.-F. Berggren, Superlattice Microstruct. **3**, 497 (1987).
3. F. Brinkop, W. Hansen, J.P. Kotthaus and K. Ploog, Phys. Rev. B**37**, 6547 (1988).
4. K-F Berggren and D.J. Newson, Semicond. Sci. Technol. **1**, 327 (1986).
5. T. J. Thornton, M. Pepper, H. Ahmed, D. Andews and G. J. Davies, Phys. Rev. Lett. **56**, 1198 (1986).
6. H. van Houten, C.W. Beenakker, B. J. van Wees and J.E. Mooij, Surf. Sci. **196**, 144 (1988).
7. T. Hiramoto, K. Hirakawa, Y. Iye and T. Ikoma, Appl. Phys. Lett., **54**, 2103 (1989).

8. M.J. Kearney and P.N. Butcher, J. Phys.C: Solid State Phys., **21**, 2539 (1988).
9. G.Y. Hu and R.F.O' Connell, J. Phys.: Condens. Matter, **2**, 5335 (1990).
10. J.R. Gao, C. de Graaf, J. Caro, S. Radelaar, M. Offenberg, V. Lauer, J. Singleton, T.J.B.M. Janssen and J.A.A.J. Perenboom, Phys. Rev. B**41**, 12315 (1990).
11. B.L. Al'tshuler and A.G. Aronov, JETP Lett. **33**, 499 (1981).
12. J.R. Gao, J. Caro, A.H. Verbruggen, S. Radelaar and J. Middelhoek, Phys. Rev. B**40**, 11676 (1989)

Self-Consistent Screening, Single Particle Energy and Plasmon Excitation in a Quasi-One-Dimensional Electronic System

V. Shikin, D. Heitmann, and T. Demel*

Max-Planck-Institut für Festkörperforschung, Heisenbergstr. 1, W-7000 Stuttgart 80, Fed. Rep. of Germany
***Permanent address: Institute of Solid State Physics, Academy of Sciences USSR, 142432 Cernogolovka, USSR**

Abstract. A consistent theory is discussed for the self-consistent screening, the single-particle energy and the dipole plasmon frequency in a one-dimensional electronic system (1DES) with a parabolic confining potential. Comparison with experiments allows a consistent characterisation of the parameters of the 1DES.

Currently there is a great interest in the investigation of quantum-confined quasi-one-dimensional electronic systems (1DES). Different arrangements have been proposed and realized to study such systems in particular starting from layered 2DES in AlGaAs-GaAs heterostructures (see e.g. Refs. 1-7 and further references therein). However, the correct characterization of such 1DES, i.e., the determination of the exact shape of the confining potential, of the energy spectrum, and of the width for the electron channel of 1DES, is a difficult task. The most accurate understanding so far comes from self-consistent numerical bandstructure calculations [8]. However, also the evaluation of such calculations is limited by a limited information on the input parameters. The difficulties to describe currently available 1DES in GaAs systems arises in particular from the strong influence of self-consistent screening. For example, in the split-gate configuration considered by Laux *et al.* [8] the external potential (without electrons) is nearly parabolic and characterized by a confinement energy and subband spacing $\hbar\Omega_0 = 5.6$ meV. The one-particle subband spacing decreases, due to self-consistent screening, to $\hbar\omega_0 = 3$ meV if the 40 nm wide channel is filled with only one electron per 15 nm along the wire. It is evident that this screening is one of the most basic characterisitics of an individual electron channel. So far, no information on this screening has been extracted from existing experimental data. In this paper we will dicuss this screening based on an analytical solution for a parabolic external potential.[9] The screened potential can be determined from magnetic depopulation of the 1D subbands in a dc Shubnikov-de Haas (SdH) experiment.[1] We will combine these results with information from a totally independent experiment, i.e., the dipole plasma excitation. We find good agreement which shows the consistency of our treatment.

A self-consistent characterization of 1D electron channels is possible if the external confining potential has a parabolic form.

$$V(x) = V_0 + \frac{1}{2}\ Kx^2 \tag{1}$$

Here V_0 and K are the phenomenological parameters which characterize the potential. The channel is oriented along the y-direction and a very strong confinement is assumed in the z-direction. If this external potential is 'filled' with electrons, self-consistent screening occurs. It has been shown in Ref. 9 that for the case $N_l a \gg 1$ (N_l is the carrier-density

Springer Series in Solid-State Sciences, Vol. 101
High Magnetic Fields in Semiconductor Physics III Editor: G. Landwehr

per length , $2a$ is the width of the channel) the effective (screened) one-particle potential $v(x)$ has the form

$$v(x) = v_0 + \frac{1}{2}\ kx^2 \tag{2}$$

with the important relation

$$k = K - \frac{4e^2 N_l}{\kappa a^2}. \tag{3}$$

Here κ is the dielectric constant of the surrounding dielectric, for the renormalized value of v_0, see Ref. 9. Integrating the 1D density of states over all occupied 1D subbands one finds for L occupied subbands and $L \gg 1$:

$$N_l = \frac{4}{3\pi}\left(\frac{2m^*\omega_0}{\hbar}\right)^{\frac{1}{2}} L^{\frac{3}{2}} \tag{4}$$

where $\omega_0^2 = k/m^*$ characterizes the one-particle energy in the screened potential (2). Information on N_l and ω_0 can be extracted from SdH type of dc measurements which detect the magnetic depopulation of the 1D subbands in a perpendicular magnetic field.[1] Taking into account the determinations for the relative position of the Fermi energy, $\delta\mu$

$$L\hbar\omega_0 = \delta\mu \quad and \quad \delta\mu = \frac{1}{2}ka^2 \tag{5}$$

we can determine the width 2a of the electron channel.

$$a^3 = \frac{3\pi}{2}\frac{\hbar^2 N_l}{km^*} = \frac{3\pi}{2}\frac{e^2 N_l b}{k\kappa} \tag{6}$$

Here $b = \kappa\hbar^2/m^* e^2$ is the Bohr radius. Combining (3) and (6) we can determine K, i.e. the parameter that characterizes the unscreened external potential

$$K = \frac{4e^2 N_l}{\kappa a^2}(1 + \frac{3\pi b}{4a}). \tag{7}$$

The screening can now be characterized by the parameter γ

$$\gamma = 1 - k/K = 1/(1 + \frac{3\pi b}{4a}). \tag{8}$$

In particular we have full screening, $\gamma \to 1$, $k \to 0$, if $b/a \ll 1$ and no screening, $\gamma \to 0$, $k \to K$, if $b/a \gg 1$.

Independent experimental information on quantum wires can be obtained from FIR experiments which excite the dipole plasmon oscillation perpendicular to the wire. It has been shown that for a parabolic confinement the dipole plasma mode frequency is exactly (Ref.9, see also Ref. 11 for a quantum mechanical treatment)

$$\omega_p^2 = \Omega_0^2 = K/m^*, \tag{9}$$

thus only governed by the curvature of the unscreened potential, K, and not by the actual screened potential, k ! Note that the evaluation of (3) in Ref.9 from static dc properties and of (9) in Ref.10 from dynamic properties is based on the same footing. We can thus combine experimental FIR data with the SdH data without any unknown parameter and check the consistency of our approach. In the following table we listed the experimental SdH data and FIR data which were measured on deep mesa etched quantum wires (Refs. 5 and 6) and quantum wires defined by split gate configurations (Ref.4). The data from the dc SdH experiment are the experimental gate voltage V_g, the width $2a$, the linear density N_l and the screened one-particle energy ω_0, whereas $\Omega_0^2 = K/m^*$ is the calculated unscreened frequency with K from (7). For comparison we give the experimental plasmon frequency, ω_p. Further we have calculated the screening parameter γ.

Ref.	V_g (meV)	N_l 10^6cm^{-1}	$2a$ nm	$\hbar\omega_0$ (meV)	$\hbar\Omega_0$ (meV)	$\hbar\omega_p$ (meV)	γ
4	- 550	6.1	220	1.6	5.5	5.0	0.91
4	- 660	3.7	160	2.0	6.0	6.0	0.89
5	-	10.8	390	1.0	4.1	4.0	0.94
5	-	18.5	226	1.5	6.4	5.8*	0.95

(* The sample discussed in the last row is actually a two-layered wire structure. The dc transport characterizes a single layer. Thus the high frequency optical plasmon mode of the two-layered wire is increased by $\sqrt{2}$. We have thus included in the table the experimental value devided by $\sqrt{2}$, i.e., 8.2 meV/$\sqrt{2}$= 5.8 meV.) From the table we observe that there is a good agreement of Ω_0, i.e. the unscreened potential parameter that was determined from the dc SdH measurements, with the plasmon frequency ω_p measured in the FIR experiment. This close agreement demonstrates the validity of assuming a harmonic oscillator confinement and the consistency of our analytic approach in Ref.9, which is the basis to determine via (3) the self-consistent screening in parabolic 1D wires. The screening parameter γ of about 0.9 for all samples demonstrates that the confining potential for currently available quantum wires in GaAs heterostructures is strongly determined by the screening properties of the 1DES.

References

1. K.F. Berggren, T.J. Thornton, D.J. Newson, and M. Pepper, *Phys. Rev. Lett.* **57**, 1769 (1986).
2. W. Hansen, M. Horst, J.P. Kotthaus, U. Merkt, Ch. Sikorski, and K. Ploog, *Phys. Rev. Lett.* **58**, 2586 (1988).
3. J. Alsmeier, Ch. Sikorski, and U. Merkt, *Phys. Rev.* **B37**, 4314 (1988).
4. F. Brinkop, W. Hansen, J. Kotthaus, and K. Ploog, *Phys. Rev.* **B37**, 6547 (1988).
5. T. Demel, D. Heitmann, P. Grambow, and K. Ploog, *Appl. Phys. Lett.* **53**, 2176 (1988).
6. T. Demel, D. Heitmann, P. Grambow, and K. Ploog, *Phys. Rev.* **B38**, 12732 (1988).
7. B.J. van Wees, H. van Houten, C.W.J. Beenaker, J.G. Williamson, L.P. Kouwenhoven, D. van der Marel, and C. Foxon, *Phys. Rev. Lett.* **60**, 848 (1988).
8. S. Laux, D. Frank, and F. Stern, *Surf. Sci.* **196**, 101 (1988).
9. V. Shikin, T. Demel, and D. Heitmann, *ZhETF* **96** (1989) 1406.
10. V. Shikin, T. Demel, and D. Heitmann, *Surf. Sci.* **229**, 276 (1990).
11. L. Brey, N. Johnson, and P. Halperin, *Phys. Rev.* **B40**, 10647 (1989).

Magnetic Field Dependence of Aperiodic Conductance Fluctuations in Narrow GaAs/AlGaAs Wires

Y. Ochiai[1], *K. Ishibashi*[2], *M. Mizuno*[1], *M. Kawabe*[1], *Y. Aoyagi*[2], *K. Gamo*[3], *and S. Namba*[2;3]

[1]Institute of Materials Science, University of Tsukuba, Tsukuba, Ibaraki 305, Japan
[2]Frontier Research Program, Institute of Physical and Chemical Research, Wako, Saitama 351-01, Japan
[3]Department of Electrical Engineering, Faculty of Engineering Science, Osaka University, Toyonaka, Osaka 560, Japan

The magnetic field dependence of aperiodic fluctuations has been studied by means of analysing the Fourier transform of the magnetoresistance in narrow GaAs/AlGaAs wires. Near the critical field of $\omega_c\tau=1$, the decay of aperiodic fluctuations, which is observed on the low frequency side of the Fourier spectra, shifts to low frequencies. The shift can be explained by reduction of the effective area for the interference.

1. Introduction

Low temperature magnetoresistance (MR) in narrow GaAs/AlGaAs wires has been measured in order to clarify the quantum effects in a mesoscopic system. The magnetic field dependence of aperiodic fluctuations in the MR has been studied by means of analysing the Fourier transform. Near the critical field of $\omega_c\tau=1$, the decay of aperiodic fluctuations, which are observed on the low frequency side of the Fourier spectra, shift to low frequencies. Here, ω_c is cycrotron frequency and τ is the scattering time in the absence of a magnetic field. The shift can be explained by a reduction of the effective area for the interference.

Quantum intereference effects in mesoscopic systems have been observed in magnetoresistance (MR) measurements for small conductors where the sample dimension is smaller than the phase coherent length l_p. In high magnetic fields, suppresion of quantum interference oscillations due to the Aharonov-Bohm (AB) effect was observed in the MR for a ring of high-mobility GaAs/AlGaAs heterostructures. [1,2] It was also observed that the period of AB oscillation shifts to lower frequencies in the Fourier spectrum of the MR. The suppression and the shift have been expected from the formation of edge states[3] in the ring and caused by the absence of backscattering between edge states because the outer edge states do not enclose a flux while inner edge states are not coupled to the outer ones. Recently a similar low frequency shift of the decay part in the Fourier spectrum has been observed in a wire of a GaAs/AlGaAs heterostructure.[4] The decay part in the spectrum is believed to correspond to aperiodic fluctuations in mesoscopic system. Stone suggests that aperiodic fluctuations in a small conductor are due to the coherent interference of electron waves.[5] Altshuler[6] and Lee and Stone[7] predict that the fluctuations in the conductance have an universal magnitude which is order e^2/h for

Springer Series in Solid-State Sciences, Vol. 101
High Magnetic Fields in Semiconductor Physics III Editor: G. Landwehr

conductors with the wire dimension less than l_p. Therefore, an oscillation component of apeoridic fluctuations in the MR is considered to come from a coherent interference between electron waves on different propagating parts in a wire. We discuss the aperiodic fluctuations of the MR in GaAs/AlGaAs heterostructure wires.

2. Experiments

The samples used in this study are narrow wires with GaAs/AlGaAs double and single heterostructures grown by the molecular beam epitaxy technique and made by using electron beam lithography and dry etching technique. The double heterostructure sample (named DHW-08) has a 800 nm-thick non-doped GaAs buffer layer, a 10 nm-thick cap layer, two 2 nm-thick spacers, and a 10 nm-thick GaAs layer sandwiched between two Si-doped AlGaAs layers of 60 nm-thickness. The single heterostructure sample (named SHW-02) has a non-doped GaAs buffer layer followed by a 6 nm-thick non-doped AlGaAs spacer layer and the 50 nm-thick Si-doped AlGaAs layer. The lithographical width and the length of those wires were about 700 nm and 3000 nm for DHW-08 or 1700 nm for SHW-02, respectively. The actual width for electrical conduction is less and about 400 nm for the two samples because of the existence of a surface depletion layer.[8] The mean free paths of DHW-08 and SHW-02 are less than 150 and 430 nm, respectively. Mobilities are estimated to be about 8600 $cm^2/Vsec$ for DHW-08 and 25000 $cm^2/Vsec$ for SHW-02 at 4.2 K, respectively. The MR measurement at low temperature was carried out under magnetic fields up to 8.5 T in a ^3He-^4He dilution refrigerator.

3. Results and Discussion

Universal conductance fluctuations can be observed in the MR with an amplitude of the order e^2/h for the two samples. While DHW-08 shows MR fluctuations up to 8 T, such fluctuations in SHW-02 cannot be detected in higher fields above 2 T because of the appearence of Shubnikov-de Haas oscillations. In the low temperature MR below 4.2 K, the Onsager-Kubo symmetry relation has been satified in SHW-02. Figure 1 shows evidence on non-local quantum interference over the wire. This indicates that l_p in SHW-02 should be longer than the wire length 1700 nm at low temperatures. As for DHW-08, magnetofingerprints were observed down to 50 mK and l_p was estimated to be comparable to the wire length in a previous study.[4]

Fourier spectra of the MR in the two samples were analysed in various ranges of the magnetic field and are relevant to the area enclosed by a pair of electron trajectories. Also it will be useful to image the actual conduction path in a wire. If the oscillation is caused by the AB effect in each conduction path in the wire, the period of ΔB in a certain oscillation component of the fluctuations is connected with the enclosed area S by the relation $\Delta BS=h/e$. With increasing magnetic fields we have observed a low frequency shift of the decay part of the fluctuations in the spectrum for DHW-08. As discussed previously,[4] the shift means the reduction of effective area for the quantum interference by the magnetic field. A similar shift has

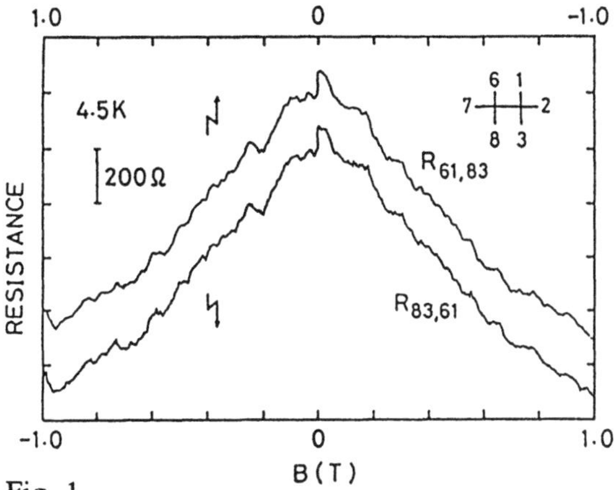

Fig. 1
$R_{61,83}$ and $R_{83,61}$ versus magnetic field at 4.5 K for SHW-02. The inset shows the electrode configuration.

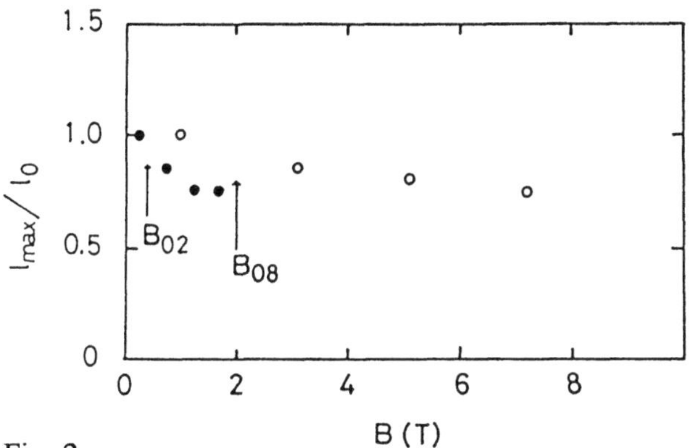

Fig. 2
Magnetic field dependence of l_{max}/l_0 at 80 mK for DHW-08(open circle) and at 1.2 K for SHW-02(closed circle). B_{02} and B_{08} stand for the critical fields of SHW-02 and DHW-08, respectively. Magnetic fields in the figure indicate the mean value in the range of the field using Fourier transform.

been obtained in the case of SHW-02. We have already defined a maximum length l_{max} which corresponds to the maximum area where the quantum interference can be successfully performed. Here l_{max} is determined by the extent of the effective conduction length or width when those are shorter than l_p. l_{max}/l_0 of the two samples are plotted in Fig. 2, where l_0 is the effective conduction length at the lowest field range in the Fourier transformation. It can be found that the decrease of l_{max} occurs near a critical field which is defined by the relation $\omega_c\tau=1$. The critical field is shown by an arrow in the figure.

In the ring experiment[1] the low frequency shift in AB oscillations has been observed and explained by the formation of edge channels in a high magnetic field. It is considered that the shift occurs when twice the cyclotron radius becomes smaller than the width of the ring. However, in quasi-ballistic conductors, it is a little difficult to explain the shift with only a relation between cyclotron radius and the width . Although the two wires have nearly the same width, there exists a clear difference in the shift as shown in Fig. 2. We consider SHW-02 to be nearly ballistic but DHW-08 to be not so. Therefore, we must take into account not only the width but also the mean free path in above criterion. Since the shift comes from a reduction of effective area of the interference, new conduction channels should appear at high fields. One explanation of the shift is expected to be a formation of edge channels.

Fourier transforms of the low temperature MR have been analysed to study the quantum interference effect in GaAs/AlGaAs narrow wires. We have found that the decay of aperiodic fluctuations in the Fourier spectrum shifts to low frequencies near the critical field of the relation $\omega_c\tau=1$. The shift in the spectrum can be explained to be due to reduction of the effective area for the interference. We consider that the reduction comes from the formation of certain conduction channels which do not affect quantum interference.

Acknowledgements

The low temperature MR was measured at the Cryogenics Center in University of Tsukuba.

References

1. G.Timp, et al.,Phys.Rev.B39 (1989) 6227.
2. C.J.B.Ford,et al.,Appl.Phys.Lett.54 (1989) 21.
3. M.Büttiker,Phys.Rev.B38 (1988) 9375.
4. M.Mizuno,et al.,Jpn.J.Appl.Phys.28 (1989) L1025.
5. A.D.Stone, Phys.Rev.Lett.54 (1985) 2692.
6. P.L.Altshuler, JETP Lett. 41 (1985) 648.
7. P.A.Lee and A.D.Stone, Phys.Rev.Lett. 55 (1985) 1622.
8. K.Ishibashi,et al., Solid State Commun. 64 (1987) 573.

Thermal Transport in Free-Standing GaAs Wires in a High Magnetic Field

A. Potts[1], *J. Singleton*[2], *T.J.B.M. Janssen*[2], *M.J. Kelly*[1,3], *C.G. Smith*[1], *D.G. Hasko*[1], *D.C. Peacock*[1,3], *J.E.F. Frost*[1], *D.A. Ritchie*[1], *G.A.C. Jones*[1], *J.R. Cleaver*[1], and *H. Ahmed*[1]

[1]Cavendish Laboratory, University of Cambridge, Madingley Road, Cambridge CB3 0HE, UK
[2]High Field Magnet Laboratory and Research Institute for Materials, University of Nijmegen, 6525 ED Nijmegen, The Netherlands
[3]G.E.C. Hirst Research Centre, East Lane, Wembley HA9 7PP, UK

Abstract. We have studied electron heating effects in free-standing GaAs wires subjected to magnetic fields of up to 20 T, applied both parallel and perpendicular to the long axis of the wires: the purpose of these measurements is to determine the contribution of one-dimensional (1D) phonons to the thermal conductivity of the wires.

The study of electronic transport in quasi one dimensional (1D) ballistic channels fabricated on semiconductor heterostructures has yielded much interesting physics over the past two years (see *e.g.* [1]). Free-standing wires can exhibit many of the same electronic properties, but additionally have laterally-confined (*i.e.* 1D) lattice vibrational modes. The physical reasons for studying such systems include: *i)* phonons are Bosons, rather than Fermions; *ii)* 1D phonon localisation, due to *mass disorder* in the wires, may occur; *iii)* there will be competing electron and 1D phonon contributions to thermal conduction [2,3].

In order to study 1D phonons, free-standing wires of a size producing lateral confinement must be fabricated (*i.e.* wavelength $\lambda \sim$ wire width). As the dominant λ of the acoustic phonon black-body spectrum is $\sim (100/T)$ nm $\sim$ 0.5 μm at 0.2 K [2], the wire width must be small and the temperature T low. The wires in this work were constructed by depositing a mask on top of an n-GaAs epilayer ($n = 10^{17}\mathrm{cm}^{-3}$). An etch was then used to remove the GaAs around the mask, leaving a triangular cross-section wire (see inset to Fig. 1a) [4]. Generally about 30 wires in parallel, with $\sim$ 0.5 μm sides, and $\sim$ 3 μm long, were constructed on one sample. The measurements described below were performed in a ^{3}He cryostat (0.4–1.2 K) with the sample *in vacuo*: the latter condition avoids thermal short–circuiting of the wires, and means that the heat-flow models simplify to 1D equations. Magnetic field (B) was provided by a 20 T Bitter coil.

The dimensionality of the lattice modes is revealed by temperature dependence of the thermal conductivity κ, which is proportional to T^n in n dimensions at low T. The first step in finding κ was to measure the wire conductivity σ as a function of T using a small ac-current (0.3 nA/wire); this enables one to use σ to measure the electron temperature. Between 0.4 and 4 K $\sigma(T)$

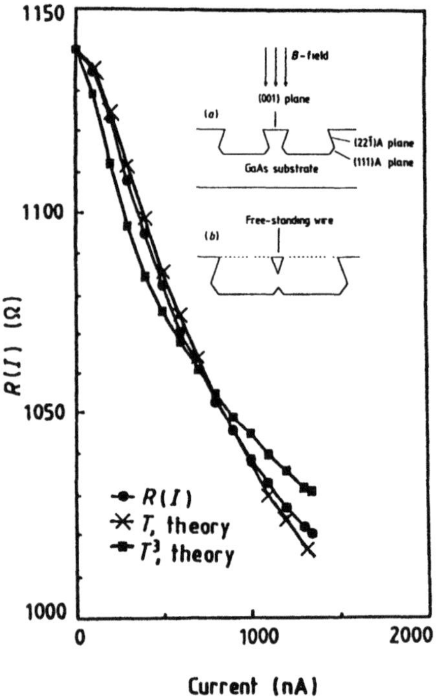

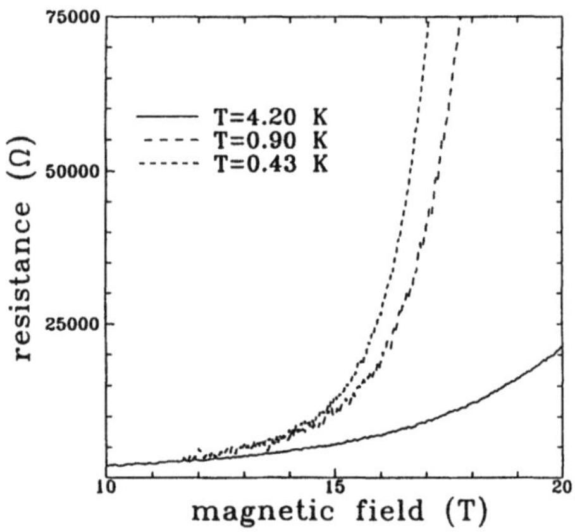

Fig.1: Left-hand side: resistance versus heating current for 27 wires in parallel at $T = 0.47$ K; both theoretical fits and data are shown. Inset: schematic of sample fabrication. Right-hand side: resistance of 27 wires in parallel versus B.

is dominated by 1D weak localisation and 3D electron-electron interactions, leading to the well-defined T-dependence $\sigma(T) = A + B\log(T)$ [4]; the same measurements showed that surface depletion leads to the electronic conduction occurring only within an approximately circular section of diameter ~ 125 nm at the wire centre. The second step was to heat the wires using a dc-current I (~ 50 nA per wire); at the same time, the small ac current was used to monitor σ. Typical results are shown in Fig. 1a for $B = 0$; as the wire heats up, its resistance falls. In order to extract κ, the heat generation and loss mechanisms must be considered. The wires are anchored at each end to the remaining epilayer, which acts as a heat sink; hence the temperature in the wire, and σ, κ, and the rate of heat generation vary along its length. The equation; $\frac{\mathrm{d}}{\mathrm{d}x}(\kappa A \frac{\mathrm{d}T}{\mathrm{d}x}) = \frac{1}{\sigma(x)} I^2$ must be solved; *i.e.* extra heat leaving region = heat generated within region. We assume that κ is given by either ΛT^3 (bulk phonons) or ηT (1D phonons, electrons), and solve the equation, using the known T-dependence of σ. Details of the method adopted to solve the equation are given in [3]; at $B = 0$, the best fit (see Fig. 1a) gives $\kappa = \eta T$. From these results, the Lorentz number $\kappa/\sigma T$ was found to be $2.1 \times 10^{-8}\mathrm{W\Omega K^{-2}}$, close to the theoretical value for electronic conduction, $\kappa/\sigma T = 2.44 \times 10^{-8}\mathrm{W\Omega K^{-2}}$: *i.e.* the electron thermal conductivity (which is also proportional to T) dominates the phonon contribution. This is thought to be a consequence of the electron-phonon scattering length being comparable to the wire length, so that the hot electrons remove the heat from the wire themselves, rather than passing it to the phonons.

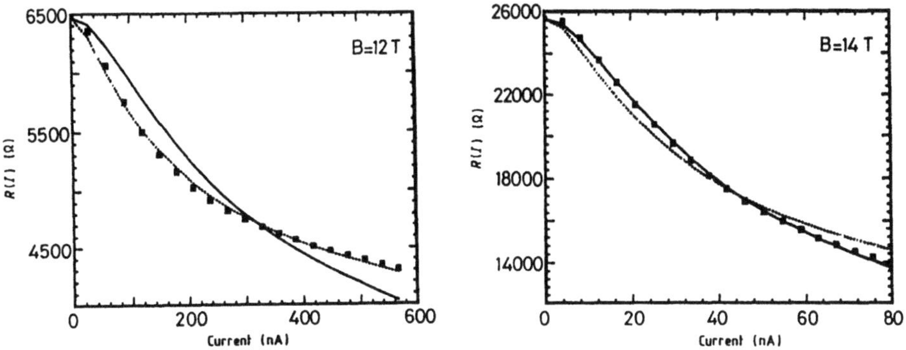

Fig.2: Resistance versus heating current for 27 wires in parallel for $T = 0.595$ K and two different fields. Data are squares; solid line $\kappa = \eta T$ fit; dotted line $\kappa = \Lambda T^3$ fit.

In order to suppress the electronic contribution to the thermal conductivity, a magnetic field B was used to localise the electrons [2,3,4]. The effect of B on the n-GaAs wires will be to confine the impurity wavefunctions in a plane perpendicular to B, so that they become ellipsoidal. Because the impurities within the wire are quasi-randomly arranged, the electrons will have to diffuse or hop perpendicular to B in order for a current to flow [5]; the conduction is no longer in an impurity band and the system undergoes a metal-insulator transition [5]. The resistance of the wires was measured as a function of B and T using small ac currents (voltage across wires < 70 μV); typical data are shown in Fig. 1b. The combination of large B and low T suppresses σ by orders of magnitude, indicating that the electronic contribution to κ will be lowered by a similar amount. The electron heating experiments were then repeated in magnetic fields, and κ calculated as above.

Strong suppression of the electronic conduction occurs for $B \simeq 10$ T (the metal-insulator transition occurs at a critical field of ≈ 11.8 T for B perpendicular to the wire [6]), and we shall concentrate on this region. Three separate regimes in the behaviour of κ are observed: *i)* for $B \leq 13$ T, $\kappa = \Lambda T^3$ appears to be a much better fit to the data (Fig. 2), both in terms of residuals and the heating current independence of the parameter Λ; *ii)* for 14 T $\leq B \leq$ 15 T, the behaviour changes, and $\kappa = \eta T$ is the better fit (Fig. 2), again both in terms of residuals and heating current independence; *iii)* for $B \geq 16$ T, neither ηT nor ΛT^3 adequately fits the data. In all cases ($B \geq 10$ T) $\kappa/\sigma T$ considerably exceeds the Lorentz number; *i.e.* the thermal conductivity is primarily due to phonons.

Order of magnitude estimates of the phonon mean free paths l may be calculated from the values of Λ and η in the field range 10-13 T. If the phonons are 3D then $l \sim 5$ nm; if they are 1D, then $N_B l \sim 7$ μm at 1 K, where N_B is the number of activated branches. The latter is comparable to the wire width; as 1D phonons cannot undergo boundary scattering [2,3], this implies scattering from wire thickness variations. The high quality of the T^3 (3D) fits in this region implies that the phonon temperature is well defined at all points in the wire, consistent with the short l, which could be interpreted as evidence for phonon localisation, as there is no other obvious reason why 3D phonons should be so strongly scattered at $T \leq 1$ K. However, these results should be treated with some caution, as although $\kappa = \Lambda T^3$ fits the data well for $B \leq 13$ T, it does not give values of Λ which are independent of T, unlike the $B = 0$ data. The transition to $\kappa = \eta T$ for $B \geq 14$ T and the poor fit at 16 T are not yet understood; it is likely that the necessary assumption of this work, that the electronic conduction is metallic and diffusive in nature, is breaking down.

In summary, we have used a high magnetic field to suppress electronic thermal conduction in free-standing n-GaAs wires. Thermal conductivity due to phonons in the wires has been observed, and there is some evidence that phonon localisation is occurring. More experiments at lower and higher temperatures and fields are under way to clarify these points.

This work was supported by the European Community, and by FOM and NWO (Netherlands) and SERC (UK).

References

[1] H. van Houten, C.W.J. Beenakker and B.J. van Wees, in; *Semiconductors and Semimetals*, M.A. Reed, ed. (Academic Press N.Y. 1990)

[2] M.J. Kelly, J. Phys. C **15**, L969 (1982).

[3] A. Potts, M.J. Kelly, C.G. Smith, D.G. Hasko, J.R.A. Cleaver, H. Ahmed, D.C. Peacock, D.A. Ritchie, J.E.F. Frost and G.A.C. Jones, J. Phys. Condens. Matter **2**, 1817 (1990).

[4] A. Potts, D.G. Hasko, J.R.A. Cleaver, C.G. Smith, H. Ahmed, M.J. Kelly, J.E.F. Frost, G.A.C. Jones, D.C. Peacock and D.A. Ritchie, J. Phys. Condens. Matter **2**, 1807 (1990);

[5] L. Altshuler and A.G. Aronov, Pis'ma Zh. Eksp. Teor. Fiz. **37**, 349 (1983) [JETP Lett. **37**,410 (1983)]; A.L. Efros and B.I. Shklovskii, J. Phys. C **8**, 149 (1975).

[6] A. Potts, J. Singleton, T.J.B.M. Janssen and M.J. Kelly (to be published).

Quantum Dots in High Magnetic Fields

U. Merkt

Institut für Angewandte Physik, Universität Hamburg,
W-2000 Hamburg 36, Fed. Rep. of Germany

Abstract. By high resolution technologies quasi-two-dimensional electron gases near semiconductor interfaces can be confined laterally to dots of diameters below 100 nm. These quantum dots contain only few electrons and exhibit discrete energy spectra. Here, the present state of our work on InSb is described.

1. Introduction

Advances of lithography and etching techniques make it possible to laterally confine quasi-two-dimensional (2D) electron systems in semiconductors into dots of diameters below 100 nm [1]. The spectrum of electrons in such dots is totally discrete, i.e., there is no free motion with continuous dispersion but we have a quasi-zero-dimensional (0D) system. Commonly, one speaks of strictly 2D or 1D electron systems if only the lowest 2D or 1D subband is occupied and addresses these situations as electric quantum limits [2]. If we take over this definition for 0D quantum dots by saying that only the lowest level is populated, we have to confine just one or two electrons in a dot. This means that one must realize one of the ultimate limits set to the miniaturization of electronic devices.

2. Fabrication

The idea of our structures on InSb is sketched in Fig. 1. We have NiCr as a Schottky barrier that pins the Fermi energy at the NiCr/InSb interface above the valence band edge [3-5]. Only underneath the narrow regions between the NiCr mesh mobile inversion electrons are induced by a gate voltage V_g. There is virtually no tunneling between adjacent dots since the barrier height and distance

Springer Series in Solid-State Sciences, Vol. 101
High Magnetic Fields in Semiconductor Physics III Editor: G. Landwehr

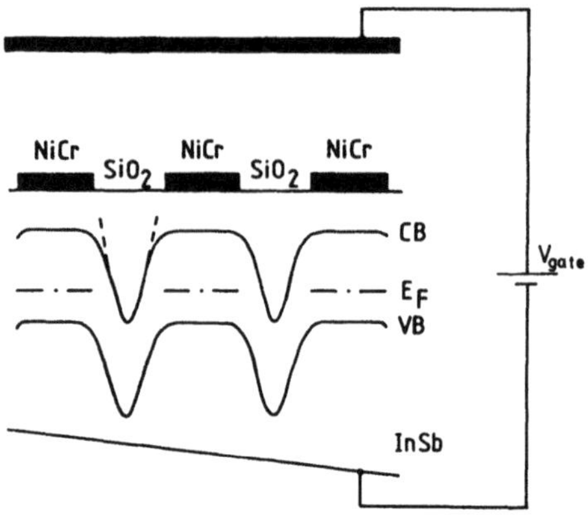

Fig. 1 Schematic cross section of the microstructured field-effect device with its lateral band structure.

between dots is of order of the band gap energy ($E_g \doteq 236$ meV) and grating constant (a=250 nm) of the dot array, respectively. In the vicinity of a minimum of the conduction band edge, the lateral potential may be approximated by a harmonic oscillator. Provided this well is narrow enough, we have discrete states or 0D quantum dots.

The desired structures are obtained by holographic lithography whose setup is depicted in Fig. 2. After the first exposure, the sample is rotated by an angle of 90^0 and the photoresist is exposed for a second time. After development there is a sinusoidal resist pattern as visualized in Fig. 2 for one lateral direction. The resist is further removed by dry etching in an oxygen plasma resulting in a periodic array of resist dots. After the fabrication of a SiO_2 insulator a homogeneous top gate is evaporated. A micrograph of a monitor sample without SiO_2 insulator and top gate is shown in Fig. 3.

The dots can be charged via field effect without direct contacts to the inversion electrons since the InSb substrate has a finite resistivity in the megaohm regime even at liquid helium temperatures. A threshold voltage V_t is determined from the onset of absorption and the voltage difference $\Delta V_g = V_g - V_t$ is used as measure of the number of electrons in a dot. In the spectroscopic experiments, the transmittance of normally incident radiation of an optically pumped far-infrared laser is recorded at liquid helium temperatures. A magnetic field applied perpendicularly to the samples tunes the dot modes in order to obtain resonant absorption for the

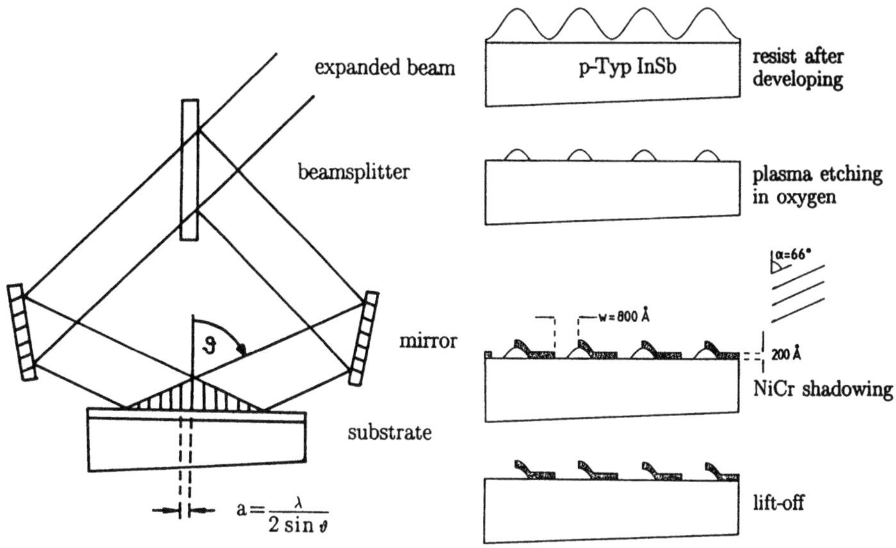

Fig. 2 Setup of the holographic lithography and principal preparation steps for quantum dot devices.

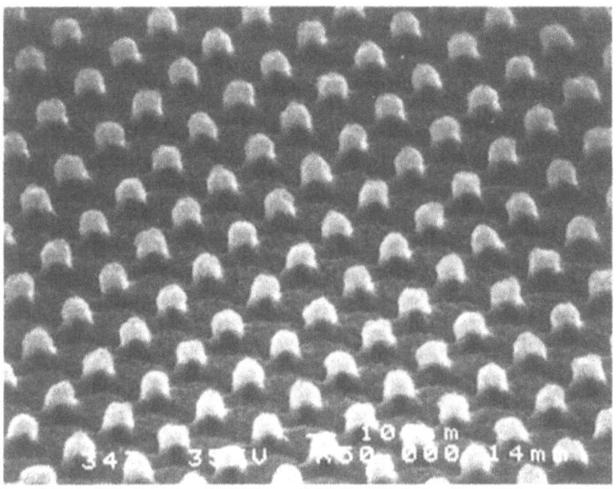

Fig. 3 Resist dots shadowed with gold for contrast enhancement. The marker is 100 nm long.

fixed laser frequencies. Simultaneously, it provides the Zeeman splitting of the atomic-like states.

3. Theoretical

Following the course of the conduction band in Fig. 1 near a minimum, we describe the lateral potential by a two-dimensional oscillator well. Its single particle eigenenergies in a magnetic field along the z direction perpendicular to the dots are given by the expression [3]

$$E_{nm} = E_{i=0} + (2n + |m| + 1)\hbar \sqrt{\left(\frac{\omega_c}{2}\right)^2 + \omega_0^2} + \frac{\hbar\omega_c}{2} m. \qquad (1)$$

We presume that the motion in z direction is frozen out into the lowest 2D subband (i=0) and ignore electron spin. The lateral motion is described by the radial n=0,1,... and the azimuthal m=0,1,... quantum number. The magnetic field strength B enters via the cyclotron frequency $\omega_c = eB/m^*$. There are only two allowed dipole transitions which are excited with the two circular light polarizations ± and which have frequencies [3]

$$\omega_\pm = \sqrt{\left(\frac{\omega_c}{2}\right)^2 + \omega_0^2} \pm \frac{\omega_c}{2} \quad . \qquad (2)$$

The recently formulated generalized Kohn theorem [6] states that in case of harmonic confinement the electron-electron interaction does not influence the spectrum. To be more precise: In the presence or absence of a magnetic field the dipole resonance frequencies of an electron system with harmonic confinement are independent of the electron number as well as of the particular form of the electron-electron interaction. Always the transition frequencies are given by the single-electron values of Eq.(2).

The relation between electron number n_0 and electronic radius R of a dot has been calculated by Chaplik [7] in the Thomas-Fermi approximation:

$$n_0 = \frac{a^* R^3}{l_0^4} \sum_{j=1}^{\infty} \frac{8R}{\lambda_j^3 \, (2R + \lambda_j a^*)} \quad . \qquad (3)$$

In this equation we have the effective Bohr radius a* and the oscillator length $l_0 = (\hbar/m^*\omega_0)^{1/2}$. The constants λ_j are the zeros of the Bessel function J_0. Like for real atoms, we expect the Thomas-Fermi result to provide a good description for higher electron numbers.

4. Experiments

Spectra for various laser energies and gate voltages are shown in Fig. 4 for linearly polarized light [3]. Spectra for energy 10.4 meV resemble cyclotron resonances of a homogeneous 2D gas but the resonance magnetic fields are shifted considerably ($\Delta B \simeq 0.4$ T) to lower values. For energy 7.6 meV, we no longer observe a distinct resonance maximum at finite fields but a monotonic decrease when the magnetic field is increased. This is in fact expected from the classical conductivities

$$\sigma_{\pm}(\omega) = \frac{e^2 n_0 \ \tau / m^*}{1 + (\omega_0^2/\omega - \omega \pm \omega_c)^2 \ \tau^2} \tag{4}$$

when the quantization frequency coincides with the laser frequency ($\omega_0 \simeq \omega$). For the energy 3.2 meV, we observe distinct but weak resonances at B $\simeq$ 1.5 T. These are the ω_- resonances of Eq.(2) which are characteristic of a system confined in both lateral directions [3-5].

In higher magnetic fields, the observed ω_+ resonance develops into the 2D cyclotron resonance [2] as is in

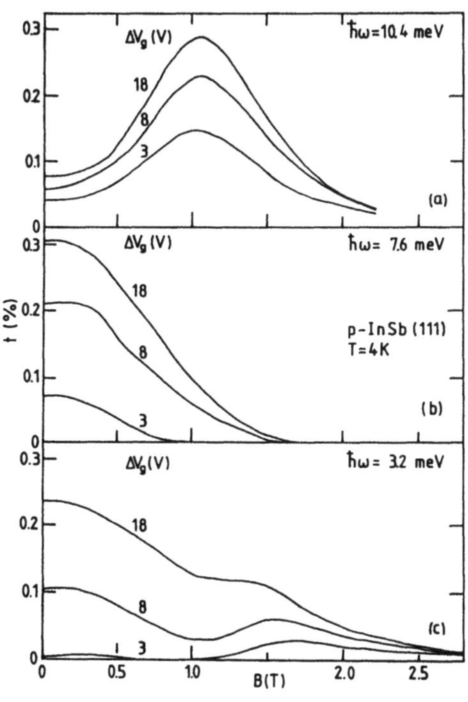

Fig. 4 Far-infrared spectra of quantum dots for three laser frequencies ω and three gate voltages ΔV_g.

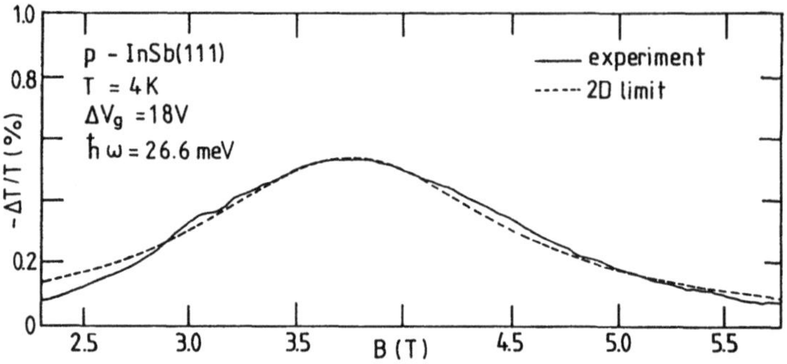

Fig. 5 Laser spectrum (solid line) and its theoretical description (dashed line) for a higher photon energy and magnetic fields where the cyclotron radius is much less than the dot radius.

fact expected when the cyclotron length $(\hbar/eB)^{1/2}$ becomes much less than the dot diameter R. Therefore, we rely on the ω_+ resonance in high magnetic fields ($\omega_c \gg \omega_0$) when we determine the electron number. We also utilize the fact that the far-infrared wavelengths by far exceed the grating constants of the dot array ($\lambda \gg a$). This means that the incident wave does not distinguish between electrons of individual dots but probes an average areal density n_0/a^2. Then the Fresnel formula for a thin conductive layer on a semiconductor of dielectric constant ϵ applies and relates the electron number to the transmittance for circular light polarization:

$$-\frac{\Delta T}{T} \simeq \frac{2/Y_0}{1+\sqrt{\epsilon}+\sigma_{\square}/Y_0} \cdot \frac{e^2 n_0 \tau/m^*}{1+(\omega-\omega_c)^2\tau^2} \cdot \frac{1}{a^2} \quad . \tag{5}$$

In this equation, we have the wave admittance of vacuum $Y_0=(\mu_0 c)^{-1}$ and the sheet conductivity $\sigma_{\square}\sim 0.1\ \Omega^{-1}$ of the two metal gates of our samples. Figure 5 shows an experimental spectrum again for the sample of Fig. 4 (solid line) at a higher laser energy 26.6 meV together with its theoretical description (dashed line). This spectrum of the ω_+ resonance is already very similar to the one of 2D cyclotron resonance as a consequence of the relatively high resonance magnetic field strength.

The experimental results obtained for the sample whose spectra we discussed are summarized in Fig. 6. The Zeeman splitting of the resonance frequency is shown together with the theoretical curves of the effective mass

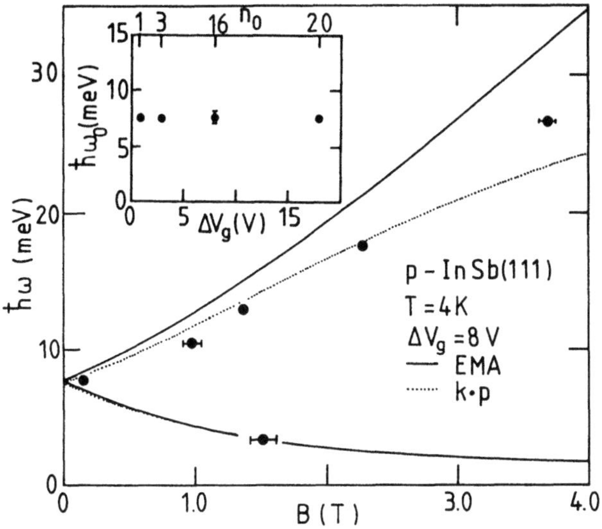

Fig. 6 Zeeman splitting of the resonance frequency. The inset gives the dependence of the quantization energy $\hbar\omega_0$ on gate voltage ΔV_g and electron number n_0.

approximation (EMA) given in Eq.(2) (solid lines) and of a simple k·p-approach (dotted lines) that seems to provide a better description [5]. The inset depicts the quantization energy in the absence of a magnetic field extrapolated from the Zeeman splitting. It is given as a function of gate voltage as well as of electron number. The electron number saturates at higher gate voltages and we could not induce more than 20 inversion electrons per dot in this particular sample. In all of our samples we observe the independence of the excitation energy on electron number and the saturation. The independence of the energy of the electron number we explain by the generalized Kohn theorem, i.e., we take it as strong evidence for harmonic confinement. The saturation seems to be a consequence of the NiCr/InSb Schottky barrier in which further electrons become bound.

5. Conclusions and Perspectives

Far-infrared spectroscopy of electronic states in InSb quantum dots demonstrates strong lateral quantization by an approximately harmonic well. The far-infrared radiation only couples to the center of mass motion in strictly harmonic potentials and one measures the bare

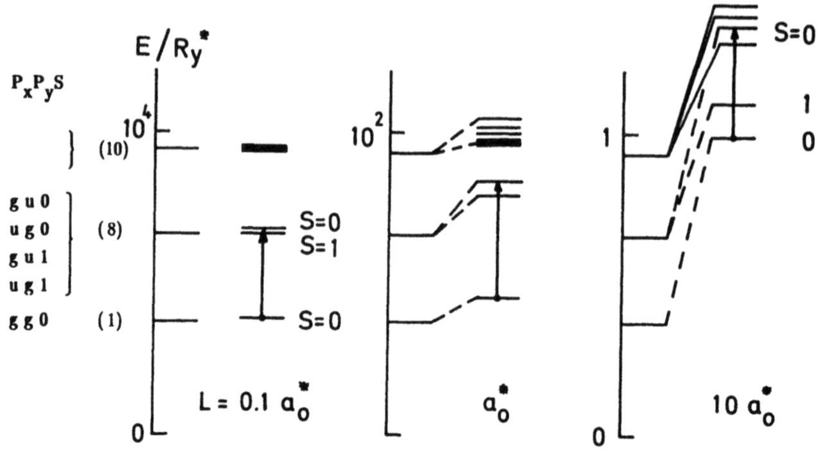

Fig. 7 Theoretical eigenenergies of two electrons in a square well potential of infinitely high barriers for three distinct lateral widths L.

frequency independent of the electron number. This way no insight is gained into the intrinsic motion.

As a foretaste of what interesting effects expect us when we find routes out of the dilemma of far-infrared spectroscopy of electrons in a harmonic potential, let us discuss the simple example of two electrons in a quadratic box with lateral length L and infinitely high walls. In Fig. 7 we show the lowest lying states of a more detailed theoretical study of Bryant [8] that are classified according to their parities P_x, P_y, and their total spin S. Degeneracies are given in parentheses. The scales of length and energy are given by the effective Rydberg constant Ry* and Bohr radius a*, respectively. The eigenenergies of independent electrons are given directly at the respective energy scales. In small dots (L=0.1 a*) the exchange splitting of the first excited level is only minor and the frequency of the optically allowed transition ($\Delta S=0$) is hardly changed when the electrons interact. In dots of intermediate size (L=a*) there is stronger increase of the eigenenergies due to Coulomb repulsion as well as larger exchange splitting. Corresponding dot sizes are realized on InSb. In still larger dots (L=10 a*) a complete rearrangement of levels starts. Exciting possibilities are also offered by the spin-orbit interaction which couples singlets and triplets thus possibly leading to metastable dot states.

Recently, deviations from harmonic confinement have been observed in far-infrared spectra of deep mesa etched dots on GaAs by Demel, Grambow, Heitmann, and Ploog [9]. Higher modes and line splittings are observed and interpreted in a collective picture as manifestations of resonant nonlocal interaction. Deviations from harmonic confinement also have been observed by Lorke, Kotthaus, and Ploog [10] in field-effect confined dots. In their voltage tunable devices the transition from isolated to coupled dots can be controlled and in addition to the modes characteristic of isolated dots new modes appear which are interpreted as edge magnetoplasmons.

An alternative very promising route to the intrinsic electronic structure is provided by transport measurements [11]. Magnetocapacitance signals also exhibit plentiful fine structure in the density of states and demonstrate the existence of the fractional quantum Hall effect [12] at low electron numbers. Electron transport through a finite chain of GaAs dots has been tuned by a split-gate configuration and the observed conductance has been discussed in terms of gaps and bands containing the corresponding number of fifteen states [13]. These few examples may serve as additional illustrations that the miniaturization of electronically active devices has reached the domain of lateral dimensions where we must employ pictures known from atomic and molecular physics.

Acknowledgements

I thank Ch. Sikorski and P. Junker for their collaboration, A.G. Aronov, G.W. Bryant, A.V. Chaplik, R. Gerhardts, D. Heitmann, J.P. Kotthaus, F. Peeters, W. Que, and U. Rößler for valuable discussions and the Deutsche Forschungsgemeinschaft as well as the Volkswagenstiftung for financial support.

References

1. "Nanostructures Physics and Fabrication", ed. by M.A. Reed and W.P. Kirk (Academic Press, Boston, 1989).
2. T. Ando, A.B. Fowler, and F. Stern, Rev. Mod. Phys. 54, 437 (1982).
3. Ch. Sikorski and U. Merkt, Phys. Rev. Lett. 62, 2164 (1989); 64, 3100 (1990).

4. U. Merkt, Ch. Sikorski, and J. Alsmeier, in "Spectroscopy of Semiconductor Microstructures", ed. by G. Fasol, A. Fasolino, and P. Lugli (Plenum Press, New York, 1989), pp. 89-114.
5. U. Merkt and Ch. Sikorski, Semicond. Sci. Technol. 5, S182 (1990).
6. L. Brey, N.F. Johnson, and B.I. Halperin, Phys. Rev. B40, 10 647 (1989).
7. A.V. Chaplik, Pis'ma Zh. Eksp. Teor. Fiz. 50, 38 (1989) [JETP Lett. 50, 44 (1989)].
8. G.W. Bryant, Phys. Rev. Lett. 59, 1140 (1987).
9. T. Demel, D. Heitmann, P. Grambow, and K. Ploog, Phys. Rev. Lett. 64, 788 (1990).
10. A. Lorke, J.P. Kotthaus, and K. Ploog, Phys. Rev. Lett. 64, 2559 (1990).
11. M.A. Reed, J.N. Randall, R.J. Aggarwal, R.J. Matyi, T.M. Moore, and A.E. Wetsel, Phys. Rev. Lett. 60, 535 (1988).
12. W. Hansen, T.P. Smith, III, K.Y. Lee, J.M. Hong, and C.M. Knoedler, Appl. Phys. Lett. 56, 168 (1990).
13. L.P. Kouwenhoven, F.W.J. Hekking, B.J. van Wees, C.J.P.M. Harmans, C.E. Timmering, C.T. Foxon, Phys. Rev. Lett. 65, 361 (1990).

Far-Infrared Transmission of Voltage-Tunable GaAs-(Ga,Al)As Quantum Dots in High Magnetic Fields

N.K. Patel[1], *T.J.B.M. Janssen*[2], *J. Singleton*[2], *M. Pepper*[1], *H. Ahmed*[1], *D.G. Hasko*[1], *R.J. Brown*[1], *J.A.A.J. Perenboom*[2], *G.A.C. Jones*[1], *J.E.F. Frost*[1], *D.C. Peacock*[1], *and D.A. Jones*[1]

[1]Cavendish Laboratory, University of Cambridge, Madingley Road, Cambridge CB3 0HE, UK

[2]High Field Magnet Laboratory and Research Institute for Materials, University of Nijmegen, 6525 ED Nijmegen, The Netherlands

Abstract. We report magneto-capacitance and far-infrared (FIR) magneto-transmission studies of an array of GaAs-(Ga,Al)As quantum dots. With zero gate voltage (V_g) applied, the FIR magneto-transmission spectrum shows the cyclotron resonance (CR) of the 2DEG, and a 2D-magnetoplasmon. On applying a negative V_g a resonance associated with the dots is observed. The response of the dots can be modelled within either a Maxwell-Garnett or an edge-magnetoplasmon model, to reveal that the effective radius of the dots can be varied by $\sim$ 50% using reasonable values of V_g.

There is a great deal of current interest in the properties of quantum dot systems, realised on high mobility heterojunctions using either metal gates [1, 2] or deep-mesa-etching [3, 4]. The laterally-confined states of such systems have been studied using capacitance [5] or optical spectroscopy [1-4]; in the former case the states are not measured directly, whilst in the latter, the number of carriers in each dot, vital in the interpretation of the spectra, must be inferred from the strength of optical transitions [2]. In this paper we have studied an array of dots, which have been specially constructed to allow *simultaneous* measurements of capacitance and transmission; the dots are relatively large (1 μm) allowing classical effects in the bounded 2DEG in each dot to be examined as a function of V_g and carrier density.

A schematic of the structure is shown in Fig. 1a. The array of dots was fabricated on a GaAs-(Ga,Al)As heterojunction, which contains a high-mobility 2DEG close to the interface and a Si δ-doped layer 200 nm below; this acts as a transparent back-contact for the capacitance measurements. A mesa with Ni-AuGe contacts connected to both 2DEG and delta-doped layer was constructed on top of the heterojunction and the substrate was wedged by 2°, in order to avoid interference effects. A lattice (unit cell 2.25 $\times$ 2.5 μm; area 4 mm^2) of pillars of high-resolution negative resist 0.5μm high was defined on the mesa using electron-beam lithography; the pillars had diameters $\approx$ 1.0 μm. Finally, a 5 nm thick transparent NiCr gate was evaporated over the central region of the lithographed area. The gate will not be continuous over the sides of the pillars, so that no gate voltage (V_g) will appear on the pillars. Therefore, on

Springer Series in Solid-State Sciences, Vol. 101
High Magnetic Fields in Semiconductor Physics III Editor: G. Landwehr

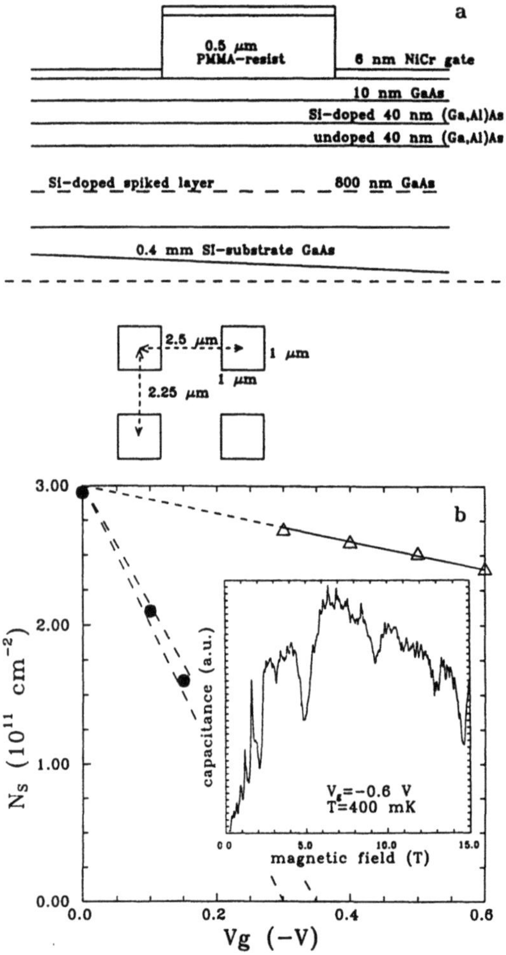

Fig.1: a) Schematic of structure. b) Inset: capacitance versus magnetic field; main figure; N_s versus V_g for regions between dots (•) and in the dots (△).

applying a negative V_g, the areas of 2DEG between the pillars are depleted, leaving "dots" of bounded 2DEG underneath. Far-infrared (FIR) transmission and magneto-capacitance measurements were carried out with the sample at 400 mK. An optically-pumped FIR laser, ($\sim$ 100 lines between 1228 and 33 μm) was used as FIR source; magnetic fields were provided by a 15 T Bitter coil. Magneto-capacitance data are shown in the inset part of Fig. 1b; clear Shubnikov-de Haas oscillations (SdHo), from which a 2D carrier density N_s can be extracted, are visible [5]. For $V_g > -0.25$ V, a set of SdHo was observed with a fundamental field which decreased rapidly as V_g became more negative. These SdHo result from the regions between the dots, depleted by the negative V_g. The N_s values from these data are shown in Fig. 1b as filled circles; an extrapolation to $N_s = 0$ indicates that the dots become isolated at -0.3 V $> V_g > -0.35$ V. For $V_g < -0.3$V, a second set of SdHo, due to the

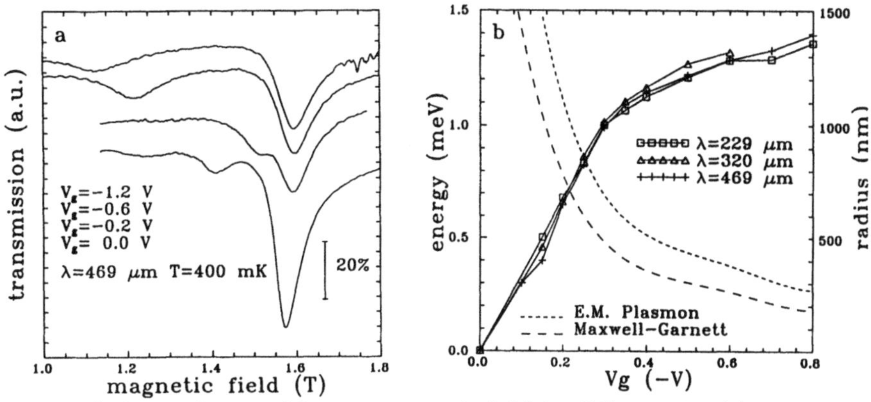

Fig.2: a) Transmission of dots vs magnetic field for different V_g. b) Dot plasmon energy (data, left-hand scale) and radius (dashed lines, right-hand scale) vs V_g.

2DEG in each dot, is visible (inset Fig. 1b). Values of N_s extracted from these oscillations are shown as triangles in Fig. 1b.

Typical magneto-transmission spectra are shown in Fig. 2a for a wavelength of 469 μm. The prominent feature at ~ 1.6 T is the cyclotron resonance (CR) of the 2DEG, which is present for all V_g, as the gate does not cover the entire mesa. At $V_g = 0$ there is a small absorption dip at ~ 1.4 T. This is a 2D magnetoplasmon, which can couple to the FIR via the periodic metal gate, and occurs at a frequency $\omega^2 = (\omega_c^2 + \Omega_q^2)$; here ω_c is the cyclotron frequency and $\Omega_q^2 = \pi N_s e^2 q\phi(q)/2m^* \varepsilon_0 \epsilon$. The wavevector is given by $q = 2n\pi/a$, with a the periodicity of the gate and n an integer; $\phi(q)$ takes account of the screening of the gate, and is 0.23 for $n = 1$ in our structure; $\epsilon = 6.9$ is the mean dielectric constant of GaAs and vacuum. Inserting the parameters for our structure, we obtain $\Omega_q = 2.1 \times 10^{12}$ s^{-1}, compared with an experimental value of 1.9×10^{12} s^{-1}.

At small negative V_g the 2D-plasmon disappears and for $V_g < -0.15$ V another resonance appears on the low-field side of the CR, moving to lower fields for more negative V_g; this is due to the dots. In contrast to the 2D case, the energy spectrum is given by (assuming a parabolic potential [6, 7]) $\omega = [(\omega_c/2)^2 + \omega_o^2]^{0.5} + (\omega_c/2)$, where ω_o is the characteristic frequency of the dots. In Fig. 2b $\hbar\omega_o$ is plotted against V_g for three laser energies. No energy dependence was observed, indicating that the potential in the dot is roughly field-independent and that the method of summing the cyclotron and dot energies has some validity for a real potential [6]. At all laser energies ω_o shows the same V_g dependence; a steep rise in the region with $0 > V_g > -0.3$ V, then a reduced slope after the dots are isolated for $V_g < -0.3$ V.

The dependence of ω_o vs. V_g can be modelled using two approaches for a bounded 2DEG. The first is the classical Maxwell-Garnett (MG) theory for *depolarisation* in arrays of 2D metallic disks [3]; ω_o is given by $\omega_o^2 = (\pi/8) N_s e^2/2m^* \varepsilon_0 \epsilon R$, where ϵ is defined as above, and R is the radius of the dots. The dashed curve in Fig. 2b gives the radius calculated using N_s from the capacitance data and ω_o^2 from the FIR-transmission curves using this approach. FETTER [8, 9, 10] has derived a model for *edge-magnetoplasmons* in a bounded 2DEG, resulting in $\omega_o^2 = (2/(g(q)+2))N_s e^2 \phi(q)/2m^* \varepsilon_0 \epsilon R$ with $\phi(q)$, R and ϵ as above, and $g(q)$ another screening function dependent on R and q; (*e.g.* $g(q) = 0.78$ for $R = 500$ nm). Using the same ϵ and $\phi(q)$ as above and varying R to fit the data produces the dotted curve in Fig. 2b.

The two models give values of R which are within ~ 30 % of each other for $V_g < -0.3$ V (*i.e.* isolated dots). Within this regime, it appears that the radius of the dot can be varied by ~ 50 % by the action of the gate. The radius extracted from the MG model is 400–450 nm at the point at which the dots first become isolated at $V_g \approx -0.35$ V; *i.e.* the theory gives a value just below the physical radius of the dots (500 nm; a slightly reduced size "seen" by the electron is expected from the depletion fields in GaAs [11]). Therefore, within the approximation used to extract the energies, it seems that the MG approach is probably a more reliable means of deducing the radius of the dots. However, a more rigorous calculation of $g(q)$ and $\phi(q)$ may well bring the edge magnetoplasmon approach into closer agreement. In the region for $V_g > -0.3$ V the dots are not isolated and so the models are of limited value.

In summary, the experiments show that simultaneous measurement of capacitance and FIR-transmission is possible. 2D-magneto-plasmons are observed in the low V_g region, which can be modelled if screening is incorporated. When the dots are isolated their response can be modelled within theories of the bounded 2DEG, to reveal that the dot radii can be easily tuned over a range of ~ 50 %.

This work was supported by the European Community, and by FOM and NWO (Netherlands) and SERC (UK).

References

[1] C.T. Lui, K. Nakamura and D.C. Tsui, Appl. Phys. Lett. **55**, 168 (1989).

[2] Ch. Sikorski and U. Merkt, Phys. Rev. Lett. **62**, 2164 (1989).

[3] S.J. Allen Jr., H.L. Störmer and J.C.M. Hwang, Phys. Rev. B **28**, 4875 (1983).

[4] D. Demel, D. Heitmann, P.G. Grambow and K. Ploog, Phys. Rev. Lett. **64**, 788 (1990).

[5] T.P. Smith III, K.Y. Lee, C.M. Knoedler, J.M. Hong and D.P. Kern, Phys. Rev. B **38**, 2172 (1988).

[6] F.M. Peeters (to be published).

[7] R.B. Dingle, Proc. R. Soc. London, Ser. A **211**, 500 (1952).

[8] A.L. Fetter, Phys. Rev. B **32**, 7676 (1985).

[9] J.W. Wu, P. Hawaylak and J.J. Quinn, Phys. Rev. Lett. **55**, 879 (1985).

[10] W. Que and G. Kirczenow, Phys. Rev. B **39**, 5998 (1989).

[11] D.A. Wharam, U. Ekenberg, M. Pepper, D.G. Hasko, H. Ahmed, J.E.F. Frost, D.A. Ritchie, D.C. Peacock and G.A.C. Jones, Phys. Rev. B **39**, 6283 (1989).

Magneto-Optical Spectrum of a Quantum Dot

*F. Geerinckx, F.M. Peeters, and J.T. Devreese**

Department of Physics, University of Antwerp (UIA), Universiteitsplein 1, B-2610 Antwerpen, Belgium
***Also at: University of Antwerp (RUCA), B-2020 Antwerpen, Belgium, and Eindhoven University of Technology, NL-5600 MB Eindhoven, The Netherlands**

The energy levels of electrons confined to a circular quantum dot with hard walls are calculated in the presence of a perpendicular magnetic field. The results are compared with the case of parabolic confinement. Important differences in the transition energies of the magneto-optical spectrum are: *i)* in the hard-wall case there are many transitions possible with different energies, whereas in the parabolic case only two transition energies are allowed. For the hard wall case, however, only a small number of them have sufficient oscillator strength to be observable; *ii)* the transition energies have a discontinuous behavior as a function of the magnetic field, and *iii)* with increasing magnetic field the hard-wall transition energies approach the two-dimensional results much faster than for the soft-wall case.

Recently it has become possible experimentally to fabricate semiconductor devices in which the electrons are confined in all spatial directions. In these so-called *quantum dots* a finite number of electrons form a quasi-zero-dimensional (Q0D) electron gas [1-4].

Applying state-of-the-art nanolithography techniques, quantum dots have been fabricated by electrostatic confinement of a two-dimensional electron gas. The confining potential in such quantum dots is probably well represented by a quadratic potential. Alternative techniques, however, based on etching techniques, fabricate columns [5] that contain a Q0D electron gas. It is expected that this technique provides more well defined confinement walls, which, in the extreme limit, may be represented by hard walls.

In this contribution we study the energy spectrum of non-interacting electrons confined to a circular quantum dot with hard walls placed in a perpendicular magnetic field. The magneto-optical spectrum has a much richer structure than for a soft-wall potential as approximated by a quadratic potential [6,7].

Recently it was proven by one of us [8] that for a quadratic confining potential the transition energies are independent of the electron-electron interaction and the number of electrons in the quantum dot. In the present paper we show that this is no longer the case for a hard wall confining potential.

Neglecting the effect of spin, the single-particle hamiltonian for an electron with effective mass m can be written as

$$H = \frac{1}{2m}\left(\vec{p} - \frac{e}{c}\vec{A}\right)^2 + V(r),$$

where $V(r) = +\infty$ for $r > R_0$ and zero elsewhere, is the hard-wall confining potential in the (x, y)-direction; $r = \sqrt{x^2 + y^2}$ and R_0 is the radius of the quantum dot. The one-electron energy levels are given by

$$E_{nl} = \hbar\omega_c\left(\alpha_{nl} + \frac{l + |l|}{2} + \frac{1}{2}\right),$$

where $\omega_c = eB/mc$ is the cyclotron frequency and $\{\alpha_{nl}; n = 0, 1, 2, \ldots\}$ are determined

Springer Series in Solid-State Sciences, Vol. 101
High Magnetic Fields in Semiconductor Physics III Editor: G. Landwehr

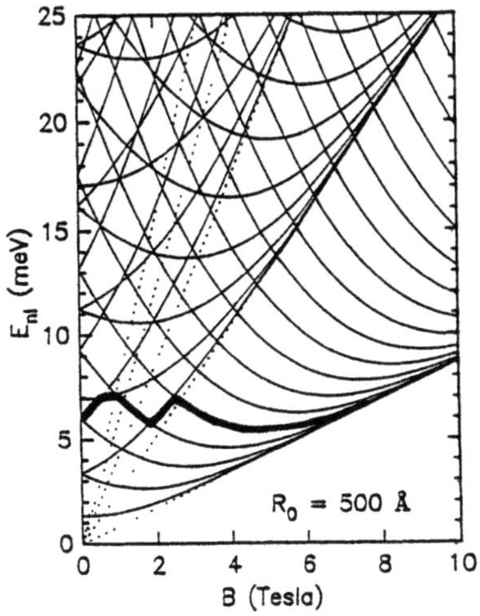

Fig. 1. Energy levels as a function of magnetic field for the hard-wall potential. The Fermi energy for a 10 electron system is indicated by the thick curve.

by the boundary condition $\Psi_{nl}(R_0, \phi) = 0$; l is the angular quantum number. In the limit of large magnetic fields, or equivalently, large quantum dot radius, $\alpha_{nl} \to n$, and the energy spectrum equals the one of a two dimensional electron in a magnetic field. In the general case however, α_{nl} will have to be calculated numerically. At zero magnetic field, the α_{nl} are related to the zeroes of the Bessel function $J_{|l|}(x)$ and the energy levels are twofold degenerate for $l \neq 0$. Fig. 1 shows the energy levels E_{nl} of a quantum dot of radius $R_0 = 500$ Å made out of GaAs ($m/m_0 = 0.065$). At high magnetic field, the energy levels degenerate into the ideal two dimensional Landau levels. At low magnetic fields the spectrum is quite complicated.

The normalized wave function, corresponding to the eigenenergy E_{nl}, is given by

$$\Psi_{nl}(r,\phi) = C_{nl}^{-\frac{1}{2}} \frac{e^{il\phi}}{\sqrt{2\pi}} \left(\frac{m\omega_c}{\hbar}\right)^{\frac{1}{2}} \left(\frac{r^2}{2l_B^2}\right)^{|l|/2} e^{-r^2/4l_B^2} {}_1\mathrm{F}_1(-\alpha_{nl}, 1+|l|, \frac{r^2}{2l_B^2})$$

with C_{nl} the normalization factor and $l_B = \sqrt{\hbar c/eB}$ the magnetic length. Fig. 2 shows the real part of the normalized wave function for two sets of quantum numbers. It is interesting to note that, because $\alpha_{nl} = \alpha_{n|l|}$, the dependence of the wave function on the sign of l lies solely in the angular part. The energy for a negative l-value $E_{nl} = \hbar\omega_c(\alpha_{nl}+\frac{1}{2})$ will always be lower than for its positive counterpart: $E_{nl} = \hbar\omega_c(\alpha_{nl}+l+\frac{1}{2})$. This follows from the fact that for negative l-values, the classical motion of the electron is *counterclockwise*, whereas for positive l-values it moves *clockwise*, counting for an energy difference of $|l|\hbar\omega_c$.

The Fermi energy E_F of a system of N non-interacting electrons at zero temperature is equal to the energy of the $(N/2)^{\mathrm{th}}$ level. Since the electronic spectrum is quite complex, the Fermi energy is not a smooth function of the magnetic field. For sufficient large magnetic fields however, the Fermi energy approaches $\hbar\omega_c/2$, i.e. the energy of the lowest Landau level in the ideal 2D case. The Fermi energy for a system of 10 electrons is indicated in Fig. 1 by the thick curve.

The interaction between an applied oscillating electric field $\vec{E}$ and the electric dipole moments of the electrons are governed by the transition amplitudes $A_{nl}^{n'l'} = \langle\Psi_{nl}|re^{\pm i\phi}|\Psi_{n'l'}\rangle$, with associated oscillator strength $f_{nl}^{n'l'} = (2m/\hbar)\omega_{nl}^{n'l'}|A_{nl}^{n'l'}|^2$, where $\omega_{nl}^{n'l'} = (E_{n'l'} - E_{nl})/\hbar$ is the transition frequency. Performing the integration over ϕ we find a rigorous selection rule on the possible $l-$values: $\Delta l = l' - l = \pm 1$. No rigorous selection rule is found on the allowed n quantum numbers!

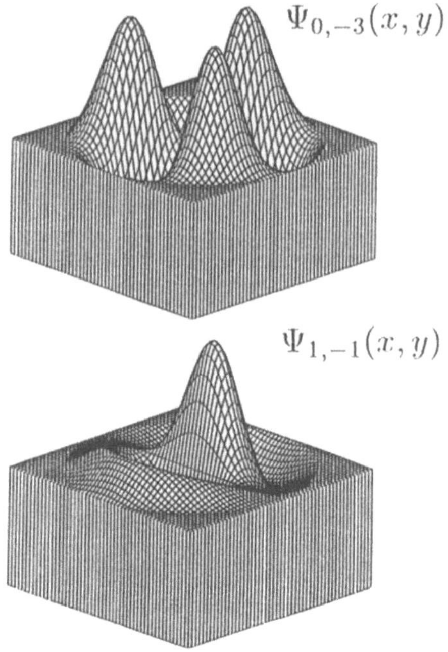

Fig. 2. Real part of the normalized wave function for two sets of quantum numbers, shown on the square $[-R_0, R_0] \times [-R_0, R_0]$ for the hard wall confinement case.

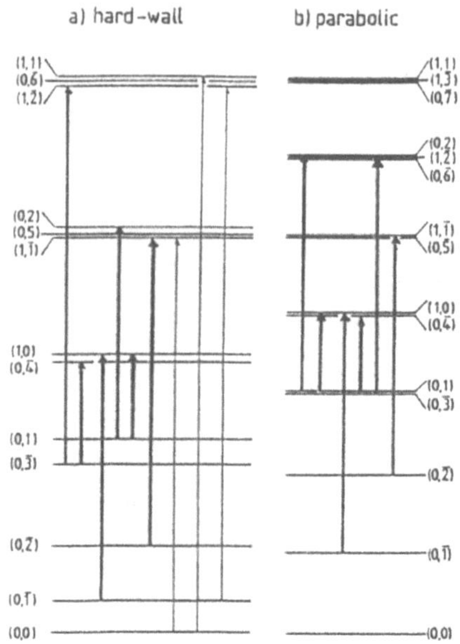

Fig. 3. Allowed magneto-optical transitions at $B = 2$ Tesla for (a) a hard-wall potential and (b) harmonic potential. In (a) transitions depicted by thin lines have transition amplitudes that are at least one order of magnitude smaller than those depicted by thick lines.

It is instructive to compare this results with the ones for a quadratic confining potential. Here the same selection rule on possible $l-$values is found, but also an additional constaint on the possible $n \to n'$ transitions: $\Delta n = 0, 1$ appears. The corresponding transition energies in this case are $\Delta E_{\pm} = \frac{1}{2}\hbar\omega \pm \frac{1}{2}\hbar\omega_c$. Thus, only two transition energies are possible, a result recovered by the hard-wall model only as a limiting case for large magnetic fields.

In Fig. 3 we compare the allowed transitions for the hard-wall model and the harmonic potential model. Although there are many more transitions possible in the hard-wall model, only a few of them have sufficient oscillator strength to be observable experimentally. The transitions with the largest oscillator strengths are given by the thick lines, while the thin lines correspond to transitions which have an oscillator strength at least an order of magnitude smaller.

The allowed transitions as a function of the magnetic field for a system of 10 electrons are shown in Fig 4. As in Fig. 3 the thick curves correspond to the most important transitions while the thin curves have oscillator strengths which are at least an order of magnitude smaller. A quadratic confinement potential with a comparable radius leads to two transition energies which are given by the two dashed curves.

In conclusion, we have studied theoretically the electronic states, the energy levels and the optical transitions of a collection of non-interacting electrons in a single

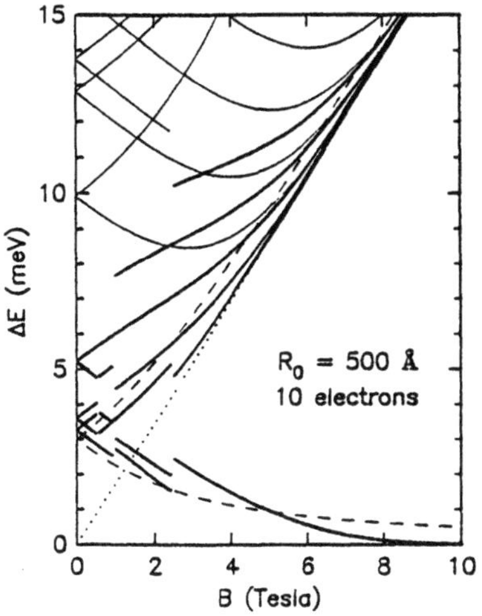

Fig. 4: Magneto-optical transition energies as a function of magnetic field for a hard-wall potential. Thin curves have a transition amplitude which is an order of magnitude smaller than those represented by the thick curves. Dashed lines are the transition energies for a quadratic confining potential. For reference, the dotted line is the cyclotron energy $\hbar\omega_c$.

quantum dot with hard-wall confinement. The results are compared to the ones of an harmonic confining potential. We found that for intermediate magnetic fields the cyclotron resonance line is split-up into several lines (five in the case of ten electrons in a quantum dot) for a hard-wall confining potential and shows jumps as a function of the magnetic field which are related to the irregular behavior of the Fermi energy. Furthermore we found that with increasing magnetic field the resonance frequency approaches the two-dimensional result much faster than for a soft-wall confining potential.

Acknowledgement. One of us (F.M.P.) is supported by the Belgian National Science Foundation. Discussions with F. Brosens are greatfully acknowledged.

REFERENCES

[1] S.J. Allen, Jr., H.L. Störmer, J.C.M. Hwang, Phys. Rev. **B28**, 4875 (1983)

[2] M.A. Reed, R.T. Bate, K. Bradshaw, W.M. Duncan, W.R. Frensley, J.W. Lee, H.D. Shih, J. Vac. Sci. Technol. **B4**, 348 (1986)

[3] W. Hansen, M. Horst, J.P. Kotthaus, U. Merkt, Ch. Sikorski, Phys. Rev. Lett. **58**, 2586 (1987)

[4] M.A. Reed, J.N. Randall, R.J. Aggarwal, R.J. Matyi, T.M. Moore, A.E. Wetsel, Phys. Rev. Lett. **60**, 535 (1988)

[5] K. Kash, A. Scherer, J.M. Worlock, H.G. Craighead and M.C. Tamargo, Appl. Phys. Lett. **49**, 1043 (1986)

[6] Ch. Sikorski, U. Merkt, Phys. Rev. Lett. **62**, 2164 (1989)

[7] C.T. Liu, K. Nakamura, D.C. Tsui, K. Ismail, D.A. Antoniadis, H.I. Smith, Appl. Phys. Lett. **55**, 168 (1989)

[8] F.M. Peeters, to be published in Phys. Rev. **B42** (1990)

RPA-Calculation of Magnetoplasmons in Quantum Dots

V. Gudmundsson[1] *and R.R. Gerhardts*[2]

[1]Science Institute, University of Iceland,
Dunhaga 3, IS-107 Reykjavik, Iceland

[2]Max-Planck-Institut für Festkörperforschung, Heisenbergstr. 1,
W-7000 Stuttgart 80, Fed. Rep. of Germany

Abstract. We investigate the longitudinal collective oscillations of a two-dimensional interacting electron gas (2DEG) confined to a disk in a perpendicular constant magnetic field. We can reproduce the exact results for the magnetoplasmon dispersion relation known for the parabolic confinement of the electrons. The effects of non-parabolic confinement and its relevance to experimental results are discussed.

1. Introduction

The measured magnetoplasmon dispersion of a 2DEG confined to a circular geometry [1-3] is well described by $\Delta E_{\pm} = \frac{1}{2}\sqrt{(\hbar\omega_c)^2 + 4(\hbar\omega_0)^2} \pm \frac{1}{2}\hbar\omega_c$, i.e. the transition energies of a single parabolically confined electron subject to a magnetic field, where ω_c is the cyclotron frequency and ω_0 is the frequency of the motion in the bare confining potential. For large disks ω_0 has been interpreted as a plasmonic frequency [1,3], and for small "quantum dots" it has been related to a parabolic confinement potential [2,3].

Recently it has been realized that this dispersion relation is exact for any number of interacting electrons in a constant magnetic field and a parabolic confinement potential if the external electric field that is radiated into the system is a dipole field, since then the Hamiltonian is exactly separable into parts containing either only the center of mass or the relative coordinates and the dipole field couples only to the center of mass motion of the whole system [4].

These results immediately raise some questions that shall be investigated here: Can this simple dispersion be reproduced in a model of Hartree interacting electrons that respond linearly to an external dipole field, even when screening effects are quite important [5]? What happens when the electrons are not confined parabolically and when the external electric field is not a dipole field?

2. The Model

Calculations in the classical regime for a 2DEG confined to a disk or a dot, have been published before [6]. Also collective excitations induced by Coulomb coupling between the different dots in a periodic array have been considered in the random phase approximation (RPA), however, with a crude phenomenological model for the individual dots [7]. Here we will present a fully quantum mechanical theory for a single quantum dot. The electron gas in the dot is considered to be exactly two-dimensional, held on a disk of radius $R \approx 1000 \mathring{A}$ in the $x-y$-plane by a confining potential $V_{conf}(r)$ with a perpendicular homogeneous magnetic field B. The Coulomb interaction between the electrons is taken into account in the Hartree approximation. The circular gauge is used for the vector potential $\vec{A}$ and the electronic states are thus labelled by the Landau band and the angular momentum quantum number, n and M, respectively. A neutral-

Springer Series in Solid-State Sciences, Vol. 101
High Magnetic Fields in Semiconductor Physics III Editor: G. Landwehr

izing background charge is not included explicitly. A similar ground state calculation [5] has revealed large magnetic-field dependent oscillations in the non-linear screening properties of the system.

In order to investigate the collective oscillations, the linear response to an external *longitudinal* (non-propagating) time dependent electric field $\vec{E}^{ext} = -\vec{\nabla}\phi^{ext}$ is considered. The self-consistent potential is then $\phi^{sc} = \phi^{ext} + \phi^{ind}$, where ϕ^{ind} is the induced potential, that is determined by Poisson's equation from the induced electron density δn_s. The induced density is in turn found from the density-density response function χ^{2D} and ϕ^{sc}, so we have then an integral equation for ϕ^{sc} that symbolically can be written as $f^{sc} = f^{ext} + \alpha\chi^{2D} f^{sc}$, with $f = q\phi$. Here q is the Fourier wave vector in the electron plane and $\alpha = 2\pi e^2/\kappa$. The RPA response function χ^{2D} is constructed using the energy spectrum and the wavefunctions of the Hartree interacting ground state, thus guaranteeing that the Coulomb interaction is treated on equal footing both in the ground and the excited state.

Due to the circular symmetry of the system there exist angular and radial eigenmodes for the plasmons. The external potential is therefore chosen of the form $\phi^{ext}(\vec{q}) = \phi^{ext}(q)\exp(-iN_p\varphi_q)$ with $N_p = \pm 1$ representing circular polarization. The only transitions that contribute to χ^{2D} are then of the type $M \to M + N_p$. $\phi^{ext}(q)$ is chosen as $\phi^{ext}(Q) = Q^{|N_P|}\exp(-\beta Q^2)$ with $Q = \sqrt{2}l$. The parameter β can then be varied to attain the dipole limit ($\beta \to \infty$) or to focus ϕ^{ext} on the inside of the system ($\beta \to 0$).

The spectrum of the collective oscillations could now be obtained from the condition $det(1 - \alpha\chi^{2D}) = 0$, but since we are interested in the dipole active modes it is more convenient (and closer to the experimental procedure) to select ϕ^{ext} as mentioned before and to solve the integral equation for ϕ^{sc}. Then we calculate the power absorption $P(\omega)$ from the Joule heating, adding a small constant imaginary part $i\eta = i\hbar\omega_c/30$ to ω.

3. Results

We assume 30 electrons in a dot and GaAs parameters, i.e. $\kappa = 12.4$ and $m = 0.067m_0$ and calculate first the power absorption $P(\omega)$ for a parabolic confinement $V_{conf}(r) = ar^2$, with $a = 50 \cdot 10^6[meV/\AA^2]$. ϕ^{ext} is chosen as a "near dipole" field ($\beta = 40.0$) and the dispersion is shown in Fig.1 together with the exact results $\Delta E_\pm$. The small inset in Fig.1 shows $P(\omega)$ for the two polarizations $N_p = \pm 1$ as a function of ω/ω_c for one filling factor $\nu = 2.0$. We see that the results of the present consistent approximation of Coulomb interaction effects in groundstate and response agree remarkably well with the exact results. This might be surprising, since the *screened* confining potential is quite different from V_{conf}. The results of Sikorski and Merkt [2] can be explained by this simple model, emphasizing the special role of the dipole field and the parabolic confinement, that make the strong screening properties of the 2DEG unimportant here. The calculated dispersion becomes also independent of the number of electrons in this case as has been seen in experiments.

If the dipole approximation correctly describes the incident light, the additional modes seen in recent experiments [3] must be due to deviations from the strictly parabolic confinement. In the dot arrays investigated experimentally, deviations from the circular symmetry may be relevant. Here we consider circular symmetric corrections to the parabolic confinement, which couple center of mass and internal motion and thus can lead to the excitation of additional modes and possibly to hybridization of different center of mass modes with anticrossings. In Fig.2 the confining potential is $V_{conf}(r) = br^2 + cr^4$, with $b = 40.0 \cdot 10^{-6}[meV/\AA^2]$ and $c = 10.0 \cdot 10^{-12}[meV/\AA^4]$ such that $V_{conf}(1000\AA) = 50meV$ as before to guarantee a similar size of the system ($R = 1000\AA$).

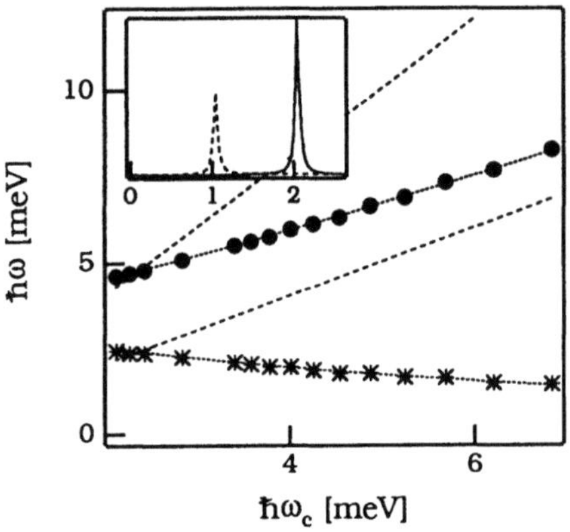

Fig.1: The spectrum of the collective oscillations in the case of a $V_{conf} = ar^2$ confinement and an external dipole field. •: the $N_p = -1$ and ∗: the $N_p = +1$ mode from the calculation. The dashed lines are $\hbar\omega_c$ and $2\hbar\omega_c$. The dotted lines are ΔE_+ and ΔE_-. The inset shows the power absorption $P(\omega)$ for $N_p = +1$ (dotted) and $N_p = -1$ (solid) as a function of ω/ω_c, $\nu = 2.0$.

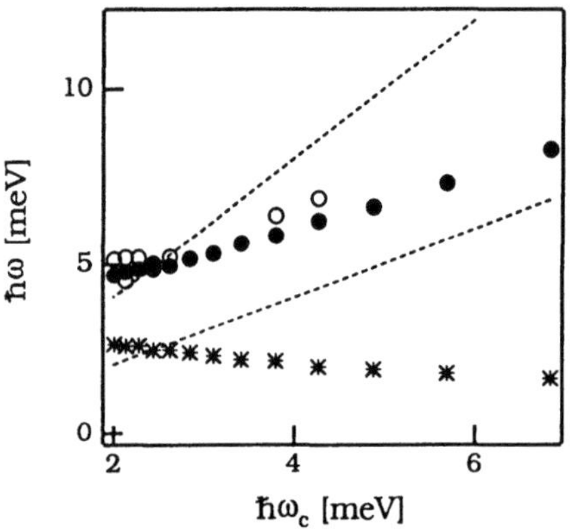

Fig.2: The spectrum of the collective oscillations in the case of a $V_{conf}(r) = br^2 + cr^4$ confinement and an external dipole field. •: a strong $N_p = -1$ mode, ∘: a weak $N_p = -1$ mode and ∗: the $N_p = +1$ mode from the calculation. The dashed lines are $\hbar\omega_c$ and $2\hbar\omega_c$.

The results are quite similar to the results for the parabolic confinement, except now the $N_p = -1$ branch does partly have a parallel branch that becomes stronger and moves further away from the main $N_p = -1$ branch if the weight of the r^4 term is increased.

In Fig.3 the confining potential is $V_{conf}(r) = br^2 + dr^6$, with $b = 40.0 \cdot 10^{-6}[meV/\AA^2]$ and $d = 10.0 \cdot 10^{-18}[meV/\AA^6]$. Again we see an additional upper branch that is almost parallel to the main branch. Also there is some kind of a more complicated splitting where the $N_p = -1$ branch crosses the $\omega = 2\omega_c$ line, reminiscent of what is seen in the experimental results of Demel et al. [3]. To obtain a better resolution of this region, we will extend our present numerical procedure, which can take into account only a few Landau levels (2 to 10), to smaller magnetic fields. The fact that the two upper modes in the experiment are running almost parallel excludes the possibility that the higher one is a higher $N_p = +1$ mode, which will be excited if the external electric field is not homogeneous over the size of the dot [6]. We have checked this by repeating the calculation with a lower β.

In summary, we have shown that the Hartree approximation for the groundstate together with the RPA for the response yields an excellent description of the far infrared response of a quantum dot with parabolic confinement potential. Moreover, our calculations indicate that the results of Demel et al. [3] may be due to deviations from the simple parabolic confinement of individual dots, which must not necessarily destroy

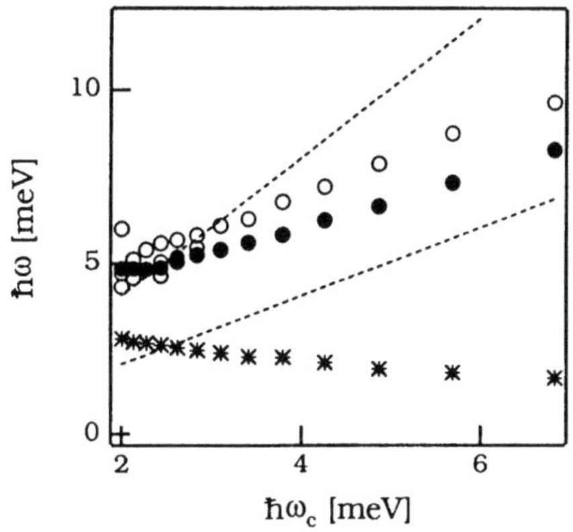

Fig.3: The spectrum of the collective oscillations in the case of a $V_{conf}(r) = br^2 + dr^6$ confinement and an external dipole field. •: a strong $N_p = -1$ mode, o: a weak $N_p = -1$ mode and $*$: the $N_p = +1$ mode from the calculation. The dashed lines are $\hbar\omega_c$ and $2\hbar\omega_c$.

the circular symmetry, but couple center of mass and internal collective motion. The full details of our calculation togehter with further results will be published elsewhere.

Acknowledgement

This research was supported in part by the Icelandic Natural Science Foundation.

References

[1] S.J. Allen, H.L. Störmer and J.C.M. Hwang, *Phys. Rev.* **B28**, 4875 (1983)

[2] Ch. Sikorski and U. Merkt, *Phys. Rev. Lett.* **62**, 2164 (1989)

[3] T. Demel, D. Heitmann, P. Grambow and K. Ploog, *Phys. Rev. Lett.* **64**, 788 (1990)

[4] P.A. Maksym and T. Chakraborty, *Phys. Rev. Lett.* **65**, 108 (1990)

[5] V. Gudmundsson, *Solid State Comm.* **74**, 63 (1990)

[6] D.C Glattli et. al., *Phys. Rev. Lett.* **54**, 1710 (1985). V.A. Volkov and S.A. Mikhailov, *Sov. Phys. JEPT* **67**, 1639 (1988). A.L. Fetter, *Phys. Rev.* **B33**, 5221 (1986)

[7] W.Que and G. Kirczenow, *Phys. Rev.* **B38**, 3614 (1988)

Microwave Conductivity of Laterally Confined Electron Systems in AlGaAs/GaAs Heterostructures

F. Brinkop[1], *C. Dahl*[1], *J.P. Kotthaus*[1], *G. Weimann*[2], *and W. Schlapp*[3]

[1]Sektion Physik, Universität München, W-8000 München 22, Fed. Rep. of Germany

[2]Walter Schottky Institut, TU München, Am Coulombwall, W-8046 Garching, Fed. Rep. of Germany

[3]Forschungsinstitut der Deutschen Bundespost Telekom, W-6100 Darmstadt, Fed. Rep. of Germany

Abstract. Lateral confinement phenomena of two-dimensional electron systems (2DES) are investigated at microwave frequencies. In a one-dimensional lateral superlattice we observe oscillations caused by an oscillatory Landau band conductivity in the magnetoconductivity parallel to the superlattice stripes at frequencies ω where $\omega\tau>1$ and $\omega/\omega_c\leq 1$. These oscillations have previously been observed only in the DC magnetoresistance. In experiments on an array of 50 μm electron discs we study dimensional resonances at moderate magnetic fields and determine the range of validity of different theoretical approximations.

1. Introduction

Lateral confinement of a 2DES to lengths which are comparable to length scales imposed, e.g., by a magnetic field, is expected to yield novel features in transport as well as in the elementary excitation spectrum. Here we study such phenomena at microwave frequencies in the regime $\omega\tau>1$, where τ denotes the momentum relaxation time. In a 1D lateral superlattice with period a we observe oscillations in the magnetoconductivity below the onset of Shubnikov-de Haas (SdH) oscillations. These oscillations have so far been reported only in DC experiments [1,2]. Their period is determined by the ratio of the classical cyclotron radius R_c and the superlattice period. In arrays of 2D electron discs we investigate dimensional resonances at magnetic fields below the magnetic quantum limit. Such resonances have first been studied in the far-infrared regime [3] and were explained in terms of the depolarization of a thin ellipsoid. Here we demonstrate that a truly 2D calculation of the modes [4] leads to a much better agreement with the experimental data than the ellipsoidal model. The experiment also yields insight into the penetration depth of edge-magnetoplasmons.

2. Sample Preparation and Experimental Setup

The samples are prepared on AlGaAs/GaAs heterostructures containing a high mobility 2DES at a single AlGaAs/GaAs interface grown by molecular beam epitaxy. To generate a 1D superlattice we define a photoresist grating with a period of a=300 nm on the surface by holographic illumination. A thin NiCr-layer evaporated on top of the photoresist grating serves as a gate electrode. The application of a gate voltage leads to a periodic density modulation via the modulated photoresist thickness [5]. In order to perform static transport measurements In/Ag ohmic contacts are alloyed in the four corners of the sample in the van der Pauw-geometry prior to the photoresist coverage. The sample with an array of heterojunction discs is produced by contact lithography and wet mesa etching. The diameter of the discs is 50 μm and their period 70 μm.

The dynamic conductivity is measured in transmission. As a microwave source we use either an Impatt diode working at f=34 GHz or a scalar network analyser which is tuned between 30 GHz and about 200 GHz. In the former case the transmitted radiation is detected with a helium-

Springer Series in Solid-State Sciences, Vol. 101
High Magnetic Fields in Semiconductor Physics III Editor: G. Landwehr

cooled bolometer using standard lock-in techniques whereas the scalar network analyser employs heterodyne detection. In both cases the microwave radiation propagates through a rectangular waveguide (Ka-band), operating in the fundamental mode around 34 GHz.

3. Bandconductivity in a 1D Superlattice

In high mobility AlGaAs/GaAs heterostructures with weak 1D periodic density modulation magnetoresistance oscillations below the onset of SdH oscillations were observed [1,2]. These oscillations can be explained in a classical [6] and a quantum mechanical [7] picture. Quantum mechanically, the oscillations are caused by an oscillatory band conductivity in the direction of the superlattice stripes. The periodic potential lifts the degeneracy of the Landau levels and leads to Landau bands with a nonvanishing dispersion and a finite width, which is controlled by the ratio of the classical cyclotron radius and the superlattice period. A variation of the magnetic field causes oscillations of the bandwidth and hence in the parallel conductivity which have first been observed in the magnetoresistance component ρ_{xx} perpendicular to the superlattice stripes. The oscillations are much weaker and phase shifted in in the ρ_{yy} component. They are periodic in 1/B and yield maxima in ρ_{xx} whenever the classical cyclotron diameter is

$$2R_c = (m+\phi)a, \qquad m=1,2,3,.... \tag{1}$$

where ϕ is a phase factor of about 0.25.

The oscillations appear at magnetic fields where $\omega_c\tau>1$ but low enough to neglect quantization of the Hall resistance. In this field regime one has $|\rho_{xy}|=B/en_s$, $|\rho_{xy}|\gg|\rho_{xx}|\approx|\rho_{yy}|$ and thus

$$\sigma_{yy} = \rho_{xx}/(\rho_{xx}\rho_{yy} - \rho_{xy}\rho_{yx}) \approx \rho_{xx}\, n_s^2 e^2/B^2 \ . \tag{2}$$

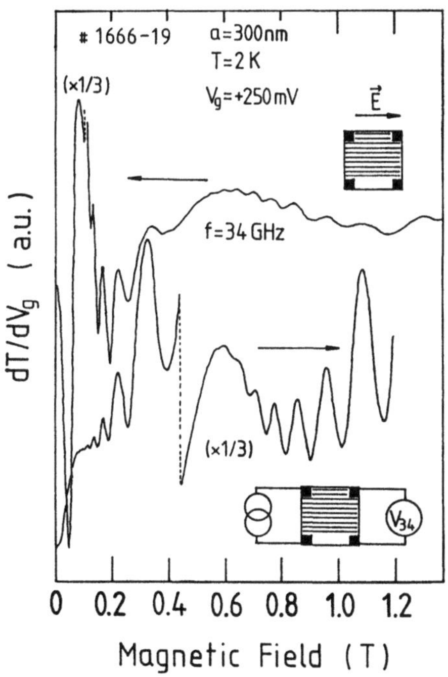

Fig. 1: Derivative of the microwave transmission with respect to the gate voltage dT/dV_g and four point magnetoresistance versus magnetic field at T=2 K. The parameters of the homogeneous sample are μ=830000 cm^2/Vs at T=4.2 K and $n_s=2.5*10^{11}$ cm^{-2}.

In DC transport experiments the dominant effect of the superlattice is therefore seen in the magnetoresistance ρ_{xx}. By contrast the microwave transmission through the sample directly reflects the conductivity in the polarization direction of the microwave field, which in our experiment is orientated in the direction of the superlattice stripes, i.e., we directly observe σ_{yy}. Fig.1 shows a comparison of a DC transport and a microwave transmission experiment of the same sample. In both measurements the bandconductivity oscillations are clearly observable below B=0.6 T. The large signal in dT/dV_g for B<0.1 T is caused by the cyclotron resonance. The cyclotron resonance position for f=34 GHz is B≈ 0.09T. It is not possible to estimate the phase factor in eq.(1) since the measurement of a derivative yields an additional phase shift. We also measure the conductivity components σ_{xx} and σ_{xy}. No oscillations in addition to SdH oscillations are observed in the perpendicular dynamic conductivity σ_{xx}. The dynamic Hall conductivity σ_{xy} is measured in a crossed waveguide arrangement. A value of $\omega\tau\approx 5$ is estimated from a theoretical fit of the transmitted signal. We thus demonstrate that the oscillatory band conductivity in a 1D lateral superlattice is present at frequencies where $\omega\tau>1$ but $\omega/\omega_c \leq 1$.

4. Dimensional Resonances in Electron Discs

As a second example of a confinement phenomenon at high frequencies we study 2D discs with a diameter of 50 μm. To describe the dynamic behaviour of confined electron systems we recall that the absorbed power is $P=(1/2)\,\mathrm{Re}(\sigma_{xx})_{eff}\,|E_0|^2$ where σ_{eff} describes the response to the external microwave field rather than to the total field. In ellipsoidal geometry and with a spatially constant external field only the lowest mode with a homogeneous depolarization field is excited. In this case the effective conductivity tensor is easily calculated by solving the classical equations of motion assuming Drude relaxation [8], which yields in the limit $\omega\tau \gg 1$ for the resonance position as a function of the magnetic field

$$\omega_\pm = \pm\omega_c/2 + \sqrt{\omega_c^2/4 + \omega_o^2} \quad \text{with} \quad \omega_o^2 = \frac{3\pi n_s e^2}{8m^*\bar{\varepsilon}\varepsilon_o d} \quad . \qquad \text{(3a, 3b)}$$

Here ω_o is the resonance frequency at zero magnetic field [9] and d the diameter of the disc.

Fig.2a shows the transmitted power versus the magnetic field at various frequencies at T=4.2 K. With increasing frequency the resonance positions first shift to lower magnetic fields whereas for frequencies above 120 GHz they shift to higher fields. The signal at 33.8 GHz, measured at T=2 K, also exhibits SdH-oscillations from which the electron density $n_s=(3.7\pm0.2)*10^{11}$ cm^{-2} is obtained. With $\omega_o/2\pi=110$ GHz and $m^*=0.07$ eq.(3a) yields a rough fit for the resonance positions (fig.2b). From ω_o we deduce $\bar{\varepsilon}=8.5$ which is higher than than the value of the half space $\bar{\varepsilon}=(1/2)(\varepsilon_{GaAs}+1)=6.8$. However, the positions of the low frequency resonances are significantly higher than eq.(3a) predicts [10]. Instead, the resonance positions at high magnetic field agree with the calculations for a strictly 2D disc [6], which yield

$$\omega_- = \frac{n_s e}{\pi\bar{\varepsilon}\varepsilon_o B d}\left[\ln\frac{d}{\ell} - \Psi\left(\frac{3}{2}\right) + 1\right], \qquad (4)$$

where Ψ is the digamma function and $\ell = n_s m^*/2\bar{\varepsilon}\varepsilon_o B^2$ is the extension of the charge associated with the edge excitation. Eq.(4) is valid only for $\ell \ll d$. Fig.2b reveals that at fields lower than about 0.7 T, corresponding to $\ell/d\approx 0.07$, the magnetic field dependence of eq.(4) begins to deviate from the experimental resonance positions. Qualitatively, the increase in the resonance frequency as compared to a thin ellipsoid arises from the confinement of the edge exitation to the length ℓ. By contrast, the ellipsoidal model implies that polarization charge is distributed over the entire disc, leading to a comparatively lower frequency.

We have also investigated the response of 2D squares with an area of 120*120 μm^2 on a sample from the same wafer. Here a fit according to eq.(3a) yields $\omega_o/2\pi=68$ GHz. This value is in close agreement with eq.(3b) using $n_s/\bar{\varepsilon}$ from the disc data and d=135μm, which is the diameter of a disc with the same area as the squares. For a more quantitative understanding a

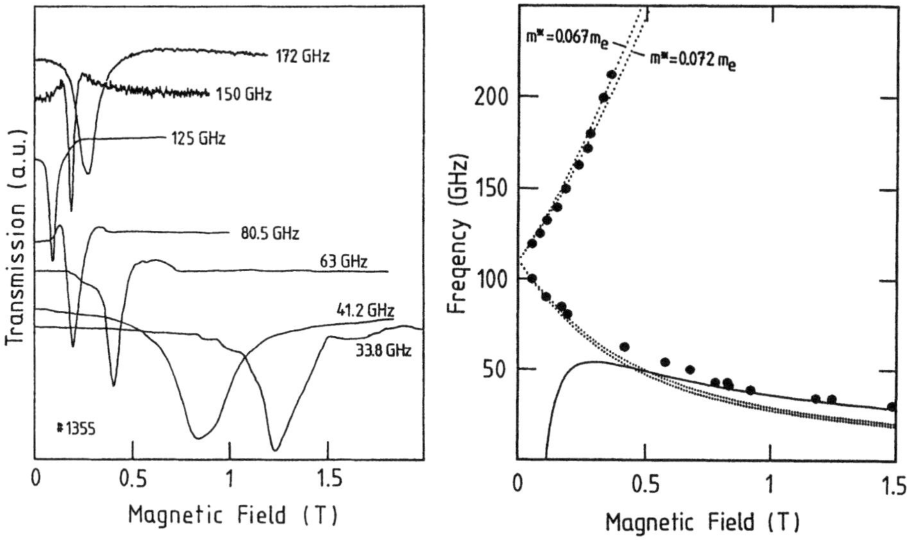

Fig.2: a) Transmitted microwave signal of 50 μm electron discs as a function of the magnetic field for various frequencies at T=4.2 K. The 33.8 GHz signal was taken at 2 K. The DC mobility of the sample is μ=550000 cm^2/Vs. The signal strengths are scaled individually.

b) Frequency versus magnetic field of the resonances displayed in a). The dotted lines are fits to the ellipsoidal model, eq.(3a). The solid line is the 2D theory, eq.(4), with $n_s/\bar{\varepsilon}$ taken from the extrapolated ω_0.

calculation of the classical dimensional resonances in finite 2D-systems for arbitrary magnetic field would be highly desirable.

We gratefully acknowledge financial support by the Deutsche Forschungsgemeinschaft and the ESPRIT Basic Research Action.

References

[1] D. Weiss, K. v. Klitzing, K. Ploog, and G. Weimann, Europhys. Lett. **8**, 179 (1989)
[2] R.W. Winkler, J.P. Kotthaus, and K. Ploog, Phys. Rev. Lett. **62**, 1177 (1989)
[3] S.J. Allen, Jr., H.L. Störmer, and J.C.M. Hwang, Phys. Rev. B **28**, 4875 (1983)
[4] V.A. Volkov and S.A. Mikhailov, Zh. Eksp. Teor. Fiz. **94**, 217 (1988) [Sov. Phys. JETP 67, 1639 (1988)]
[5] F. Brinkop, W. Hansen, J.P. Kotthaus, and K. Ploog, Phys. Rev. B **37**, 6547 (1988)
[6] C.W.J. Beenakker, Phys. Rev. Lett. **62**, 2020 (1989)
[7] R.R. Gerhardts, D. Weiss, and K. v. Klitzing, Phys. Rev. Lett. **62**, 1173 (1989)
[8] B.A. Wilson, S.J. Allen, Jr., and D.C.Tsui, Phys. Rev. B **24**, 5887 (1981)
[9] R.P. Leavitt and J.W. Little, Phys. Rev. B **34**, 2450 (1986)
[10] To a lesser extent this is also true for the data shown in Ref. 3

Part IV

Magneto-Transport in 2D Systems

Magnetotransport in a Two-Dimensional Electron Gas Subject to a Weak Superlattice Potential

R.R. Gerhardts, D. Pfannkuche, D. Weiss, and U. Wulf

Max-Planck-Institut für Festkörperforschung, Heisenbergstr. 1, W-7000 Stuttgart 80, Fed. Rep. of Germany

We present experimental and theoretical results on commensurability oscillations of the magnetoresistivity in high-mobility two-dimensional electron systems subject to a weak superlattice potential. Experimentally, the band conductivity of a sample with a holographically produced square-grid potential is shown to be considerably smaller than that of the same sample with a similar linear grating potential of the same lattice constant. This is explained by a magnetotransport theory with due consideration of collision broadening effects and the peculiar subband splitting of Landau levels resulting in a Hofstadter-type energy spectrum.

1. Introduction

With modern techniques of microstructuring, it is possible to prepare quasi two-dimensional electron gases (2DEGs) which are periodically density-modulated in one or both lateral directions on a submicron scale. In gated structures, the density modulation can be tuned from a weakly modulated 2DEG to a situation with a periodic array of areas which are completely depleted of electrons. Interesting new optical and transport properties have been observed on such systems, especially in the presence of a perpendicular magnetic field.

For strongly modulated systems with a grid of depleted spots in a multiply connected 2DEG, magnetotransport anomalies such as a quenching of the Hall effect and Aharonov-Bohm oscillations of the longitudinal resistance have recently been reported [1]. In the present work, we discuss the opposite limit of a weak lateral superlattice potential imposed on the 2DEG, which leads to a novel type of magnetoresistance oscillations first detected by Weiss et al. [2] on unidirectionally modulated samples with high mobility.

The 2DEG with a weak periodic modulation in one lateral direction shows, in addition to the Shubnikov-de Haas oscillations at high values of the perpendicular magnetic field $\vec{B}$, low-field oscillations which are also periodic in $1/B$. These oscillations originate from the commensurability of the two relevant length scales in the system, the cyclotron radius R_c of electrons at the Fermi energy and the period a of the modulation, and are by now well understood [3–8]. The oscillations of the resistivity $\rho_\perp$, measured when the current flows perpendicular to the equipotentials of the modulation potential, are closely related to the classical guiding center drift of cyclotron orbits in the periodic electric field due to the modulation. These have been first explained in terms of a 'band conductivity' within a quantum mechanical picture assuming *ad hoc* a constant relaxation time τ [3,4], but can also be obtained from a quasi-classical argument [6]. The oscillations of the resistivity component $\rho_{||}$, measured when current and equipotentials are parallel, are, on the other hand, due to quantum oscillations of the scattering rate, originating from the peculiar oscillating bandwidth of Landau levels (LLs) which are broadened by the superlattice potential, [7,8] and cannot be explained within the classical picture [6].

In the present article we extend the previous work to the case of a periodic modulation of the 2DEG in both lateral directions. The most striking experimental result is that the additional modulation in the second lateral direction leads to a drastic suppression of the band conductivity. We demonstrate that, once the peculiar single-electron energy spectrum for this situation is taken into account correctly, this suppression is explained by straightforward application of conventional transport theory.

The two-dimensional motion of electrons in the presence of both a 2D superlattice potential of period a [$a = (a_x a_y)^{\frac{1}{2}}$ for rectangular symmetry with periods a_x and a_y] and a perpendicular magnetic field $\vec{B}$ introducing the magnetic length $l = (c\hbar/eB)^{\frac{1}{2}}$ leads to intricate commensurability problems. As a function of B, a complicated self-similar energy spectrum (Hofstadter's butterfly) has been obtained in the two complementary, but mathematically equivalent limits of, first, a strong lattice potential and a weak magnetic field in the tight-binding approximation [9,10] and of, second, a weak periodic perturbation in a Landau quantized 2DEG [11–13]. In the second case, which has been discussed in the context of the quantized Hall effect, one finds that each LL splits into p subbands if

$$Ba^2/\Phi_0 \equiv a^2/2\pi l^2 = p/q \; , \tag{1}$$

i.e., if the flux Ba^2 per unit cell is a rational multiple of the flux quantum $\Phi_0 = hc/e$. Each of these subbands is predicted [12,13] to contribute an integer multiple (which may be large [13] for large q) of e^2/h to the Hall conductivity σ_{xy}, in such a manner that the total contribution per LL (and per spin) sums to e^2/h. Up to now, this subband splitting could not be verified experimentally.

We want to demonstrate that this subband splitting is the key for the understanding of the suppression of the band conductivity observed in experiment.

2. Experimental facts

In Fig.1 we summarize typical results of a series of experiments in which a grid modulation with square-lattice symmetry was created in two steps by holographic illumination [2,3] exploiting the persistent photoconductivity effect in Si-doped $Al_xGa_{1-x}As$ at low temperatures. In the first step the samples were illuminated with an interference line pattern produced by a split laser beam reflected from two mirrors as sketched in the inset of Fig.1a. Thus, a linear grating pattern (parallel to y-direction) with modulation in x-direction was created and the resistivity components $\rho_{xx}^{(1d)} = \rho_\perp$ (Fig.1b) and $\rho_{yy}^{(1d)} = \rho_\parallel$ (Fig.1a) for the unidirectionally modulated (1dm) systems were measured. The second illumination (which always increased the density of the 2DEG somewhat) was performed after a rotation of the sample by 90°. The resistivities $\rho_{xx}^{(2d)} = \rho_{yy}^{(2d)}$ of the bidirectionally modulated (2dm) samples show oscillations which, at small magnetic fields ($B \leq 0.6T$), are very similar to and in phase with the weak oscillations of $\rho_{yy}^{(1d)}$ (Fig.1a). On the other hand, they are much smaller than and 90° out of phase with the large-amplitude oscillations of $\rho_{xx}^{(1d)}$ in the corresponding 1dm situation. The data shown in Fig.1a were obtained from a sample with mobility $1.4 \cdot 10^6 cm^2/Vs$ and electron density $N_s = 5.1 \cdot 10^{11} cm^{-2}$ after the second illumination, those of Fig.1b from a sample with mobility $1.2 \cdot 10^6 cm^2/Vs$ and $N_s = 3.7 \cdot 10^{11} cm^{-2}$.

As has been shown previously [2,3], 1dm samples exhibit, in addition to the usual Shubnikov-de Haas (SdH) oscillations at higher magnetic fields ($B \geq 0.6T$ in Fig.1), characteristic low field oscillations, also periodic in $1/B$, with minima of $\rho_{xx}^{(1d)}$ and maxima of $\rho_{yy}^{(1d)}$ at B-values for which the cyclotron radius $R_c = l^2 k_F$ of the electrons at the Fermi energy ($E_F = \hbar^2 k_F^2/2m$) satisfies the commensurability condition

$$2R_c = a(\lambda - \frac{1}{4}), \quad \lambda = 1, 2, \ldots \; . \tag{2}$$

Since $k_F^2 = 2\pi N_s$, the period $\Delta(1/B) = (a/\Phi_0)(2N_s/\pi)^{-\frac{1}{2}}$ results.

The band conductivity $\Delta\sigma_{yy}$, which is responsible for the large-amplitude oscillations of $\rho_{xx}^{(1d)}$ [3,4,7,8], is apparently strongly suppressed in the 2dm case.

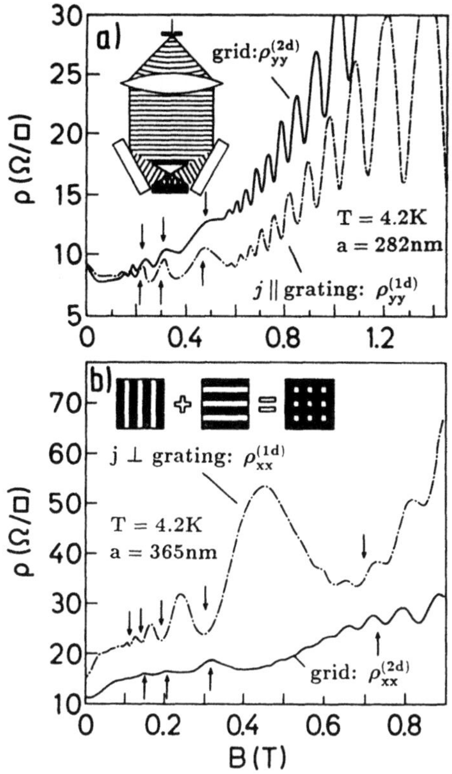

Fig. 1: Magnetoresistance in a grating (with modulation in x-direction) and a grid potential for two periods and samples. The insets sketch the creation of the potential by in situ holographic illumination (a), and the resulting pattern (b). The arrows indicate the flat band situations defined by Eq.(2), where the band conductivity contribution has minima [the second illumination always increases the electron density]. The grid potential, created as superposition of two gratings at right angles, suppresses the band conductivity in high-mobility samples, and the oscillations due to the scattering rate (with maxima at the arrow positions) dominate.

3. Model and energy spectrum

All the magnetotransport oscillations observed on modulated 2DEGs can be understood within a quantum mechanical approach using the simplest approximations for both the modulation potential and the collision broadening. Following the philosophy of Ref.8, we describe the modulation by a potential

$$V(x,y) = V_x \cos Kx + V_y \cos Ky, \tag{3}$$

with the same period $a = 2\pi/K$ in both lateral directions. Since for the high-mobility samples of Fig.1 the mean free path ($\sim 10\mu m$) is much larger than a ($\sim 300nm$) and than the cyclotron radius (for $B > 0.1T$), we include the effect of the weak modulation in the energy spectrum. We start from the Landau quantized, unmodulated system with energy eigenvalues $E_n = \hbar\omega_c(n+\frac{1}{2})$ and eigenstates $|n,k_y\rangle$ with wavefunctions [14] $\psi_{nk_y}(x,y) \propto \exp(ik_y y)\phi_n(x-x_0)$, where $\omega_c = eB/mc$ is the cyclotron frequency, $x_0 = -l^2 k_y$ the center coordinate and ϕ_n an oscillator function, and we calculate the effect of the modulation potential (3) in first order perturbation theory, neglecting the coupling of LLs. For the 1dm case ($V_y = 0$), this approximation has been shown [7,8] to be extremely good for the parameter values of interest ($V_x \sim 0.3meV$, $a \sim 300nm$, $E_F \sim 10meV$) and for $B > 0.1T$. The potential $V_x \cos Kx$ then is diagonal in the basis $|nk_y\rangle$,

$$\langle n,k_y'|V_x \cos Kx|n,k_y\rangle = \delta_{k_y',k_y} V_x \mathcal{L}_n \cos Kx_0 \ , \tag{4}$$

with $\mathcal{L}_n = \exp(-\frac{1}{2}X)L_n(X)$, where $X = \frac{1}{2}l^2K^2$ and L_n is a Laguerre polynomial. This leads to broadened Landau bands with energies $E_n(k_y) = E_n + \mathcal{L}_n V_x \cos Kx_0$, i.e., an oscillatory dependence of the bandwidth on the quantum number n [3,8]. The zeroes of the $L_n(X)$ yield condition (2) with $R_c = l(2n+1)^{\frac{1}{2}}$ for flat bands, which occur at energy values $E_\lambda = \frac{1}{8}(a/l)^2\hbar\omega_c(\lambda - \frac{1}{4})^2$ for $\lambda = 1, 2, ...$ (here the asymptotic form of $L_n(X)$ for large n was used).

With an additional modulation in y-direction, these Landau bands split into subbands, since the matrix elements

$$\langle nk_y'|V_y \cos Ky|nk_y\rangle = \frac{1}{2}V_y\mathcal{L}_n(\delta_{k_y',k_y+K} + \delta_{k_y',k_y-K}) \tag{5}$$

couple Landau states with center coordinates differing by integer multiples of l^2K. For the following it is important to note that, in the case of a square lattice, which is exclusively considered here, all potential matrix elements in the n-th LL have the common factor $\mathcal{L}_n$. As a consequence, the condition for flat bands is the same for the 2dm ($V_y = V_x$) and for the corresponding 1dm ($V_y = 0$) case. Moreover, since $\mathcal{L}_n$ is the only n-dependent factor of the matrix elements, the internal energy structure of all the LLs is the same, $E_{n;\alpha} = \hbar\omega_c(n+\frac{1}{2}) + \mathcal{L}_n\,\epsilon_\alpha$, and the eigenstates are of the form $|n;\alpha\rangle = \sum_{\lambda=-\infty}^{\infty} c_\lambda(\alpha)|n, k_y + \lambda K\rangle$ with coefficients $c_\lambda(\alpha)$ satisfying Harper's equation [10,12,13]

$$\left\{V_x \cos\left(Kl^2[k_y + \lambda K]\right) - \epsilon_\alpha\right\} c_\lambda(\alpha) + \frac{1}{2}V_y\left[c_{\lambda+1}(\alpha) + c_{\lambda-1}(\alpha)\right] = 0\,. \tag{6}$$

If the commensurability condition (1) holds, the eigenvalue problem is periodic in the sense that $c_{\lambda+p}(\alpha)$ satisfies the same equation as $c_\lambda(\alpha)$. This suggests a Bloch-type ansatz $c_\lambda(\alpha) \equiv c_\lambda(k_x, k_y; j) = \exp(i\lambda Kl^2k_x)u_\lambda(k_x, k_y; j)$ with p-periodic $u_\lambda(\vec{k}; j) = u_{\lambda+p}(\vec{k}; j)$. The latter are the components of the eigenvectors of a $p \times p$ effective Hamiltonian matrix $H(\vec{k}) = \mathcal{L}_n h^{(p)}(\vec{k})$ given (for $p > 2$) by

$$h^{(p)}_{\lambda,\lambda'}(\vec{k}) = \begin{cases} V_x \cos(Kl^2[k_y + \lambda K]) & \text{if } \lambda{=}\lambda'{=}1, ..., p, \\ \frac{1}{2}V_y \exp(iKl^2k_x) & \text{if } \lambda{=}\lambda' - 1{=}1, ..., p-1 \text{ or } \lambda{=}p, \lambda'{=}1, \\ \frac{1}{2}V_y \exp(-iKl^2k_x) & \text{if } \lambda{=}\lambda'{+}1{=}2, ..., p \text{ or } \lambda{=}1, \lambda'{=}p, \\ 0 & \text{otherwise.} \end{cases} \tag{7}$$

The eigenvalues $\epsilon(\vec{k}; j)$ ($j = 1, ..., p$) of $h^{(p)}(\vec{k})$ yield p energy subbands $E_{nj}(\vec{k}) = E_n + \mathcal{L}_n\epsilon(\vec{k}; j)$ (per LL) defined on the magnetic Brillouin zone (MBZ) $|k_x| \leq \pi/aq$, $|k_y| \leq \pi/a$ [12,13]. If the allowed values of $\epsilon(\vec{k}; j)$ are plotted versus the ratio q/p of Eq.(1), i.e. versus the inverse number of flux quanta per unit cell of the superlattice, one obtains a self-similar pattern [12], known as Hofstadter's butterfly [10]. For an integer value $p = p_0$ and $q = 1$, one gets p_0 bands. For a rational value p/q close to p_0, one gets p bands which group themselves into p_0 clusters close to the p_0 bands of the integer case.

Contrary to the unmodulated case, the velocity operator now has nonzero intra-LL matrix elements which are diagonal in $\vec{k}$ but not in the subband quantum number j. It can be shown from systematic first order perturbation theory for the eigenstates that the matrix elements of the velocity operators can be calculated from the $p \times p$ matrices [12]

$$v^{(p)}_\mu(\vec{k}) = \hbar^{-1}\,\partial H(\vec{k})/\partial k_\mu\,, \tag{8}$$

where $\mu = x, y$. For the following it is important to note that, e.g. the matrix $v^{(p)}_y(\vec{k})$ is independent of the modulation amplitude V_y and depends only on the modulation in x-direction. It can be shown [13] that the energy eigenvalues $E_{nj}(\vec{k})$ ($j = 1, ..., p$) are q-fold degenerate in the MBZ. In the limit $V_y \to 0$, the $E_{nj}(\vec{k})$ become independent of

k_x and the p subbands merge into a single band, $E_{nj}(\vec{k}) \to E_n + V_x\mathcal{L}_n \cos Kx_0$, where $0 \leq x_0 \leq qa$. For further details we refer to Refs.12,13.

4. Collision broadening and conductivities

To describe collision broadening effects, we follow recent work for the 1dm case [7,8,15] and introduce into the Green's function $G^-_{n\alpha}(E) = [E - E_{n\alpha} - \Sigma^-(E)]^{-1}$ a quantum-number independent self-energy (a cut off $n \leq 2E_F/\hbar\omega_c$ is implied)

$$\Sigma^-(E) = \Gamma_0^2 \sum_{n;j} (l^2/2\pi) \int d^2k \, G^-_{n,\vec{k};j}(E) \, , \tag{9}$$

which in the absence of modulation ($V_x = V_y = 0$) reduces to the self-consistent Born approximation (SCBA) for randomly distributed short-range scatterers [16], with $\Gamma_0^2 = \frac{1}{2\pi}\hbar\omega_c \cdot \hbar/\tau$ and τ the corresponding life time for zero magnetic field. With the spectral function $A_{n\alpha}(E) = \frac{1}{\pi}\mathrm{Im}G^-_{n\alpha}(E)$ this yields for the density of states (DOS) $D(E) = 2\sum_n D_n(E) = \mathrm{Im}[\Sigma^-(E)/(\pi l \Gamma_0)^2]$, where $D_n(E) = (2\pi)^{-2}\sum_j \int d^2k \, A_{n\alpha}(E)$ is the DOS of the n-th LL and one spin direction.

As a consequence of the oscillatory width of the modulation-broadened LLs amplitude oscillations of the Landau quantized DOS result. For the 1dm case, these have previously been calculated and observed in magnetocapacitance experiments [15]. A typical result for the DOS in the 2dm case is shown in Fig.2. For a comparison of the 1dm ($V_y = 0$) and the 2dm ($V_y = V_x$) case, Fig.3a shows $D_n(E)$ for $B = 0.23T$ and $a = 300nm$, which means $p = 5$ and $q = 1$, and for both small ($\Gamma_0 \ll V_x\mathcal{L}_n$) and large collision broadening [17]. A general result, previously found from CPA-calculations in the strong-modulation tight-binding limit [18], is that, in the presence of collision broadening, the fine structure due to the many subbands resulting from large p (and large q) values is smeared out and the DOS appears as a continuous function of the magnetic field, in spite of the highly singular B-dependence of the energy spectrum.

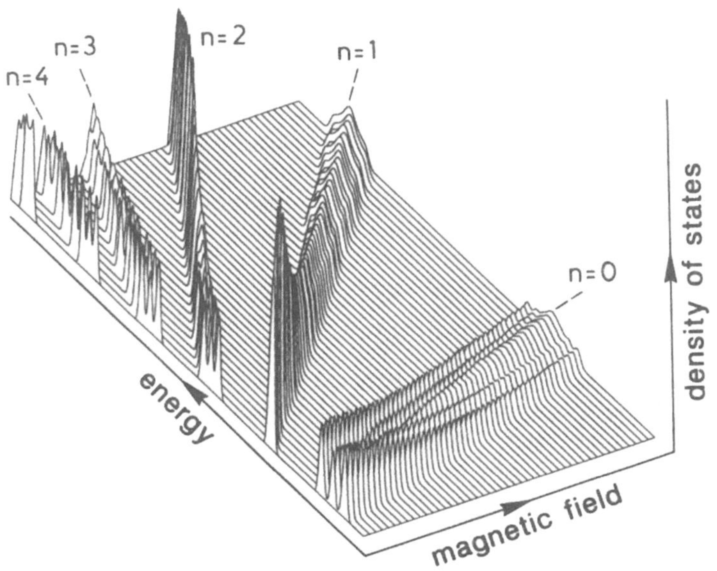

Fig. 2: Density of states for a grid modulation with $V_y = V_x = 0.25meV$ and $\Gamma_0 = 0.035\sqrt{B[T]}\ meV$, for $0.5T \leq B \leq 1.65T$ and $0 \leq E \leq 5meV$ [17].

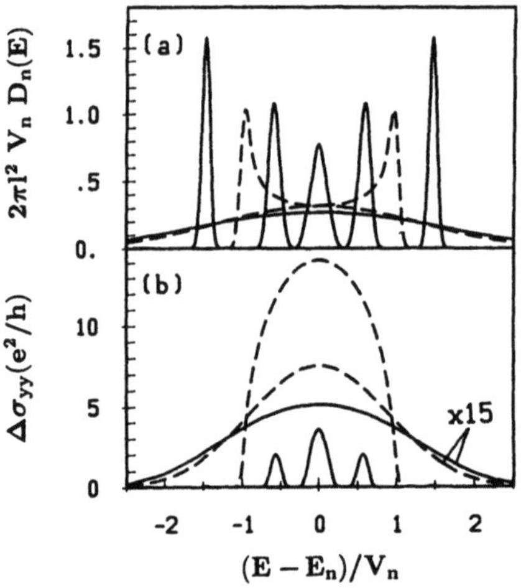

Fig. 3: Calculated density of states $D_n(E)$ (a) and band conductivity $\Delta\sigma_{yy}(E)$ (b) for one Landau level and two values of the collision broadening, $\Gamma_0/V_n = 1.0$ and 0.05. Solid (dashed) curves are for a grid (grating) potential with $V_x\mathcal{L}_n{=}V_y\mathcal{L}_n{=}V_n$ ($V_x\mathcal{L}_n{=}V_n$, $V_y{=}0$) and $p/q{=}5$. For $\Gamma_0/V_n{=}1.0$ the internal bandstructure is not resolved, $D_n(E)$ and $\Delta\sigma_{yy}(E)$ (here $15\times\Delta\sigma_{yy}(E)$ is plotted) are similar for grid and grating. For $\Gamma_0/V_n{=}0.05$, the resolved subband splitting dramatically reduces $\Delta\sigma_{yy}$ for the grid [with only tiny contributions from the narrow outer bands (near ± 1.5)].

Our calculation of conductivities is based on Kubo's formulas [16,19], which in the approximation consistent with Eq.(9) read [19,8] $\sigma_{\mu\mu} = \int dE\, [-f'(E)]\sigma_{\mu\mu}(E)$, with f' the derivative of the Fermi function and

$$\sigma_{\mu\mu}(E) = \frac{e^2\hbar}{2\pi}\int d^2k \sum_{n,n'}\sum_{j,j'} |\langle n';\alpha'|v_\mu|n;\alpha\rangle|^2 A_{n\alpha}(E)A_{n'\alpha'}(E)\ , \tag{10}$$

where $\alpha = (\vec{k};j)$ and $\alpha' = (\vec{k};j')$. We distinguish two contributions, $\sigma_{\mu\mu}(E) = \sigma^{sc}_{\mu\mu}(E) + \Delta\sigma_{\mu\mu}(E)$, a band conductivity $\Delta\sigma_{\mu\mu}(E)$ arising from intra-LL contributions ($n' = n$) which diverges in the absence of random scatterers and vanishes for the unmodulated system, and an inter-LL ($n' \neq n$) contribution $\sigma^{sc}_{\mu\mu}(E)$, which arises from scattering and is the only contribution in the unmodulated case. We will consider here only the situation were both collision broadening ($\sim \Gamma_0$) and modulation broadening ($\sim V_x\mathcal{L}_n$) are much smaller than $\hbar\omega_c$, so that the resistivities are given by $\rho_{xx} \approx \sigma_{yy}/\sigma^2_{yx}$ and $\rho_{yy} \approx \sigma_{xx}/\sigma^2_{yx}$ with $\sigma_{yx} \approx e^2N_s/m\omega_c$.

4.1 Band conductivity

Since the intra-LL ($n' = n$) velocity matrix elements are proportional to the modulation potential, one might replace $E_{n\alpha}$ by E_n in the spectral functions, in order to calculate $\Delta\sigma_{\mu\mu}(E)$ to lowest order in the modulation ($V_x\mathcal{L}_n, V_y\mathcal{L}_n \ll \hbar\omega_c$),

$$A_{n\alpha}(E)A_{n\alpha'}(E) \approx [A_n(E)]^2 \approx (\pi\Gamma_0)^{-1}\delta(E - E_n). \tag{11}$$

In view of the SCBA result $A_n(E) = (\pi\Gamma_0)^{-1}[1-(E-E_n)^2/4\Gamma_0^2]^{\frac{1}{2}}$ the second approximation in Eq.(11) is reasonable and relates the collision broadening of the LLs to a constant relaxation time $\tau = \hbar/\Gamma_0$. As a consequence of the first approximation in Eq.(11), the sum over j and j' in $\Delta\sigma_{\mu\mu}(E)$ can be expressed by the trace in the p-dimensional space and thus easily be evaluated, for example,

$$\int d^2k \sum_{j,j'} \left|\langle n;\vec{k},j'|v_y|n;\vec{k},j\rangle\right|^2 = \int d^2k \; Tr^{(p)} \left[v_y^{(p)}(\vec{k})\right]^2$$

$$= \frac{2\pi}{aq}\int_0^{aq}\frac{dx_0}{l^2}\left[\frac{1}{\hbar}\mathcal{L}_n V_x K l^2 \sin(Kx_0)\right]^2 = \pi\left[\mathcal{L}_n V_x K l/\hbar\right]^2 \; . \tag{12}$$

The resulting $\Delta\sigma_{yy}$ is independent of V_y and equals exactly the result for a unidirectional modulation in x-direction. In the interesting range of temperatures, where k_BT is larger than $\hbar\omega_c$ but smaller than the energy separation $\Delta_\lambda \approx \frac{1}{4}m\omega_c^2a^2(\lambda-\frac{1}{4})$ of adjacent flat bands, we thus recover the result [4,3,5,8]

$$\Delta\sigma_{yy} \approx \frac{e^2}{h}\frac{V_x^2}{\Gamma_0\hbar\omega_c}\frac{4}{ak_F}\cos^2\left(2\pi\frac{R_c}{a}-\frac{\pi}{4}\right). \tag{13}$$

Extending Beenakker's [6] quasi classical calculation to the 2dm case, one finds exactly the same result: the calculated band conductivity $\Delta\sigma_{yy}$ is independent of the modulation in y-direction, in sharp disagreement with the experiment.

To understand the suppression of the band conductivity observed in experiment, we must take the peculiar subband splitting of the Hofstadter-type energy spectrum seriously. From the mobility at zero magnetic field, one can estimate [8] that, in the experiments shown in Fig.1, the collision broadening is small enough to resolve the gross features of the Hofstadter spectrum, if one is not near a flat band situation. If the splitting of the subbands j and j' is resolved, the corresponding spectral functions do not overlap, and thus the non-diagonal matrix elements of the velocity between these subbands do not contribute to $\Delta\sigma_{\mu\mu}(E)$. Then the band conductivity of the 2dm system is considerably smaller than that of the corresponding 1dm system, as is visualized for a typical situation by the numerical results in Fig.3b.

4.2 Scattering conductivity

We now turn to the inter-LL contribution $\sigma_{\mu\mu}^{sc}$ to the conductivity. Here we neglect the effect of the modulation on the velocity matrix elements, i.e., small corrections of the order $V_x\mathcal{L}_n/\hbar\omega_c$ [7,8]. Then, the modulation affects $\sigma_{\mu\mu}^{sc}(E)$ only via the self-energy. To leading order in $|\Sigma^-/\hbar\omega_c|$, one gets from Eqs.(9) and (10) $\sigma_{\mu\mu}^{sc}(E) = (e^2/\hbar)\sum_n(2n+1)$ $[2\pi l^2\Gamma_0 D_n(E)]^2$. For zero modulation, this is a well known result [16]. In the interesting temperature range, $\hbar\omega_c < k_BT < \Delta_\lambda$, the Fermi function in the energy integral defining $\sigma_{\mu\mu}^{sc}$ can be replaced by the first term of the expansion $f'(E) = f'(E_n)+f''(E_n)(E-E_n)+...$, so that $\sigma_{\mu\mu}^{sc}$ can be expressed in terms of an effective scattering rate $\tilde{\Gamma}_n$ defined by

$$\tilde{\Gamma}_n = 2\pi\int dE\;[2\pi l^2\Gamma_0 D_n(E)]^2 \;, \tag{14}$$

which oscillates as a function of n with minima for flat bands [20]. For $k_BT \ll \Delta_\lambda$ the sum over n is easily performed to yield the Drude-type result

$$\sigma_{xx}^{sc} = \sigma_{yy}^{sc} = (e^2N_s/m\omega_c^2)\;\tilde{\Gamma}_{n_F}/\hbar \;, \tag{15}$$

where $n_F = E_F/\hbar\omega_c$ and $E_F = N_s/D_0$ with the zero-B DOS $D_0 = m/\pi\hbar^2$ has been inserted. It is obvious from Eq.(14) that $\sigma_{\mu\mu}^{sc}$ becomes maximum if the Landau bands at the Fermi energy become flat (and the peaks of the DOS become high near $E = E_F$). In this situation the *band conductivity* becomes minimum, since the intra-LL velocity matrix elements approach zero. The numerical results shown in Fig.4 demonstrate this antiphase

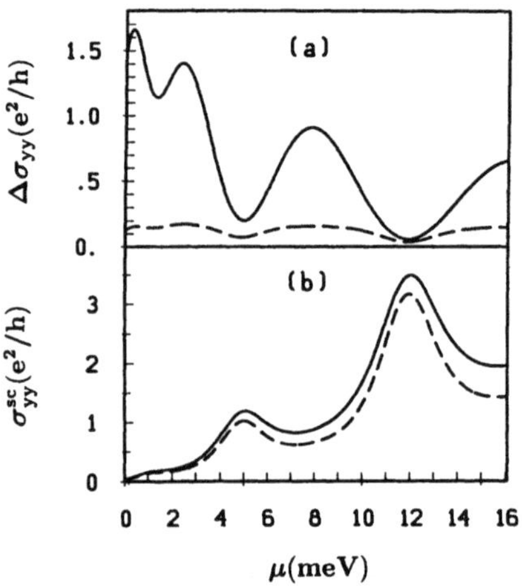

Fig. 4: Calculated band- (a) and scattering (b) contribution to the conductivity versus chemical potential for temperature $T = 5K$, $B = 0.23T$, $a = 300nm$ (p/q=5), and $\Gamma_0 = 0.02meV$. Solid (dashed) curves are for grid (grating) potential with V_x=V_y= $0.25meV$ ($V_x = 0.25meV$, $V_y = 0$).

behavior of the different contributions to the conductivity, and also the suppression of the band conductivity in the case of a grid modulation.

5. Summary and perspective

We have presented a straightforward quantum transport theory which explains all the novel magnetoresistance oscillations observed on 2D electron systems in weak lateral superlattices [2]. Among these, only the large-amplitude 'band conductivity' oscillations, observed in unidirectionally modulated systems when the current flows in the modulation direction, can be understood within a quasi classical approach [6], together with the accompanying strong positive magnetoresistance at $B < 0.1T$ [22]. The weaker antiphase oscillations, observed when the current flows in the other direction and determined by the 'scattering conductivity', as well as the suppression of the band conductivity in bidirectionally modulated systems reflect properties of the peculiar, quantized energy spectrum, which persist at elevated temperatures where the individual SdH oscillations are not resolved.

Our theory predicts that at very low temperatures the novel oscillations appear as amplitude modulations of the SdH oscillations. For a 2D grid modulation, the magnitude of the band conductivity depends on the values of V_x and Γ_0 in a complicated manner and may dominate the scattering conductivity or not. Experimentally, both situations are possible, too, as is shown in Fig.5 for a gated sample, which allows to tune the amplitude of the modulation. The electron density was calculated from the SdH period and increases with increasing gate voltage from top to bottom. From the position of the arrows, indicating the flat-band condition Eq.(2), and from the heights of the SdH maxima, we conclude that the importance of the band conductivity decreases from top to bottom, but we know little about the magnitude of the built-in superlattice potential. The fact that the SdH peaks show no internal structure, although this should be resolved due to both the small collision broadening ($\Gamma_0 \sim 50\mu eV$ at $B = 0.5T$) and the low temperature ($k_BT \approx 4\mu eV$), indicates that mesoscopic fluctuations of the electrostatic potential ($\sim 0.1mV$) due to

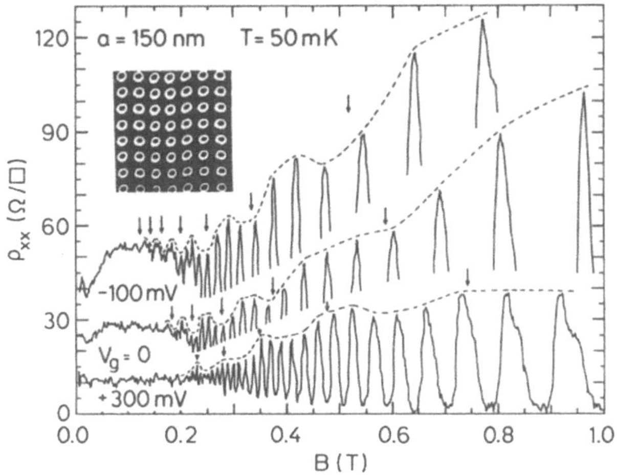

Fig. 5: Magnetoresistivity at temperature $T = 50mK$ of a gated microstructured sample with a grid (period $a = 150nm$) defined by electron lithography and subsequent etching. The inset shows the array of photoresist spots before the metal gate was deposited. Curves are for gate voltages $V_g = -100$, 0, and $+300mV$, with electron densities $N_s = 1.7$, 2.2, and $3.5 \cdot 10^{11} cm^{-2}$ and with mobilities $\mu = 1.0$, 1.2, and $1.5 \cdot 10^6 cm^2/Vs$, respectively. Arrows indicate the B-values satisfying Eq.(2).

inhomogeneities and nonlinear screening effects become here important [23]. This raises a variety of new and very interesting questions for future research.

We are grateful to G. Weimann, Walter Schottky Institut München, for providing us with high mobility sample material and to A. Menschig, IV Physikalisches Institut der Universität Stuttgart, for performing the electron beam lithography. The work was supported in part by the Bundesministerium für Forschung und Technologie, Germany (under Grant No. NT-2718-C).

References

[1] C.G. Smith, M. Pepper, R. Newbury, H. Ahmed, D.G. Hasko, D.C. Peacock, J.E.F. Frost, D.A. Ritchie, G.A.C. Jones, and G. Hill, J. Phys.: Condensed Matter **2**, 3405 (1990)

[2] D. Weiss, K. v.Klitzing, K. Ploog, and G. Weimann, Europhys. Lett. **8**, 179 (1989); see also in *High Magnetic Fields in Semiconductor Physics II*, edited by G. Landwehr, Springer Series in Solid-State Sciences Vol. **87** (Springer-Verlag, Berlin 1989), p. 357

[3] R.R. Gerhardts, D. Weiss, and K. v.Klitzing, Phys. Rev. Lett. **62**, 1173 (1989)

[4] R.W. Winkler, J.P. Kotthaus, and K. Ploog, Phys. Rev. Lett. **62**, 1177 (1989)

[5] R.R Gerhardts, in *Science and Engineering of One- and Zero-Dimensional Semiconductors*, ed. by S.P. Beaumont and C.M. Sotomajor Torres (Plenum Press, New York, 1990), p. 231

[6] C.W.J. Beenakker, Phys. Rev. Lett. **62**, 2020 (1989)

[7] R.R. Gerhardts and C. Zhang, Phys. Rev. Lett. **64**, 1473 (1990); Surf. Sci. **229**, 92 (1990)

[8] C. Zhang and R.R. Gerhardts, Phys. Rev. B **41**, 12 850 (1990)

[9] M.Ya. Azbel', Sov. Phys. JETP **19**, 634 (1964); [Zh. Eksp. Teor. Fiz. **46**, 929 (1964)]

[10] R.D. Hofstadter, Phys. Rev. B **14**, 2239 (1976)

[11] A. Rauh, phys. stat. sol. (b) **69**, K9 (1975)

[12] For a recent review see: D.J. Thouless, in *The Quantum Hall Effect*, edited by R.E. Prange and S.M. Girvin (Springer-Verlag, New York, 1987)

[13] N.A. Usov, Sov. Phys. JETP **67**, 2565 (1988), [Zh. Eksp. Teor. Fiz. **94**, 305 (1988)]

[14] The sample is considered as a part of an infinitely extended system. The plane waves are normalized on a length L_y, the ϕ_n on the total x-axis. To count states correctly, x_0 is considered within a length L_x taken as a suitable (B-dependent) multiple of the period a. Finally, the limit L_x, $L_y \to \infty$ is taken.

[15] D. Weiss, C. Zhang, R.R. Gerhardts, K. v.Klitzing, and G. Weimann, Phys. Rev. B **39**, 13020 (1989)

[16] T. Ando, A.B. Fowler, and F. Stern, Rev. Mod. Phys. **54**, 437 (1982)

[17] For computational convenience, the plots are calculated using the approximation $A_{n\alpha}(E) = (2\pi\Gamma^2)^{-\frac{1}{2}} \exp[-(E - E_{n\alpha})^2/2\Gamma^2]$.

[18] G. Czycholl and W. Ponischowski, Z. Physik B.- Cond. Matter **73**, 343 (1988)

[19] See, e.g., R.R. Gerhardts, Z. Phys. B **22**, 327 (1975)

[20] If one replaces, in the spirit of Eq.(11), the factor $[...]^2$ in Eq.(14) by $2l^2\Gamma_0 D_n(E)$, one gets the n-independent scattering rate $\tilde{\Gamma}_n = 2\Gamma_0$, and σ_{xx}^{sc} reduces to Eq.(9) of Ref.21, where Γ_0 is written as $(N_I U_0^2/2\pi l^2)/\Gamma$. One thus misses the leading order contribution to the oscillations of $\rho_{yy}^{(1d)}$, retaining higher order contributions with small amplitudes of the order $(V_x \mathcal{L} n/k_B T)^2$.

[21] P. Vasilopoulos and F.M. Peeters, Phys. Rev. Lett. **63**, 2120 (1989)

[22] P. Streda and A.H MacDonald, Phys. Rev. B **41**, 11 892 (1990)

[23] U. Wulf, V. Gudmundsson, and R.R. Gerhardts, Phys. Rev. B **38**, 4218 (1988); V. Gudmundsson and R.R. Gerhardts, Phys. Rev. B **35**, 8005 (1987)

2D Electrons in a Tilted Magnetic Field: Effect of the Spin-Orbit Interaction

Yu.A. Bychkov[1], *V.I. Mel'nikov*,[1], *and E.I. Rashba*[2]

[1]L.D. Landau Institute for Theoretical Physics, SU-117940 Moscow, USSR
[2]The European Branch of the Landau Institute at ISI Foundation,
Villa Gualino, Viale S. Sevevo 65, I-10133 Torino, Italy

Abstract. Spin-orbit (SO) interaction introduces changes in the spectrum of 2D electrons in a perpendicular magnetic field $B_\perp$, resulting in the appearance of specific beats in oscillatory phenomena. We investigate the effect of tilting of a magnetic field on the energy spectrum and show that a moderate parallel field $B_\parallel$, whose Zeeman energy does not exceed the cyclotron quantum, strongly reduces the effect of the SO interaction. As a result, the beats become suppressed, their period increases.

1. Introduction

In crystals lacking an inversion center, energy bands are split by spin-orbit interaction. For 2D electrons this splitting may be caused both by the absence of an inversion center in the bulk and by the asymmetry of a confining layer. It is manisfested in several phenomena [1], most convincingly in the Shubnikov-de Haas (SdH) oscillations [2-5]. The splitting must result in the appearance of beats in all oscillatory phenomena. Such beats, well known in 3D [6,7] and recently observed for 2D electrons in InAs [3a] and $In_xGa_{1-x}As$ [4] quantum wells, allow a reliable measurement of the SO splitting at the Fermi level. In a tilted field a dramatic suppression of the beats has been observed by the group of Fang and Stiles when the tilt angle θ increased [3b]. This phenomenon has been attributed to spin dependence of the scattering rate. We show that the θ dependence of the spectrum implies a new mechanism of beat suppression.

2. A Normal Field - Energy Spectrum and Beats

The model SO interaction is written as (for more details see [8]):

$$H_{SO} = \alpha\ [\vec{\sigma}\ x\ k]\vec{\nu}, \tag{1}$$

where $\vec{\sigma}$, k and $\vec{\nu}$ are the Pauli matrices, quasimomentum and unit vector normal to the 2D layer, respectively. The energy spectrum is $E(k) = k^2/2m \pm \alpha k$. The parameters α and $\Delta = m\alpha^2/2$ $(\hbar = 1)$ may be considered as SO coupling constants. Typically, $\alpha = 10^{-10} - 10^{-9}$ eV·cm. The spectrum for $\mathbf{B}\|\vec{\nu}$ is [1b]

$$\epsilon_0 = \delta, \; \epsilon^{\pm}\sqrt{\delta^2+\gamma^2 n}, \qquad n\geq 1, \tag{2}$$

where ϵ_n are in units of the cyclotron frequency $\omega_c = eB_{\perp}/mc$, $\gamma = (4\Delta/\omega_c)^{1/2}$, $\delta = 1/2 - \beta$, $\beta = gm/4m_0$, m_0 is the free electron mass and g is a g-factor. The spectrum consists of two non-equidistant ladders. The separation $\omega_{\pm}$ (B,ϵ) between levels in each ladder depends on the spin index ±. In the limit $\zeta \gg \omega_c$, where ζ is the Fermi energy, SdH oscillations may be described by a semiclassical theory, and then the contribution of each ladder is $\sigma_{\pm} \propto \cos\,(2\pi\zeta/\omega_{\pm} + \varphi_{\pm})$. Here $\omega_{\pm}(B) = \omega_{\pm}(B,\zeta)$, and $\varphi_{\pm}$ is the phase. For a spin independent scattering the terms σ_+ and σ_- are completely cancelled if at the Fermi level the equality $2\epsilon_n^+ = \epsilon_{n+s}^- + \epsilon_{n+s+1}^-$ holds. Such an arrangement of levels will be termed: a built-in configuration. The cancellation gives rise to beats in the SdH oscillations, the integer $s \geq 0$ enumerates to the nodes of the beats. The above criterion of cancellation is equivalent to

$$(s + 1/2)^2 = (1-2\beta)^2 + (\Delta_{sp}/\omega_c)^2, \tag{3}$$

where $\Delta_{sp} = 4(\Delta\zeta)^{1/2}$ is the spin splitting at the Fermi level. For $s \gg 1$ Eq.(3) coincides with the criterion used in [4], and the number of oscillations between two nodes is $\zeta/\Delta_{sp} \approx \frac{1}{4}\sqrt{\zeta/\Delta}$. This factor is closely connected to the difference $2\sqrt{\Delta/\zeta\omega_c}$ in the level spacing in the two ladders. The condition $\zeta \gg \Delta_{sp}$ is the criterion for observation of semiclassical beats.

3. A Tilted Field

In this case ($\theta \neq 0$) eigenvalues of the infinite system

$$(\epsilon-n-\frac{1}{2}-\beta)\,a_n = \beta\tan\theta b_n + \gamma\sqrt{n+1}b_{n+1},$$

$$(\epsilon-n-\frac{1}{2}+\beta)\,b_n = \beta\tan\theta a_n + \gamma\sqrt{n}a_{n-1}$$

cannot be found in an explicit form. Even in the semiclassical limit the system is equivalent to a generalized Mathieu equation (cf.[9]). The origin of this complication is the fact that the parameter $n \gg 1$ actually enters through the factor $\gamma^2 n \sim (\Delta_{sp}/\omega_c)^2 \sim s$, which is not large [cf.(3)]. Below are expounded some results of a numerical investigation of (3) for the parameter values typical for InAs based alloys ($m \approx 0.04m_0$, $\beta \approx -0.1$, $\alpha \approx 0.9\cdot 10^{-9}eV\cdot$cm, carrier concentration $\approx 10_{12}$cm^{-2}). Values of $B_{\perp}$ for different built-in configurations depend strongly on $B_{\parallel}$. This dependence is shown in Fig.1 for a last node, $s = 1$ (the number of energy levels below ζ is $l \leq 25$). It is seen that $B_{\perp}$ is doubled when the ratio of the Zeeman to the cyclotron frequencies $\omega_Z/\omega_c = 2\;|\beta|\;B/B_{\perp} \approx 0.5$; for larger values of B the dependence $B_{\perp}(B)$ is close to linear. The increase in the magnitude of $B_{\perp}$ with θ has been reported in [4]. Fig.2 shows the spacing D between adja-

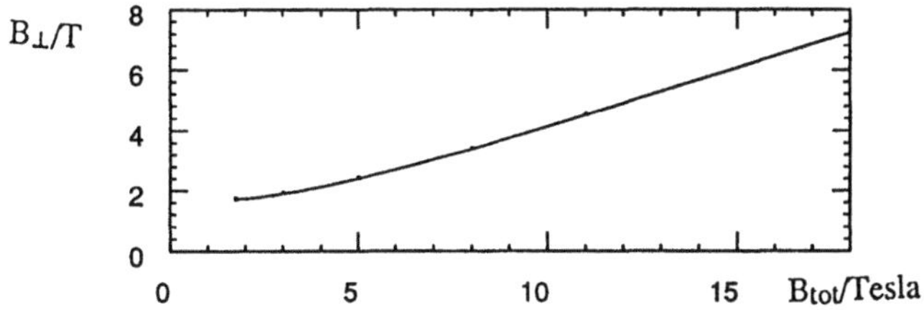

Fig.1. The built-in field $B_\perp$, corresponding to the last node of beat patterns, plotted versus the total field B (in Tesla). The parameter values are given in the text.

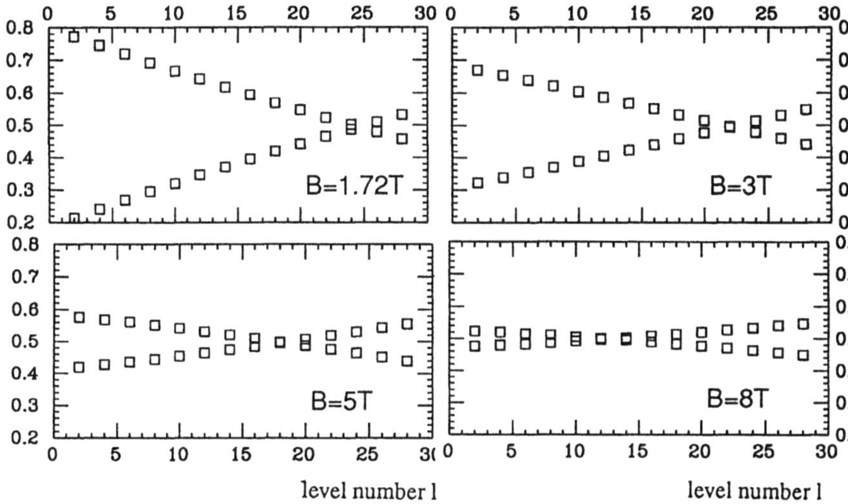

Fig.2. The spacings $D^{\pm}$ plotted versus l. For every B the magnitude of $B_\perp$ corresponds to the last built-in configuration, $D^+ = D^-$, at the Fermi level (Fig.1).

cent levels (in units of ω_c) as a function of l, the level number. Adjacent levels belong to different ladders, and the two families, D^+ and D^-, correspond to left and right hand adjacent levels. For every B the data is shown for a field $B_\perp(B)$ [cf.Fig.1], corresponding to a built-in configuration ($D_l^+=D_l^-$) at the Fermi level. The spacings in the ladders are $D_l^+ + D_l^-$ and $D_{l-1}^+ + D_l^-$. Hence, the difference in spacings is $D_l^+ - D_{l-1}^+ \approx dD_l^+/dl$. It follows from Fig.2 that with increasing B (or the angle θ) the spectrum changes drastically: most affected are the fine properties of it responsible for beats. Indeed, the derivatives D' diminish, and the spacings between levels in the two ladders become closer and nearly coincide at $B \geq 8T$. Since beats are caused by a difference in these spacings, they are suppressed, and the conductivity cannot be described in the framework of semi-classical phenomenology. The period of the beats estimated in this way increases as $(D')^{-1}$. The field $B \approx 8T$, for which D' is strongly reduced, is in reasonable agreement with the field $\approx 6T$ in which beats disappear [3b]. The above analysis is relevant for strong fields, i.e., for the last nodes. In weak fields, i.e., for $s \gg 1$, the beats are expected in a wider region of θ values.

Summary

The beats in SdH oscillations arise due to a difference in the level spacing in the two ladders corresponding to the different spin states. In a tilted field the effect of SO interaction is reduced, and this difference diminishes. Hence, the beats are suppressed, the last nodes disappear at a moderate magnitude of the ratio ω_Z/ω_c of the Zeeman and cyclotron frequencies.

References

[1a] Yu.A. Bychkov, E.I. Rashba, Usp. Fiz. Nauk **146**, 531 (1985)

[1b] E.I. Rashba, Fiz. Tverd. Tela (Leningrad) **2**, 1224 (1960) (Sov. Phys.- Solid State **2**, 1109)

[2] H.L. Störmer, Z. Schlesinger, A. Chang, D.C. Tsui, A.C. Gossard and W. Wiegmann, Phys. Rev. Lett. **51**, 126 (1983)

[3a] L.L. Chang, E.E. Mendez, N.J. Kawai and L. Esaki, Surf. Sci. **113**, 306 (1983)

[3b] J. Luo, H. Munekata, F.F. Fang and P.J. Stiles, Phys. Rev. **B41**, 7685 (1990)

[4] B. Das, D.C. Miller, S. Datta et al., Phys. Rev. **B39**, 1411 (1989)

[5] R. Wollrab, R. Sizmann, F. Koch et al., Semicond. Sci. Technol. **4**, 491 (1989)

[6] E.I. Rashba, I.I. Boiko, FTT **3**, 1277 (1961)

[7] D. Seiler, in: Landau Level Spectroscopy, ed. by G. Landwehr and E.I. Rashba (North-Holland, Amsterdam), 1031 (1990)

[8] E.I. Rashba, V.I. Sheka, ibid, 133

[9] Yu.A. Bychkov, V.I. Mel'nikov, E.I. Rashba, Zk. Eksp. Teor. Fiz. **98**, 717-726 (1990) (Sov. Phys. JETP **71 (2)**, 401-405 (1990))

Evidence for Negative Sign of the Thermodynamic Density of States of 2D Electrons in Si Inversion Layers

S.V. Kravchenko, V.M. Pudalov, D.A. Rinberg, and S.G. Semenchinsky

Institute for the Metrological Service, Andreevskaya nab. 2, SU-117334 Moscow, USSR

Using the floating gate technique, we found that for a two-dimensional electron gas in a quantizing magnetic field the rate of increase of the chemical potential with magnetic field can be much higher than the maximum value for a noninteracting gas. This corresponds to a negative sign of the thermodynamic density of states. The effect can be explained by the interactions between electrons and allows us to estimate the energy of the electron-electron interaction in Si inversion layers.

Recently, by measuring the magnetocapacitance of Si-MOSFET's, we have observed [1] that in the integer QHE regime the influence of the electron-electron interactions on the thermodynamic density of states is so drastic that the latter may become negative. This phenomenon was predicted theoretically by Efros [2]. However the ac measurements performed in Ref.1 are not absolutely reliable evidence for the negative sign of the thermodynamic density of states as one needs to assume a constant effective sample area (i.e. independent of a magnetic field) and to use parameters which cannot be determined in the present experiment.

In the present work we studied the influence of the electron-electron interactions directly on the chemical potential, μ, of a 2D electron gas using the so-called "floating gate technique" [3]. This method is free of any additional assumptions. The results confirm that the thermodynamic density of states is negative at partial filling of the energy levels in sufficiently strong magnetic fields. A comparison of our data with calculations allows us to estimate the energy of the electron-electron interaction in Si inversion layers in a magnetic field.

The experiments were performed on high-mobility Si MOSFET's. The data represented below were obtained with the sample Si 2-16B (the maximum zero magnetic field mobility of the electrons was 3 m^2/Vs at $T = 1.5$ K). The gate was connected to a direct current amplifier with an input resistance greater than 10^{14} Ω. After setting the gate voltage, V_g, the gate was isolated from the voltage source and the variations of V_g, proportional to the variations of μ, were measured directly using the amplifier.

In high magnetic fields the electron energy spectrum in a 2D system is completely quantized, and the positions of the energy level centers in Si inversion layers can be represented by the expression

$$E_i^0 = \hbar\Omega_c\left[\frac{1}{2}+N\right] \pm \frac{1}{2}g^0\mu_B H \pm \frac{1}{2}\Delta E_v^0, \qquad (1)$$

where Ω_c is the cyclotron frequency, N is the Landau quantum number, g^0 is the g-factor assumed to be 2 (see Ref.4), μ_B is the Bohr magneton, and ΔE_v^0 is the valley splitting. Magnetic field dependence of ΔE_v^0 is absent [5] or negligibly weak [3,4]. Thus the maximum rate of increase of the chemical potential with magnetic field for a noninteracting gas is

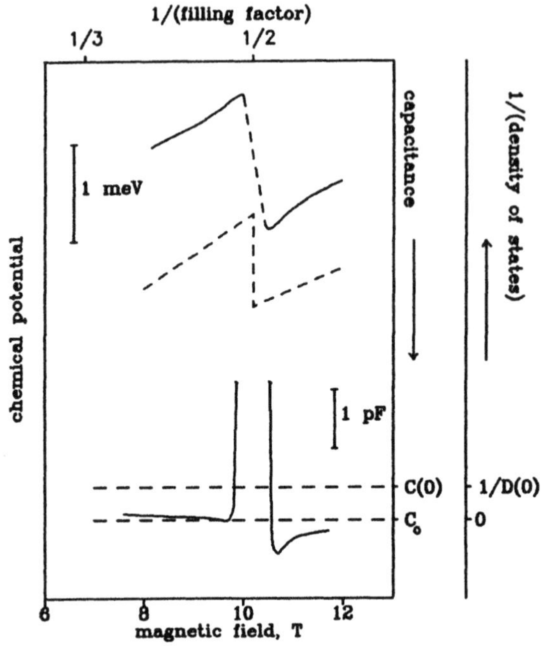

Fig.1 The chemical potential (upper curve) and the magnetocapacitance (lower curve) as functions of H $\propto \nu^{-1}$. T = 1.3 K and n_s= $5\cdot 10^{15}$ m^{-2}; C(0) and D(0) are zero-field capacitance and density of states respectively; C_0 is the capacitance at $\partial n_s/\partial\mu\to\infty$. The dashed curve shows the calculated$^0\mu$(H) dependence for noninteracting gas at T = 0 and in the absence of disorder.

$$\left[\frac{\partial\mu}{\partial H}\right]^0_{max} = \frac{e\hbar}{m^*c}\left[\frac{1}{2}+N\right] \pm \frac{1}{2}g^0\mu_B, \tag{2}$$

where + and - correspond to antiparallel and parallel spin directions respectively, e is the elementary charge, and m^* is the effective mass. The lowest value of m^* is 0.19 m_0 where m_0 is the free-electron mass [6,7]. Hence, for the zero Landau level

$$\left[\frac{\partial\mu}{\partial H}\right]^0_{max} \le \frac{1}{2}\left[\frac{e\hbar}{0.19m_0c} \pm g^0\mu_B\right]. \tag{3}$$

Nonzero temperature and disorder lead to a broadening of the energy levels and therefore to a decrease of $\partial\mu/\partial$H.

Fig.1 shows the chemical potential as a function of H (upper curve). The abrupt drop uf μ in the vicinity of H $\simeq$ 10.2 T corresponds to the filling factor ν = 2 (ν = $2\pi n_s l_H^2$, where l_H is the magnetic length). The excitation of eddy currents [8] makes it impossible to determine μ at filling factors close to ν = 2 and therefore μ(H) was interpolated in this region (dashed line). The dashed curve is the calculated μ(H) for a noninteracting 2D elctron gas at T = 0 and in the absence of disorder; ΔE_v^0 was chosen to be 0.2 meV [5]. One can see that at ν < 2 the slope of measured μ(H) dependence is nearly twice as high as the maximum slope for the noninteracting gas. Capacitance measurements analogous to those described in Ref.1 show that just in this region of filling factor the inverse thermodynamic density

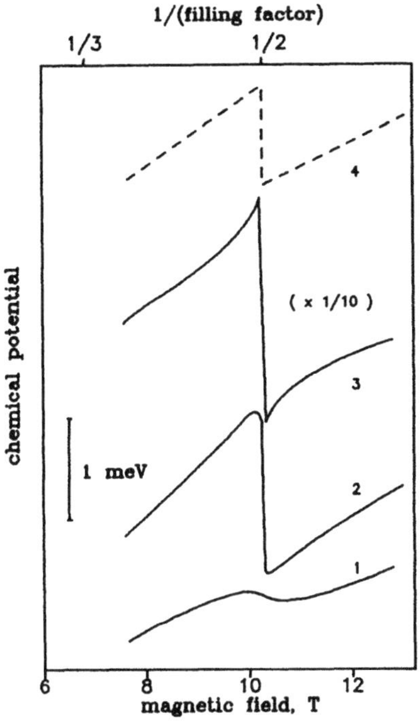

Fig.2 $\mu(H)$ calculated for T = 1.5 K, $n_s = 5\cdot10^{15}m^{-2}$ and α = 0 (curve 1), α = 0.06 (curve 2) and α = 0.782 (curve 3; vertical scale for this curve is ten times lower than that for other curves). Dashed curve (4) corresponds to $\alpha = T = \Gamma = 0$. The curves are shifted vertically for clarity.

of states, $1/D \equiv [\partial n_s/\partial\mu]^{-1}$ becomes negative (lower curve in Fig.1). At $2 < \nu < 3$, where $[\partial n_s/\partial\mu]^{-1}$ is close to zero, the measured slope is nearly equal to the calculated maximum one. At T = 4 K the rate of increase of the chemical potential with magnetic field is always less than $[\partial\mu/\partial H]^0_{max}$ (at least at $H \leq 12$ T), and the thermodynamic density of states obtained from the capacitance measurements is always positive. Note that the measured drop of μ at $\nu = 2$ is considerably higher than the calculated one for a noninteracting gas. This g-factor enhancement has been observed previously by many authors (Ref.4, 5 and references therein).

The too high slope of the $\mu(H)$ dependence can be explained by the electron-electron interaction. The energy (per particle) of interacting electrons belonging to the same energy level at $\{\nu\} \leq 1/2$ is nearly equal to [9,10]

$$H_{ee} = -\frac{\alpha e^2}{(\epsilon l_H)}\{\nu\}^{\frac{1}{2}}, \tag{4}$$

where α is a positive dimensionless constant equal to 0.782 for the classical Coulomb interaction [9,10], ϵ is the dielectric constant, and $\{\nu\}$ is the filling factor for the considered energy level: $\{\nu\} \equiv \nu - \mathrm{INT}(\nu)$. At increasing magnetic field the filling of the upper level diminishes and therefore the rate of increase of μ should be higher[11] than that for a noninteracting gas. The condition $\partial\mu/\partial H > [\partial\mu/\partial H]^0_{max}$ corresponds to a negative sign of the density of states.

Using Equ.(4) and taking into account the electron-hole symmetry we calculated dependences $\mu(H)$ for different α (to take account of the influence of scattering we used the conventional Gaussian formula for the one-particle density of states; the level broadening, Γ, was chosen to be 0.1 meV at $H \simeq 10$ T in accordance with the experimental results of Ref.3). The results are shown in Fig.2. Approximate agreement between calculated and measured slopes of $\mu(H)$ dependences and between energy splittings at $\nu = 2$ can be obtained by taking the parameter α to be 0.06, which indicates that the energy of the electron-electron interaction is around one order of magnitude less than the classical interaction energy. The slopes of the measured $\mu(H)$ dependence at $\nu < 2$ and $\nu > 2$ are respectively slightly higher and lower than corresponding slopes of the $\mu(H)$ calculated for $\nu = 0.06$. This is believed to be due to the slightly asymmetric shape of the density of states.

It is worth mentioning that the unexpectedly high slope of the $\mu(H)$ dependence was also noticed recently for a 2D electron gas in a GaAs/AlGaAs heterostructure at $\nu < 2$ [12]. However, the effect was not explained in that paper.

In summary, by measuring the chemical potential of 2D elctron gas in a strong magnetic field, we have obtained independent evidence for the negative sign of the thermodynamic density of states at partial fillings of energy levels. The results allow us to estimate the energy of the electron-electron interaction in Si inversion layers, which appears to be about an order of magnitude less than the classical Coulomb interaction.

REFERENCES

1. S.V. Kravchenko, V.M. Pudalov and S.G. Semenchinsky, Phys. Lett. A141, 71 (1989)
2. A.L. Efros, Solid State Commun. 65, 1281 (1988)
3. V.M. Pudalov, S.G. Semenchinsky and V.S. Edelman, Sov. Phys. JETP 62, 1079 (1985)
4. T. Ando, A.B. Fowler and F. Stern, Rev. Mod. Phys. 54, 437 (1982)
5. I.V. Kukushkin, S.V. Meshkov and V.B. Timofeev, Usp. Fiz. Nauk 155, 219 (1988)
6. G. Abstreiter, J.P. Kotthaus, J.F. Koch and G. Dorda, Phys. Rev. B14, 2480 (1976)
7. U. Kunze and G. Lautz, Surf. Sci. 142, 314 (1984)
8. V.M. Pudalov, S.G. Semenchinsky and V.S. Edelman, Solid State Commun. 51, 713 (1984)
9. D. Levesque, J.J. Weis and A.H. McDonald, Phys. Rev. B30, 1056 (1984)
10. P.K. Lam and S.M. Girvin, Phys. Rev. B30, 473 (1984)
11. A.H. McDonald, H.C.A. Oji and K.L. Liu, Phys. Rev. B34, 2681 (1986)
12. E. Böckenhoff and K. von Klitzing, Proceedings of the 8th Conference on Electronic Properties of 2D Systems, Grenoble, p. 297 (1989)

Electronic Transport in Regular Doping Structures

F. Koch

Technische Universität München, Physik-Department E16,
W-8046 Garching, Fed. Rep. of Germany

We consider dopant atoms incorporated in crystalline semiconductors in an ordered array. For structures in which donors have been condensed into periodically ordered sheets, we discuss some recent transport and cyclotron resonance experiments. There is evidence from transport that partial ordering within the plane of dopant atoms is also possible.

I. Introduction

Traditionally dopants have been introduced into a semiconducting material by indiffusion, implantation and by incorporation from the melt. The result is a random, statistically distributed assembly spread out over dimensions of typically 1 μm. Although the implantation technique allows lateral resolution and control of depth to a fraction of this length, the thermal annealing which is necessary to achieve electrical activation causes significant diffusion broadening.

Basic functions of the dopant atoms are to provide mobile carriers and to create potential variations that can be controlled by applied voltages. In device applications transport of the charges plays an essential role in both active and passive elements of the structure. Whether dealing with the channel of a MES device, the p-n junctions of a transistor or simply the contact regions of a LED, the conductivity is reduced by scattering of the mobile carriers by the randomly fluctuating impurity ion potentials.

In this discussion we consider dopant atoms arranged in an ordered lattice such as to reduce or totally eliminate the scattering. In a periodic structure electrons become eigenstates diffracting coherently from the ordered array of ionic potentials. Even a partial, liquid-like ordering can provide for improved transport by reducing the structure factor in the Born scattering description. Ordered impurities also provide free carriers and are actually much better for sharply defined potential variations.

The ultimate challenge to the growth techniques will be to learn how to build the impurities into a three-dimensionally ordered lattice. That is a tall order for future work. For the present it is possible to approximate the ideal in a number of ways. In particular, vertical ordering that makes use of the epitaxial growth techniques has been developed to a fine art. We consider some examples in the next Sec.II. There follows in Sec.III a discussion of possible ways to

Springer Series in Solid-State Sciences, Vol. 101
High Magnetic Fields in Semiconductor Physics III Editor: G. Landwehr

achieve partial ordering in a planar arrays of impurity atoms.

II. Properties of Vertically Ordered Impurities

Using MBE or related growth techniques it has proved possible to achieve atomically-sharp confined sheets of impurity atoms. These can be assembled into a periodic stack. As a result the impurities of the random system in Fig.1 appear condensed into sheets. We may expect dramatically altered transport properties, in particular a conductivity σ which differs for motion of charge in the planes and perpendicular to them. For the present discussion we consider dopant sheets with a density $N = 1/a^2$, where a is the average in-plane spacing. The planes are separated by the distance b. The quotient b/a is called the aspect ratio. Layers for which $N^* \geq 0.13/(a_B{}^2)$ will be degenerate and metallically conducting. In this expression a_B is the effective Bohr radius.

Among the various possible ordered systems of this type there are several that have been studied in recent years. Ref./1/ has introduced the so-called δ-layers for which $r \to \infty$ and $N > N^*$. Electrons are in quantized subbands, each with a distinct and different density distribution about the sheet of positive ions. In Fig.2 is shown the binding potential, subband levels and charge densities for a typical δ-layer. The reader is reminded about the role of oscillatory magnetotransport measurements in determining the charge densities and energy levels. This has become a classic approach to characterizing the δ-layer. Magnetotransport in parallel /2/ and perpendicular /3/ fields, as well as quantum-Hall-effect /4/ experiments, have been used to study these layers.

The periodically repeated δ-layer in Fig.1 raises a number of new and basic questions about transport in an ordered system of impurities. There are two distinct limits that need to be explored. We refer to these as the δ-layer ($r \to \infty$) and the δ-minibands that are expected when $r \gtrsim 1$.

In the first of these, electrical conduction will be confined to the plane. There is a $\sigma_{\parallel}$ only. The relevant question is whether the condensation of donors into a plane has achieved a parallel conductivity greater or less than

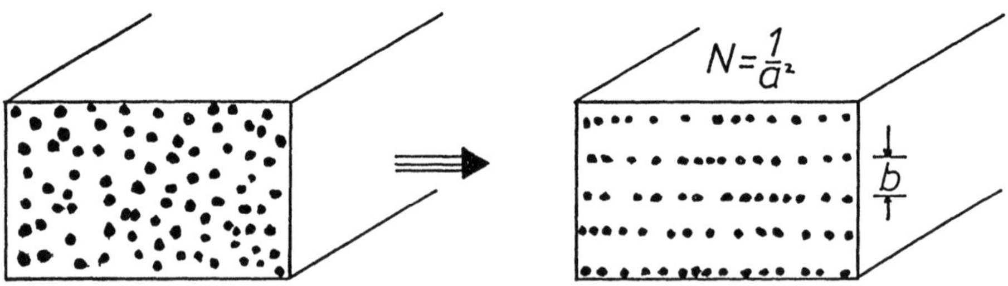

Fig.1 By means of epitaxial growth dopant atoms can be ordered in a periodic series of sheets with density $N = 1/a^2$, space distance b apart.

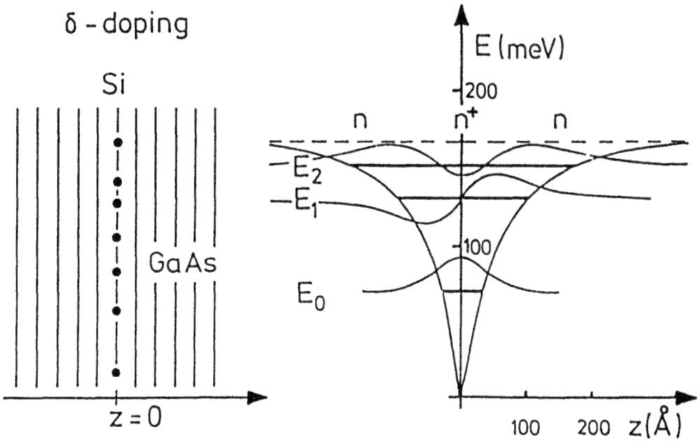

Fig.2 Potential well, energy levels and wavefunctions calculated for GaAs donors with density $N = 4.1 \times 10^{12}\ cm^{-2}$ in a single sheet of width 6Å. The parabolic approximations for the bandstructure has been used. Exchange and correlations has been ignored.

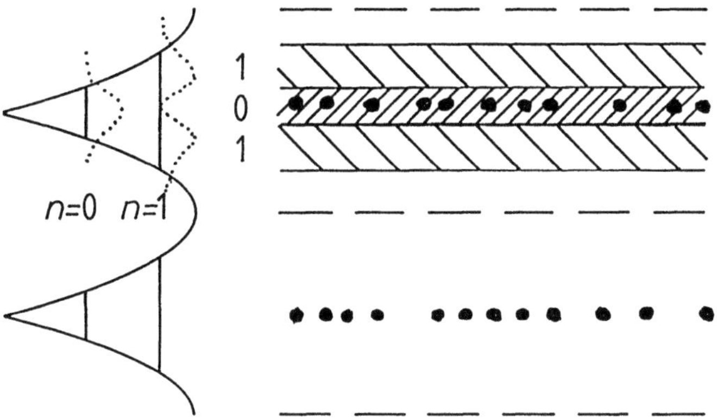

Fig.3 For uncoupled, neighboring δ-layers with two occupied subbands (n = 0 and 1) the parallel conductivity is the sum of contributions from a carrier (n = 0) strongly interacting with the donors and a carrier (n = 1) which is well screened and mobing at considerable distance from the donors. Is a conductivity greater than that for homogenous, random doping possible?

that of the random distribution of scatterers. Having condensed the dopants may have increased the density into a degenerate, metallic limit. It leads, as Fig.3 shows, to groups of distinctly different carriers. The n = o ground state subband is closely bound to the ions. Its mobility, is low and moreover must be expected to depend on the exact degree of confinement of the ions. Any spreading in the $\hat{z}$-distribution will sensitively influence the conductivity of

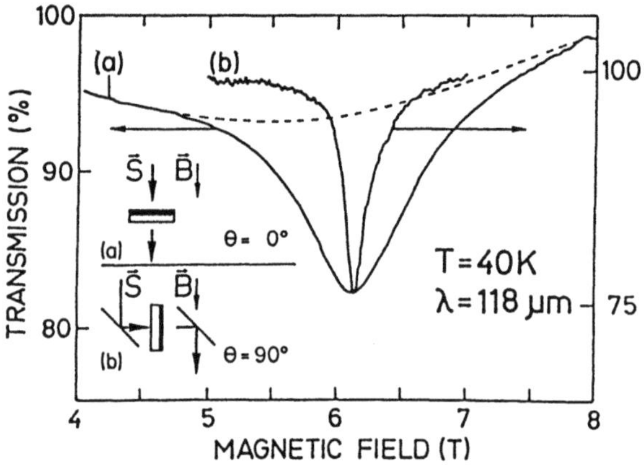

Fig.4 Cyclotron resonance of thermally ionized electrons measured with the magnetic field perpendicular (a) and parallel (b) to the doping planes of a sample with $N = 2.2 \times 10^{11}$ cm^{-2}, r = 4.6. The narrow line comes from high mobility electrons moving between the doping planes in a structure like Fig3.

these carriers. Ordering within the plane of the ionic charge will enhance this contribution to $\sigma_{\parallel}$. There is a second group of carriers (the n = 1 subband) which is both physically remote from the ions and that sees only the screened, potential fluctuations. Whether or not on balance there is an increased conductivity remains to be explored. It is clear from the figure that one must carefully account for the bounding regions between neighboring potential wells. Finite overlap allows a $\sigma_{\perp}$ and redistributes the charges. Just where the optimum σ occurs as the aspect ratio r is varied is an open question.

Planar structures provide distinctly different magnetotransport properties than homogeneously doped materials. These structures have open spaces, regions in which there are no dopant atoms at all. The layer-doped semiconductor is divided into doped sheets and undoped empty spaces, much like the holes in Swiss cheese. Charges confined to moving in the holes only by a magnetic field, will be observed with a very high mobility. This is the principle explored in ref./5/. The cyclotron line from carriers moving between the doping sheets shows up as a narrow and distinctly different line (Fig.4). The experiments are done in the dilute limit of doping ($N = 2.2 \times 10^{11}$ cm^{-2}, r = 4.6). Some of the electrons thermally excited from their parent atoms are confined by the parallel magnetic field to the "doping holes" between the planes. These high mobility electrons generate the sharp line in Fig.4.

Conduction in a δ-miniband system for which there is $N > N^*$ and r tends to a value near one, is characterized by having both a $\sigma_{\parallel}$ and $\sigma_{\perp}$. The quantity of particular interest in this case is $\sigma_{\perp}$. It depends on the minibanding effect in the $\hat{z}$-direction. This increases with $r \to 1$. At the same time the sensitivity to potential fluctuations in the plane will increase. For the periodic part of the potential there is no

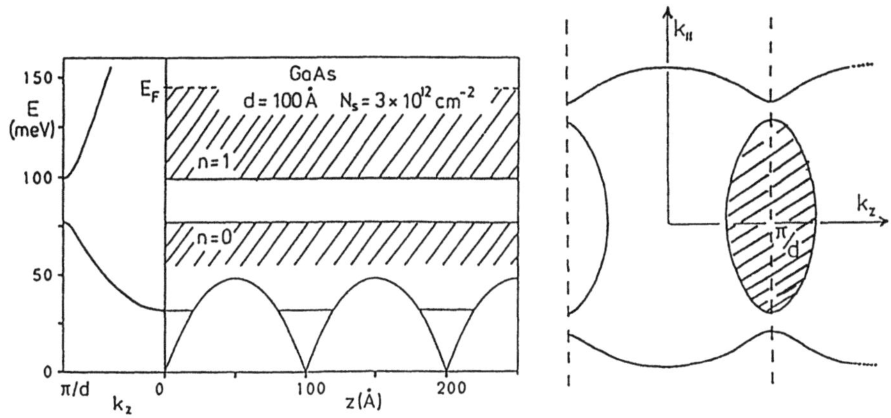

Fig.5 Energy-band structure $\epsilon_F \sim 140$ meV consists of two distinct sheets in the $\mathbf{k}_{\parallel}$, $\mathbf{k}_z$ plane.

scattering. It remains to find the optimal conditions with regard to r. Finally, the question must be asked if this optimal $\sigma_{\perp}$ is then really better than if random doping had been used. There is reason to believe that it will be so.

To illustrate the uniquely different electronic properties of a periodically ordered miniband system we calculate self-consistently the case of $N = 3 \times 10^{12}$ cm^{-2} spaced 100 Å apart. The aspect ratio is 1.7. The resulting mini-bandstructure in ϵ vs.k_z appears in Fig.5. It has been calculated for GaAs using a parabolic dispersion law and ignoring exchange and correlation effects. The Fermi energy is fixed to accommodate correctly all the 2×10^{18} electrons per cm^3. The carriers are found to occupy two subbands. The Fermi surface in Fig.5 consists of two separate sheets in the first and second Brillouin zones.

III. Transport and Possible Lateral Ordering

Vertical structuring of the impurity density is easily achieved by epitaxial growth. Lateral order is a more subtle effect and it is not easily obtained during the growth procedure. Nevertheless, some degree of local ordering can result, if not during growth then by subsequent treatment of the layers. The conductivity with or without magnetic field reacts sensitively when there is coherent scattering.

For oscillatory magneto-transport in metallic δ-layers, regular placement of impurities is expected to increase the mean free path. However, a comparison of a calculated mean free path assuming a totally random distribution of scatterers, with that observed in the experiments cannot be done with convincing precision. There is evidence for changes in mean free path for magnetotransport experiments when done under pressure. In ref./6/ the oscillatory amplitude for each of the periods is monitored as pressure is applied in increments. Fig.6 repeats this data here. The peaks represent

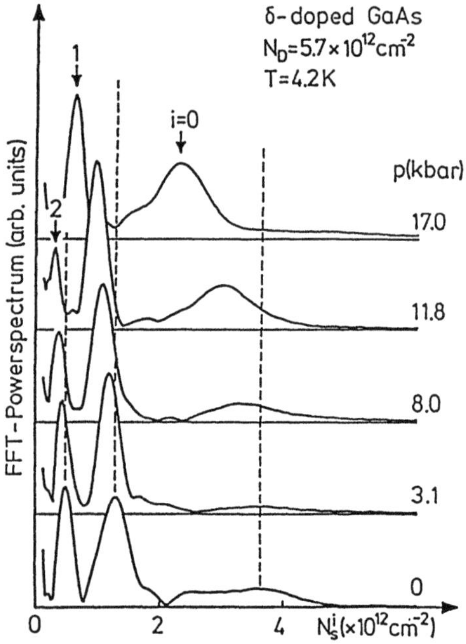

Fig.6 Fourier transform spectra of a δ-doped GaAs samples with applied pressure. When carriers disappear and N_s^i decreases there is a dramatic increase in the peak amplitude for $i = 0$.

periods in the oscillations and give the number of electrons for a given subband. Amplitude is an indication of the oscillation strength. We note the drastic increase of amplitude that comes with the disappearance of some of the mobile charges. The effect is most pronounced for the = o subband.

Along with the changes in the oscillation spectrum, the conductivity at B = 0 also changes. The number of carriers is decreased and the mass rises under pressure. When both of these are accounted for there remains an increased average mobility for the carriers.

The explanation that we propose for the pressure experiments is related to that which has been used to account for HgSe(Fe) data /7/. In the latter, a fraction of the statistically random Fe-sites exists as charged scatterers while others are neutral. The Fermi energy is pinned at the Fe impurity level. It is argued the charged sites because of Coulomb energy consideration, are arranged as close as possible to a lattice. This explains the low scattering rate observed in the experiments. For the present case, the reduction of free carriers is a consequence of DX-center occupation. When a pressure value is reached for which the DX-center energy equals the Fermi energy a fraction of the positive Si-donor ions are neutralized. By choosing to neutralize always those centers with lowest electronic Coulomb energy, which is linked with irregularly positioned close pairs or other configurations, one ends up with a structure which appears partially ordered.

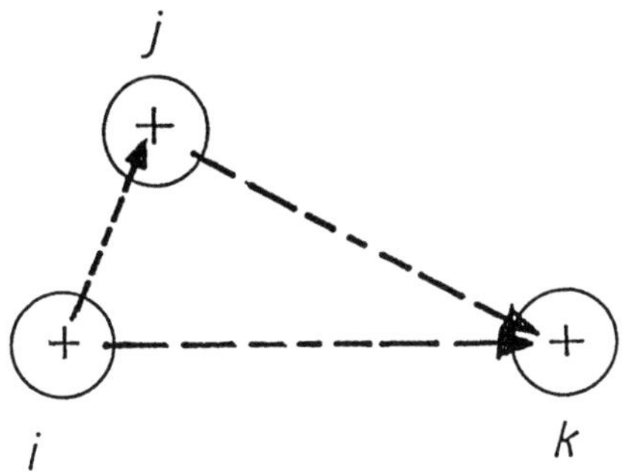

Fig.7 Schirmacher's triangle. An electron tunneling directly from **i** to **k** will interfere with the tunneling amplitude reaching **k** via the intermediate site **j**.

The same principle applies for electronic passivation of donors with atomic hydrogen. For experiments with δ-layer doped Si it has been found that the resistance of the layer first decreases when hydrogen from a plasma source is introduced /8/. The effect observed in these experiments is too strong to be caused solely by a partial ordering of the donor ions, but this could be one contribution.

Magnetic field experiments are suited to detect coherence in transport phenomena because of the influence of the field on the phase. In particular, whenever the coherent motion of the particle generates an area through which magnetic flux penetrates, there is expected to be an important effect. An example of this is the transport via hopping in a sheet of donor impurities. The magnetoresistance has been studied experimentally in refs./9,10/.

The theory in refs./11,12/ describes how for hopping transport in the variable range limit there occurs destructive interference. As shown in Fig.7, an electron starting at site **i** may tunnel directly to final site **k**. It may also do so by coherently tunneling first through site **j** and then appear at **k**. Doing so it has experienced a phase shift which on the average leads to destructive interference. The tunneling resistance is increased. The area generated by the alternate paths is a triangle. Its value is the area per impurity a^2. It is argued that a perpendicular field that satisfies $a^2B \sim \phi_o$, where ϕ_o is the flux quantum, destroys the interference. The result is a negative magnetoresistance such as observed in a number of systems where planar hopping transport occurs /10,13,14/.

The interference effect can account for at most a 50% decrease of resistance. It is now certain that the large effects of the field that are generally found have yet another origin. It has been suggested /13,15/ that the B-field also induces a lowering of resistance in variable range hopping, because it increases the density of states at the Fermi energy. The arguement considers that wave-function shrinkage reduces the overlap and thus lowers the impurity band width. As the impurity levels coalesce energetically it becomes easier to reach a given site. This effect is counteracted by the exponentially increasing resistance that comes from the tunneling matrix element. The complex

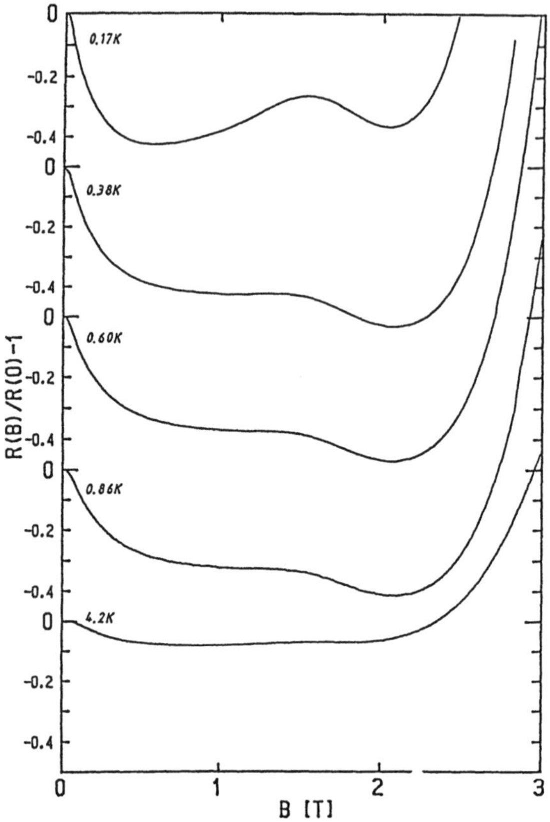

Fig.8 Magnetoresistance curves observed for a doping layer in GaAs ($N = 0.8 \times 10^{11}$ cm^{-2}, $r = 2.7$) for different temperatures. The negative resistance charge has contributions from the interference effect that exists because of the triangular paths in Fig.7.

interrelationship of all these mechanisms needs to be explored in further work. Nevertheless, it is clear that a magnetic field is a sensitive probe of coherency aspects in planar transport. Fig.8 shows an example of the nonmonotonic resistance variations caused by an perpendicular magnetic field. In this data the exponentially strong increase in resistance above ~2.5 T is well understood. The negative effect and its variation with field remain to be described in detail.

Lateral ordering in a planar system of dopants is a challenge to growth technology. To achieve a perfectly ordered array of dopant atoms may not be an easy task, but it should also not be impossible. Following the lead provided by the growth of AlAs wires on a slightly tilted GaAs surface, we question if a similar approach could achieve a system of impurities aligned in rows. Such "graphoepitaxy" requires freezing the dopant atoms in position after a diffusion step. Realizing that doping in a given atomic plane involves densities of 1-10 atomic percent, it is not unreasonable to look for interactions, Coulomb forces or even substrate-mediated influences, that would lead to a self-organizing system. Why should not impurity atoms at sufficient density on a substrate form a lattice that can be frozen in place? This would be the case in the best of all worlds, but even partial order as in a liquid would greatly modify the transport scattering. It's something to think about. Observing the effects related to transport and magnetotransport in ordered dopant structures would be a great thrill.

Acknowledgements

A number of students have contributed to the ideas and experimental results that are contained in this article. Recently I have much enjoyed the interaction with theory colleagues Shklovskij, Raikh and Schirmacher.

References

/1/ A. Zrenner, H. Reisinger, F. Koch and K. Ploog, Proc. 17th Int. Conf. Phys. Semicond. (San Francisco 1984), ed. J.D. Chadi and W.A. Harrison, Springer Berlin 1985, p.325

/2/ A. Zrenner, H. Reisinger, F. Koch, K. Ploog and J.C. Maan, Phys. Rev. B33, 5607 (1986)

/3/ A. Zrenner and F. Koch, Surface Sci. 196, 671 (1988)

/4/ A. Zrenner, F. Koch, J. Leotin, M. Goiran and K. Ploog, Semicond. Sci. Technol. 3, 1132 (1988)

/5/ H. Sigg, K. Ploog, Qui-yi Ye and F. Koch, Phys. Rev. Lett. 64, 1951 (1990)

/6/ A. Zrenner, F. Koch, R. Williams, R.A. Stradling, K. Ploog and G. Weimann, Semicond. Sci. Technol. 3, 1203 (1988)

/7/ W. Dobrowolski, K. Dybko, C. Skierbiszweski, T. Suski, E. Litwin-Staszweska, S. Miotkowska, J. Kossut and A. Mycielski, Proc. 19th Int. Conf. Phys. of Semicond. (Warszawa 1988), ed. W. Zawadzki, Inst. of Physics 1988, p. 1247

/8/ R. Hollenbach, Diplomathesis, Tech. Univ. München (1988) and to be publ.

/9/ Qiu-yi Ye, A. Zrenner, F. Koch and K. Ploog, Semicond. Sci. Technol. 4, 16 (1989)

/10/ Qiu-yi Ye, B.I. Shklovskii, A. Zrenner, F. Koch and K. Ploog, Phys. Rev. B41, 8477 (1990)

/11/ V.L. Nguyen, B.Z. Spivak and B.I. Shklovskii, Zh. Eksp. Teor. Fiz. 89, 1777 (1985)

/12/ W. Schirmacher, Phys. Rev. B41, 2461 (1990)

/13/ Yi-ben Xia, E. Bangert and G. Landwehr, Phys. Status Solidi, B144, 601 (1987)

/14/ H. Sigg, K. v. Klitzing, M. Hauser and K. Ploog, Inst. Phys. Conf. Ser. 95, 11 (1988)

/15/ M. Raikh, Sol. State Commun. (to be publ.)

High Magnetic Field Transport in II–VI Heterostructures

R.N. Bicknell-Tassius, S. Scholl, C. Becker, and G. Landwehr

Physikalisches Institut der Universität Würzburg,
W-8700 Würzburg, Fed. Rep. of Germany

In the present work we report the results of magneto-transport measurements on some Hg–based II–VI semiconductor epitaxial layers grown by molecular beam epitaxy. The transport measurement were carried out at temperatures in the range 0.4 - 4.2 K in magnetic fields up to 10.0 T. Further, we point out the necessity of using multicarrier models for data interpretation and show finally some Shubnikov-de-Haas results on samples with high mobility carriers.

1 Introduction

The family of II–VI semiconductors includes materials with a wide range of properties (i.e. band gap, effective mass, etc.). On the one hand there are the Hg-based materials, which are semimetals, and on the other hand there the Zn-based materials with room temperature band gaps as large as 3.8 eV. Due to this large range of band gaps, the II–VI compound semiconductors are quite interesting for potential commercial applications and fundamental research. The use molecular beam epitaxy (MBE)to grow epitaxial layers of these materials has been recently widely employed in an effort to improve the the properties of these materials. Due to the low growth temperature during the MBE growth process, it is hoped that better control of the electrical properties can be obtained. Earlier growth techniques, which typically used higher growth temperatures, have been frustrated due to "self–compensation" effects. This has meant that reliable substitutional doping these materials has not been readibly achievable. Much of the recent II–VI MBE effort has been focussed on the growth of the narrow gap material HgCdTe. This is due to the fact that HgCdTe is presently the material of choice in the fabrication of infrared focal–plane arrays.

Van der Pauw Hall transport measurements are often employed as an initial characterization technique on epitaxial samples, because these measurements can be done quite quickly and easily. The quality of the sample usually is then characterized by the measured mobility and carrier concentration at low temperatures.

In the past also Shubnikov–de–Haasoscillationshave been reported for II–VI samples [1]. We have observed Shubnikov–de–Haas oscillations on several of our samples, allowing us a deeper look at the properties of those samples.

Springer Series in Solid-State Sciences, Vol. 101
High Magnetic Fields in Semiconductor Physics III Editor: G. Landwehr

2 Single carrier model failure and mobility-conductivity analysis

In recent years most transport data of MBE grown II–VI–semiconductor layers and heterostructures was interpreted using a classical one–carrier model. As is commonly known the transport phenomena are described in terms of a carrier concentration n and usually a Hall mobility μ when using a single carrier model. These two parameters are obtained from the measured values of the specific resistivity $\rho_{xx}(B=0)$ and of the Hall coefficient $R_H(B)$ at one given value of B using the formulae

$$n = -\frac{1}{eR_H(B)} \tag{1}$$

$$\mu = -\frac{R_H(B)}{\rho_{xx}(0)}. \tag{2}$$

The results are normally plotted in the form of carrier concentration or mobility as a function of temperature (e.g. [7]). In the valid range of the single–carrier model n and μ should be independent of magnetic field. If this evaluation is done for different values of magnetic field the calculated parameters of most mercury–containing samples appear to be a function of field, in some samples one or even two changes from n-type to p-type or vice versa occur (see fig. 1).

This clearly shows that such a simple calculation method should not be used for those samples. For bulk samples of HgTe, multiple–carrier models using up to four different carrier types at low temperatures have been employed[4, 5, 6], where the parameters were determined by fitting. As an alternative method we use the mobility-conductivity-analysis instead of a multi–carrier fit [3].

The mobility-conductivity-analysis, which was described originally by Beck [2], uses a classical model based on a solution of the Boltzmann equation that allows conduction in multiple bands, energy dependent relaxation times and nonparabolic bands. Within this model the components of the conductivity tensor are expressed in terms of a conductivity density $s(\mu)$ in the following way

$$\sigma_{xx} = \int_{-\infty}^{\infty} \frac{s(\mu)d\mu}{1+\mu^2B^2} \tag{3}$$

$$\sigma_{xy} = \int_{-\infty}^{\infty} \frac{\mu B s(\mu)d\mu}{1+\mu^2B^2}. \tag{4}$$

In principle the conductivty density now can be calculated by reversing these transformations. Practically this reversing can be done only using a small (in our case 7) number of field values due to numerical diffculties and due to the fact that σ_{xx} and σ_{xy} are known only at a finite number of magnetic field

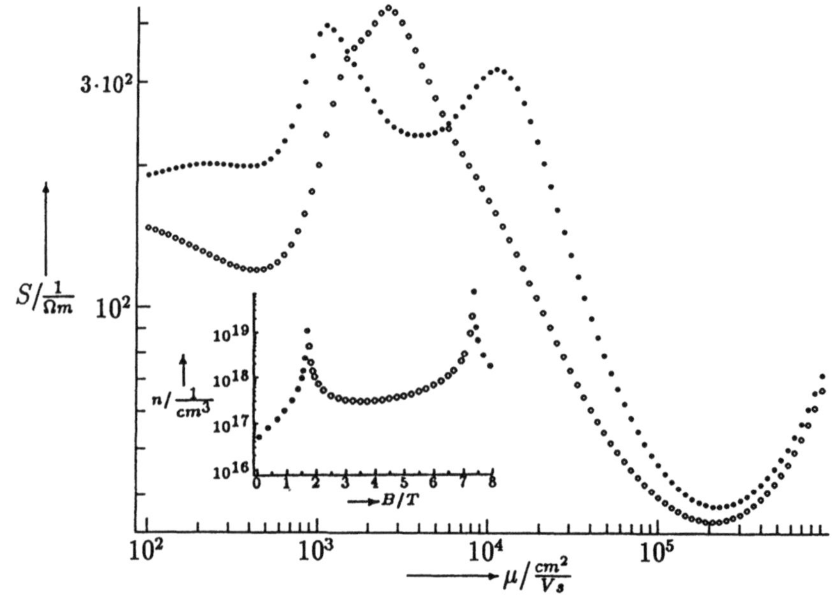

Figure 1: Mobility-conductivity-spectrum of a HgTe sample at 0.4 K, o holes, • electrons. Inset shows the same data evaluated in the single–carrier model (o p-type, • n-type).

values with a certain measurement error. As a consequence the calculation yields in general only an envelope $S(\mu)$, which can be interpreted as the maximum conductivity density at a given mobility that does not contradict the measured data. This envelope is still useful, because there are maxima at mobilities where real carriers in terms of the classical model exist. If quantum mechanical effects like magnetic freezeout occur, there may be artifact carrier types in the spectrum, also the data must be taken from field regions where no significant oscillations are visible. Therefore and also because of the possible effects of sample inhomogenities on van der Pauw measurements a cautious interpretation is required. Figure 1 shows the conductivity-mobility-spectrum for the sample with the two type changes. As you expect from the two changes, there are three types of carriers present.

3 Single layer samples showing Shubnikov–de–Haas oscillations

As already mentioned one parameter that is typically used to optimize the growth process is the mobility of the carriers present at low temperatures. To demonstrate that it is not always safe to regard the presence of high mobility carriers as a quality criterium we show the results on a $4\mu m$ thick

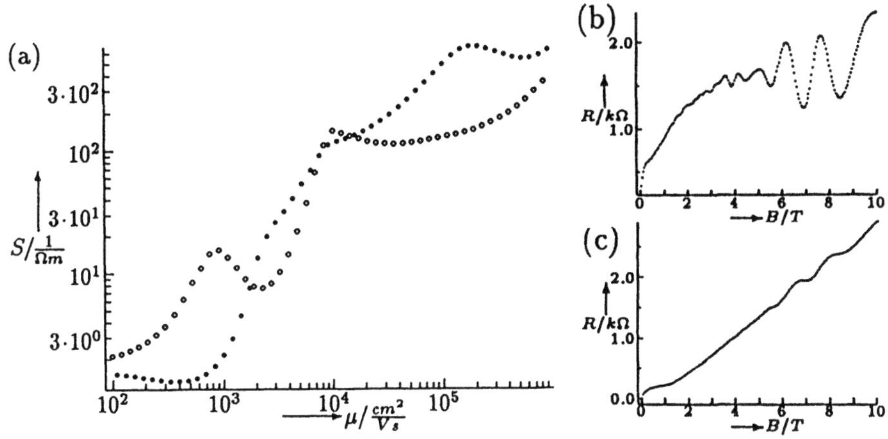

Figure 2: (a) Mobility-conductivity spectrum of a high mobility single layer HgCdTe sample (van der Pauw configuration), T = 0.4 K, o holes, • electrons, the maxima close to $10^4 cm^2/Vs$ are due to magnetic freezeout.
(b) SdH oscillations measured on a Hall bar made from this sample, T = 0.4 K.
(c) Hall effect of this Hall bar, T = 0.4 K.

$Hg_{0.88}Cd_{0.12}Te$ layer. We have obtained similar results on HgTe epilayers. The mobility–conductivity analysis of measurements in van der Pauw geometry shows n–type carriers with a mobility high enough to expect Shubnikov–de–Haas oscillations (fig.2a).

Because the van der Pauw measurements show no evidence of oscillations a Hall bar was made from the sample. The results on this Hall bar show strong oscillations and some structure in the Hall effect that looks like Hall plateaus (fig. 2b,c). This indicates that van der Pauw measurements are useful to get some qualitative results, but for a more detailed characterization this method should not be employed.

The magnetoresistance at 45° tilt angle between sample and magnetic field showed that all found periods changed approximately in the way one would expect for two–dimensional carriers.

For a more detailed analysis of the oscillation structure the second derivative of the resistance with respect to the field was recorded by field modulation (inset of fig. 3). The maxima in the second derivative were numbered with integers and plotted against inverse field. This plot shows three sections where the points are on a line (fig. 3).

From these results we conclude that the carriers responsible for the oscillation are two–dimensional carriers and that three subbands are occupied. The properties which can be estimated are listed in table 1. The concentrations listed there are not individual subband occupations at zero field, but the number of carriers responsible for each part of the oscillation. The step-

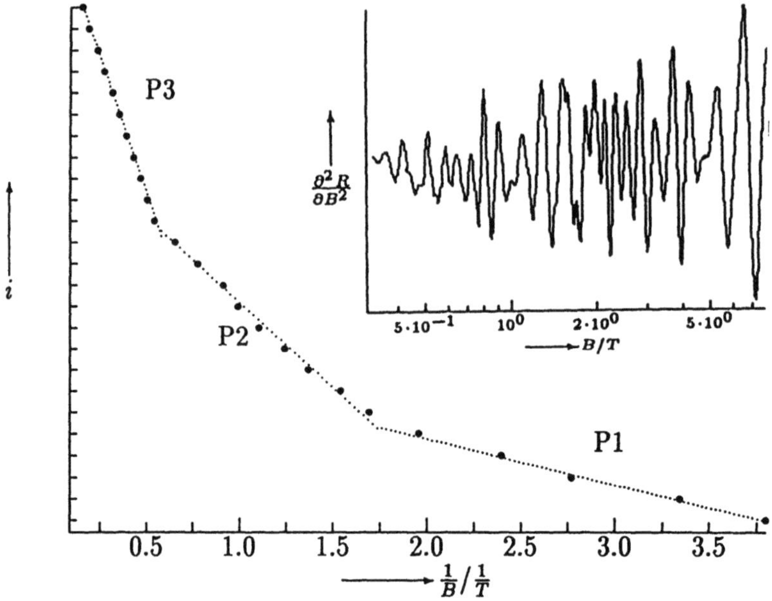

Figure 3: Number of oscillation maxima of the HgCdTe sample against inverse field, inset shows oscillation pattern in arbitrary units as functions of field, T = 1.5 K

Table 1: Parameters of the Shubnikov-de-Haas oscillations, the values preceeded by "≈" could be estimated only due to noise

Nummer	$P/\frac{1}{T}$	$\frac{m^*}{m_0}$	T_D/K	$n_{2D}/\frac{1}{cm^2}$
1	$(4.5 \pm 0.7)\cdot 10^{-1}$	≈ 0.02	≈ 2	$(1.1 \pm 0.2)\cdot 10^{11}$
2	$(1.3 \pm 0.3)\cdot 10^{-1}$	≈ 0.02	≈ 2	$(3.7 \pm 0.7)\cdot 10^{11}$
3	$(3.8 \pm 0.6)\cdot 10^{-2}$	0.3 ± 0.2	—	$(1.3 \pm 0.2)\cdot 10^{12}$

like structures in the Hall effect seem to be quantized Hall effect steps, but the measured resistance does not agree with the expected quantized values. This is probably due to conduction by additional nonoscillating low mobility carriers. A mobility-conductivity analysis performed on the low field part of these measurements where no oscillations are visible ($B < 1$ T), yields concentrations of the high mobility carriers that agree within the uncertainty of the evaluation with the Shubnikov-de-Haas results. Whether the carriers reside close to the surface of the sample or near the growth interface to the CdTe buffer will be a matter of further investigation of our samples. From other groups there are results which indicate that those carriers reside close to the buffer layer [13].

4 Superlattice with hole mobilities close to $10^5 cm^2/Vs$

In most II-VI samples carriers with the highest mobility are electrons. In one of our $HgTe/Hg_{0.11}Cd_{0.89}Te$ superlattices the mobility-conductivity analysis of the van der Pauw data (see fig. 4a) shows an enormously high hole mobility. Subsequent Hall bar measurements confirmed this high mobility, but no evidence for Shubnikov-de-Haas oscillations was found in the magnetoresistance. The second derivative of the magnetoresistance (fig. 4b) shows nice oscillations. By tilted field measurements the oscillating carriers were found to be two–dimensional. Because the oscillation pattern looks as if only one occupied subband is involved the data was fitted to the oscillation equation [1]

$$\frac{\Delta\rho}{\rho_0} = \sum_{R=1}^{\infty} \frac{5}{2} \left(\frac{RP}{2B}\right)^{1/2} \frac{\beta T m' \cos(R\pi m' g^*/2)}{\sinh(R\beta T m'/B)} e^{-R\beta T_D m'/B} \cos\left[2\pi\left(\frac{R}{PB} - \frac{1}{8} - R\gamma\right)\right], \tag{5}$$

but additionally the effect of the field modulation was included. This fit allowed the parameters listed in table 2 to be determined.

Table 2: Shubnikov-de-Haas parameters of the superlattice with high mobility holes

$P/\frac{1}{T}$	$\frac{m^*}{m_0}$	T_D/K	g^*	γ
0.127 ± 0.020	0.03 ± 0.01	12 ± 3	25 ± 5	0.18 ± 0.05

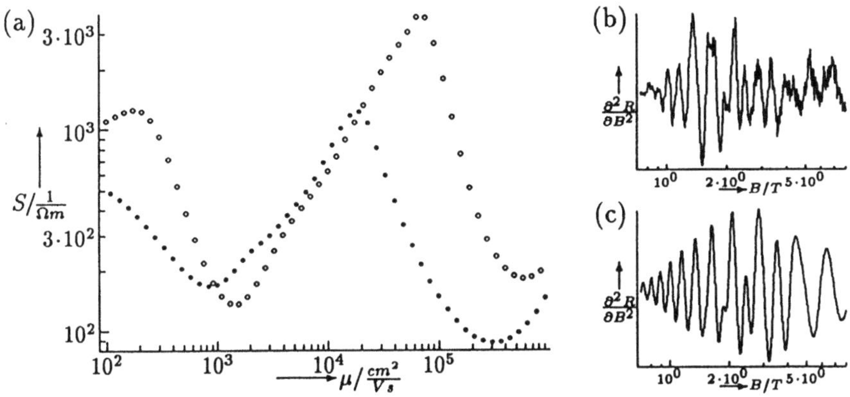

Figure 4: (a) Mobility-conductivity spectrum of superlattice with high mobility holes (van der Pauw method), T = 0.4 K.
(b) Second derivative of the magnetoresistance showing SdH oscillations, T = 1.5 K.
(c) Numeric simulation of the oscillation,
(b) and (c) show second derivative in arbitrary units versus field.

Out of these parameters the expected form of the oscillation has been calculated (fig. 4c), at least in the lower field region the agreement is quite good. In fields higher than 2 T there is some difference, probably due to the band structure.

5 Conclusion

We were able to demonstrate that a single carrier evaluation of transport data should not be used for semimetallic Hg–based samples and a multi–carrier–type model like the mobility–conductivity–analysis should be used instead. Moreover, we showed that detailed Hall bar measurements are essential for correct understanding of the conduction in Hg–based MBE-samples at low temperatures especially if the samples seem to contain high mobility carriers.

6 ACKNOWLEDGEMENTS

This project was supported by the Bundesministerium für Forschung und Technologie (Bonn).

References

[1] D.G. Seiler, A.E. Stephens *The Shubnikov-de-Haas Effect in Semiconductors : A Comprehensive Review of Experimental Results* to be published in Modern Problems in Condensed Matter Science : Landau Level Spectroscopy, edited by G. Landwehr and E.I. Rashba

[2] W.A. Beck, J.R. Anderson, J. Appl. Phys. **62**, 541 (1987)

[3] W.A. Beck, F. Crowne, J.R. Anderson, M. Gorska, Z. Dziuba, J. Vac. Sci. Technol. **A 6**, 2772 (1988)

[4] Z. Dziuba, Phys. stat. sol. (b) **117**, 531 (1983)

[5] Z. Dziuba, Phys. stat. sol. (b) **118**, 319 (1983)

[6] J.R. Meyer, C.A. Hoffman, F.J. Bartoli, J.M. Perez, J.E. Furneaux, R.J. Wagner, R.J. Koestner, M.W. Goodwin, J. Vac. Sci. Technol **A 6**, 2775 (1988)

[7] K.A. Harris, S. Hwang, Y. Lansari, J.W. Cook Jr., J.F. Schetzina, M. Chu, J. Vac. Sci. Technol. **A 5**, 3085 (1987)

[8] J.N. Schulman, Yia-Chung Chang, J. Vac. Sci. Technol. **A 4**, 2114 (1986)

[9] M.A. Berding, S. Krishnamurthy, A. Sher, A.-B. Chen, J. Vac. Sci. Technol. **A 5**, 3014 (1987)

[10] N.P. Ong, J.K. Moyle, J. Bajaj, J.T. Cheung, J. Vac. Sci. Technol. **A 5**, 3079 (1987)

[11] D.G. Seiler, G.B. Ward, R.J. Justice, R.J. Koestner, M.W. Goodwin, M.A. Kinch, J.R. Meyer, J. Appl. Phys. **66**, 303 (1989)

[12] C.A. Hoffman, J.R. Meyer, F.J. Bartoli, J.W. Han, J.W. Cook Jr., J.F. Schetzina, J.N. Schulman Phys. Rev. B **39**, 5208 (1989)

[13] P.R. Emtage, T.A. Temofonte, A.J. Noreika, C.F. Seiler, Appl. Phys. Lett. **54**, 2015 (1989)

Density of States of a Two Dimensional Electron Gas in High Magnetic Fields

S. Das Sarma

Center for Theoretical Physics and
Center for Superconductivity Research, Department of Physics,
University of Maryland, College Park, MD 20742, USA

Abstract. We develop a theory for the electronic density of states of a weakly disordered two dimensional electron gas in the presence of a strong external magnetic field oriented normal to the electron layer. The density of states is calculated using the self-consistent Born approximation for the electron-impurity scattering, retaining Landau level coupling in the theory. The electron-impurity scattering potential is calculated in a non-linear screening approximation where scattering and screening self-consistently determine each other. Screening is treated in the random-phase-approximation by retaining the bubble diagrams and the polarizability is obtained by solving the vertex function within the ladder approximation (which is consistent with the self-energy being treated in the single-site approximation). The resultant level broadening and the electronic density of states cannot easily be characterized by a single parameter such as the zero-field mobility which uniquely characterizes the usual short-range approximation extensively used in the literature. We find that the density of states calculated from this non-linear, self-consistent screening theory is, in general, much smoother and flatter, and the Landau level broadening much larger than that implied in the short-range approximation. The density of states depends on the actual impurity distribution (and, not just on the zero-field mobility) in the system and the level broadening and screening are oscillatory function of the chemical potential. We conclude that in many experimental situations the short-range approximation is even qualitatively wrong.

I. INTRODUCTION

When a strong external magnetic field (B) is applied normal to a two-dimensional electron gas (2DEG), the system becomes quantized into a series of Landau levels (denoted by the index N = 0, 1, 2, ... throughout this paper) whose energy E_N, in the non-interacting situation, is given by ($\hbar$ = 1 throughout)

$$E_N = (N+\tfrac{1}{2})\omega_c \,, \tag{1}$$

Springer Series in Solid-State Sciences, Vol. 101
High Magnetic Fields in Semiconductor Physics III Editor: G. Landwehr

where,

$$\omega_c = eB/mc \tag{2}$$

is the cyclotron frequency and m the electron (band) effective mass. Each Landau level, denoted by the discrete index N, is highly degenerate with a macroscopic degeneracy given by $(2\pi\ell^2)^{-1}$ per spin per unit area where $\ell = (c/eB)^{1/2}$ is the Landau radius or the magnetic length. Thus, for a given two-dimensional electron density N_s at T = 0, a certain number n of Landau levels will be occupied by electrons where n is the largest integer satisfying the inequality condition

$$N_s \leq (n/2\pi\ell^2) , \tag{3}$$

with the equality being valid when n levels are exactly filled and (n+1) th level completely empty. In general, at T = 0 the chemical potential or the Fermi energy (E_F) will be in the (n-1) th Landau level (n = 1, 2, 3, ...)

$$E_F = (n-\tfrac{1}{2})\omega_c , \tag{4}$$

except in the situations (of measure zero) whence n Landau levels are exactly full and, $E_F = n\omega_c$. Thus, in a constant magnetic field, the Fermi energy remains locked in individual Landau levels as N_s increases (decreases) and then jumps to the next higher (lower) Landau level as the next integer satisfies the inequality Eq. (3). The situation is similar when B is changed at a constant N_s -- E_F remains locked in an individual level and then jumps to the next lower (higher) Landau level as B is increased (decreased) beyond the next critical value determined by Eq. (3).

The density of states (DOS) of the non-interacting 2DEG in the presence of an external normal magnetic field is easily seen to be a series of delta functions at the quantized Landau energies

$$D(E) = (2\pi\ell^2)^{-1} \sum_{N=0} \delta(E-E_N) \tag{5}$$

where the delta function sharpness reflects the complete quantization of the system as indicated by Eq. (1). Throughout this work we assume the electrons to be spinless fermions, neglecting the electron spin degeneracy completely. (In a strong magnetic field, the spin degeneracy is lifted due to Zeeman splitting which we are assuming to be very large.) Note that the delta-function singular DOS would imply rather singular thermodynamic properties of a 2DEG in the presence of an external magnetic field.

The above rather simple-minded and well-known theoretical picture for a 2DEG in the presence of an external magnetic field does not hold for real 2DEG systems where all indications are that the DOS is a smooth function of energy (and, E_F certainly moves continuously through the Landau levels as quantum Hall effect experiments definitively demonstrate). The DOS of a real 2DEG, while being a weakly oscillatory function of energy, does not have the strong delta function singularities of the non-interacting theory. For example, measured thermodynamic properties of a 2DEG in the presence of a strong external magnetic field such as the magnetic susceptibility and the specific heat are rather weak oscillatory functions of B (or, N_s) implying a smooth (and, small) variation in $D(E_F)$ as a function of E_F. Model calculations based on experimentally measured specific heat or magnetic susceptibility show the DOS to be much smoother and broader than one would intuitively expect.

In real systems, one expects disorder arising from random impurities in the system to have a smoothening effect on the singular D(E) given by Eq. (5). In particular, scattering by the random impurities should broaden the DOS D(E) with a characteristic broadening parameter $\Gamma_N(E)$ determined by the strength of the impurity scattering potential associated with the random disorder. In the weak disorder case, one expects standard ensemble-averaged diagrammatic perturbation theory to be valid in calculating the broadened DOS. Such a calculation was carried out a long time ago by Ando and Uemura[1] for a model of randomly distributed "short-range" (actually, zero-range) scatterers, neglecting the coupling between Landau levels. Within the self-consistent Born approximation (SCBA) and assuming that the scattering potential associated with the electron-impurity interaction is a point-like delta function in real space, Ando and Uemura[1] obtained the following expression for the DOS in the strong-field limit:

$$D(E) = (2\pi\ell^2)^{-1} \sum_{N=0} (\pi\Gamma)^{-1} \left\{1-\left[\frac{E-E_N}{\Gamma}\right]^2\right\}^{1/2} , \qquad (6)$$

where the constant level broadening Γ which, in this short-range approximation, is independent of the Landau-level index N (and, energy E) depends only on the strength of the impurity scattering potential and is, therefore, completely determined by the <u>zero-field</u> electronic mobility (μ_o) of the system:

$$\Gamma = (2\omega_c/\pi\tau)^{1/2} \equiv \Gamma_{SR} , \qquad (7)$$

where the scattering time τ is extracted from the mobility (μ_o) of the system at zero magnetic field:

$$\tau = m\mu_o/e . \qquad (8)$$

We shall refer to the results (6) - (8) as the short-range approximation,[1] denoted by the subscript SR in Γ_{SR} given in Eq. (7). The approximations made in deriving these equations are many: (1) Weak disorder, so that self-consistent Born approximation (SCBA) involving only single impurity scattering is adequate; (2) the strong field limit so that there is no coupling between Landau levels which implies that one must have $\omega_c >> \Gamma$ with the overlap between Landau levels negligibly small; (3) zero-range electron-impurity interaction which can be characterized by a delta-function point scattering potential in real space.

The work presented in this paper goes beyond the above set of approximations and develops a more realistic theory of strong field DOS in a 2DEG. (The Feynman diagrams for our theory are shown in Fig. 1a-1d.) In particular, we relax approximations (2) and (3) listed above while still maintaining (1). Thus, we work within the SCBA assuming the impurity disorder to be weak so that the single-site scattering approximation (Fig. 1a and 1b) is adequate. But we take into account the fact that the impurities in a real 2DEG are not point delta function scatterers - they are actually charged impurities which interact with an electron via the (screened) long-range Coulomb interaction (Fig. 1c and 1d). We also include Landau level coupling in our theory so that the strong-field ($\omega_c >> \Gamma$) restriction is relaxed.

In fact, it is easy to see that in a strong magnetic field scattering and screening must be obtained self-consistently where each determines (and, is determined by) the other. The level broadening Γ is obviously determined by the (screened electron-impurity) scattering potential. But, electronic screening itself is determined by the DOS which, in turn, is determined by Γ. Thus, screening and the scattering potential mutually determine each other and must be obtained self-consistently. One particularly spectacular aspect of this self-consistency is its strong intrinsic dependence on the position of the chemical potential E_F. For example, when E_F is at the middle of a broadened Landau level, $E_F \approx (N+\frac{1}{2})\hbar\omega_c$, the system is like a "metal" because there are empty available electronic states at E_F for the electrons to make transitions to. This system obviously screens very strongly, making the effective electron-impurity interaction weak and short-ranged, giving rise to a small Γ. On the other hand, when E_F is near the edges of a Landau level (and, the magnetic field is large enough so that there is an energy gap in the DOS between the neighboring levels), the system is like an "insulator" with no free electronic states available infinitesimally above the Fermi level. Screening is, consequently, weak and the effective electron-impurity interaction is long-ranged and strong, giving rise to a large Γ. Thus, $\Gamma(E_F)$ (and, therefore the DOS and screening) show strong oscillatory behavior as a function of the Fermi level position. There have been many experimental verifications of this oscillatory behavior of level broadening, DOS and screening as a function of E_F in a 2DEG under a strong external magnetic field. Clearly, self-consistency plays a strong role in this oscillatory behavior. In real systems, any possible Landau level overlap substantially modifies the screening and the non-linear screening-scattering self-consistency becomes more complicated in the presence of Landau level overlap.

We should emphasize that the self-consistent screening effect makes the short-range impurity scattering model a bad approximation in the strong-field situation. While the impurity potential is short-ranged when the Fermi level is at the center of the Landau level, it is long-ranged when the Fermi level lies at the Landau level edges. This suggests that disorder, in the presence of a magnetic field, cannot be characterized by the zero-field mobility. In fact, the qualitative discussion given above indicates that the strong-field broadening should, in general, be larger than the corresponding short-range result. This is precisely the experimental observation - thermodynamic measurements using three different kinds of experimental techniques, namely, the de Haas-van Alphen magnetization studies, the specific heat and the magneto-capacitance measurements all indicate the DOS to be much broader than the short-range result. The general experimental consensus has been that the strong-field DOS of a 2DEG is, in general, much broader and flatter than that implied by the short-range SCBA-based scattering theoretic model with substantially more (than that given by the short-range theory) DOS in between the Landau levels. We should mention that there have been earlier attempts[4] at the self-consistent screening theory of strong-field level-broadening along the same line being discussed here, but these attempts did not include actual calculations of the complete DOS and left out the Landau level coupling effect.

II. THEORY

In this review we omit the theoretical details which can be found in the literature.[2,3] The basic idea is to treat the elastic scattering potential not as an arbitrary model potential (as is done in the short-range approximation) but as a bare Coulomb potential which needs to be screened by the polarizable 2DEG. We emphasize that as a theory there is nothing wrong with the short-range approximation, it just so happens that it is not a good model for the modulation-doped 2DEG occuring in high-mobility GaAs heterostructures. The impurity scattering associated with disorder in high mobility GaAs heterostructures is known to arise from charged Coulomb centers. Even in an ideal ultra-pure GaAs heterostructure, the dopants (on the $Al_xGa_{1-x}As$ side) that produce the 2DEG via modulation doping are positively charged and scatter the electrons elastically (albeit rather weakly, because they are positioned rather far away from the 2DEG). It is well-known,[5] that in the absence of any external magnetic field, the screened impurity potential due to these distant

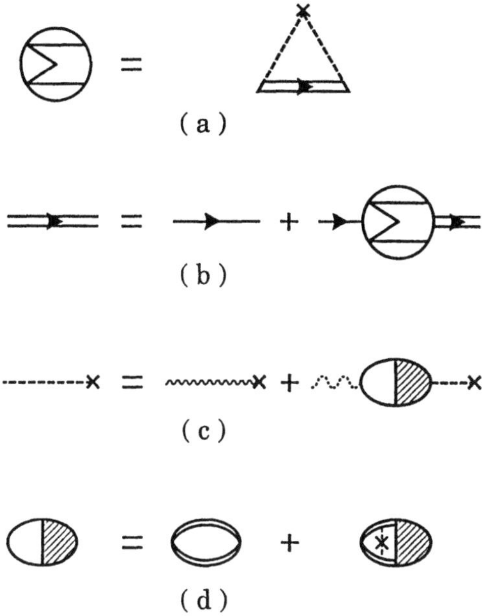

Fig. 1: (a) - (d) Show the Feynman diagrams used in our theory for the electron- impurity scattering (a and b) and screening (c and d) respectively.

dopants is long-ranged and most of the electronic scattering is forward scattering which contributes to level broadening, but not to electrical resistance. Thus, even though the mobility is greatly enhanced by modulation doping, the level broadening (which depends on the single-particle scattering time τ_s, rather than the transport relaxation time τ_t) may not equivalently be reduced. In fact, in high-mobility modulation-doped heterostructures, it is now well accepted[5] that $\tau_t >> \tau_s$, because the long-range impurity potential due to the distant dopants contribute mostly to τ_s^{-1} and <u>not</u> to τ_t^{-1}. Clearly, in such a situation the short-range approximation is not only quantitatively poor but is a qualitatively inadequate model. The self-consistent screening theory addresses this particular issue in high magnetic fields by calculating level broadening, screening and the DOS within one non-linear self-consistent scheme as shown in Fig. 1.

The model used for our calculation is that of an ideal (i.e. zero thickness) 2DEG of 2D electron density N_s interacting at T = 0 with a random, planar distribution of charged Coulomb impurity centers of 2D charge density N_i, with a length parameter 'a' denoting the separation between the planes of confinement of the 2DEG and the random impurities. Thus, our model is completely defined by three independent parameters N_s, N_i, and a. (We note that, even though for the sake of simplicity, we take these parameters as independent, in reality they are not completely independent because of the charge neutrality and Poisson's equation requirements.) More complicated models involving three dimensional impurity distribution and quasi-two dimensionality of the electron layer can easily be incorporated in the calculation, but we refrain from doing that because this requires more unknown parameters with no particular improvement in our physical understanding of the problem. Note that our minimal model (with three parameters N_s, N_i, and a) is already substantially more complicated than the short-range approximation which involves only one independent parameter, namely, the zero-field mobility of the system. All the parameters in our calculation are chosen appropriate for GaAs-based 2DEG, but the theory is, of course, of general validity to any modulation-doped system.

In Fig. 1(a)-(d) we show the many-body Feynman diagrams used in our self-consistent electron-impurity scattering calculation. As noted earlier, scattering depends on screening (cf. Fig. 1(a)) through the scattering potential whereas screening (Fig. 1(c) and (d)) depends on scattering through the renormalized propagator via Dyson's equation (Fig. 1(b)) which determines the electronic dielectric function (Fig. 1(c)) through the RPA, and, the polarizability (Fig. 1(d)) function through the ladder vertex equation. We retain Landau level coupling in our calculation. In addition to the diagrams shown in Fig. 1(a)-(d), one must also calculate the chemical potential self-consistently by requiring that the total density of electrons equal N_s. We do not provide the detailed equations and formulae here,[2,3] except to note that the calculation is made quite complicated by the fact that one needs to solve self-consistently a set of two coupled integral equations (Figs. 1(a) and (d)) along with two coupled algebraic equations (Figs. 1(b) and (c)) in the presence of Landau level coupling. Depending on the situation (eg. the location of the chemical potential) the number of iterations required for self-consistency is between 50 and 500.

III. RESULTS AND DISCUSSIONS

From our discussions it is clear that the self-consistent DOS of a 2DEG in the presence of an external magnetic field is a complicated non-linear function of a number of variables - in addition to depending on the magnetic field B, D(E) depends on the 2D electron density N_s, the impurity density N_i and the effective spacer thickness or the electron-impurity separation a (note that if we relax our delta-doping model for the charged impurities, the impurity configuration would have to be described by more variables). Thus,

D(E) in our theory is an explicit function of four variables B, N_s, N_i, and a. The corresponding situation for the short-range theory is simpler with N_s, N_i, and a entering the DOS only through the zero-field mobility μ_o which is experimentally measured. The free-electron result is, of course, a series of delta functions located at the Landau energies, dependent only on the magnetic field. Since the short-range scattering result for the DOS is widely (in fact, almost exclusively) used, we will show our calculated self-consistent results along with the corresponding short-range results for the sake of comparison. As one would see, in many experimentally relevant situations the self-consistent DOS is strikingly different from the short-range result and, as emphasized in the introduction, in general, the self-consistent DOS is broader and flatter than the short-range result, being in closer agreement with the measured DOS. All our results are calculated for a 2DEG confined in a GaAs heterostructure with the impurity plane separated by a distance a from the 2D electron plane.

In Fig. 2 we show our calculated self-consistent DOS, D(E), with a = 50Å, μ_o = 80,000 cm^2/Vs and N_s = 2x10^{11}cm^{-2} for three values of the magnetic field B = 2.6(a), 3.4(b), and 4.4(c) Tesla. The self-consistent DOS is shown by the solid lines whereas the dashed lines give the short-range result and the Fermi level (i.e. the chemical potential) E_F is also shown. The self-consistent DOS is much broader than the short-range results, and, for B = 2.6T (Fig. 2(a)), where Landau level overlap is very significant, the self-consistent DOS, in sharp contrast to the oscillatory short-range result, is almost flat, being qualitatively similar to the zero-field non-interacting two-dimensional density of states which is a constant. At higher fields (Figs. 2(b) and (c)), the self-consistent DOS shows much weaker oscillations as a function of E than does the short-range result.

In general, the zero-field mobility is not a good parameter in characterizing the strong-field DOS. This is most clearly seen from Fig. 3 where we depict the calculated self-consistent DOS for μ_o = 400,000 cm^2/Vs, N_s = 2x10^{11}cm^{-2}, B = 3.4T with three different values of the electron-impurity separation a = 0 (dot-dash line), 100Å (heavy dash), and 150Å (solid) and compare it with the short-range result (light dash) which is uniquely determined by the zero-field mobility. In varying the spacer thickness parameter a (while keeping μ_o fixed) we obviously had to adjust the impurity concentration N_i appropriately. It is obvious from Fig. 3 that the self-consistent DOS is not uniquely determined by the zero-field mobility μ_o, but is a function of both a and N_i (or, more generally, of the actual impurity distribution in the sample). We emphasize that each curve in Fig. 3 represents the calculated DOS of a 2DEG with identical values of N_s

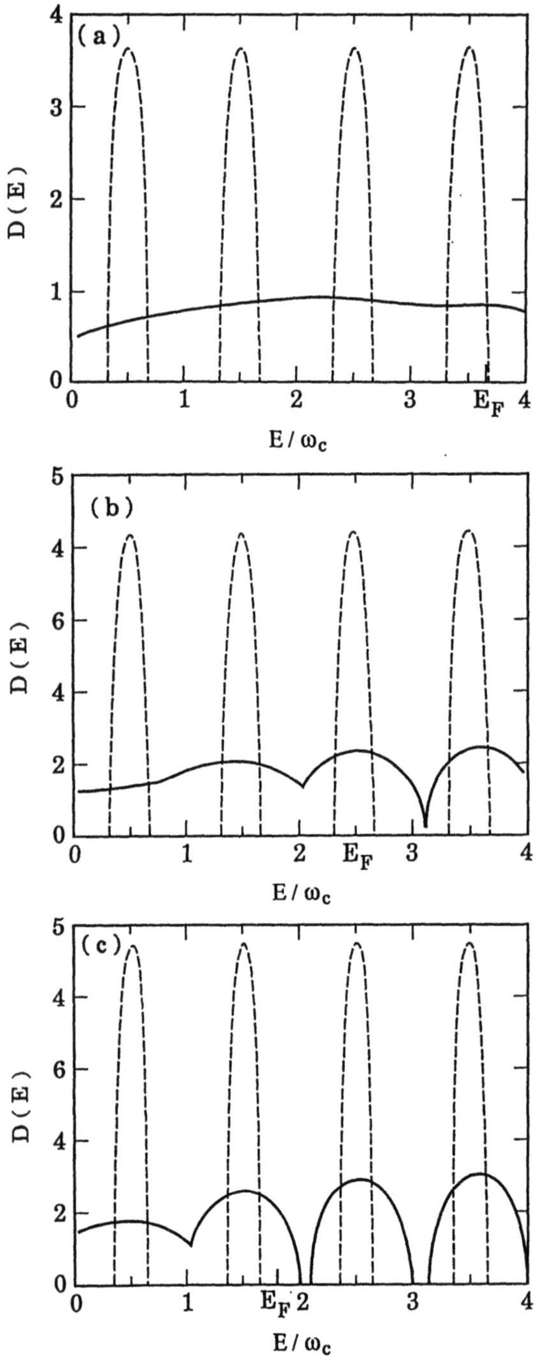

Fig. 2: Shows the DOS D(E) with a = 50Å, μ_o = 80,000cm^2/V.s, N_s = 2x10^{11}cm^{-2} for magnetic field values B = 2.6T (a), 3.4T (b), and 4.4T (c). Dashed lines are the results for the short-range model.

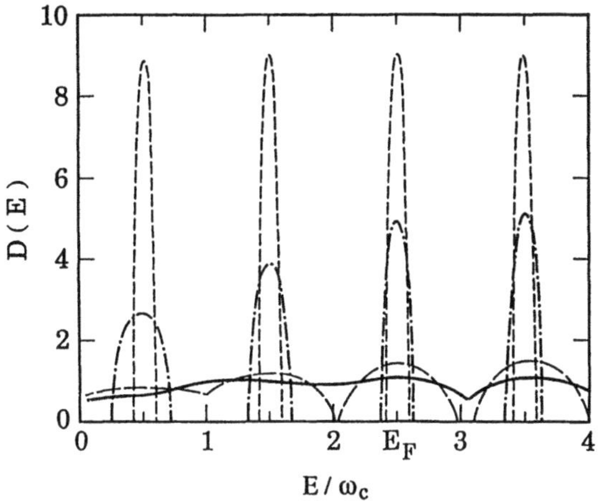

Fig. 3: Shows the DOS D(E) with fixed μ_o = 4000,000cm^{-2}/V.s, N_s = 2x10^{11}cm^{-2}, and B = 3.4T, but with variable a = 0 (dot-dashed line), 100Å (short dashed line) and 150Å (solid line). Short-range model results are shown by long dashed line.

and μ_o, but with different impurity distributions (i.e. N_i and a in our model).

From these results and from many other sets[3] of such results that we have (which are not being shown here), we form the following general conclusions about our calculated self-consistent DOS:

(1) In general, self-consistent D(E) could be very different from the short-range result -- in particular self-consistent DOS is broader and flatter than the short-range result.

(2) The zero-field mobility μ_o does not uniquely parameterize the self-consistent DOS which depends sensitively on the actual impurity distribution in a modulation-doped structure.

(3) For a fixed value of μ_o, the self-consistent DOS gets closer to the short-range result as the spacer thickness a (or, the effective electron-impurity separation) decreases (implying an increase of N_i).

(4) For a fixed value of the separation parameter a, the self-consistent DOS gets closer to the short-range result as the zero-field mobility μ_o increases (implying a decrease of N_i).

(5) For a fixed value of the impurity density N_i, the self-consistent DOS gets closer to the short-range result as a decreases (implying a decrease of μ_o).

(6) In any given situation, self-consistent D(E) becomes qualitatively similar to the short-range result for E > > E_F.

(7) In general, the self-consistent DOS is qualitatively closer to the short-range result for higher fields.

(8) Finally, knowing D(E) does not uniquely give $D(E_F) = D(E=E_F)$ because there is an implicit dependence of the DOS on the chemical potential (to be discussed below).

The above qualitative remarks about our calculated self-consistent DOS are all in agreement with the experimental conclusions from various thermodynamic (eg. magnetization, capacitance, and specific heat), transport and optical measurements. We should point out that a direct comparison between our self-consistent theory and experimental results is difficult because one does not, in general, know the detailed impurity distribution in an experimental sample, and, as emphasized above, the self-consistent DOS depends sensitively on the details of the impurity distribution, rather than on the mobility μ_o itself.

In comparing with experimental results one often needs the DOS at the Fermi energy (chemical potential), $D(E_F)$, rather than the full D(E). Normally, as, for example, in the short-range model, $D(E_F)$ is uniquely determined by D(E), and, is given by $D(E_F) \equiv D(E=E_F)$ from Eq. (6). But, in the self-consistent model there is an implicit dependence of the DOS on the chemical potential (in addition, to the explicit dependence through energy) because screening depends on where the Fermi level lies (i.e. whether E_F is at the middle of the Landau level leading to strong screening or whether it is at the edge of a Landau level leading to weak screening). Thus the self-consistent DOS, D(E), is a function of E_F through the non-linear screening, i.e. $D(E) \equiv D(E;E_F)$. This implicit dependence of the DOS on the actual position of the Fermi level is a novel feature of the screening-level broadening self-consistency being studied here. In the short-range theory knowing D(E) completely defines $D(E_F)$, but the same is not true in the self-consistent theory!

In Fig. 4 we show our calculated self-consistent $D(E_F) \equiv D(E=E_F;E_F)$ as a function of the applied magnetic field B for $N_s = 2\text{x}10^{11}\text{cm}^{-2}$; $\mu_o = 30{,}000\ \text{cm}^2/\text{V.sec.}$, and $a = 0\text{Å}$. The self-consistent result is shown by the solid line whereas the short-range result is shown as the dashed line.

In Fig. 5 we show our calculated DOS (Fig. 5(a)) and level broadening (Fig. 5(b)) as a function of the electron density N_s for fixed values of B = 4T, $a = 0\text{Å}$, $\mu_o = 30{,}000\ \text{cm}^2/\text{V.s.}$ In Fig. 5(a) the self-consistent and the short-range results are shown by the solid and the dashed lines respectively whereas in Fig. 5(b) the self-consistent results for various Landau levels are shown (the short-range Γ is a constant of value around $\Gamma/\omega_c \approx 0.1$).

Results shown above clearly demonstrate the main qualitative features of the self-consistent DOS vis-à-vis the widely used short-range result. In general, the self-consistent broadening is substantially (by factors of 1.5 to 100) larger than the short-range broadening, and, consequently, the self-consistent DOS is

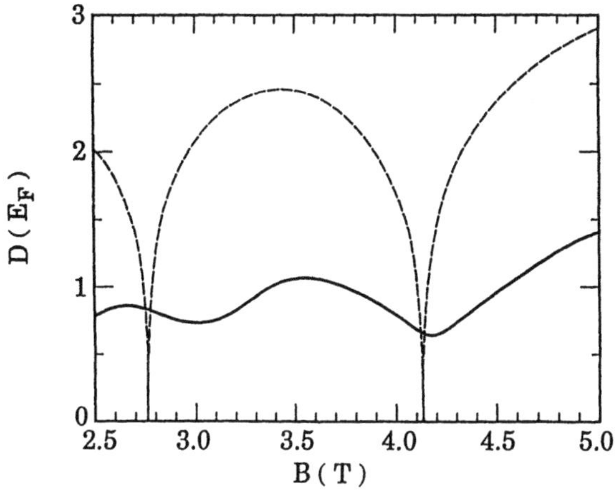

Fig. 4: Shows the DOS $D(E_F)$ as a function of magnetic field. The dashed line is short-range model (SRM) result.

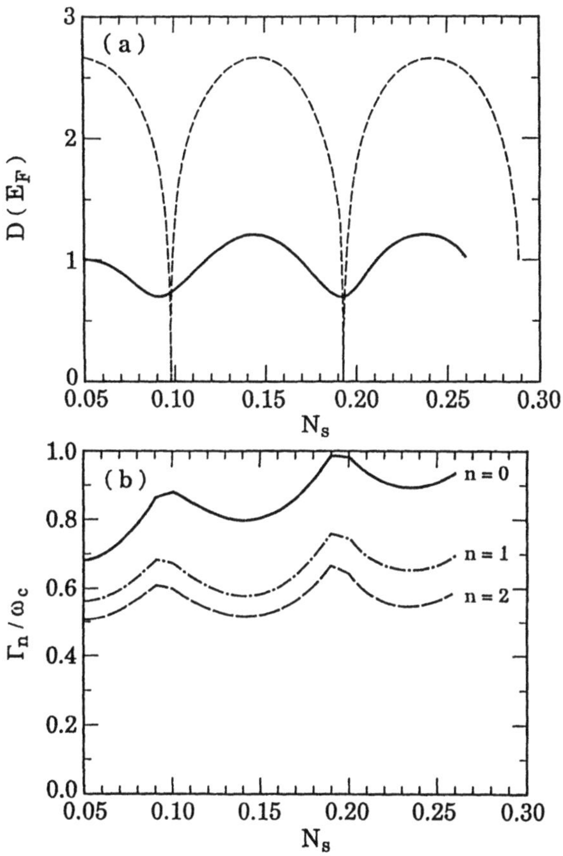

Fig. 5: Shows (a) the DOS $D(E_F)$, and, (b) the Landau level broadening Γ_n, as a function of the electron density N_s.

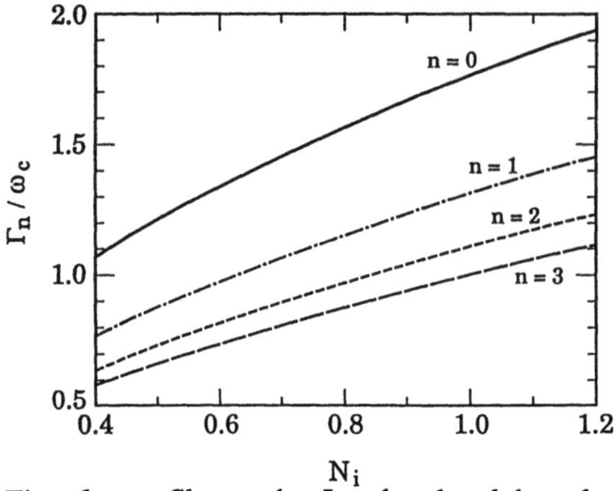

Fig. 6: Shows the Landau level broadening Γ_n as a function of the impurity density N_i at B = 3.5T (a), and, B = 4.2T (b).

substantially flatter than the short-range DOS. For example, the relative variation (i.e. the difference between the peaks and the valleys) in the DOS is about 10-40% in the self-consistent theory and is about 100-300% in the short-range theory. Many experimental papers in the last ten years have concluded that in modulation-doped GaAs heterostructures, the strong-field DOS is substantially flatter (and, the broadening significantly larger) than the results of the short-range theory. We believe that the self-consistent screening theory of the DOS as developed in this paper is a possible explanation for this discrepancy.

Finally, in Fig. 6 we show the calculated self-consistent Landau level broadening Γ_n for fixed values of a = 50Å and $N_s = 2\text{x}10^{11}\text{cm}^{-2}$ as a function of the impurity density N_i for two different values of the magnetic field: B = 3.5T (Fig. 6(a)) and 4.2T (Fig. 6(b)). In the short-range theory, $\Gamma \propto N_i^{1/2}$ because the zero-field mobility μ_o is inversely proportional to N_i (for fixed a and N_s) in the single-site approximation. In general, the self-consistent level-broadening does not have a simple power law dependence on the impurity density N_i because the self-consistent theory depends sensitively on a number of parameters such as E_F, a, N_s, and B. The dependence on N_i in Fig. 6 is, however, closer to a linear power law than a square root one.

The above representative results for the calculated DOS give one a good flavor of the qualitative features of the self-consistent theory. Obviously, there being a large number of parameters and variables in the problem (viz. N_s, N_i, a, μ_o, E_F, B, n, E) a single paper, no matter how comprehensive, cannot really provide an exhaustive set of results for all possible experimental situations. We have, therefore, emphasized here generic features of the self-consistent DOS, emphasizing, in particular, its differences with the simple (and, extensively used) short-range theory. The self-consistent theory is in much better agreement with the experimental results, not only quantitatively but also qualitatively. But, in some sense one pays a price for this experimental agreement - one cannot characterize the self-consistent DOS by a single suitable parameter in contrast to the short-range theory where zero field mobility completely and uniquely define the strong-field DOS. Thus, the simplicity of the short-range theory is lost and, in order to compare with experimental results, one must now know the actual impurity distribution and different impurity distributions having the same zero-field mobility may have strikingly different strong-field DOS (cf. Fig. 3). This dependence of the DOS and the level-broadening on the details of the impurity distribution and the inadequacy of μ_o as a unique parameter defining the strong-field DOS have been repeatedly discussed in the experimental literature.

We conclude this section by discussing the approximations, the limitations, and the shortcomings of the self-consistent theory. We emphasize that the theory manifestly assumes a weak scattering limit so that the single-site-approximation within the SCBA is adequate to calculate the electron-impurity self-energy. This approximation is totally uncritical and, within our theory, we have no way of improving upon the SCBA or, for that matter, even to estimate the error introduced by this approximation. Secondly, we neglect all effects of electron correlation (i.e. electron-electron interaction) beyond the random-phase-approximation (RPA) for screening (i.e. except for the Coulomb screening introduced by the polarization bubble diagrams). The use of RPA for screening (which, however, is calculated self-consistently in the electron-impurity interaction by keeping the SCBA self-energy diagrams and the ladder vertex corrections) is also a critical approximation of the theory that we do not know how to improve upon. Thus, both of these important approximations, namely the weak-disorder SCBA and the screening RPA, are essential approximations of our theory that cannot be relaxed. We also make some additional non-essential approximations such as assuming delta doping for the impurity configuration so that we can parametrize the impurity distribution by only two parameters the impurity density N_i and the electron-impurity separation a, and, assuming the electron layer to be perfectly two dimensional. We believe that these non-essential approximations have small effects on our calculated numerical results.

IV. CONCLUSIONS

In this paper we review a theory[2,3] for the density of states of a weakly disordered two dimensional electron gas in the presence of a strong external magnetic field by treating scattering and screening self-consistently. Consistent with many different experimental observations, our calculated DOS is, in general, much smoother and flatter than that implied by the short-range approximation where the transport relaxation time entering the zero-field mobility completely determines the Landau level broadening and the single-particle DOS. We find that, in addition to self-consistent screening effect (which is strongly dependent on the exact location of the chemical potential with respect to Landau level edges), the DOS may be strongly affected by Landau level overlap particularly at weaker fields. Our calculated DOS cannot be parametrized by any single parameter and, in particular, zero-field mobility is qualitatively inadequate in parametrizing the strong-field DOS -- we show (Fig. 3) that the same zero-field mobility could produce drastically different DOS depending on the exact impurity configuration. This is understandable since self-consistent screening depends on the details of the impurity configuration and different impurity configurations could produce the same zero-field mobility. In our model, the impurity configuration is parametrized by two parameters N_i and a, and, therefore, the DOS depends on N_i, a, N_s, and B. In real systems, one may need several more parameters to characterize the impurity distribution, consequently making the DOS dependent on more independent parameters. Thus, one unfortunate consequence of the self-consistent screening theory is the loss of simplicity of the short-range approximation where a single parameter (namely, the zero-field mobility) uniquely determines the strong-field DOS.

ACKNOWLEDGEMENT

The theory for the strong-field density of states reviewed here has been developed by the author in close collaboration with Dr. X.C. Xie. The author thanks Dr. X.C. Xie as well as Dr. Q.P. Li for discussions and collaboration. This work is supported by the United States Office of Naval Research (US-ONR), the United States Army Research Office (US-ARO), the National Science Foundation (NSF), and the United States Department of Defense (US-DOD).

REFERENCES

1. T. Ando and Y. Uemura, J. Phys. Soc. Jpn. 36, 959 (1974).
2. S. Das Sarma and X.C. Xie, Phys. Rev. Lett. 61, 738 (1988) and J. App. Phys. 54, 5465 (1988); Qiang Li, X.C. Xie, and S. Das Sarma, Phys. Rev. B40, 1381 (1989).
3. X.C. Xie, Q.P. Li, and S. Das Sarma, Phys. Rev. B (to be published).
4. S. Das Sarma, Sol. State Commun. 36, 357 (1980); S. Das Sarma, Phys. Rev. B23, 4529 (1981); R. Lassnig and E. Gornik, Sol. State Commun. 47, 959 (1983); Y. Murayama and T. Ando, Phys. Rev. B35, 2252 (1987).
5. S. Das Sarma and F. Stern, Phys. Rev. B15, 8442 (1985).

Penrose Lattice in High Magnetic Fields

T. Hatakeyama

Department of Physics, Faculty of Science,
University of Tokyo, Bunkyo-ku, Tokyo 113, Japan

Abstract. The one-electron energy spectrum of a Penrose lattice in high magnetic fields is studied with a tight-binding Hamiltonian. We show the following results characteristic of a Penrose lattice. (1) The magnetic field dependence of the energy spectrum shows a butterfly shape caused by Landau quantization. (2) The magnetic field dependence of the energy spectrum is characterized by two periods whose ratio is equal to the golden mean $(1+\sqrt{5})/2$, and repeats quasiperiodically. The origin of this quasiperiodic behavior of the energy spectrum has been clarified analytically.

1 Introduction

The electronic structures of quasicrystals have received much attention recently [1]. In particular the electronic structure of a one-dimensional quasicrystal made by a Fibonacci sequence is studied extensively. It is believed that the energy spectrum of a one-dimensional quasicrystal comprises the Cantor set and the states are neither "extended" nor "localized" [2].

For a two-dimensional quasicrystal, Kohmoto and Sutherland reported electronic structure calculations of a Penrose Lattice shown in Figure 1 [3]. They pointed out the occurrence of a central peak with zero width at zero energy in the density of states, and showed that the states which form this peak in the density of states are exactly localized at the distinct sites. The present author and H. Kamimura first calculated the electronic structures of the Penrose lattice in a magnetic field and showed that exactly localized states exist even in a magnetic field and form a central peak at zero energy in the density of states [4].

It is known that the energy spectrum of electrons in a magnetic field in a periodic potential shows complex and exotic structures. Hofstadter showed that the energy spectrum of electrons in a square lattice in a magnetic field is crucially influenced by the rationality or irrationality of the number of the magnetic flux quanta in the cross section of the unit cell perpendicular to the magnetic field [5]. In a rational magnetic field the energy spectrum is divided into a finite number of the Landau levels, while in an irrational field the spectrum comprises the Cantor sets, and the field dependence of the energy spectrum shows a butterfly shape.

In the case of the Penrose lattice in a magnetic field, the situation is expected to become more interesting, because the ratio of areas of fat and thin

Springer Series in Solid-State Sciences, Vol. 101
High Magnetic Fields in Semiconductor Physics III Editor: G. Landwehr

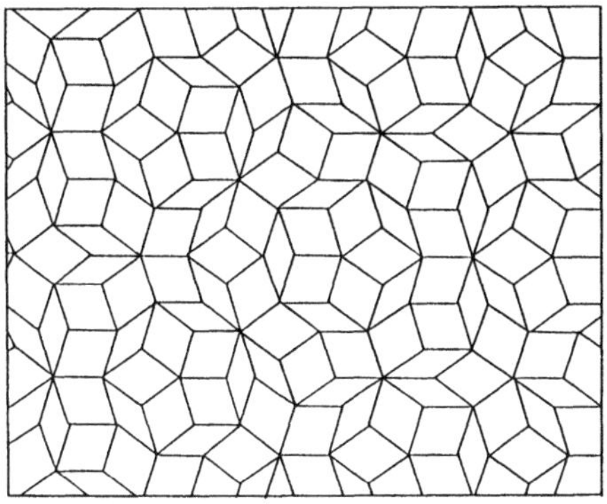

Figure 1. Penrose lattice. It comprises fat and thin rhombuses.

rhombuses from which the Penrose lattice is constructed is the irrational golden mean

$$\tau \equiv \frac{(1+\sqrt{5})}{2}, \tag{1}$$

and thus the ratio of magnetic fluxes passing through a fat rhombus to those through a thin rhombus cannot be rational. That is, if the magnetic flux which passes through thin rhombuses is rational with regard to the flux quantum (hc/e), the flux through fat rhombuses is irrational, or vice versa. Furthermore the ratio of the numbers of fat and thin rhombuses is also the golden mean. In this proceedings, we investigate the effect of this irrationality on the energy spectrum, and show that it causes the quasiperiodic field dependence of the energy spectrum.

2 Numerical Results

The tight-binding Hamiltonian for the Penrose lattice is given by

$$\mathcal{H} = \sum_{\langle i,j \rangle} |i\rangle t_{ij} \langle j|, \tag{2}$$

where $|i\rangle$ denotes the wavefunction on an i-th lattice site, and t_{ij} is the transfer integral between i-th and j-th lattice sites.

In the absence of a magnetic field t_{ij} is taken to be unity between neighboring lattice sites and to be zero for others. When the magnetic field is applied perpendicularly to the Penrose lattice, the transfer integral t_{ij} is given by

$$t_{ij} = \exp\left[i\frac{e}{\hbar c}\mathbf{H}\cdot\frac{\mathbf{r}_i \times \mathbf{r}_j}{2}\right], \tag{3}$$

with the vector potential $\mathbf{A} = \mathbf{H} \times \mathbf{r}/2$. In general, the transfer integral t_{ij} is a complex number,

$$t_{ij} = \exp[i2\pi\phi_{ij}], \tag{4}$$

where phase ϕ_{ij} is determined by the gauge.

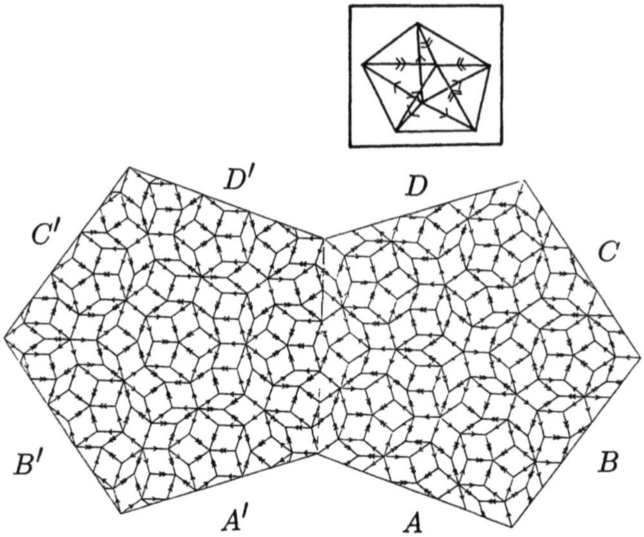

Figure 2. The developed edgeless Penrose lattice. The lines A', B', C' and D' are the same as the lines A, B, C and D respectively. This Penrose lattice is made by deflating the seed as shown in the inset. The seed consists of five fat rhombuses bent along the shorter diagonal line.

To diagonalize the Hamiltonian (2) numerically we adopt an edgeless Penrose lattice shown in Figure 2 instead of the infinite Penrose lattice. This lattice is made by "deflating" the pentagonal seed shown in the inset of Figure 2 [6, 7]. This seed is composed of five fat rhombuses bent along the shorter diagonal line. The pentagonal seed can be constructed by connecting these five rhombuses under the "matching rule" of Penrose tiling.

Figure 3 shows the calculated energy E as a function of the magnetic field $\tilde{H}$, where $\tilde{H}$ represents a dimensionless magnetic field defined by $\tilde{H} = eHa^2\sin(2\pi/5)/hc$, with $a^2\sin(2\pi/5)$ being the area of a fat rhombus. The number of sites in the Penrose lattice is 4937. The size dependence of the electronic structure is checked by changing the number of sites.

From the magnetic field dependence of energy levels we notice that it exhibits a self-similar pattern which is similar to the butterfly shape of the energy spectrum in a square lattice pointed out by Hofstadter. However, a remarkable difference between Landau levels in quasiperiodic and periodic systems is the ratio of two successive repeated periods in their field dependence. In a Penrose Lattice it is exactly the golden mean τ. Furthermore we have extended the calculation of the energy spectrum up to $\tilde{H} \simeq 20$ and the result is shown in Figure 4. By seeing Figure 4, we notice that the periods in which similar energy spectra are repeated form the Fibonacci sequence of $\tau 1 \tau 1 \tau \ldots$type, in which τ and 1 correspond to larger and shorter periods, respectively, where the n-th period of Fibonacci sequence $q(n)$ is given by a formula

$$q(n) = 1 + \tau^{-1}(\lfloor (n+1)/\tau + 1/2 \rfloor - \lfloor n/\tau + 1/2 \rfloor), \qquad (5)$$

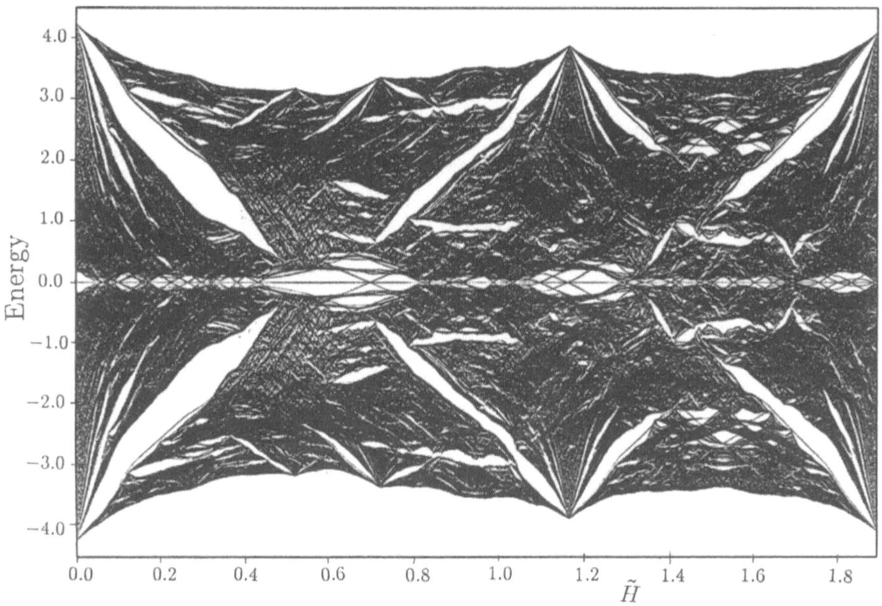

Figure 3. The field dependence of the energy spectrum for the Penrose lattice, where $\tilde{H}$ is taken from zero to 2.0.

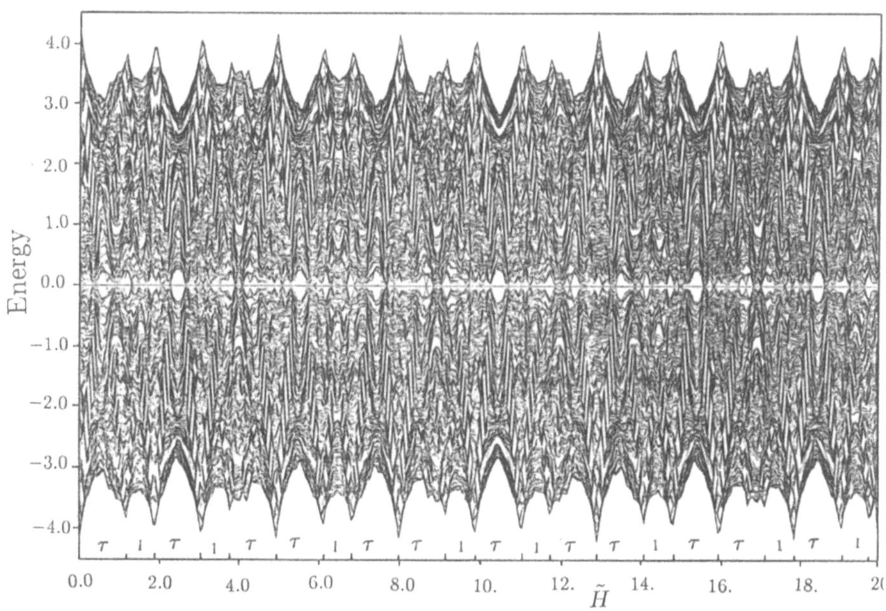

Figure 4. The field dependence of the energy spectrum for the Penrose lattice where $\tilde{H}$ is taken from zero to 20.

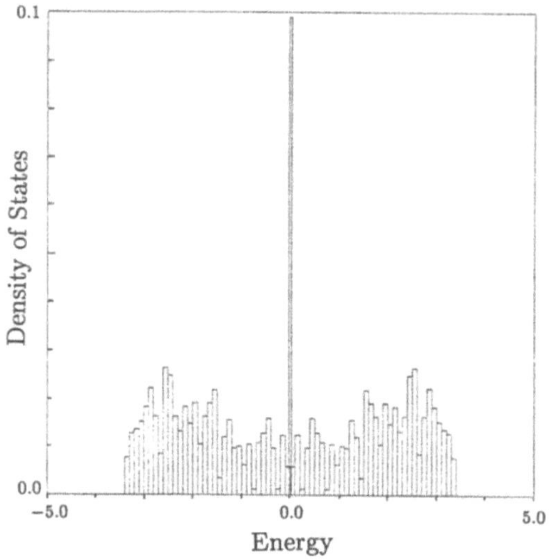

Figure 5. The calculated density of state, for the Penrose lattice consisting of 10256 sites in the magnetic field $\tilde{H} = 0.723$.

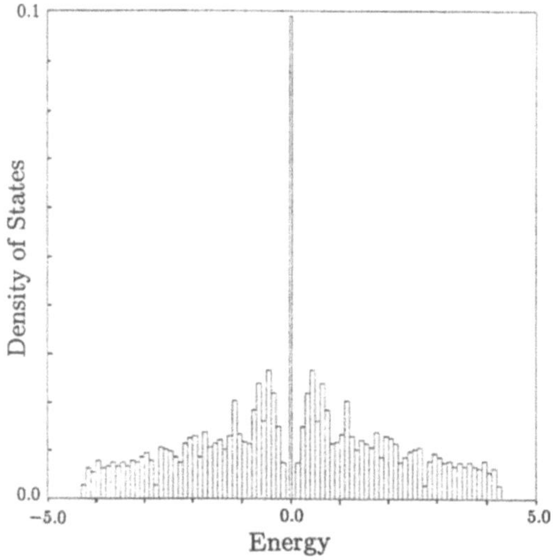

Figure 6. The calculated density of states per one site for the Penrose lattice consisting of 10256 sites in the absence of magnetic field.

where $\lfloor p \rfloor$ represents an integer largest but not larger than p [8, 9]. In the following section we investigate this quasiperiodicity of the field dependence of the energy spectrum analytically.

Finally we comment on the localized states at zero energy. Figure 5 shows the calculated density of states per site for an edgeless Penrose lattice of 10256

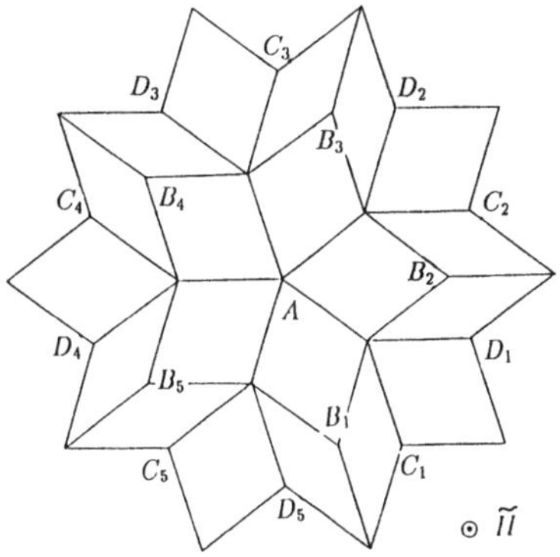

Figure 7. The representation of one of zero-energy states. This zero-energy state has an amplitude only at the sites A, B_i, C_i and D_i with $i = 1$ to 5.

sites in the magnetic field $\tilde{H} = 0.723$. In Figure 6 the density of states in the absence of a magnetic fields is shown for comparison. A central peak with zero width at zero energy still occurs even in a magnetic field irrespective of its strength. It comprises the localized states at the distinct sites. The wavefunction of one of these states can be derived analytically as follows.

In Figure 7 the basic pattern of the Penrose lattice is given. From the numerical results we notice that the wavefunction for one of zero-energy states has an amplitude only at the sites A, B_i, C_i and D_i with $i = 1$ to 5. We assume that this wavefunction is expressed in the following form with a five-fold symmetry,

$$\phi_{\mathrm{ring}} = c_{\mathrm{A}}|\mathrm{A}\rangle + c_{\mathrm{B}}\sum_{i=1}^{5}|\mathrm{B}_i\rangle + c_{\mathrm{C}}\sum_{i=1}^{5}|\mathrm{C}_i\rangle + c_{\mathrm{D}}\sum_{i=1}^{5}|\mathrm{D}_i\rangle \tag{6}$$

where c_{A}, c_{B}, c_{C} and c_{D} denote the amplitudes at sites of A, B, C and D types, respectively.

Since for a zero-energy state

$$\mathcal{H}\phi_{\mathrm{ring}} = 0, \tag{7}$$

we obtain

$$\left.\begin{array}{l} c_{\mathrm{C}}e^{i\pi\tau\tilde{H}} + c_{\mathrm{D}}e^{-i\pi\tau\tilde{H}} = 0, \\ c_{\mathrm{B}} + c_{\mathrm{C}}e^{-i\pi\tau\tilde{H}} + c_{\mathrm{D}}e^{i\pi\tau\tilde{H}} = 0, \\ c_{\mathrm{A}} + c_{\mathrm{B}}e^{i\pi\tilde{H}} + c_{\mathrm{B}}e^{-i\pi\tilde{H}} + c_{\mathrm{C}}e^{i\pi\tau^{-1}\tilde{H}} + c_{\mathrm{D}}e^{-i\pi\tau^{-1}\tilde{H}} = 0. \end{array}\right\} \tag{8}$$

The exponential factors in the above equations come from transfer integrals. By solving the above equations, we obtain the solution,

$$\begin{cases} c_A = -4\cos(\pi\tilde{H})\sin(2\pi\tau\tilde{H}) + 2\sin(\pi\tilde{H}), \\ c_B = 2\sin(2\pi\tau\tilde{H}), \\ c_C = \exp\left[-i\pi\left(\tau\tilde{H} + \frac{1}{2}\right)\right], \\ c_D = \exp\left[i\pi\left(\tau\tilde{H} + \frac{1}{2}\right)\right]. \end{cases} \tag{9}$$

3 Proof of Quasiperiodicity in the Magnetic Field Dependence of the Energy Spectrum

Let us consider a finite system of a Penrose lattice similar to an edgeless Penrose lattice, which consists of a_{2m} fat rhombuses and a_{2m-1} thin rhombuses where a_n is the n-th Fibonacci number generated by a recursion relation,

$$a_{n+1} = a_n + a_{n-1}, \quad (a_0 = 1, a_1 = 1). \tag{10}$$

We note that the ratio of the number of fat to thin rhombuses in the Penrose lattice is the golden mean, and that the ratio a_n/a_{n-1} is in a good approximation considered as the golden mean. In fact, the following relation holds between a_n and a_{n-1}:

$$\frac{a_n}{\tau} = a_{n-1} + (-1)^n \frac{1}{\tau^{n+1}}. \tag{11}$$

This relation is derived by using the relation (10) and

$$\tau^2 = \tau + 1. \tag{12}$$

The total magnetic flux which penetrates the whole system is quantized;

$$a_{2m}\phi_{\text{fat}} + a_{2m-1}\phi_{\text{thin}} = n, \tag{13}$$

where n is the number of a flux quantum, and ϕ_{fat} and ϕ_{thin} denote the magnetic flux divided by a flux quantum which penetrate fat and thin rhombuses, respectively.

We note that although phase ϕ_{ij} in the equation (4) depends on the gauge, the following equations

$$\begin{aligned} \oint_{\text{fat}} \phi_{ij} &= \phi_{\text{fat}} + n, \\ \oint_{\text{thin}} \phi_{ij} &= \phi_{\text{thin}} + m, \end{aligned} \tag{14}$$

are independent of gauge, where $\oint_{\text{fat}} \phi_{ij}$ and $\oint_{\text{thin}} \phi_{ij}$ are the sum of phase ϕ_{ij} for a closed loop along the four sides of a fat and a thin rhombus, respectively. The energy spectrum of the tight-binding Hamiltonian of this system (2) is a periodic function of ϕ_{fat} and ϕ_{thin} under the condition (13)

$$E(\phi_{\text{fat}} + n, \phi_{\text{thin}} + m) = E(\phi_{\text{fat}}, \phi_{\text{thin}}), \tag{15}$$

where $E(\phi_{\text{fat}}, \phi_{\text{thin}})$ denotes the energy spectrum characterized by ϕ_{fat} and ϕ_{thin}. We can rewrite the equation (13) in the following way:

$$\begin{aligned} \phi_{\text{fat}} &= na_{2m-2} - sa_{2m-1}, \\ \phi_{\text{thin}} &= -na_{2m-1} + sa_{2m}, \end{aligned} \tag{16}$$

with identity

$$a_{2m}a_{2m-2} - a_{2m-1}^2 = 1, \tag{17}$$

where s is an arbitrary parameter. From the equation (15), we have

$$E(na_{2m-2} - sa_{2m-1}, -na_{2m-1} + sa_{2m}) = E(-sa_{2m-1}, sa_{2m}) \equiv \tilde{E}(s) \tag{18}$$

Since a_{2m} and a_{2m-1} are integers, $\tilde{E}(s)$ is the periodic function of s whose period is unity;

$$\tilde{E}(s+m) = \tilde{E}(s). \tag{19}$$

In a uniform magnetic field we have,

$$\phi_{\text{thin}} = \frac{1}{\tau}\phi_{\text{fat}}, \tag{20}$$

because the ratio of the areas of fat to thin rhombus is τ. In this condition n in the equation (13) is considered as the strength of the magnetic field.

From the equations (16) and (20) we obtain

$$s = \frac{n}{\tau} \tag{21}$$

Thus the energy spectrum of the magnetic field strength n is,

$$\tilde{E}\left(\frac{n}{\tau}\right). \tag{22}$$

The energy spectrum $E(n/\tau)$ has a quasiperiod m_l in the sense that

$$\lim_{l\to\infty} \sup_{n\in Z} \left\| \tilde{E}\left(\frac{n+m_l}{\tau}\right) - \tilde{E}\left(\frac{n}{\tau}\right) \right\| = 0, \tag{23}$$

where

$$m_l \in \{ka_l + \lfloor k/\tau + 1/2 \rfloor a_{l-1} | k \in Z, k > 0\}. \tag{24}$$

We note that a quasiperiod m_l consists of two periods $a_{l+1} = a_l + a_{l-1}$ and a_l, the ratio of which becomes τ as $l \to \infty$. The n-th period $q(n)$ is given by

$$q(n) = a_l + \{\lfloor (n+1)/\tau + 1/2 \rfloor - \lfloor n/\tau + 1/2 \rfloor\} a_{l-1}, \tag{25}$$

which is essentially the same as the equation (5).

Then the equation (23) is proved as follows; using the the equation (11) we obtain the relation

$$\frac{ka_l + \lfloor k/\tau + 1/2 \rfloor a_{l-1}}{\tau}$$
$$= ka_{l-1} + \lfloor k/\tau + 1/2 \rfloor a_{l-2} + (-1)^l \frac{k/\tau - \lfloor k/\tau + 1/2 \rfloor}{\tau^l}. \tag{26}$$

From the equations (26) and (19), we obtain

$$\left\| \tilde{E}\left(\frac{n+m_l}{\tau}\right) - \tilde{E}\left(\frac{n}{\tau}\right) \right\|$$
$$= \left\| \tilde{E}\left(\frac{n}{\tau} + (-1)^l \frac{k/\tau - \lfloor k/\tau + 1/2 \rfloor}{\tau^l}\right) - \tilde{E}\left(\frac{n}{\tau}\right) \right\|. \tag{27}$$

Thus we have

$$\left| (-1)^l \frac{k/\tau - \lfloor k/\tau + 1/2 \rfloor}{\tau^l} \right| < \frac{1}{2\tau^l} \overset{l \to \infty}{\longrightarrow} 0. \tag{28}$$

Since $\tilde{E}(s)$ is a periodic continuous function of s, we note that

$$\lim_{h \to 0} \sup_{s \in R} ||\tilde{E}(s+h) - \tilde{E}(s)|| = 0. \tag{29}$$

Thus, from the equations (27), (28) and (29), we conclude that the equation (23) holds.

4 Conclusions

We have calculated the energy spectrum of the Penrose lattice in a magnetic field and found that the spectrum splits into Landau levels in the presence of a magnetic field. It has been shown that the density of states has a central peak with zero width at zero energy and this corresponds to the exactly localized states. The field dependence of the energy spectrum has a self-similarity; it has a butterfly shape and repeats quasiperiodically. This quasiperiodic magnetic field dependence of the energy spectrum has been clarified mathematically.

Acknowledgment—We would like to thank Professor Hiroshi Kamimura and Professor Hideo Aoki. The numerical calculations were done on the HITAC S-810/20 at the Computer Center, University of Tokyo. This work was supported by a Grant-in-Aid from the Ministry of Education, Science and Culture.

References

[1] D. Schechtman, I. Blech, D. Gratias and J. W. Cahn: Phys. Rev. Lett. **53**, (1984) 1951.

[2] M. Kohmoto, L. P. Kadanoff and C. Tang: Phys. Rev. Lett. **50** (1983) 1870.

[3] M. Kohmoto and B. Sutherland: Phys. Rev. Lett. **56**, (1986) 2740.

[4] T. Hatakeyama and H. Kamimura: Solid State Commun. **62** (1987) 79.

[5] D.R. Hofstadter: Phys. Rev. **B14**, (1976) 2239.

[6] R. Penrose: Bull. Inst. Math. Appl. **10**, (1974) 266.

[7] M. Gardner: Sci. Am. **236**, No.1, (1977) 110.

[8] N.G. de Bruijn: Ned. Akad. Weten. Proc. Ser. **A84**, (1981) 39.

[9] N.G. de Bruijn: Ned. Akad. Weten. Proc. Ser. **A84**, (1981) 53.

Semi-Metallic Behaviour in GaSb/InAs Heterojunctions

R.W. Martin, M. Lakrimi, S.K. Haywood, R.J. Nicholas, N.J. Mason, and P.J. Walker

Clarendon Laboratory, University of Oxford, Parks Road, Oxford OX1 3PU, UK

Abstract: We have performed magnetotransport measurements on a series of GaSb/InAs double heterojunctions grown by MOVPE, with the InAs thickness ranging from 100 to 2000 Å. The samples exhibit coexisting two-dimensional holes located at the interface, in the GaSb, and higher mobility two-dimensional electrons in the InAs. In addition to extracting the hole and electron densities from the gradients of the Hall resistance, these are also estimated, together with the respective mobilities, by the use of a two carrier fit. We present a systematic study of the two carrier densities as a function of the InAs layer thickness.

1 Introduction:

With the conduction band of InAs lying below the valence band of GaSb, it is possible to switch the energy gap, E_g, of a quantum well or superlattice from positive (semiconductor) to negative (semimetal) by varying the InAs layer thickness [1]. We have recently exploited the smallness of E_g to grow GaSb/InAs and InSb/InAs superlattices with enhanced absorption in the region of 10 μm on $< 111 >$-oriented GaAs substrates using atmospheric MOVPE [2]. This type II structure is also interesting for magnetotransport measurements. At very small thicknesses, the system can be made insulating when the Fermi energy lies above the valence band of the GaSb. As the InAs thickness is increased, it is estimated that a semiconductor-semimetal transition occurs in the vicinity of 100 Å [1]. Beyond this thickness, electrons from the valence band of the GaSb transfer intrinsically into the conduction band of the InAs. In the majority of structures studied to date, it is necessary to invoke a considerable amount of extrinsic doping, say donor states at the interface, to account for large electron concentrations [3]. In the present publication, we probe the semiconductor-semimetal transition and report on Shubnikov-de Haas and Hall measurements performed on GaSb/InAs double heterojunctions (DHET) closer to intrinsic behaviour. At high magnetic fields, the Hall gradient appears compensated and the Hall plateaus are dependent upon the hole and electron filling factors. We have also characterised different growth switching sequences at the interfaces between GaSb and InAs.

2 Growth and magnetotransport measurements:

Since both anion and cation change across a binary interface such as GaSb/InAs, many different growth switching sequences are possible. Furthermore, we found that the introduction of growth pauses and different switching sequences at the interface yielded interfaces with superior transport properties. Our study of the different possibilities

Springer Series in Solid-State Sciences, Vol. 101
High Magnetic Fields in Semiconductor Physics III Editor: G. Landwehr

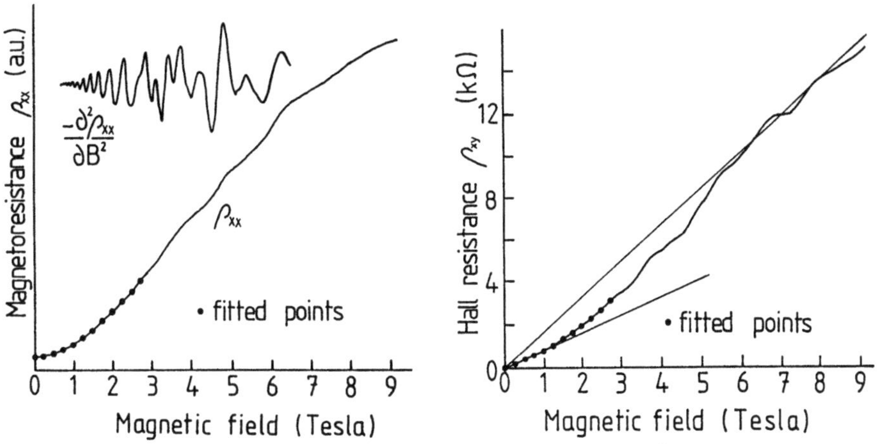

Figure 1: Typical ρ_{xx} and ρ_{xy} recordings at 700 mK for a 500 Å DHET. The dots represent the fits to ρ_{xx} and $\rho_x y$ using a two carrier model.

revealed that it is important to avoid the formation of Ga antisite defects on the GaSb side of the interface which dominate the electrical properties of bulk GaSb.

We found that the "InSb" MBE-type interface [4] gives poor electrical results when adapted to MOVPE growth. This interface sequence consists of Ga-off, Sb-off, In-on and As-on. We have used a different "InSb" type interface consisting of Ga-off, In-on, Sb-off and As-on to good effect. A "GaAs" type interface (Sb-off, Ga-off, As-on and In-on) produced low mobilities and very high carrier concentrations. As for the growth pauses, we bracket the optimum pause between 0 and 3 seconds since 0 and 5 seconds pauses gave poor electrical mobilities. We achieved high mobilities, up to 80,000 cm^2/Vs, in our DHETs using tertiarybutylarsine (TBAs) as the arsenic source. A detailed description of the optimisation of the growth with respect to III-V ratios, temperature and growth pauses/times can be found in [5,6] and the comparison between a few interface types is to be reported elsewhere.

The samples consist of a single layer of InAs sandwiched between layers of p-type GaSb, with InAs thickness varying from 100 to 2000 Å. The growth sequence is described in reference [5]. We performed SdH and Hall measurements using a resistive magnet up to 9 T in the temperature range of 0.5 to 4.2 K and also a ^{3}He cryostat down to 350 mK with fields up to 14.5 Tesla. A typical recording of the magnetoresistance ρ_{xx} and the Hall resistance ρ_{xy}, at 700 mK, is shown in figure 1 for a sample with an InAs thickness of 500 Å. The hole and electron densities are extracted from the slopes of the Hall resistance and also from the Fourier transform of the doubly differentiated ρ_{xx}, i.e. $\partial^2\rho_{xx}/\partial B^2$. We used a classical two carrier model to fit the low field ρ_{xx} and ρ_{xy} to estimate the densities, N_e and N_h, and mobilities, μ_e and μ_h, of the electrons and holes respectively [7].

In figure 1, we have drawn two lines to illustrate the slope changeover in ρ_{xy} with increasing magnetic field. A two carrier fit to ρ_{xy} shows that $\mu_e \gg \mu_h$ so that at very small magnetic field intensities, the Hall resistance it is inversely proportional to N_e, while at high magnetic fields such that $\mu B \gg 1$, it is well approximated by $1/(e(N_e - N_h))$, [7]. The slopes shown correspond to carrier densities of 7.3×10^{11} and 3.7×10^{11} cm^{-2} at low and high fields respectively; thus the hole density is 3.6×10^{11} cm^{-2}.

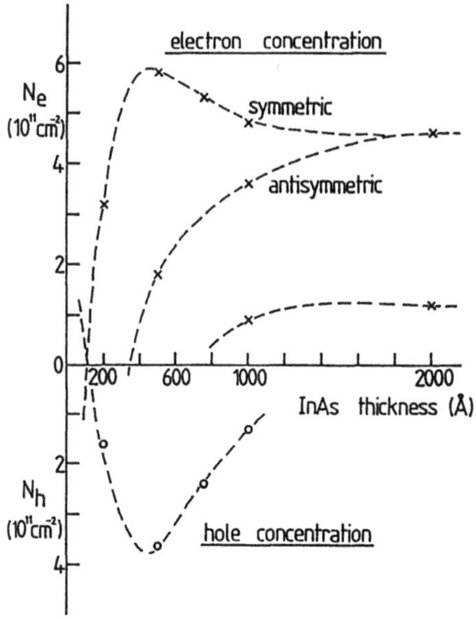

Figure 2: Hole and electron density variations as a function of the InAs layer thickness.

The difference in carrier concentration, $N_e - N_h$, is equal to $3.7\times10^{11}\ cm^{-2}$ constitutes strong evidence for the presence of interface and donor states.

For a Hall plateau with an integer filling factor to appear in a two carrier system, the Fermi energy level must simultaneously be in the localised regions for both the electrons and holes and consequently the plateaux appear compensated and given by $R_{QHE} = (h/e^2)/(\nu_e - \nu_h)^{-1}$. This behaviour has been reported by other groups [3,8]. However, the samples studied here are closer to intrinsic behaviour than has previously been seen. The plateau at 7 Tesla has a Hall resistance of 11.7 $k\Omega$ which does not correspond to an integer filling factor. The existence of plateau like behaviour suggests that the Fermi energy may be in the localised regions for the electrons with $\nu_e = 4$ but not for the holes. We estimate $\nu_h = 1.8$ and the coresponding hole density as $3.1\times10^{11}\ cm^{-2}$. Because of the intrinsic nature of our samples, the hole and electron concentrations vary self-consistently with magnetic field. Assuming a fixed net charge density of $3.7\times10^{11}\ cm^{-2}$, it follows that $N_e = 6.8 \times 10^{11}\ cm^{-2}$ at the plateau, hence yielding a filling factor of 4 as suggested.

We have then curve fitted the low field ρ_{xx} and ρ_{xy}. As illustrated in figure 1 by the dotted line, a good fit is found using $7.1\times10^{11}\ cm^{-2}$, $3.7\times10^{11}\ cm^{-2}$, 80000 cm^2/Vs and 5700 cm^2/Vs for N_e, N_h, μ_e and μ_h, respectively. These values are in good agreement with the above estimates.

Finally, we discuss the variation of the hole and electron densities as a function of the thickness of the InAs layer, as shown in figure 2. To a first approximation and at small thicknesses, the variation of the hole concentration mirrors that of the electrons as the structure is close to intrinsic. The difference between the two concentrations is made up extrinsically and increases with increasing layer thickness. Below 400 Å only one peak is found in the Fourier transform of $\partial^2\rho_{xx}/\partial B^2$ which accounts for the total electron density, which is thus in a single occupied level. As the InAs thickness is increased, the single quantum well breaks into two heterojunctions, one at each interface, and for sufficiently thick InAs layers, they contain an equal amount of carriers.

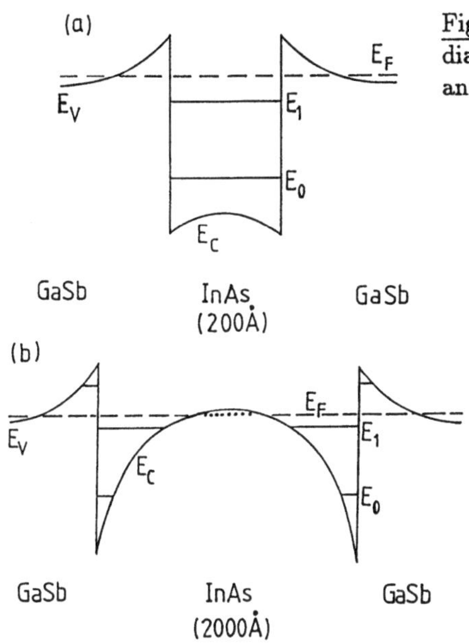

Figure 3: Sketch of the band energy diagram of the GaSb/InAs structure for an InAs layer of a) 200 Å and b) 2000 Å.

For weakly coupled heterojunctions, as found at thicknesses of order 1000 Å, the lowest states form symmetric and anti-symmetric combinations. These are resolved in the Fourier analyses as two different carrier concentrations, while the width of the transform for the widest well (2000 Å) allows us to estimate that the two interfaces are the same to within 5%. A higher subband is also occupied for InAs thicknesses above 800 Å. The various level populations are shown in figure 3.

When figure 2 is turned upside down, it bears a striking resemblance to the famous Lennard-Jones potential for two atoms of hydrogen brought closer by confinement; there results a strong repulsion at short distances as the electron wavefunctions overlap. This is precisely what happens as our InAs layer is thinned, the repulsive and attractive terms corresponding to the bonding E^+ and antibonding E^- states respectively. As the DHET is thinned the overlap between the two wavefunctions becomes large enough to split the levels into E^+ and E^- and give maximum binding at around 500 Å. These states then become the symmetric and antisymmetric states of the quantum well potential at small thicknesses until finally the DHET turns insulating around 100 Å.

3 Conclusion:

In conclusion, we have clearly shown the semiconductor-semimetallic transition as the InAs layer thickness is increased and sketched the changeover of the structure from a single quantum well, at small thickness, to a double heterojunction. We have discussed the compensation of the plateaux values in terms of the filling factors of the holes and electrons. Finally, the use of TBAs enhanced our electrical characteristics and the best interface is found to be that of "InSb" in the case of MOVPE grown structures.

4 References:

1. L.L. Chang, N. Kawai, G.A. Sai-Halasz, R. Ludeke & L. Esaki, Appl. Phys. Lett. 35, 939-942 (1980) and see also E.E. Mendez, L.L. Chang, C.-A. Chang, L.F. Alexander & L. Esaki, Surf. Sci. 142, 215-219 (1984).
2. M. Lakrimi, N.J. Mason, R.J. Nicholas, G.B. Stringfellow, G. Summers & P.J. Walker, submitted to J. Cryst. Growth.
3. E.E. Mendez, L.Esaki & L.L. Chang, Phys. Rev. Lett. 55, 2216-2219 (1985).
4. G. Tuttle, H. Kroemer & J. English, J. Appl. Phys. 67, 3032-3037 (1990).
5. S.K. Haywood, R.W. Martin, N.J. Mason & P.J. Walker, submitted to J. Cryst. Growth.
6. S.K. Haywood, R.W. Martin, N.J. Mason & P.J. Walker, J. Cryst. Growth 97, 489 (1989).
7. R.A. Smith, *Semiconductors*, Cambridge University Press, Cambridge, England (1978), pp114-115.
8. J. Beerens, G. Gregoris, J.C. Portal, E.E. Mendez, L.L. Chang & L. Esaki, Phys. Rev. B36, 4742-4747 (1987).

Evidence for Exchange Enhancement of the Cyclotron Energy from Quantum Lifetime Measurements of a Two Dimensional Electron Gas

M. Hayne[1], *A. Usher*[1], *J.J. Harris*[2], *and C.T. Foxon*[2]

[1]Department of Physics, Exeter University, Stocker Road, Exeter EX4 4QL, UK

[2]Philips Research Laboratories, Cross Oak Lane, Redhill, Surrey RH1 5HA, UK

Abstract. We report magnetoresistance measurements of the temperature dependence of the quantum lifetime of three GaAs/GaAlAs single heterojunctions which show remarkable evidence for an exchange enhancement of the Landau level separation. Furthermore, our results indicate the critical importance of the range of scattering in these systems.

1. Introduction

The distinction between the momentum relaxation time τ and the quantum lifetime τ_q is now well established. In contrast to the situation for silicon MOSFETs, where the two times are approximately equal [1], low temperature magnetoresistance studies of the two dimensional electron gas (2DEG) in GaAs/GaAlAs single heterojunctions reveal a τ_q which is between 4 and 60 times smaller than τ [2]-[6]. These contrasting results indicate that the dominant scattering mechanism at low temperatures in the latter system is long range scattering from impurities which are spatially separated from the 2DEG [7].

Theoretical studies by Ando [8], and Isihara and Smrcka [9] have considered the effect of short range scatterers on the 2DEG formed in the inversion layer of a silicon MOSFET, calculating the Landau level broadening self-consistently and deriving an expression for the amplitude of the Shubnikov-de Haas oscillations $\Delta\rho_{xx}$ at intermediate magnetic fields ($\omega_c\tau_q<2$). Coleridge *et al.* [4] pointed out that for heterojunctions the distinction between τ and τ_q is needed, and have shown that in the intermediate magnetic field regime the expression for $\Delta\rho_{xx}$ is

$$\Delta\rho_{xx} = 4\rho_0 \frac{\chi}{\sinh\chi} \exp\left(-\frac{\pi}{\omega_c \tau_q}\right). \tag{1}$$

Where the $\chi/\sinh\chi$ (with $\chi = 2\pi^2 kT/\hbar\omega_c$) term accounts for the effect of finite temperature and ρ_0, the zero field resistivity, is given by $\rho_0 = m^*/ne^2\tau$. Ando's central assumption of short range scattering limits the validity of (1) to the field range where the cyclotron diameter of the electrons near the Fermi energy is large compared with the range of the scattering.

The samples used in this study (details of which are given in table 1) were grown by MBE at Philips Research Laboratories, Redhill. Measurements were made between 0.3K and 4.2K on each sample, and at different electron concentrations obtained by persistent photoexcitation. The quantum lifetime was then found from $-\pi m^*/e\tau_q$, the slope of $\ln(\Delta\rho_{xx}\sinh\chi/4\chi)$ plotted against $1/B$ as shown in figure 1. The intercept is equal to $\ln(\rho_0)$.

Table 1: Sample details and results. The last column gives the enhancement of the Landau level splitting as discussed in the text.

SAMPLE spacer	n x10^{-11} (cm^{-2})	μ x 10^{-6} (cm^2/Vs)	τ (ps)	τ_q (ps)	Enhance-ment (%)
G580	1.26	2.44	94.5	2.25	30
(800Å)	1.77	3.04	118	2.6	30
G578	2.05	1.34	51.9	1.7	30
(400Å)	2.75	1.11	43.0	1.8	30
	2.98	1.13	43.8	3.2	30
G419	3.71	1.16	44.9	1.6	25
(200Å)	4.01	1.24	48.0	2.5	10
	4.40	1.36	52.7	3.75	10

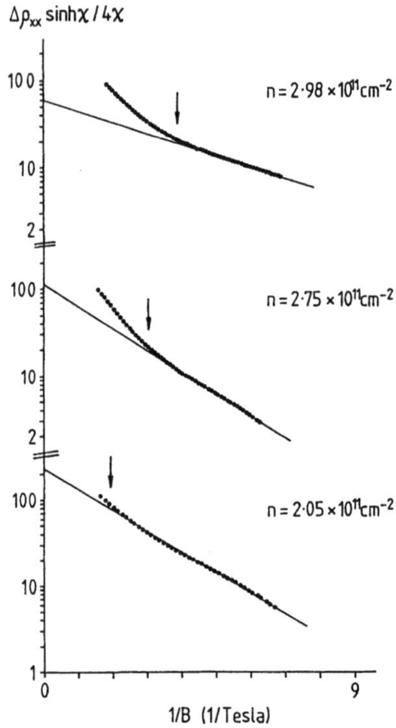

Figure 1: Plot of ln($\Delta\rho_{xx}$sinh$\chi/4\chi$) vs. 1/B for three electron concentrations in G578 at 0.5K. The straight line regions occur at low field where the cyclotron radius is large compared with the range of scattering.

2. Results and discussion

Analysis of the data using ω_c=eB/m* with m*=0.068m_0 was found to yield the unlikely result that τ_q *increased* as the temperature was increased between 0.3K and 1.3K (see figure 2). Since the zero field resistance, and thus the momentum relaxation time was constant in this temperature range, it is reasonable to assume that the quantum lifetime should also be temperature independent. The analysis was therefore

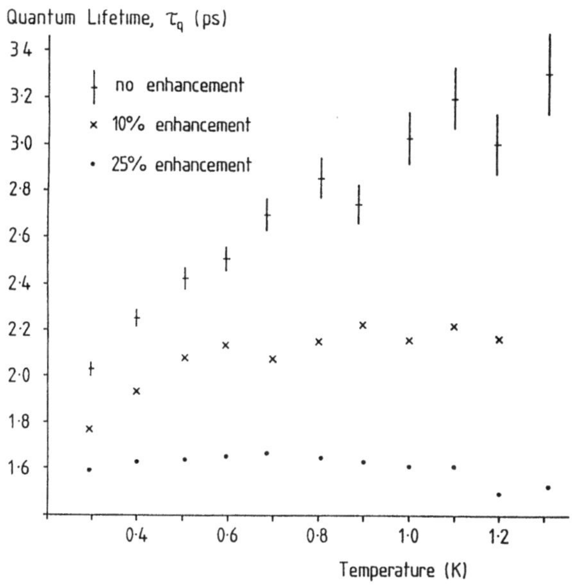

Figure 2: The temperature dependence of τ_q using the usual Landau level splitting, and splittings enhanced by 10% and 25%. For this sample (G419 at $n=3.71\times10^{11}$ cm^{-2}) an enhancement of 25% restores the expected temperature independence.

repeated replacing the usual Landau level separation by an exchange enhanced value, and the anomalous temperature dependence of the quantum lifetime was removed. The enhancement required was 20 to 30%, decreasing at higher electron concentrations due to increased scattering in the narrow spacer layer sample, and to the onset of inter-subband scattering as the second subband became occupied (see table 1). Above 1.3K oscillations are only resolved at higher fields, where the samples have already passed from the short to the long range scattering regime, and so (2) is no longer valid. By 4.2K the zero field resistance rises significantly, and so the assumption of a temperature independent τ_q would also not be valid.

An enhancement of the g-factor in silicon MOSFETs was first noticed by Fang and Stiles [10] and it was attributed to exchange interactions by Janak [11] and Ando and Uemura [12]. The effect has also been firmly established in GaAs/GaAlAs heterojunctions by activation and tilted field measurements [13] of the spin splitting of Landau levels. More recently activation measurements of the Landau level splitting at a filling factor of $\upsilon=2$ demonstrated the presence of an exchange enhancement of about 30% [14]. These results can equivalently be interpreted as the observation of large wavevector magnetic excitons in the case of enhancement of the Landau level splitting, and spin wave excitations for g-factor enhancement experiments. After taking disorder into account good agreement was found between the activation experiments of reference 14 and the theory of magnetic excitons [15]. The results presented here are the first reported evidence that the exchange enhancement of the Landau level separation previously only seen at $\upsilon=2$ is present even at filling factors as high as $\upsilon=100$.

Equation 1 was obtained for the intermediate field range where $\omega_c\tau_q<2$, but subsequent experiments have shown it to be valid up to $\omega_c\tau_q=4$ [4],[6]. It can be seen from figure 1 that our data produces a substantial straight line region, with the expected deviations from this behaviour at high fields. We do not, however, attribute these deviations to a non-validity of (1) as a result of a critical $\omega_c\tau_q$. The $\chi/\sinh\chi$ term in (1) should cause the deviation field to have a linear temperature dependence. This is not the case (except for sample G580 at its lowest electron concentration); at fixed electron concentration the deviation is at constant field despite a four-fold increase in temperature.

We interpret the high field deviation as the transition between short and long range scattering, when the size of the cyclotron orbit becomes small compared with the range of potential fluctuations in the sample; in agreement with studies of magnetoconductivity peaks [16]. As the electron concentration is increased the remote ionised impurities, which are the main source of scattering, are screened out and the deviation moves to lower field or larger cyclotron orbit (figure 1). Further demonstrating the crucial role played by screening, the results of table 1 show that τ_q tends to increase with electron concentration, corroborating the results of other authors [2],[5],[7]. In the case of G580 at its lowest electron concentration, τ_q has the lowest value of all the samples studied, and so the $\omega_c\tau_q$ limit is reached before deviations due to scattering range can occur. This is reflected in the approximately linear temperature dependence of the deviation field seen only at this electron concentration.

3. Conclusions

We have performed a systematic study of the quantum lifetime of electrons in a 2DEG formed at the interface of a GaAs/GaAlAs heterojunction. This has shown the important role that the relative length scales of the cyclotron orbit and the scattering mechanisms play in these systems. Temperature dependence studies have provided startling new evidence for an exchange enhancement of the cyclotron energy of up to 30% in some samples, present even at fields below 0.2T.

The authors would like to thank Dr. R.J. Nicholas for useful discussions.

References

1. F.F. Fang, A.B. Fowler and A. Hartstein Phys. Rev. **B16** 4446 (1977)
2. M.A. Paalanen, D.C. Tsui and J.C.M. Hwang Phys. Rev. Lett. **51** 2226 (1983)
3. J.P. Harrang, R.J. Higgins, R.K. Goodall, P.R. Jay, M. Laviron and P. Delescluse Phys. Rev. **B32** 8126 (1985)
4. P.T. Coleridge, R. Stoner and R. Fletcher Phys. Rev. **B39** 1120 (1989)
5. R.G. Mani and J.R. Anderson Phys. Rev. **B37** 4299 (1988)
6. M. Lakrimi PhD. thesis Sussex University (1988)
7. S. Das Sarma and F. Stern Phys. Rev. **B32** 8442 (1985)
8. T. Ando J. Phys. Soc. Jap. **37** 1233 (1974)
9. A. Isihara and L. Smrcka J Phys. **C19** 6777 (1986)
10. F.F. Fang and P.J. Stiles Phys. Rev. **174** 823 (1968)
11. J.F. Janak Phys. Rev. **178** 1416 (1969)
12. T. Ando and Y. Uemura J. Phys. Soc. Jap. **37** 1044 (1974)
13. For example; R.J. Nicholas, R.J. Haug, K v. Klitzing and G. Weimann Phys. Rev. **B37** 1294 (1988)
14. A. Usher, R.J. Nicholas, J.J. Harris and C.T. Foxon Phys. Rev. **B41** 1129 (1990)
15. C. Kallin and B.I. Halperin Phys. Rev. **B30** 5655 (1984)
    C. Kallin and B.I. Halperin Phys. Rev. **B31** 3635 (1985)
16. M. Lakrimi, A.D.C. Grassie, K.M. Hutchings, J.J. Harris and C.T. Foxon. To be published.

Determination of Spin Splitting for Two-Dimensional Carriers with Strong Spin-Orbit Coupling from the Quantum Transport Phenomena

S.I. Dorozhkin

Max-Planck-Institut für Festkörperforschung, Heisenbergstr. 1, W-7000 Stuttgart 80, Fed. Rep. of Germany and Institute of Solid State Physics, Academy of Sciences of the USSR,, Chernogolovka, Moscow District, SU-142432 Moscow, USSR

Abstract. The dependence of the Shubnikov-de Haas oscillation beating on the magnetic field tilting and the substrate bias is experimentally investigated in the p-channels of Si (110) MOSFETs. The results indicate the existence of a g-factor anisotropy for the magnetic field orientations relative to both the channel plane and the crystal axes. The first result is explained in terms of the size quantization for systems with a four-fold degenerate valence band. The dependence on the substrate bias is related to the variation of the zero-field spin splitting with the modified confining potential.

1. Introduction

In the absence of an inversion symmetry a spin-orbit coupling lifts the spin degeneracy already at zero magnetic field. As a result two branches appear in the energy spectrum. From the simple quasiclassical consideration in the case of a small splitting an existence of two close Fermi surfaces may result in a beating of the Shubnikov-de Haas oscillations. For two-dimensional systems this effect was first observed [1] and investigated in detail [2] in the p-channels of Si (110) MOSFETs. The quantitative model [2,3] of the effect enables a determination of both the zero-field spin splitting and the g-factor of the two-dimensional carriers from the node positions and their shift in a tilted magnetic field. Later results on the beating effect in a number of other materials can be quantitatively described in terms of this model [3]. This model is based on the energy spectrum proposed by Bychkov and Rashba [4].

In this paper we present experimental results on the dependence of the beating pattern in the Si MOSFETs on the magnetic field tilting. These results provide clear evidence that the magnetic field component H_x, parallel to the channel, has a much smaller influence on the energy of magnetic levels than it could be expected from the Zeeman splitting in the perpendicular field. Moreover this influence depends on the orientation of H_x relative to the crystal axes. These effects look like an anisotropy of the g-factor. A simple consideration of the size quantization in systems with a four-fold degenerate valence band provides a qualitative explanation of the weak effect of H_x on the energy spectrum and predicts the similar result for a number of materials.

Another experimental result reported here is the observation of the effect of a substrate bias on the node position.

2. Experimental results

The measurements were carried out on the two-dimensional p-channels in the MOSFETs produced on the Si (110) surface. The carrier peak mobility was about 3000 cm^2/Vs. The magnetoresistance was measured as a function of the magnetic field at different tilting angles. A magnetic field component parallel to the channel plane was aligned either along [1-10] or [001] crystal axes. Typical experimental traces are presented in Fig. 1 as a function of the perpendicular magnetic field component H_z for the case of a carrier concentration $n_s=2.2\times10^{12}$ cm^{-2}. The beating effect appears as a minimum in the oscillation amplitudes (a node) and a simultaneous change of the parity of the filling factors corresponding to the magnetoresistance minima. The magnetic field tilting in the [001] direction results in an increase of the filling factor N corresponding to the node in the beating pattern. Tilting in the [1-10] direction decreases the N value. For the data in Fig.1 we estimate the N values as following. Curve 1: N=19.5, curve 2: N=21, curve 3: N=22.5. The results for different tilting angles and carrier concentrations are summarized in Fig. 2. Fig. 3 demonstrates the dependence of the node position on the substrate bias. The positive bias shifts the node to the higher filling factors and the value of this shift for U_{sb}=+20 V is of the order of unity.

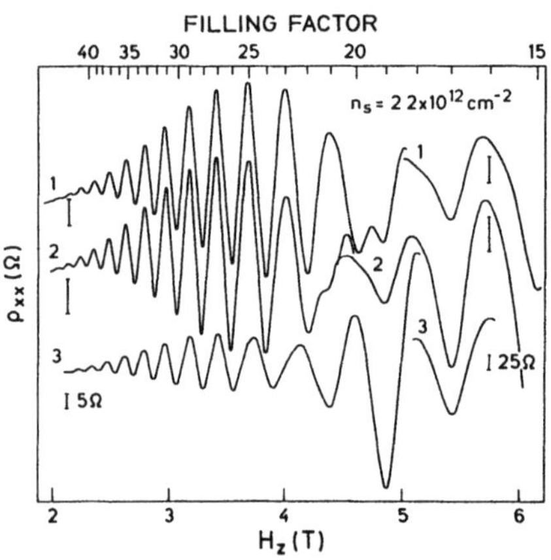

FIG.1 Shubnikov-de Haas oscillations in the Si (110) p-channels in perpendicular (curve 2) and tilted (curves 1 and 3) magnetic fields as a function of a perpendicular field component H_z. Corresponding filling factors are also shown. Tilting angle $\theta=45^0$. 1-$H_x \parallel$[1-10], 3-$H_x \parallel$[001]. The scale for the magnetoresistance is shown near the curves by marks corresponding either to 5Ω at smaller fields or to 25Ω at higher fields. Temperature T=1.4 K.

3. Discussion

In our case the observation of the one node only does not allow independent determination of the zero-field splitting and the g-factor. Nevertheless some quantitative estimations of these values can be done [2,5]. There are two important results in Fig. 2. The first one is a weak dependence of the node position on the H_x/H value when this value is not close to unity. The second one is a different sign of the node shift for different H_x orientations. The first result can be interpreted only in terms of the anisotropic g-factor for the following reason. At high perpendicular magnetic field when the spin-orbit effects are not of importance the spin splitting exceeds one half of the cyclotron splitting which follows from the observation of the magnetoresistance minima at odd filling factors. In this case for the isotropic g-factor the tilting of the magnetic field should result in drastic changes of the node position which is inconsistent with the experimental observations. In Ref. [5] from similar results an estimation was obtained that $g_z/g_x>2.5$, where g_z and g_x are the g-factor values related to the magnetic field components H_z and H_x respectively. This anisotropy can be explained in terms of the size quantization for the four-fold degenerate valence band corresponding to the angular momentum J=3/2. As a result of the size quantization the energy states near the bottom of the lowest subband are formed by heavy holes with the momentum projection $m_J=\pm 3/2$. It is easy to show that in the first order of the perturbation theory the parallel magnetic field H_x does not influence the energy of these states which is equivalent to $g_x=0$. This result is quite general and should be valid for a number of materials (for example, p-channels in GaAs and Ge) in which the heavy hole branch corresponds to $m_J=\pm 3/2$. For a finite wave vector the classification of the states mentioned above is not strictly valid which would result in a nonvanishing g_x value

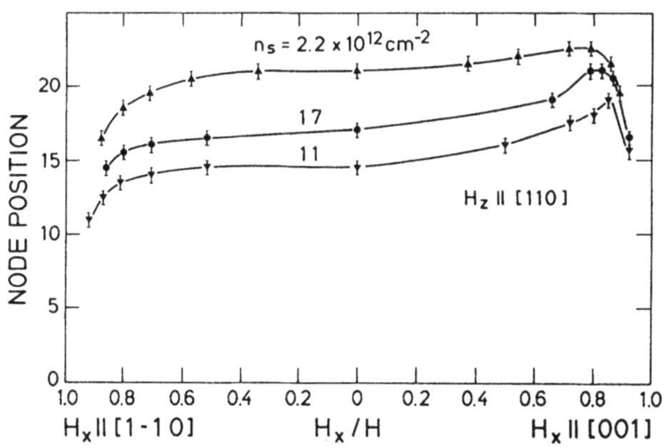

FIG.2. Dependence of the filling factor corresponding to the node position on the relative value H_x/H of the magnetic field component H_x parallel to the channel for the two H_x orientations and different carrier concentrations. Here H is a total magnetic field. Corresponding concentrations are shown near the curves.

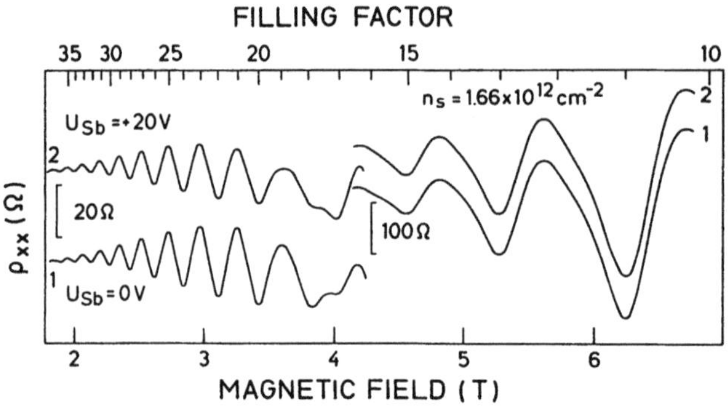

FIG.3. Effect of the substrate bias U_{sb} on the node position. The curves 1 (U_{sb}=0 V) and 2 (U_{sb}=+20 V) were measured in perpendicular magnetic field at different values of the gate voltage adjusted to produce the same carrier concentration $n_s=1.66\times10^{12}$ cm^{-2}. T=1.4 K.

and nonzero node shift. The dependence of the node shift on the H_x orientation we relate to the crystal anisotropy of the g- factor. An explanation of this effect cannot be done in a simple way and needs rather complicated calculations.

The effect of the substrate bias on the node position can easily be understood qualitatively. The zero-field spin splitting can exist only in the absence of the inversion symmetry. In the Si p-channels the asymmetry of the confining potential is the only origin of this effect. The potential drop near the Si-SiO_2 interface is much sharper than that in between the channel and the bulk of the sample originating from the gate electric field. An application of the positive substrate bias reduces the potential asymmetry which leads to the decrease of the zero-field splitting. As a result the node position should shift to larger filling factors [2].

The author gratefully acknowledges a grant from the Alexander von Humboldt Foundation.

References

1. K. von Klitzing, G. Landwehr, and G. Dorda, Solid State Commun. **14**, 387 (1974).
2. S.I. Dorozhkin and E.B. Ol'shanetskii, JETP Lett. **46**, 502 (1988).
3. S.I. Dorozhkin, Phys. Rev. **B 41**, 3235 (1990).
4. Yu.A. Bychkov and E.I. Rashba, JETP Lett. **39**, 78 (1984).
5. S.I. Dorozhkin, Solid State Commun. **72**, 211 (1989).

Magnetoresistance and Hall Effect Measurements on Molecular-Beam Epitaxy-Grown ZnSe-on-GaAs Epilayers in High Magnetic Fields

P. Kempf[1], *M. von Ortenberg*[1], *T. Marshall*[2], *and J. Gaines*[2]

[1]Institut für Halbleiterphysik und Optik,
Hochmagnetfeldanlage der Technischen Universität Braunschweig,
Mendelssohnstr. 3, W-3300 Braunschweig, Fed. Rep. of Germany

Abstract. Magnetoresistance and Hall effect measurements were performed on MBE-grown ZnSe-on-GaAs-epilayers with donor concentrations on the insulating side of the metal-insulator transition in high magnetic fields and temperatures from 1.8 K to 300 K. In the hopping regime, a temperature and concentration dependent negative magnetoresistance effect is observed. The results are compared to the Yosida-Toyozawa localised-spin model and analysed using a semiempirical formula first introduced by Khosla and Fischer.

1. Introduction

In recent years, a lot of work has been done to understand the formation and the properties of an impurity band in n-conducting bulk ZnSe [1] - [4]. Molecular Beam Epitaxy (MBE) is now capable of producing high-quality epilayers. Thus, new interest has been raised in research and applications. On In-doped ZnSe epitaxial layers on GaAs substrates, transverse DC-magnetoresistance and Hall effect measurements were performed in a temperature range from 1.8 K to 300 K and in magnetic fields up to 17 Tesla. The data were taken on cleaved-square van der Pauw samples, with indium contacts in the corners. This paper extends previous work of T. Marshall and J. Gaines [5] on the same set of samples. Their results are summarised in table 1 and show that all samples have a donor concentration of less than the critical value of about $4.4 \cdot 10^{17} cm^{-3}$ for the MI-transition. At low temperatures, all samples obey Mott's $T^{1/4}$-law for variable-range hopping.

Table 1: Donor and acceptor concentrations of the examined samples after [5]. The value N_D for sample 371 is estimated in accordance to the systematic behavior of T_M.

Sample this work	in [5]	N_D ($10^{17} cm^{-3}$)	N_A ($10^{17} cm^{-3}$)
273	1	3.52	0.44
308	3	2.67	0.07
309	4	3.32	0.00
310	5	2.69	0.10
337	6	2.13	0.25
371	/	4.1	

Springer Series in Solid-State Sciences, Vol. 101
High Magnetic Fields in Semiconductor Physics III Editor: G. Landwehr

2. Magnetoresistance Results

Magnetoresistance in n-type bulk-ZnSe has been studied before by [2], [4],and [6], but mostly for compensated samples and lower magnetic fields. To our knowledge, high-magnetic-field resistance studies on ZnSe have not been performed so far. Investigations similar to ours have been made by Khosla and Fischer [7] for In-doped CdS samples mostly on the metallic side of the MI transition.

Fig. 1 shows the relative magnetoresistance normalised to the value at B = 0 Tesla, $\Delta\rho/\rho(0) = (\rho(B) - \rho(0))/\rho(0)$, for two representative samples. Typically, negative magnetoresistance (NMR) starts in the hopping conduction regime between 30 and 60 K and increases with decreasing temperature. Some

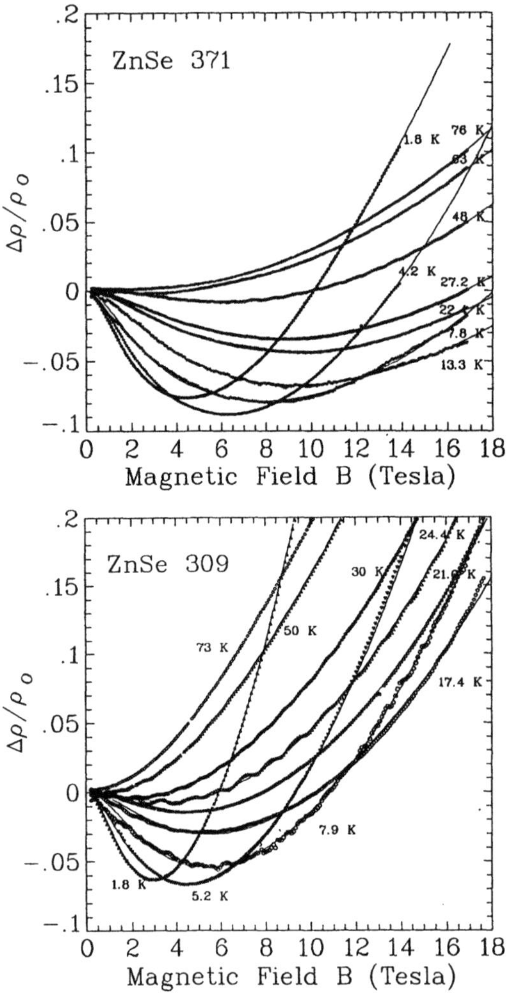

Fig. 1: Magnetoresistance results for a) sample 371 ($N_D \approx 4.1 \cdot 10^{17} cm^{-3}$ and b) sample 309 ($N_D \approx 3.32 \cdot 10^{17} cm^{-3}$). Symbols denote the data, lines are the fits.

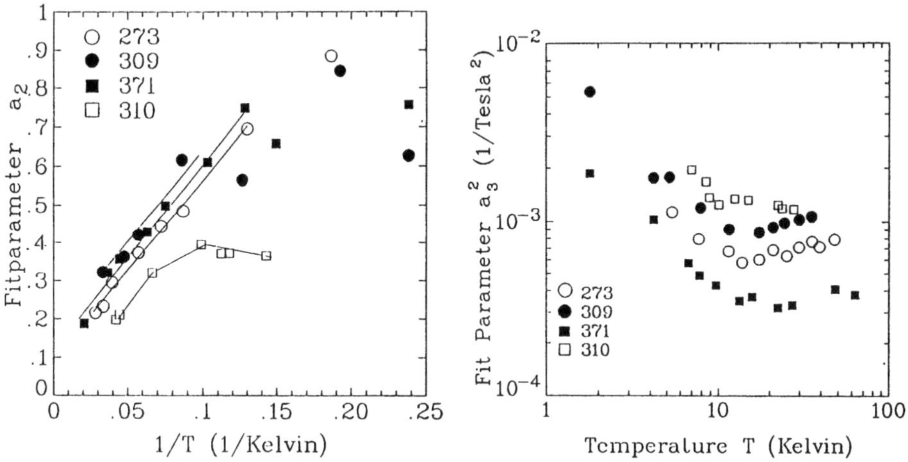

Fig. 2: Dependence of the fit parameters a) a_2, and b) a_3^2 on the temperature. The lines are to guide the eye only.

of the samples show a maximum of $-\Delta\rho/\rho(0)$ in the range 2 - 8 K. In accordance to [4], NMR is more pronounced for samples with a higher donor concentration.

To analyse our data, we tried to apply the localised-spin-model by Yosida [8] and Toyozawa [9]. In this theory, NMR is proportional to the squared magnetisation of the localised spin system: $\Delta\rho/\rho(0) \sim M^2$ where $M^2 \sim \tanh^2(\mu B/kT) \sim B^2$ for low magnetic fields and tends to saturation for high magnetic fields. This model explains some qualitative features of our data, e.g. the increase of NMR with increasing donor concentration and decreasing temperature. However, the $\Delta\rho/\rho(0)$ for our data already deviate for low fields ($B < 0.5$ Tesla) from the predicted quadratic dependence, long before the minimum of the magnetoresistance is reached. We conclude that the Yosida-Toyozawa model is no good description as far as our measurements are concerned. That does not mean that the concept of localised magnetic moments is not valid. A semiempirical formula based on this concept has been successfully used to model magnetoresistance data on CdS [7]:

$$\frac{\Delta\rho}{\rho(0)} = -a_1 \ln(1 + a_2^2 B^2) + \frac{a_3^2 B^2}{1 + a_4^2 B^2}$$

This expression gives good fits to our data for the samples with $N_D > 2.7 \cdot 10^{17} cm^{-3}$. Its positive component is based on a model by Sondheimer and Wilson [10] where conduction in two bands of differing conductivites (i.e. conduction and impurity band) is assumed in which the magnetic field induces a change in the relative populations. In our case, the fit parameter a_4 can be neglected for temperatures above 150 K and below 70 K, indicating transport in the conduction and impurity band, respectively, alone. The negative component is extracted from calculations of Appelbaum [11] using third-order expansion of the s-d exchange Hamiltonian, which accounts for the scattering of conduction electrons on localised impurities. The parameters a_2 and a_3 show a pronounced temperature dependence which is summarised in fig. 2. a_2 should be proportional to $1/T$ after [7], but a logarithmic dependence on $T^{-\alpha}$ where $\alpha \approx 0.25 - 0.5$ cannot be excluded within the accuracy of our data. a_3 which accounts for the positive component shows a minimum in the hopping conduction regime around 15 K. For the samples with lower donor concentration, a much more pronounced positive component of the magnetoresistance is observed which cannot be described by a simple B^2 dependence.

3. Hall Effect Results

The Hall effect measurements in van der Pauw geometry were taken in opposite magnetic field directions in order to eliminate magnetoresistance effects and currents were reversed to eliminate thermal voltages. Despite that, reliable measurements appeared to be increasingly difficult for decreasing temperatures. An example for the behavior of the Hall coefficient is given in fig. 3. The small inset shows the Hall coefficient for low magnetic fields and for B = 17 Tesla. The Hall coefficient for low magnetic fields shows a temperature dependence typical for impurity conduction (see e.g. [12]). At high magnetic fields, the Hall coefficient decreases and tends to saturation. This behavior is most pronounced around the Hall maximum and may be connected to two-band behavior. At the lowest temperatures, in the variable range hopping regime, R_H increases with increasing magnetic field.

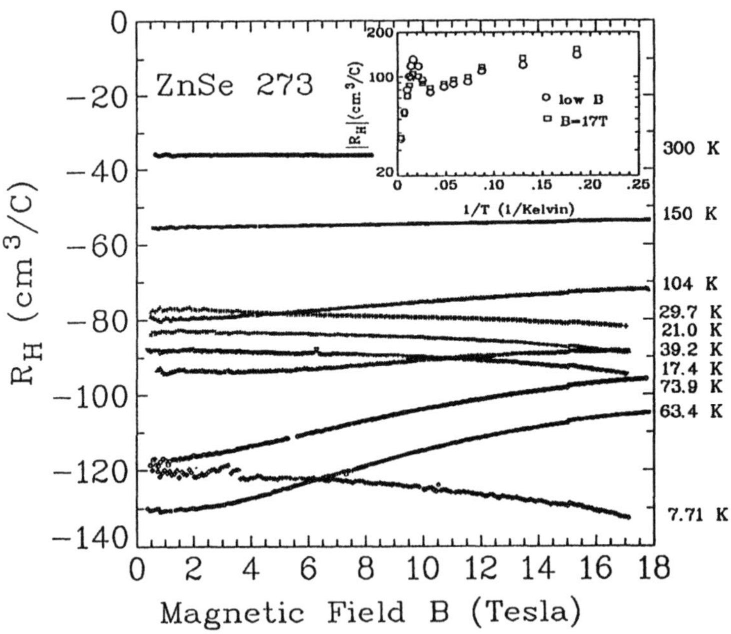

Fig. 3: Dependence of the Hall coefficient R_H for sample 273 on temperature and magnetic field. The inset compares the values of the Hall coefficient for low magnetic field and B = 17 Tesla.

4. Discussion

Magnetoresistance and Hall effect measurements on ZnSe-on-GaAs epilayers with donor concentrations close to the insulating side of the MI-transition were performed with the following results:

1. In the hopping regime, pronounced negative magnetoresistance effects are observed whose magnitude increase with increasing donor concentrations. Excellent fits can be achieved by using a semiempirical formula by Khosla and Fischer [7] which allows the analysis of the magnetoresistivity in terms of a positive and negative component.
2. The analysis of the positive component of the resistance after [7] and of the high field behavior of the Hall coefficient consistently shows two-band behavior in the temperature range between 150 K and 50 K which is more pronounced for the lower doped samples. For lower temperatures, where the Hall effect becomes anomalous, an increase of the coefficient with increasing field can be tentatively linked to magnetic freeze-out effects. A more detailed analysis will be the subject of future work.

Acknowledgments. We are very much indebted to K. Dybko and W. Dobrowolski for their helpful assistance in the numerical calculations and to H. Simontowski and R. Hoffmann for their technical support in the High Magnetic Field Facility in Braunschweig.

References

[1] B.R. Sethi, P.C. Mathur, J. Appl. Phys. **49**, 3618 (1978)
[2] P.C. Mathur, B.R. Sethi, O.P. Sharma, J. Appl. Phys. **50**, 4463 (1979)
[3] D.D. Neodeoglo, phys. stat. sol. (b) **80**, 369 (1977)
[4] G.N. Ivanova, D.D. Neodeoglo, A.V. Simashkevich, I.N. Timchenko, phys. stat. sol. (b) **103**, 643 (1981)
[5] T. Marshall, J. Gaines, Appl. Phys. Lett. **56**, 2669 (1990)
[6] V.A. Kasiyan, D.D. Neodeoglo, A.V. Simashkevich, I.N. Timchenko, phys. stat. sol. (b) **139**, 559 (1987)
[7] R.P. Khosla, J.R. Fischer, Phys. Rev. B **2**, 4084 (1970)
[8] K. Yosida, Phys. Rev. **107**, 396 (1957)
[9] Y. Toyozawa, J. Phys. Soc. Japan **17**, 986 (1962)
[10] E.H. Sondheimer, A.H. Wilson, Proc. Roy. soc. **A310**, 435 (1947)
[11] J.A. Appelbaum, Phys. Rev **154**, 633 (1967)
[12] E.H. Putley, *The Hall Effect in Semiconductor Physics*, Dover, NY 1968

Part V

Magneto-Transport in 3D Systems

High Magnetic Field Effects in Semimagnetic Semiconductors

M. von Ortenberg

Institut für Halbleiterphysik und Optik,
Hochmagnetfeldanlage der Technischen Universität Braunschweig,
Mendelssohnstr. 3, W-3300 Braunschweig, Fed. Rep. of Germany

Abstract. Among Semimagnetic Semiconductors the Fe-based modifications are of special interest, since the Fe^{2+}-level is degenerate with the conduction band. This fact gives the possibility of a pinning of the Fermi energy, so that the "Three Dimensional Analogue of the Quantum Hall Effect" is observed. Detailed analysis of the experimental data reveals that nonstoechiometry of the compound seriously affects the carrier system. Cyclotron measurements in fields up to 150 Tesla with magnetic field induced hot carrier effects are presented.

1. Introduction

Semimagnetic Semiconductors are a new group of materials with tunable energy bandstructure [1-3]. This tunablity is achieved via the exchange interaction of the quasi-free band electrons with localized electrons of statistically substituted paramagnetic ions in the mixed crystal system. After the initially favoured Mn-doped mixed crystals nowadays iron-based materials attract increasing interest. Specially activity is focused on HgSe:Fe where the Fe-dopant manifests not only as a paramagnetic center but also as a donor degenerate with the conduction band [4]. The latter results in the "Three Dimensional Analogue of the Quantum Hall-Effect" [5].

2. Transport Properties of Fe-based HgSe

The energy bandstructure of HgSe is of the inverted zincblende type, so that an effective negative energy gap is obtained for this zero-gap material. Whereas Mn as paramagnetic dopant manifests essentially by the exchange interaction within the range of investigated conduction or valence band states, for Fe we have to consider the donor properties of the Fe^{2+}-state about 210 meV above the conduction band edge as visualized in Fig. 1 [4]. There seems to be only negligible hybridization between the 3d-Fe^{2+} and the s/p-states of the conduction band. The most striking effect is the pinning of the Fermi energy for iron concentrations $n_{Fe} > 5{*}10^{18}$ cm^{-3} in the mixed valence regime, where Fe^{2+} and Fe^{3+} coexist. This leads to various interesting effects [4]. Experimentally a dramatic increase of the carrier mobility is observed for samples in the mixed valence regime with respect to samples with a Fermi energy below the Fe^{2+}-level. To explain this increase *J. Mycielski* created the concept of the space charge superlattice of ionized donors to suppress long range scattering [6]. This model was later modified by *Kossut*

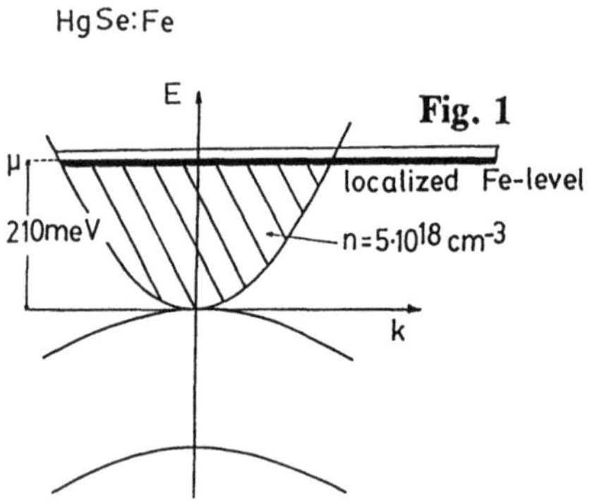

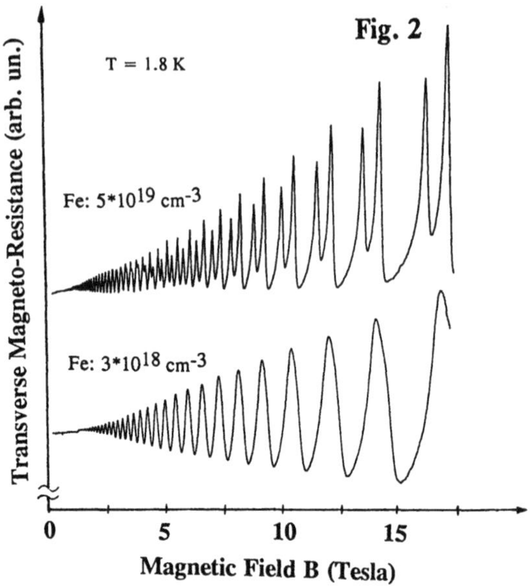

Fig. 1 The energy band structure of HgSe:Fe is of the inverted zincblende type with a Fe^{2+}-donor level about 210 meV above the conduction band edge.

Fig. 2 The transverse magneto-resistance of HgSe:Fe for the two different Fe-concentrations of $n_{Fe} = 5*10^{19}$ cm^{-3} and $n_{Fe} = 3*10^{18} cm^{-3}$ in the upper and lower part of the figure respectively. After [9].

assuming only short range correlation of the Fe^{3+}-scatterers [3]. We present now experimental data, which cannot be explained within those existing models. In Fig. 2 we have reproduced the data of the transverse magneto-resistance for two different samples with nominal Fe-doping of $5*10^{19}$ cm^{-3} and $3*10^{19}$ cm^{-3} for the upper and lower trace respectively. Whereas the upper data are recorded for material well within the mixed valence regime having a Fermi energy of 210 meV, the lower one should have an Fermi energy of only about 100 meV. As a matter of fact, both samples exhibit in the Shubnikov-de Haas effect exactly the same period indicating a carrier concentration of $4.5*10^{18}$ cm^{-3}. The Dingle temperature and spin splitting of both materials are obviously quite different. From the view point of the existing theories we could explain the data in the following way: whereas the $5*10^{19}$ cm^{-3}-sample exhibits definitely the behaviour of a Fermi-level pinned material in the mixed valence regime, the $3*10^{18} cm^{-3}$-sample shows for all but the concentration of quasi-free carriers the properties of a material having only Fe^{3+}- and no Fe^{2+}-ions. The lacking $1.5*10^{18}$ cm^{-3} or more quasi-free carriers to fill up the conduction band to the measured concentration of $4.5*10^{18}$ cm^{-3} with the Fermi level just below the Fe^{2+}-level are provided by Se-vacancies of the slightly nonstoechiometric compound. The measurement of the Hall-effect as represented in Fig. 3 together with the transverse magneto-resistance, however, shows definitely that the Fermi energy is pinned to the Fe^{2+}-level and carrier

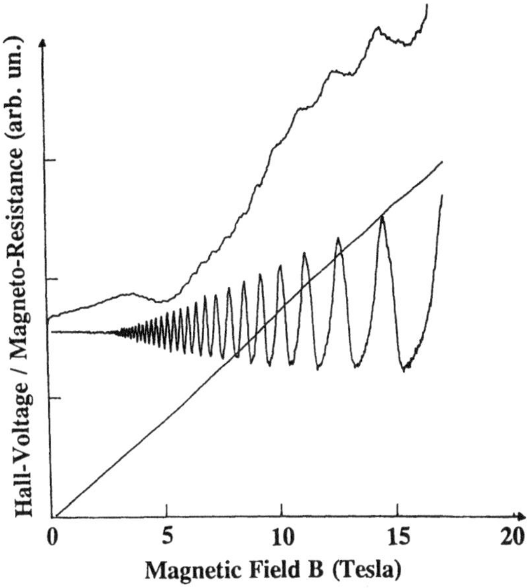

Fig. 3 The modulation of the Hall-voltage (upper part after suppression of the linear part) and magneto-resistance of a sample with $n_{Fe} = 3*10^{18}$ cm^{-3} are in phase. After [9].

fluctuations occur between the localized and delocalized states. The observed oscillations in the Hall-effect are with respect to magnitude and phase correlation to the Shubnikov-de Haas modulation of the transverse magneto-resistance similar to those of the samples with higher Fe-concentration and therefore definitely within the mixed-valence regime. Please note that the maxima of the oscillations of Hall-effect and magneto-resitance coincide on the magnetic field axis. The broad minimum in the Hall-data around 5 Tesla is an artifact and due to a minor nonlinearity of solenoid current and magnetic field. The above results prove that the existence of carrier fluctuations and thus the possibility for a correlation of the Fe^{3+}-ions in the mixed valence regime is not a sufficient condition to explain the exorbitant mobility increase and decrease of the Dingle temperature for carrier concentrations higher than $5*10^{18}$cm^{-3}. Both in experimental investigations and theory substantially more effort is necessary to clear up the situation.

Oscillations in the Hall-effect are not always due to a modulation of the concentration of quasi-free carriers. A careful analysis of the experimental data has to be performed, to ensure that no admixture of the transverse magnetoresistance is present. Even after definite confirmation, that the observed structures are inherent to the Hall effect, the physical origin may differ. So *Mani et al.* have observed oscillatory behaviour in the Hall effect due to scattering effects [7]. Analysis of the phase shift with respect to the Shubnikov-de Haas oscillations in the transverse magneto-resistance gives, however, an unambiguous criterion for the physical mechanism involved. In the data of Fig. 3 scattering effects can be definitely excluded because of the zero-phase shift of Shubnikov-de Haas oscillations. We

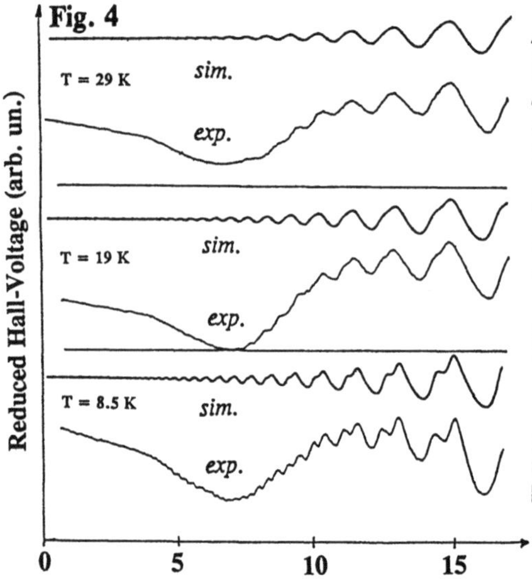

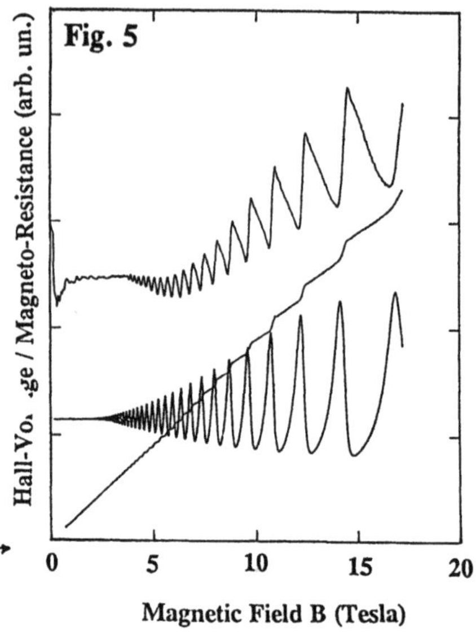

Fig. 4 Direct comparison of experimental Hall voltage after suppression of the linear part and the simulation. After [9].

Fig. 5 The modulation of the Hall-voltage (upper part after suppression of the linear part) of $Hg_{0.88}Fe_{0.12}Se$ has a phase shift of π with respect to the magneto-resistance.

have also studied the corresponding temperature dependence of the oscillations in the Hall-effect as shown in Fig. 4. With increasing temperature the spin splitting is decreased as expected and the oscillations smear out. The model of carrier fluctuations between delocalized and localized levels can simulate also for high temperature very well the experimental data as shown in this figure.

Of special interest is the development of the HgSe:Fe system from simple Fe-doping at relatively small concentrations < 10^{20} cm^{-3} to an actual mixed crystal compound $Hg_{1-x}Fe_xSe$. In Fig. 5 we have plotted the corresponding data of the transverse magnetoresistance and the Hall effect for a mixed crystal with the composition parameter $x = 0.12$ corresponding to an iron concentration of $n_{Fe} = 3.5*10^{21}$ cm^{-3}. Please note that now only 1.5 per mille of the Fe^{2+}-ions are ionized to Fe^{3+}. The spin-splitting is now so large, that at low temperature the coincidence of the spin levels of adjacent Landau states is encountered as verified by temperature dependent measurements. The Dingle temperature is still of very high quality. There is a very pronounced modulation of the Hall-voltage about ten times larger as for $n_{Fe} \approx 10^{19}$ cm^{-3}. The striking result is, however, that there is a phaseshift of π between the Shubnikov-de Haas oscillations and the modulation of the Hall-voltage. This indicates that the dominant physical mechanism involved is not the carrier fluctuation as observed for lower Fe-concentrations. Similar results were obtained for a composition parameter of $x = 0.1$.

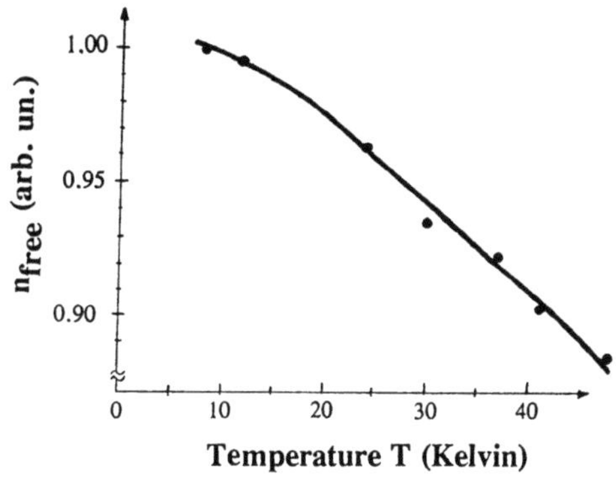

Fig. 6 The concentration of quasi-free carriers in $Hg_{0.88}Fe_{0.12}Se$ decreases with increasing temperature.

The study of the temperature dependence of the Hall effect for x = 0.12 shows that the effective concentration of quasi-free carriers decreases with increasing temperature as shown in Fig. 6. There is no model to explain this effect so far.

3. Magnetooptical Investigations in the Mixed Valence Regime of HgSe:Fe

Whereas magneto-transport measurements result in detailed information on the carrier system with respect to the Fermi energy as normalizing quantity, no direct information on the energy splitting and the cyclotron mass of the material is derived in this way in contrast to magneto-optical measurements. Due to the high carrier concentration in HgSe:Fe the experimental magneto-optical investigation is not straightforward, since radiation of the submillimeter range encounters a negative dielectric function and thus total reflection. Because of this fact the strip-line method provides the most suitable experimental setup [8]. In Fig. 7 we have plotted the strip-line spectra on $Hg_{0.99}Mn_{0.01}Se$:Fe for three different radiation wavelengths as a function of the magnetic field up to 30 Tesla [9]. The dominant structure is independent of the applied radiation frequency and reflects the optical Shubnikov-de Haas effect as nonresonant quantum effect. The exchange-interaction-enhanced spin splitting of the sharp, frequency-dependent resonance line between 20 and 30 Tesla is due to the paramagnetic resonance of the Mn^{2+} embedded into the high-mobility matrix of HgSe:Fe. The rather obscured structure below 3 Tesla can be interpreted as the low field combined spin-flip transition [10]. Due to the special "parallel" configuration of the strip-line system no cyclotron resonance is observed [8].

To measure the cyclotron resonance directly, evidently a direct transmission experiment in Farady configuration has to be performed. To obtain a detectable transmission even for a sample thickness of only 1 μm we have to apply radition beyond the plasma edge at 87 meV. Such high radition energies, however, require corresponding high magnetic field intensities in the megagauss regime for the study of the cyclotron resonance. In Fig. 8 we have plotted the experimental data of the direct transmission of 9.6 μm wavelength radition through a HgSe:Fe

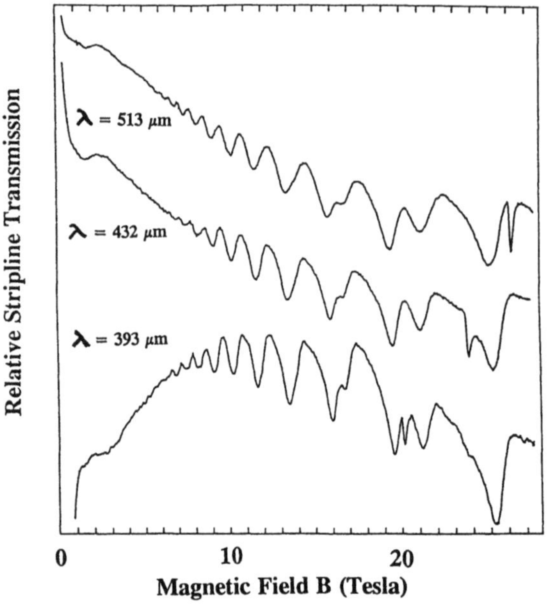

Fig. 7 The strip-line transmission on Hg(Mn)Se:Fe for three different radiation wavelengths. After [11].

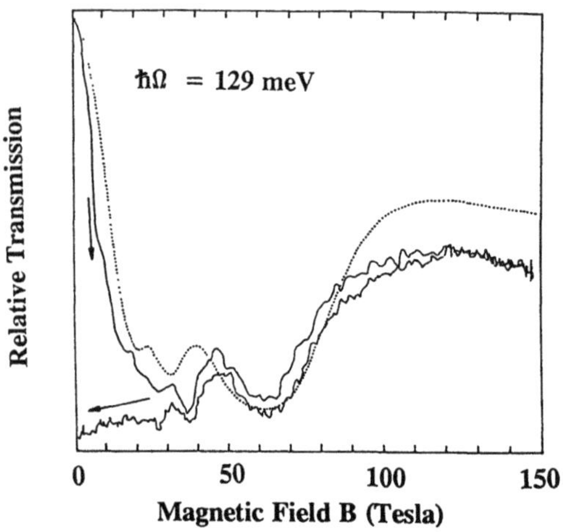

Fig. 8 The megagauss cyclotron resonance on HgSe:Fe for up and down sweep by the solid curves in direct comparison with the simulation given by the dotted curve. After [11].

sample of about 1 μm thickness for up and down sweep by the solid curves [11]. A pronounced multi-resonance spectrum is recorded with the cyclotron resonance at about 60 Tesla. The low field resonances can be attributed to harmonic transitions. Please note the strong decrease of transmission at zero field after the magnetic field pulse. Due to the strongly induced eddy currents in the sample the carriers are actually under "hot carrier" condition and have changed the dispersion of the material. This fact is also reflected by the large line width of the cyclotron resonance which exceeds by far the value expected from the Dingle temperature derived from magneto-transport experiments. The resulting cyclotron mass has a value of $m_c = 0.55*m_0$.

4. Prospects

Due to the high mobility and the high average carrier concentration HgSe:Fe is an excellent host material to study the influence of additional perturbations. In this way can for $Hg_{1-x}Mn_xSe$:Fe the exchange interaction between the Mn^{2+} and the quasi free cariers be studied in detail. Other dopants as Co, Ni, Zn etc. are also promising candidates. For future applications, however, the combination with Q2D systems is of great interest. So the transition from the 3D to the Q2D quantum Hall-effect should be observable [5]. We propose now an additional new Q2D system for HgSe:Fe: the "scattering super lattice". Since the carrier mobility of the Fermi-level pinned system depends slightly, but definitely on n_{Fe} [12], we can produce a superlattice having in the different periodic layers the same energy band structure and the same Fermi energy but different carrier lifetime as visualized schematically in Fig. 9. Formally we can attribute to the different layers a complex hamiltonian, whose imaginary part describes the scattering properties. Since the real part is essentially the same for all the layers, the periodic super-lattice potential is purely imaginary. The investigation of such a superlattice from both experimental and theoretical point of view is a challenge for the future.

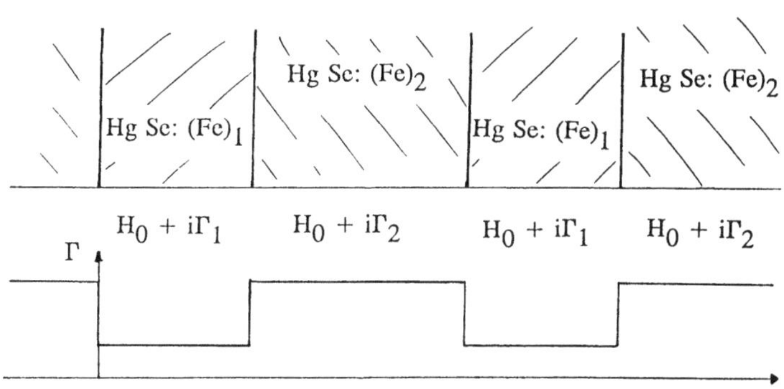

Fig. 9 The "scattering super lattice" with purely imaginary super-lattice potential Γ.

Acknowledgement. The author likes to express his gratitude to the many coworkers in the international cooperation between Braunschweig, Leuven, Tokyo, and Warsaw involved in this investigation. Only within this network the present study was possible.

5. References

[1] R.R. Galazka and J. Kossut, Lecture Notes in Physics **133**: *Narrow Gap Semiconductors and Applications*, ed. W. Zawadzki, Springer Verlag 1980, p. 245

[2] N.B. Brandt and V.V. Moshchalkov, Advances in Physics **33**, 193 (1984)

[3] J. Kossut, Semiconductors and Semimetals **25**, 183 (1988)

[4] A. Mycielski, MRS Symp. Proc. **89**, 159 (1987)

[5] M. von Ortenberg, O. Portugall, W. Dobrowolski, A. Mycielski, R.R. Galazka, and F. Herlach, J. Phys. C: Solid State Phys. **21**, 5393 (1988)

[6] J. Mycielski, Solid State Commun. **60**, 165 (1986)

[7] R.G. Mani, J.R. Anderson, W.B. Johnson, Proc. of the Intnl. Conf. *High Magnetic Fields in Semiconductors Physics*, Würzburg 1986, Springer Series in Solid-State Sciences **71**, p. 436 (1987)

[8] M. von Ortenberg, in *Infrared and Millimeter Waves*, ed. K.J. Button, Vol III, p. 275 (1980)

[9] I. Laue, O. Portugall, M. von Ortenberg, and W. Dobrowolski, this volume

[10] W. Dobrowolski, M. von Ortenberg, J. Thielemann, and R.R. Galazka, Phys. Rev. Lett. **47**, 541 (1981)

[11] O. Portugall, H. Yokoi, S. Takeyama, L. van Bockstal, K. Buchholz-Stepputtis, M. von Ortenberg, N. Miura, F. Herlach, and W. Dobrowolski, Proc. of the *20th Intnl. Conf. on the Physics of Semiconductors*, Thessaloniki 1990, p. 2287

[12] F.S. Pool, J. Kossut, U. Debska, R. Reifenberger, Phys. Rev. **B35**, 3900 (1987)

The 3D Analogue of the Quantum Hall Effect in HgSe:Fe and Its Temperature Dependence

I. Laue[1], *O. Portugall*[1], *M. von Ortenberg*[1], *and W. Dobrowolski*[2]

[1]Institut für Halbleiterphysik und Optik,
Hochmagnetfeldanlage der Technischen Universität Braunschweig,
Mendelssohnstr. 3, W-3300 Braunschweig, Fed. Rep. of Germany

[2]Institute of Physics, Polish Academy of Sciences,
Al. Lotnikow 32/46, PL-02-668 Warsaw, Poland

Abstract. We present temperature dependent measurements of oscillations in the Hall resistance in $Hg_{1-x}Fe_xSe$ referred as "the 3D analogue of the quantum Hall effect" [1] in magnetic fields up to 18 Tesla and temperatures between 4.2 and 50 K. Highly doped material ($x \approx 0.1$) exhibits a strong modulation of the Hall resistance which cannot be explained by fluctuations of the carrier concentration in a system with a pinned Fermi energy. The measurements are compared with simultaneously recorded Shubnikov-de Haas-data.

1. Introduction

HgSe:Fe is a diluted magnetic semiconductor with a localized resonant donor state degenerate with the conduction band [2]. For sufficiently high donor concentrations the Fermi energy gets pinned to the localized state, resulting in a mixed valence system of coexisting ionized and unionized donors.

In recent years several phenomena arising from this behavior were studied. The application of an external magnetic field for example, causes fluctuations of the concentration of the quasi-free charge carriers. These fluctuations can be observed as a modulation of the Hall resistance, which is in phase with the oscillations of the transverse magnetoresistance. This effect is referred as "the 3D analogue of the quantum Hall effect" (3D QHE) [1] because the oscillations are the limiting case of superpositions of the plateau-like 2D-QHE- structures for each of the k_z-quantum numbers in bulk material.

2. Measurements on weakly doped material

We studied the temperature dependence of the Hall oscillations in a sample with an iron concentration of 5×10^{19} cm^{-3} (Fig.1). The curves were obtained by measuring the Hall voltage and subtracting a voltage proportional to the current through the magnet coil. The oscillations have an amplitude of 0.8% of the Hall voltage at 4.2 K. We observed no phase shift between these oscillations and the oscillations of ρ_{xx}. The Hall data could be compared with numerical computations of the carrier concentration under the boundary condition of a fixed chemical potential [3].

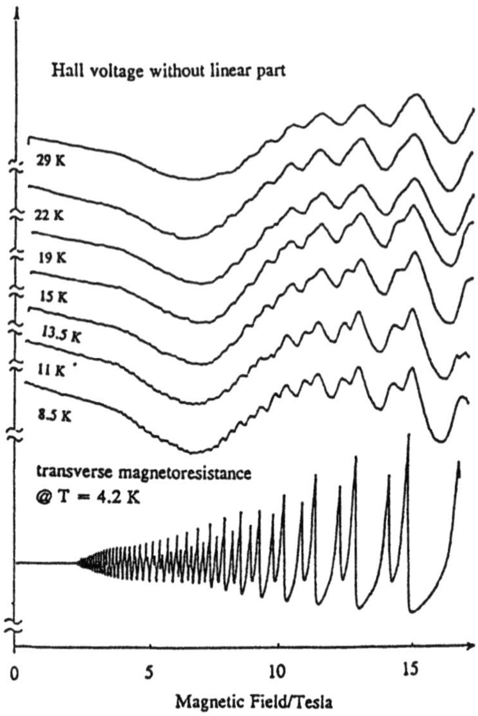

Fig. 1: The temperature dependence of the Hall effect in weakly doped $Hg_{1-x}Fe_xSe$ (x=0.003), compared with a Shubnikov-de Haas-measurement at liquid helium temperature.

We observed a monotoneous increase of the oscillation half-width with increasing temperature which is essentially caused by the smoothing of the Fermi-Dirac- distribution. On the other hand the amplitude of the oscillations decreases monotoneously with rising temperature. The broad minimum of the curves between 5 and 10 Tesla is caused by nonlinearities of the Bitter coil. The Hall resistance and thus the carrier concentration was independent from temperature in the considered temperature region.

Measurements on a sample with an iron concentration of 2×10^{19} cm^{-3} showed comparable results.

3. Measurements on highly doped material

We also made Hall measurements on highly doped $Hg_{1-x}Fe_xSe$ (x = 0.10 ... 0.12 according to an iron concentration of 2 ... 2.4×10^{21} cm^{-3}). We observed a strong modulation of the Hall voltage, which is, however, shifted in phase in such a way that a minimum of the Hall resistance corresponds to a maximum of the transverse magnetoresistance. For this fact these oscillations cannot be explained by fluctuations of the carrier concentration due to the pinning of the Fermi energy. On the other hand the large amplitude of the modulation (about 5% of the Hall voltage at T=4.2 K) excludes scattering effects as an essential cause [4]. The observed oscillations were reproducible with other samples and symmetric in both directions of the

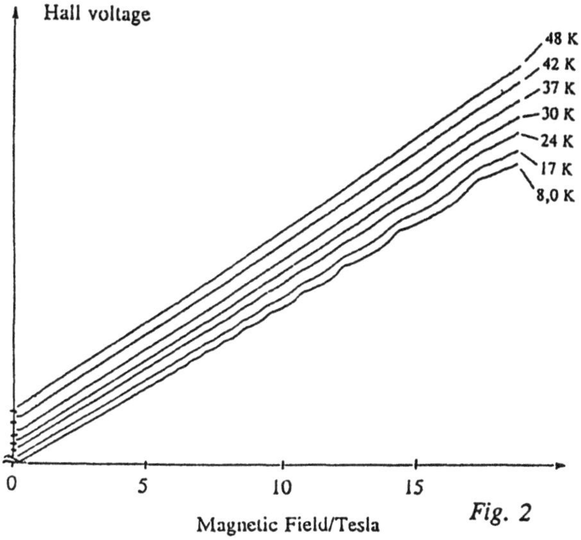

Fig. 2: The temperature dependence of the Hall effect in $Hg_{1-x}Fe_xSe$ with x=0.12. The origin of the Hall voltage was slightly shifted for each temperature for better comparison.

magnetic field. One possible mechanism causing these oscillations might be due to the direct charge transport j_{loc} between localized states, which cannot be neglected for such high donor concentrations. The current density j_{loc} may be considered in a good approximation as being independent of the magnetic field B, while the charge transport by delocalized electrons j_{del} is dependent on B through ρ_{xx}. The boundary condition of a constant total current density $j_{loc} + j_{del}$ leads to a minimum of j_{del} at a maximum of ρ_{xx}. This mechanism might explain the observed phase shift between ρ_{xx} and $U_{Hall} = R_{Hall} \times I_{del}$. The amplitude of this modulation is difficult to estimate because we have no information on the grade of homogenity of the spatial donor distribution in the host lattice which has much influence on the hopping probability between localized states.

Fig. 2 shows the temperature dependence of the Hall effect in $Hg_{1-x}Fe_xSe$ for x=0.12. For higher temperatures the modulation vanishes and leads to the classical limit. We observe a slight increase of the Hall resistance with increasing temperatures indicating a decrease in the carrier concentration of about 10% for temperatures up to 40 K. Shubnikov-de Haas- measurements (Fig. 3) confirm this result. The magnetic field positions of the oscillations are shifted to lower magnetic field intensities with increasing temperature.

It is remarkable that a spin-splitting of the Shubnikov-de Haas-oscillations becomes observable for higher temperatures (arrows). At liquid helium temperature the spin-splitting $g^*\mu_B B$ is equal to an integer multiple of the Landau level splitting $\hbar\omega_c$. A numerical simulation of this behavior with a Landau-level scheme had no satisfying

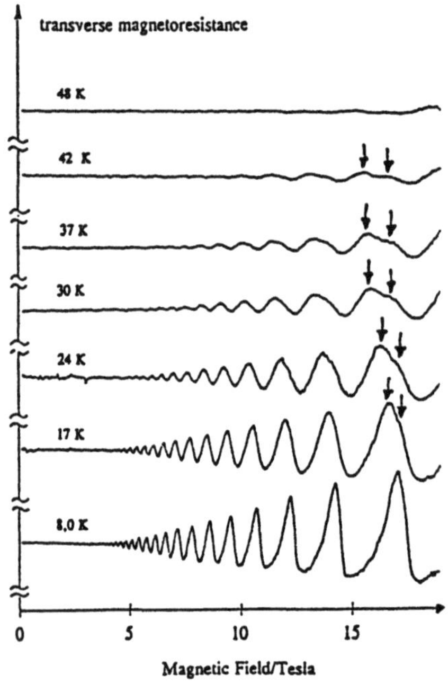

Fig. 3: The temperature dependence of the Shubnikov-de Haas-effect in $Hg_{1-x}Fe_xSe$ with x=0.12. The arrows indicate a spin-splitting of the oscillations for higher temperatures.

results because too many band parameters for this mixed crystal system (x=0.12) are not yet known (e.g. band gap, spin-orbit-splitting, direct Fe-Fe exchange interaction).

On the other hand we could fit well a Landau-level scheme using the Kossut-Gałazka-model [5] to our magnetotransport measurements of more weakly doped material ($x \approx 0.001$). These simulations result in values for the exchange interaction matrix elements α=-1.8 eV ±25% and β=+3.0 eV ±25%. Fitting procedures for different iron concentrations show best results, if all donors are considered to contribute to the exchange interaction in the same way. Therefore the common neglection of the unionized Fe^{2+} donors which are assumed to be magnetically inactive clearly does not hold.

4. Summary and prospects

The oscillations of the Hall resistance in $Hg_{1-x}Fe_xSe$ referred as the 3D analogue of the quantum Hall effect [1] could be measured reproducibly in samples with an iron concentration between 10^{19} and 10^{20} cm^{-3} (x = 0.0006 ... 0.006) in fields up to 18 Tesla. Their temperature dependence was studied and compared with numerical simulations.

Furthermore we have investigated magnetotransport in highly doped material ($x \approx 0.1$) and have observed a strong modulation of the Hall resistance which cannot be explained to be caused by fluctuations of the carrier concentration due to the pinning of the chemical potential.

Whereas the carrier concentration of the more weakly doped material ($x \approx 0.001$) seems to be independent from temperatures up to 50 K, the carrier concentration in the highly doped samples decreases about 10% between 4.2 K and 40 K.

The challenge for the future is to produce quasi-2D HgSe:Fe samples by "hot wall epitaxy" to investigate the dependence of the Hall oscillations on the layer thickness. The transition between the 2D and the 3D case of the quantum Hall effect will be examined.

Our measurements have also shown, that it is worthwile to pay more attention to the highly doped HgSe:Fe samples in future, which were somehow neglected in most of the common publications about this material.

References

[1] M von Ortenberg, O Portugall, W Dobrowolski, R R Galazka, F Herlach, J Phys C **21**, 5393 (1988)

[2] A Mycielski, P Dzwonkowski, B Kowalski, B A Orlowski, M Dobrowolska, W Dobrowolski, J M Baranowski, J Phys C **19**, 3605 (1986)

[3] I Laue, O Portugall, M von Ortenberg, W Dobrowolski, to be published in Acta Physica Polonica

[4] R G Mani, J R Anderson, W B Johnson, in: Proc Int Conf on High Magnetic Fields in Semicond Phys 1986, Springer Series in Solid State Physics **71**, Springer, Berlin, 1987.

[5] J Kossut, in: Semimetals and Semiconductors **25**, Academic Press, 1988.

Influence of Localization on the Hall Effect in Narrow-Gap, Bulk Semiconductors

R.G. Mani and J.R. Anderson

Department of Physics, University of Maryland,
College Park, MD 20742, USA

Abstract. Our transport study of the narrow gap, bulk (3D) semiconductors $Hg_{1-x}Cd_xTe$ and InSb reveals incipient, non-quantized "Hall plateaus" which coincide with the minima of Shubnikov-de Haas oscillations, analogous to the quantum Hall effect in 2-dimensional systems. We attribute this effect to the existence of quasi-mobility gaps at the bottom of each Landau level, originating from localization due to shallow hydrogenic donors and disorder.

1. Introduction

Although Shubnikov-de Haas (SdH) oscillations of the magnetoresistance in 3D systems have been explained,concurrent oscillatory deviations in the Hall effect, from the classical behavior ($R_{xy} \sim H$), are still not well understood [1]. The effect of localization upon the Hall effect in 3D semiconductor systems has attracted limited attention although localization is the key factor which influences the width of the integral quantum Hall plateaus in 2D systems [2]. Finite width of the quantum Hall plateaus and dissipationless current flow in 2D systems has been attributed to the existence of localized states in the tails of the 2D Landau bands, which creates a mobility gap at the Fermi level for near integral filling factors. In contrast, the simple, ideal 3D Landau level (LL) spectrum does not allow for mobility gaps between Landau levels since each LL extends to infinite energy. Thus, dissipationless transport is not expected in 3D semiconductor systems. Also, quantization of R_{xy} to combinations of fundamental constants would not occur in 3D systems since parallel conduction in the 3rd dimension introduces a geometrical factor, the sample thickness, which varies from sample to sample.

2. Experiment

In our transport studies, the four terminal resistance, R_{xx}, and the Hall effect were measured with a constant dc current, I, applied to Hall bar type samples in the transverse configuration, $I \perp H$, at low temperatures, $0.5K < T < 4.2K$. Here, we report results for an InSb sample and a $Hg_{1-x}Cd_xTe$ sample, HCT_A ($x=0.206\pm0.004$). In Fig. 1, we plot the low-temperature transport data for the InSb sample, n(4.2K) =

Springer Series in Solid-State Sciences, Vol. 101
High Magnetic Fields in Semiconductor Physics III Editor: G. Landwehr

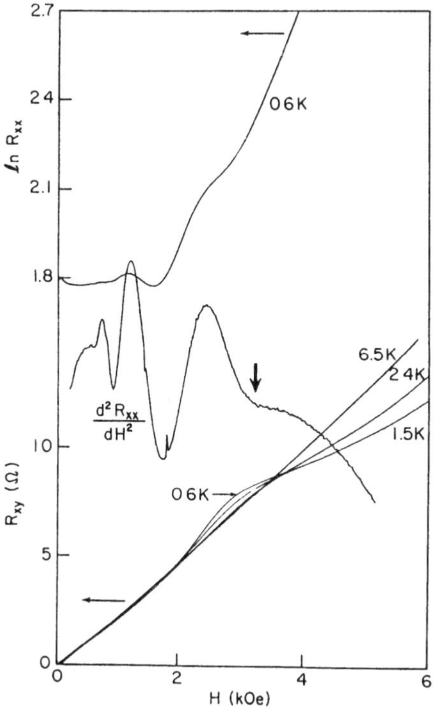

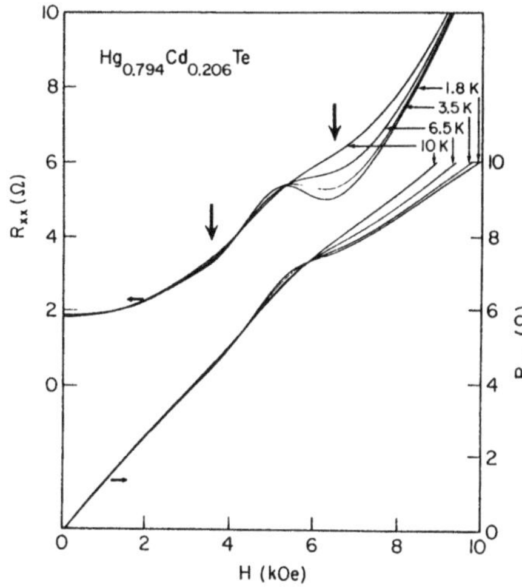

Fig. 2) The temperature dependance of R_{xx} and R_{xy} are shown for HCT_A. Note the temperature dependence of the SdH minimum (heavy arrows).

Fig. 1) As R_{xx} exhibits weak SdH oscillations in InSb (top), the oscillatory part, d^2R_{xx}/dH^2, of R_{xx} is shown in the center. The heavy arrow indicates the last SdH minimum. R_{xy} shows a "plateau" at the lowest temperatures (bottom).

$5x10^{14}$ cm^{-3} and $\mu(4.2K)$ = 125,000 cm^2/Vs. Fig. 1 (top) shows that R_{xx} increases rapidly *vs* H with weak SdH oscillations superimposed upon the background. The SdH oscillations were enhanced using standard field modulation techniques and the results are also shown in Fig. 1 (center). The Hall effect, shown in Fig. 1 (bottom), exhibits classical behavior for T > 6.5K and develops inflections with decreasing temperatures which result in a "Hall plateau" for H ~ 3.5kOe. Note that the "Hall plateau" coincides with the last SdH minimum but $R_{xy} \neq h/e^2$ as for a $\nu=1$ quantum Hall plateau in a 2D system [2]. Finally, magneto-optical studies on the same sample indicate impurity cyclotron resonance (ICR) in addition to free electron cyclotron resonance *at fields as low as* H ~ 2.5kOe for $\hbar\omega$ = 3meV while SdH and Hall oscillations persist to ~ 4kOe [3].

Fig. 2 shows the transport data for HCT_A, n(4.2K) = $1.1x10^{15}$cm^{-3} and $\mu(4.2K)$ = 300,000cm^2/Vs. The data suggest large R_{xy} oscillations as in the other sample. Also, the SdH oscillations show anomalous behavior *vs* T: The last minimum gets deeper with decreasing T while the height of the last maximum is roughly independent of T. The standard picture of 3D, I ⊥ H, SdH oscillations associates peaks in R_{xx} with an

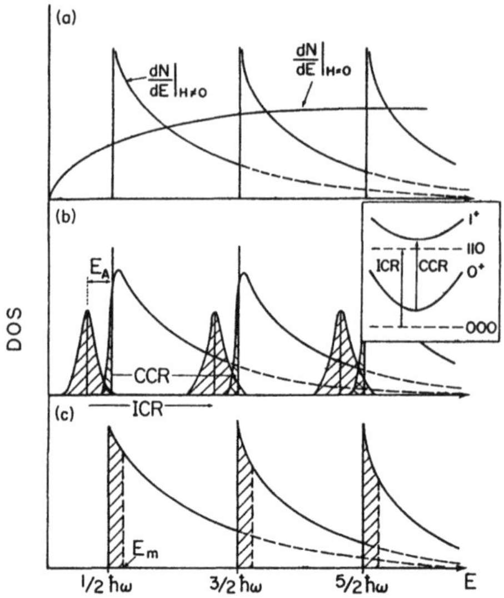

Fig. 3) The ideal 3D - DOS is shown in A. The inset (B) shows transitions associated with ICR and CCR. The model DOS used to simulate the Hall effect is shown in C.

enhancement in the scattering, $R_{xx} \sim \sigma_{xx}/\sigma_{xy}^2$, as the Fermi level, E_F, sweeps through the singularity in the density-of-states (DOS) at the bottom of each Landau level [1]. Thus, the peak SdH amplitude is usually expected to increase as the Fermi function "sharpens-up" with decreasing T. However, our observation of a deeper SdH minimum with decreasing T suggests that scattering is suppressed for particular filling factors, i.e., bands of localized states which do not carry current sweep across the Fermi level *vs* H and the lack of current carrying states at E_F suppresses dissipation.

An illustration of our model for these transport effects is shown in Fig. 3. The ideal, disorder-free, 3D, DOS for H=0 ($dN/dE_{H=0} \sim E^{1/2}$) and $H \neq 0$ ($dN/dE_{H\neq 0}$) are shown in Fig. 3A while the effects of impurities and disorder are illustrated in the rest of the figure. Previous magneto-optical studies of InSb and $Hg_{1-x}Cd_xTe$ have revealed transitions between donor bound states (inset Fig. 3B) associated with the N=0 and the N=1 Landau levels, i.e., ICR, in addition to the free-electron cyclotron-resonance (CCR) [4]. The transition-energy between donor bound states of different Landau levels is only slightly greater than the Landau level separation, $\hbar\omega_c$, since the hydrogenic binding energy $R^* << \hbar\omega_c$ in these narrow gap systems [5]. For our transport studies, these results imply the existence of a reservoir of quasi-localized states below each Landau level (see Fig. 3B). Spatial fluctuations of the bandedge due to disorder would also effectively localize a fraction of the zero-transverse-momentum, i.e., $\hbar k_z \sim 0$, states at the bottom of each Landau level (see Fig. 3B). We assume that these quasi-localized states do not contribute to R_{xy}. These points suggest that our transport results may be simulated using the model DOS shown in Fig. 3C in which each Landau level includes a mobility edge,

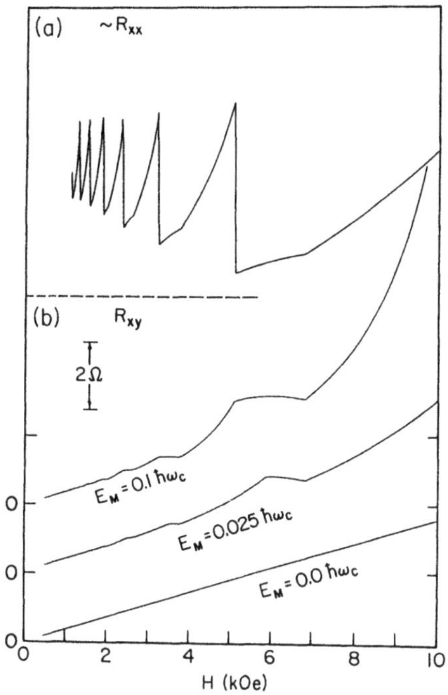

Fig. 4) SdH (R_{xx}) oscillations are shown in A. In B, R_{xy} is shown for three values of E_M (see fig. 3C).

E_m, to separate the quasi-localized states from the current carrying states. As in the percolation picture of the QHE, we have also assumed that current carrying states "speed up" and exactly compensate for the lost current carrying capability due to localization [2]. Thus, in our picture, $R_{xy} = H/e\ \Sigma n_{ex}{}^{i} S^{i}$. Here, $n_{ex}{}^{i}$ is the number of extended states below E_F associated with the i'th Landau level, S^i is the "speed-up factor" which assures that the plateau resistance is independent of $E_m/\hbar\omega$, and E_F satisfies charge neutrality by counting the extended states and quasi-localized states at each value of H. The calculated DOS at E_F, neglecting finite temperature effects, which is ~ R_{xx}, exhibits SdH oscillations *vs* H (Fig. 4, top). Fig. 4 also shows that R_{xy} develops "plateaus" *vs* H which become wider as $E_m/\hbar\omega_c$ is increased. This behavior is similar to the observed correlation between enhanced quantum Hall plateau width and increased disorder (localization) in 2D systems [2]. However, the R_{xy} "plateaus" *are not* quantized to $h/\nu e^2$ as in the 2D quantum Hall effect. Finally, minimal scattering occurs when R_{xy} develops "plateaus". However, R_{xx} does not vanish since current carrying states associated with lower LL's occur at E_F even when E_F is pinned in the mobility gap of a particular LL.

3. References

1) see refs. 1-6 in R. G. Mani, Phys. Rev. B 41, 7922 (1990).
2) see The Quantum Hall Effect, ed. R. Prange and S. Girvin, (Springer-Verlag, New York, 1987).
3) J. B. Choi, Ph.D Thesis, Univ. of Maryland, 1989 (unpublished).
4) W. S. Boyle and A. Brailsford, Phys. Rev. 107, 903 (1957).
5) For InSb ($Hg_{0.8}Cd_{0.2}Te$), $R^* \sim 0.7$meV (0.3meV).

Magnetotransport Investigation of HgSe:Ce in High Magnetic Fields

R. Bogaerts[1], *I. Deckers*[1], *B. Momont*[1], *G. Pitsi*[1], *F. Herlach*[1], *M. von Ortenberg*[2], *and W. Dobrowolski*[3]

[1]Laboratorium voor Lage Temperaturen en Hoge-Veldenfysica, KU Leuven, Celestijnenlaan 200 D, B-3030 Leuven, Belgium

[2]Institut für Halbleiterphysik und Optik, Hochmagnetfeldanlage der Technischen Universität Braunschweig, Mendelssohnstr. 3, W-3300 Braunschweig, Fed. Rep. of Germany

[3]Institute of Physics, Polish Academy of Science, PL-02-669 Warsaw, Poland

Abstract. Oscillations of the transverse magnetoresistance and the Hall voltage in HgSe:Ce are studied at temperatures between 1.2 K and 77 K, in magnetic fields up to 35 T. From the Shubnikov-de Haas measurements band structure parameters are calculated and compared to the values for HgSe. Due to scattering effects SdH-like oscillations are observed in the Hall voltage. Resistivity measurements show a minimum in the temperature dependence.

1 Introduction

Diluted magnetic II-VI compounds containing elements of the iron group have been intensively studied. The exchange interaction between the mobile charge carriers and the 3d electrons of the magnetic ions accounts for the strong semimagnetic character of $Hg_{1-x}Mn_xTe$ and $Hg_{1-x}Mn_xSe$, which causes an anomalous temperature dependence of the SdH oscillations [1].

The study of HgSe:Fe has revealed new properties due to the degeneracy of the Fe donor level and conduction band states. In the mixed valency region of Fe ions, large values of the mobility and low Dingle temperatures are observed [2]. These indicate a reduction in efficiency of ionized impurity scattering as well as scattering caused by spin fluctuations (Kondo effect).

Using the rare earth element cerium as a dopant, a new diluted magnetic compound is obtained. By means of magnetoresistance, Hall voltage and resistivity measurements, the interaction between the conduction electrons and the magnetic impurities is studied. In particular, we concentrate on scattering phenomena introduced by the impurities and on the influence of the exchange interaction on the band structure.

2 The Shubnikov-de Haas effect

Samples of n-type HgSe:Ce doped with cerium ($n_{Ce} = 10^{19}\mathrm{cm}^{-3}$) were grown by the Bridgman method at the Institute of Physics, Warsaw. The magneto-

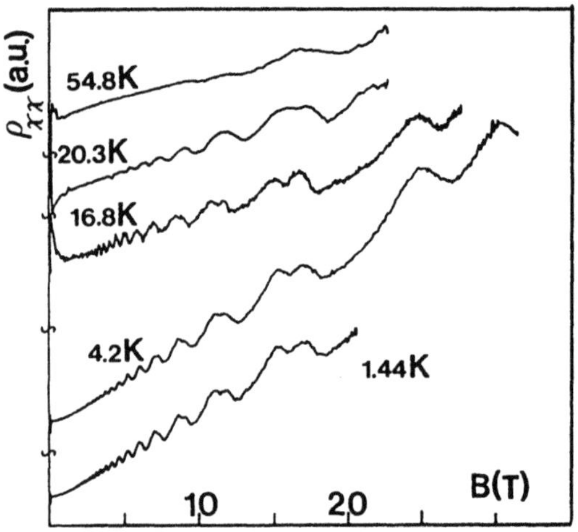

Figure 1: Temperature dependence of the SdH oscillations for sample C1.

transport measurements were carried out in pulsed magnetic fields up to 35 T (sample C1) and DC fields up to 17 T (sample C2). The magnetic field was applied in a [100] direction, perpendicular to the current.

Fig. 1 shows the temperature dependence of the SdH oscillations. No semimagnetic effects, such as a non-monotonic temperature dependence of the amplitude or a shift of the spin splitting, are observed. This justifies the analysis of the SdH data with the formula for ordinary semiconductors [3], and the application of the unmodified Pidgeon-Brown model to fit the oscillation maxima. The parameters obtained are listed in Table 1.

The carrier concentrations derived from the period of the oscillations agree within 1 % with the values calculated from the high field Hall constant. Calculation of the mobility temperature, $T_\mu = e\hbar/2\pi k_B m^* \mu$, allows the non-thermal broadening to be separated into two contributions: one due to inhomogeneities and the other one to electronic scattering. The non-thermal broadening is found to be caused primarily by inhomogeneities.

A spin splitting of the Landau levels is clearly observed in both samples C1 and C2 for the oscillation peak at 16 T and 12 T, respectively. This maximum is due to the level with Landau quantum number 2, as determined in a fit of the oscillation peaks to Landau levels for HgSe (Fig. 2). The g-factors are therefore calculated using the equation given by Ponomarev et al. [4],

$$\frac{H_2^+}{H_2^-} = \frac{\sum_{M=0}^{2}(M^{\frac{1}{2}} + (M+\nu)^{\frac{1}{2}})}{\sum_{M=1}^{2}(M^{\frac{1}{2}} + (M-\nu)^{\frac{1}{2}})}, \quad \nu = \frac{m^*|g|}{2m_0}. \tag{1}$$

The Fermi energies derived from the fit are found to be consistent with the theoretical values, calculated from the Kane model.

The obtained sample parameters for HgSe:Ce agree with the values for HgSe determined in [3] and [5], and show the same dependence on carrier concentration.

Table 1: Sample Parameters.

	n (cm^{-3})	m^*/m_0	T_D(K)	T_μ(K)	g	$\varepsilon_F^{th.}$(meV)	ε_F^{fit}(meV)
C1	1.37 10^{18}	0.0648	8.6	0.8	5.9	107	108
C2	0.96 10^{18}	0.0412	13.0	1.5	10	86	88

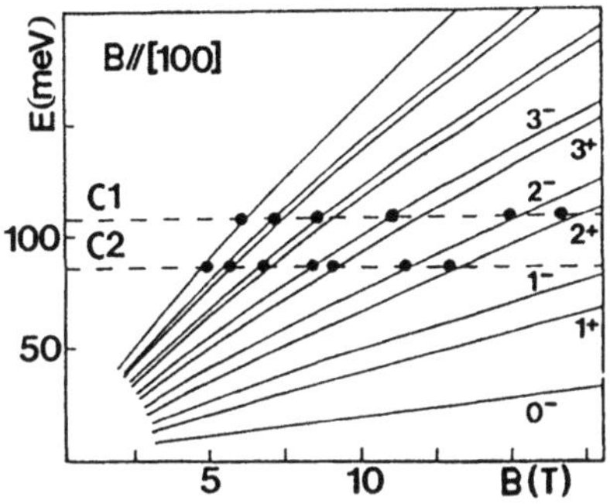

Figure 2: Fit of the oscillation peaks to Landau levels derived from the PB model for HgSe. Band parameters are taken from [5].

3 Effects of electron scattering

Fig. 3 shows a typical Hall voltage trace. After subtraction of the linear part the oscillations are found to be shifted in phase relative to the magnetoresistance oscillations (Fig. 4). The shift, which is approximately π in both samples, has been observed in several semiconductors and is characteristic of scattering [3]. In our measurements, however, it is remarkable that the amplitudes of Hall and SdH oscillations have the same order of magnitude, while scattering gives only a second order contribution to the Hall voltage.

Temperature dependent resistivity measurements were carried out to obtain more experimental information on scattering phenomena in HgSe:Ce. The results are shown in Fig. 5. In the 50-110 K region, a decrease of the resistivity is observed with decreasing temperature, comparable to measurements on HgSe

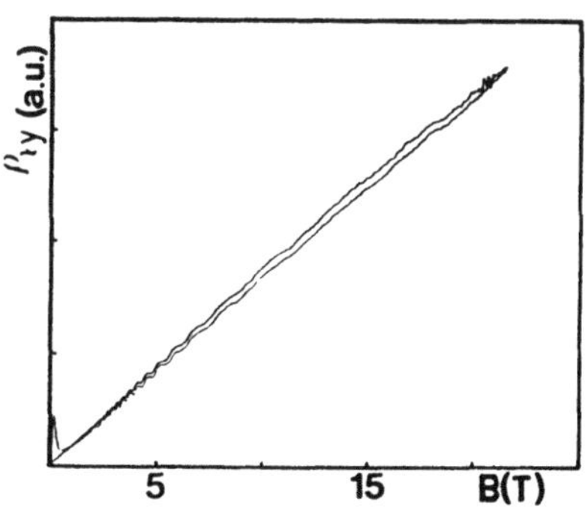

Figure 3: Typical Hall voltage trace for sample C1.

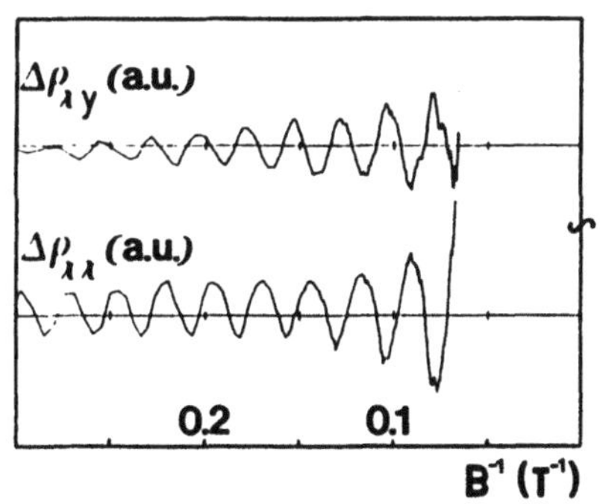

Figure 4: Phase shift between ρ_{xx} and ρ_{xy} .

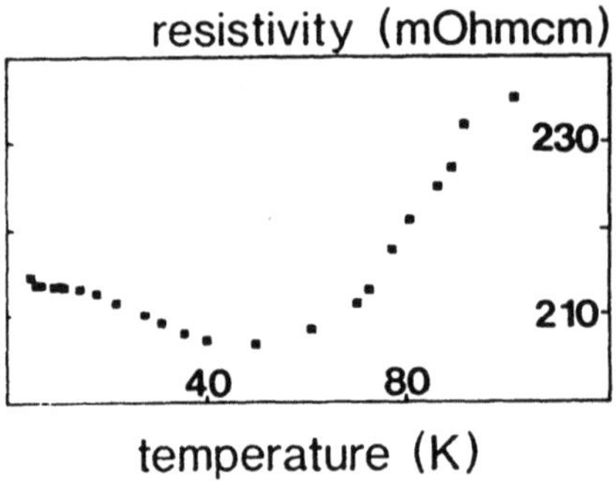

Figure 5: Resistivity of sample C2 as a function of temperature.

[6]. At about 50 K, a minimum occurs and at lower temperatures the resistivity increases. We attribute this to the Kondo effect, introduced by the presence of the Ce impurities. Samples with different Ce doping should be grown to study the minimum in the resistivity with respect to impurity concentration.

4 Conclusions

The first magnetotransport measurements on HgSe:Ce are reported. At an impurity concentration of $10^{19}cm^{-3}$, no influence of the exchange interaction on the conduction band is detected. Band parameters calculated from the SdH effect are comparable to those derived for HgSe.

Remarkable scattering effects, probably due to the Ce ions, are observed in Hall voltage and resistivity measurements. The Kondo effect is considered as an explanation for the resistance minimum.

Acknowledgements

The measurements at Braunschweig were carried out by B. Momont under sponsorship of the ERASMUS program. R. Bogaerts is *Aspirant* and G. Pitsi is a *Research Associate* of the Belgian NFSR.

References

[1] J. Kossut, Semiconductors and Semimetals, Vol. 25, Ed. R. K. Willardson and A. C. Beer, Academic Press, New York (p. 183).

[2] Z. Wilamowski, Acta Phys. Pol. A **A77**, 133 (1990).

[3] A. I. Ponomarev, G.A. Potapov, and I. M. Tsidil'kovskii, Sov. Phys. Semicond. **11**, 24 (1977).

[4] A. I. Ponomarev, G.A. Potapov, G. I. Kharus, and I. M. Tsidil'kovskii, Sov. Phys. Semicond. **13**, 502 (1979).

[5] R. R. Galazka, W. Dobrowolski, and J. C. Thuillier, phys. stat. sol. (b) **98**, 97 (1980).

[6] B. A. Lombos, E. Y. M. Lee, A. L. Kipling and R. W. Krawczyniuk, J. Phys. Chem. Solids **36**, 1193, (1975).

Spin Dependent Scattering in $(Cd_{1-x-y}Zn_xMn_y)_3As_2$ from the Shubnikov–de Haas Waveshape Analysis

W. Lubczyński[1], *J. Cisowski*[1], *J. Kossut*[2], *and J.C. Portal*[3]

[1]Department of Solid State Physics, Polish Academy of Sciences, Ul. Kawalca 3, PL-41-800 Zabrze, Poland

[2]Institute of Physics, Polish Academy of Sciences, Al. Lotników 32/46, PL-02-668 Warsaw, Poland

[3]CNRS-SNCI, BP 166X, F-38042 Grenoble Cedex, France and INSA-CNRS, Av. de Rangueil, F-31077 Toulouse Cedex, France

Abstract. The study of the Shubnikov-de Haas effect (SdH) in $(Cd_{1-x-y}Zn_xMn_y)_3As_2$ with x=0.03 and y=0.006 has been performed in high magnetic fields up to 25 T. From the detailed waveshape analysis of SdH oscillations, the contribution of spin dependent scattering of the conduction electrons on magnetic ions to the total relaxation time has been determined and compared with the theoretical prediction derived for this purpose.

1. Introduction

The material studied in this work, i.e. $(Cd_{1-x-y}Zn_xMn_y)_3As_2$ (CZMA) belongs to the family of diluted magnetic (semimagnetic) semiconductors (DMS) with a small energy gap for low Zn and Mn contents. The first study of the quantum transport in CZMA [1] has shown that the Landau spin subband energies in this system are very strongly influenced by the exchange interaction between the band electrons and localised magnetic moments of Mn ions. Additionally, the existence of spin dependent scattering in CZMA has been also observed [1]. In order to study this phenomenon in a greater detail, we have performed the SdH effect measurements in high fields and we have undertaken an attempt to describe it theoretically with complexities of the band structure of CZMA taken into account.

2. Theory of spin dependent scattering

The presence of localised magnetic moments in DMS not only modifies their band structure [2], but also gives rise to an additional scattering mechanism of the carriers. The contribution of this mechanism to the total relaxation time can be observed in the difference between Dingle temperatures T_D^+ for spin-up- and T_D^- for spin-down electrons deduced from the SdH effect [3].

In order to analyse this process theoretically, we have assumed that we deal with the system of non-interacting magnetic atoms, randomly distributed in the host crystal. We have taken the electron wave function suitable for a narrow-gap material in the presence of an external magnetic field in the form derived in [4], with coefficients corrected for the exchange interaction. The perturbing Hamiltonian H_S, which

Springer Series in Solid-State Sciences, Vol. 101
High Magnetic Fields in Semiconductor Physics III Editor: G. Landwehr

gives rise to the electron scattering by the magnetic impurities, is taken as

$$H_S = J(\vec{r})\cdot(\vec{S}\cdot\vec{\sigma} - y\cdot\sigma_z\cdot\langle S_z\rangle) \qquad (1)$$

where $\vec{S}$ and $\vec{\sigma}$ are the spin operators of Mn ions and the band electrons, respectively, $J(\vec{r})$ is an exchange constant, S_z and σ_z are the components of $\vec{S}$ and $\vec{\sigma}$ in the direction of the applied magnetic field and y denotes the Mn content. In order to derive the relaxation time due to the spin dependent scattering (in the first Born approximation), we have calculated the square of the matrix element of H_S between the initial $\phi(N,k_z,j)$ and final $\phi(N',k_z',j')$ electron states. Symbols N and k_z denote the Landau number and the wave vector component parallel to the external magnetic field,respectively, and j is the z-component of the total momentum, becoming +1/2 for spin-up- or -1/2 for spin-down electrons, which, in what follows, will be labelled "+" and "-". Assuming that $J(\vec{r})$ is short-ranged (which is reasonable because the wave function of d-electrons of Mn ions are well localised) and summing over all possible final states, the following formula for the relaxation time of electrons at the Fermi surface can be obtained

$$\left(\frac{1}{\tau}\right)_{N,j} = A\cdot\left(\langle S_z^2\rangle-(2y-y^2)\langle S_z\rangle^2\right) + B\cdot\langle S^+S^-\rangle + C\cdot\langle S^-S^+\rangle \qquad (2)$$

where $S^{\pm}=S_x \pm iS_y$ and $\langle\ldots\rangle$ denotes the thermal average. The values of A, B and C depend on the electron wave functions and density of states calculated at the Fermi level, the exchange integrals $\alpha=\langle S|J|S\rangle$ and $\beta=\langle X|J|X\rangle=\langle Y|J|Y\rangle=\langle Z|J|Z\rangle$ [2] (S,X,Y,Z represent Kohn-Luttinger amplitudes of which the electron wave functions are composed) and a number of magnetic ions in the unit volume. The form of these dependences is given in detail in [5]. The contribution of this type of scattering to the total Dingle temperature is related to the the relaxation time τ_j as [6]

$$T_D^{\pm} = \frac{\hbar}{2\pi k_B\tau_{\pm}} \qquad (3)$$

where k_B denotes the Boltzmann constant. In Eq.(3) we have dropped the index N, because the dependence of τ_j on N directly corresponds to the dependence on magnetic field. The above calculations of the electron relaxation time due to scattering by magnetic ions lean on the approximation that the scattering process is elastic. It is not strictly true for spin-flip transitions, which are here allowed, since the electron wave function is a mixture of spin-up- and down functions [4]. In the case of spin-flip transition, the energy difference between the initial and final states is of the order of the Zeeman splitting of d-electrons in Mn ions. However, for the magnetic fields considered, this energy is smaller than the Landau level broadening Γ ($\Gamma=\pi k_B\overline{T}_D$, where $\overline{T}_D$ is the average of the Dingle temperatures for spin-up- and spin-down electrons), corresponding to $\overline{T}_D \cong 10$ K as found for our material. Therefore, the inelasticity of the scattering can be neglected.

3. Results and Discussion

Fig.1 presents the SdH oscillations for one of the samples studied, after removal of a weak monotonic background. A node at about 3.5 T at 1.6 K is clearly visible. In order to analyse

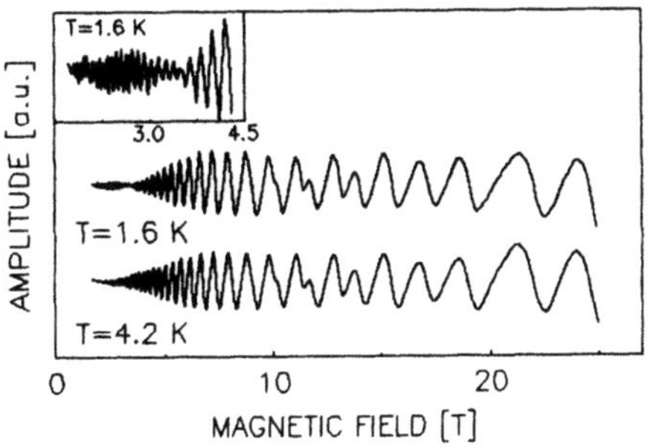

Fig.1. SdH oscillations for a sample of CZMA with x=0.03, y=0.006 and with the electron concentration of $4.05\times10^{18}cm^{-3}$. The insert presents the low-field oscillations in greater detail.

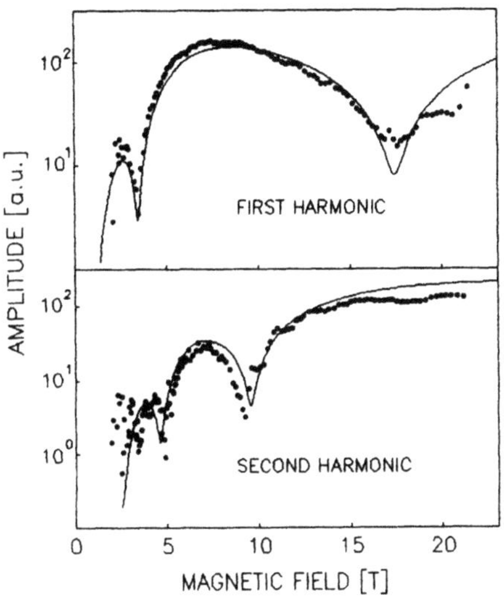

Fig.2. The amplitude of the first and second harmonics of the SdH oscillations from Fig.1 for T=1.6 K, obtained from the fitting procedure (circles). The solid lines represent the theoretical dependences [3] calculated using Eq.(3) with the exchange interaction constants $N_0\alpha$=-0.4 eV and $N_0\beta$=2.14 eV (where N_0 is the number of cations per unit volume).

the experimental data, the least-squares fitting technique [3] was used to determine the amplitudes and phases of the harmonic content for each SdH trace recorded. As can be seen in Fig.2, the first harmonic shows pronounced minima and their depth is related to the difference in T_D^+ and T_D^- (for $T_D^+ = T_D^-$, the amplitude would vanish entirely). The obtained amplitudes of the first and second harmonics as a function of magnetic field are compared with the theoretical dependences derived for the case of spin dependent scattering [3,7] - see Fig.2. In these calculations we have applied the band structure model mentioned earlier and magnetization data obtained from independent measurements on the same material. Values of the sp-d exchange interaction constants $N_0\alpha$ and $N_0\beta$ have been independently determined in the standard way by fitting the theoretical model of the band structure to the experimental g-factor data [2]. A good agreement between positions of calculated and observed minima of SdH harmonics (see Fig.2), which depend sensitively on α and β, is indicative of the consistency of this analysis. In description of the magnetic system of Mn ions in CZMA we have used the results of Ref.[8].

Appearence of minima of the first and second harmonic amplitudes is connected with the fact that g-factor of free

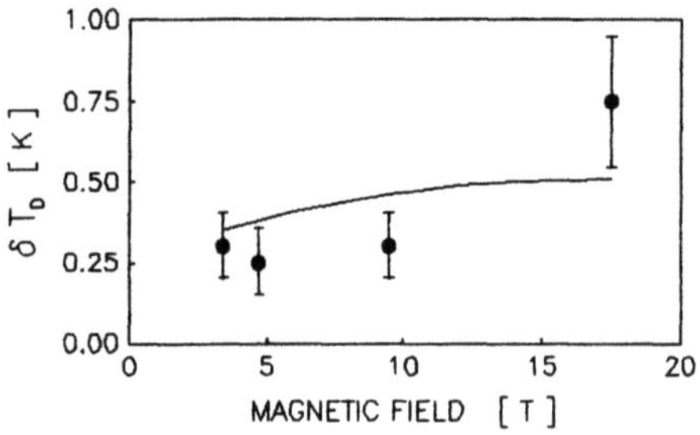

Fig.3. The difference Dingle temperature at 1.6 K as a function of magnetic field for CZMA. Points represent data obtained from the analysis of the experimental results and the solid line - the theoretical prediction calculated on the basis of Eq.(3).

carriers in DMS strongly depends on magnetic field and temperature [2]. In the presence of spin dependent scattering, the depths of these minima are controled by values $\delta T_D = (T_D^+ - T_D^-)/2$.

Direct comparison between theory and experiment is shown in Fig.3. It can be seen that the presented theoretical model renders values of the difference Dingle temperature δT_D in good agreement with observations. It should be emphasised that no fitting parameters have been used in these estimates.

References

1. W.Lubczyński, J.Cisowski, J.C.Portal and W.Żdanowicz, Acta Phys.Polon.A **73**, 509 (1988)
2. J.Kossut,in *Semiconductors and Semimetals*, ed.by J.K.Furdyna and J.Kossut, (Academic Press,Boston 1988),vol.**25**,p.183
3. M.Vaziri and R.Reifenberger, Phys.Rev.B **32**, 3921 (1985)
4. P.Kacman and W.Zawadzki, Phys.Stat.Sol.(b) **47**, 629 (1971)
5. W.Lubczyński, J.Cisowski, J.Kossut and J.C.Portal, submitted for publication
6. L.Roth and P.Argyres,in *Semiconductors and Semimetals*, ed.by R.K.Willardson and A.C.Beer (Academic Press, New York 1966), vol.1,p.159
7. R.J.Higgins and D.H.Lowndes, in *Electrons at the Fermi Surface*, ed.by M.Springford (Cambridge University Press, 1980), p.393
8. C.J.M.Denissen, H.Nishihara,J.C.van Gool and W.J.M.de Jonge, Phys.Rev.B **33**,7637 (1986)

High-Electric-Field Transport in $Hg_{0.8}Cd_{0.2}Te$ Under a Large Quantising Magnetic Field*

C.K. Sarkar[1], *P.P. Basu*[2], *and D. Chattopadhyay*[2]

[1]Department of Electronics and Telecommunication Engineering, Jadavpur University, Calcutta-700032, India

[2]Institute of Radio Physics and Electronics, 92, Acharya Prafulla Chandra Road, Calcutta-700009, India

Abstract : The energy relaxation times τ_e of hot electrons in HgCdTe are calculated for extreme magnetic quantum limit (MQL) at 1.5 K assuming interaction with acoustic phonons. Relevant complexities of band structure and scattering rates are included. The calculated values of τ_e assuming thermal phonons are found to be several times higher than the experimental values for magnetic flux densities (B) of 4T and 6T. Inclusion of nonequilibrium phonons gives a satisfactory agreement between the theoretical and the experimental values of τ_e. The model also yields the velocity-field relationship agreeing with the experimental results.

We give here a theoretical analysis of the energy loss rate of hot electrons in HgCdTe at 1.5 K in MQL and compare the results with the experimental data recently reported by NIMTZ and STADLER/1/. The motivation is to obtain an insight into the basic hot-carrier kinetics in this technologically important material at low temperatures.

Our model includes scattering by acoustic phonons in the nonparabolic conduction band /2/, the free-carrier screening, nonequipartition of the phonons, and the broadening of the Landau level due to electron impurity interactions /3/. We assume a nondegenerate carrier distribution having an electron temperature T_e. This assumption is justified by the specific heat data /1/ for T_e lying between 2.5 K and 3.5 K. For a carrier concentration of about $10^{14}cm^{-3}$, as found in the experimental samples, the application of the magnetic field coupled with band-nonparabolicity makes the degeneracy effect negligible, the Fermi level lying about $2k_BT_e$ below the band-edge.

The expressions for the average power loss per electron in the presence of a heating electric field parallel to B due to deformation potential acoustic and piezoelectric scattering for thermal phonons are derived following the techniques of Refs. /4-6/. The quantity τ_e is obtained from the total power loss P using the relationship

$$P=(1/2)k_B (T_e - T_L) / \tau_e \tag{1}$$

where k_B is the Boltzmann constant and T_L is the lattice temperature. The factor 1/2 on the right-hand side of (1) appears due to the one translational degree of freedom due to Landau quantisation.

Numerical results are obtained with the material parameters given by DORNHAUS and NIMTZ /7/: Effective mass ratio=0.0065, band gap=70meV, mass density = 7.654 x 10^3 kg m^{-3}, Lande's 'g' factor = 90 (for B=4T) and

* This paper was presented at the 1988 Würzburg Conference and inadvertently not included in the proceedings

Springer Series in Solid-State Sciences, Vol. 101
High Magnetic Fields in Semiconductor Physics III Editor: G. Landwehr

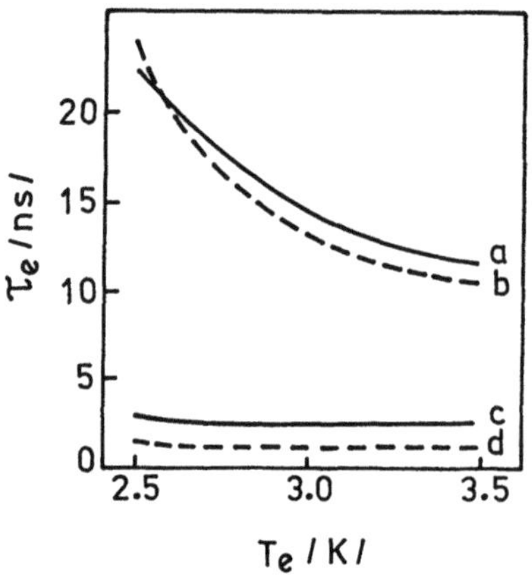

Fig. 1. Plots of τ_e with T_e at T_L=1.5K. The solid and the dashed curves are for B=4T and 6T, respectively. a, b : experimental values; c,d : theoretical values assuming thermal phonons.

80 (for B=6T), acoustic velocity = 3.017 x 10^3 m s^{-1}, acoustic deformation potential = 9 eV, piezoelectric modulus = 0.0335 C m^{-2}, piezoelectric velocity = 1.948 x 10^3 m s^{-1}, dielectric constant=18. The calculated results are shown in Fig.1. The experimental values of τ_e are found to be much higher than the theoretical values. Thus, considering thermal phonons, the experimental results on τ_e cannot be explained by including the complexities of the band structure and the scattering rates.

To explain the experimental results the nonequilibrium acoustic phonons which slow down the carrier cooling rate /8/ are considered. We find that the deformation potential acoustic phonons are primarily responsible for the energy loss of hot electrons, piezoelectric scattering contributing only 30% to the total scattering. In the analysis with nonequilibrium phonons, the piezoelectric scattering is neglected to avoid unmanageable expressions. This simplification is particularly permissible since the phonon life time τ_p is not precisely known owing to the complicated nature of the phonon relaxation mechanism /9,10/.

The ocupation number of nonequilibrium phonons is obtained by equating the rate of phonon generation due to hot electrons to the rate of phonon decay /8/. The power loss per electron, interacting with nonequilibrium phonons, then yields values of τ_e provided τ_p is known.

We fit the experimental results on τ_e here by adjusting the values of τ_p, and give the required values of τ_p in Table 1. Such values of τ_p are not unreasonable in view of the discussion given in Refs./10,11/.JAY-GERIN /9/, in his calculations of the phonon drag contribution to the thermoelectric power in GaAs at low temperatures, also used values of τ_p of this order.

The contribution of the boundary scattering to τ_p can be roughly taken as d/2s where d is the shortest transverse dimension of the sample and s is the sound velocity. For the given sample this yields for τ_p a value of 100 ns which falls in the range of values of τ_p found in our analysis. However, as the value of τ_p may be significantly enhanced due to the mismatch factor with boundaries /12/, processes other than boundary scattering, such as, phonon-phonon interaction seems to control the phonon decay in this situation. With increasing heating of the electron gas phonon-phonon interaction gets stronger which accounts for the decrease of τ_p with increasing T_e for a given B.

Table 1. Phonon life times

B/T/	T_e/K/	τ_p/ns/
4	2.5	62.4
4	3.0	41.0
4	3.5	32.1
6	2.5	111.5
6	3.0	66.3
6	3.5	55.0

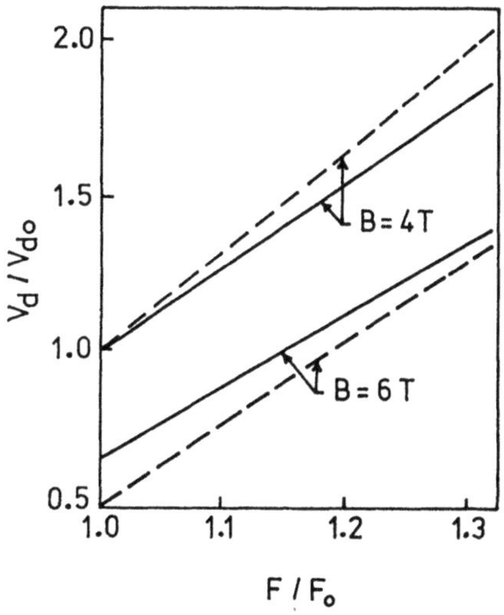

Fig.2. Plots of V_d/V_{do} with F/F_o. The solid and the dashed curves give respectively the theoretical and the experimental values.

The velocity-field characteristics are obtained considering additionally the ionised impurity scattering and using the values of τ_p that fit the theoretical values of τ_e to the experiments. The results are displayed in Fig.2. To facilitate comparison with experiments, the electric field F is normalised by the value F_o corresponding to T_e = 2.5K when the system begins to behave like a non-degenarate gas /1/. The drift velocity V_d is normalised by V_{do}, the value it attains at B=4T and T_e=2.5K. The normalisation allows a comparison of the theoretical values of V_d/V_{do} with the experimental values /1/ of the corresponding current ratios : the possible uncertainty in the value of carrier concentration does not affect the comparison.

The enhancement of the mobility due to weakening of the ionised impurity scattering with the increasing field is offset by the acoustic phonon scattering the effect of which increases with the carrier heating. Therefore, the drift velocity increases almost linearly with the electric field, which agrees satisfactorily with experiments.

In conclusion, we believe that the nonequilibrium acoustic phonons control the cooling of hot carriers in HgCdTe in MQL at low temperatures. Our model accounts for the experimental results on τ_e as well as the electric-field dependence of the drift velocity.

References

1. G. Nimtz, J.P. Stadler : Physica B 134, 359 (1985); also J.P. Stadler : Ph.D. Thesis, University of Cologne (1985)
2. U.P. Phadke, S. Sharma : J. Phys. Chem. Solids 36, 1 (1975)
3. L.M. Roth, P.N. Argyres : Semiconductors and Semimetals, ed. by R.C. Willardson, A.C. Beer, Vol.1 (Academic, New York 1966) p.159
4. H. Kahlert, G. Bauer : Phys. Rev. B 7, 2670 (1973)
5. G. Bauer, H. Kahlert, P. Kocevar : Phys. Rev. B 11, 968 (1975)
6. I.I. Pinchuk, Phys. Stat. Sol. (b) 97, 355 (1980)
7. R. Dornhaus, G. Nimtz : Springer Tracts in Modern Physics, ed. by G. Hohler, Vol.98 (Springer, Berlin 1983) p.119
8. E.M. Conwell : Solid State Physics, ed. by F. Seitz, D. Turnbull, H. Ehrenreich, Suppl.9 (Academic, New York 1967) p.127
9. J.P. Jay-Gerin : Phys. Rev. B 12, 1418 (1975)
10. B.A. Aronzon, E.Z. Meilikhov : Fiz. Tekh. Poluprovodn. 13, 974 (1979)/Engl. Transl. Sov. Phys. Semicond. 13, 568 (1979)/
11. B.M. Askerov : Fiz. Tekh. Poluprovodn. 16, 2083 (1982) / Engl. Transl. Sov. Phys. Semicond. 16, 1346 (1982)/
12. P. Bordone, C. Jacoboni, P. Lugli, L. Reggiani, K. Kocevar : J. Appl. Phys. 61, 1460 (1987)

Magnetotransport in n-GaAs and n-$Al_xGa_{1-x}As$ in High Magnetic Fields Under Hydrostatic Pressure

J.C. Portal[1], *A. Kadri*[2], *E. Ranz*[1], *and K. Zitouni*[2]

[1]CNRS/INSA, F-31077 Toulouse Cedex, France and
CNRS/SNCI, F-38042 Grenoble Cedex, France
[2]MOE-Lab., University of Oran-Es-Sénia, 31100-Oran, Algeria

Abstract. Magnetotransport experiments under hydrostatic pressure performed before and after successive 1.4eV-light-excitations on barely metallic n-GaAs and n-$Al_xGa_{1-x}As$ (x<0.32) samples, indicate a strong influence of deep defect states near the metal-insulator transition. In n-GaAs, evidence is given for pressure-induced electrical activity of the EL2 defect metastable configuration which is found to act as an acceptor-like resonant level strongly affecting both the magnetic freezeout and the nearest-neighbor hopping conduction regimes. In n-$Al_xGa_{1-x}As$, the light-induced transformation of the deep DX-trap into its metastable effective-mass-like configuration was used to monitor the concentrations of shallow donors in a wide range across the Mott density. High magnetic fields were then used to demonstrate a non-hydrogenic behavior of the Shallow-donor-like Γ-states as a result of an anticrossing interaction with higher lying resonant impurity states.

1. Introduction

Shallow donor-like impurity states are well known to play a dominant role in the magnetic field-induced metal-insulator transition (MIT) in barely metallic n-type III-V materials, as a result of impurity wavefunction shrinkage [1]. This magnetic field-induced MIT has been studied for a variety of III-V compounds:InSb[2-3], InAs[4,5], InP[6,7], GaAs[8,9], and evidence has been given for correlated hopping conduction on the insulating side of the transition which turns out to be of the variable range type at very low temperatures.

In addition to these shallow states, much experimental evidence [5,10-12] has shown the existence of deep levels forming non-Γ-like localized resonances which, when driven into the fundamental gap under pressure, are found to strongly affect the behavior of shallow donors near the magnetic field-induced MIT. At this threshold pressure, an abrupt increase of the binding energy of the shallow donors and a simultaneous

Springer Series in Solid-State Sciences, Vol. 101
High Magnetic Fields in Semiconductor Physics III Editor: G. Landwehr

decrease of their concentration are usually observed not only indicating that both shallow and deep states arise from the same impurity centers, but also suggesting that they are two possible configurations excluding each other.

Quite similar features were also demonstrated in far-infrared transmission spectroscopy [13] for shallow and deep DX-related energy levels in n-$Al_xGa_{1-x}As$ with almost the same behavior as in the case of InSb [10-12] where the localized resonance also exhibit a DX-like behavior[14].

In this work, magnetotransport experiments under pressure are used to demonstrate the influence near the MIT of EL2 defect states in barely metallic n-GaAs and the DX-centers in n-$Al_xGa_{1-x}As$

2. Experimental techniques

n-GaAs samples for this study were cut from lightly Si-doped bulk crystals grown by the liquid encapsulated Czochralski technique. At P=1bar and T=4.2K, n_H and μ_H were $1.7\times10^{16}cm^{-3}$ and $920cm^2/Vs$ for sample 1 and $1.9\times10^{16}cm^{-3}$ and $850cm^2/Vs$ for sample 2 indicating that both samples are barely metallic with N_d-N_a above the Mott density $n_c \simeq 1.6\times10^{16}cm^{-3}$. n-$Al_xGa_{1-x}As$ samples were Si-doped at $2\times10^{17}cm^{-3}$ for x=0.16 and $2\times10^{18}cm^{-3}$ for x=0.32. These samples were grown by molecular beam epitaxy on undoped semi-insulating GaAs substrate and AlGaAs buffer layer At P=1bar and T=77K, earlier studies of the persistent photoconductivity effect (PPC)[15] indicated that n_H and μ_H were $1.8\times10^{17}cm^{-3}$ and $1200cm^2/Vs$ for x=0.16 and $1.15\times10^{15}cm^{-3}$ and $300cm^2/Vs$ for x=0.32. Only the latter sample showed PPC at 77K with $n_H=5\times10^{17}cm^{-3}$ and $\mu_H=900cm^2/Vs$.

R_H and $\rho_\perp$ measurements were performed in magnetic fields up to 20T, under hydrostatic pressures up to 15kbar, for two current and two magnetic field directions in order to eliminate background signals which can strongly affect both R_H and $\rho_\perp$ values during magnetic freezeout(MFO). Care was also taken to avoid any electric field effect and the measurements were always taken in the ohmic behavior region.

3. EL2-Defect States Effects in n-GaAs

3.1 The Magnetic Field-Induced MIT in the Dark

The data in the dark shown in Figs.1(a)&1(b) indicate clearly a magnetic field-induced MIT as a result of MFO effects at

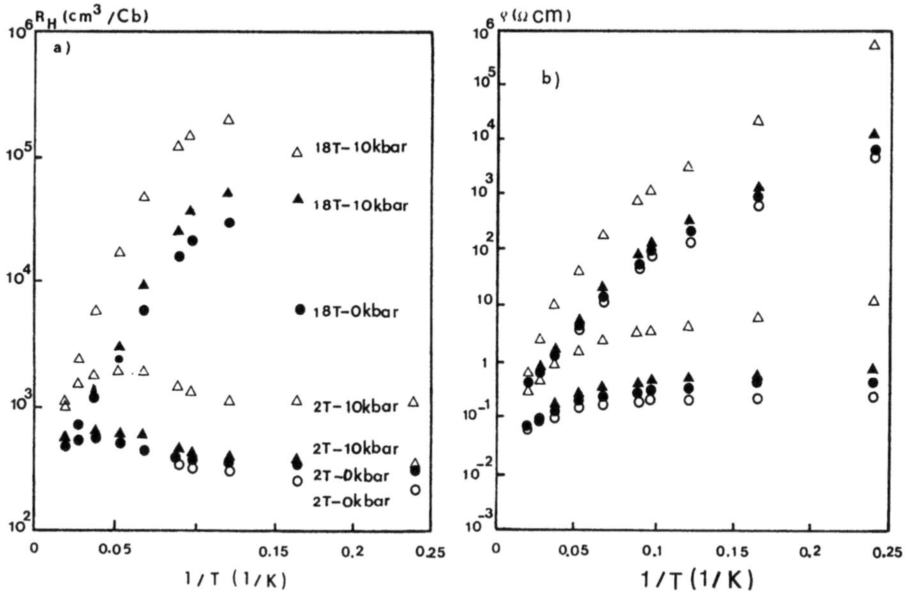

Fig.1: (a)- Hall coefficient & (b)- transverse magnetoresistivity versus 1/T at ≠B and P :- Black symbols stand for data in the dark, -open symbols stand for data after saturated illumination.

T>10K, followed by a hopping conduction at lower temperatures where $\rho_{\perp}$ and R_H show a characteristic behavior [2,3] (a much smaller activation for ρ and a strong decrease of R_H until it becomes scarcely detectable in a noise background).

The results of analysis are summarized in Fig.2 for MFO effects and in Fig.3 for the hopping conduction (ε_3). As can be seen in Fig.2, the binding energy (E_b) of the shallow donors could only be observed above a threshold field (B_c=6T) which appears to be correlated with the actual donor concentration through the Mott criterion in the presence of magnetic field. At higher fields ($B>B_c$), E_b values are found much lower than the predicted hydrogenic-like behavior[16], and as usually observed at $N_d>n_c$[2,5,7], impurity wavefunction overlap[17] can account for E_b reduction.The data in Fig.3 indicate clearly a correlated nearest-neighbor type of hopping conduction, for ε_3 increases with B while the magnetoresistivity pre-exponential factor (ρ_{30}) behaves as $B^{1/2}$(18) starting at B values (≃ 4T) much lower than the predicted critical value (B ≃ 9T) as observed in other III-V materials (2,3,6).

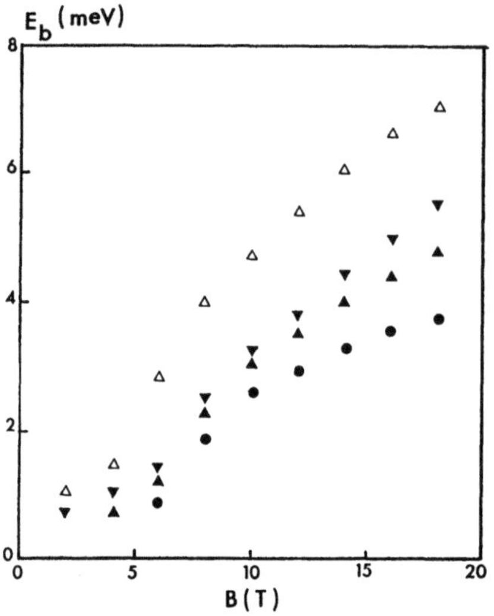

Fig.2: Binding energy of the shallow donors (E_b) versus B (●)-dark,1 bar; (▲)-Dark, 10kbar (▼)-Dark,11.5kbar; (△)- after satureted illumination at 10 kbar.

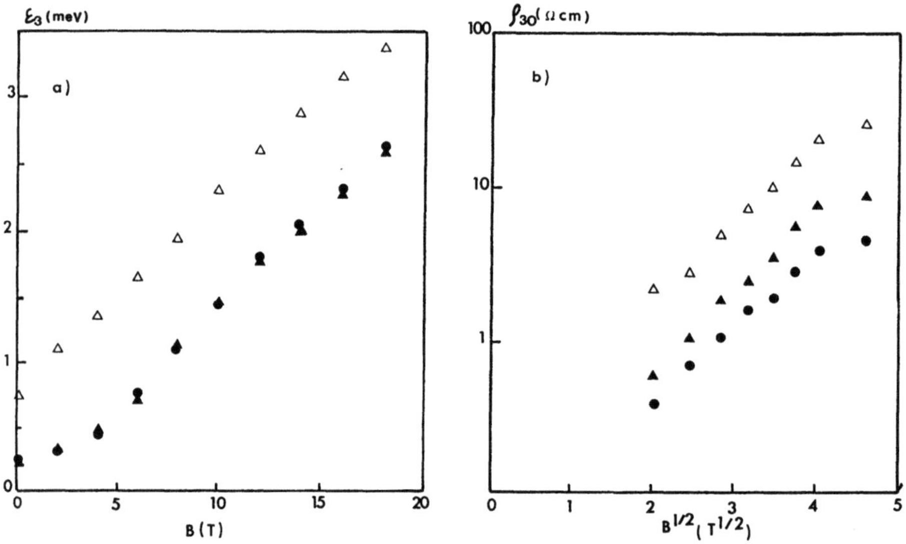

Fig.3: (a)- Activation energy (ε_3) of the hopping conduction versus B, (b)- magnetoresistance pre-exponential term (ρ_{30}) versus $B^{1/2}$. The symbols have the same meaning as in Fig.2.

3.2 Effects of Hydrostatic Pressure

As shown in Figs.1(a)&1(b), hydrostatic pressure enhances both the MFO effects and the hopping conduction. As expected, E_b of the shallow donors increases with increasing pressure. However, as can be seen in Fig.2, the pressure coefficient of E_b increases well above the hydrogenic behavior at high pressures and magnetic fields. Such effects, observed earlier in magnetooptical studies under pressure on pure GaAs[19], indicate an increasing contribution from a localized impurity potential as when non Γ-like deep resonances are driven into the fundamental gap under pressure. The origin of the localized resonances will be discussed in Section 3.3 below. At $T<10K$, pressure is found to shift the $B^{1/2}$ behavior of ρ_{30} to higher values, without practically affecting ε_3, suggesting a main effect of pressure on the percolation paths on which may critically depend the constant term in the pre-exponential factor [18].

3.3 Light-induced Metastability Effects

Illuminating the samples at 4.2K with the LED is found to induce persistent changes of both n_H and μ_H. As can be seen in Figs.1(a)&1(b), these effects strongly depend on pressure. At P=0kbar, a slight persistent increase of n_H and μ_H is observed depending on the exposure time to the LED light, and these effects saturate at $\simeq$ 30% higher than the dark data. At P=10kbar, illumination with the LED is in turn found to induce a strong persistent decrease of both n_H and μ_H again depending on the light exposure time, and also saturating. In contrast to those at P=1bar, the persistency effects at high P are found to affect strongly both ε_1 and ε_3 regimes as shown in Fig.2 & 3.

The recovery of the initial characteristics could be obtained by warming the samples at $T>120K$, indicating in this way a reversible metastable behavior as usually observed when deep defects states are involved. By the fact that in the pressure range investigated ($P\leq15$kbar) no contribution could be expected from the DX-related energy levels [20], the metastability effects must be attributed to the EL2 defect states which were shown in recent optical experiments under pressure[21] to play an important role not only in semi-insulating GaAs but also in lightly doped n-type GaAs as well.

The analysis of our data shows that at high pressures, the persistency effects reveal a light-induced decrease of the net donor concentration (N_d-N_a decreases from its initial dark value to about $6x10^{15}cm^{-3}$ after saturated illumination) and a simultaneous increase of the concentration of charged centers as indicated by the strong mobility decrease. According to Baj et al.[21], P higher than 3kbar can induce not only a very efficient optical activity of the metastable state but also an additional absorption indicating a change of the charge state of EL2 defect from neutral to negatively charged as a result of trapping an extra-electron in its metastable configuration. Our results strongly support this conclusion and as suggested by Von Bardeleben[22], the EL2 metastable state acts as a resonant acceptor-like level separated by a potential barrier from both conduction band states and shallow donor states in almost the same way as the deep DX-level in direct-band-gap $Al_xGa_{1-x}As$. As is the case for DX-levels, a deepening of E_b (see §3.2 above) is observed when the EL2 metastable state is driven into the energy gap under both pressure and magnetic field. The light-induced drastic enhancement of both E_b in ε_1 regime and the localization parameters on the insulating side of the transition (ε_3 regime) can be well accounted for by a decrease of both wave-function overlap and correlations as a result of a decrease of the concentration of shallow donors when most of EL2 defect centers are photoquenched in their metastable configuration.

3.4 Electric Field Induced Metastability effects

In addition to pure thermal recovery processes, we have observed also electric field recovery processes. I-V measurements under pressure indicate an electric field-induced persistent free carrier increase confirmed by R_H measurements. At T<15K, the threshold electric field is ≃ 500mV/cm and decreases with increasing temperature to ≃ 100mV/cm at 77K, indicating a thermally-assisted effect. However, in contrast to pure thermal recovery processes, the electric field is found to induce new metastability effects under pressure. As shown in Fig.4, in the dark and before any illumination applied, the application of a weak electric field at 11.5kbar is found to persistently increase n_H twice with respect to its initial value, indicating an additional persistent ionization of EL2 metastable state.

These effects not only confirm the doubly occupancy of the EL2 metastable state as shown in §3.3 , but also suggest a "shallow-deep-bistability" behavior, for a persistent increase

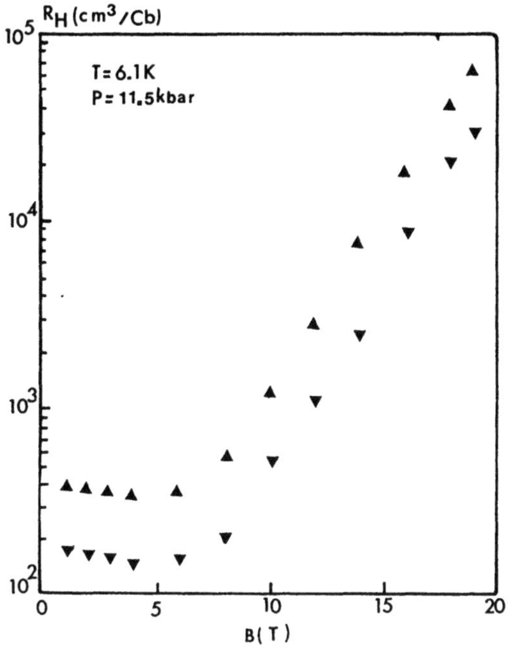

Fig.4: Dark RH versus B (▲) before and (▼) after the application of an electric field.

of shallow donor states is observed as is the case of DX-related states in n-$Al_xGa_{1-x}As$[13]. Striking weak-electric-field effects were also observed around the shallow-deep-levels crossover in n-$Al_xGa_{1-x}As$[23], with almost the same features as described here, except that no additional ionization of deep DX-level was detected.

4. DX-Centers Effects in n-$Al_xGa_{1-x}As$

4.1 Light-controlled deep-shallow-DX-levels transformation

MFO experiments were performed in this case by cycling the samples in the following way: they were first slowly cooled down to 4.2K in the dark, at P=1bar for x=0.32 and P=13.5kbar for x=0.16. Then R_H and ρ measurements were performed by slowly warming up the samples in the dark up to 77K, well below the critical temperature (120K), to prevent any thermal electron recapture by the deep DX level. Next, to monitor conduction-band populations, successive illuminations were applied at 77K, photoionizing the deep DX level into its metastable shallow-like configuration.The samples were then cooled down to 4.2K for the next measurements. Such a cycle was repeated with the accumulated light exposure until complete photoionization of the deep DX level occurred.

As expected, in the absence of any illumination, both samples were highly resistive at 4.2K indicating almost complete carrier trapping into the deep DX level. Because of a strong decrease of carrier mobilities at low temperatures and the onset of a hopping-type conduction, meaningful results could be obtained only over a temperature range typically >30K. In this range, our low B data are in good agreement with those reported earlier by Theis [13]. Both samples in the dark show a well defined R_H thermal activation indicating carrier trapping into a shallow level already shifted from Γ-conduction-band-edge at B=0T. As first shown by Theis [13], this shallow level is the effective-mass-like configuration of the DX center. This is in deed confirmed by illuminating the samples, for a monotonic de crease of the binding energy (E_b) of metastable shallow states is observed as their concentration is continuously increased by photoionizing the deep DX-level.

4.2 Nonhydrogenic Behavior of the Shallow Donors

Both samples show an increase of R_H and $\rho_\perp$ with increasing B characteristic of MFO effects due to an increase of E_b of the shallow donors as shown in Fig.5. The light-induced increase of metastable shallow states concentrations results in a shift of E_b to lower energies at all B strengths, until an Insulator-Metal Transition set in close to Mott critical density ≃ $5x10^{16}cm^{-3}$. Further illumination led to the appearance of Shubnikov-de-Haas (SdH) oscillations characteristic of a metallic behavior. In both samples, Hall concentrations are found very close to those derived from the period of the SdH oscillations excluding any contribution from higher lying conduction bands.

The influence of higher lying states appears through nonhydrogenic behavior of metastable shallow state as clearly indicated by E_b versus B variation shown in Fig.5. Indeed, distinct features are observed for each of the investigated samples. In the sample with x=0.16 under 13.5kbar, E_b increases with B first as observed for hydrogenic levels up to about 15T, then much more strongly at higher B. Although this effect is diminished by increasing the concentrations of metastable shallow states , it could be observed up to the MIT. The same nonhydrogenic behavior is observed in the other sample with

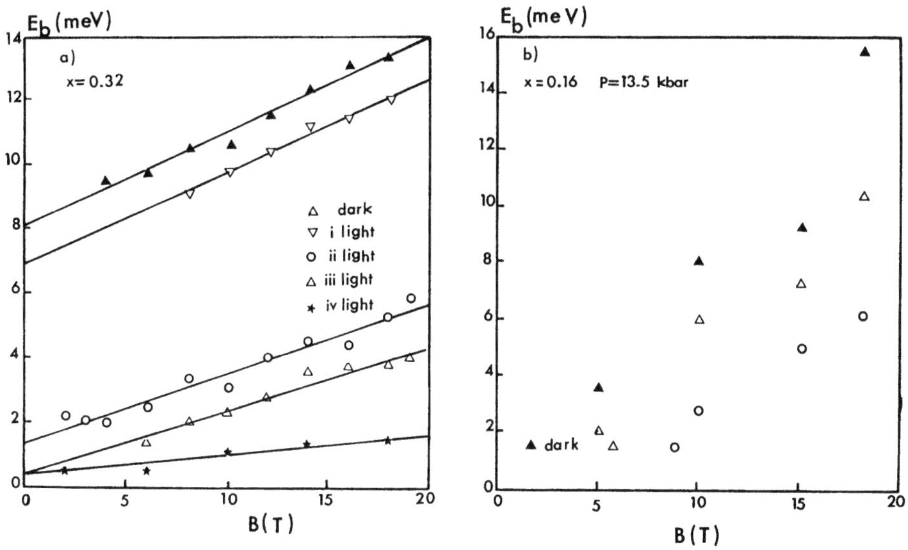

Fig.5: Variations of the binding energy of the shallow donors versus B for different light exposure times (see text).

x=0.32 starting at low B followed by an approximately linear variation of E_b versus B at a rate of about 0.4meV/T in the dark. This rate is very close to the shift of the lowest Landau level (≃0.55meV/T) and suggests an anticrossing interaction effect with higher lying resonant impurity states as observed in other III-V compounds [5,10-12]. Both results confirm early FIR data [13] where an anticrossing interaction between $\Gamma(A_1)$ and $L(A_1)$ states was assumed to explain the rapid increase of the 1s-2p transition at alloy compositions x>0.25. Our data indicate that this resonance holds at x≃0.32.

Acknowledgements

We acknowledge Drs B.Goutiers, D.Lavielle, R.Sirvin for technical assistance, Dr N.Chand from AT&T Bell Labs. for AlGaAs samples, Dr B.Gil from GES-USTL for GaAs samples, Dr C.Grattepain and Dr Huber from Thomson-France for SIMS facilities and technical discussions. We are grateful to EEC, NATO and Conseil Régional Midi Pyrénées for financial support. This work was also supported by Ministère des Affaires Etrangères-CCST/DPRST-Algerian Ministry for Higher Education.

References

[1] Y.Yafet, R.W.Keyes, E.N.Adams, J.Phys.Chem.Solids 1, 137 (1956)

[2] J.L.Robert, A.Raymond, R.Aulombard, C.Bousquet, Phil.Mag.B42, 1003 (1980) [3] M.Abdulgader, R.Mansfield, P.Fozooni, in Proceedings of the International Conference on High Magnetic Fields in Semiconductor Physics, Wurzburg -FRG -1986, edited by G. Landwehr (Springer ,Berlin,1986) Springer Ser. Solid- State Sci. Vol.71, p.381.

[4] L.A.Kaufman, L.J.Neuringer, Phys. Rev. B 2, 1840 (1970)

[5] A.Kadri,R.Aulombard,K.Zitouni,M.Baj,L.Konczewicz,Phys.Rev. B31, 8013(1985) 85)

[6] S.Aboudy, R.Mansfield, P.Fozooni, in Proceedings of the International Conference on High Magnetic Fields in Semiconductor Physics,Wurzburg-FRG -1986, edited by G. Landwehr (Springer,Berlin,1986),Springer Ser Solid- State Sci Vol.71, p.518.

[7] A.Kadri, K.Zitouni, L.Konczewicz, R.Aulombard, Phys.Rev.B 35(12),6260 (1987)

[8] P.Debray, M.Sanquer, R.Tourbot, G.Beuchet, D.Huet, in Proceedings of the International Symposium on GaAs and Related Compounds,Heraklion-Greece- 1987, edited by A.Christou in I.O.P.,vol.91 (3), 263 (1988)

[9] M.C.Maliepard, M.Pepper, R.Newbury, J.E.F.Frost, D.C.Peacock, D.A.Ritchie, G.A.C.Jones, Phys. Rev. B 39, 1430 (1987)

[10] S. Porowski,L. Konczewicz, A. Raymond, R. Aulombard, J.L. Robert, M.Baj, in Proc. Int. Conf. on the Application of High Magnetic Fields in semic. Physics, Grenoble 1982, ed. by G.Landwehr (Springer, Berlin, 1983), p.357

[11] Z Wasilewski, A.M.Davidson, R.A.Stradling, S.Porowski, in Proc. Int. Conf on the Application of High Magnetic Fields in semicond. Physics, Grenoble 1982, ed. by G.Landwehr (Springer, Berlin, 1983), p.233

[12] L.C.Brunel, S.Huant, M.Baj, W.Trzeciakowski, Phys.Rev.B33, 6863 (1986)

[13] T.N. Theis, in Proceedings of the 3rd Intern. Conf.onShallow Impurities in Semiconductors (Linköping-Sweden-1988) edited by B.Monemar in IOP, 95,p.307 (1990)

[14] L.Dmowski, M.Baj, P.Ioannides, R.Piotrzkowski, Phys.Rev.B26, 4495 (1982)

[15] B.Goutiers, G.Gregoris, D.Lavielle, J.C.Portal, N.Chand, Appl.Phys.Lett. 55 1124 (1989).

[16] D.M.Larsen, J. Phys. Chem. 29, 271 (1968)

[17] G.Pearson and L.Bardeen, Phys. Rev 75, 865 (1949)

[18] B.I.Shklovskii and A.L.Efros, in Electronic Properties of Doped Semicond. ed. by M.Cardona (Springer, Berlin, 1984)
[19] Z.Wasilewski and R.A.Stradling, Semic.Sci.Technol.1, 264 (1986)
[20] D.K.Maude, J.C.Portal, L.Dmowski, T.Foster, L.Eaves, M.Nathan, M.Heiblum, J.J.Harris, R.B.Beall, Phys. Rev. Lett. 59(7), 815 (1987).
[21] M.Baj and P.Dreszer, Phys. Rev. B 35(14), 10470 (1989)
[22] H.J. von Bardeleben, Phys. Rev. B 40(18), 12546 (1989)
[23] J.C.Portal, P.Basmaji, R.L.Aulombard, P.Gibart, D.Gauthier, D.Lavielle, A.Celeste, J.F.Carlin, J.M. Sallese, in Proc. 19th ICPS-Warsaw-1988,ed. by Institute of Physics (PAS),p.1067 (1988)

AC Conduction in n-Type InSb in the Magnetic Freeze-Out Region

S. Abboudy[1], *R. Mansfield*[2], *and P. Fozooni*[3]

[1]Physics Department, Faculty of Science, Alexandria University, Alexandria, Egypt
[2]Physics Department, University of Brunei, Brunei
[3]Physics Department, Royal Holloway and Bedford New College, University of London, Egham, Surrey TW20 0EX, UK

Measurements of the longitudinal alternating current (a.c.) conductivity of n-type InSb samples (carrier concentration from 6×10^{14} to 2×10^{15} cm^{-3}) in the frequency range of 10^2 to 10^5 Hz were made at temperatures down to 0.04 K and in magnetic fields up to 70 kG.

The investigated samples are metallic-like in zero magnetic field. A strong magnetic field ($> H_c$) is applied to shrink the donor wavefunctions so that samples are located on the insulator side. H_c here is the critical value of the magnetic field at which metal-insulator transition occurs. The real part of the conductivity σ_{ac} was found to vary as w^s where w is the angular frequency. The exponent s is approaching unity at low temperatures and high fields but decreasing as the temperature T is increased. At the lowest temperatures σ was independent of T but at higher T the temperature dependence is stronger than the linear dependence predicted by the simple pair approximation theory. Data are interpreted in terms of multiple hopping of electrons which becomes important at high temperatures and/or low frequencies. The temperature independent value of $\sigma(w)$ was found to diverge as the critical value of the magnetic field H_c is approached which is associated with the onset of the metallic-like behaviour.

1. Introduction

At low temperatures transport in doped and compensated semiconductors occurs due to discontinuous "hopping" jumps of charges between well defined localized states. In a.c. hopping conduction it is necessary to transfer electrons between pairs of sites, while in the direct current (d.c.) one a continuous percolation path between the electrodes has to be formed for the current to perpetuate. This suggests the approach of a simple pair approximation. The basic assumption of the pioneering work of Pollak and Geballe [1] is that the a.c. conductivity is due to electrons that tunnel back and forth between two donors, (randomly situated in space and energy), and not among larger groups of donors. This has been referred to as the quantum-mechanical tunnelling (QMT) or two-site model.

It is generally accepted to describe the frequency dependence of the a.c. conductivity $\sigma_{ac}(w,T)$ as

$$\sigma_{ac}(w,T) = \sigma(w,T) - \sigma(0,T) = A\,w^s \tag{1}$$

where $\sigma(w,T)$ is the frequency-dependent conductivity measured under the a.c. electric field, $\sigma(0,T)$ is the d.c. limit of the conductivity, w is the angular frequency and A and s are constants.

Efros [2] has used the Austin and Mott [3] one-electron approximation in which the energy of an electron on site i is independent of the occupation numbers of all other sites. An equation of the form

$$\sigma(w)=(\pi^4/12)\; e^2\; K_B\; T\; g_0^{\,2}(a/2)^5\; w\; \ln^4(\nu_{ph}/w) \tag{2}$$

Springer Series in Solid-State Sciences, Vol. 101
High Magnetic Fields in Semiconductor Physics III Editor: G. Landwehr

is obtained. K_B is the Boltzmann constant, a is the localization length and $\nu_{ph} \sim 10^{12}$ Hz. Pollak [4] was the first to consider the effect of intersite correlation between electrons on neighbouring sites. Subsequently, Efros [2]; Long [5] and recently Efros and Shklovskii [6] have dealt with this problem. It is suggested that the intersite Coulomb interaction affects just the close neighbours, the Coulomb interaction is assumed to be screened out for remote neighbours. For this case the conductivity takes the form

$$\sigma(w) = (\pi^3 e^4/3\epsilon)(a/2)^4\ w \ln^3(\nu_{ph}/w)\ g_0^2 \tag{3}$$

where ϵ is the dielectric constant.

In n-InSb materials, the effective Bohr radius is about 650 Å, this is very large compared to hydrogen atom. The wavefunctions associated with donor sites extend over large distances in the host crystal. A metal-insulator (MI) transition in such systems can be induced by changing the spatial extent between donors using a reasonable magnetic field. This technique has been used by many authors [7&8]. According to Ishida and Otsuka [9], for moderate compensation ratio the MI-transition occurs at a critical (threshold) value H_c, which can be deduced from the formula

$$(N_D\ a_{\parallel}\ a_{\perp}^2)^{1/3} = 0.26 \tag{4}$$

where N_D is the donor concentration, $a_{\parallel}$ and $a_{\perp}$ are the Yafet, Keyes and Adams parameters [10].

In the present study, three n-InSb specimens of parameters given in the table were examined. The magnetic field was kept parallel to the electric field and the longitudinal conductivity was measured (in the frequency range of 10^2-10^5 Hz) as a function of temperature at fields > H_c in steps of 5 kG and up to 70 kG and temperatures down to 0.04 K. The d.c. limit was also measured under the same conditions of the field and temperature.

Parameters of the n-type InSb samples.

Sample	$(N_D - N_A)$ cm^{-3}	H_c kG
5714	5.7×10^{14}	8
9914	9.9×10^{14}	18
2015	2.0×10^{15}	32

2. Results and Discussion

The temperature dependence of the conductivity of sample 5714 at H=60 kG is shown in Fig. 1. One would observe the following: (a) In the higher temperature range, a sharp exponential drop in conductivity occurs and this is due to the freeze-out of electrons from the conduction band onto the donor sites. (b) As the temperature is reduced, the decrease in $\sigma(w)$ is less pronounced. In this region, the conductivity is realized to be in the hopping region. (c) At much lower temperatures, depending on the magnetic field, $\sigma(w)$ becomes independent of temperature and weakly dependent on magnetic field for higher values of the field.

The excess a.c. conductivity, deduced from equation (1), due to hopping processes is obtained at different temperatures and magnetic fields. The frequency dependence of $\sigma_{ac}(w)$ for sample 5714 is displayed in Fig. 2(a) at H=40 kG. It is obvious that the conductivity is following a power law of the form $\sigma \sim w^s$. The exponent s estimated from measurements at frequencies 1 kHz-

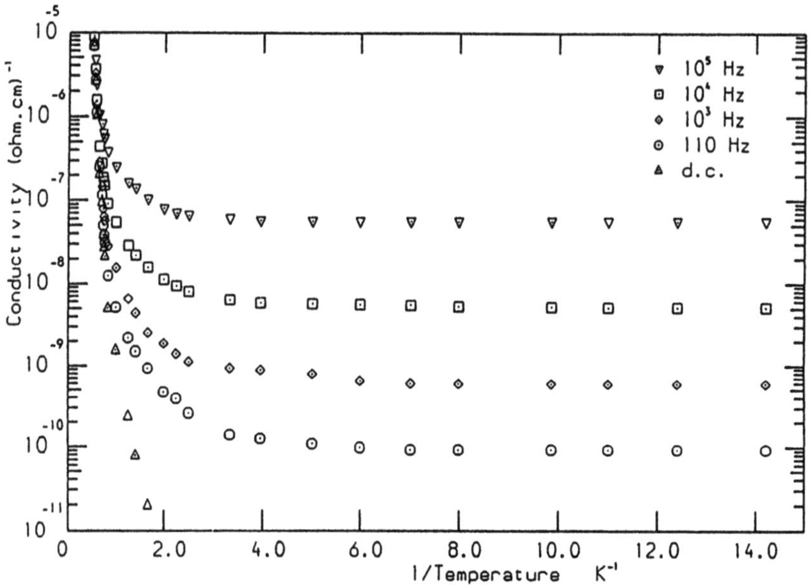

Fig. 1 The temperature dependence of the conductivity of sample 5714 at H=60 kG.

100 kHz was found to be temperature dependent, increasing with decreasing T and approaching unity at the lowest temperatures. This is shown explicitly in Fig. 2(b) at H=50 and 40 kG. The variation of s with T is also observed for the other samples and data are demonstrated in Fig. 3.

According to the two-site model, s is given by $s = 1-4/\ln(\nu_{ph}/w)$ which implies that s is temperature independent, ($s \sim 0.8$ for moderate frequency). For lower values of $s < 0.8$, one has to consider unphysically small values of ν_{ph} varying with T. Furthermore, the QMT model predicts that s should increase as w is decreased. No such increase in s (as w is lowered) is observed. s is indeed, frequency independent or slowly an increasing function of w. These discrepancies between experimental data and the theory might be considered as evidence against the QMT model.

The QMT model predicts that $\sigma(w)$ is varying only linearly with T. Such temperature dependence seems to be weaker than the experimental observations. On the other hand, at low temperatures, $\sigma(w)$, is temperature independent. Finally, the temperature dependence of $\sigma(w)$ at lower frequencies seems to be more pronounced than that at higher frequencies. The possibility, then, of using the simple pair approximation as being the dominant mechanism of conduction cannot be totally justified in the present study. Generally, the temperature and frequency dependence of $\sigma_{ac}(w,T)$ can be expressed as

$$\sigma_{ac}(w,T) \sim w^s\, T^d \tag{5}$$

Values of the exponent d are obtained from plots of log σ_{ac} vs. log T at different fields and frequencies. For sample 5714, at H=50 kG and frequencies in the range of 10^2-10^5 Hz, d takes values in the range of 3.9 down to 2.0. Similar behaviour is observed at different fields for this sample and the other two samples. d was found to be greater than unity and increases as the frequency is decreased. A similar trend was observed when the actual measured values of $\sigma(w,T)$ were used. According to Pollak [11] as T is increased, a cluster of three, four etc. centres can be formed instead of the simple pair

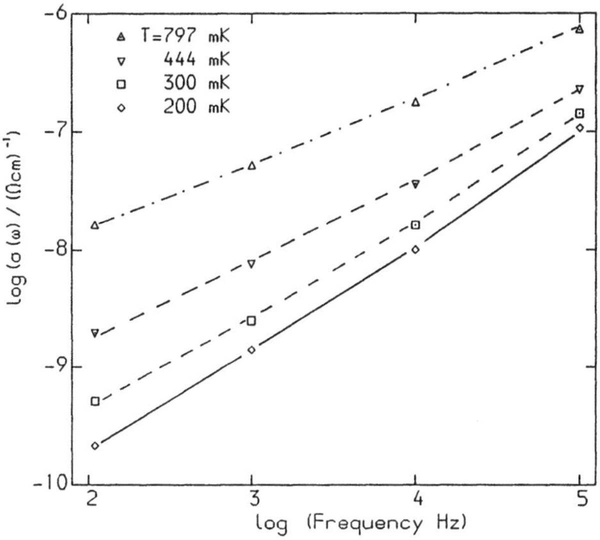

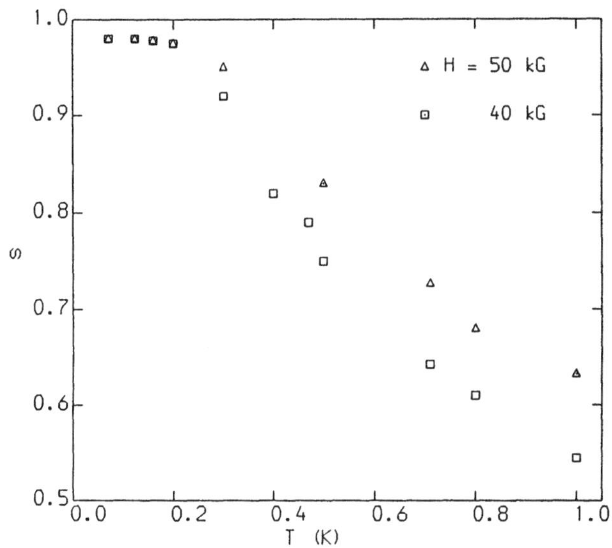

Fig. 2 (a) The frequency dependence of $\sigma_{ac}(\omega)$ of sample 5714 at H=40 kG. (b) †he temperature dependence of the exponent "s" at fields 40 and 50 kG.

and the electron can hop twice, thrice or more in one half-period of the a.c. field depending on the size of the cluster leading to an increase in the conductivity. The likelihood of such multiplets increases with increasing $r_w = a/2 \ln(\nu_{ph}/w)$ as well as T. Such multiplet hopping conduction weakens the frequency dependence but in the meanwhile strengthens the temperature dependence. An increased dependence of the conductivity on r_w means a decreasing dependence on w, the multiple hops then will result in a decreased frequency dependence in addition to the increased temperature dependence.

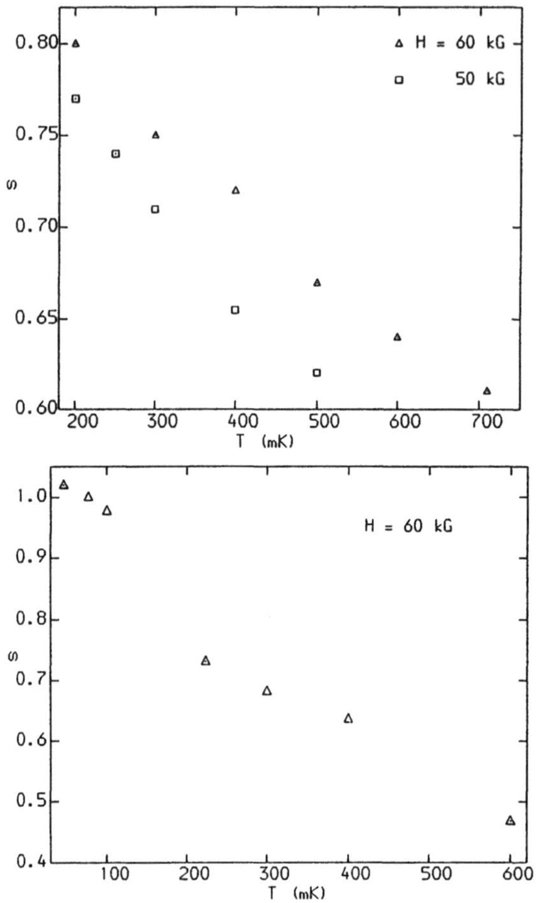

Fig. 3 The temperature dependence of the exponent "s" of (a) sample 9914 and (b) sample 2015.

Following Pollak's arguments then multiple hopping becomes more important at lower frequencies and high temperatures. This might explain the higher values of d at lower frequencies as well as the excess in its value above unity.

The temperature-independent values of σ (w) at different fields and frequencies for samples 5714 and 9914 are plotted in Fig. 4 as a function of H. It is obvious that the dependence at high fields is weak, however, as the field is reduced a pronounced change in σ (w) is observed and the a.c. conductivity strongly depends on H. At high magnetic fields, the wavefunctions are expected to be strongly (highly) localized, and as the field is reduced towards lower values, a metallic-like behaviour is expected at fields $< H_c$.

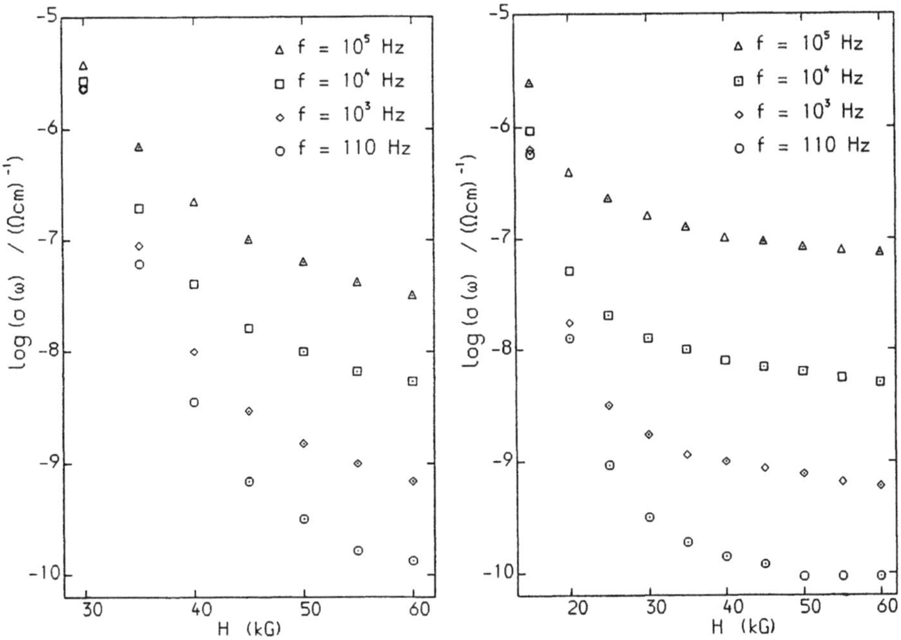

Fig. 4 The magnetic field dependence of the lowest temperature limit of $\sigma(\omega)$ of (a) sample 5714 and (b) sample 9914.

References

[1] Pollak, M. and Geballe, T.H., Phys. Rev., 122, 1742 (1961).
[2] Efros, A.L., Phil. Mag. B, 43, 829 (1981).
[3] Austin, I.G. and Mott, N.F., Adv. Phys., 18, 41 (1969).
[4] Pollak, M. Phil. Mag., 23, 519 (1971).
[5] Long, A.R., Adv. Phys., 31, 553 (1982).
[6] Efros, A.L. and Shklovskii, B.I., "*Electron-Electron Interactions in Disordered Systems*", Ed. Efros and Pollak, Elsevier Pub., 409 (1985).
[7] Mansfield, R. Abdul-Gader, M. and Fozooni, P., Solid State Elect., 28, 109 (1985).
[8] Abdul-Gader, M., Mansfield, R. and Fozooni, P., "*Application of High Magnetic Fields in Semiconductor Physics*", Springer Verlag, Ed. Landwehr, 71, 381 (1987).
[9] Ishida, S. and Otsuka, E., J. Phys. Soc. Japan, 43, 14 (1977).
[10] Yafet, Y., Keyes, R.W. and Adams, E.N., J. Chem. Solids, 1, 37 (1956).
[11] Pollak, M., Phys. Rev. A, 138, 1822 (1965).

Negative Magnetoresistance of InP in the Hopping Region

S. Abboudy[1], *R. Mansfield*[2], *and Lim Chee Ming*[2]

[1]Physics Department, Faculty of Science, Alexandria University, Alexandria, Egypt

[2]University of Brunei, Brunei

A characteristic feature of hopping conduction is a very large positive magnetoresistance caused by the field reducing the overlap of the impurity wave functions and giving rise to a resistivity dependent exponentially on H^2. There is however evidence of an initial negative magnetoresistance in GaAs, InP and Ge which reaches a maximum of up to 10% in a field of approximately 10 kG and then decreases rapidly as the effect is swamped by the positive effect. This effect occurs when the zero field resistivity, due to variable range hopping, varies with temperature as $\rho = \rho_0 \exp(T_0/T)^x$ where x depends on the energy dependence of the density of states at the Fermi energy and the samples are in the localized or non-metallic region. Benzaquen et al. [1] have interpreted their results on GaAs using the mechanism proposed by Fukuyama and Yosida [2] who assume the field causes a Zeeman splitting of impurity states with a majority of carriers in the upper Zeeman state. This results in an increase in overlap of wave functions with neighbouring states. The relative magnetoresistance then becomes

$$\frac{\rho_H - \rho_0}{\rho_0} = \left\{[(\cosh(a_1 H))^{-1} - 1] + \{\exp(K_s H^2) - 1\}\right\} \quad (1)$$

The term in the square bracket is the negative magnetoresistance where a_1 is given by

$$a_1 = (x\beta d/2)[g^*\mu_B/(E_c - E_F)](T_0/T)^x \quad (2)$$

Here $x = 1/4$, d = dimension of the system = 3, $\beta \cong 1$, g^* = effective g-factor, μ_B = Bohr magneton, E_F= Fermi energy, E_c = mobility edge. Although $E_c - E_F$ increases with magnetic field it is assumed that this field dependence is weaker than the Zeeman shift $g^*\mu_B H/2$. The second term in (1) is the positive magnetoresistance and for VRH and constant density of states

$$K_s = t^1\frac{e^2a^4}{c^2\hbar^2}\left(\frac{T_0}{T}\right)^{3/4}, \quad t^1 = .0025 \quad (a = \text{donor radius}) \quad (3)$$

Benzaquen et al [1] were able to reproduce the general features of their results using (1) but consistency between the theoretical and experimental values of a_1 and K_s was poor. In the case of a_1 Benzaquen (private communication) used an effective Bohr magneton $e\hbar/2m^*$ with $m^*/m = 0.068$. However the Zeeman energy splitting is usually expressed as $g^*\mu_B H$ where the g^* factor includes the band parameters such as spin-orbit coupling and μ_B is the normal Bohr magneton $e\hbar/2m$. For GaAs $g^* = -0.44$ [3] and when a_1 is calculated

Springer Series in Solid-State Sciences, Vol. 101
High Magnetic Fields in Semiconductor Physics III Editor: G. Landwehr

using this value of g^* and $\mu_B = e\hbar/2m$ the agreement between the values of a_1 obtained from experiment and theory is very much improved.

This paper considers measurements made of the negative magnetoresistance on lightly doped samples of InP. Comprehensive measurements have been made of the resistivity of three samples of n-type InP having a doping level such that $N_D^{1/3}a = 0.19$ to 0.16. The results for the zero field and positive magnetoresistance have been discussed by Mansfield et al. [4]. Nearest neighbour hopping (NNH) is observed in the temperature range 4 to 2 K and variable range hopping (VRH) below 2 K. Good numerical agreement is obtained between theory and experiment in the NNH region and also in the VRH region where the resistivity varies as $\exp(T_0/T)^{1/4}$ which is predicted for a constant density of states at the Fermi level. An estimate of K_s can be made from a plot of $\ln(\rho(H)/\rho(0))$ against H^2. Such a plot is only linear from 15 kG when the PMR dominates the NMR to 30 kG which is the weak magnetic field limit given by $H < 2H_c = 2N_D^{1/3}c\hbar/ae$. The asymptotic slope over this restricted region will be greater than K_s given by (3). In fact the value of t_1 derived by this method was greater than the theoretical value by a factor two.

The interpretation of the negative magnetoresistance is complicated because there are several mechanisms which cause a variation of resistance with field. Apart from the two competing processes given in (1) there has recently been a series of papers suggesting an additional process [5,6,7]. This arises because phase coherence in certain paths containing closed loops increase the resistance. The effect of a magnetic field passing through such loops destroys this phase coherence and reduces the resistance. Entin-Wohlman et al. have provided graphs of the magnetoconductivity $(\sigma_H - \sigma_0)/\sigma_0$ as a function of the flux $\Phi_M = HR_M^{3/2}\xi^{1/2}$ through an effective area $R_M^{3/2}\xi^{1/2}$ where R_M = hopping distance and ξ is the localization length. For the InP samples described here the concentration is low so that the localization length should in this case be close to the donor radius 84×10^{-8} cm. The localization length which should be used here is $\xi = a|1 - n/n_c|^{-\beta}$ where $\beta \cong 1$ and n and n_c are the concentration of impurities and critical concentration for the M-I transition. An estimate of the hopping length can be derived from T_0 and ξ using for example the expression $R_M \cong 0.4\xi(T_0/T)^{1/4}$ given by Sharafam et al. [8]. This predicts a value of $R_M \cong 6\xi$. However the doping level is such that the distance between donors is $R_D \cong 3\xi$ hence the hop length using this formula is approximately $2R_D$. This appears unrealistically low and since there is some uncertainty in the numerical constant in the formula for R_M a value of $R_M = 0.6\xi(T_0/T)^{1/4}$ has been used to calculate the magnetoconductance.
The table gives details of the samples together with values of a_1 derived using (2) and values of K_s obtained from the asymptotic slopes of the zero field $\ln(\rho_H/\rho_0)$ versus H^2 plots. The contribution to the magnetoresistance arising from the PMR (1), the Zeeman effect (2) and the orbital effect (3) were derived for each magnetic field for which measurements had been made and the difference d_i between the theoretical and experimental values were calculated. The sum Σd_i for each point in the negative

Table

Sample	N_D-N_A $10^{15}cm^{-3}$	T_O 10^4K	ε_3 meV	a_1 $10^{-4}G^{-1}$ theory	data fit	K_s $10^{-9}G^{-2}$ H^2 plot	data fit
2	5.9	3.3	0.76	5.05	8.98	2.76	2.65
3	7.8	3.7	0.56	6.5	9.8	2.47	2.47

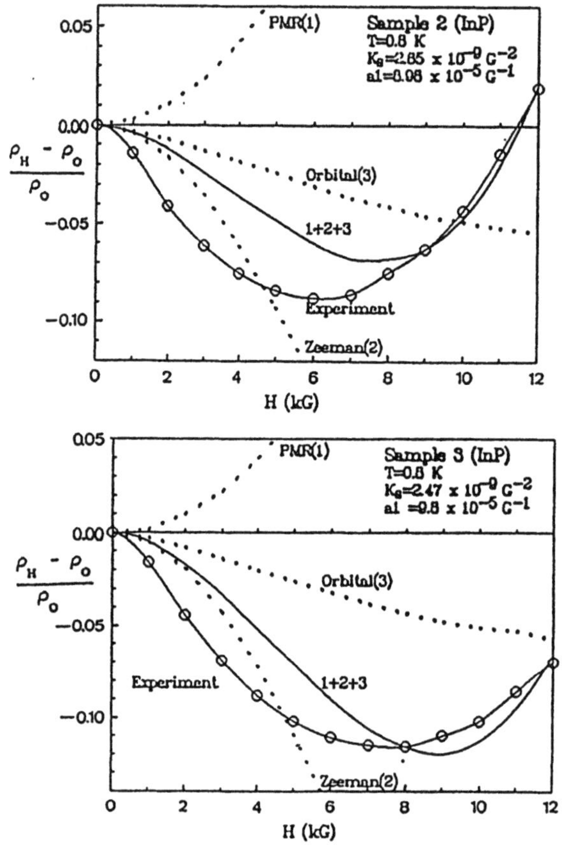

Figure 1. Negative Magnetoresistance of two samples of InP with magnetic field perpendicular to the current. The dotted curves show contributions from 1 positive magnetoresistance, 2 the Zeeman effect and 3 the orbital magnetoresistance [7]. The sum of these three contributions is shown by the continuous curve 1+2+3. The circles show the experimental points and the continuous line through them is a cubic spline fitted to the points.

magnetoresistance region was then calculated and the parameters a_1 and K_s were varied to minimize this sum to achieve the best fit. Figure 1 shows the best fit to the data and the required values of a_1 and K_s are given in the Table. The general features of the results are reproduced with values of K_s in reasonable agreement with its high field values but agreement between a_1 required to

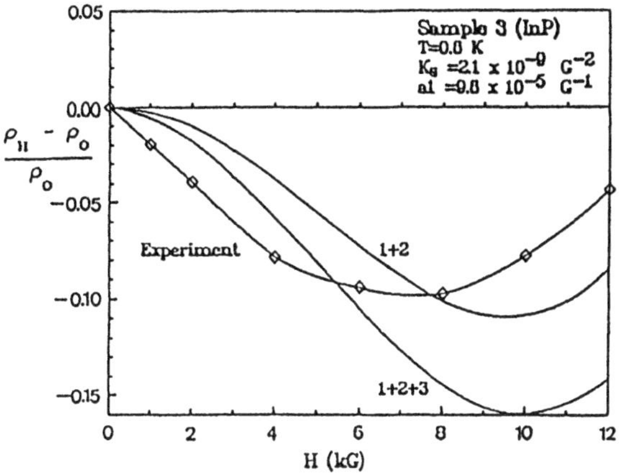

Figure 2. Magnetoresistance of sample 3 with magnetic field parallel to the current. The experimental points are in better agreement with 1+2 suggesting that the orbital contribution in this case is small or negligible.

fit the data and theory is not so good. There are, however, several terms in the expression for a_1 whose values are uncertain such as $E_c-E_F = \varepsilon_3$, $\beta \cong 1$ and the value of T_0. Figure 1 also shows that the two terms (1) and (2) increase very rapidly with field and are both large compared to the measured magnetoresistance so that the agreement could be improved with a relatively small magnetic field dependence of these parameters. The contribution due to the orbital effect is small compared to the other two contributions even though an artificially large hop length $R_H \cong 9\xi$ has been used. It should be noted however that the main discrepancy between theory and experiment occurs in the low field region and that a substantial increase in the low field orbital contribution is obtained by increasing R_H still further. In fact a good fit is obtained if R_H is increased to 15ξ with a corresponding decrease in a_1 which would then become closer to its theoretical value given in the Table.

Figure 2 shows the longitudinal magnetoresistance with magnetic field parallel to the current. Also shown are 1+2 which is the sum of the contributions due to PMR with the appropriate value of K_s derived from measurements in the longitudinal configuration and the Zeeman effect and also 1+2+3 with the addition of the orbital magnetoresistance. A comparison between the experimental curve and the two theoretical curves show that the agreement with 1+2 is much better than with 1+2+3 and suggests that since the Zeeman effect is expected to be isotropic the difference between the transverse and longitudinal magnetoresistance is due to the absence or much reduced orbital effect.

References

1. M.Benzaquen, D.Walsh and K.Mazuruk, Phys.Rev.**B38**, 10933 (1988)
2. H.Fukuyama and K.Yosida, J.Phys.Soc. Japan **46** , 102 (1979)
3. C.Herman, G.Lampel and C.Weisbuch, Proc. of the 13th Int. Conf. on the Phys. of Semiconductors, Paris, 130 (1976)
4. R.Mansfield, S.Abboudy and P.Fozooni, Phil. Mag. **B57**,667 (1988)
5. V.I.Nguyen, E.Z.Spivak and B.I.Shklovskii, JETP Lett.**41**, 42 (1985) and Sov.Phys.-JETP **62**, 1021 (1985)
6. U.Sivan, O.Entin-Wohlman and Y.Imry, Phys.Rev.Lett. **60**, 1566 (1988)
7. O.Entin-Wohlman, Y.Imry and U.Sivan, Phys.Rev. **B40**, 8342 (1989)
8. W.N.Shafarman, D.W.Koon and T.G.Castner, Phys.Rev. **B40** 1216 (1989)

Part VI

Magneto-Optics, Mainly in Quantumm Well Structures

Magneto-Optic and Quantum Transport Studies of MBE InSb and InAs

R.A. Stradling

Physics Department and Interdisciplinary Research Centre in Semiconductor Materials, Imperial College, London SW7 2BZ, UK

Abstract. The MBE growth and doping of the narrow gap materials InSb and InAs are studied. Magneto-optical experiments are extremely informative in the case of high purity films of InAs showing the presence both of a low mobility accumulation layer and a high mobility bulk region and enabling the galvanomagnetic properties to be interpreted in terms of parallel conductance involving the two regions. Silicon can be incorporated fully as a donor up to concentrations of $3x10^{18}$ cm^{-3} in the case of InSb and $6x10^{19}$ cm^{-3} with InAs. The control of silicon doping is utilised to prepare high-concentration spike-doped samples and nipi structures showing enhanced subband gap absorption, large optical non-linearities and increased carrier lifetimes. Shubnikov-de Haas measurements are undertaken with the spike-doped samples to determine the subband occupancy and to investigate possible dopant diffusion.

1. Introduction

Only a small fraction of the effort expended on improving the technology of epitaxial GaAs has been directed at the narrow-gap III-V materials. In order to create high-quality low-dimensional structures of materials such as InSb and InAs, it is therefore necessary to develop first the appropriate growth and doping techniques. At the last Wurzburg Conference the Imperial College Group reported the observation of quantum confinement effects in MBE InSb samples slab-doped with silicon [1]. Silicon is known to be an amphoteric dopant in InSb but nevertheless doping during MBE growth is now sufficiently under control [2] that n-type spike-doped structures can be grown with high mobility and little dopant diffusion. This paper discusses similar growth and doping studies of InAs and then reviews quantum transport and optical measurements with selectively-doped InSb and InAs structures.

Springer Series in Solid-State Sciences, Vol. 101
High Magnetic Fields in Semiconductor Physics III Editor: G. Landwehr

2. MBE Growth and Doping Studies of InAs

InAs is generally thought to have markedly inferior material quality compared with GaAs and several other semiconductors. Very recently however linked electrical and magneto-optical studies were reported for some very high mobility epitaxial films grown by MBE on GaAs substrates [3].

The Hall mobility derived from the product of the conductivity σ and the Hall coefficient R was of the order of 100,000 cm^2/Vs at 77K for films of greater than 5μm thickness and was found to decrease rapidly with decreasing film thickness. It is tempting to interpret this result as demonstrating the increasing importance of scattering from the high concentration of misfit dislocations known to exist close to the interface as the films became thinner. However the magneto-optical studies showed that this interpretation was incorrect. A broad but strong absorption line together with a number of sharp lines were observed in experiments with a far-infrared laser system (fig. 1). The two strongest of the sharp lines could be immediately identified from earlier magneto-optical experiments [4] as arising from cyclotron resonance from the free carriers in the bulk of the InAs film and as the 'impurity-shifted cyclotron resonance' (i.e. the $1s\text{-}2p_+$ shallow donor line). The broad line moved to higher magnetic field with the characteristic $1/\theta$ dependence for cyclotron resonance of a two dimensional electron gas (2DEG) while the sharp lines stayed at the same field position. Earlier electrical measurements [5] had shown the InAs suffers from electron accumulation at the surface and the 2DEG giving the broad cyclotron resonance line is thought to originate from a surface accumulation layer. Dislocation scattering was clearly unimportant as the line widths of both the bulk cyclotron resonance and donor impurity lines were independent of film thickness. The apparent fall-off in the Hall mobility with decreasing film thickness can now be interpreted as arising from the increasing proportion of the current flowing through the low mobility surface skin and does not represent any degradation in the bulk quality of the material. The line widths determined for the bulk cyclotron resonance were slightly narrower than the cyclotron resonance lines reported earlier for high purity VPE films of thickness of the order of 20μm where

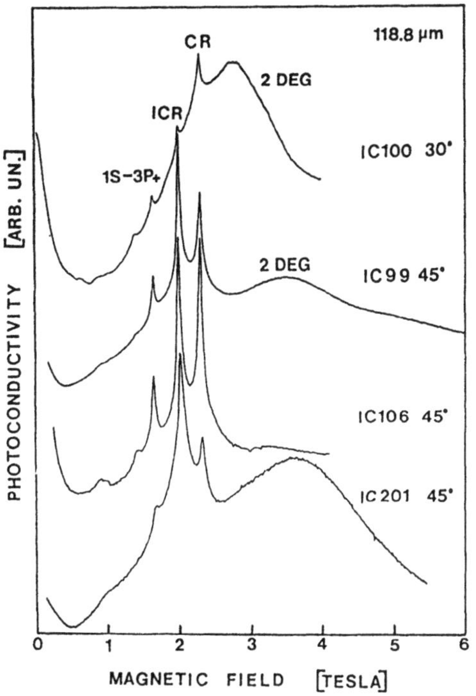

Fig. 1 shows the far-infrared magneto-optical spectrum for a number of high purity InAs samples taken with a laser wavelength of 118 µm. The lower three recordings were taken with the sample tilted at 45° with respect to the magnetic field causing the broad line due to the 2DEG to be displaced upwards in field.

the measured Hall mobility at 77K was 180,000 cm^2/Vs [6]. On the basis of the magneto-optical measurements the mobility in the bulk of the MBE films was shown to be 250,000 cm^2/Vs at 77K with a shallow donor concentration of about 5x10^{14} cm^{-3}. A detailed fit of the thickness and magnetic field dependencies of the galvanomagnetic properties was consistent with these figures [3]. For example the transverse magnetoresistance is very sensitive to the presence of layered or parallel conductance. Fig. 2 shows the results of magnetoresistance measurements for a number of samples of differing thicknesses together with the theoretical variation calculated assuming the thickness independent electrical parameters quoted above for the bulk and 2DEG regions. Also included in the figure is the result quoted in ref. [7] for a much thicker vpe

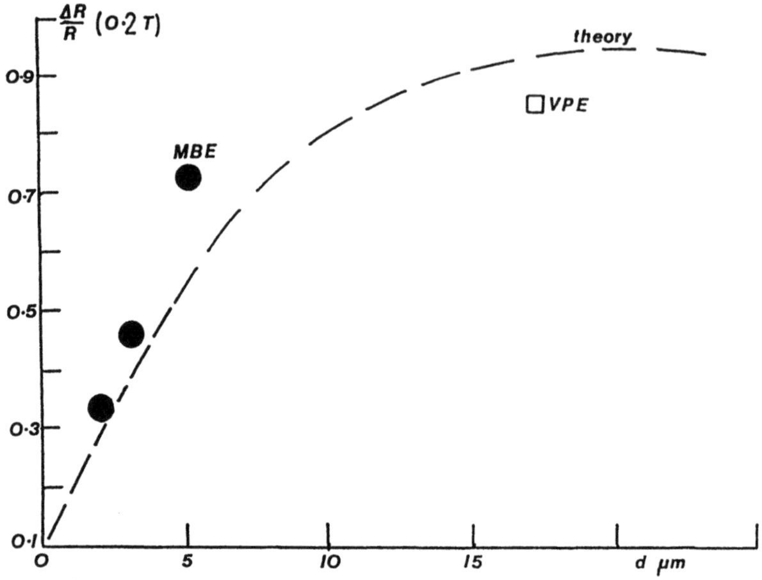

Fig. 2 shows the transverse magnetoresistance measured at a field of 0.2T as a function of thickness for a series of InAs epilayers together with the theoretical prediction for the electrical parameters quoted in the text.

sample of similar electrical parameters. It can be seen that the simple two-carrier theory provides a good fit to experiment.

The identification of two of the lower-field (high energy) magneto-optical lines initially caused some puzzlement. The excited states of the transitions involved were clearly associated with the N=1 Landau levels as the lines tracked with the cyclotron resonance line but the energies were greater than the difference between the ground state of the shallow donors and the N=1 Landau level. There were also significant qualitative differences in character between these two lines and the lines associated with shallow donors. In particular these two lines persisted to bias electric fields which caused the impact ionisation of the shallow donor states. A second difference can be seen in the results shown in figure 3. The two anomalous lines show a doublet structure which is not seen with the other donor lines whose final state is associated with the N=1 Landau level. As the two anomalous lines extrapolate back to much higher energy than the shallow donor energy

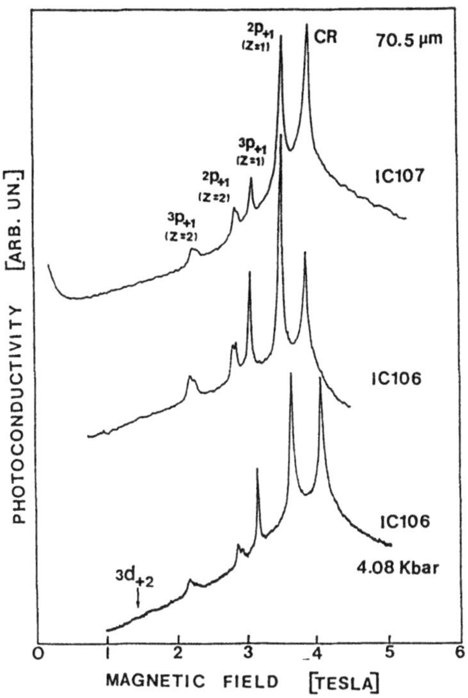

Fig. 3 shows magneto-optical spectra for several high-purity InAs epilayers taken at a laser wavelength of 70.6 μm. The two lines seen at fields of about 2.2 and 3T are thought to arise from singly ionised double (Z=2) donors.

in zero magnetic field (12 cm^{-1}), they could be 1s - $2p_+$ lines from two impurities whose chemical shifts are much greater than the effective Rydberg for Z=1 donors. However a much more likely interpretation is that the two lines are 1s - $2p_+$ and 1 - $3p_+$ transitions for singly ionised state of double (Z=2) donors. The theoretical field dependence for these lines agrees extremely well with experiment. The doublet structure could arise from the non-cubic local potential around the impurity site, or from the presence of two different (Z=2) donors. One plausible candidate for a double donor is the arsenic antisite centre although any substitutional impurity having a valence differing by two from the host, or an interstitial could also be responsible for the observed spectral lines.

It has proved possible to dope InAs with silicon donors to much higher concentrations than with either InSb or GaAs. Concentrations of carriers up to $6x10^{19}cm^3$

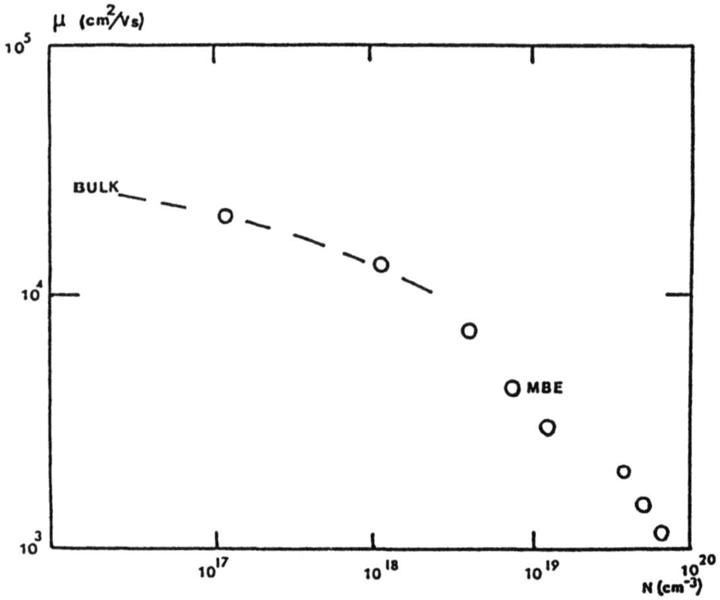

Fig. 4 shows the variation of Hall mobility against carrier concentration for MBE silicon-doped InAs

have been achieved without dropping below the bulk mobilities expected for this concentration (see fig. 4). The reason for the attainment of much higher donor concentrations with InAs than with the other III-Vs may be associated with the nature of the amphoteric native defects [8]. Alone amongst the III-V compounds this defect level is located well-up into the conduction band rather than being localised in the gap. According to the ideas developed by Walukiewicz [8], the defect will therefore act to stabilise the silicon atoms on the indium sites until the Fermi energy becomes substantially greater than the energy of the amphoteric defect.

3. Developments with Selectively-Doped InSb and InAs

The Shubnikov-de Haas effect has now been studied with a considerable number of atomic-plane (or spike doped) samples [9]. Fourier analysis of the rather complicated peaked structure in the magnetoresistance shows that many subbands are occupied. The very high doping levels achievable in the case of InAs give rise to the occupancy of at least eight subbands in a sample with a

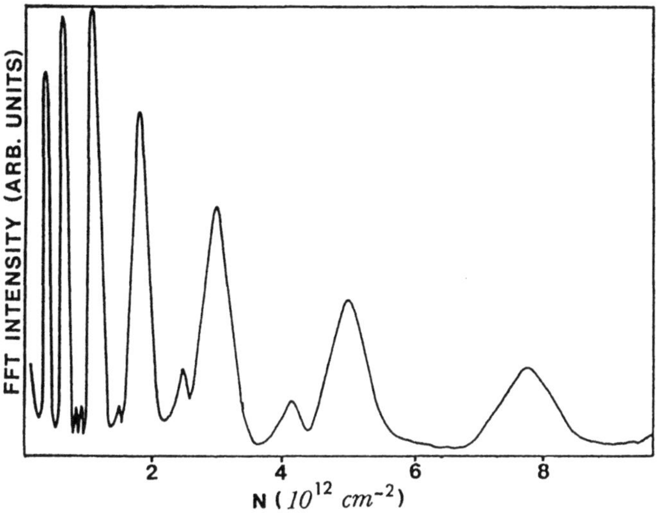

Fig 5 shows the Fourier analysis of the Shubnikov-de Haas effect observed with a high concentration spike-doped sample of InAs. At least eight subbands are occupied.

planar doping level of $2x10^{13}$ cm^{-2} (see fig. 5). The relative occupancy of the i=0 subband with respect to higher order subbands provides a sensitive test for diffusion of the dopant away from the doping plane as the population of the i=0 subband falls with increasing well width and the carriers are redistributed into higher subbands. With InSb it has been possible to obtain excellent agreement between the populations derived from the Shubnikov-de Haas measurements and those calculated self-consistently [10] - see table 1. This agreement shows that the silicon is fully activated as a donor and diffuses less than about 6nm for planar doping levels up to $4x10^{12}$ cm^{-2}. If this concentration of donor atoms remained located within a monolayer, the equivalent bulk doping level of $1.2x10^{20}cm^{-3}$ would exceed the maximum bulk doping level achieved with MBE by well over an order of magnitude. Even if the silicon were redistributed over a distance equivalent to the depth resolution associated with the Shubnikov-de Haas effect (6nm), the equivalent bulk density ($7x10^{18}$ cm^{-3}) remains slightly higher than that achieved in the bulk doping experiments with full activation of the silicon doping ($3x10^{18}$ cm^{-3}) [2]. It therefore would appear that higher concentrations are achievable with atomic plane

Table 1. A Comparison of the Carrier Concentrations for Spike-doped InSb and InAs deduced from Fourier Analysis of Shubnikov-de Haas Data with Those Predicted by Self-Consistent Calculations

Sample	InSb 126		InAs 148		InAs 212	
	exp. (10^{12} cm^{-2})	theory	exp. (10^{12} cm^{-2})	theo.	exp. (10^{13} cm^{-2})	theo.
i=0	1.05	1.09	2.21	2.64	0.77	1.04
i=1	0.51	0.50	1.19	1.18	0.49	0.46
i=2	0.28	0.27	0.68	0.63	0.29	0.25
i=3	0.18	0.14	0.34	0.31	0.18	0.13
i=4	0.10	0.07	0.2	0.15	0.10	0.07
i=5	0.03	0.03	0.06	0.06	0.06	0.03
i=6	0.01	0.01	0.02	0.02	0.03	0.02
Total						
Sh-deH	2.2	2.12	4.7	5.00	1.90	2.00
Hall	2.5	-	5.2	-	1.5	-
Growth	2.47	-	4.9	-	2.0	-

doping than in bulk growth but further experiments with yet higher doping densities are needed to confirm this result.

In the case of InAs the occupation of the i=0 subband is always some 25-30% lower than that predicted by self-consistent calculations (see table 1). Furthermore the ratio of the occupancy of the i=0 to i=1 subband is substantially less than two whereas with InSb this ratio is about two. This ratio is only weakly dependent on the total doping concentration and for different materials should increase slowly with increasing ratio of the effective mass to the dielectric constant (m^*/ϵ); i.e. with decreasing Bohr radius for the hydrogenic donors. Thus the ratio of the two lowest subbands for InAs should be significantly greater than two and the deviation between the experimental values and the self-consistent calculations could be indicative of substantial dopant movement during growth.

The control of growth and doping achieved with these narrow-gap materials have enabled us to grow doping (or"*nipi*") superlattices with novel and potentially useful optical, electro-optical and electronic properties with both InSb and InAs. The electrostatic origin of the nipi superlattice potential means that proportionally larger shifts in the effective bandedge can be engineered in narrow gap semiconductors than with GaAs, and the smaller conduction band masses increase the electron wavefunction spread in the superlattice direction. The resultant enhancement in the optical absorption coefficients for radiation below the bulk bandedge have been demonstrated as well as the associated optical non-linearities and increased lifetimes [11,12].

4. Conclusions

Farinfrared magnetospectroscopy and quantum transport measurments have been extremely useful in characterising thin MBE films of InAs and InSb. Good control of doping has been achieved enabling good-quality spike-doped and nipi structures to be grown.

5. Acknowledgements

This paper has reviewed the work of a large number of staff at Imperial College. The major contributions from the growth team of R. Droopad, I.Ferguson, A. d'Oliveira, S.D.Parker, R.L. Williams; from E. Skuras, M. Williams and R.L.Williams on the transport measurements; from S.N. Holmes and P.D. Wang on the magneto-optical studies and E.J. Johnson and A. Mackinnon for theoretical assistance are all greatfully acknowledged. This work was supported by the Science & engineering Research Council through the Low Dimensional Structure Initiative.

6. References

[1] R. Droopad, S.D. Parker, E. Skuras, R.L. Williams, R.A. Stradling, R.B. Beall & J.J. Harris Vol. **87** Springer Series in Solid State Sciences p199 (1989).

[2] S.D. Parker, R.L. Williams, R. Droopad, R.A. Stradling, K.W.J.Barnham, S.N.Holmes, J.Laverty, C.C.Phillips, E.Skuras, R.Thomas, X.Zhang, A.Staton-Bevan & D.W.Pashley
Semicond. Sci. & Tech. **4** 663 (1989)

[3] S.N. Holmes, R.A.Stradling, P.D. Wang, R.Droopad, S.D.Parker & R.L.Williams
Semicond. Sci. & Tech. 4 303 (1989)

[4] C.M. Sotomayor-Torres & R.A.Stradling
Semicond. Sci. & Tech. 2 323 (1986)

[5] H.A.Washburn, J.R.Sites & H.H.Wieder
J.App. Phys. **50** 4872 (1979)

[6] H.Reisenger, H.Schaber & R.E.Doezema
Phys. Rev. **B24** 5960 (1981)

[7] H.H.Wieder
App. Phys. Lett. **25** 206 (1974)

[8] W. Walukiewicz
MRS Symposium Proceedings **Vol 104** 483 (1988)

[9] R.L.Williams, E.Skuras, R.A.Stradling, R.Droopad, S.N. Holmes & S.D. Parker
Semicond. Sci. & Tech. 5 S338 (1990)

[10} E.A.Johnson & A. Mackinnon
Semicond. Sci & Tech. **6** S189 (1990).

[11] C.C.Phillips
App. Phys. Lett. **56** 151 (1990)

[12] C. Hodge, C.C.Phillips, R.H.Thomas, S.D.Parker, R.L.Williams & R.Droopad.
Semicond. Sci & Tech. **5** S319 (1990)

Magneto-Optical Studies of Impurities in Coupled and Perturbed Quantum Wells

R. Ranganathan, B.S. Yoo, J.-P. Cheng, and B.D. McCombe

Department of Physics and Astronomy and
Center for Electronic and Electro-optic Materials,
State University of New York at Buffalo, Buffalo, NY 14260, USA

Abstract. Recent far infrared magneto-spectroscopic studies of shallow donor impurities in GaAs/AlGaAs multiple-quantum-well (MQW) structures are described. Three examples are given: 1) Photothermal ionization spectroscopy of high excited states in well-center-doped QWs; 2) The effects of wave function penetration through narrow barriers on the electronic states of shallow donor impurities in coupled double quantum wells; and 3) The effects of electric fields (Stark effect) on shallow donors in wide quantum wells.

1. Introduction

The importance of impurities and defects in determining the electronic and optical behavior of semiconductors is well-recognized. With semiconductor heterostructures for many applications, as well as basic physical studies, accurate placement of impurities and the control of doping density (selective doping) play an important role, and it is therefore important to determine the effects of the confining structures on the electronic states of the impurities.

There has been significant progress in understanding the effects of confinement on shallow (hydrogenic) impurities in simple quantum-well (QW) structures over the past few years, both theoretically, through variational calculations in the envelope function approximation, and experimentally, through the use of various spectroscopic techniques [1]. In this paper we describe some recent far infrared magnetospectroscopic work directed at investigating high excited states of shallow donor impurities doped in the centers of GaAs wells in GaAs/AlGaAs MQW structures, the effects of coupling of pairs of wells in coupled-double-quantum-well (CDQW) structures on the impurity states, and the effects of external perturbations (large electric fields) on the electronic states of the shallow impurities. Results illustrate the dramatic effects that confinement, coupling, and external fields have on the electronic states of the shallow donors.

2. Theoretical Background

In the effective-mass approximation the Hamiltonian for a shallow donor in a MQW structure in the presence of a uniform external magnetic field, B,normal to the layers is given by [2,3] (in dimensionless units)

$$H = -\nabla^2 - 2/r + \gamma L_z + \gamma^2\rho^2/4 + V_W(z). \qquad (1)$$

Springer Series in Solid-State Sciences, Vol. 101
High Magnetic Fields in Semiconductor Physics III Editor: G. Landwehr

Here $\gamma = \hbar\omega_c/2R^*$ is a dimensionless measure of the magnetic field, ω_c is the cyclotron frequency, eB/m^*c, with m^* the effective mass, and R^* is the effective rydberg ($R^* = m^*e^4/2\hbar^2\epsilon^2 = 5.83$ meV for GaAs). The distance $r = [\rho^2 + (z-z_i)^2]^{1/2}$ is the position of the electron measured with respect to the impurity ion located at $\rho=0$, $z=z_i$, and $\rho = [x^2 + y^2]^{1/2}$. The projection of the orbital angular momentum along the magnetic field, L_z, is a constant of the motion. The unit of energy in Eq. (1) is the effective rydberg, and the unit of length is the effective Bohr radius ($a^* = \hbar^2\epsilon/e^2m^* = 98.7$ Å for GaAs).

The MQW potential, V_W, is given by [3]

$$V = \begin{cases} 0, & -L_W/2 + n(L_W + L_B) < z < L_W/2 + n(L_W + L_B), \\ V_o, & L_W/2 + n(L_W + L_B) < z < -L_W/2 + (n+1)(L_W + L_B), \end{cases} \quad (2)$$

where L_W is the well width, L_B is the barrier width, and n is an integer. The barrier height, V_o, is approximately $0.6\Delta E_g$, where ΔE_g is the band gap difference between the AlGaAsand GaAs.

Solutions of the Schrödingerequation with the Hamiltonian of Eq. (1) for a single, isolated impurity have been obtained through the use of a variational approach with trial (envelope) wave functions expanded in gaussion basis sets and with appropriate matching conditions at the interfaces for the wave functons [2,3]. The simpler case of an impurity in an isolated QWwas treated earlier by a similar approach [4]. The results of the variational calculations for the ground and low-lying excited states of an isolated donor are consistent with physical intuition. For an impurity at a well center the binding and transition energies increase with increasing well width. For constant well width the binding and transition energies decrease as the impurity ion is moved from the well center to an edge, and decrease even further as the impurity is moved into the barrier (with the electron remaining in the well).

To date calculations have concentrated primarily on impurity states in a magnetic field associated with the lowest two Landau levels. Recently, calculations for higher excited states have been carried out for a strictly two-dimensional (2D) system [5], and the correspondence between the low magnetic field and the high magnetic field quantum numbers established for this case. This correspondence appears to contradict the well-established correspondence for shallow donors in bulk (3D) semiconductors [6]. Experimental results and theoretical considerations are discussed in Sect. 4.1., and it is shown that there is no contradiction, and that there is a natural evolution from the three-dimensional to the two-dimensional case through the quasi-2D situation realized in QWs.

3. Experimental Considerations

The samples used in these experiments were grown by molecular beam epitaxy (MBE) at Cornell University. Typical sample structures consist of a 2000Å GaAs buffer, followed by a 1500Å cladding layer of $Al_xGa_{1-x}As$, followed by the MQWstructure, another 2000Å $Al_xGa_{1-x}As$ cladding layer, and, finally, a

100Å GaAs cap. Quantum well widths were between 125Å and 500Å, and barrier widths were 125Å, except for the 500Å well-width sample, which had 200Å barriers. All samples discussed here were doped over the central 1/3 of the GaAs wells with Si donors. Magnetotransmission data were obtained with two separate Fourier Transform Spectrometers, a SPECAC Model 40.000 polarizing interferometer, and a BOMEMDA3.02 spectrometer, in conjunction with a 9T superconducting magnet system, light-pipe, condensing-cone optics, and a Ge:Ga photoconductive detector. The photoconductivity spectra were obtained with an audio frequency, capacitive-coupling technique [7].

4. Results and Discussion

4.1. Higher Excited States

Experimental results for the low-lying transitions, 1s-2p($m=\pm1$) (bulk notation), in well-center-doped samples are in good agreement with variational calculations [1] over a wide range of well widths. However, little work has been reported for higher excited states, particularly states that are associated with higher Landau levels. MacDonald and Ritchie [5] have recently considered such states theoretically for strictly two-dimensional systems and have established unambiguous correspondences between the low-field and high-field quantum numbers for this situation. However, these correspondences are in apparent contradiction with well-established correspondences for three-dimensional hydrogenic impurities in a magnetic field.

In the low field limit the appropriate quantum numbers describing the 3D hydrogenic impurity are (n,ℓ,m), where n is the principal quantum number, ℓ is the orbital angular momentum quantum number, and m is the azimuthal quantum number. The wave function of such a state has ($n-\ell-1$) nodal spheres and ($\ell-|m|$) nodal cones along the z-direction. Parity under the transformation $z \to -z$ is even or odd according to whether $(-1)^{\ell-m}$ is +1 or -1, respectively. In lowest order the problem of a hydrogenic impurity strongly confined in the center of a quantum well can be treated as a 2D hydrogenic impurity plus an associated subband structure in the z-direction. The appropriate quantum numbers for this system are ($\tilde{n},m,i$), where i is the subband index, and $\tilde{n}$ is the principle quantum number of the strictly 2D hydrogen atom. The wave functions in this case have ($\tilde{n}-|m|-1$) nodal cylinders and (i-1) nodal planes along the z-direction. Parity along z is even or odd according to whether $(-1)^{i-1}$ is +1 or -1, respectively. Confinement along the z-direction condenses the nodal cones into planes parallel to the x-y plane. Thus the correspondence between the 3D and 2D limiting cases at very low magnetic field is easily established as

$$n-\ell-1 \to \tilde{n}-|m|-1\,,$$
and
$$\ell-|m| \to i-1. \tag{3}$$

In simple physical terms the confinement removes some of the degeneracies of the 3D states by raising states with larger wave function extent in the z-direction to higher energies. Three dimensional states with $|m|=\ell$ have the smallest z-extent among states having the same n and ℓ; thus these states belong to the ground subband in the quasi-2D system. For example, with the rules of Eq.(3) 1s, 2s, 2p(m=±1), 3s, 3p(m=±1) all become associated with the lowest confinement subband, while 2p(m=0), 3p(m=0), 3d(m=±1), 4p(m=0) become associated with the second subband , etc. As magnetic field is increased the detailed evolution of the low field states will depend on the energies of the confinement subbands, the impurity states associated with them, and the interaction (or lack thereof) of these states via an interaction Hamiltonian given by the difference between the 2D and 3D Coulomb terms ($H_1 = [e^2/\epsilon][(1/\rho)-(1/r)]$. States of the same parity and m are coupled by H_1. In the limit of very large confinement, $R^* << \hbar\omega_c << \Delta E_i$, where ΔE_i is the separation between the lowest pair of subbands , unperturbed states do not cross at any reasonable magnetic field, and the correspondence between low magnetic field ($\tilde{n}$,m,i) and high field (N,m,i) states is unambiguous [5]. Here N is the Landau quantum number. Making use of the concept of nodal surface conservation, the correspondence is given by

$$N \rightarrow (\tilde{n}-1) + (m - |m|)/2. \qquad (4)$$

This is identical to the results for the strictly 2D case [5]. It is clear from Eqs. (4) that the (3,1,1) and (4,1,1) quasi-2D low field states become associated with the N = 2, and N = 3 Landau levels, respectively. Also, from Eq.(3) these two quasi-2D states come from the bulk 3p(m=+1) and 4p(m=+1) 3D states, respectively.

Experimental results obtained from photoconductivity spectra on two well-center-doped MQW samples (sheet density per well = 4 x 10^9 cm^{-2} and 5 x 10^9 cm^{-2}, for the 125Å and 150Å well-width samples, respectively) are shown in Fig. 1. The frequency positions of the various lines are shown as a function of magnetic field. The assignment of the lowest two transitions is unambiguous. The transitions are labeled in the bulk notation; in the strong confinement limit these are the (1,0,1) → (2,-1,1) and (2,1,1) transitions, respectively. The solid lines are the results of an approximate expression, transition energy = E[1s-2p(m=-1)] + $k(\hbar\omega_c)$ + Δ_k, in which the 1s-2p(m=-1) transition energy is taken from Ref.6, k is an integer, and Δ_k is a parameter giving the transition energy differences at zero field. It is clear that the agreement is very good.

To identify the higher energy transitions it is necessary to examine the selection rules. For the infrared electric field polarized in the plane of the wells the selection rules are $\Delta m = \pm 1$, $\Delta i = 0$. Since the transition energies are much less than the separation between the lowest two subbands, the above considerations lead to the conclusion that these are the 1s → 3p(m=+1), and the 1s → 4p(m=+1) transitions in the conventional notation ((1,0,1) → (3,1,1) and (4,1,1)). Thus it is clear from the slopes that the 3p(m=+1) state is associated with the N=2 Landau level and the 4p(m=+1) state is associated

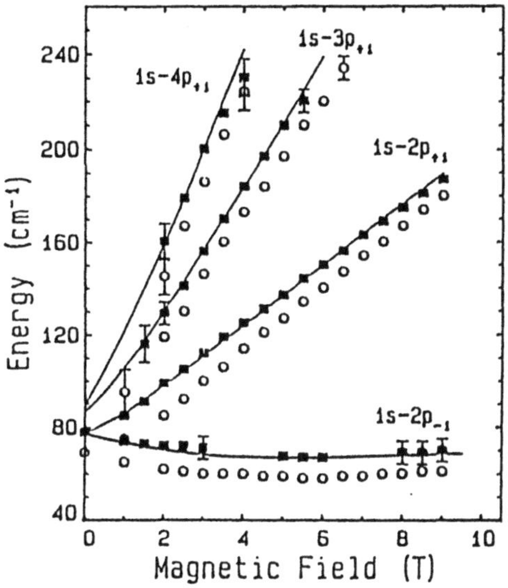

Fig. 1. Field dependence of transition energies for two well-center-doped samples. ■ - 125Å wells; o - 150Å wells.

with the N=3 Landau level, in agreement with the correspondences established for the strictly 2D case [5].

Further detailed studies of the correspondences and the physical reasons underlying the differences between the 3D and quasi-2D results will be presented elsewhere [8].

4.2. Donor Impurities in Coupled Double Quantum Wells

A coupled double quantum well (CDQW)consists of a pair of quantum wells separated by a narrow barrier region. The wavefunction penetration through the narrow barrier leads to a splitting of the degenerate (in the case of isolated wells) ground state into symmetric (lower) and antisymmetric (higher) states. Excitons and electric field effects have been investigated in such structures without intentional doping [9,10], but there has been no published work on impurity effects in CDQWs.

In coupled wells the wave function is more spread out than for isolated wells of the same width. For the case of interest in which the impurities are located in the centers of the wells, the peaks of the symmetric state wave function are displaced toward the narrow barrier region, and away from the impurity ions (Fig. 2). Qualitatively, this leads to a reduction in the binding (and transition) energies compared to the isolated well case, since both the attractive potential energy and the kinetic energy of the electron are reduced. This situation has been investigated experimentally through magneto-transmission studies of the ground state (m=0) to first excited state (m=+1) transitions for several samples with different interwell barrier widths. The samples studied consist of twenty pairs of GaAs wells, each 170Å wide with

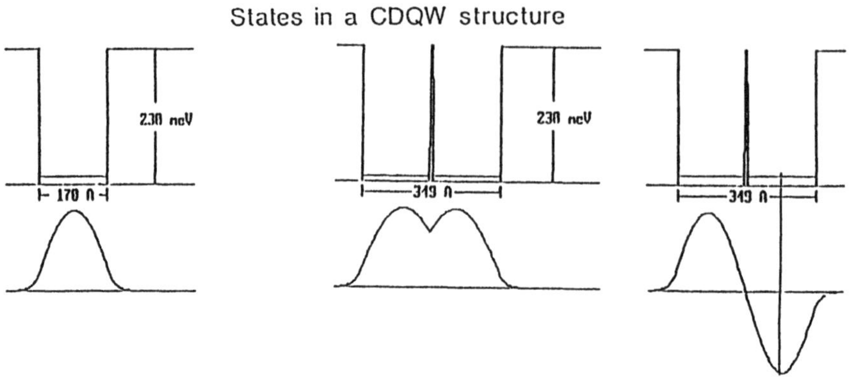

Fig. 2. Energy levels and wave functions for a single QW and CDQW structure.

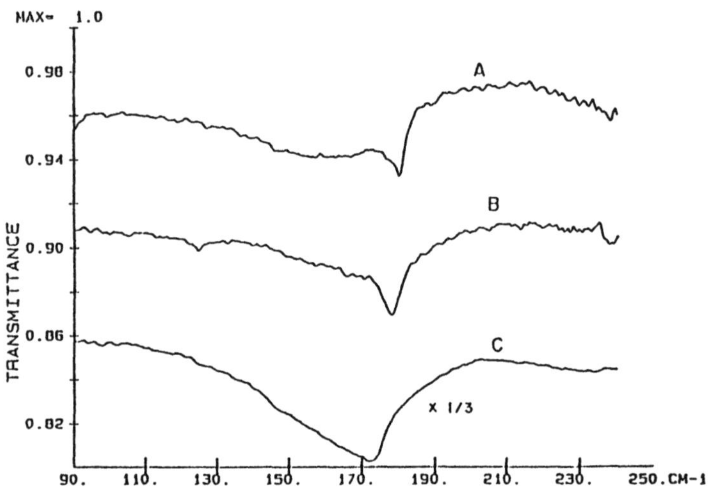

Fig. 3. Magnetotransmission spectra for 3 CDQW samples. The narrow barrier widths are: A - 48Å; B - 18Å; C - 9Å.

adjacent pairs separated by 125Å barriers. The central one third of each well is doped with Si donors at 1×10^{16} cm^{-3} for the two widest barrier samples, and 3×10^{16} cm^{-3} for the narrowest barrier sample. The narrow barrier widths for the three samples are: A - 48Å, B - 18Å, and C - 9Å. Results of magnetotransmission measurements at 9T for the ground state (m=0) to first excited state (m=+1) transition for the three samples are shown in Fig. 3. The traces are the ratios of the transmission spectra at 9T to that at zero field. The coupling effect is manifest as a decrease in the transition energy as the narrow barrier width decreases. The magnetic field dependence of the transmission minima for the three samples is shown in Fig. 4. The 48Å barrier of sample A produces minimal coupling, and the wells behave roughly as isolated wells. Coupling effects become clearly measureable at a barrier width

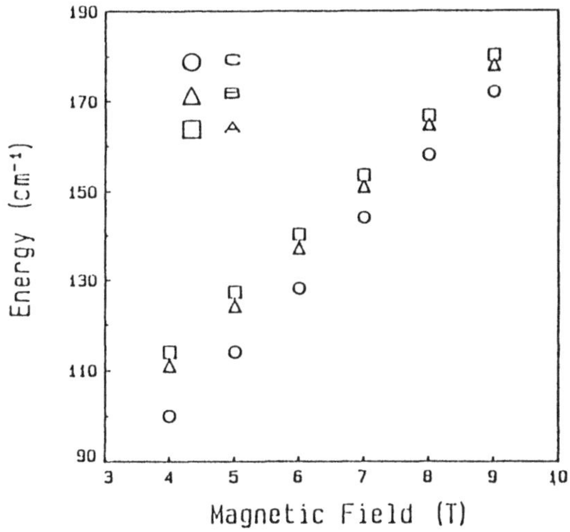

Fig. 4. Frequency vs field of transmission minima for 3 samples of Fig. 3.

of 18Å, for which there is a downward shift of 3 cm^{-1}; and a much larger shift of 10 cm^{-1} is observed for sample C. Very recent variational calculations show good agreement with these results [11].

4.3. Stark Effect of Confined Impurities

The donor impurity transitions can be used as the basis for photoconductive detectors in the far IR. The peak energy in the response curve can be tailored simply by adjusting the well width of the structure. However, it is more desirable to have a method of tuning the peak in the response continuously in situ. The use of an externally applied electric field (Stark tuning) provides such a possibility. In bulk semiconductors such Stark tuning cannot be realized in a practical sense since the donors are field ionized at small values of electric field. In quantum confined structures, however, ionization is prevented by the confining barriers, permitting the possibility of applying large electric fields to neutral donors. The evolution of the impurity ground state in the presence of an electric field in a quantum well has been calculated by Brum et al.[12]; however the effect on the impurity excited states has not been calculated, and no experiments have been reported. The qualitative effects of an applied electric field can be easily seen. In the absence of an electric field donors at the center of the well have binding and transition energies determined by the square quantum well confining potential. The z-component of the electronic wave (envelope) function is peaked at the position of the impurity ion in this case leading to increased binding and transition energies compared with the bulk. In the presence of an electric field the peak of the z-component of the electronic wave function is displaced toward one of the barriers away from the impurity ion; this causes a decrease in the potential energy, and an increase in the kinetic energy, resulting in a markedly decreased binding energy.

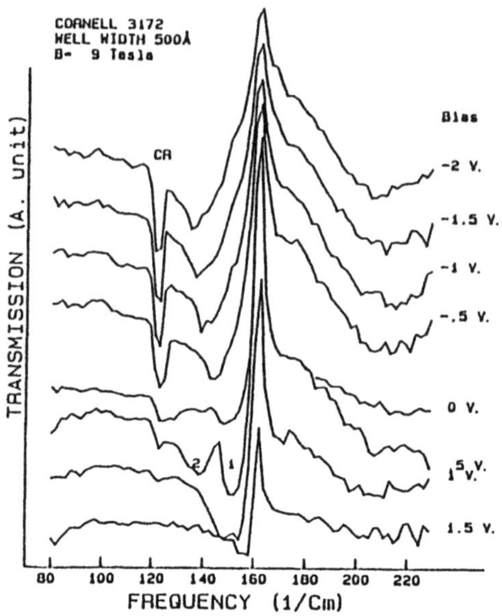

Fig. 5. Magnetotransmission spectra obtained as described in the text.

In the present work a MQW structure with wide GaAs (approximately 500Å) wells as the "i" region of a p-i-n diode has been investigated by far IR magneto-spectroscopy. The structure consists of 20 such wells doped over the central one-third with Si donors at 1 x 10^{16} cm^{-3} separated by 200Å $Al_{0.3}Ga_{0.7}As$ barriers sandwiched between p^+ $Al_{0.3}Ga_{0.7}As$ and n^+ GaAs contact regions. The structure is mesa-etched, and approprite contacts are provided to apply external bias. In order to avoid leakage problems a modulation technique was devised. With this technique a square wave voltage at 1 kHz is applied between the p^+ and n^+ contacts. The upper part of the square wave (p^+ contact) is adjusted to cancel the built-in potential. This creates zero electric field during this part of the cycle. During this half of the cycle an in-situ red LED is pulsed for 170 μsec to ensure neutralization of the donors in the wells. The lower part of the square wave modulation is adjusted to provide a desired (uniform as long as the donors are neutral) electric field across the multiple quantum well structure. The transmitted far IR intensity is detected, amplified, and demodulated by a lock-in amplifier referenced to the modulation voltage. This produces a dc output voltage proportional to the difference between the far IR intensity transmitted through the sample with zero electric field and that transmitted at some desired value of electric field.

Differential transmission data taken at 9T in this manner are shown in Fig. 5. The bias labels on the traces are the voltages of the lower half of the square wave modulation. The sharp positive-going feature at 162 cm^{-1} is the ground state (m=0) to first excited state (m=+1) transition of donors at the centers of the wide quantum wells with zero electric field. As electric field is increased, three negative-going features become apparent. For biases more negative than about 1V a confined free electron cyclotron resonance (CR) is observed at 121 cm^{-1}, whose intensity increases with increasing electric field (more negative bias). The intensity increases because of the reduced binding energy and the concomitant increase in the thermally excited free carrier density. The other two lines are both donor-impurity-related. Line 1 is the ground state (m=0) to first excited state (m=+1) transition of donors near the centers of the wells. At the maximum electric field applied it is shifted down in frequency by 27 cm^{-1} in comparison to the zero electric field case. The observed shift corresponds to a 75% shift in the transition frequency at zero magnetic field. Line 2, which moves very rapidly with increasing electric field and merges into CR at about zero volts bias, is not well-understood at present.

For comparison a variational calculation of the electric field effects on the ground state to first excited state transition energy for a single donor located at the center of a GaAs well with infinite barriers has been carried out [13]. At zero magnetic field the calculated shift in transition frequency is 25 cm^{-1} at the maximum electric field of 2.3 x 10^4 V/cm, in good agreement with the experimental results.

Acknowledgements This work was supported by ONR and ONR/SDIO under the MFEL program. The authors are indebted to W. Schaff, J. Ralston and G. Wicks for the expert MBE growth of the samples.

References

1. See A.A. Reeder, J-M. Mercy and B.D. McCombe, IEEE J. Quant. Elec., QE-24, 1690 (1988) for a review.
2. R.L. Greene and P. Lane, Phys. Rev. B34, 8639 (1986).
3. P. Lane and R.L. Greene, Phys. Rev. B33, 5871 (1986).
4. R. Greene and K. K. Bajaj, Phys. Rev. B31, 913 (1985).
5. A.H. MacDonald and D.S. Ritchie, Phys. Rev. B33, 8336 (1985).
6. C.J. Armistead, R.A. Stradling and Z. Wasclewski, Semicond. Sci. Technol. 4, 557 (1989).
7. J-M. Mercy, N.C. Jarosik, B.D. McCombe, J. Ralston and G. Wicks, J. Vac. Sci. Technol. B4, 1011 (1986).
8. J.-P. Cheng and B.D. McCombe, Phys. Rev. B, Rapid Commun., submitted.
9. Y.J. Chen, E.S. Koteles, B.S. Elman and Armiento, Phys. Rev. B36, 4562 (1987).
10. S. Charbonneau, M.L.W. Thewalt, E.S. Koteles and B.S. Elman, Phys. Rev. B38, 6287 (1988).
11. R. Ranganathan, B.D. McCombe, Y. Zhang, N. Nguyen, M. Rustgi and W. Schaff, to be published.
12. J.A. Brum, C. Priester and G. Allan, Phys. Rev. B32, 2378 (1985).
13. B.S. Yoo, B.D. McCombe, L. He and W. Schaff, to be published.

Magneto-Luminescence and Magneto-Luminescence Excitation Spectroscopy in Strained Layer Heterostructures

R. Küchler[1], *P. Hiergeist*[1], *G. Abstreiter*[1], *J.-P. Reithmaier*[2,a], *H. Riechert*[2], *and R. Lösch*[3]

[1]Walter Schottky Institut, TU München, Am Coulombwall, W-8046 Garching, Fed. Rep. of Germany
[2]Siemens AG, W-8000 München, ([a]temporary address)
[3]Forschingsinstitut der Deutschen Bundespost Telekom, W-6100 Darmstadt, Fed. Rep. of Germany

1. Introduction

Magneto-luminescence (PL) and magneto-luminescence excitation (PLE) spectroscopy are powerful tools to examine optical properties of semiconductor quantum wells [1,2]. The main effect of a magnetic field applied perpendicular to a quantum well is the splitting of the subbands into Landau levels. In interband magneto optical studies the reduced mass of electron and hole determines the observed splittings. However, exciton effects have to be taken into account. [3-5].

Two different types of samples are studied. In the first part we concentrate on undoped strained layer $In_yGa_{1-y}As$ multi-quantum-wells with various In concentrations and well widths. The barriers are wide enough to ensure that the wells are isolated from each other. The single sided remote doped quantum-well which is discussed in the last part of this paper consists of an 14 nm wide $In_yGa_{1-y}As$ layer with a In content of 20%. All samples were grown by molecular beam epitaxy and characterized by photoluminescence, electrical transport measurements and transmission electron microscopy.

The samples are placed in the center of a superconducting magnet ($B < 15.5$ T). The sample temperature can be varied from 0.3 K to 300 K. The magnetic field is orientated normal to the epitaxial layers. An Ar-ion laser or a dye- or Ti-sapphire laser beam is directed to the sample surface via glass fiber technique. A triple grating spectrometer with multichannel detection is used to record the luminescence spectra. The luminescence excitation spectra are measured with a 1m double grating spectrometer and a cooled InGaAs photomultiplier.

2. Experimental results and discussion

Fig.1 shows typical luminescence excitation spectra of a $In_yGa_{1-y}As$/GaAs MQW with 12 nm wide wells and In content of

Springer Series in Solid-State Sciences, Vol. 101
High Magnetic Fields in Semiconductor Physics III Editor: G. Landwehr

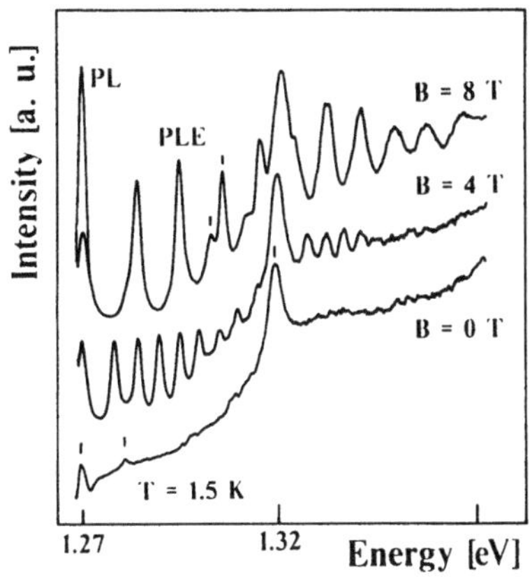

Fig. 1 Luminescence excitation spectra for different magnetic fields. At B = 0 T the marked transitions correspond to e1-hh1, e2-hh1 and e1-lh1. At B = 8 T the satellite transitions are marked. Also shown is the luminescence signal.

24% for different magnetic fields. Without magnetic field discrete transitions can be observed, which are attributed to the interband transitions e1-hh1, e1-hh2 and e2-hh1. All transitions are in excellent agreement with theoretical calculations using the sample parameters and a conduction band offset of 0.6 [6].

With increasing magnetic field the spectrum splits into discrete Landau levels (Fig. 1). Two sets of Landau levels are observed with linear dependence in magnetic field. In excitation spectroscopy we expect the allowed transitions (diagonal transitions, DT, $\delta n = 0$) to dominate the spectrum [7]. From the splitting between the 0-0 transition and the intersection of the higher transitions (n-n) extrapolated to B=0 T, we deduce the exciton binding energy [8,9]. The results agree well with the values extracted from the separation of the exciton line and the continuum at B=0 T. From the slope of the DT we find the reduced mass of the Landau transitions.

At high magnetic fields satellite transitions can be observed. We attribute these to off-diagonal transitions (ODT) with $\delta n=1$. This additional information allows to deduce electron and hole mass independently. Using this electron mass we can also derive the hole mass of the lh-transition.

In table 1 (#1-5) we summarize the measured values for different samples. Most obvious feature is the dependence of the exciton binding energy on well width.

In contrast to luminescence of undoped quantum wells, doped quantum wells give broad lines with a high energy cut-off, determined by the Fermi energy as shown in Fig. 2a. This lineshape is believed to be a consequence of indirect transitions between electrons with wave vectors from 0 up to k_F and holes close to k=0 [10], which is confirmed by PLE experiments [11]. In a magnetic field this broad line splits into Landau levels (Fig. 2b,c). At low temperatures and low excitation intensities electrons in Landau levels in the energy range up to E_F will mainly recombine with holes in the lowest

Table 1: Exciton binding energy and reduced mass for various samples

#	L/nm	In/%	E_{exc}/meV (B=0 T)	E_{exc}/meV (B>0 T)		m^*/m_0	m_c/m_0	m_h/m_0
1	5	24	10.6	8.8		0.046	---	---
2	8	24	9.8	9.2		0.046	0.064	0.16
3	10	12	7.3	7.2		0.045	0.062	0.17
4	12	24	7.0	7.8	hh	0.045	0.062	0.16
			---	6.3	lh	0.055	0.062	0.46
5	15	12	6.4	6.4		0.044	---	---
6	14	20	---	---		0.046	0,062	0,18

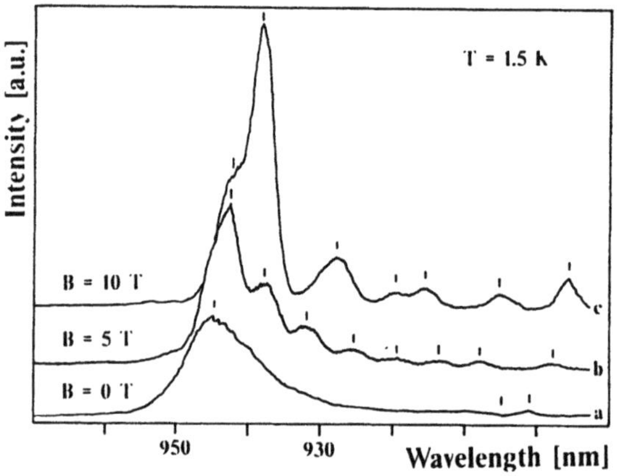

Fig. 2 Luminescence spectra for different magnetic fields. At B = 0 T the cut-off due to the Fermi edge and the transition from the second electron subband are marked.

hole Landau level. By increasing the sample temperature and thereby thermally populating higher hole Landau levels DT become more and more intense and dominate the spectrum at temperatures above 40 K, as shown in Fig. 3. Comparing DT and ODT we can derive the effective mass of electrons and holes, listed in table 1 (#6).

As for the undoped samples at low magnetic fields the higher Landau levels shift almost linearly with magnetic field, but in the high-field region nonparabolicity effects are much more pronounced. In contrast, the 0-0 transition does not follow a simple linear or quadratic dependence on magnetic field. At several magnetic fields an additional line rises at the low energy side of the 0-0 transition. This line gains in intensity and a clear anticrossing behaviour can be recorded. The crossing points scale with B^{-1}. At these magnetic fields the higher transitions deviate from linear behaviour.

As the Fermi energy is only a few meV below the second electron subband, we detect also luminescence from transitions between thermally activated electrons in this band with holes in the lowest subband. These transitions are very efficient due to increased overlap of the wavefunctions of electrons and holes. The intensity of this

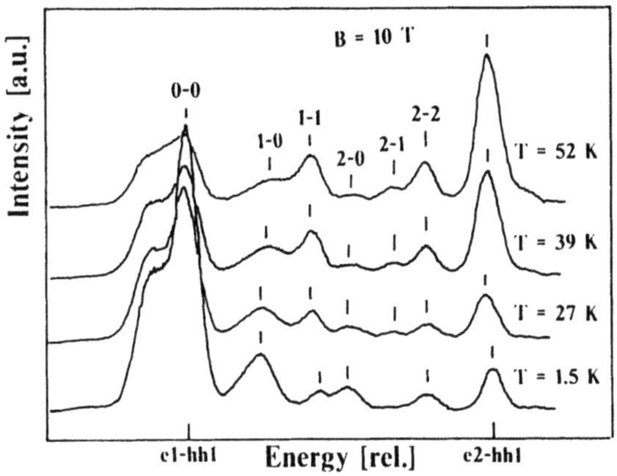

Fig. 3 Luminescence spectra for different temperatures at B = 10 T. The individual spectra are shifted to match the 0-0 transition.

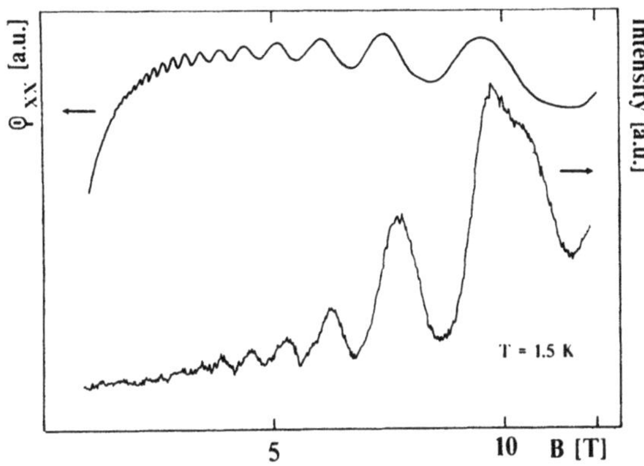

Fig. 4 Comparison between the intensity oscillation of the transition involving the second subband and simultaneously performed Shubnikov-de Haas measurement.

transition undergoes periodic oscillations with increasing magnetic field. These oscillations compare very well with simultaniously performed Shubnikov-de Haas measurements as shown in Fig. 4. ("optical SdH" [12]). Similar oscillations can be derived from the 0-0 transition, which also exists in samples without second populated subband.

References

1. J.M. Worlock, A.C. Maciel, A. Petrou, C.H. Perry, R.L. Aggarwal, M. Smith, A.C. Gossard and W. Wiegmann, Surf. Sci. 142, 486 (1984)
2. D.C. Rogers, J. Singleton, R. J. Nicholas, C.T. Foxon and K. Woodbridge, Phys. Rev. B34, 4002 (1986)
3. O. Akimoto and H. Hasegawa, J. Phys. Soc. of J., 22, 181 (1967)
4. P.C. Makado and N.C. McGill, J. Phys. C19, 873 (1986)
5. G. Duggan, Phys. Rev. B37, 2759 (1988)

6. J.-P. Reithmaier, R. Höger, H. Riechert, A. Heberle, G. Abstreiter and G. Weimann, Appl. Phys. Lett, 56, 536 (1990)
7. S.K. Lyo, E.D. Jones and J.F. Klem, Phys. Rev. Lett. 61, 2265 (1988)
8. K. J. Moore, G. Duggan, K. Woodbridge, C. Roberts, Phys. Rev. B41, 1090 (1990)
9. D.D. Smith, M. Dutta, X.C. Liu, A.F. Terzis, A. Petrou, M.W. Cole and P.G. Newman, Phys. Rev. B40, 1470 (1989)
10. S.K. Lyo and E.D. Jones, Phys. Rev. B38, 4113 (1988)
11. O. Heinrich, to be published
12. W. Chen, M. Fritze, A.V. Nurmiko, D. Ackley, C. Colvard and H. Lee, Phys. Rev. Lett. 64, 2434 (1990)

Magneto-Excitons in CdTe/(CdMn)Te Quantum Wells

W. Ossau

Physikalisches Institut der Universität Würzburg, Am Hubland, W-8700 Würzburg, Fed. Rep. of Germany

We have performed low temperature photoluminescence and photoluminescence excitation measurements on CdTe/(CdMn)Te single quantum wells in magnetic fields up to 9.5 T. The analysis of the temperature and magnetic field dependence of the splitting of the heavy-hole exciton in Faraday configuration shows that the splitting of the exciton in the nonmagnetic layer is caused by exchange interaction with the Mn^{++} ions in the barrier. From the asymmetric behaviour of the splitting of the heavy-hole excitonic recombination we identify a Type-I to Type-II transition of heterostructures with small manganese concentration in the barrier. The asymmetric splitting furthermore allows to determine the valence band offset to 20 ± 5 % of the total bandgap discontinuity. In addition we deduce from the finestructure of the spectra the binding energy of the heavy-hole exciton as a function of the well width.

1. INTRODUCTION

During the past decade the study of two dimensional semiconductor systems, like quantum wells and superlattices, has attracted great attention [1]. Heterostructures have been investigated mainly with III-V semiconductors, such as GaAs and (AlGa)As, where the lattice constants are nearly the same. In recent years, quantum well structures and superlattices composed of II-VI compounds such as CdTe and $Cd_{1-x}Mn_xTe$ have been grown by MBE [2]. In CdTe-(CdMn)Te heterostructures, there are besides the bandgap two important features that depend on the Mn content - the lattice constant and the magnetic properties of the alloy.

$Cd_{1-x}Mn_xTe$ belongs to a group of materials referred to as semimagnetic or dilute magnetic semiconductors (DMSs). The interest in DMS materials stems from the unique magnetic and magnetooptic properties they exhibit due to the presence of Mn. Manganese ions (Mn^{++}) possess a total spin of S=5/2 and interact among themselves via an antiferromagnetic exchange interaction [3]. Furthermore, the Mn^{++}-ions interact via spin exchange with conduction band electrons and valence band holes, resulting in large g-factors, strongly circularly polarized band edge luminescence, giant Faraday rotation and other interesting phenomena that are new to semiconductor physics.

The second important parameter varying with x is the difference in the lattice constants of the two types of layers $\Delta a(x) = 0.149 \cdot x$ Å [4], resulting in strained layer heterostructures. This lattice mismatch is accommodated by elastic strain rather than by the generation of misfit dislocations, provided that the layer thicknesses are less than some critical value (300 nm, [5]). The biaxial stress applied on the CdTe layer can be decomposed in a hydrostatic part, which increases the bandgap and an uniaxial strain which splits the valence band in separated light-and heavy-hole states [5]. In our particular case the expected hydrostatic shift is close to the value of the uniaxial splitting.

Springer Series in Solid-State Sciences, Vol. 101
High Magnetic Fields in Semiconductor Physics III Editor: G. Landwehr

The distribution of the total bandgap discontinuity at the interface between the conduction and valence band is not known at present. The published studies without exception are performed on multiple quantum wells or superlattices. In some papers an agreement with the common anion rule, i.e. a valence band offset that is zero or small (6 to 9 %) can be found [6,7,8]. On the other hand some publications claim a valence band offset that is about 40 % of the total discontinuity [9,10]. Latter values are in agreement with recent theoretical calculations that yield a significant valence band offset of 33 % [11] or even 70 % [12]. The great deviation of the reported values may be caused by interface roughnesses, manganese contamination of the well or the unknown strain conditions of the heterostructures. All cited publications have required subband calculations for the determination of the band edge discontinuities. This method is rather indirect and depends strongly on the used parameters for the well width, the manganese concentration, the mass values for electrons and holes and their degree of nonparabolicity. The method we used in this publication is more direct and doesn't depend on the knowledge of the mentioned parameters except the manganese concentration of the barriers.

2. EXPERIMENTAL RESULTS AND DISCUSSION

The single quantum well structures investigated are grown on (100)-oriented (CdZn)Te substrates by conventional molecular beam epitaxy. The Mn concentrations of the samples discussed vary between $x \approx 0.07$ and $x \approx 0.26$. Our study is concentrated on single quantum well structures that are grown between thick (CdMn)Te barriers. Therefore we deal with a defined strain situation. All the quantum wells studied are completely strained, whereas the barriers are unstrained. This brings up a great advantage for the determination of the valence band offset with these single quantum well structures compared to multiple quantum wells or superlattices where the stress is divided up somehow between well and barrier. The optical properties of the quantum wells were investigated using low temperature photoluminescence **PL** and photoluminescence excitation spectroscopy **PLE** by standard techniques.

In a magnetic field all ground-state excitonic lines split into two components with circular polarization $\sigma+$ and $\sigma-$. In Faraday configuration the luminescence is emitted by dipole allowed $\Delta m = \pm 1$ recombinations of electrons ($J = 1/2$) with light-holes ($J_Z = \pm 1/2$) or heavy-holes ($J_Z = \pm 3/2$) respectively [2].

Fig. 1 shows spectra of four quantum wells with different well widths. The luminescence correlated with the well is indicated by **QW**. Due to the quantum-size effect we observe a blue-shift of the peak position relative to that of CdTe with decreasing well width. The linewidth is less than 10 meV even for the sample with the thinnest well ($L_Z = 4$ nm), indicating very abrupt heterointerfaces with a variation of ΔL_Z less than 0.2 nm. The luminescence near 790 nm is emitted by the subtrate and not troublesome for the analysis of the data.

To elucidate further the origin of the observed photoluminescence, we have studied spectra as a function of sample temperature and magnetic field strength. All of the obtained spectra show a fine structure that can be seen in figure 2. For zero magnetic fields the main luminescence splits into two components, a high and a low energy peak. The relative intensity of these peaks is a function of well width and sample temperature. The lower energy line that is more pronounced in thin quantum wells disappears by raising the sample temperature from 1.8 to 20 K, whereas the intensity of the higher energy peak increases. In narrow quantum wells the high energy component exceeds the lower energy peak in intensity not until the sample temperature is raised to 10 K.

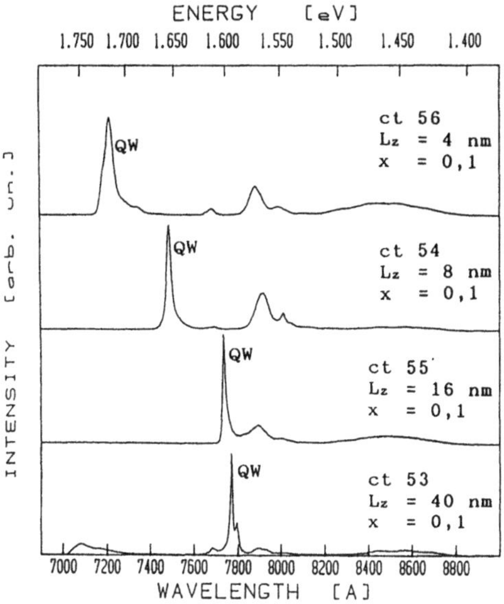

Fig. 1 Photoluminescence spectra of four samples. The lines correlated with the quantum wells are marked with QW.

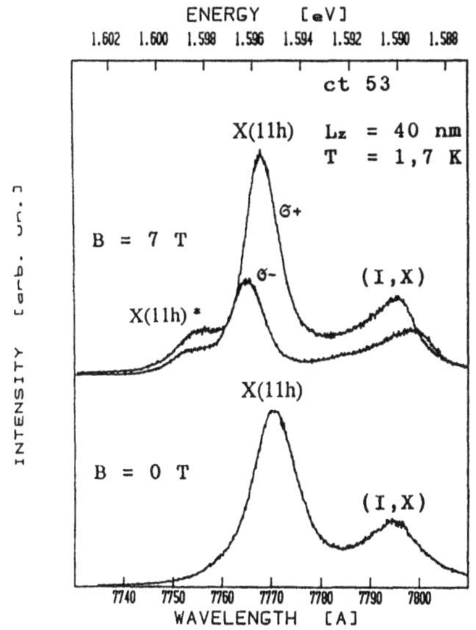

Fig. 2 Influence of the magnetic field on the photoluminescence spectra.

A similar temperature dependence was obtained by Zhang et al. [13]. They interpreted the high energy component as the free exciton recombination and the low energy component as the recombination of excitons bound to potential wells caused by random strain and compositional fluctuations at the heterointerfaces. The energy distance of the lines observed by Zhang and coworkers was approximately 25 meV. In our case the emission lines are separated by about 7 meV only for all samples we have studied. This energy is similar to the localization energy of excitons bound to acceptors in CdTe [14]. Therefore the lower energy line may be interpreted as an exciton bound to an impurity. The higher energy line stems from the recombination of an electron with a heavy-hole in the first subband respectively [X(11h)].

Figure 2 also shows two luminescence spectra obtained at a magnetic field of 7 T for σ^+ and σ^- polarization. Both lines X(11h) and (I,X) split in two subcomponents. In general our results obtained for the splitting of the excitonic recombination in magnetic fields can be summarized in two main features. For quantum well widths larger than the exciton diameter (approximately 12 nm) we observe a small splitting of the subcomponents. If the quantum well width becomes smaller than the exciton diameter the Zeeman-splitting is drastically enhanced. For all well widths the splitting is a nonlinear function of the magnetic field strength. In addition on the high energy side of the free exciton luminescence new lines can be observed with increasing field strength. These lines are attributed to the excited states of the heavy-hole exciton (2S and 3S). From the analysis of these lines we determined the binding energy of the heavy-hole exciton. The binding energy increases with decreasing well width. From our data we obtain 16.3 meV for a quantum well with L_Z = 18 nm and 23.7 meV for a wellwidth of 6.5 nm. Details of these studies will be published elsewhere [15].

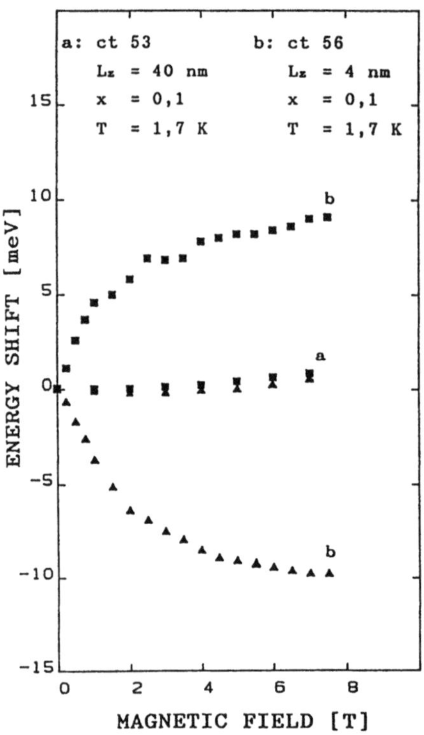

Fig. 3 Magnetic field dependence of the PLE peak energies for two samples. A) Sample with wide quantum well L_Z = 40 nm. B) Sample with a narrow quantum well L_Z = 4 nm. The manganese concentration is 10 % for both structures.

In Fig. 3 we have plotted the peak positions of the luminescence lines of the ground state of the heavy-hole exciton for two different quantum wells. One with L_Z = 4 nm and one with a 40 nm well width. The manganese concentration for both quantum wells is about 10 % . For quantum well widths larger than the exciton diameter (about 12 nm) we observe a small Zeeman-splitting and a small diamagnetic shift of the centre of gravity. If the well width becomes smaller than the exciton diameter the Zeeman splitting is drastically enhanced. The observed spin-splittings obtained at low temperatures show a strong increase with field strength up to 4 T and then saturate. This splitting behaviour is a result of the sp-d exchange interaction of the spin moments of the exciton (s-d interaction for electrons and p-d interaction for holes) with those of the Mn^{++} ions within its orbit and similar to that of free excitons in bulk (CdMn)Te.

We neglect the intrinsic g-factors ($g^* \approx 1$) and apply the empirical expression for the magnetization of bulk (CdMn)Te. The observed photon energies for transitions $|3/2, \pm 3/2\rangle \rightarrow |1/2, \pm 1/2\rangle$, this means recombinations of heavy-holes with electrons, is according to Gaj et al. [16]:

$$E_{\sigma\pm} = E(B=0) \pm \frac{1}{2} (\beta-\alpha)\ N_0\ x\ \langle S_z \rangle \quad , \qquad (1)$$

where α and β are exchange constants for conduction band and valence band respectively, x is the manganese concentration, N_0 is the number of cations per unit volume and $\langle S_Z \rangle$ is the averaged spin per Mn site:

$$\langle S_Z \rangle = S_0\, B_{5/2}\, [5\, \mu_B\, B \,/\, k_B\, (T + T_{AF})] \ ,$$

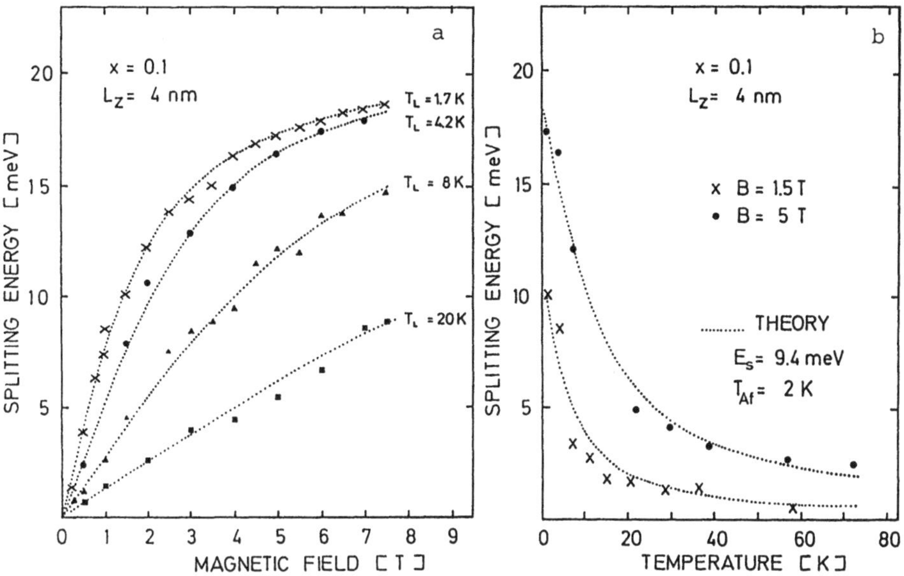

Fig. 4 Energy difference of the σ^+ and σ^- subcomponents as a function of magnetic field and temperature. Fig. 4a shows the splitting for different temperatures as a function of magnetic field. Fig. 4b shows the splitting for two magnetic fields as a function of the sample temperature. The dotted lines are calculated with equation (1) using E_s = 9.4 meV and T_{AF} = 2.0 K.

where $B_{5/2}$ is a standard Brillouin function for S=5/2, μ_B is the Bohr magneton, k_B is the Boltzmann constant and T the temperature. S_0 and T_{AF} are fitting parameters which vary strongly with x and more weakly with the temperature. For bulk (CdMn)Te the obtained saturation values $S_0(x)$ are less than S=5/2 and suggest that a fraction of manganese ions forms pairs and/or clusters with strong antiferromagnetic ordering [16]. The remaining ions align in the magnetic field according to a Brillouin function with T_{AF} greater than zero, reflecting an antiferromagnetic type of interaction among those ions.

In Fig. 4a we have plotted the energy difference of the peak positions of the PLE maxima of the heavy-hole exciton in a thin quantum well as a function of magnetic field and sample temperature. The observed spin-splitting obtained at low temperatures shows a strong increase with field strength up to 4 T and then saturates. By increasing the sample temperature the splitting decreases. For temperatures greater than 5 K no saturation is observed at 8 T. This splitting behaviour is a result of the sp-d exchange interaction of the spin moments of the exciton (s-d interaction for electrons and p-d interaction for holes) with those of the Mn^{++} ions within its orbit. Figure 4b shows the energy difference of the peak positions as a function of temperature for two magnetic field strengths. The splitting energy decreases with increasing temperature as is expected from equation (1). Even for temperatures higher than 60 K the splitting caused by the exchange interaction is larger than the splitting taking into account the intrinsic g-factors of bulk CdTe only [14].

In Fig. 4 the dotted lines are calculated with (1), where we have assumed for $E_s = S_0(\beta-\alpha)N_0x$ = 9.4 meV and for T_{AF} = 2.0 K. Within the experimental error we get excellent agreement between the experimental data and the calculated

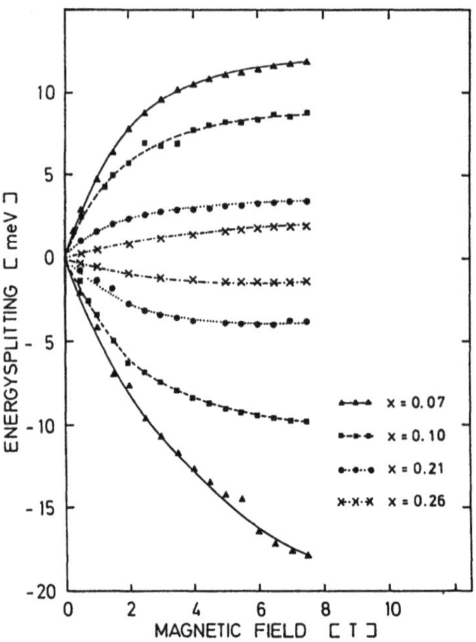

Fig. 5 Splitting of the PLE maxima of the heavy hole exciton for samples with different manganese concentration of the barrier as a function of the magnetic field strength.

values. This result clearly demonstrates that the spin-splitting of the heavy-hole exciton of the non-magnetic CdTe layer is strongly influenced by the paramagnetic behaviour of the DMS barrier material.

To demonstrate this influence further, we have studied the splitting of the heavy-hole exciton as a function of magnetic field strength for quantum wells with well widths of 8 $\pm$ 0.8 nm and manganese concentrations varying between 0.07 and 0.26. In Fig. 5 it is to be seen that the splitting of the subcomponents of the heavy-hole exciton increases if the manganese concentration decreases. If the energy splitting is caused by the interaction of the exciton with the semimagnetic layer via the parts of the wavefunction penetrating into the barrier this result unambigously shows that there has to be a significant natural valence band offset in CdTe/(CdMn)Te-quntum wells. Applying the parameters S_0 and T_{AF} we have obtained by PLE measurements [17] for the barrier material, the magnetization of a (CdMn)Te layer with $x = 0.26$ is twice that of a layer with $x = 0.07$ at sample temperatures $T = 1.7$ K and field strength $B = 7$ T. The splitting of the heavy-hole exciton for a quantum well with the small x value for the barrier, however, is more than eight times stronger than that of the exciton with the greater magnetization of the barrier material. This indicates that the QW-exciton in the sample with $x = 0.26$ has a strong confinement. If the potential well for the heavy-hole would be caused by the strain alone the change in the barrier concentration from $x = 0.07$ to $x = 0.26$ should rise the potential height approximately by 3.3 meV at most. Subband calculations prove that this small change in barrier height is not able to produce the drastic change in the confinement observed even for small barrier heights.

In Fig. 5 it is also to be seen that the splitting of excitons confined within low barriers is asymmetric. This behaviour is not observed for the (CdMn)Te barrier layer where the observed Zeeman-splitting remains symmetric. If the splitting of the QW-excitons is proportional to the part of the wavefunction interacting with the barrier, an asymmetric splitting can only be caused by a different penetration

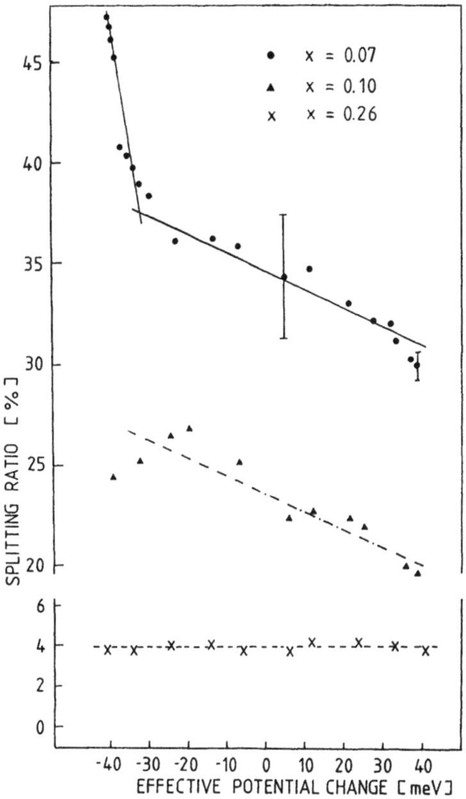

Fig. 6 Ratio between the splitting of the quantum well and barrier exciton as a function of the effective potential height.

for spin-up and spin-down states. In an external magnetic field the band edges of the (CdMn)Te barriers are split. On account of the large exchange constant for the valence band this splitting is more pronounced for hole-states. The potential height of the m_j = -3/2 spin-state increases with increasing field strength, whereas the potential well of the m_j = +3/2 spin-state becomes smaller. This obviously results in a stronger confinement of the -3/2-heavy-hole. The +3/2-heavy hole, however, caused by the decreasing potential height, is less confined. In the framework of penetrating wavefunctions this means an enlarged interaction of the +3/2-spin-state with the magnetic moments of the Mn^{++}-ions, whereas the interaction of the -3/2-spin-state is reduced. Experimentally this is confirmed by an enhanced red-shift of the σ^+- and a reduced blue-shift of the σ^--subcomponent of the heavy-hole PLE-maxima.

Similar asymmetric splittings of QW-exciton states are observed in the ZnSe/(ZnFe)Se-system and explained by a spin-dependent type-I to type-II transition of the heterostructure [18]. A type-I to type-II transition in CdTe/(CdMn)Te-quantum wells may be observed for small manganese concentrations, where the natural valence band offset is comparable with the change of potential well by the applied magnetic field.

To demonstrate this, we have plotted in Fig. 6 the ratio of the Zeeman-splitting between the heavy-hole QW-exciton and the barrier heavy-hole exciton as a function of the effective quantum well potential for quantum wells with $L_z = 8 \pm 0.8$ nm and different manganese concentrations of the barriers. The effective potentials are deduced from PLE-measurements on the barrier material.

The splitting ratios of the σ^--components belong to positive, that of the σ^+-components to negative effective potentials. Figure 6 shows two important facts. Firstly, if the barrier manganese concentration is high enough a change of the potential height of the quantum wells with magnetic field does not influence the subband state of the heavy-hole. This means that a potential change of $\pm$ 40 meV is small compared with the natural valence band offset. If the valence band offset is caused by strain only, the height of the potential well for the heavy-hole in a QW with 26% manganese content of the barriers should be 4.4 meV at most.

The second important fact that can be deduced from Figure 6 is the clear indication for a type-I to type-II transition for the QW with only 7 % manganese concentration of the barriers. The splitting ratio shows a clear dependence on the effective potential. For increasing potential height the splitting ratio (amount of penetrating wavefunction) becomes smaller. For negative effective potentials the spin-state is less confined as can be seen from the increasing splitting ratio. Furthermore we observe for an effective potential of -30 meV a drastic change of the splitting ratio. This increase of the splitting ratio is interpreted with the type-I to type-II transition. If we identify the effective potential where we observe the kink in the curve with the potential necessary to compensate the natural band offset, we deduce for the system CdTe/(CdMn)Te a valence band offset of 20 $\pm$ 5 % of the total band-gap discontinuity. The uncertainty mainly comes from the fact that it is not known how the hydrostatic part of the biaxial stress is divided up between conduction and valence bands. The valence band offset determined here is in excellent agreement with that we have deduced from the investigation of the anisotropy of the exchange interaction in CdTe/(CdMn)Te quantum wells [19].

Acknowledgements

The author would like to thank Prof. G. Landwehr for helpful discussions, the stimulating interest in the progress of this work and the critical reading of this manuscript. Further I would like to thank R.N. Bicknell-Tassius, A. Waag and S. Schmeußer for providing the high quality samples without which this work would not have been possible. The author is obliged to the following collaborators for the contributions discussed in this paper: B. Kuhn, A. Krämer, S. Fischer and E. Vornberger. Finally I would like to thank the Deutsche Forschungsgemeinschaft for the financial support of this work.

References:

[1] L. Esaki, IEEE Trans. Quantum Electron. QE-22 (1985) 1611.
[2] W. Ossau, S. Fischer and R.N. Bicknell-Tassius, J. Crystal Growth 101 (1990) 905 and references therein
[3] R.R. Galazka, S. Nagata and P.H. Keesom, Phys. Rev. B22 (1980) 3344
[4] N. Bottka, J. Stankiewicz, W. Giriat, J. Appl. Phys. 52 (1981) 4189
[5] N. Magnea ,F. Dal'bo, C. Fontaine, A. Million, J.P. Gaillard, L.S. Dang, Y. Merle d'Aubigné and S. Tatarenko, J. Crystal Growth 81 (1987) 501
[6] S.K. Chang, A.V. Nurmikko, J.W. Wu, L.A. Kolodziejski and R.L. Gunshor, Phys. Rev. B 37 (1988) 1191
[7] T.J. Gregory, C.R. Hilton, J.E. Nicholls, W.E. Hagston, J.J.Davies, B. Lunn and D.E. Ashenford, Journal of Crystal Growth 101 (1990) 584
[8] R.L. Harper, R.N. Bicknell, D.K. Blanks, N.C. Giles, J.F. Schetzina, Y.R. Lee and A.K. Ramdas, J. appl. Phys. 65 (1989) 624

[9] A. Wasiela, Y. Merle d'Aubigné, J.E. Nicholls, D.E. Ashenford and B. Lunn, Solid State Communications 76 (1990) 263
[10] W. Heimbrodt, O. Goede, H.E. Gumlich, H. Hoffmann, U. Stutenbäumer, B. Lunn and D.E. Ashenford, Journal of Luminescence, in print
[11] J. Tersoff, Phys. Rev. Lett. 56 (1986) 2755
[12] N.E. Christensen, I. Gorczyca, O.B. Christensen, U. Schmid and M. Cardona, Journal of Crystal Growth 101 (1990) 318
[13] X.C. Zhang, S.K. Chang, A.V. Nurmikko, L.A. Kolodziejski, R.L. Gunshor and . Datta, Phys. Rev. B31 (1985) 4056
[14] W. Ossau, T.A. Kuhn and R.N. Bicknell-Tassius Journal of Crystal Growth 101 (1990) 135
[15] B. Kuhn, W. Ossau, to be published
[16] J.A. Gaj, R. Planel and F. Fishman, Solid State Commun. 29 (1979) 435
[17] A. Krämer, Diplomathesis, University Würzburg, unpublished
[18] X. Liu, A. Petrou, J. Warnock, B.T. Jonker, G.A. Prinz and J.J.Krebs, Phys. Rev. Letters 63 (1989) 2280
[19] E. Vornberger, W. Ossau, A. Waag, R.N. Bicknell-Tassius and G. Landwehr, Proceedings of the 20th Int. Conf. on the Physics of Semiconductors 2 (Aug. 1990) 1569, Thessaloniki, Greece

Free and Bound Magnetic Polarons in CdTe/(Cd,Mn)Te Quantum Wells

D.R. Yakovlev[1;*], *W. Ossau*[1], *G. Landwehr*[1], *R.N. Bicknell-Tassius*[1], *A. Waag*[1], *and I.N. Uraltsev*[2]

[1]Physikalisches Institut der Universität Würzburg, Am Hubland, W-8700 Würzburg, Fed. Rep. of Germany

[2]A.F. Ioffe Physico-Technical Institute, Academy of Sciences of the USSR, 194021 St. Petersburg, USSR

*Permanent address: A.F. Ioffe Physico-Technical Institute Academy of Sciences of the USSR, 194021 St. Petersburg, USSR

We report the first experimental observation of free magnetic polarons in diluted magnetic semiconductor quantum wells. $CdTe/Cd_{0.78}Mn_{0.22}Te$ multiple quantum well structures with quantum well widths up to 100 Å were investigated by reflection, photoluminescence and photoluminescence excitation spectroscopy at low temperatures and in magnetic fields. Free and bound magnetic polaron energies were determined by the suppression of polaron formation with increasing temperature and in external magnetic fields.

1. INTRODUCTION

The application of molecular-beam epitaxy (MBE) has recently led to the successful growth of high quality CdTe/(Cd,Mn)Te multiple quantum wells (MQW) /1/. In semimagnetic semiconductors the exchange interaction between the spins of electrons and holes and the localized magnetic ions gives rise to large Zeeman splittings and to magnetic polaron effects. Two types of magnetic polarons have been identified in semimagnetic semiconductors : bound magnetic polarons (BMP), where the carriers are localized at impurities (3,4) and free magnetic polarons (FMP) consisting of carriers trapped by their own, self-consistently generated exchange potential /2/. To our knowledge the free magnetic polaron has not been observed in QWs and there is only limited evidence for its existence in bulk materials /3,5/. Theoretical model calculations of FMP in quantum wells have been performed /6,7/. The specific feature of this polaron is that an exciton localized in the nonmagnetic CdTe QW forms a polaron due to the interaction with magnetic ions in the (Cd,Mn)Te barrier layers.

2. EXPERIMENTAL DETAILS

$CdTe/Cd_{0.78}Mn_{0.22}Te$ MQW structures were grown at the University of Würzburg by MBE technique on (100)-oriented CdTe substrates at T=230°C. The structure studied in detail consists of 20 alternating layers of CdTe (L_z = 30Å) and $Cd_{0.78}Mn_{0.22}Te$ (L_b = 120Å). A dye laser tunable over the spectral range between 1.73 ÷ 1.87 eV was used for photoluminescence (PL) and excitation spectroscopy. The magnetic fields up to 6 T were applied perpendicular to the QW layers and luminescence was analysed in the Faraday geometry by a 1 m Jobin-Yvon spectrometer with a cooled photomultiplier. The reflection spectra were measured in p-polarization at an incidence angle of 45°. This method has been described in detail in a previous publication /8/.

3. RESULTS AND DISCUSSION

The reflection spectrum at T = 1.6 K is represented in Fig.1 by the curve drawn with a dash-dotted line. The resonance frequency $\hbar\Omega_0$ of the free n=1

Springer Series in Solid-State Sciences, Vol. 101
High Magnetic Fields in Semiconductor Physics III Editor: G. Landwehr

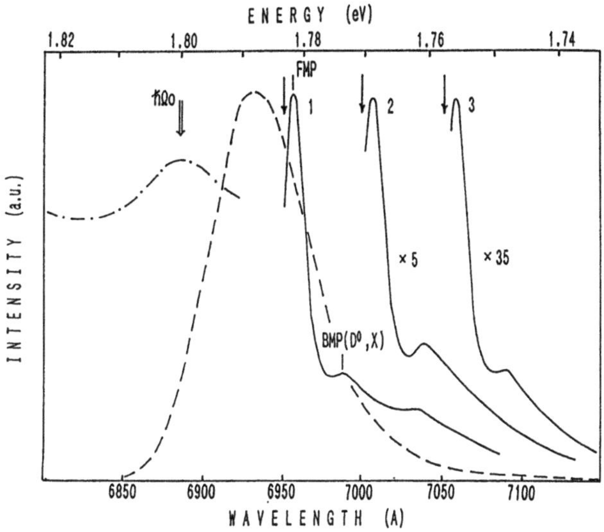

FIG.1 Reflection (dash-dotted) and photoluminescence (dashed) spectra of $CdTe/Cd_{0.78}Mn_{0.22}Te$ MQW (Lz=30Å; Lb=120Å) at T=1.6K. ħΩo-resonance frequency of n=1 HH exciton. Solid curves 1,2,3 are the luminescence spectra under selective excitation (excitation energies are shown by arrows).

heavy-hole (HH) exciton was determined by the fitting procedure described in /8/. The n=1 light-hole exciton is shifted by 25 meV to higher energies by strain and quantum confinement effects. The broad photoluminescence band obtained under nonresonant excitation ($\hbar\Omega_{exc} > \hbar\Omega_o$) has a halfwidth 20 meV (Fig.1, dashed curve) and a maximum shifted by 12 meV to lower energies. We attribute this band to the recombination of localized excitons. The width of the spectrum of localized states is caused by fluctuations of the QW width by one or two monolayers with different lateral dimensions of the islands /9/.

Selective excitation in the contour of the PL band ($\hbar\Omega_{exc} < \hbar\Omega_o$), i.e. excitation of excitons in localized states, leads to a dramatic narrowing of the luminescence band and appearence of a three line structure (Fig.1, solid curves 1,2,3). The strongest peak in this structure is 2 meV below the excitation energy and the second peak has an energy difference of 9 meV. The relative intensities and energy separations of these peaks from $\hbar\Omega_{exc}$ are almost independent of the excitation energy. The luminescence band narrowing under selective excitation is the consequence of the absence of energy relaxation between localized states within the exciton lifetime. From this fact we conclude the absence of spatial migration of the excitons as well. The excitons recombine free or bound at an impurity, at the sites of the QW where they were created by selective excitation. A similar situation has been investigated in Cd(S,Se) alloys /10/.

The observed energy shift of 2 meV is less than any binding energy of excitons at impurities in CdTe. Therefore we associate this peak with a free magnetic polaron in the QW which has been predicted theoretically /6,7/. The peak with a binding energy of 9 meV is attributed to bound magnetic polarons for excitons bound at neutral donors (BMP(D^0,X)). The BMP energy contains the magnetic polaron energy and the exciton binding energy at an impurity center, which is increased due to quantum confinement in QWs.

It is well known that magnetic polarons can be destroyed: either in an external magnetic field due to the alignment of Mn^{2+} spins or by a temperature increase due to a decrease of the Mn^{2+} magnetic susceptibility. The Zeeman pattern of magnetic polarons under selective excitation is rather

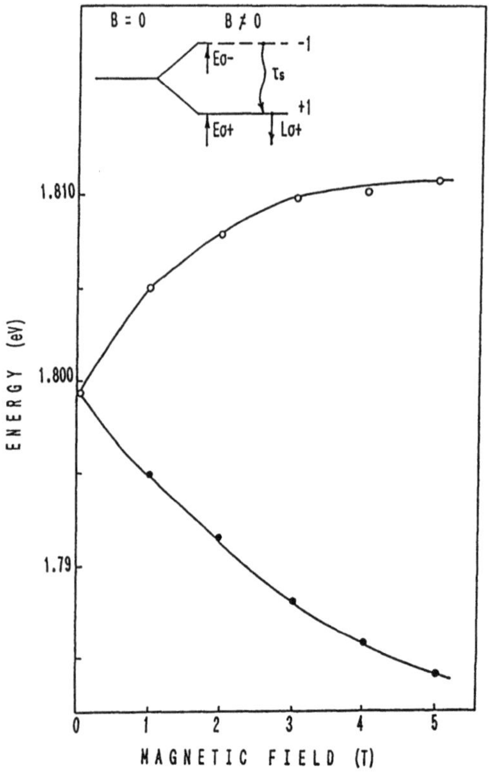

FIG.2 Exciton peak energies in PL excitation spectra as a function of magnetic field at T=1.6K. Excitation was circularly polarized (○ -σ-, ● - σ+) luminescence was analysed in σ+ polarization. The optical Zeeman transitions are shown in the insert; τ_s is the spin relaxation time.

complicated as one can see in Fig.3. Therefore we would like to describe in succession the magneto-optical data under nonresonant and selective excitation.

The Zeeman splitting of the free excitons was measured directly with photoluminescence excitation spectroscopy in Faraday geometry. In Fig.2 the energy position of the n=1 HH exciton peaks are shown as a function of a magnetic field in both excitation polarizations. The giant Zeeman splitting is due to strong exciton exchange interaction with the Mn^{2+} ions and corresponds to an effective g-factor of approximately $g^* \approx 140$. A schematic diagram of the optical transitions involved is shown in the insert. We define the polarization of the excitation light so that σ+ excitation (Eσ+) populates the lower Zeeman level which results in the σ+ luminescence (Lσ+). Excitons created by σ- excitation (Eσ-) on the upper Zeeman level also take part in σ+ luminescence due to fast spin relaxation processes τ_s.

The selectively excited spectra represent a different Zeeman pattern. Fig.3 shows the lower branch of the FMP Zeeman doublet and both branches of BMPs. The lower branches of the FMP and the BMP(D^0,X) are parallel, so we conclude that the upper branches must be similiar. The value of the Zeeman splittings are the same as observed for free excitons (see Fig.2). However, using selective excitation we found a strong asymetry of the Zeeman doublet. The upper branch shifts only by 3 meV and the main part of the Zeeman splitting is in the shift of the lower branch. The second difference between selectively and nonresonantly excited spectra is the inversion of the polarization. The upper branch is populated by σ+ excitation contrary to the free exciton case (Fig.2,3). It should be noted that the Zeeman pattern under selective excitation is independent of the excitation energy.

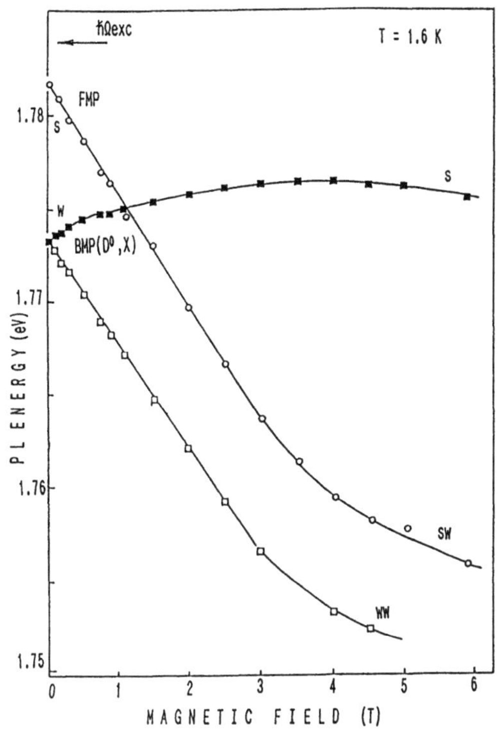

FIG.3 PL peak energies of FMP (○) and BMP(D^0,X) (□,■) as a function of magnetic field under selective excitation $\hbar\Omega exc = \hbar\Omega o - 16.5$ meV. Excitation polarization: full signs - σ+ open signs - σ-. PL was analysed in σ+.

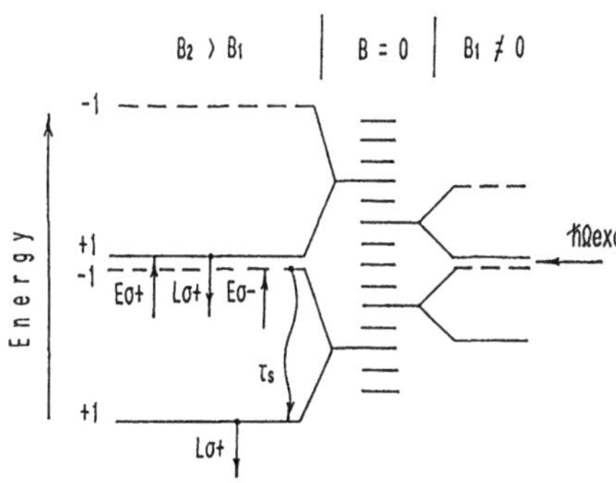

FIG.4 Schematic diagram of optical Zeeman transitions of excitons under selective excitation in the band of localized states. E stands for excitation and L for luminescence. τ_s denotes the spin relaxation process.

Fig.4 represents the model of selective excitation in the spectrum of localized states based on the absence of energy relaxation. At zero magnetic field the levels in the band of localized states are shown in the middle of the scheme. An arrow denotes the energy of the excitation light which in the absence of a magnetic field resonantly excites one exciton level only. In a magnetic field due to the Zeeman splitting different states are excited. As can be seen on the right and left sides of Fig.4 the lower level of one state and the upper level of the other state are resonant with the exciting energy. By changing the magnetic field the excitation passes through the spectrum of localized states. All observed unusual experimental behaviour can be explained in the framework of the model by this "scanning" process.

The energy positions of the PL peaks excited by σ+ polarized light are independent of the magnetic field due to the "scanning" process. These peaks are shifted from the excitation energy by the energy of free or bound magnetic polarons. The σ- excited peaks are shifted to lower energies by the full value of the Zeeman splitting. This leads to strong asymetry of the Zeeman doublet under selective excitation which was found experimentally (see Fig.3). The "scanning" process in the model of selective excitation also explains the inversion of polarization which is easy to be seen if one compares Fig.4 with the insert of Fig.2.

As discussed above in the model of selective excitation the upper branch of the Zeeman doublet should show no shift in a magnetic field. Therefore the energy changes of this branch are caused by the modification of the BMP binding energy. The diamagnetic shift of the exciton in CdTe is negligible small (0.2 meV at 3 T) and the 3 meV shift of the upper Zeeman branch of the BMP(D^0,X) to high photon energies in Fig.3 demonstrates the suppression of magnetic polarons in external magnetic fields.

The luminescence peak energies of FMPs and BMP(D^0,X)s under selective excitation conditions were measured as a function of temperature in the interval 1.6 - 30 K. The peak of the FMP shifts by 2 meV to higher energies and at 30 K its maximum cannot be resolved in the wing of the excitation line. The shift of the BMP(D^0,X) is 2,5 meV at 30 K. The high energy shifts of the polaron lines with increasing temperature give evidence for the suppression of magnetic polaron formation due to a decrease of the magnetic susceptibility in the barrier layers.

Summarizing the data of magnetic field- and temperature dependences, we have determined the polaron energies of 2 meV for the FMP and 3 meV for the BMP(D^0,X).

In conclusion, we have reported the first experimental observation of the theoretically predicted free magnetic polaron formation in CdTe/(Cd,Mn)Te MQWs. The polaron energies of free and bound magnetic polarons have been determined in quantum wells for the first time.

Acknowledgement - The authors wish to thank T.A.Kuhn for his help. D.R.Yakovlev acknowledges support of the Alexander von Humboldt Foundation. The project was sponsored by the Bundesministerium für Forschung und Technologie, Bonn.

REFERENCES

1. W.Ossau, S.Fischer, R.N.Bicknell-Tassius,J.of Crystal Growth 101, 905 (1990)
2. P.A.Wolff, In: Semiconductors and Semimetals, Vol.25, p.413, ed. by J.K.Furdyna and J.Kossut, Academic Press, London 1988
3. A.Golnik, J.Ginter and J.A.Gaj, J. Phys.C: Solid State Phys. 16, 6073 (1983)
4. A.V.Nurmikko, X.C.Zhang, S.K.Chang, L.A.Kolodziejski, R.L.Gunshor and S.Datta, J. of Luminescence 34, 89 (1985)
5. D.Heiman, P.Becla, R.Kershaw, D.Ridgley, K.Dwight, A.Wold and R.R.Galazka, Phys. Rev. B 34, 3961 (1986)
6. C.E.T.Goncales da Silva, Phys. Rev. B 33, 2923 (1986)
7. J.W.Wu, A.V.Nurmikko and J.J.Quinn, Phys. Rev. B 34, 1080 (1986)
8. E.L.Ivchenko, V.P.Kochereshko, P.S.Kop'ev, V.A.Kosobukin, I.N.Uraltsev and D.R.Yakovlev, Solid State Commun. 70, 529 (1989)
9. P.S.Kop'ev, I.N.Uraltsev, Al.L.Efros, D.R.Yakovlev and A.V.Vinokurova, Sov. Phys. Semicond. 22, 259 (1988)
10. S.Permogorov, A.Reznitskii, S.Verbin, G.O.Müller, P.Flögel and M.Nikiforova, Phys. Stat. Sol. (b) 113, 589 (1982)

Magnetoluminescence of Optically Oriented Excitons in GaAs/AlGaAs Superlattices

E.L. Ivchenko, V.P. Kochereshko, I.N. Uraltsev, and D.R. Yakovlev

A.F. Ioffe Physico-Technical Institute, Academy of Sciences of the USSR, SU-194021 St. Petersburg, USSR

Abstract. Magnetic field effects on the optically oriented angular momenta of excitons and spins of electrons have been studied in $GaAs/Al_{.35}Ga_{.65}As$ short-period superlattices to separate the radiative recombination of excitons either generated directly under optical excitation or created from nongeminate photoelectrons and photoholes. Anisotropy of the g-factor in the electron miniband has been measured for the first time from the effect of tilted magnetic fields on the luminescence polarization.

1. Introduction

Recent observations [1] of polarized luminescence in short-period superlattices (SLs) due to optical spin orientation of electrons has stimulated polarized magnetoluminescence measurements to study the electronic structure of SLs. The effect of a longitudinal magnetic field (Faraday configuration) on a suppression of the electron spin relaxation has been found in GaAs/AlGaAs Sls to be a measure of both the **k**-linear spin-splitting of electron minibands and the momentum relaxation time of photoelectrons [2]. The relative contribution of extended and localized states of electrons to the SL luminescence has been evaluated from an analysis of depolarization shapes induced by a magnetic field in the Voigt configuration [3].

In this paper, we present a comparision of the magnetic field effects on the polarization induced by optical orientation of electron spins or angular momenta of excitons in $GaAs/Al_{.35}Ga_{.65}As$ SLs with period range d=30-150 Å. Finding a distinct difference in the effects we are able to reveal contributions to the luminescence both from excitons generated directly under optical excitation or from excitons formed from uncorrelated electron and hole pairs. Analyzing the tilted magnetic field effect on the electron spin orientation we find anisotropy of the electronic g-factor in SLs with d<100 Å.

2. Transverse and Longitudinal Magnetic Field

Fig.1 displays the heavy-hole exciton and electron-acceptor luminescence, shown by a solid line, taken from the 40 Å-period SL excited along the SL axis by circulary polarized light at the energy of some meV higher than that of the exciton resonance, $\hbar\omega_0$, evaluated from the reflection spectrum [4]. The degree of circular polarization, P_{cir}, of the luminescence (dotted line) governed by the lifetime and spin relaxation time

Springer Series in Solid-State Sciences, Vol. 101
High Magnetic Fields in Semiconductor Physics III Editor: G. Landwehr

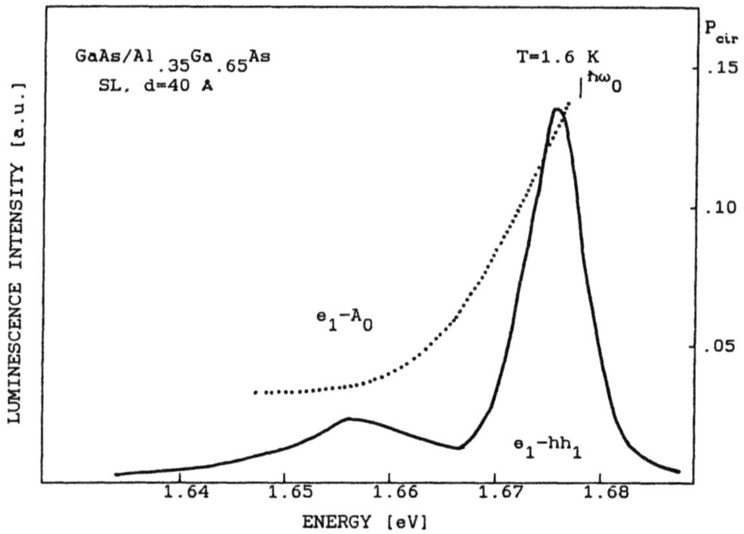

Fig.1. Photoluminescence spectrum (solid) and polarization of luminescence (dotted) taken from the 40 Å-period GaAs/AlGaAs superlattice.

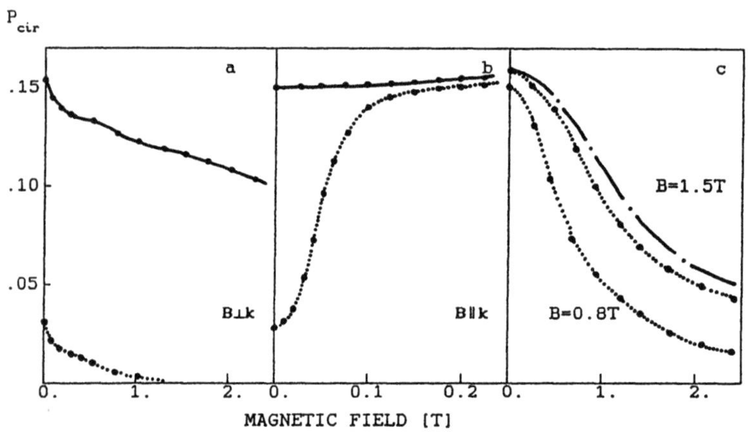

Fig.2. Circular polarization taken at the heavy-hole exciton (solid) and the e_1-A (dotted) peak as a function of magnetic field: (a)-transverse, (b)-longitudinal and (c)-transverse at $B_{\parallel}$=0.8 and 1.5T. Dashed-dotted curve is calculated with $g_{\parallel}=g_{\perp}$.

of electrons shows the spectral dependence which is expected if different recombination processes contribute to the SL luminescence.

The application of a transverse magnetic field **B⊥k** produces a precession of the oriented miniband electrons and consequently depolarizes the recombination light. The depolarization shape taken at the e_1-A_0 luminescence peak, as shown in Fig.2a by a dotted line, can be described by a sum of two Lorentzians. One of them with the halfwidth Γ of 0.7 T is characteristic of free electrons and the other with Γ=0.05 T is due to electron states localized along the SL axis [3]. Contributions to the polarization from optically oriented spins of extended and localized electrons one can identify also on the depolarization curve taken at the exciton peak, as shown in Fig.2a by a solid line. This fact evidences that the exciton luminescence is contributed by exciton recombination of uncorrelated electron and hole pairs. The dominant contribution to the polarization, which is influenced only slightly by a transverse magnetic field, is assigned to excitons generated with oriented momenta under optical excitation. The small value of the g-factor, which one has to propose for an explanation of a weak effect of transverse magnetic field, is not unexpected for the heavy-hole exciton in GaAs/AlGaAs SLs [5].

Strong support of the above assignment is obtained from an analysis of the longitudinal magnetic field effects on the polarization, P_{cir}. Since the precession mechanism has been found to dominate in electron spin relaxation in SLs at low temperatures [2], an increase of P_{cir} is expected in the presence of a longitudinal magnetic field if a contribution of the oriented electron spins to polarization dominates. A suppression of the electron spin relaxation occurs when the cyclotron frequency of electrons becomes higher than the inverse of the electron momentum relaxation time. In contrast, no remarkable effect of a longitudinal magnetic field on P_{cir} of oriented excitons is expected. The precession mechanism of the angular momentum relaxation of excitons is ineffective because the **k**-linear spin-splitting of the heavy-hole exciton band is smaller than that of the electron miniband by a m_e/m_{hh} factor. The experimental findings shown in Fig.2b are consistent with our predictions: the strong increase of P_{cir} is observed for the e_1-A_0 peak, where polarization is governed by optically oriented spins of electrons, and no effect of a longitudinal magnetic field on P_{cir} is found at the exciton peak where the contribution of the oriented-momentum excitons dominates.

The effects of a magnetic field in the Faraday and Voigt configuration one can use to study exciton and electron dynamics and localization in two-dimensional systems by measuring the effects at different spectral positions of the exciton luminescence line.

3. Tilted Magnetic Field

Strong effects of a tilted magnetic field are expected on the optically oriented spin of electrons that is influenced by both transverse and longitudinal magnetic fields. The application of a tilted field produces a combined effect of suppression of the electron spin relaxation resulting from the longitudinal component of a field and of the electron spin precession with Larmor frequency, $\Omega=(\mu_B/\hbar)\cdot[(g_\parallel B_\parallel)^2+(g_\perp B_\perp)^2]^{1/2}$, here $g_\parallel, g_\perp$ is the electron g-factor parallel and perpendicular to the SL axis. The depolarization curve is shown to be independent on the electron spin lifetime and it governed by the longitudinal field effect, $\mu_B g_\parallel B_\parallel$, if the electron spin relaxation is suppressed.

Fig.2c shows the polarization of the e_1-A_0 peak as a function of a transverse magnetic field in the presence of a longitudinal magnetic field of 0.8 or 1.5 T. One can see an about two-fold increase of the halfwidth as the longitudinal magnetic field becomes higher. The halfwidth of the latter curve is attributed to a complete suppression of spin relaxation which is governed by the ratio $g_\parallel/g_\perp$. The depolarization curve calculated for the latter case with $g_\parallel=g_\perp$, shown by a dashed- dotted line, is wider than the experimental one. The dotted curves are result of a fitting with $g_\parallel=0.8g_\perp$. Measuring this difference as a function of the SL period we found that it diminishes at d>100 Å, when the coupling between quantum wells quenches and extended Bloch states transform into localized quantum well eigenstates [4].

Isotropic g-factors of electrons in bulk GaAs and AlGaAs alloys calculated in the framework of the $\mathbf{k}\cdot\mathbf{P}$ perturbation theory in a three-band model are found to be in good agreement with experimental data [6]. Assuming the validity of the theory for short-period SLs one might assign the observed g-factor anisotropy to a strong effect of the SL valence band anisotropy.

References

1. E.L.Ivchenko, P.S.Kop'ev, V.P.Kochereshko, I.N.Uraltsev, D.R.Yakovlev, Sov.Phys. JETP Let. **47**,407 (1988)
2. I.N.Uraltsev, E.L.Ivchenko, P.S.Kop'ev, V.P.Kochereshko, D.R.Yakovlev, Phys.Stat.Sol. b**150**,673 (1988)
3. I.N.Uraltsev, V.P.Kochereshko, P.S.Kop'ev, A.M.Vasiliev, D.R.Yakovlev, Surface Science. **232**, 143 (1990)
4. E.L.Ivchenko, V.P.Kochereshko, P.S.Kop'ev, V.A.Kosobukin, I.N.Uraltsev, D.R.Yakovlev,Solid State Commun .**70**,529 (1989)
5. G.W.Bauer,T.Ando, Phys.Rev.B**37**, 3130 (1988)
6. C.Hermann, C.Weisbuch Phys.Rev.B**15**, 823 (1977)

Determination of the Hole Effective Magnetic Moment in Quantum Wells in a Parallel Magnetic Field

A. Fasolino[1], *G. Platero*[2], *M. Potemski*[3,4], *J.C. Maan*[4], *K. Ploog*[5], *and G. Weimann*[6]

[1]SISSA, Strada Costiera 11, I-34100 Trieste, Italy
[2]Instituto de Ciencia de Materiales (CSIC), Universidad Autonoma de Madrid, Cantoblanco, E-28049 Madrid, Spain
[3]Institute of Physics, Polish Academy of Sciences, PL-02-668 Warsaw, Poland
[4]Max-Planck-Institut FKF-HML, BP 166X, F-38042 Grenoble, France
[5]Max-Planck-Institut für Festkörperforschung, Heisenbergstr. 1 W-7000 Stuttgart 80, Fed. Rep. of Germany
[6]Walter Schottky Institut, TU München, Am Coulombwall, W-8046 Garching, Fed. Rep. of Germany

Abstract. We present a theoretical and experimental study of interband magneto-optics in GaAs-AlGaAs quantum wells in a magnetic field parallel to the layers. This study yields a precise and direct determination of the parameter κ, describing the hole effective magnetic moment. Experimentally, for thin wells, both transitions from heavy and light hole states experience a diamagnetic shift but *only* those deriving from light hole states are split by the field. The observed light hole splitting is shown to depend solely on the parameter κ as far as the well width is less than the magnetic length. From a comparison between theory and experiments, we obtain for GaAs, a value of $\kappa = 0.7$ which is substantially smaller than the value commonly assumed in the literature.

The magneto-optics of GaAs quantum wells (QW's) in a magnetic field along the growth direction ($B_\perp$) has been thoroughly studied both experimentally and theoretically [1]. Much less has been done in the parallel field configuration ($B_\parallel$); in $B_\parallel$ the electronic cyclotron motion occurs in the plane containing the growth direction, thus being affected by the confining potential barriers. The degeneracy of bulk Landau levels with respect to the position of the orbit centre is then destroyed by the superposed magnetic and confinement potentials and levels become spatially dispersive [2-4].

We observe experimentally that the parallel field in thin QW's has the peculiar effect of splitting only the electronic and light hole levels while the heavy hole levels are not split up to high fields.

This behaviour can be explained theoretically as follows. For narrow wells, i.e. for well widths smaller than the magnetic length $l_m = \sqrt{\hbar/eB}$, confinement effects largely dominate over the effect of the magnetic field and the pure heavy and light hole character of the valence subbands at $B = 0, k_\parallel = 0$ is only slightly affected by the field. It can be shown [5] that such pure heavy $J_x = \pm 3/2$ hole states (x being the growth direction) are not split by the field while the pure light hole states $J_x = \pm 1/2$ are split by an amount $2\kappa B$ where κ is the effective hole magnetic moment introduced by Luttinger [6] in analogy to that of electrons g^*. The experimentally observed light hole splitting allows a direct determination of the parameter κ. The present analysis yields a value $\kappa = 0.7$ much smaller than that obtained by magneto-reflectance in bulk GaAs ($\kappa = 1.2$) [7]. There, the information on this parameter cannot be singled out and the observables

Springer Series in Solid-State Sciences, Vol. 101
High Magnetic Fields in Semiconductor Physics III Editor: G. Landwehr

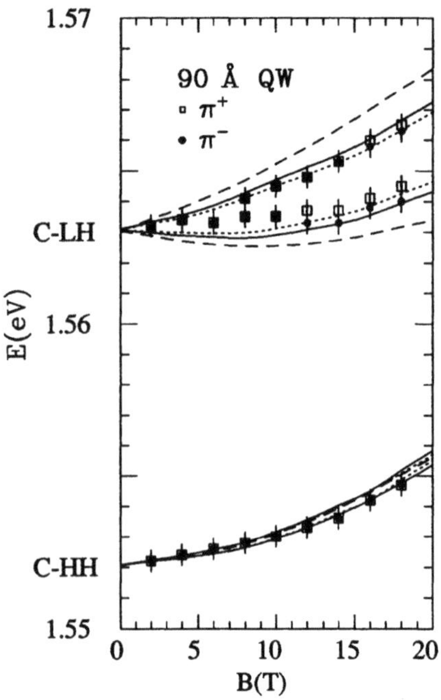

Fig.1. Interband transition energies $C_1 - HH_1$ and $C_1 - LH_1$ for the 90 Å GaAs QW versus $B_\parallel$ in Tesla. Dots and squares show the experimental behaviour for both linear polarizations of the light. Solid and dotted lines are the theoretical results for the parameter $\kappa = 0.7$ for the two polarizations (they differ by $g^*\mu_B B$); the dashed lines are calculated for $\kappa = 1.2$ (for clarity the electronic spin splitting is not included in this case). The calculated $C_1 - HH_1$ and $C_1 - LH_1$ transitions are down-shifted by 11.2 and 15.5 meV respectively for $\kappa = .7$ and by 11.7 and 16.15 meV for $\kappa = 1.2$.

are dominated by the other Luttinger parameters $\gamma_1, \gamma_2, \gamma_3$ describing the anisotropic effective hole masses.

We have performed excitation spectroscopy experiments on high quality GaAs undoped QW's (90, 140 Å) sandwiched between $Ga_{.74}Al_{.26}As$ barriers in $B_\parallel$, up to 18 Tesla. Experiments are performed in the Voigt configuration with the two linear polarizations of the incident light (π_+, π_-). The lowest excitonic energy transition from the first heavy hole subband ($HH_1 \rightarrow C_1$) does not split appreciably up to the highest value of the field. Conversely, the next observed transition, corresponding to the first light hole subband ($LH_1 \rightarrow C_1$), splits into two peaks. This splitting increases at first and appears to saturate at high magnetic fields as shown in Figs. 1. and 2. The spin splitting versus $B_\parallel$ deviates from linearity at higher fields for thinner QW's as shown in Fig. 2. In Figs. 1 and 2 the experimental results are compared with the theoretical curves, calculated within the envelope function formalism with the method described in Ref. [4]. We do not include $\mathbf{k} \cdot \mathbf{p}$ coupling between conduction and valence bands; non-parabolicity is approximately taken into account through a heavier electronic effective mass $m^*(E) = m^*(1 + 2\frac{E-E_{gap}}{E_{gap}})$. The valence band is described by the Luttinger Hamiltonian which accounts for the mixing of the heavy and light hole levels. The calculated transitions $HH_1 \rightarrow C_1$ and $LH_1 \rightarrow C_1$ shown in fig.1 for both polarizations of the light are obtained with the value of $\kappa = 0.7$ that we propose here. The agreement with experiments is remarkable. However, the simple approach of treating separately the electron and hole spin does not account for the observed behaviour of the experiments in the two polarizations of the light; within this picture each light hole level should make transitions at $B \neq 0$ to the electronic levels of opposite spin depending on the polarization of the light. Experiments show that only the lowest energy $LH_1 \rightarrow C_1$ transition is split in the two polarizations of the light. The dashed line is instead calculated with $\kappa = 1.2$. The discrepancy between theory and experiment for the latter value strongly indicates that the hole magnetic moment is overestimated by this value. The results

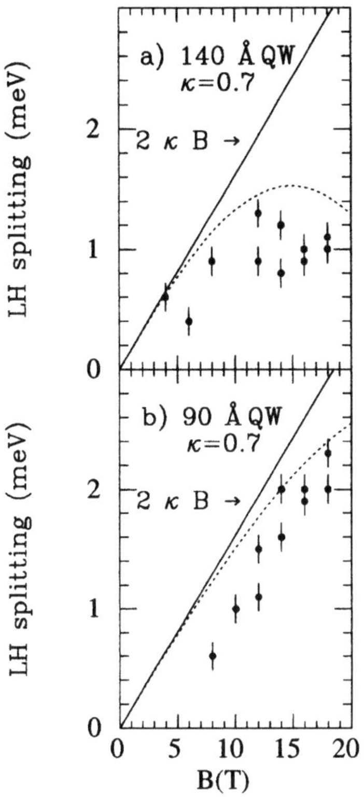

Fig.2. First light hole (LH_1) splitting versus $B_{\|}$ for the 140 $\AA$ (panel a) and 90$\AA$ (panel b) GaAs QW's. Dotted lines show the theoretical results for $\kappa = 0.7$ and dots are the measured splittings in both light polarizations. The solid line, with slope equal to $2\kappa B$, represents the expected behaviour for pure light hole states (see text).

for the 140 $\AA$ sample are similar and lead to the same conclusion. Excitonic effects are not included in our theory. In comparing our results with experiments we have rigidly shifted for all fields the calculated transitions at $k = 0$ by an amount representing the binding energies for heavy and light hole excitons. The values we have used are slightly bigger than those reported in Ref. [1, 8].

In Fig 2 the calculated light hole splitting is compared with the experimental values. The solid line represents the slope $2\kappa B$ which would be expected for pure light hole states. We see that both theoretical and experimental results deviate from this slope when the magnetic field increases, i.e. the magnetic energy starts to be comparable to the confinement energy. Clearly this occurs at higher fields for thin wells. This deviation parallels the deviation of the wavefunction from a pure light hole character. In the limit of very large wells this behaviour is lost because the magnetic energy is bigger than the confinement energy even at very low fields and the hole states have a mixed character as in bulk [6]. As shown in Fig 1. also the heavy hole states are split at high fields where they start to acquire a mixed character. This splitting is seen experimentally only for the 140 $\AA$ QW (0.4 meV at 18 T).

In conclusion, we have studied the effect of a parallel magnetic field on the optical transitions in thin QW's. From a comparison of excitation spectroscopy and envelope function calculations we extract a new value for the Luttinger parameter κ. The behaviour of the interband transitions with field gives direct information also on the mixing of the hole magnetic levels.

Acknowledgments Calculations are performed as part of the SISSA-CINECA joint project, sponsored by the Italian Ministry of Scientific Research. One of us (G. P.) is indebted to the Comision Interministerial de Ciencia y Tecnologia of Spain under contract MAT88-0116-C02-02.

[1] see e. g. G.E.W. Bauer, T. Ando, Phys. Rev. **B37**, 3130 (1988) and references therein.
[2] J.C. Maan, in *Festkörperprobleme*, ed. P.Grosse (Vieweg, Braunschweig, 1987), Vol.27, p.137.
[3] G. Platero, M. Altarelli, Phys.Rev. **B39**,3758 (1988).
[4] A. Fasolino, M. Altarelli, *Proc. of the 19th International Conference on the Physics of Semiconductor*, ed. W. Zawadszki, Inst. Phys. Polish Academy of Sciences, Wroclaw 1988, p.361; A. Fasolino to be published.
[5] A. Fasolino, G. Platero, M. Potemski, J.C.Maan, K. Ploog, G. Weimann, to be published.
[6] J.M. Luttinger, Phys. Rev. **102**, 1030 (1956).
[7] K. Hess, D. Bimberg, N.O. Lipari, J.U. Fischbach and M. Altarelli, *Proc. of the 13th International Conference on the Physics of Semiconductors*, F.G. Fumi ed., Rome 1976, p. 142.
[8] L.C. Andreani,A. Pasquarello, Europhys. Lett.,**6**,259 (1988).

Resonant Raman Scattering in High Magnetic Fields

T. Ruf, R.T. Phillips, F. Iikawa**, and M. Cardona*

Max-Planck-Institut für Festkörperforschung, Heisenbergstr. 1
W-7000 Stuttgart 80, Fed. Rep. of Germany
***Permanent address: Department of Physics, University of Exeter, Exeter EX4 4QL, UK**
****On leave from: Instituto de Fisica, UNICAMP, Cidade Universitaria-13081, Campinas, SP, Brazil**

Abstract. The efficiency of higher order Raman scattering by LO phonons in GaAs exhibits large enhancements in a magnetic field when resonances with interband transitions occur. Scattering by up to four LO phonons was systematically investigated. Features observed for different circular polarizations and orders of scattering are discussed.

1. Introduction

In a magnetic field the Stokes LO phonon Raman intensity of GaAs or InP exhibits oscillations due to Landau level structure. Magneto-Raman profiles, obtained from measuring the Raman intensity vs. magnetic field for fixed excitation energies, yield information on interband magneto-optical transitions and electron-phonon interactions in such magnetically quantized systems. [1-7] The effect of valence band mixing as well as corrections for Coulomb effects have to be taken into account in order to arrive at a consistent assignment of the observed fan lines. [2,4,5,7] In InP information on resonant magneto-polarons can be extracted. [2] Several Raman terms lead to resonant enhancement of the scattering intensity in incoming or outgoing resonance. When both, incident and scattered photons, match with interband magneto-optical transitions, both denominators in the expression for the Raman efficiency become singular and double resonance occurs. This situation was achieved in GaAs by tuning valence Landau levels such that their separation equals the energy of the LO phonon and choosing the excitation energy appropriately. [3-5] Since magneto-Raman fan plots reflect bulk electronic structure, the band dispersion (nonparabolicity) and electron effective mass of GaAs could be experimentally determined up to 300 meV above the E_0 fundamental gap. [1,7] In a magnetic field the efficiency for multiphonon Raman scattering is significantly increased with overtones up to 9 LO being easily observed. [6] In the following we present investigations on the magneto-resonant behavior of multiphonon-scattering in GaAs.

2. Results and Discussion

Exemplary magneto-Raman profiles which were recorded at the spectral positions of 2 LO- and 4 LO-scattering are given in Fig. 1. The spectra, taken in $\bar{z}(\sigma^+,\sigma^+)z$- and $\bar{z}(\sigma^-,\sigma^-)z$-backscattering configurations using circularly polarized light, exhibit strong selection rules with the peaks being mutually exclusive for the two geometries. Pronounced resonances occur even at rather low magnetic fields. The large difference in magnetic field between

Springer Series in Solid-State Sciences, Vol. 101
High Magnetic Fields in Semiconductor Physics III Editor: G. Landwehr

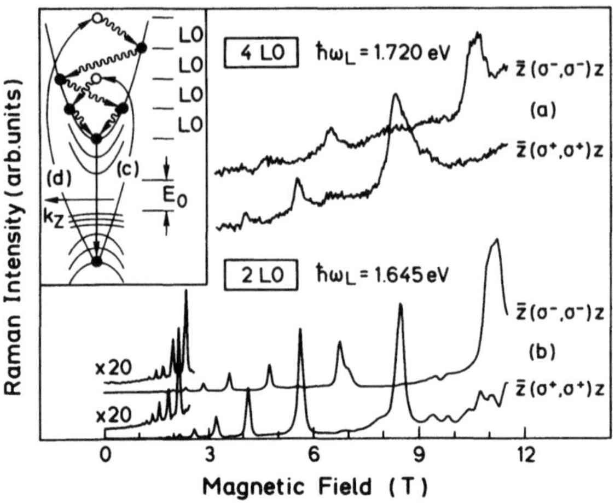

Fig. 1: Magneto-Raman profiles from (a) 4 LO- and (b) 2 LO-scattering in GaAs (001). The difference in excitation energies amounts to 2 LO. The inset shows possible Raman processes which lead to the same outgoing resonances for both orders.

associated features in the two configurations indicate that valence levels of light mass are involved in the interband magneto-optical transitions which are resonantly enhanced. [2] The difference in the energies of the incident photons between the 4 LO- (Fig. 1a) and the 2 LO-profiles (Fig. 1b) amounts to about twice the zone center LO phonon energy in GaAs (37 meV). The two types of profiles exhibit the same resonances in the respective scattering geometries in 2 LO- as well as in 4 LO-scattering when the excitation energy is changed appropriately. This suggests that the enhancement of the Raman efficiency occurs for both cases in outgoing resonance, after all phonons have been emitted. In the inset of Fig. 1, possible Raman processes for 2 LO- (Fig. 1c) and 4 LO-scattering (Fig. 1d) are shown. In these processes, which may also occur with phonon scattering in the light mass valence levels, outgoing resonance takes place after all phonons have been emitted in a cascade process mediated by intra-level and intra-band Fröhlich scattering. Theoretical evidence for such processes to be important has been given in Refs. [8] and [9]. Further investigations which take the dispersion and mixing of Landau levels into account are needed in order to arrive at a satisfying description for the microscopic processes of multiphonon emission in III-V-semiconductors in a magnetic field.

Figure 2 gives resonance positions from magneto-Raman profiles for 2 LO-scattering. The fan lines for both, $\bar{z}(\sigma^+,\sigma^+)z$- (dots) and $\bar{z}(\sigma^-,\sigma^-)z$-configurations (crosses) converge at an energy of $E_0 + 2\,\mathrm{LO}$ when the magnetic field goes to zero. This identifies the observed resonances to be of outgoing character. The fan plot in Fig. 2 illustrates that the resonances observed in the two scattering configurations are mutually exclusive. By comparison with interband magneto-optical transitions calculated with a modified Pidgeon-Brown-type 8×8 $\vec{k}\cdot\vec{p}$-Hamiltonian, [2,5] the fan lines can be assigned. In $\bar{z}(\sigma^+,\sigma^+)z$, transitions between the $|n, V, -\frac{3}{2}\rangle$-admixtures of light mass valence levels and $|n, C, \downarrow\rangle$ electron states are observed. The range of Landau quantum numbers n for these transitions in Fig. 2 extends from $1 \le n \le 10$. In $\bar{z}(\sigma^-,\sigma^-)z$-configuration the resonant recombination occurs between $|n, V, +\frac{3}{2}\rangle$ light mass and $|n, C, \uparrow\rangle$ electron levels. Such transitions are observed for $0 \le n \le 10$. For excitation energies above 1.67 eV the character of the resonances changes. Fan lines from 2 LO-scattering far above the E_0 fundamental gap have been assigned to intermediate resonances involving heavy mass valence levels and were used to determine the nonparabolicity of the conduction band in GaAs. [7] In scattering configurations with complementary circular polarizations ($\bar{z}(\sigma^\pm,\sigma^\mp)z$), all

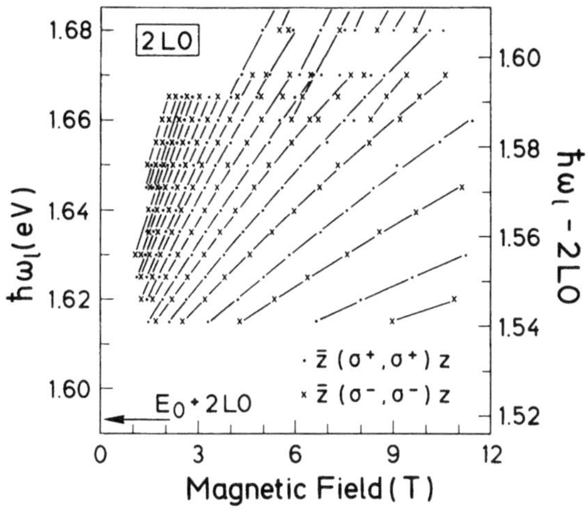

Fig. 2: Fan plot of magneto-Raman resonances in 2 LO-scattering. The fan lines can be assigned to interband magneto-optical transitions involving light mass valence levels. Outgoing resonances are found.

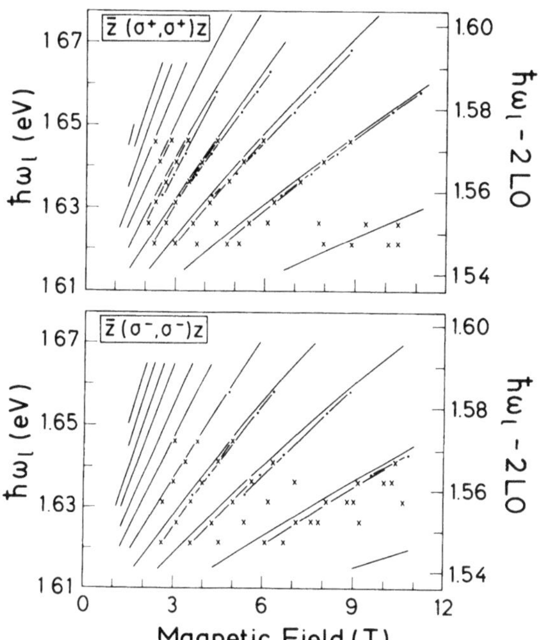

Fig. 3: Fan plots of magneto-Raman resonances in 2 LO- (solid lines), 3 LO- (dots), and 4 LO-scattering (crosses). The transition energies were corrected by 0 LO, 1 LO, and 2 LO, respectively, to make the similarities visible.

resonances which occur in separate geometries in Fig. 2 are observed at the same time. This finding is similar to the case of 1 LO-scattering in InP, however, with complementary and equal circular polarizations being exchanged. [2]

The common features of màgneto-Raman resonances and the similarities found in the various orders of multiphonon scattering which we investigated are highlighted in Fig. 3. The solid lines stand for the 2 LO-results from Fig. 2. For these fan lines the energy scales given to the left and right side of the figure apply. Dots which are connected by lines as a guide to the eye are experimental data from 3 LO-scattering. For these data the energy of one LO phonon was subtracted from the energy of the exciting photons. Crosses, also connected by guiding lines, represent resonances found in 4 LO-magneto-Raman profiles. The energy of two LO phonons was subtracted from these transitions prior to plotting. The synopsis of fan lines from second to fourth order Raman scattering shows that indeed always the same magneto-optical interband transitions are observed in the multiphonon magneto-Raman profiles. The resonant enhancement has outgoing character occuring after the emission of all participating phonons as was conjectured before (Fig. 1). This feature presumably extends to still higher orders of scattering. Note, however, that for GaAs there is a distinct difference between the cases of multiphonon and one-phonon scattering. Whereas valence levels of light mass participate in the resonances presented here, heavy mass levels in outgoing resonance are involved in the latter. [3-5] In contrast to that one-phonon as well as multiphonon magneto-Raman resonances in InP arise from interband transitions where light mass valence levels contribute.

Acknowledgments

Thanks are due to H. Hirt, M. Siemers, and P. Wurster for expert technical help. R. T. Phillips acknowledges support from the Royal Society and the Deutsche Forschungsgemeinschaft. F. Iikawa acknowledges support from the Alexander von Humboldt Foundation. We would like to thank S. I. Goubarev for a critical reading of the manuscript.

References

1. G. Ambrazevičius, M. Cardona, and R. Merlin, Phys. Rev. Lett. **59**, 700 (1987).
2. T. Ruf, R. T. Phillips, A. Cantarero, G. Ambrazevičius, M. Cardona, J. Schmitz, and U. Rössler, Phys. Rev. B **39**, 13378 (1989).
3. T. Ruf, C. Trallero-Giner, R. T. Phillips, and M. Cardona, Solid State Commun. **72**, 67 (1989).
4. C. Trallero-Giner, T. Ruf, and M. Cardona, Phys. Rev. B **41**, 3028 (1990).
5. T. Ruf, R. T. Phillips, C. Trallero-Giner, and M. Cardona, Phys. Rev. B **41**, 3039 (1990).
6. T. Ruf and M. Cardona, Phys. Rev. Lett. **63**, 2288 (1989).
7. T. Ruf and M. Cardona, Phys. Rev. B **41**, 10747 (1990).
8. V. I. Belitskii, A. V. Gol'tsev, I. G. Lang, and S. T. Pavlov, Zh. Eksp. Teor. Fiz. **86**, 272 (1984) [Sov. Phys. JETP **59**, 155 (1984)].
9. V. I. Belitsky, A. V. Goltsev, I. G. Lang, and S. T. Pavlov, phys. stat. sol. (b) **122**, 581 (1984).

Magnetooptical Investigations on S and P Excitons in Multiple Quantum Wells

J. Engbring and C.D. Ludwig

Institut für Physik, Universität Dortmund,
W-4600 Dortmund 50, Fed. Rep. of Germany

1. Introduction

Magnetooptical investigations have gained increasing interest as a powerful tool for the determination of electronic parameters of heterostructures [1,2,3].

For a complete description of excitons in GaAs/AlGaAs quantum wells and their complex behavior in magnetic fields, valence band mixing must be taken into account [4,5]. However, in strained QW-samples an uniaxial stress in growth direction increases the heavy hole - light hole splitting and the interaction of the hh band and lh band decreases [6]. If hh and lh levels are clearly separated, the interpretation of optical spectra is much simplified. Due to the decreased valence band mixing it is possible to describe the magneto-exciton in a simplified model.

Starting with a two-band model, we have derived an empirical formula based on semiclassical arguments. This analytical formula describes the behaviour of excitons in weak and in strong magnetic fields.

The formula is tested by comparison with magnetoexcitons, which we have investigated by two-photon spectroscopy (P excitons) and one-photon spectroscopy (S excitons) on the same sample. Two-photon measurements in quantum wells were performed for the first time by Fröhlich et al. [7]. The authors interpreted their data without taking into account excitonic effects. The combination of these two experimental techniques allows us to measure 16 excitons in magnetic fields up to 6.5 T. The experimental data of P excitons allow to determine the valence band mass and the conduction band mass separately.

2. Experiment

For the two-photon measurements we use a tunable, pulsed dye-laser ($\hbar\omega$=1.44-1.55eV, $\tau \approx$60ps, P$\approx 10^{12}$W/m^2) and a pulsed high power CO_2-laser ($\hbar\omega$=0.117eV, $\tau \approx$250ps, P$\approx 10^{13}$W/m^2). Details of the experimental setup are given in ref. [8].
The laser beams have to be synchronized and focussed on the sample to be in perfect overlap. The simultaneous absorption of two

photons is detected by monitoring the subsequent luminescence of the excitonic ground state. The sample is mounted in a helium-cryostat with a superconducting split-coil and kept at a temperature of 2.2K. The experiments are performed in Faraday-configuration (B‖k).

Our sample is a $GaAs/Al_{0.33}Ga_{0.67}As$ multiple quantum well structure of 66 GaAs layers of 98Å between barriers of 98Å, which is glued on BaF_2 as sample holder. Due to the different expansion coefficients of GaAs and BaF_2 our sample is strained ($P \approx 0.3$ GPa) [6]. In this contribution we will only be concerned with the exciton of the first heavy hole subband and the first conduction subband (hh1-cb1 exciton). In our spectra the lh1-cb1 1S exciton is also observed.

Since the energy of the dye-photon is close to the gap energy, the two-photon absorption can be interpreted as a two-step process. In the first step a dye-photon generates a virtual (nS) exciton (m=0). In a second step a CO_2-photon induces a transition according to the selection rule $\Delta m = \pm 1$ depending on its polarisation ($\sigma^{\pm}$). Therefore transitions are induced, if the sum of the energies of the CO_2-photon and the dye-photon is resonant with the energy of a P exciton state. In fact, our spectra show nearly no dependence on the polarization of the dye-laser but a drastic dependence on the polarisation of the CO_2-laser.

3. Fit of Magnetoexciton

For the analysis of the experimental data we propose a semiempirical formula (eq. (1)).The limits for low and high magnetic fields are well described by this formula. In real quantum well layers excitons are

$$E = E_{gap} - E_{ex}/2 + E_B - \sqrt{E_{ex}^2/4 + E_{ex}\,(E_B - E_{ze})} \tag{1}$$

$$E_{ex} = E_{ryd}/(n-t_d)^2 \tag{2}$$

$$E_{ze} = \mu_B B\left[\left(\frac{1}{m_{cb}} - \gamma_1\right) m - 2\varkappa J_z\right] \quad ; \quad |J_z| = 3/2 \tag{3}$$

$$E_B = \frac{\hbar e}{m_e}\left[\frac{1}{m_{vb}}(n_{vb}+0.5) + \frac{1}{m_{cb}}(n_{cb}+0.5)\right] B \tag{4}$$

$$\text{with} \quad 1/m_i = 1/m_i^0\left[1 + \alpha_i \frac{4 e m_i^0}{\hbar^3}(n_i+0.5)\,B\right] \tag{5}$$

$$\text{and} \quad E_i = \frac{\hbar^2 k^2}{2 m_i^0 m_e} + \alpha_i k^4 \quad ; \quad i = vb,\ cb \tag{6}$$

expected to behave somewhat intermediate between the pure two-dimensional and the three-dimensional case. We take into account this fact by introducing a parameter t_d in eq. (2). Since the bands are nonparabolic, we allow the effective masses to be dependent on

Table 1: Exciton-labels and corresponding Landau-labels, polarization dependence for dye- and CO_2-laser

exciton-label (n,m)	Landau-label (n_{vb}/n_{cb})	dye-polarization	CO_2-polarization
1S (m = 0)	0 / 0	$\sigma^\pm$	/
2S (m = 0)	1 / 1	$\sigma^\pm$	/
$2P^+$ (m = +1)	0 / 1	$\sigma^\pm$	σ^+
$2P^-$ (m = -1)	1 / 0	$\sigma^\pm$	σ^-

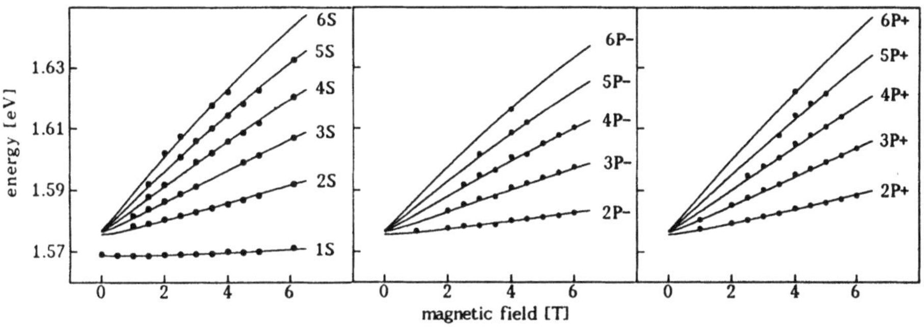

Fig. 1: Data of magnetooptical measurements, solid lines are calculated from formula (1) with parameters given in Table 2

the magnetic field (eq. (5)). The parameters α_i are chosen according to eq. (6).

For high magnetic fields Landau-quantum numbers n_{vb} and n_{cb} can be assigned to each peak. Now each peak corresponds to an exciton with quantum numbers n and m for the low field limit [9]. The relation of these quantum numbers can also be found by comparison of the selection rules for the interband Landau transitions with those for the magnetoexcitons. These relations are given in Table 1. We are now able to assign a pair of these quantum numbers to each resonance in our spectra. The numbers were inserted into formula (1). A least squares fit, which takes into account all measured magnetoexcitons, leads to the six fit parameters in Table 2.

In Figure 1 the measured exciton energies are plotted versus the magnetic field. The solid lines in Figure 1 represent the magneto-excitons calculated from eq. (1) . The agreement is good, if we take into account, that only six fit parameters were necessary to describe all 16 magnetoexcitons. We therefore conclude, that the fitted values of the parameters are fairly accurate.

Table 2: Parameters used in formula (1), left side: fit parameters ; right side: parameters from literature

E_{gap} [eV]	1.577	±0.0002	E_{ryd} [meV]	4.2	a)
t_d	0.3	±0.04	γ_1	6.85	ref.[10]
m^0_{vb}	0.069	±0.001	$\varkappa$	1.2	ref.[10]
α_{cb} [eVÅ^4]	800	±1300			
m^0_{vb}	0.155	±0.004			
α_{vb} [eVÅ^4]	-10100	±1200			

a) $E_{ryd} = 13.6\text{eV}\ \mu/\varepsilon^2$; $\mu = (1/m^0_{cb} + 1/m^0_{vb})^{-1}$; $\varepsilon = 12.53$ (ref.[10])

4. Acknowledgements

We thank Dr. G. Khitrova (Optical Science Center, University of Arizona, Tucson, USA) for the high quality MQW samples. Discussions with Prof. Dr. D. Fröhlich and financial support by the Deutsche Forschungsgemeinschaft are gratefully acknowledged.

References

1. J.C. Maan, G.Belle, A. Fasolino, M. Altarelli, and K. Ploog, Phys. Rev. B**30**, 2253 (1984).
2. W. Ossau, B. Jäkel, and E. Bangert, High Magnetic Fields in Semiconductor Physics, ed. G. Landwehr, Sol. State Sci. **71**, 213, (Springer 1987).
3. F. Ancilotto and A. Fasolino, Superl. and Microstr. **3**,187 (1987).
4. G.E.W. Bauer and T. Ando, Phys. Rev. B**38**, 6015, (1988).
5. L. J. Sham, High Magnetic Fields in Semiconductor Physics II, ed. G. Landwehr, Sol. State Sci. **87**, 232, (Springer 1989).
6. M. Schlierkamp, R. Wille, K.Greipel, U. Rössler, W. Schlapp, and G. Weimann, Phys. Rev. B**40**, 3077, (1989).
7. D. Fröhlich, R. Wille, W. Schlapp, and G. Weimann, Phys. Rev. Lett. **61**, 1878, (1988).
8. D. Fröhlich, Ch. Neumann, B. Uebbing, and R. Wille, Phys. Stat. Sol. (b), **159**, 297, (1990).
9. A. H. MacDonald and D. S. Ritchie, Phys. Rev. B**33**, 8336, (1986).
10. Landolt-Börnstein, ed. O. Madelung, Group III, Band 17a, (Springer, 1982).

Magnetic Levels in Fibonacci Superlattices and Temperature Dependence of Spin Relaxation in Quantum Wells at High Magnetic Fields

J.C. Maan[1], *V. Chitta*[1], *D. Toet*[1], *M. Potemski*[1;x], *and K. Ploog*[2]

[1]Max-Planck-Institut für Festkörperforschung, Hochfeld-Magnetlabor, BP 166X, F-38042 Grenoble Cedex, France

[2]Max-Planck-Institut für Festkörperforschung, Heisenbergstr. 1, W-7000 Stuttgart 80, Fed. Rep. of Germany

On leave from: Institute of Physics, Polish Academy of Sciences, PL-02-668 Warsaw, Poland

Abstract. New experimental results on magnetic levels in Fibonacci superlattices and on spin relaxation in quantum wells at high magnetic fields are presented. In the case of the Fibonacci superlattices a unique self-similarity in the shape of magneto-optical spectra is demonstrated. In the case of spin-relaxation in quantum wells, a strong temperature dependence of the rate of the relaxation of conduction band spin character of excitonic states which decreases with temperature as $T^{-3/4}$ has been observed.

1. Introduction

In this paper we will describe some recent experimental results on magnetic levels in quasiperiodic Fibonacci superlattices and on the temperature dependence of the relaxation of the magnetic moment of photoexcited carriers in quantum wells. Both subjects have in common that they make use of interband magneto-optical absorption at high magnetic fields, but the physical problems addressed are quite different.

The Fibonacci superlattice studied is a structure consisting of $Ga_{0.8}Al_{0.2}As$ barriers 1.12 nm thick which separates wells of 1.69 or 2x1.69 nm. The different well thicknesses are arranged according to a Fibonacci sequence (see below). In such a structure the density of states of the magnetic levels is studied when a magnetic field is applied parallel to the layers. Since the magnetic length, $l=(\hbar/eB)^{1/2}$ at fields up to 20T, is about 8 nm and the barriers are thin, the carriers orbit through the barriers and the effect of the non-periodicity can be studied. This level structure has previously been studied theoretically [1], here we want to present the first experimental results.

In the experiment on spin relaxation in quantum wells, the magnetic field is applied perpendicular to the layers. In this case electrons and holes can orbit in the plane of the layers and the continuous two-

Springer Series in Solid-State Sciences, Vol. 101
High Magnetic Fields in Semiconductor Physics III   Editor: G. Landwehr

dimensional density of states at zero field is split into sharp and discrete (i.e. separated by gaps) Landau levels. For the conduction band these Landau levels are equidistant and almost linear field dependent, whereas each of them is split in two sublevels, one for each component of the electron magnetic moment. In an earlier paper [2], it has been shown that the relaxation of the electron magnetic moment (spin) is greatly affected by the discrete nature of the density of states. In particular it was possible to establish that photocreated electrons relax preferentially within the Landau level ladder between states with the same spin orientation, and that the spin-flip time may be substantially larger than the time for radiative recombination. In this paper we present new data on the temperature and field dependence of this phenomenon.

2. Magnetic levels in Fibonacci superlattices.

The structure studied consists of a series of $Ga_{0.8}Al_{0.2}As$ barriers, designated "a" with thickness 1.12nm, which separate GaAs wells, designated "b" or "bb" with thickness 1.69 nm respectively 3.38 nm. The sequence of layers is arranged according to a Fibonacci series. In the Fibonacci series [3], the nth generation w_n of some entity consists of the sum of the two previous generations w_{n-1} and w_{n-2}. In the case of numbers and starting with w_1=1 and w_2=1, this leads to a series 1,1,2,3,5,8,13,21,34,... etc. The ratio w_n/w_{n-1} approaches $\left(1+\sqrt{5}\right)/2$, the golden mean. In an analogous manner, we can construct a string with "a"s and "b"s where w_n is the concatenation of w_{n-2} and w_{n-1} for n is odd and w_{n-1} and w_{n-2} for n is even. Starting with w_1="a" and w_2="b" we find: "a", "b", "ab", "abb", "ababb", "ababbabb", etc. The number of the elements in each generation being equal to the Fibonacci number of that generation. The actual superlattice consisted of 233 layers, arranged according to this sequence, corresponding to generation w_{13}.

We want to investigate the level structure when a magnetic field is applied parallel and perpendicular to the layers. The parallel field energy spectrum of Fibonacci SL with very similar parameters has been studied theoretically before [1]. When z is the SL growth direction, V(z) describes the consecutive potential of the barriers and the wells, and is non-periodic for the Fibonacci SL. For a magnetic field along the growth axis, (carriers orbiting in the layers) the Hamiltonian for the in-plane motion is given by:

$$H_{x,y}=-\frac{\hbar^2}{2m}\frac{d^2}{dx^2}+\frac{e^2B^2x'^2}{2m}]e^{ik_y y}\Phi(x)=E_{x,y}e^{ik_y y}\Phi(x) \qquad (1a)$$

with $x=x'+\hbar k_y/eB$, and $E_{x,y}=E_N=(N+1/2)\hbar\omega_c$, with ω_c the cyclotron frequency, for all orbit center coordinates, $\hbar k_y/eB$. For the z-motion:

Figure 1

Eigenvalue spectrum for the Fibonacci SL as described in the text (bars) and the density of states (solid line). For comparison the density of states for a corresponding periodic SL is also shown.

$$H_z = \left[-\frac{\hbar^2}{2m}\frac{d^2}{dz^2}+V(z)\right]\Psi(z)=E_i\Psi(z) \qquad (1b)$$

with E_i a series of eigenvalues, which is the solution of 1b. The full eigenvalue spectrum is therefore the sum of $E_i + E_N$. and consists of a set of regular Landau levels associated with each of the eigenvalues E_i. For a periodic SL V(z) is periodic and the solutions of 1b can be found with the Kronig-Penney model and E_i becomes $E(k_z)$, with k_z the SL wavevector.

In Fig. 1 we show the eigenvalue spectrum of 1b for the Fibonacci sample described before, and the corresponding density of states at zero magnetic field (for which $E_{x,y}$ reduces to $k^2{}_x$ and $k^2{}_y$) and compare this with the density of states (DOS) of a periodic sample with comparable layer thicknesses. It can be seen that the DOS of the periodic and the non-periodic sample are in fact rather similar, which shows that the non-periodic sample can be seen as a weakly disordered superlattice. The effect of the disorder is in fact the appearance of gaps in the eigenvalue spectrum of E_i (the bars in the lower part of the figure).

For a magnetic field parallel to the layers we cannot separate the z-motion from the x,y motion as done in eq. 1. Instead, the Hamiltonian is given by:

$$\left[-\frac{\hbar^2}{2m}\frac{d^2}{dz'^2} + \frac{e^2B^2z'^2}{2m} + V(z)\right]e^{ik_y y}e^{ik_x x}\Psi(z)=$$

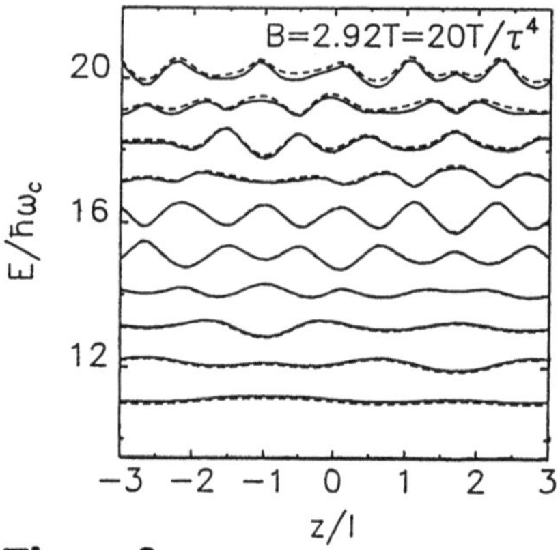

Figure 2

Magnetic levels for the Fibonacci SL described in the text as a function of the position of the cyclotron orbit center for a field of 20T (solid lines) and of 2.92T (dashed lines). In units of the cyclotron energy and magnetic length the results are the same.

$$\left(E+\frac{\hbar^2 k^2_x}{2m}\right)e^{ik_y y}e^{ik_x x}\Psi(z) \qquad (2)$$

with $z'=z+\frac{\hbar k_y}{eB}$.

I.e. plane waves in the x,y direction but an eigenvalue spectrum in the z-direction which depends on $\hbar k_{y/eB}$, the position of the cyclotron orbit center with respect to the non-periodic potential. Furthermore, since the magnetic length varies with the magnetic field for a fixed potential V(z), the resulting DOS becomes strongly field dependent. In Fig. 2 this center orbit dispersion is shown for the actual superlattice for k_x=0 and B=20T. It can be seen the solutions of eq.2 resemble that of a Landau level ladder with more or less equidistant levels, but broadened as a consequence of the non-translational invariant potential. Nevertheless since the SL is only weakly disordered, the broadening is always less than the Landau level separation.

In Fig. 3 and 4 we show measured interband excitation spectra for the FSL with a field parallel and perpendicular to the layers. The anisotropy in the level structure as previously described can indeed be observed. For a field perpendicular to the layers no clear Landau level ladder is observed, whereas in a parallel field more or less equidistant and linearly field dependent transitions can be distinguished. However it is quite notable that the intensity of each of these transitions varies in an irregular manner. Some transitions which are pronounced at some fields are much more broadened at somewhat different field values.

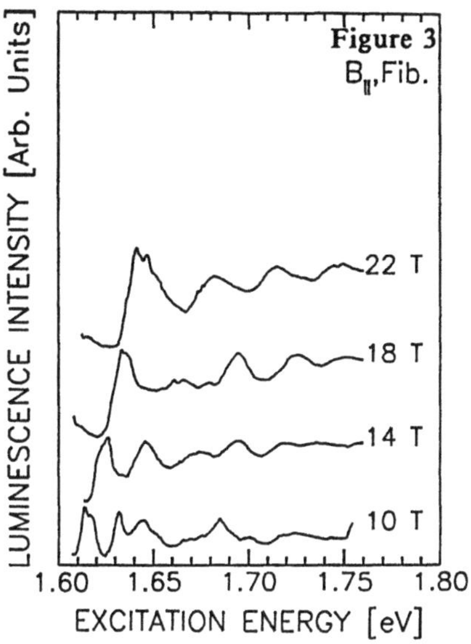

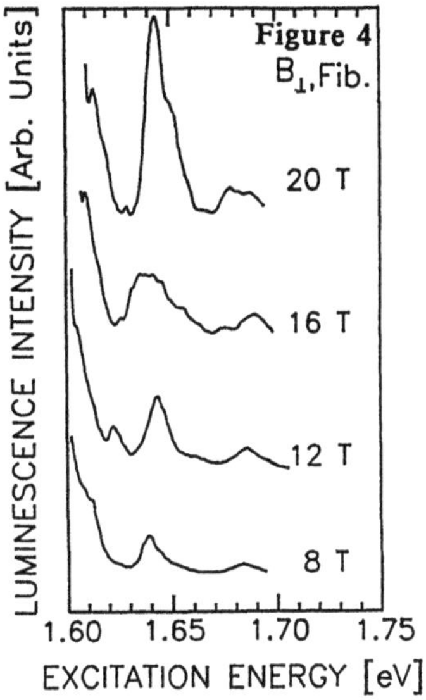

Figure 3

Excitation spectra of the Fibonacci SL for the field parallel to the layers at different magnetic fields.

Figure 4

As figure 3 with a field perpendicular to the layers.

This fact directly reflects the different scale on which the electron "sees" the non-periodicity as the magnetic length is changing with field.

A pertinent prediction of previous theoretical work [1] was that the density of states, D(E), when expressed in units of the cyclotron energy is self similar for magnetic field values related by τ^{2n} (n=..-2,1,0,1,2..). I.e. $D(E/\hbar\omega_c{}^a)=D(E/\hbar\omega_c{}^b)$ when $\omega_c{}^a/\omega_c{}^b$. We therefore scaled the experimental data at different fields by dividing the energy scale by the appropriate cyclotron energy at each field and compared the shape of the spectra by calculating the maximum of the cross-correlation function defined as:

$$\text{Max } p_{B_0,B}=\int(I_{B_0}(E)-\bar{I}_{B_0})(I_B(E-\Delta E)-\bar{I}_B)dE \qquad (3)$$

In this equation I_B is the luminescence intensity at (reduced) energy E for a spectrum at field B, I_{B_0} is the same at some particular value B_0. ΔE is a slight shift to align the spectra and which is adjusted to give a maximum. $\bar{I}$ represents the "average intensity of the spectrum. The maximum of this function thus obtained as a function of B using some

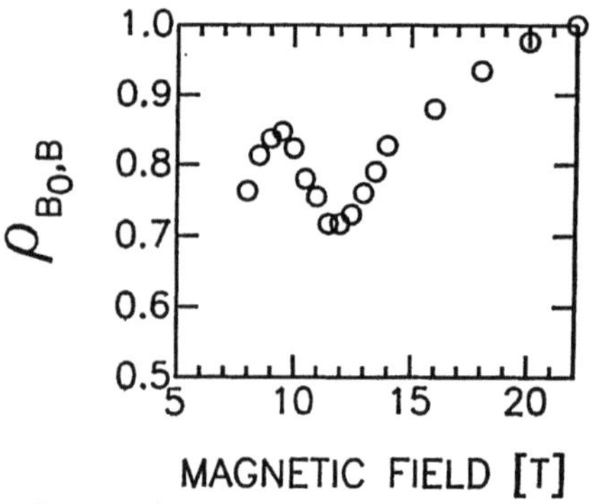

Figure 5

Cross-correlation between spectra (equation 3) at B_0=22T and at a field B as a function of B.

arbitrary value for the reference field B_0, 22T in this particular example, is shown in fig. 5. For B close to B_0 $\rho_{B_0,B}$ approaches unity, since the cross-correlation of a spectrum with itself is unity. As B decreases, ρ_{B,B_0} decreases and rises again and shows a clear maximum at around 9.5 T. This curve shows that the spectra at 9.5 and 22T resemble each other most, and are much more similar than those between 22T and 12T (the minimum in Fig. 5) for example. The theoretical predicted value for the ratio of the field values for which the spectra are most similar is τ^2=2.62, whereas experimentally 2.3 is found. Many possible explanations can be invoked to explain such differences because the detailed shape of the spectra depends on many effects neglected in this simple analysis. For instance only transition at k_x=0 are considered, the hole levels are neglected, excitonic effects are not accounted for etc. Nevertheless it is remarkable that indeed similar spectra are found at field values related to each other by a value which is very close to the theoretical one.

Self-similar properties in artificial structures with in-built geometrical self-similarity, like our SL have been studied before in other systems [4-9]. Experimentally, for instance spectra of acoustic phonons in Fibonacci SL or X-ray scattering [7,8] have shown fractal properties. I.e. the finer details of the spectrum are small replicas of the entire spectrum. Similar effects have been observed of the critical current versus critical field curves in superconducting fractal networks [9]. However the self similarity observed here is fundamentally different because the shape of the entire spectrum itself shows the self-similar property at particular values of an external parameter, the field. This is quite different from the observation of a spectrum, in a spectrum in a spectrum etc. The other experiments reflect in some sense a sort of interference between a length (X-ray or phonon wavelength, magnetic length) and the geometry of the structure, seen at a certain scale. Therefore the position of each "interference maximum" depends on the particular geometry of the structure. For the self similarity of the magnetic levels this is not the case because the particular parameters of the structure do not determine any particular resonance, and -in principle- for any given value of B_0 one can find other field values at $\tau^{2n}B_0$ with a self similar density of states. In other words the value of B_0 has no relation to geometry of the structure.

3. Spin conservation of photocreated carriers at high magnetic fields.

In quantum wells with a magnetic field applied perpendicular to the layers, both electron and valence subbands are split in discrete (i.e. separated by gaps with zero density of states) levels. For the electrons the quantised orbital motion leads to more or less equidistant linear field dependent levels, each of this levels split into two component due the quantisation of the magnetic moment (Zeeman effect). For the hole levels such a separate quantisation of orbital motion and of the magnetic moment is not possible because of the spin-orbit coupled nature of those states. Furthermore due to band-mixing effects, the levels are not equidistant nor linear field dependent. Selection rules allow transitions between some particular hole levels and electron Landau levels with a well defined spin orientation, and different transition are allowed for different polarizations of the helicity (σ^+ and σ^-) of the exciting light for light propagation along the field direction (Faraday configuration). As in any optical experiment on undoped semiconductor samples, the excitation of an electron from the full valence, is associated with the creation of a positively charged hole. Therefore the lowest excited state of the system is not the free hole and the free electron, but rather the electron bound to its own hole by Coulomb interaction; a state called an exciton. This Coulomb interaction makes that interband transitions can in principle not be described in terms of pure band states [10-12]. This fact is equally true in a magnetic field. However it is usually assumed that at sufficiently high magnetic fields where the magnetic energies exceed substantially the Coulomb correction, the transitions can still be labeled and identified from the free interband transitions. Indeed, a comparison between theory an experiment [12,13] has shown that this is true for GaAs QW for fields larger than ~10T. It is therefore possible to draw a transition energy scheme (fig.6) for the various possible interband transitions allowed in both polarizations of the light.

We assume that the actual energies will be lowered by excitonic interactions but that the state (i.e. the character of its wavefunction) can still be characterized by some corresponding hole level (indicated in the figure by h, which is generic for any hole state) and electron

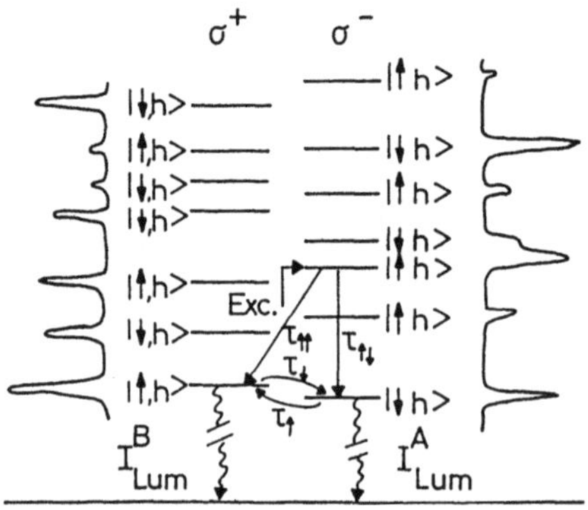

Figure 6

Schematic representation of possible excitonic transition in a quantum well in a perpendicular magnetic field. h designates an arbitrary hole state and ↑ or ↑ the electron magnetic moment. The definition of the relevant relaxation rates as used in the text are defined.

Landau level, the latter having a Landau level index and a well defined orientation of the magnetic moment (indicated with arrows). This level scheme would be a series of delta functions separated by gaps for perfect samples and would look very much like an atomic spectrum. Excitation in a particular state with light of the appropriate energy can take place as indicated in the figure. Carriers in this excited state may relax down the ladder and end up in one of the two lowest states which have respectively a spin-up and a spin-down character for the CB component of the wavefunction. Finally the electron can recombine with the hole and the system returns to the ground state.

In a conventional luminescence experiment one sees at low temperatures and under low excitation only one luminescence peak corresponding to the lowest possible energy of the system before recombination. The reason for this is that usually the relaxation lifetime between the excited states is much longer than the radiative recombination lifetime. At high magnetic fields in quantum wells we observed a clear luminescence from the higher lying state B (see fig.6) although $\Delta E_{AB} >> kT$ and this observation implies that the recombination from B to the ground state is faster that the relaxation from B to A. Excitation spectroscopy experiments can then be performed on each of the luminescence peak A and B and the results are shown in Fig.7. It can be seen that, although the spectral position of each peak in the excitation spectra in a given polarization is independent whether peak A or B is observed, some excited states are more intense when observing A and some others when observing B. From a comparison with the calculated energy levels we have concluded that the common property which governs the preferential relaxation is the conduction band spin [2]. I.e

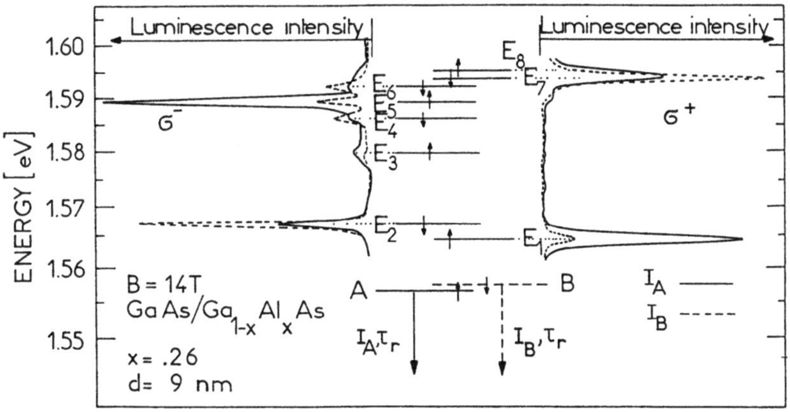

Figure 7

Luminescence intensity as a function of the energy of the σ^- (left) and σ^+ (right) polarized exciting light in a field of 14T. A and B designate the energy position of the two components of the heavy hole spin split ground state. The luminescence of A (B) is σ^- (σ^+) polarized. The solid lines show the excitation spectra taken by measuring the intensity of A , and the dashed lines by measuring the intensity of B. The arrows label the orientation of the conduction band magnetic moment involved in the transition as extracted from the comparison between theory and experiments [2].

carriers in excitonic states with a mainly CB spin-up character relax down to states having the same CB spin character. Such effects are absent at low fields and in samples with lower quality (broader lines) and our conclusion was that this effect is a consequence of a bottleneck for the electron spin relaxation when states become completely discrete.

In order to obtain more experimental information about the spin relaxation mechanism we have studied the above described effect as a function of the lattice temperature. Fig.8 shows the measured ratio of the luminescence intensity of peak B with respect to peak A when exciting in a state which preferentially relaxes to A (transition E_1 in Fig.7, crosses) or to B (transition E_2, circles) as a function of the temperature. The dashed line is $\exp[-\Delta E_{AB}/kT)$ with ΔE_{AB} the splitting between A and B. If all carriers would thermalise to state A before recombination (no bottlenecks nor preferential relaxation) this would be the expected temperature dependence of I_B/I_A. For excited states with preferential relaxation to A this is indeed more or less also the measured dependence. For excitation in a state which relaxes to B, strong deviations from the thermalised behaviour occur, which is just the effect described before. In fact since I_B/I_A is larger than one at

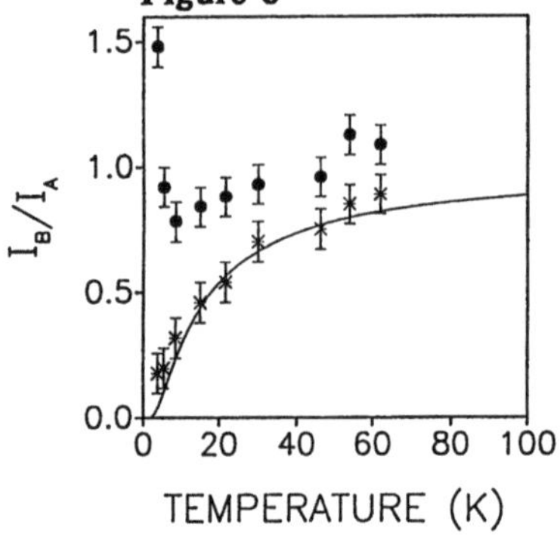

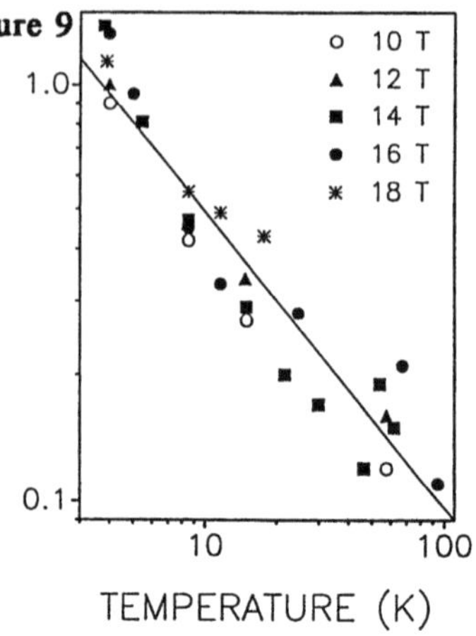

Figure 8

Ratio of the luminescence intensities from peak B and A, for excitation in a state with the same electron spin as A (E_1 in Fig. 7, crosses) and in a state with the same spin as B (E_2, circles) as a function of temperature.

Figure 9

Temperature dependence the ratio of the spin flip time between A and B (τ_s) and the recombination lifetime (τ_r). The drawn line is a $T^{-3/4}$ power dependence.

low temperature the preferential relaxation is sufficiently large to create a population inversion between A and B. With increasing temperatures the two experimental results approach each other, i.e. the tendency for preferential relaxation disappears. This disappearance of the effect is a direct consequence of the temperature dependence of the relaxation mechanism itself and is not just due to thermal occupation.

In order to obtain information about the relevant times we can calculate with a rate equation model I_B/I_A for excitation in a particular state:

$$I_B/I_A = \frac{\frac{\tau_\downarrow}{\tau_r} + \left(1 + \frac{\tau_{\uparrow\uparrow}^{i}}{\tau_{\uparrow\downarrow}}\right)\frac{\tau_\downarrow}{\tau_\uparrow}}{1 + \frac{\tau_{\uparrow\uparrow}^{i}}{\tau_{\uparrow\downarrow}}\left(1 + \frac{\tau_\downarrow}{\tau_r}\right)} \tag{4}$$

For the definition of the various times see Fig. 6. In equation (4) $\tau_\downarrow/\tau_r$ is the ratio between the relaxation time from B to A, $\tau_\downarrow$, with respect to the radiative recombination time τ_r from B or A. This ratio does not depend on which state the carriers have been originally excited. $\tau_{\uparrow\uparrow}/\tau_{\uparrow\downarrow}$ is the ratio between the relaxation times times from a particular excited state (i) to one of the two lowest states B or A while conserving spin, with respect to that while flipping the spin (see also fig.6). This latter ratio depends on the nature of the excited state. $\tau_{\downarrow}/\tau_{\uparrow}$ describes the ratio between up and down processes between A and B and is responsible for the tendency of thermalisation between A and B. If τ_r is much longer than $\tau_{\downarrow}$, the photocreated carriers will stay long enough in A and B to establish a quasi-thermal equilibrium. In this case the ratio I_B/I_A will be the same as the ratio of the occupation of A and B and will be given by $\exp(-\Delta E_{AB}/kT)$. It is reasonable to assume that the ratio of the thermalisation times $\tau_{\downarrow}/\tau_{\uparrow}$ is therefore given by $\exp(-\Delta E_{AB}/kT)$ also when τ_r is shorter. In this case the occupancy of A and B will of course not be given $\exp(-\Delta E_{AB}/kT)$ even when the ratio of the thermalisation times can be described that way. With this assumption Eq. (4) shows that the observation of preferential relaxation depends on two ratio's of relaxation and recombination times and from a single experiment these numbers can therefore not be determined. However, exploiting the fact that $\tau_{\downarrow}/\tau_r$ is the same whatever state is excited, it is possible to determine lower bounds for this ratio.

In fig.9 we show the temperature dependence of τ_s/τ_r where $\tau_s^{-1}=\tau_{\downarrow}^{-1}+\tau_{\uparrow}^{-1}$, i.e. the ratio of the spin relaxation times between A and B with respect to the recombination lifetime. It can be seen that this ratio is independent of the magnetic field and has a temperature dependence proportional to $T^{-3/4}$. A similar strong temperature dependence is also found for $\tau_{\uparrow\downarrow}{}^{i}/\tau_{\uparrow\uparrow}$, i.e. the probability for spin flip to that of spin conservation. Since this latter ratio is a property of the excited state it shows experimentally a magnetic field dependence which is probably related to the relative positions of intermediate states which may be invoked in the relaxation.

Both τ_s/τ_r and $\tau_{\uparrow\downarrow}/\tau_{\uparrow\uparrow}$ describe the efficiency of energy relaxation with a simultaneous spin flip with respect to that while conserving the same magnetic moment. It is reasonable to assume that the strong temperature dependence of these ratio's comes mainly from the temperature dependence of the spin flip mechanism and therefore the decrease of these numbers with temperature shows that the spin flip becomes more and more efficient. In bulk semiconductors spin relaxation mechanisms have experimentally been extensively studied; often using using similar optical pumping techniques as the one described here [14]. Theoretically four main mechanisms have been identified; k-linear terms in the conduction band [15], non-pure character of conduction band states due to spin orbit interaction [16], hyperfine interaction with magnetic nuclei of the host lattice [17] and exchange interaction between electrons and holes [18]. For bulk p-type semiconductors at zero field the temperature dependence of these mechanisms have been calculated by Fishman and Lampel [19] and are given by T^{-3},T^{-2},$T^{-1/2}$ and $T^{-1/2}$ respectively. The latter two dependences are most similar to the one we measured here. The calculated mechanism for hyperfine interactions seems to be orders of magnitude less efficient than that of exchange [19]. We therefore think that this latter mechanism could also be the most important one here. However since the interest in spin relaxation phenomena in two dimensional systems is relatively recent [2,20-22], there does not yet exist a theory which is directly applicable to our data, and any conclusion about possible mechanisms is rather speculative at this moment.

REFERENCES

1. Y.Y.Wang and J.C.Maan, Phys. Rev. **B40**, 1955, (1989)
2. M.Potemski, J.C.Maan, A.Fasolino, K.Ploog and G.Weimann, Phys. Rev. Lett. **63**, 2409, (1989)
3. M.R.Schroeder, "Number Theory in Science and Communication", Springer Series in Information Sciences, Vol. 7, Springer Verlag, Berlin, (1986)
4. M. Kohmoto, L. P. Kadanoff and C. Tang, Phys. Rev. Lett. **50**, 1870 (1983)
5. M. Kohmoto, B. Sutherland and C. Tang, Phys. Rev. **B35**, 1020 (1987)
6. J. M. Luck and Th. M. Nieuwenhuizen, Europ. Lett. 2, 257 (1986) Th.M. Nieuwenhuizen, J. Phys. (Paris) **47**, C3-211 (1986)
7. R. Merlin, K. Bajema and R. Clarke, Phys. Rev. Lett. **55**, 1768 (1985)
8. J. Todd, R. Merlin and R. Clarke, Phys. Rev. Lett. **57**, 1157 (1986) and R. Merlin, K. Bajema, J. Nagle and K. Ploog, J.Physique, **48 C5**, 503
9. J.M.Ghez, Y.Y.Wang, R.R.Rammal, B.Pannetier, and J.Bellissard, Solid State Commun. **64**, 1291, (1987) and B.Pannetier, O. Buisson, P.Gandit, Y.Y.Wang, J.Chuassy and R.Rammal, Surface Sci. **229**, 331, (1990)
10. G.E.W.Bauer and T. Ando, Phys. Rev. **B37**, 3130, (1988).
11. S-R.E.Yang and L.J.Sham, Phys. Rev. Letters **58**, 2598 (1987).
12. F.Ancilotto, A.Fasolino, and J.C.Maan, Phys. Rev. **B38**, 1788 (1988)
13. L.Vina, G.E.W.Bauer, M.Potemski, J.C.Maan, E.E.Mendez and W.I.Wang, Surface Sci. **229**, 504, (1990).
14. Optical Orientation, Modern Problems in Condensed Matter Sciences, Vol. **8**, ed. F.Meier and B.P.Zakharchenya (North-Holland, Amsterdam, 1984).

15. M.I.D'yakonov and V.I.Perel, Zh. Eksp. Teor. Fiz. **60**, 1954 (1971); [Sov. Phys. Solid State **13**, 3023 (1972)].
16. R.J.Elliot, Phys. Rev. **B96**, 266 (1954); Y.Yafet in *Solid State Physics*, F.Seitz and D.Turnbull eds., Academic, New York 1963, Vol.14, p.1
17. A.W.Overhauser, Phys. Rev. **89**, 689, (1953)
18. G.L.Bir, A.G.Aronov, G.E.Pikus, Zh. Eksp. Teor. Fiz. **69**, 1382 (1975), [Sov. Phys, JETP **42**, 705 (1976)].
19. G.Fishman, G.Lampel, Phys. Rev. **B16**, 820 (1977).
20. A.Berg, M.Dobers, R.R.Gerhardts and K.v.Klitzing, Phys. Rev. Lett, **64**, 2563 (1990).
21. M.R.Freeman, D.D.Awschalom, J.M.Hong and L.L.Chang, Phys. Rev. Lett. **64**,2430, (1990)
22. T.Uenoyama and L.J.Sham, Phys. Rev. Lett. **64**, 3070, (1990)

Anomalous Subband Landau Level Coupling in GaAs-AlGaAs Heterojunctions

G.C. Wiggins[1], *D.R. Leadley*[1], *D.J. Barnes*[1], *R.J. Nicholas*[1], *M.A. Hopkins*[1], *J.J. Harris*[2], *and C.T. Foxon*[2]

[1]Clarendon Laboratory, University of Oxford, Parks Road, Oxford OX1 3PU, UK

[2]Philips Research Laboratories, Redhill, Surrey RH1 5HA, UK

Abstract Electric subband energy differences are measured in GaAs–GaAlAs heterojunctions by tilted field cyclotron resonance and magnetotransport as a function of carrier density and temperature. At small tilt angles we see unusual resonant subband Landau level coupling, with an overlap of the branches of the coupling rather than the expected anticrossing.

1. Introduction

The energies of confined electric subbands in heterojunctions are determined by the shape of the confining potential, which depends strongly on the depletion charge density in the GaAs and the carrier density (n_s) in the 2DEG [1,2]. Once the upper (E_1) level is populated the difference in subband energies, E_{01} can be measured from the periodicities of the Shubnikov–de Haas oscillations [3], but at lower n_s optical techniques must be used [4,5].

We have used cyclotron resonance (CR) to measure E_{01} from the resonant subband Landau level coupling (RSLC) in tilted fields, but find an anomalous behaviour which will be discussed first. The experiments were performed at 4.2K using a Fourier transform spectrometer. The samples used were conventional GaAs–GaAlAs heterojunctions with n_s between 0.85 and 11.4 $\times 10^{11}$cm^{-2}, grown at Philips Research Laboratories. G635 was slightly different with a superlatice buffer layer grown to trap impurities migrating from the substrate.

2. Anomalous Resonant Subband Landau Level Coupling

In ultra–high mobility heterojunctions RSLC exhibits totally unexpected behaviour at small tilt angles. RSLC arises from hybridization of the Landau levels and electric subbands when the magnetic field is not perpendicular to the 2DEG, usually giving rise to a split resonance, with two branches which *anticross*. Fig. 1a shows typical anticrossing behaviour, obtained at 25° tilt in G635 (μ = 3.4$\times 10^6$cm^{-2}/Vs), where E_{01} is 122 cm^{-1}. A second coupling at E_{02} is also evident. There is in fact a gap between the pinning energies of the two branches, as predicted by Merlin [6] and confirmed qualitatively by Huant *et al.* [7]. When the sample is illuminated with a red LED (Fig. 1b) the coupling energy drops to 87 cm^{-1}, the maximum splitting of the resonance is reduced and the gap between the two branches has essentially disappeared. The resonance at higher fields cascades across several couplings and approaches the 3D cyclotron energy.

At very small tilt angles (<5°), however, the anticrossing behaviour breaks down and the two branches of the coupling cross over in energy substantially. This can be clearly seen in Figure 2 where the coupling regime is studied in more detail. At intermediate angles illumination can also cause such an overlap. Typical CR traces are shown in the inset in Fig. 2. Even greater overlap was observed in G580, with μ = 2$\times 10^6$cm^2/Vs, at 3° tilt.

The low coupling energy prior to illumination demonstrates that the depletion charge is extremely low (N_{dep} ~5$\times 10^{10}$cm^{-2}) [2], and illumination reduces it even

Springer Series in Solid-State Sciences, Vol. 101
High Magnetic Fields in Semiconductor Physics III Editor: G. Landwehr

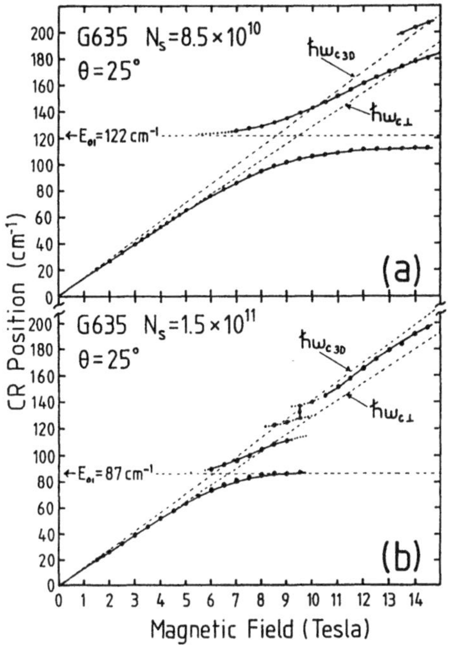

Fig. 1 CR position as a function of magnetic field at large tilt angle (a) in dark and (b) after illumination.

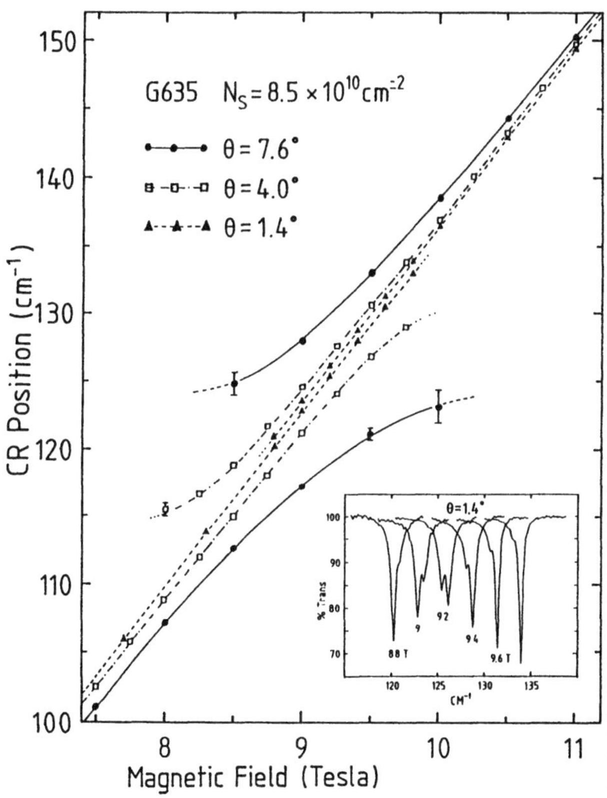

Fig. 2 CR position at small angle showing breakdown of anticrossing behaviour. Insert – typical data.

further to ~$1.5\times10^{10}cm^{-2}$. It is also clear from the reduced splitting magnitude in Fig. 1b that illumination, as well as reducing the coupling energy, significantly reduces the matrix element governing the strength of the coupling, z_{01}. These two facts suggest that the depletion charge is so low that the electrons in the 2DEG contribute substantially to the slope of the potential near the interface. Outside the 2DEG the low depletion charge results in an extremely small slope. This means that the second electric subband state is *very* large in z–extent compared to the lowest subband. The small spatial overlap results in a small z_{01}, and the small slope outside the 2DEG explains the small subband separation.

The large spatial difference between the two lowest subbands may provide an explanation of the overlapping subband coupling. It is generally accepted that RSLC does not occur at the true intersubband separation, but rather at an energy which is *depolarization shifted* [8]. In fact there are two effects [9]: a depolarization effect – tending to increase the effective coupling energy; and an exciton–like effect which tends to decrease it (though this term is very small in GaAs [10]). In first order perturbation theory, the depolarization effect depends on the overlap integral between the first and second subbands [11]. As the coupling is approached, the upper subband is strongly hybridized, and assumes a greater and greater first subband character, reducing the z–extent of the hybrid wavefunction and so increasing the spacial overlap of the wavefunctions.

If the depolarization effect were considered in second order perturbation, where the effect of hybridization on the wavefunctions is taken into account, one would expect a depolarization shift which depends on the degree of hybridization. This should lead to a *renormalization* of the depolarization shift: increasing hybridization raises the effective coupling energy higher and higher and smears out the coupling. At high tilt angles this effect is not so clearly seen because it simply reduces the size of the gap between pinning energies of the branches.

3. Measured Subband Energies

Transport measurements were performed, with the field perpendicular to the 2DEG, at 500mK for samples with narrow spacer layers [12]. n_s was varied between 4.5 and 11.7×10^{11} cm^{-2} by photoexcitation. With the upper subband occupied two series of Shubnikov– de Haas oscillations were observed from whose periodicities the carrier concentration in each subband (n_i) were deduced and E_{01} calculated. The occupation of the upper subband was typically less than 10% of that in the lower subband and could be detected for $n_1 > 2\times10^{10}$ cm^{-2}. The resulting values of E_{01} are shown in Fig. 3, together with the calculations of Stern & Das Sarma as solid lines. The dashed line represents the Fermi energy calculated for all the electrons in one subband, which divides the regions where we can use transport and optical measurements. Also shown in Fig. 3 are RSLC results, not corrected for depolarisation which may reduce the coupling energy by ~20%, for several wider spacer layer samples.

These show a larger subband splitting in the dark, where the depletion charge is ~$3.5\times10^{10}cm^{-2}$, reducing to ~$1.5\times10^{10}cm^{-2}$ on illumination. Similar measurements by Rikken *et al.* [5] who found an increase of E_{01} with n_s when the carrier concentration was varied by persistent photoconductivity, but the opposite behaviour when n_s was changed with a back–gate voltage, demonstrating that E_{01} is sensitive to the exact shape of the confining potential. Fig. 3 shows that in G580 the GaAs is exceptionally pure with N_{dep} ~$1.5\times10^{10}cm^{-2}$ in the dark. G635 would be expected to have a similar background impurity level to G580 (as the mobilities are similar and an order of magnitude larger than the other 800Å spacer sample G63), but pinning of E_F at the superlattice 0.5μm from the 2DEG increases the slope of the potential and so increases E_{01}.

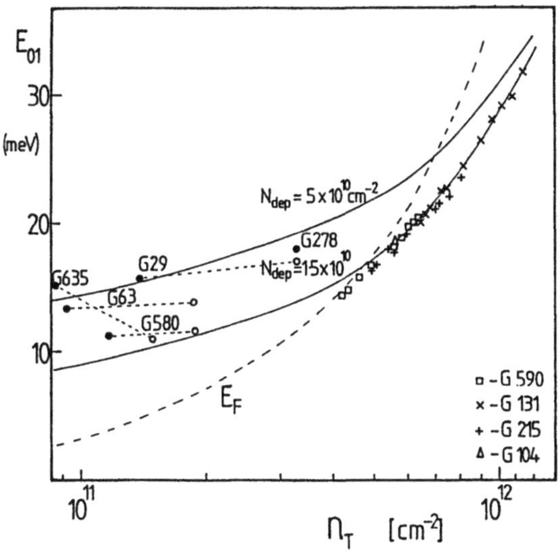

Fig. 3
Subband separation, as a function of n_s: measured by RSLC before (•) and after illumination (o); from SdH (other symbols); and calculated [ref 2] (lines).

4. Temperature Dependence of Subband Energies

We have measured the change in E_{01} with temperature by two methods:– in transport by parallel field depopulation; and optically by RSLC.

Figure 4 shows the effective mass deduced from CR at 2.2K and 80K. By comparing data with the sample flat and tilted (by ~11° so as to avoid the anomalous RSLC effect) the position of RSLC may be found. The additional features in the high temperature traces are due to the magnetophonon effect [13]. The subband coupling can be seen to move from 146.7cm^{-1} at 2.2K to 172.4cm^{-1} by 80K, ie. the energy increases with temperature at a rate of 0.33$cm^{-1}K^{-1}$.

Shubnikov-de Haas oscillations are not resolved above ~6K so we have used the diamagnetic subband energy shift in a magnetic field parallel to the 2DEG to depopulate the upper subband [3,12] and infer n_1 between 0.5K and 80K. To fit the change of n_1 with temperature requires E_{01} to increase at a rate of 0.037 $meVK^{-1}$ (ie. 0.30$cm^{-1}K^{-1}$ which agrees well with the optical result).

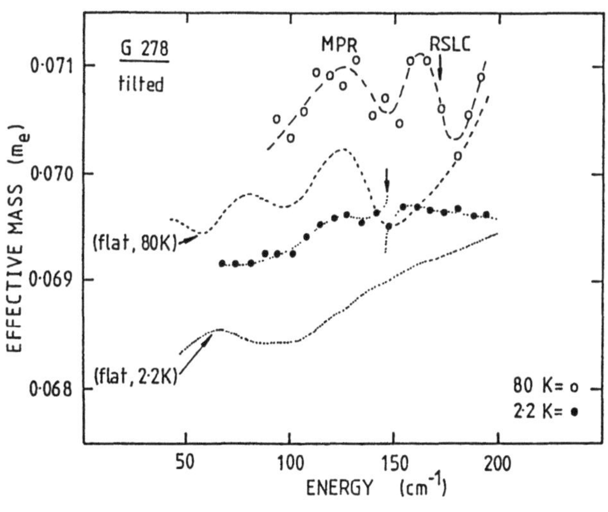

Fig. 4 Effective mass as a function of energy from CR at 2.2K and 80K. Tilted field data (circles) shows RSLC at arrows. MPR oscillations also seen in 80K data.

5. Conclusion

Strange resonant subband Landau level coupling behaviour is found in high mobility heterojunctions, violating the usual anticrossing picture. This suggests the effective intersubband energy is shifting as coupling is approached, and may be attributed to renormalization of the depolarisation shift. Existing perturbation theories fail to explain this anomalous behaviour. From measurements of E_{01} we find extremely low depletion charge densities in the best heterojunctions. Finally we measure the of increase of E_{01} with temperature.

References

The reader may find the conversion $100cm^{-1} \equiv 12.4meV$ useful!

1. T. Ando; *J.Phys.Soc.Japan* **51**, 3893 (1982)
2. S. Das Sarma and F. Stern; *Phys.Rev.B* **30**, 840 (1984).
3. J.J. Harris, D.E. Lacklison, C.T. Foxon, F.M. Selton, A.M. Suckling, R.J. Nicholas and K.W.J. Barnham; *Semicond.Sci.Technol.* **2**, 783 (1987).
4. G. Abstreiter and K. Ploog; *Phys. Rev. Lett.* **42**, 1308 (1979).
5. Rikken G.L.J.A., H. Sigg, C.J.G.M. Langerak, H.W. Myron, P.A.A.J. Perenboom and G. Weimann; *Phys. Rev. B* **34**, 5590 (1986).
6. R. Merlin; *Solid State Commun.* **64**, 99 (1987).
7. S.Huant, M.Grynberg and B.Etienne; *Solid State Commun.* **65**, 457 (1988).
8. M. Zaluzny; *Solid State Commun.* **56**, 235 (1985).
9. T. Ando; *Solid State Commun.* **21**, 133 (1977)
10. A.D.Wieck, F.Thiele, U.Merkt, K.Ploog, G.Weimann & W.Schlapp; *Phys. Rev. B* **39**, 3785 (1989).
11. S.J. Allen, D.C. Tsui and B. Vinter; *Solid State Commun.* **20**, 425 (1976).
12. D.R. Leadley, R.J. Nicholas, J.J. Harris and C.T. Foxon; Submitted to *Semicond. Sci. Technol.* (1990).
13. D.J. Barnes *et al. Proc. Wurzburg III* (1990).

Cyclotron Resonance in the 2D Hole Gas of GaAs/AlGaAs Heterostructures

G. Kalinka[1], *F. Kuchar*[1], *R. Meisels*[1], *E. Bangert*[2], *W. Heuring*[2], *G. Weimann*[3], *and W. Schlapp*[4]

[1]Institut für Festkörperphysik, Universität Wien, A-1090 Vienna, Austria and L. Boltzmann Institut für Festkörperphysik, A-1060 Vienna, Austria
[2]Universität Würzburg, Würzburg, Fed. Rep. of Germany
[3]Walter Schottky Institut, TU München, Am Coulombwall, W-8046 Garching, Fed. Rep. of Germany
[4]FTZ, W-6100 Darmstadt, Fed. Rep. of Germany

Magneto-optical spectroscopy in the far infrared is a valuable tool to study the energetic structure close to the extrema of the energy bands. In heterostructures like GaAs/AlGaAs the valence band structure is very complex due to the strong electric field near the interface and the lack of inversion symmetry. As a consequence the light hole band is split off the heavy hole band which is highest in energy. Also the spin degeneneracy is lifted for $k \neq 0$ and there exists a strong mixing between states of different spin orientation. States mainly consisting of heavy hole states with spin-up are called a-states, whereas states mainly consisting of heavy hole states (hh) with spin-down are called b-states [1,2]. Transitions may occur between states of the same spin orientation ($na \rightarrow ma,\ n,m = 1,2,...,\ kb \rightarrow lb,\ k,l = -2,-1,0,1,...$). but also between states with different spin orientation (CSR - Combined Spin flip Resonances). For the assignment of the experimentally observed absorption peaks to transition energies the energetic position of the Landau states as a function of the magnetic field was calculated.

In our transmission experiments in Faraday geometry we used a Fouriertransform spectrometer in the range of $10 - 50\,cm^{-1}$ $(1.3 - 6.5\,meV)$ and small magnetic field intervals up to 11 T. The three samples studied were single heterostructures with p_s between 2 and $7 \times 10^{11}\,cm^{-2}$ and μ between 2 and $8.4 \times 10^4\,cm^2/Vs$ at 4.2 K. The measuring temperature was 1.5 K. Fig. 1 shows spectra at high magnetic fields of the sample with the lowest hole concentration . Only the lowest Landau level $(1a)$ is occupied and the transition with the highest probability is $(2a \rightarrow 1a)$. The transition energies in the whole magnetic field range from 2 to 11 T for the sample with $p_s = 4 \times 10^{11}\,cm^{-2}$ is shown in Fig. 2. Transitions with $\Delta N = n - m = 1$ have high transition probabilities and these transitions $(2a \rightarrow 1a, 3a \rightarrow 2a, 2b \rightarrow 1b, 1b \rightarrow 0b)$ showed a strong absorption ($\sim 10\%$). Due to the fourfold symmetry of the warping of the hh band the calculation predicts strong transitions also for $\Delta N = 4$ and

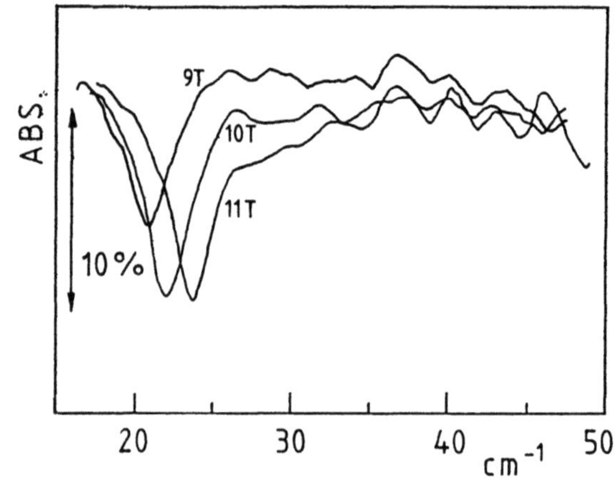

Fig. 1: Absorption spectra of a p-AlGaAs/GaAs Heterostructure with $p_s = 2.7 \times 10^{11}\,cm^{-2}$ at B=9, 10 and 11 T. The lines result from the transiton $(2a \rightarrow 1a)$

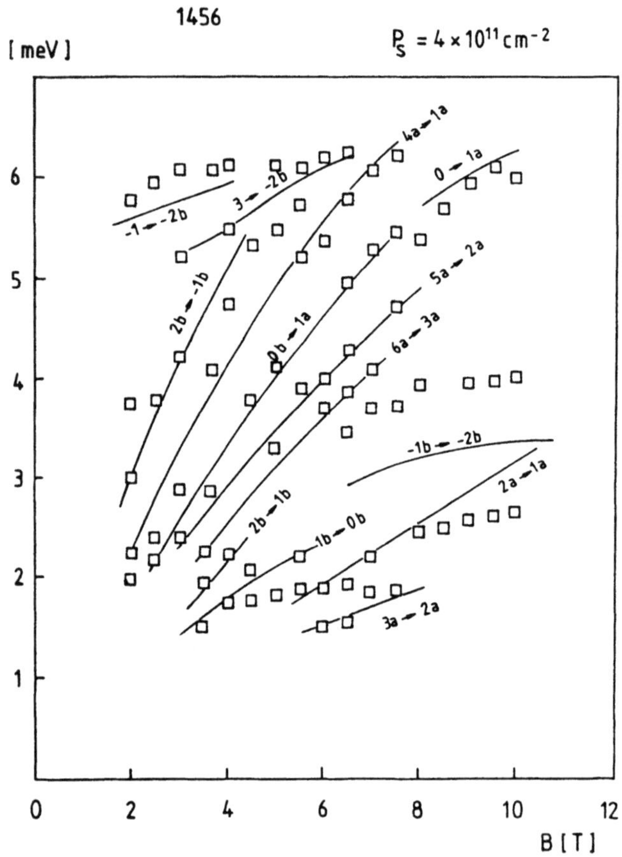

Fig. 2: Transition energies versus magnetic field, $p_s = 4 \times 10^{11}\,cm^{-2}$.

are experimentally found ($5a \to 2a$, $4a \to 1a$, $2b \to -1b$) predicted by the calculation to be the strongest one.

The magnetic field dependend mixing of the states is seen for the transition ($0b \to 1a$), a CSR from the $0b$ (spin-down) to the $1a$ (spin-up) state. The heavy hole state $0b$ changes its character in high magnetic fields containing predominantly contributions from the light hole state (0) (denoted only by numbers). At 4 meV and high magnetic field an absorption line is observed whose field dependence becomes flat with increasing magnetic field indicating the transition ($-1b \to -2b$). The reason for this behaviour is strong deviation from linearity of the ($-1b$) level. The calculated values for that transition are about 25% smaller than the observed data. Also for the transition ($2a \to 1a$) there exists a deviation from the calculated data at high magnetic fields. At low magnetic fields and high energies ($\sim 6\,meV$) we found absorption lines corresponding to transitions between heavy hole states to light hole states ($-1 \to -2b$, $3 \to -2b$). Although the calculated transition probabilities for these transitions are high the experimentally observed absorptions are very weak.

Fíg. 3 shows the data for the sample with $p_s = 7 \times 10^{11}\,cm^{-2}$. For that hole concentration the ($2a$) Landau level is occupied by holes in the available magnetic field range. Therefore the strongest absorption is the transition ($3a \to 2a$). The calculated values deviate slightly from the experimental data. The transition ($-1b \to -2b$) exhibits a deviation of 20%. A transition marked by the dotted line could not be identified. Due to the high hole concentration there are numerous CSR ($0b \to 1a$, $1b \to 2a$, $2b \to 3a$). For low magnetic fields and high energies the theory does not predict transitions from light hole states to heavy hole states but CSR with weak transition probability. Due to the vincinity of the calculated absorption lines in that region the experimentally observed transition cannot be identified.

The Fourier Transform Spectroscopy allowed to follow the absorption lines quasi-continously hardly possible in earlier laser experiments [3] and to compare the experimental data with the calculted transition energies. The observed transitions are: (a) "regular" cyclotron resonance, $\Delta N = 1$, e.g. ($2a \to 1a$) roughly corresponding to $\omega_c = eB/m_{hh}$, ($m_{hh} \sim 0.4\,m_0$), (b) $\Delta N = 4$ transitions caused by the fourfold anisotropy of the heavy hole band, (c) CSR transitions due to the mixing of the spin states and (d) $lh \to hh$ transitions. The energetic position, the slope of the lines and the transition probabilities are in good agreement with the experiment. Deviations for the lowest transitions ($-1b \to -2b, 3a \to 2a$, $2a \to 1a$) at high magnetic fields may arise because of the strong anisotropy and nonparabolicity existing in AlGaAs/GaAs heterostructures.

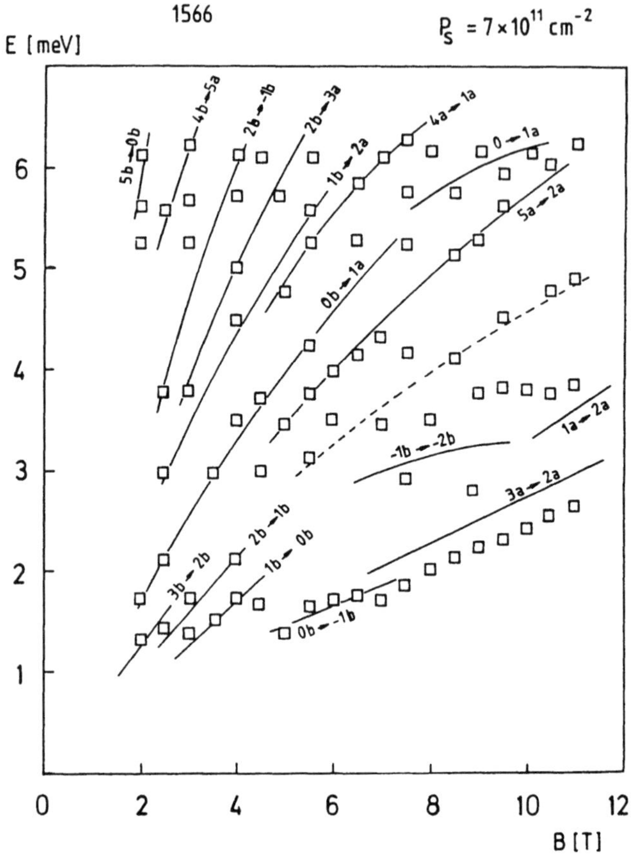

Fig. 3: Transition energies versus magnetic field, $p_s = 7 \times 10^{11}\ cm^{-2}$.

References:

[1] E. Bangert and G. Landwehr, Surf. Sci. **170**, 593 (1986) and references therein

[2] E. Bangert and G. Landwehr, Superlattices and Microstructures **1**, 363, (1985)

[3] W. Erhardt et al., Surf. Sci. **170**, 581, (1986)

Cyclotron Resonance Measurements in GaAs/AlGaAs Heterostructures in the Fractional Quantum Hall Range

M. Besson, H. Drexler, P. Graf, E. Gornik, G. Böhm, W. Ettmüller, and G. Weimann

Walter Schottky Institut, TU München, Am Coulombwall, W-8046 Garching, Fed. Rep. of Germany

Cyclotron resonance (CR) absorption in GaAs/AlGaAs heterostructures is studied in strong magnetic fields ($\leq$ 14T) as a function of temperature (0.4K $\leq$ T$\leq$ 12K) and charge density. At low temperatures a weak broadening of the cycloton resonance linewidth is observed at fractional filling factors. These structures disappear with increasing temperature.

1. Introduction

Cyclotron resonance (CR) investigations of the two-dimensional electron gas were first carried out as long as 15 years ago in Si-MOS inversion layers /1/. Most of the recent investigations are devoted to GaAs/AlGaAs heterostructures due to their high mobility and the appearance of the fractional QHE. Effects of band-nonparabolicity /2/, the 2D polaron /3/ and coupling to higher subbands (RSLC) /4/ can be studied. In this contribution we present our resent results of CR-investigations in GaAs/AlGaAs heterostructures in the integer and fractional quantum-Hall range.

2. Experimental details

The samples represented here are MBE grown, modulation doped GaAs/AlGaAs heterostructures with a free carrier density of $2.3 \cdot 10^{11}$ cm^{-2} (μ = $2.5 \cdot 10^5$ cm^2/Vs) and $1.55 \cdot 10^{11}$ cm^{-2} (μ = $1.4 \cdot 10^6$ cm^2/Vs), respectively.

The measurements are made by means of a Fourier transform spectrometer connected via a waveguide to a 17 T superconducting magnet-system. The transmitted radiation is detected with a silicon bolometer. The resolution of the spectrometer was set to 0.1 cm^{-1}. The samples are wedged at 2° in order to avoid line shape distortion of the transmission spectra due to interference effects. The free carrier density is determined by SdH measurements and can be varied by illumination with a red LED.

A Drude model is used to evaluate the cyclotron mass and the scattering rate $1/\tau$ of the transmission data /6/. In addition $1/\tau$ is determined by numerical calculations based on the self-consistent Born-approximation (SCBA) /7, 8/.

3. Experimental results

Typical cyclotron resonance transmission spectra by a large number of magnetic fields are shown in Fig 1. They clearly demonstrate the oscillation of the CR-amplitude and linewidth with filling factor, as the oscillation in linewidth results in an inverse dependence of the transmission minimum.

The evaluated scattering rate and the CR-mass are depicted in figure 2. The scattering rate $1/\tau$ shows a clear maximum for filling factor $\nu = 2$ and 4 , whereas a step is seen for $\nu = 1$. Due to a small tilt of the sample in the magnetic field $1/\tau$ increases again above 13 T due to resonant subband Landau-level coupling (RSLC). The corresponding CR-mass drops with decreasing magnetic field, then rises at $\nu = 2$ and has a maximum at about $\nu = 2.5$. A small additional structure in $1/\tau$ and the CR-mass can be seen around $\nu = 4/3$. In addition, results from a self-consistent calculation for $1/\tau$ are given in this figure. They show a good agreement with the experimental values at integer filling factors. The step at $\nu = 1$ cannot be reproduced since spin splitting is not included in this calculation.

In order to study the CR-linewidth at fractional values of filling factors in more detail we have investigated another sample with lower carrier density ($n_s = 1.55\cdot10^{11}$ cm^{-2}) and higher mobility ($\mu = 1.4\cdot10^6$ cm^2/Vs). In magnetotransport measurements carried out on this sample at 0.4K fractional effects at filling factors $\nu = 5/3$, 4/3, 2/3, 4/5 and 3/5 can be observed, thus confirming the high quality of the sample.

The evaluated CR-scattering rate for this sample is shown in Fig. 3a for 0.4 and 12K. At about 9T a RSLC- transition is observed. The scattering rate $1/\tau$ shows for filling factor $\nu = 2$ a pronounced maximum. For both temperatures $1/\tau$ shows a drastic increase at $\nu = 1$ and a sharp maximum at slightly higher filling factor. Below $\nu = 1$ a strong temperature dependence for the scattering rate is observed. For 0.4K,

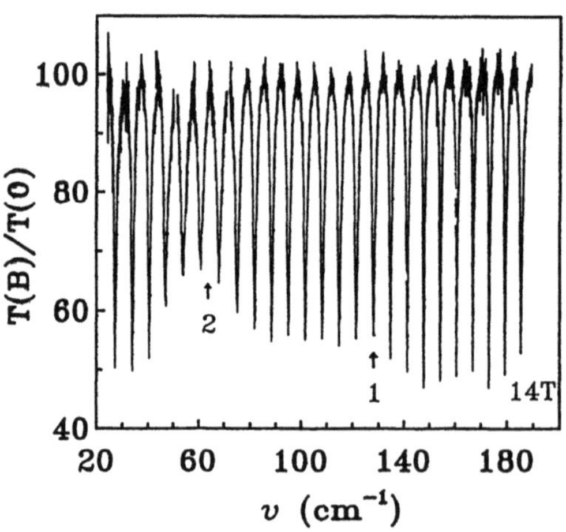

Fig 1: Normalised transmission spectra for various values of B (in steps of 0.5 T) at 1.2K for sample No. 1 ($n_s = 2.3\cdot10^{11}$ cm^{-2}, $\mu = 2.5\cdot10^5$cm^2/Vs) The arrows indicate filling factor 1 and 2.

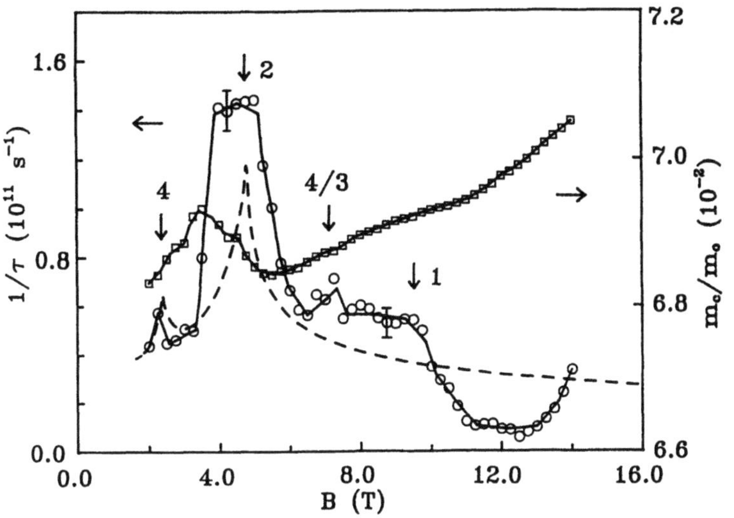

Fig 2: The evaluated CR-mass and scattering rate for sample No. 1 over magnetic field. The full lines are guides to the eye. The error bars in $1/\tau$ indicate the uncertainty of the analysis. The dashed line represents the result of the theoretical calculation (the background doping in the GaAs-layer was chosen to be $6\cdot10^{14}$ cm^{-3}).

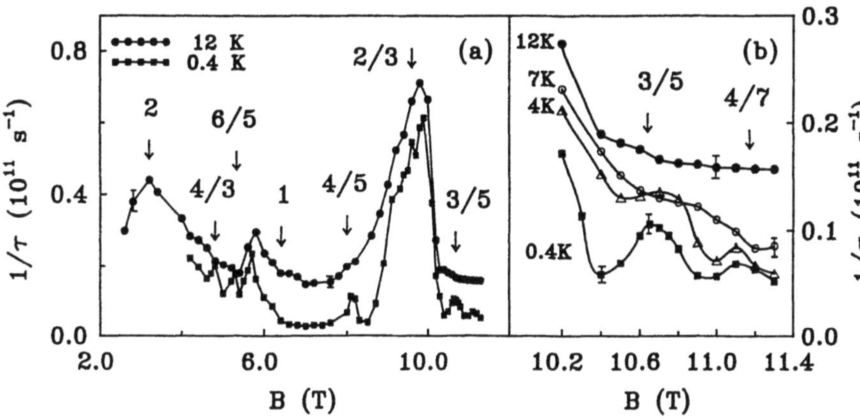

Fig. 3: (a) The evaluated scattering rate $1/\tau$ at 0.4 and 12K for sample No. 2 (n_s = $1.55\cdot10^{11}$ cm^{-2} and μ = $1.4\cdot10^{6} cm^2/Vs$). (b) the scattering rate $1/\tau$ at 0.4K, 4.0K, 7.0K and 12K in dependence of magnetic field for sample No. 2.

small but well developed maxima are found at fractional filling factors of 2/3, 4/3, 3/5, 4/5 and even 6/5 (indicated by the arrows). The filling factor υ = 2/3 has about the same maximum as the RSLC-transition and is therefore masked. To get more insight in these observed structures we have investigated fillingfactor 3/5 and 4/7 in more detail. This is shown in Fig. 3b. At the fractional state 3/5 a clear maximum in $1/\tau$ can be seen for 0.4 and 4K which disappears for higher temperatures. For ν = 4/7 the magnitude of the observed structure is of the order of the error bar.

From these observations one can propose that the occurrence of peaks in linewidth at fractional occupacy is due to oscillatory screening produced by the density of states gap formed on condensation into the ground state. The fact that fractions can only be seen up to about 1K in dc conductivity is not contradictory to the observation of a broadening in CR-linewidth for filling factor 3/5 at temperatures up to 4K. Although the fractional gap is much smaller than this experimental temperature a certain amount of the electrons will be condensed. In CR- investigations one is apparently able to observe fractions in locally condensed areas. For an observation in dc transport these fractions have to extend from contact to contact.

To confirm the consistency of the data the free carrier density of the sample was changed by illumination (not shown here). As expected the peak positions of the observable fractional filling factors shift with free carrier density.

4. Conclusion

In conclusion we have made CR-investigations in GaAs/AlGaAs heterostructures. In a low mobility sample maxima in CR-linewidth occur at filling factors 2 and 4, whereas a step can be seen at ν = 1. In a high mobility sample with low carrier density additional small maxima in linewidth are observed at fractional filling factors which disappear at higher temperatures.

5. Acknowledgement

We thank the Deutsche Forschungsgemeinschaft for sponsoring this work.

6. References

/1/ G. Abstreiter, P. Kneschaurek, J. P. Kotthaus, J. F. Koch, Phys. Rev. Lett. **32**, 104 (1974)

/2/ F. Thiele, U. Merkt, J. P. Kotthaus, G. Lommer, F. Malcher, U. Rössler, G. Weimann, Solid State Commun. **62**, 841 (1987)

/3/ W. Seidenbusch, Phys. Rev. B **36**, 1877 (1987)

/4/ Z. Schlesinger, J. C. M. Hwang, S. J. Allen Jr., Phys Rev Lett. **50**, 2098 (1983)

/5/ G. L. J. A. Rikken, H. Sigg, C. J. G. M. Landerak, H. W. Myron, J. A. A. J. Perenboom and G. Weimann, Phys Rev. B **34**, 5590 (1986)

/6/ K. W. Chiu, T. K. Lee, J. J. Quin, Surf. Sci. **58**, 182 (1976)
/7/ T. Ando, Y. Murayama, J. Phys. Soc. Jpn. **54**, 1519 (1985)
/8/ R. Lassnig, W. Seidenbusch, E. Gornik, W. Weimann: in "Landau levelwidth and cyclotron resonance in 2D Systems", Proceedings of the 18. International conference on the Physics of semiconductors, Vol. 1, o. Engstrom (ed.), Singapore, World Scientific 1987 , p 593.

Impurity Emission from GaAs/GaAlAs Heterostructures in the FIR

M. Witzany[1], *E. Gornik*[1], *W. Ettmüller*[1], *G. Böhm*[1], *G. Weimann*[1], *W. Knap*[2], *and J.L. Robert*[2]

[1]Walter Schottky Institut, TU München, Am Coulombwall, W-8046 Garching, Fed. Rep. of Germany

[2]Université des Sciences et Techniques du Languedoc, F-34360 Montpellier, France

We determined the binding energies of impurities in GaAs-AlGaAs heterostructures by means of a magnetically tunable GaAs detector at magnetic fields of up to 10T. We found strong oscillations with filling factor of the total impurity emission intensity. At the maxima we could analyze the spectra because screening is less effective.

1. Introduction

Shallow donors in confined semiconducting materials such as GaAs-GaAlAs-heterostructures are of great interest due to their effect on transport and optical properties of the material. Experimental and theoretical studies [1-8] have revealed a strong position dependence of the impurity binding energy. As suggested in [9] cyclotron resonance emission is a useful tool to measure the binding energy of impurities in a two-dimensional electron gas (2-DEG). Impurities near the 2-DEG introduce additional localized states below each Landau level. Transitions may also occur between these localized states. In heterostructures these transitions could only be seen in emission experiments [9-10] (impurity emission IE).

In this paper we report our studies of cyclotron emission by impurities in GaAs/GaAlAs heterostructures. The experiments were performed in a cryostat containing two magnets with a maximum field of 14 T/10 T. The emitted radiation is guided to a magnetically tunable high resolution GaAs detector. The detector is sensitive to FIR due to an impurity transition ($1s \rightarrow 2p^-$, $2p^0$, $2p^+$). In our case the detector magnetic field was kept constant and the sample magnetic field was tuned. Every time the emission energy coincides with one of the detector lines we observe a peak in the spectrum. Table 1 contains the sample parameters. The typical conduction band edge is shown in Fig 1. There is a weak background dopant concentration in the spacer and in the GaAs which influences the spectra.

Table 1:

Sample	Carrier Concentration	Mobility	Spacer Thickness
S1	$2.24 * 10^{11}$ cm^{-2}	250000 cm^2/Vs	250 Å
S2	$5.12 * 10^{11}$ cm^{-2}	100000 cm^2/Vs	150 Å
S3	$4.20 * 10^{11}$ cm^{-2}	755000 cm^2/Vs	150 Å

Springer Series in Solid-State Sciences, Vol. 101
High Magnetic Fields in Semiconductor Physics III Editor: G. Landwehr

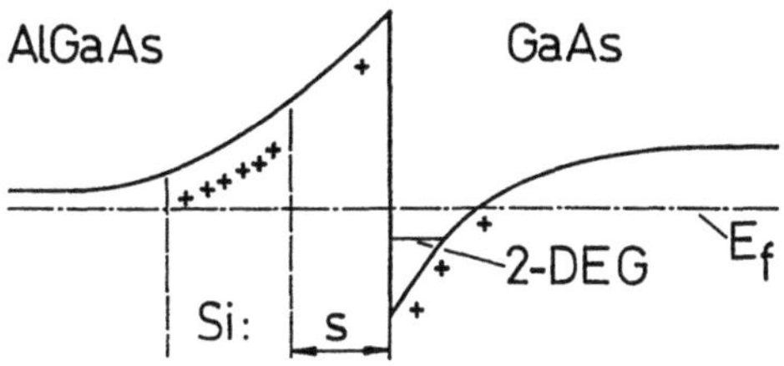

Figure 1: Conductionband edge of the used heterostructures. The spacer thickness s is 250 Å for sample 1 and 150 Å for samples 2 and 3.

2. Spectra

Fig. 2 shows a typical emission spectrum of sample 1 using a detector magnetic field of 2 T. At this magnetic field the detector is sensitive at 8 meV. The excitation field of the emitter was 3 V/cm and 15 V/cm. The high field spectrum which is five times larger than the low field spectrum was scaled to the same height as the other line. As shown in [9], for high excitation fields the emission spectrum is dominated by the cyclotron emission of free electrons, whereas it contains an impurity-shifted part for low excitation fields. The difference of the two curves represents the impurity emission spectrum. In Fig. 3 the filling factor dependence of the relative impurity emission intensity is plotted (integrated impurity emission intensity / integrated CR-emission intensity), thus avoiding any artefacts due to the excitation of electrons. We found maxima near integer filling factors for samples 1 and 2 but non for sample 3. This might be due to the high mobility of sample 3 which leads to a higher electron temperature, and therefore screening effects cannot be observed any more [11-12].

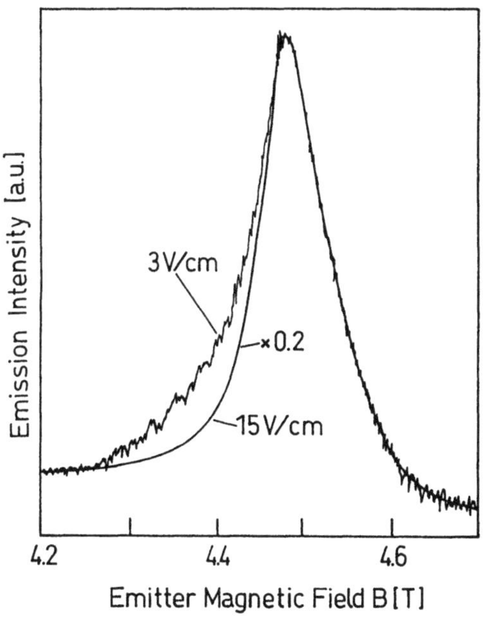

Figure 2: The typical emission spectrum for two different excitation fields. The detector is sensitive at 8 meV.
The 15 V/cm spectrum is scaled to the same height as the 3 V/cm spectrum.

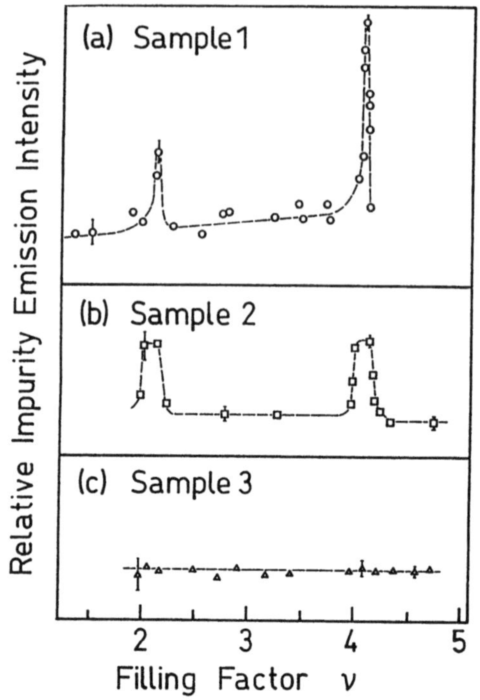

Figure 3: Filling factor dependence of the impurity emission intensity divided by the cyclotron emission intensity of free electrons.

3. Analysis of the Spectra

The observed spectra can be converted into impurity binding energies using a simple model based on a harmonic impurity potential:

$$\hbar\omega_{obs} = \sqrt{(\hbar\omega_c)^2 + E_B{}^2}$$

This assumption has been previously used in absorption and emission studies successfully. Especially for small binding energies this is a good assumption. The resulting spectra are shown in Fig. 4. They consist of a peak at about 1.5 meV to 3 meV depending on the sample and the magnetic field. In addition to this main peak we observed a long tail up to energies of 4 meV to 8 meV. With increasing magnetic field the line becomes more and more symmetric. At low magnetic fields the tail represents very high energies. Therefore this part of the emission spectrum is identified to be due to impurities located near the interface between GaAs and GaAlAs or, in the case of 8 meV, within the 2-DEG. It is interesting to note that this tail nearly disappears at high magnetic fields (magnetic freeze-out). This behaviour is due to the mechanism of excitation: At high magnetic fields the impurity states become completely localized and the efficiency of excitation disappears.

In Fig. 5 the dominant binding energy is plotted vs. the magnetic field. Within the magnetic field range measured the binding energy increases about linearly with magnetic field, which is in agreement with calculations [3]. In Sample 2 oscillations of the binding energy

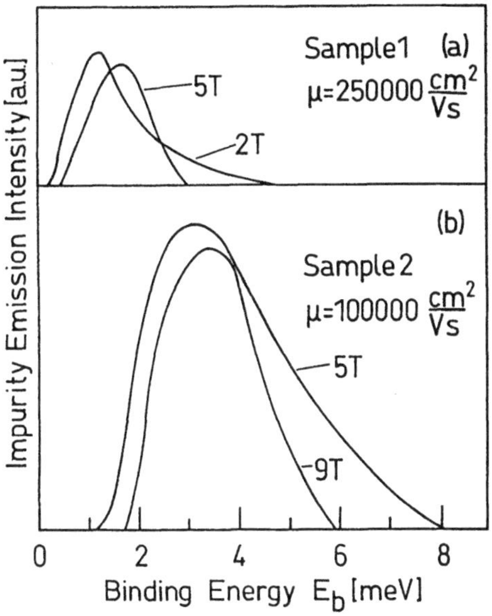

Figure 4: Transformed impurity emission spectra (impurity emission intensity vs. impurity binding energy). The spectra consist of a main peak at 1.5 meV to 4 meV and a long tail up to 4 to 8meV.

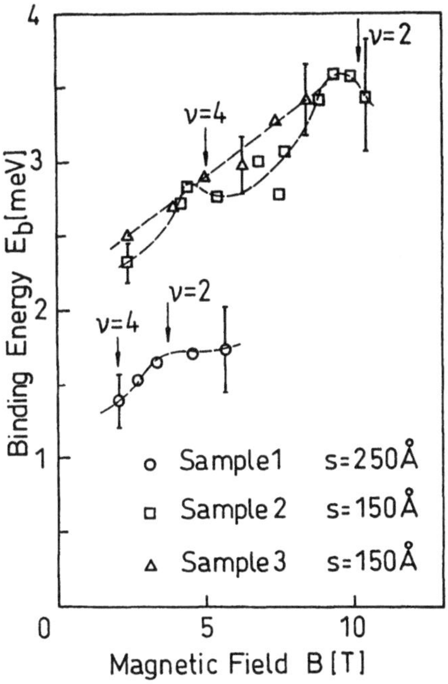

Figure 5: Peak position of the impurity emission intensity vs. magnetic field. The binding energy increases with magnetic field. The binding energy for sample 1 (250 Å spacer) is lower than for samples 2 and 3 (150 Å spacer).

are observed, which occur at the same filling factors as in Fig. 3. However these oscillations are only of the order of 0.3 meV, which is comparable to the error. The reason for the weak oscillations are screening effects, which are very weak due to the finite electron temperature in the emission experiment. For the lowest electric fields the electron temperature can be estimated to be 15 K to 20 K. The binding energies for samples 2 and 3 which have the same spacer thickness (s=150 Å) are comparable though the mobilities differ by a factor of 7.5. These binding energies are 1 meV larger than for sample 1 (s=250 Å). The dominant binding energy therefore is correlated to the spacer thickness. This is a hint that the main peak in the impurity emission spectrum is due to impurities located behind the spacer. However, the long tail extending to energies up to 8 meV is only observed in sample 2. This means that the lower mobility is mainly due to higher background doping.

4. Conclusion

We could separate the impurity emission spectrum from the spectrum of free electrons. With increasing magnetic field the binding energy of low energy impurities increases, whereas the emission of high energy impurities freezes out. We observed filling factor oscillations in impurity emission intensity and in binding energy. The main part of the impurity emission spectrum could be assigned to impurities located behind the spacer, whereas the tail is due to background doping.

5. Acknowledgment

We wish to thank the Deutsche Forschungsgemeinschaft for sponsoring this work.

6. References

[1] G. Bastard Phys Rev B **24**, (1981) 4714
[2] R.L. Greene, K.K. Bajaj Solid State Commun. **45**, (1983) 825
[3] R.L. Greene, K.K. Bajaj Phys Rev B **31**, (1985) 913
[4] R.L. Greene K.K., Bajaj Phys Rev B **34**, (1986) 951
[5] N.C. Jarosik, B.D. McCombe, B.V., Sanabrook, J. Comas, J. Ralston, G. Wicks Phys Rev Lett **54**, (1985) 1283
[6] G.Bauer (ed.): *Two-Dimensional Systems: Physics and new Devices* (Springer, 1987), p. 156
[7] S. Huant, R. Stepienski, G. Martines, V.Thierry-Mieg, B. Etienne Superlatt Microstruc **5**, (1989) 331
[8] W.Zawadski (ed.): *Narrow Gap Semiconductors Physics and Applications* (Springer, 1990), p. 160
[9] E. Gornik, W. Seidenbusch, R. Christanell, R. Lassnig, C.R. Pidgeon Surf Sci **196**, (1987) 339
[10] J.L. Robert, A. Raymond, J.Y. Mulot, C. Bousquet, J.P. André, W. Knap, M. Kubisa, W. Zawadski, E. Gornik, W. Seidenbusch, M. Witzany, Proc. 19th Int. Conf. on the Physics of Semiconductors (Warsaw), ed. W. Zawadski, 1988
[11] T. Ando, Y. Uemura J. Phys. Soc. Jap. **34**, (1971) 632 [12] T. Ando, Y. Uemura J. Phys. Soc. Jap. **31**, (1971) 331

Electron Lifetime in the Low Field Limit: A Microwave Cyclotron Resonance Study

M. Watts[1], *I. Auer*[1], *R.J. Nicholas*[1], *J.J. Harris*[2], *and C.T. Foxon*[2]

[1]Clarendon Laboratory, University of Oxford, Parks Road, Oxford, OX1 3PU, UK

[2]Philips Research Laboratories, Redhill, RH1 5HA, UK

Abstact. We have observed cyclotron resonance (CR) in several very high mobility GaAs-AlGaAS heterojunctions at magnetic fields as low as 19 mT. We deduce a single particle lifetime one to two orders of magnitude longer than the "quantum lifetime" deduced from damped Shubnikov de Haas oscillations in these and similar samples, and which is close to the DC scattering time. We also observe a weak linewidth maximum when the Larmor radius (L_0) ≈ the spacer thickness. We conclude that the CR lifetime shows no evidence of enhanced forward peaked scattering and has the same angular scattering dependence as the DC mobility.

Cyclotron resonance has been extensively studied in two dimensional electron gas systems (2DEG). Though much of this work has been on the very high quality GaAs-AlGaAs heterojunctions currently available, only magnetic fields above ≃ 0.6 Tesla have been extensively investigated. This corresponds to a wavelength of ≃ 1.2mm, the limit of most far infra red laser systems. Using microwave Gunn oscillators we have extended this wavelength range upto 18mm allowing cyclotron resonance to be observed at fields as low as 0.019 Tesla. Such fields correspond to a Larmor radius of the order of ≃ 0.2μm thus sampling potential fluctuations over a similar range.

In discussing the magneto-transport aspects of a 2DEG two characteristic times are neccessary; the first, the familiar "scattering time" τ_t, is obtainable from the Boltzman equation and determines the DC mobility. The second lifetime is that inferred from the energy uncertainty associated with the Landau level broadening. Traditionally this is referred to as the "quantum lifetime". Analytically, τ_q differs from τ_t by the omission of an angular dependent differential scattering cross section, making it sensitive to scattering through all angles. The scattering time τ_t however is most sensitive to back scattering and so underestimates the scattering in a system where forward peaked scattering dominates. Experimentally, τ_t is measured as the DC conductivity and extracted using a simple Drude formulism. The quantum lifetime τ_q is deduced by fitting an envelope to the low field Shubnikov de Haas amplitude as described by Ando's theory of magnetoconductivity (1). In Si the ratio τ_t/τ_q is measured as being very close to 1 (2), whilst in GaAs-AlGaAs heterojunctions the ratio has been found to be as large as 20 - 50 (3,4,5). Fang et al (5) and Das Sarma (6) attribute this difference to the shorter screening length in Si (with respect to a typical Fermi wavelength) causing largely isotropic scattering.

From our absorption linewidth we are able to deduce a time characteristic of the energy uncertainty between local cyclotron resonance transitions. We obtain a lifetime close to that of the DC mobilty and so a value more than an order of magnitude longer than those of magnetoconductivity measurements.

The experiments were all carried out at 1.5K in magnetic fields up to 2 Tesla. A range of conventional high mobility GaAs-AlGaAs heterojunctions were

Springer Series in Solid-State Sciences, Vol. 101
High Magnetic Fields in Semiconductor Physics III Editor: G. Landwehr

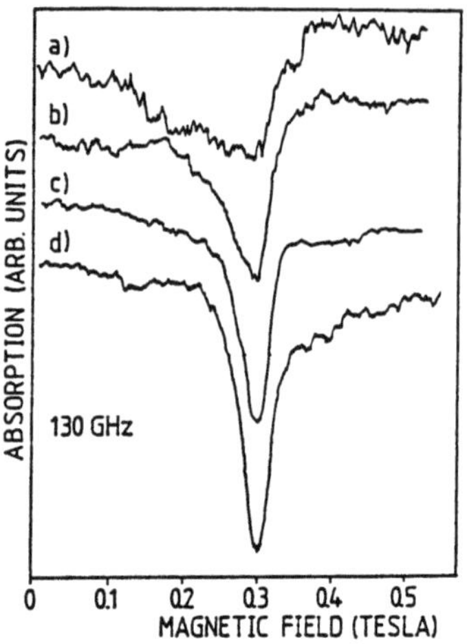

Figure 1:
Obtaining a symmetrical lineshape through cavity tuning :
(a) Complete detuning
(b) Tuned short of resonance
(c) On resonance
(d) Tuned past resonance.

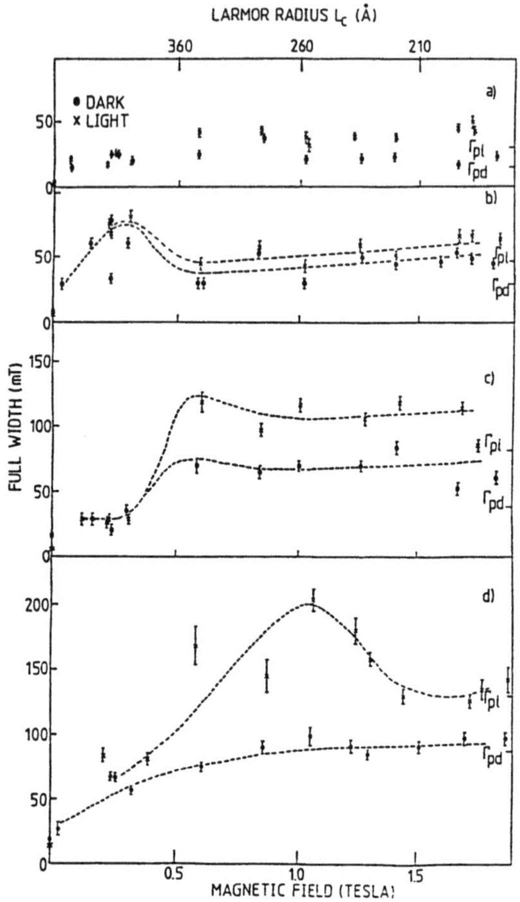

Figure 2 : Resonance linewidth (mT) versus magnetic field (T) for various spacer thicknesses: (a) 1200Å (b) 800Å (c) 400Å (d) 200Å. The zero Tesla linewidths correspond to equivalent width deduced from the DC mobility.

studied for different spacer layer thicknesses and electron concentrations. Above 0.5 Tesla a far infra red laser was used and transmission was measured conventionally. The magnetic field region below 0.5 Tesla neccessitated the use of Gunn diodes of centre frequencies 17 GHz, 35 GHz, 65 GHz and 90 GHz; this range could be extended to include 2nd and 3rd harmonics by judicious choices of waveguide. The sample was situated with the detection bolometer in a resonant tunable cavity. The tunability of the cavity enabled the adverse effects of the cavity on the resonance lineshape to be overcome; this process of tuning the lineshape is shown in figure 1. In this way the absorption corresponds to a reduction in the overall microwave energy density in the cavity at resonance.

For the 17 GHz experiment a frequency stabilized Gunn diode was used in a conventional microwave bridge circuit. A single turn coil immediately above the sample was used to introduce an additional modulated magnetic field at 115 kHz giving a differential resonance signal.

In figure 2 the full width at half maximum of the absorption curve is plotted as a function of magnetic field and Larmor radius. The experimentally measured

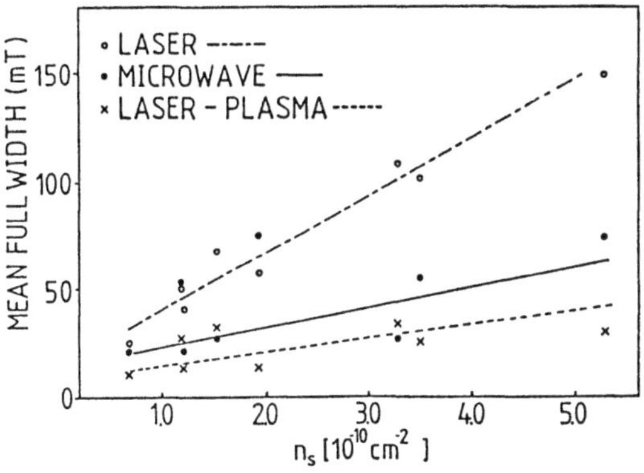

Figure 3 : Mean linewidth plotted as a function of carrier concentration n_s.

full width will only directly reflect the the true resonance broadening when the conductivity at resonance is sufficently low as allow a small absorption approximation to be made. In our high mobility samples this is not the case and the reflective properties of the 2DEG must be included in the overall dielectric response of the system (for example (7)). Effectively the linewidth of the resonance has two contributions: an inherent width due to the various scattering mechanisms plus an additional (carrier density dependent) artificial broadening. We refer to the additional term as the "plasma broadening" Γ_p. Failure to allow for Γ_p will lead to an underestimate of any characteristic time deduced. The extent of the term Γ_p in shorter wavelength experiments is well understood but its contribution in the microwave regime is less certain. We anticipate a steady increase in Γ_p with frequency saturating at around 0.5 Tesla where the sample dimensions become larger than the incident wavelength. The extent of this artificial broadening, as calculated for the shorter wavelength case, is indicated by Γ_{pd} and Γ_{pl} in figure 2 for dark and light carrier concentrations respectively.

The microwave resonances in figure 2, occuring at fields less than 0.5 Tesla, are with the exception of the 800Å spacer sample, consistently narrower than the laser linewidths. This is due to the fall off of the plasma contribution with decreasing frequency. Furthermore, the very low field measurements in fig.2 are much less sensitive to increases in carrier concentrations with illumination; whilst the higher frequency linewidths are very carrier density dependent: the plasma term in the linewidth increasing by an amount proportional to the increase in carrier density. This suggests that little plasma broadening is present in the very low field resonances. Thus from our microwave data we can extract directly a close approximation to the inherent scattering linewidth. This situation is depicted in figure 3 where the mean values of microwave linewidth, laser linewidth and plasma corrected laser linewidth are plotted as a function of carrier density.

In figure 2 for each sample there is a weak linewidth maximum at a magnetic field which is very sample dependent. This feature is clearest for the low field microwave measurements on the 800Å sample and for the 200Å sample at around 1 Tesla. The 400Å sample shows a maximum at around 0.5 Tesla but this may be complicated by the decreasing plasma contribution. We attribute these features to the characteristic range of the underlying potential becoming comensurate with the Larmor radius of the cyclotron orbit, as discussed by Prasad (8) for the case of a Gaussian potential. The extent of the potential is of the order of the spacer layer thickness, explaining both the relative spacing of these features and the absence of any feature for the 1200Å sample.

At low temperatures and magnetic fields where phonon effects are negligible, the CR linewidth is determined by the extent of potential fluctuation over the range of the Larmor radius. The origins of the underlying potential fluctuation in a modulation doped GaAs–AlGaAs heterojunction are the long range Coulomb tails from remote ionized impurities, these set up a random potential dependent on both spacer size and electron concentration. In the microwave regime where the Larmor radius is very large ($\approx$ 0.2μm for B_r = 19mT), we observe very narrow linewidths, indicating that the potential fluctuations are small over even such a large range. The extent of this fluctuation can be estimated by the energy halfwidth of the resonance divided by the Larmor radius, in this sense the linewidth is sensitive to the gradient of the underlying random potential function. For the 1200Å sample measured at 19mT the fluctuation is of the order of 0.015 meV/0.2 μm.

The mean microwave linewidths for each sample are easily expressed as a characteristic lifetime; for the (dark n_s) 200Å, 400Å, 800Å and 1200Å samples these are respectively : 14 ps, 28 ps, 12 ps and 36 ps. Calculating the DC scattering times from the appropriate mobilities yields τ_t/τ_{cr} ratios of 3, 1.5, 6.6 and 3.4. These ratios are between one and two orders of magnitude lower than those obtained deducing a quantum lifetime from SdH measurements. The enhanced ratio for the 800Å spacer is due to enhanced impurity scattering as already discussed. From the comparable values of classical and CR lifetime obtained we must conclude that the angular sensitivity of the CR lifetime to impurity scattering is similar to that of the DC scattering lifetime. Our data is in no way suggestive of a forward peaked scattering distribution as are the results obtained from magneto–conductivity measurements. Such a discrepancy suggests that different quantities are being measured. The local nature of the cyclotron resonance transition make it appropriate to interpret the lifetime obtained as a "single particle dephasing time", that is the time in which an electron can be considered as existing in a single particle momentum eigenstate. We interpret the lifetime τ_q not as a strict electron dephasing time, but as a quantity sensitive to potential fluctuations throughout the sample. In this respect the quantum lifetime τ_q is a measure of the energy uncertainty associated with the overall density of states. This quantity, whilst still a useful parameter in theoretical descriptions, is not a measure of the single particle lifetime. The spatially variant potential causes local filling factor variations throughout the sample dimensions; the subsequent phase smearing is equivalent to an ensemble averaging over all states made available to all electrons by the random scattering potential.

Acknowledgements.

We would like to thank A.J. O'Connell and R. Storey.

References.

1. T. Ando, J. Phys. Soc. Japan 37 (1974) 1233.
2. F.F. Fang et al. Phys. Rev. B 16 (1977) 4446.
3. M.A. Paalanen et al. Phys. Rev Letters 51 (1983) 2226.
4. J.P. Harrang et al. Phys. Rev. B 32 (1985) 8126.
5. F.F. Fang et al. Surf. Sci. 196 (1988) 310.
6. S. Das Sarma and F. Stern, Phys. Rev. B 32 (1985) 8442.
7. T.A. Kennedy et al. Solid State Comm. 18 (1976) 275.
8. M. Prasad and S. Fujita, Physica 91A (1978) 1.

Electron Scattering Studies in Semiconductors at High Magnetic Fields by Far-Infrared Cyclotron Resonance

H. Kobori, T. Ohyama, and E. Otsuka

College of General Education, Osaka University,
Toyonaka, Osaka 560, Japan

Extensive study of scatterings in quantizing magnetic fields has been made on carrier transport for basic semiconductors (Ge, CdS, InSb and GaAs). Probabilities of various scatterings are expressed by Cyclotron Resonance Line-Width (CRLW) obtained from far-infrared (84 - 513 μm) cyclotron resonance experiments.

1. INTRODUCTION

Half of the Cyclotron Resonance Line-Width (CRLW) represents the inverse of the carrier transport relaxation time and is connected directly with the total sum of probabilities of various scatterings. Accordingly, the cyclotron resonance method is quite useful for studying carrier scattering mechanism as well as band structure. In comparison of the cyclotron effective energy, $\hbar\omega_c$, with the thermal energy, k_BT, two extreme limits exist, where ω_c is the cyclotron angular frequency, k_B the Boltzmann constant and T the absolute temperature. One is realized when $\hbar\omega_c \ll k_BT$, which is called the classical limit, and the other one is the quantum limit, where $\hbar\omega_c \gg k_BT$. In the classical limit, or low magnetic field case, the CRLW can be explained by the semi-classical Boltzmann transport theory. Actually, the agreement between experimental and theoretical results is excellent [1]. Under such a high magnetic field that satisfies the quantum limit, the quantum effect due to the magnetic field cannot be neglected and the semi-classical method breaks down. Then quantum-statistical treatment is necessary for understanding this phenomenon. The first rigorous quantum-statistical treatment of cyclotron resonance line- shape was derived by Kawabata [2] on the basis of the general Langevin equation of Brownian motion, starting from the Kubo formula. This formalism was applied to ionized impurity scattering, assuming a bare Coulomb potential. Thereafter, many theories have come out for various scatterings; ionized impurity- [3], acoustic deformation potential- [4], acoustic piezo-electric- [5], and polar optical phonon [3,6] scatterings. At the same time, numerous experiments have been made for ionized impurity- [7], acoustic deformation potentia-l [8,9] and polar optical phonon [7] scatterings. In this paper, we present CRLW in quantizing magnetic fields for various scattering (ionized impurity-, neutral impurity-, acoustic deformation potential-, acoustic piezo-electric- and polar optical phonon scatterings) mechanism in basic semiconductors (Ge, CdS, InSb and GaAs).

2. EXPERIMENTAL PROCEDURES

Far-InfraRed (FIR) lasers, discharge and optical pumped ones, have been used for cyclotron resonance experiments. FIR light transmitted through the sample is detected by an n-InSb Putley type or Sb-doped Ge detector. Pulsed FIR transmission has been made at a repetition of 30Hz in synchronized combination with photo-excitation by a xenon flash lamp at 15 Hz. When not using the xenon flash lamp, thermal excitation was adopted. Four kinds of semiconductors were studied; Ge, CdS, InSb and GaAs. Their conductivity types and characteristics are shown in Table I. All the samples were wedge shaped and had a typical size of 5x5mm^2.

Table 1 List of samples

Sample	N_d ($\times10^{14}cm^{-3}$)	N_a ($\times10^{14}cm^{-3}$)	Growth method
pure Ge	$N_a - N_d < 10^{12}cm^{-3}$		FZ
pure CdS	—	—	—
n-InSb	2.52	2.48	—
n-GaAs	55	15	MBE

3. EXPERIMENTAL RESULTS AND DISCUSSIONS

The time averaged absorption power of the cyclotron resonance of frequency ω is proportional to the real part of circular electrical conductivity $\sigma_{+-}(\omega)$. In the quantum limit, almost all the conduction electrons populate the lowest Landau level. The distribution function is such that $f[\varepsilon_n(k_z)] \cong 0$ for $n>0$ with the energy $\varepsilon_n(k_z)$ connected with the Landau quantum number n and wave number k_z along the magnetic field. Then we have

$$\mathrm{Re}[\sigma_{+-}(\omega)] = \frac{2e^2}{m^*V}\sum_{k_z} \frac{f[\varepsilon_0(k_z)]\Gamma(\omega, \omega_c; k_z)}{[\omega-\omega_c-\Delta(\omega, \omega_c; k_z)]^2+[\Gamma(\omega, \omega_c; k_z)]^2} \tag{1}$$

Here m^* is the effective mass and V the crystal volume. The functions $\Delta(\omega,\omega_c;k_z)$ and $\Gamma(\omega,\omega_c;k_z)$ represent the peak shift and relaxation rate of the cyclotron resonance, respectively. We assume that the resonant cyclotron angular frequency ω_r is given by $\omega_r=\omega_c+\Delta(\omega,\omega_c;k_z)\cong\omega_c$. Postulating plural scatterings, the relaxation rate $\Gamma(k_z)$ is expressed by

$$\Gamma(k_z)=\sum_s \Gamma_s(k_z) \tag{2}$$

$\Gamma_s(k_z)$ is the relaxation rate of individual scatterings. We hereafter discuss only the ionized impurity-, neutral impurity-, acoustic deformation potential-, acoustic piezo-electric- and polar optical phonon scatterings. In the quantum limit, we have

$$\Gamma_s(k_z)=\sum_M \Gamma_s^{0\to M}(k_z)+\sum_N \Gamma_s^{1\to N}(k_z). \tag{3}$$

$\Gamma^{0\to M}(k_z)$ and $\Gamma^{1\to N}(k_z)$ are the relaxation rates for the scatterings with $n=0 \to n=M$ and $n=1 \to n=N$ (M,N: integer), respectively. $\Gamma^{0\to M}(k_z)$ ($\Gamma^{1\to N}(k_z)$) is corresponding to the scattering before (after) photon absorption ($n=0 \to n=1$) at cyclotron resonance. Assuming that $\hbar\omega_{AP} < \hbar\omega_c < \hbar\omega_{OP}$, where $\hbar\omega_{AP}$ and $\hbar\omega_{OP}$ are the phonon energies of acoustic and optical phonons, respectively, $0 \leq M \leq \omega_{OP}/\omega_c$ and $0 \leq N \leq \omega_{OP}/\omega_c+1$ for polar optical phonon scattering, on the other hand, M=0, N=0,1 for all scatterings except polar optical phonon scattering. When the energy dependent relaxation rate $\Gamma(k_z)$ is obtained, CRLW γ is determined. The definition of CRLW γ is

$$\gamma = \gamma_L + \gamma_R \tag{4}$$

with

$$\mathrm{Re}[\sigma_{+-}(\omega_r-\gamma_L)] = \mathrm{Re}[\sigma_{+-}(\omega_r+\gamma_R)] = \mathrm{Re}[\sigma_{+-}(\omega_r)]/2. \tag{5}$$

We have calculated the relaxation rate $\Gamma(k_z)$ and CRLW γ numerically for various scatterings.

Figure 1 shows the experimental results and theoretical curves for the temperature dependence of the half CRLW in pure Ge, in which the acoustic deformation potential scattering dominates in comparison with other scatterings. "Intra" and "inter" represent theoretical predictions for intra- ($n=0 \to n=0$ and $n=1 \to n=1$) and inter- ($n=1 \to n=0$) Landau level scatterings, respectively. We take the anisotropy of the effective mass of the conduction electron into account. As may be seen from Fig1, CRLW for intra-Landau level scattering becomes remarkably small with decreasing temperature due to the effect of inelastic scattering. The inter-Landau level scattering can be treated as quasi-elastic, since the acoustic phonon energy is small compared with the cyclotron effective energy $\hbar\omega_c$. At very low temperature, only spontaneous phonon emission is dominant and the CRLW becomes independent of temperature. The increase of CRLW with magnetic field arises from the fact that the acoustic deformation potential scattering is of short range type. Figure2 shows the temperature dependence of the half CRLW for both experimental and theoretical results in pure CdS. Designated "lg" and "tr" are corresponding to the longitudinal and transverse modes of acoustic phonons, respectively. As the piezo-coupling constant is large in CdS, acoustic piezo-electric scattering is considered to be dominant in pure CdS below room temperature. The behaviour is rather similar in its

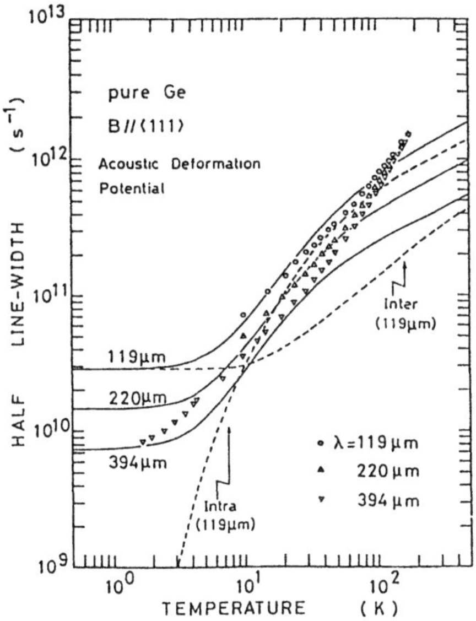

Fig.1 Temperature dependence of half CRLW by acoustic deformation potential scattering in pureGe for wavelengths of 119,220 and 394 μm at the <111> magnetic field direction.

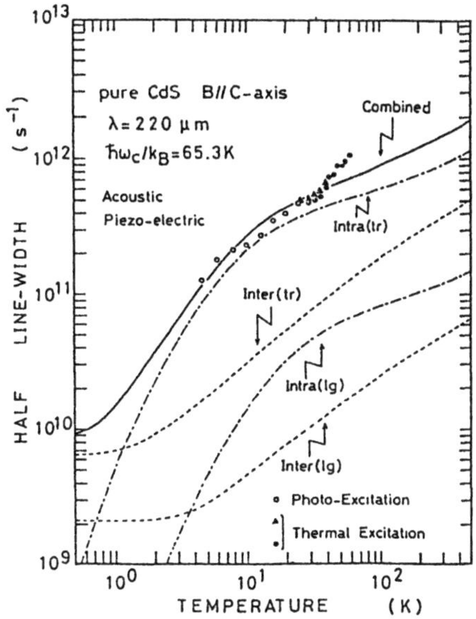

Fig.2 Temperature dependence of half CRLW by acoustic piezo-electric scattering in pure CdS for a wavelength of 220 μm; magnetic field parallels c-axis.

temperature dependence to that for acoustic deformation potential scattering. A difference shows up in the magnetic field dependence. Acoustic piezo-electric scattering is of longer range compared with acoustic deformation potential scattering. The CRLW varies slowly with magnetic field. The anisotropy of the effective mass of the conduction electron is taken into account in the calculation. Figure 3 gives the experimental and theoretical curves of temperature dependence of the half CRLW in n-InSb. The contribution of the ionized impurity scattering is considered to be dominant below 30K. With shorter wavelength, or higher magnetic field, the CRLW is found to decrease. This comes from the fact that the ionized impurity scattering is of long range type. Above 30K, the CRLW increases steeply with temperature. We thus consider the optical phonon scattering to be dominant. The optical phonon emission is practically forbidden, since the optical phonon energy $\hbar\omega_{OP}$ is larger than the cyclotron effective energy $\hbar\omega_c$. Figure4 indicates the temperature dependence of the half CRLW in n-GaAs. In the low temperature region, the CRLW is practically determined by neutral impurity-, ionized impurity- and electron-electron scatterings. On the other hand, in the high temperature region, it is dominated by acoustic deformation potential-, acoustic piezo-electric- and polar optical phonon scatterings.

The following material constants have been used:

(1) Ge $m_t^*=0.082\ m_0$, $m_l^*=1.58\ m_0$, $\rho_m=5.36$ g/cm^3, $v_s=5.94\times10^5$ cm/s and $E_{1,\perp}=18$eV.

(2) CdS $m^*_{\perp,c}=0.18\ m_0$, $m^*_{//,c}=0.20\ m_0$, $\rho_m=4.82$g/cm^3, $\kappa=8.58$, $v_s^l=4.28\times10^5$cm/s, $v_s^t=1.81\times10^5$cm/s, $(K_l^2)_{av}=3.52\times10^{-3}$ and $(K_t^2)_{av}=2.63\times10^{-2}$.

(3) n-InSb $m_e^*=0.0139\ m_0$, $\kappa=17.88$, $N_I=5.0\times10^{14}$cm^{-3}, $T_D=205$ K and $\alpha=0.087$.

[ρ_m: mass density, v_s:sound velocity, κ:dielectric constant, E_1:acoustic deformation potential constant, K:piezo-coupling constant, N_I; ionized impurity concentration, T_D: Debye temperature and α: polaron coupling constant)

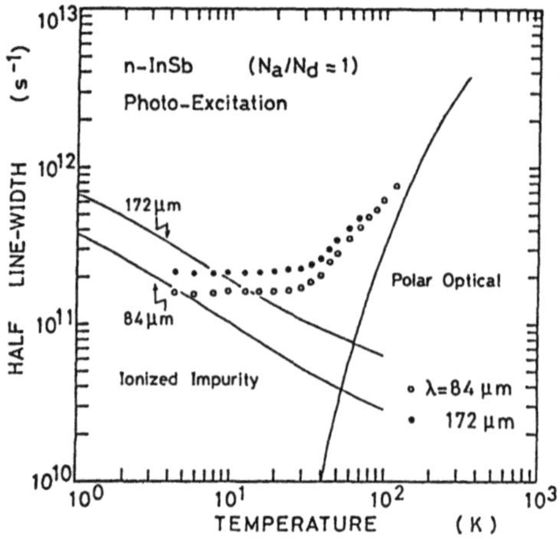

Fig.3 Temperature dependence of half CRLW by ionized impurity and polar optical phonon scatterings in n-InSb for wavelengths of 84 and 172 μm.

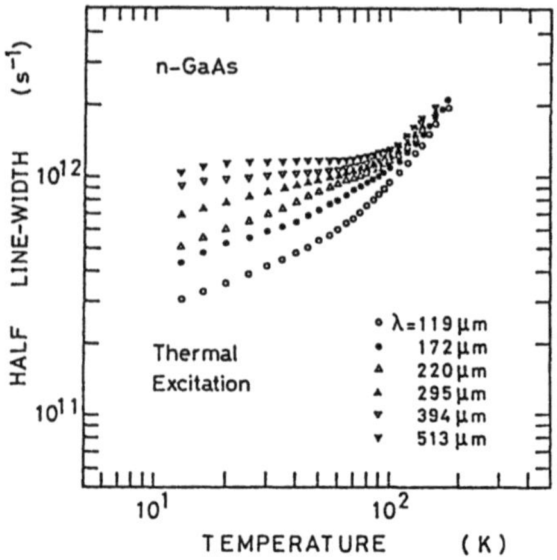

Fig.4 Temperature dependence of half CRLW by various scatterings in n-GaAs for wavelengths of 119 - 513 μm.

REFERENCES

1. E.Otsuka: Jpn.J.Appl.Phys. 25 303 (1986)
2. A.Kawabata: J.Phys.Soc.Jpn. 23 999 (1967)
3. J.V.Royen, J.D.Sitter and J.T.Devreese: Phys.Rev. B30 7154 (1984)
4. A.Suzuki and K.Dunn: Phys.Rev. B25 7754 (1982)
5. M.Saitoh and A.Kawabata: J.Phys.Soc.Jpn. 23 1006 (1967)
6. M.Saitoh: J.Phys. A16 1795 (1983)
7. O.Matsuda and E.Otsuka: J.Phys.Chem.Sol. 40 809 (1979)
8. T.Ohyama, K.Murase and E.Otsuka: J.Phys.Soc.Jpn. 25 729 (1968)
9. H.Kobori, T.Ohyama and E.Otsuka: Solid State Commun. 64 35 (1987)

Anisotropy in the Conduction Band of GaAs

H. Mayer and U. Rössler

Institut für Theoretische Physik, Universität Regensburg,
Postfach 397, W-8400 Regensburg, Fed. Rep. of Germany

The anisotropy of the cyclotron resonance and of its spin-splitting, being due to the nonparabolicity of the conduction band, is calculated systematically by performing an algebraic reduction of a 14×14 $\mathbf{k}\cdot\mathbf{p}$ Hamiltonian to a 2×2 conduction band Hamiltonian. By use of an established set of band parameters for GaAs the observed anisotropy can be quantitatively explained without any fitting of parameters.

1. Introduction

The anisotropy of the lowest s-type conduction band in cubic semiconductors is not unexpected from its overall dispersion throughout the Brillouin zone. Its detailed detection in GaAs by cyclotron resonance experiments is a challenging task, however, which has been solved only a couple of years ago [1,2]. Because the effect becomes stronger with increasing electron energy in excess of the conduction band minimum, investigations at higher magnetic fields yield stronger evidence [2]. More recently experimental results for up to 100 T were reported by Miura [3].

The theoretical concept for a quantitative understanding of the observations has existed in principle since 1966, when Ogg [4] formulated the conduction band Hamiltonian by group-theoretical considerations as an invariant expansion. Essentially this Hamiltonian was used by Golubev et al. [1] to interpret their low magnetic field data.

Our perturbation-theoretical derivation of this Hamiltonian starts from a 14×14 $\mathbf{k}\cdot\mathbf{p}$-Hamiltonian, which by partitioning, i.e. block-diagonalization, is systematically reduced to a 2×2-Hamiltonian for the conduction band [5]. In this work the weighting factors of all invariants are traced back to the band parameters of the 14×14 $\mathbf{k}\cdot\mathbf{p}$-model, which for GaAs are known with high accuracy from a wide spectrum of different experimental results [6]. Frequently, similar five-level $\mathbf{k}\cdot\mathbf{p}$-models are used [7,8], which represent simplifications of the 14×14-model as outlined in section 2. The attempt to fit the results of the five-level $\mathbf{k}\cdot\mathbf{p}$-model to the experimental data was only partially successful [2]: the authors succeeded in fitting the anisotropy of the spin-splitting but failed to fit the anisotropy in the average resonant magnetic field.

In this paper, using the 14×14 $\mathbf{k}\cdot\mathbf{p}$-model including the off-diagonal spin-orbit interaction Δ^- [9] and the parameter set of [6], we find an almost perfect interpretation of Sigg's [2] data without any adjustable parameter. This de-

Springer Series in Solid-State Sciences, Vol. 101
High Magnetic Fields in Semiconductor Physics III Editor: G. Landwehr

monstrates the importance of remote band contributions which are partially neglected in the five-level **k**·p-model of [2,8].

2. The 14×14 k·p-model

For semiconductors with open gaps like GaAs and InP the energy separations between the lowest conduction band (Γ_{6c}) on one side and the *p*-bonding valence ($\Gamma_{8v}+\Gamma_{7v}$) and *p*-antibonding conduction band states ($\Gamma_{8c}+\Gamma_{7c}$) on the other side are comparable. Therefore, the proper basis for a **k·p** expansion around the Γ-point should include at least these 14 states, whose energy separations are well known from spectroscopic data. These states are directly coupled by the three momentum matrix elements P (between Γ_{6c} and $\Gamma_{8v}+\Gamma_{7v}$), P' (between Γ_{6c} and $\Gamma_{8c}+\Gamma_{7c}$), and Q (between $\Gamma_{8v}+\Gamma_{7v}$ and $\Gamma_{8c}+\Gamma_{7c}$) [10], see Fig. 1.

Besides the **k**·p coupling within the space of the 14 basis states of our model we consider coupling to remote bands outside of this space by second order perturbation terms. This introduces the correction terms C to the conduction band mass and the remote band contributions contained in the Luttinger parameters γ_1, γ_2, γ_3. In the presence of a magnetic field we have to consider also the free electron g-factor and remote band contributions to the electron (C') and hole g-factors (κ and q). In contrast to [5,6] we extended the 14×14-model by including the off-diagonal spin-orbit coupling Δ^- between $\Gamma_{8v}+\Gamma_{7v}$ and $\Gamma_{8c}+\Gamma_{7c}$ [9] as well as the free electron contributions to the hole g-factor in the valence band part of the Hamiltonian.

We do not consider here the asymmetry induced k-linear term in the valence band [11], whose effect is negligible for GaAs in the context of this paper. Remote band contributions in the upper conduction bands, which correspond to the Luttinger parameters in the valence band, are also omitted, because so far there is no evidence of their influence. All parameters of this 14×14 **k**·p-model (see Fig. 1) are well established from a variety of different experiments and some calculations. Their values are given in Tab. 1.

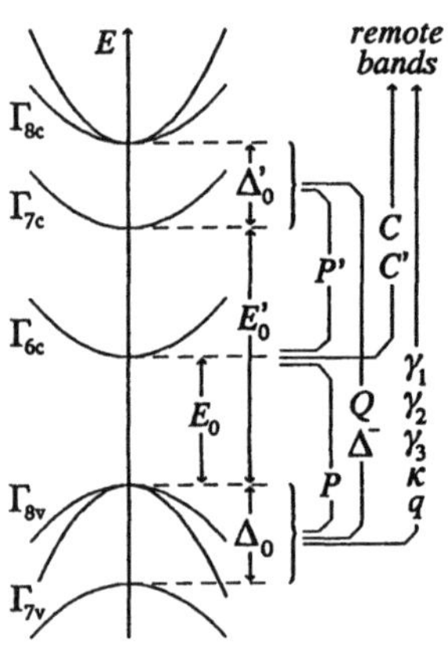

Fig. 1: The 14×14 **k**·p-model

Tab. 1: Band parameters

$E_0 = 1.519$ eV	$\gamma_1 = 6.85$	$C = -1.878$
$\Delta_0 = 0.341$ eV	$\gamma_2 = 2.10$	$C' = -0.021$
$E_0' = 4.488$ eV	$\gamma_3 = 2.90$	$P = 10.493$ eVÅ
$\Delta_0' = 0.171$ eV	$\kappa = 1.20$	$P' = i4.780$ eVÅ
$\Delta^- = -i0.050$ eV	$q = 0.01$	$Q = 8.165$ eVÅ

3. Derivation of the Conduction Band Hamiltonian

Using the 14×14-model for a numerical calculation of the conduction band Landau levels obscures the influence of the various coupling terms. In order to make the effects of the couplings on the conduction band explicit and to gain a more compact Hamiltonian for the conduction band only we follow the procedure outlined in [5] to perform an algebraic block-diagonalization of the 14×14-matrix Hamiltonian. The matrix elements of this Hamiltonian are polynomials in the components of the wave vector operator $\mathbf{k}$, which in the presence of a magentic field $\mathbf{B}$ obeys the commutation relations $\mathbf{k} \times \mathbf{k} = -i\,(e/\hbar)\mathbf{B}$. The cumbersome and tedious algebra which is necessary to obtain the 2×2 conduction band Hamiltonian was carried out using the algebraic computation expert system MACSYMA [12,13].

The resulting conduction band Hamiltonian can be formulated as a sum of invariants, represented as products of 2×2 matrices ($X^{(1)} = \mathbb{1}_{2\times 2}$, $X_\ell^{(4)} = \sigma_x, \sigma_y, \sigma_z$) and irreducible tensor components formed out of the components of the electron wave vector operator $\mathbf{k}$:

$$\mathcal{H} = \sum_{\kappa,\lambda} a_{\kappa,\lambda} \sum_{\ell} X_\ell^{(\kappa,\lambda)} \, \mathcal{K}_\ell^{(\kappa,\lambda)*} \, .$$

The superscript κ indicates the irreducible representation according to which these quantities transform.

Thus the expansion coefficients $a_{\kappa,\lambda}$ can be expressed in terms of the 14×14-model parameters. Except for terms containing Δ^- or the additional valence band g-factor contributions the present results [13] are identical to those of [5] (up to a few misprints).

4. Results and Discussion

For a given orientation of the magnetic field the conduction band Hamiltonian was diagonalized using a basis of oscillator eigenfunctions, following the standard procedure indicated in [5]. Calculations were performed with the parameters of Tab. 1 and for zero component of the wave vector parallel to the magnetic field. From the calculated energy eigenvalues the spin-splitting Δ and the anisotropy of the average magnetic field with respect to that of the (001)-direction are determined. The results are compared with the experimental data of [2], see Fig. 2.

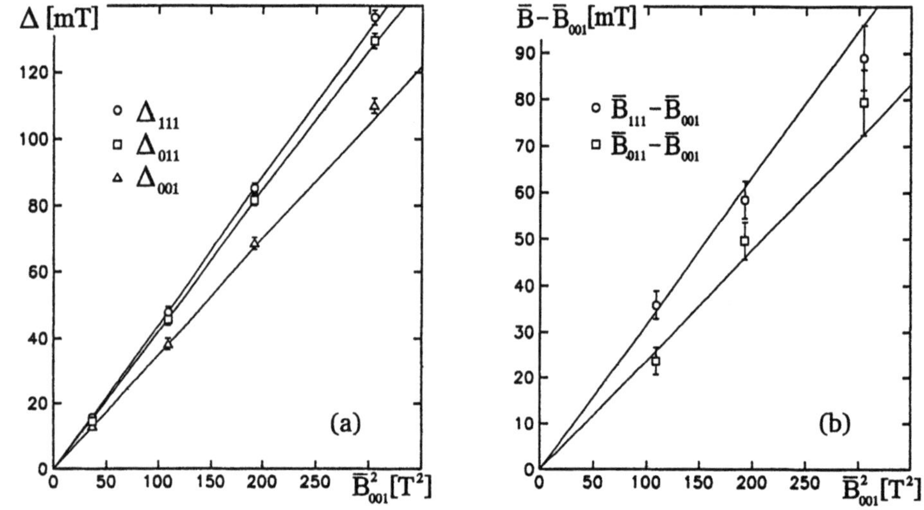

Fig. 2: Calculated spin-splittings of CR (a) and anisotropy of average resonant magnetic field (b) (solid lines) in comparison with experimental data of [2].

References

[1] V.G. Golubev, V.I. Ivanov-Omskii, I.G. Minervin, A.V. Osutin, D.G. Polyakov, Sov. Phys. JETP Lett. **40**, 896 (1984); Sov. Phys. JETP **61**, 1214 (1985)

[2] H. Sigg, J.A.A.J. Peerenboom, P. Pfeffer, W. Zawadzki, Solid State Commun. **61**, 685 (1987)

[3] N. Miura, in: *High Magnetic Fields in Semiconductor Physics II, Proc. Int. Conf. Würzburg 1988*, ed. G. Landwehr, Springer Series in Solid State Sciences 87, Springer Berlin 1989, p. 618

[4] N.R. Ogg, Proc. Phys. Soc. **89**, 431 (1966)

[5] M. Braun, U. Rössler, J. Phys. C: Solid State Physics **18**, 3365 (1985)

[6] U. Rössler, Solid State Commun. **49**, 943 (1984)

[7] C. Hermann, C. Weisbuch, Phys. Rev. **B 15**, 823 (1977)

[8] W. Zawadzki, P. Pfeffer, H. Sigg, Solid State Commun. **53**, 777 (1985)

[9] M. Cardona, N.E. Christensen, G. Fasol, Phys. Rev. **B 38**, 1806 (1988)

[10] In principle the matrix elements formed with spin-orbit split states can be different due to relativistic effects.

[11] M. Cardona, N.E. Christensen, G. Fasol, Phys. Rev. Lett. **56**, 2831 (1986)

[12] MACSYMA™ Symbolics Inc., developed at MIT, Cambridge, Mass. 1986

[13] H. Mayer, Diplomarbeit, Regensburg 1990

Zeeman Effect of the Carbon Acceptor in GaAs, II

R. Atzmüller, M. Dahl, H. Kraus, G. Schaack, E. Bangert, and W. Schmitt

Physikalisches Institut der Universität Würzburg, Am Hubland,
W-8700 Würzburg, Fed. Rep. of Germany

1 Introduction

The Zeeman splitting of shallow acceptors in semiconductors with degenerate bands has been studied experimentally as well as theoretically for more than two decades [1][2]. Two years ago we reported [2] on our examination of the ground state ($1S_{3/2}\Gamma_8$) and the first excited state ($2P_{3/2}\Gamma_8$) of the carbon acceptor in bulk GaAs. We have continued our work by investigating the splitting of the D–line ($2P_{5/2}\Gamma_8 \rightarrow 1S_{3/2}\Gamma_8$) applying a standard FIR–Fourier spectrometer. We have used the same sample as before, that is a LEC grown crystal which shows a persistent hole concentration ($p = 9 \cdot 10^{15}\, cm^{-3}$) after bleaching the EL2 absorption.

As the well resolved experimental spectra contain detailed information on the splitting of three different states each for $\vec{B} \parallel [001]$, [111] or [110] and two possible polarisations in Voigt–geometry, we were encouraged to extend Baldereschi's acceptor model to the strong magnetic field case and to develop its solutions which can be considered as practically exact ones.

2 Theory

The key idea for the solution of the shallow acceptor problem is the transformation of Luttingers kp–matrix into spherical tensor form which was accomplished first by Baldereschi et al. [1]. The effect of an external magnetic field is included by substituting $\hbar\vec{k} \rightarrow \hbar\vec{k} + e\vec{A}$, $\vec{A} = \frac{1}{2}\vec{B} \times \vec{r}$. By subsequent application of the angular momentum recoupling scheme [3] the Hamiltonian can be ordered in terms of different powers of the magnetic field dependence. So we find: (Symbols as in [1][2])

$$
\begin{aligned}
H &= H_0 + H_{lin} + H_{qua} \\
H_0 &= -\boldsymbol{k}^2 + \mu\, \boldsymbol{k}^{(2)} \cdot \boldsymbol{J}^{(2)} - \delta \sum_{m=0,\pm 4} a_m \left(\boldsymbol{k}^{(2)} \times \boldsymbol{J}^{(2)}\right)^{(4)}_m + \frac{2}{r} \\
H_{lin} &= \boldsymbol{B}^{(1)} \cdot \left[\boldsymbol{L}^{(1)} - \frac{\mu}{2}\sqrt{\frac{5}{3}}\left(\boldsymbol{L}^{(1)} \times \boldsymbol{J}^{(2)}\right)^{(1)} - \frac{\mu}{2}\sqrt{10}\left(\mathcal{L}^{(2)} \times \boldsymbol{J}^{(2)}\right)^{(1)} - \frac{2\kappa}{\gamma_1}\boldsymbol{J}^{(1)}\right] \\
&\quad + \frac{\delta}{2}\sum_{m=0,\pm 4} a_m \left\{\boldsymbol{B}^{(1)} \times \left[\left(\boldsymbol{L}^{(1)} \times \boldsymbol{J}^{(2)}\right)^{(3)} - \left(\mathcal{L}^{(2)} \times \boldsymbol{J}^{(2)}\right)^{(3)} + \sqrt{5}\left(\mathcal{L}^{(2)} \times \boldsymbol{J}^{(2)}\right)^{(4)}\right]\right\}^{(4)}_m \\
H_{qua} &= -\frac{1}{2}\boldsymbol{B}^{(0)} \cdot \boldsymbol{r}^{(0)} + \frac{\mu}{4}\sqrt{\frac{5}{3}}\boldsymbol{B}^{(0)} \cdot \boldsymbol{Q}^{(20)} + \frac{1}{4}\boldsymbol{B}^{(2)} \cdot \boldsymbol{r}^{(2)} + \frac{\mu}{4}\sqrt{\frac{5}{3}}\boldsymbol{B}^{(2)} \cdot \left[\boldsymbol{Q}^{(02)} - \sqrt{7}\boldsymbol{Q}^{(22)}\right] \\
&\quad - \frac{\delta}{4\sqrt{3}}\sum_{m=0,\pm 4} a_m \left\{\boldsymbol{B}^{(0)} \times \boldsymbol{Q}^{(24)} + \boldsymbol{B}^{(2)} \times \left[\boldsymbol{Q}^{(02)} - \frac{2}{\sqrt{7}}\boldsymbol{Q}^{(22)} + \sqrt{\frac{5}{2}}\boldsymbol{Q}^{(23)} - \sqrt{\frac{55}{14}}\boldsymbol{Q}^{(24)}\right]\right\}^{(4)}_m
\end{aligned}
$$

Springer Series in Solid-State Sciences, Vol. 101
High Magnetic Fields in Semiconductor Physics III Editor: G. Landwehr

$$Q^{(\rho\tau)} := \left(r^{(\rho)} \times J^{(2)}\right)^{(\tau)}, \quad \mathcal{L}^{(2)} := i\left(r^{(1)} \times k^{(1)}\right)^{(2)}, \quad a_{\pm 4} := 1,\; a_0 = \frac{\sqrt{70}}{5}$$

$$B^{(0)} := -\frac{1}{\sqrt{3}}B^2 \text{ and } B \text{ in units } Ry/(\varepsilon^2\gamma_1^2\mu_B).$$

All spherical tensor operators have been constructed according to Edmonds' rule [2][3]. The field independent part H_0 contains a dominating spherically symmetric term and a cubic correction proportional to δ which reduces the symmetry to that of the group O_h. When applying an external field the symmetry is further reduced to that of C_{4h} for $\vec{B} \parallel [001]$, C_{3i} for $\vec{B} \parallel [111]$ and C_{2h} for $\vec{B} \parallel [110]$.

Our eigenfunctions are constructed as follows: We first combine the four degenerate Bloch functions, which form a $J = \frac{3}{2}$ spinor quartet , and spherical harmonics Y_M^L up to $L = 8$. The latter ones constitute the angular dependence of the envelope functions. The above combinations are chosen so that we get functions (spin orbit parts) that transform according to the representation Γ_k of the given symmetry group, i.e. C_{4h}, C_{3i} or C_{2h}, respectively. Finally a total eigenfunction of symmetry Γ_k is a linear combination of all spin orbit parts of this symmetry Γ_k with a radial function as expansion coefficient for each spin orbit part.

When applying the Hamiltonian H to the total eigenfunction we obtain a system of differential equations. The number of equations equals the number of spin orbit parts involved. For the low symmetry orientation [110] we have to use 55 coupled differential equations. These we solve by what we call the matrix method, i.e. we expand each radial function into a set of about 20 basis functions and diagonalize the corresponding 1100×1100 matrix numerically. For all states considered excellent convergence was found for $L \leq 8$.

In order to achieve a unique assignment of the large number of observed transitions to initial and final states we had to use not only the selection rules (see Table 1) but also to compute the matrix elements of the dipole operator $\vec{v} = \frac{1}{\hbar}\frac{\partial H}{\partial \vec{k}}$, which is appropriate for the magnetic field case.

Table 1: Selection rules for $\vec{B} \parallel [001]$ (C_{4h}), $\vec{B} \parallel [111]$ (C_{3i}) and $\vec{B} \parallel [110]$ (C_{2h}).

C_{4h}	Γ_5	Γ_6	Γ_7	Γ_8
Γ_5	v_0	v_+	0	v_-
Γ_6	v_-	v_0	v_+	0
Γ_7	0	v_-	v_0	v_+
Γ_8	v_+	0	v_-	v_0

C_{3i}	Γ_4	Γ_5	Γ_6
Γ_4	v_0	v_+	v_-
Γ_5	v_-	v_0	v_+
Γ_6	v_+	v_-	v_0

C_{2h}	Γ_3	Γ_4
Γ_3	v_0	v_+, v_-
Γ_4	v_+, v_-	v_0

3 Discussion

The Zeeman splitting of the D–line ($156.1\,cm^{-1}$) is shown in Fig. 1.a by a series of transmission spectra for increasing magnetic fields in [001] orientation with polarization $\vec{E} \perp \vec{B}$. As the spectra for the other orientations and polarisations are resolved equally well, they are not reproduced here. The positions of the absorption lines were determined by fitting Lorentz profiles to the data. The results for the three different orientations and both polarisations are presented in Fig. 2.a–f by dashed lines. They clearly demonstrate the anisotropy of the acceptor states (compare especially Fig. 2.d and 2.f). The theoretical transition energies (full lines) are shown as well in Fig. 2 where the symbol $k \to k'$ means a transition from a Zeeman sublevel Γ_k of $2P_{5/2}\Gamma_8$ to $\Gamma_{k'}$ of

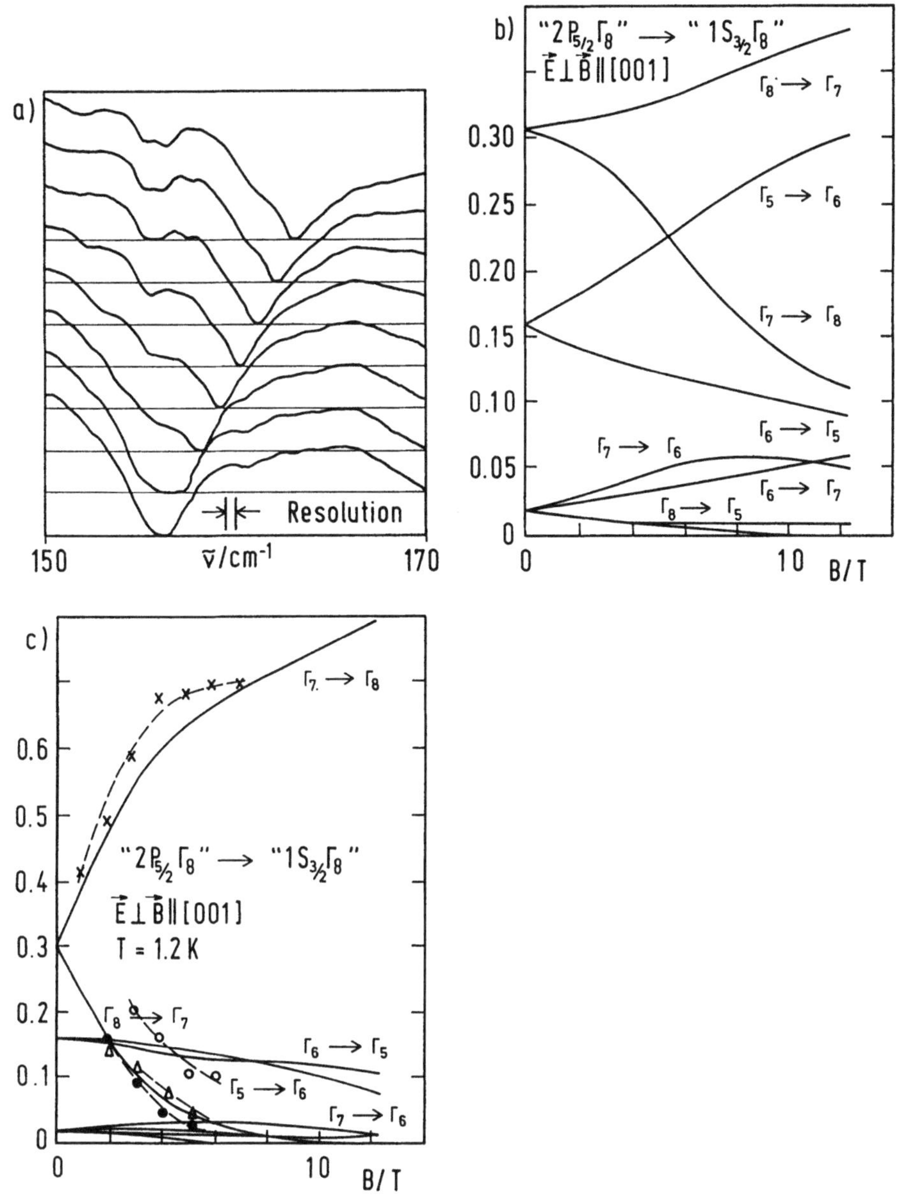

Figure 1: a) Typical transmission spectra ($\vec{E} \perp \vec{B} \parallel [001]$) for magnetic fields $B = 0, 1, 2, 3, 4, 5, 6, 7\,T$ from bottom to top at $T = 1.2\,K$.
b) Normalized dipole matrix elements $|\langle k'|v_\pm|k\rangle|^2$ (see text).
c) Normalized transition intensities ($T = 1.2\,K$): Full lines: theoretical results; dashed lines: experimentally determined absorption strengths.

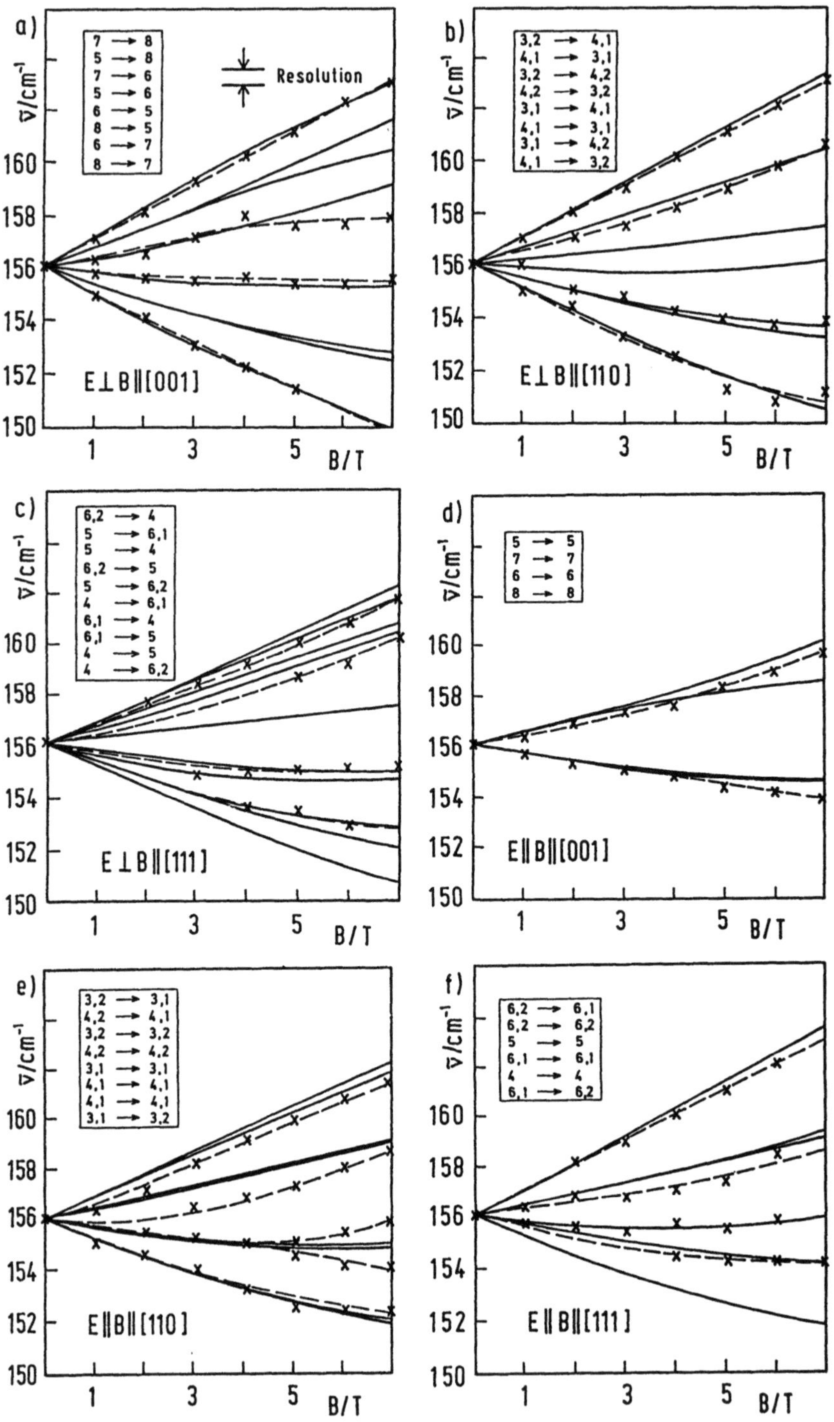

Figure 2: Field dependence of the splitting of the D–line. Comparison of theoretical (full lines) and experimental results (dashed lines).

the ground state $1S_{3/2}\Gamma_8$. The sequence of the symbols given in each plot corresponds to the sequence of the lines. For the orientation of the field [111] the representation Γ_6 occurs twice in the decomposition of O_h–Γ_8. Therefore the upper Γ_6 level is called 6,1 and the lower one is called 6,2. We apply the same ordering for the representations Γ_3 and Γ_4 of the [110] orientation.

The deviations of the calculated and the measured transition energies are remarkably small and well below the experimental resolution (0.5 cm^{-1} !) for all orientations and polarizations. This satisfying agreement is attained by exact diagonalization of the full Hamiltonian including H_{lin} and H_{qua} and cannot be achieved by simple perturbation theory. We would like to point out that some transitions that are allowed theoretically are not observed in the experiment. This can be explained quantitatively by the oscillator strengths of these transitions. Fig. 1.b presents our calculated dipole matrix elements $|\langle k'|v_\pm|k\rangle|^2$, which show a marked dependence on the magnetic field. When taking into account additionally the temperature dependent occupation of the ground state Zeeman levels, we find the normalized transition probabilities plotted in Fig. 1.c by the full lines. The observed intensities (dashed lines) follow closely the theoretical field dependence. The lines not observed just correspond to those close to the horizontal axis of Fig. 1.c.

A description of the Zeeman splitting of the G–line ($2P_{3/2}\Gamma_8 \rightarrow 1S_{3/2}\Gamma_8$) by our theory reproduces the experimental data [2] equally well. It is worth noticing that our theory is completely determined by the Luttinger parameters given below, so no further adjustable quantities enter our calculation.

We conclude our considerations by giving the g–values in Table 2. However, we must point out that the description of the splitting in terms of these g–values may lead to errors up to 100% already for 7 Tesla! Thus the g–values are of limited significance.

Table 2: g–values obtained for the Luttinger parameters $\gamma_1 = 6.65$, $\gamma_2 = 1.95$, $\gamma_3 = 2.63$, $\kappa = 1.1$; experimentally deduced values in brackets

state	$1S_{3/2}\Gamma_8$	$2P_{3/2}\Gamma_8$	$2P_{5/2}\Gamma_8$
g_1	0.21(0.22)	0.24(0.09)	−1.7(−1.3)
g_2	0.11(0.15)	0.21(0.26)	1.2(1.0)

References

[1] A. Baldereschi, N.O. Lipari, Phys. Rev. **B8**, 2697 (1973), Phys. Rev. **B9**, 1525 (1974)

[2] J. Schubert, M. Dahl, E. Bangert, in: G. Landwehr (Ed.), High Magnetic Fields in Semiconductor Physics II, Berlin, Springer–Verlag, 567 (1989)

[3] R.A. Edmonds, Angular Momentum in Quantum Mechanics, Princeton, University Press (1960)

Stimulated Interband Landau Emission due to Electromagnetic-Force Excitation of InSb at the Quantum Limit

T. Morimoto and M. Chiba

Institute of Atomic Energy, Kyoto University, Uji, Kyoto 611, Japan

It is shown that stimulated interband Landau emission due to electromagnetic-force excitation of bulk InSb is observed accompanied by a drastic saturation of the current in the quantum limit. It is also pointed out that the Landau levels indicate remarkable lowering in the vicinity of the critical current density J_C where the stimulated emission sets out.

We have recently found significant stimulated interband Landau emission, as shown in Fig.1, in bulk InSb subjected to a high magnetic field H satisfying the quantum limit condition, $\hbar\omega_{c1} > kT$, when passing a current density J as low as 20 - 50 A/cm^2 in the transverse configuration $J \perp H$ [1,2]. Here, $\omega_{c1} = eH/cm_1^*$ is the cyclotron frequency of electrons, k the Boltzmann constant, and T the temperature. This phenomenon is an inverse effect of interband magneto-optical absorption (IMO) in semiconductors [3,4], and hence the peak frequency of the emission is tunable by an applied magnetic field H through the change in the Landau levels [1,2].

The excitation of electrons and holes is achieved by the action of the $J \times H$ force, which we call Electromagnetic-Force (EMF) excitation [1,2,5], by passing a current density J ($//\hat{x}$) in quantizing, high magnetic fields H ($//\hat{z}$) through narrow-gap semiconductors, such as InSb and HgCdTe with small values of energy gap E_g (0.23 eV for InSb). Here we report the notable features connected with the excitation mechanism and the onset of stimulated emission.

Though the peak frequency ω of the emission is controlled by H, it was also found to decrease with increasing J, as shown in Fig.2. The emitting surface of the sample #M' (4.4 mm × 3.13 mm × 0.2 mm) was etched by CP-4A solution, and it did not form a Fabry-Pérot cavity. These data were taken after the sample, which was used repeatedly in the preceding experiments [1,2], was degraded due to damage suffered in the high current experiments, and hence the value of the critical current density, J_C, at which the emission sharply rises up, shifted to somewhat higher currents. Nevertheless, it still indicates the relationship between the onset of the stimulated emission and the decrease in the energy of the emitted photon that might lead to a virtual decrease of the gap energy [1,2]. However, it should be attributed to the lowering of the Landau levels due to the presence of a strong induced electric field, E_y^*, as given by $E_y^* \cong (JH/n|e|c)$; it easily reaches a value higher than 100 V/cm even at the present *cw*-operation using small current densities, in contrast to the rather small value, 10~20 V/cm, of the applied longitudinal electric field E_x [1,2,5]. Here, n and e are the concentration of electrons and the electronic charge, respectively.

Springer Series in Solid-State Sciences, Vol. 101
High Magnetic Fields in Semiconductor Physics III Editor: G. Landwehr

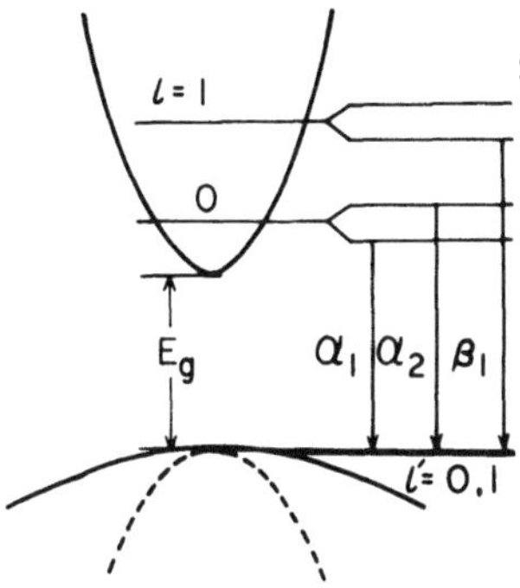

Fig.1. Landau levels of InSb at $k_z = 0$ nearby 80 K and series of interband Landau emissions. The Landau levels of heavy holes contributing mainly to the emission have not yet separated for magnetic fields up to 7 T.

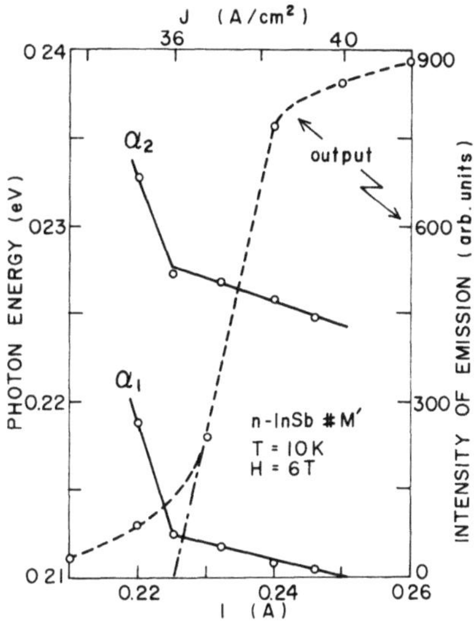

Fig.2. Current dependence of the photon energies of α_1 and α_2 emissions (solid lines), and that of the total output of the emission (dotted line) at 10 K for H = 6 T. The surfaces of the sample are chemically etched.

The peak frequency of the emission can then be written as follows for the strongest α_1 (0+ → 0') and α_2 (0- → 0') emissions (see Fig.1), neglecting the contribution from the term due to E_x :

$$\hbar\omega = E_g + \frac{1}{2}\hbar\omega_{c1}\,[1 \pm (m_1^* g/2m)] + \frac{1}{2}\hbar\omega_{c2,h} - |e|E_y^*\, l^* \ , \tag{1}$$

where g is the effective g-factor of electrons, $\omega_{c2,h}$ the cyclotron frequency of heavy holes ($m_{2,h}^* = 0.42m$) contributing mainly to the emission by EMF excitation; l^* is the mean free path of lucky electrons drifting along $-y$ direction, it is the order of the ambipolar diffusion length (~1 μm). It has been pointed out that the Landau levels are lowered by applying a strong electric field E_x perpendicular to the magnetic field H by the kinetic energy, $(1/2)(m_1^* + m_2^*)c^2(E_x/H)^2$, of drifting electron-hole pairs [6,7], but the substantial effect of annihilation of electrons (holes) due to recombination has never been included in the theoretical treatments up to now. In Eq.(1) the difficulty due to the finite lifetime of electrons is avoided by including the term with l^*. The last term of Eq.(1) represents the acquired energy of an electron travelling a distance l^* in the field E_y^* until annihilation by recombination. In fact, assuming a reasonable value of l^* of 1~2 μm, we obtain the relative lowering of Landau levels of about 20 meV as actually observed in the vicinity of J_c.

In Fig.3 the current dependence of the total output of the emission at 80 K is shown for another bulk n-InSb sample with emitting surfaces made by

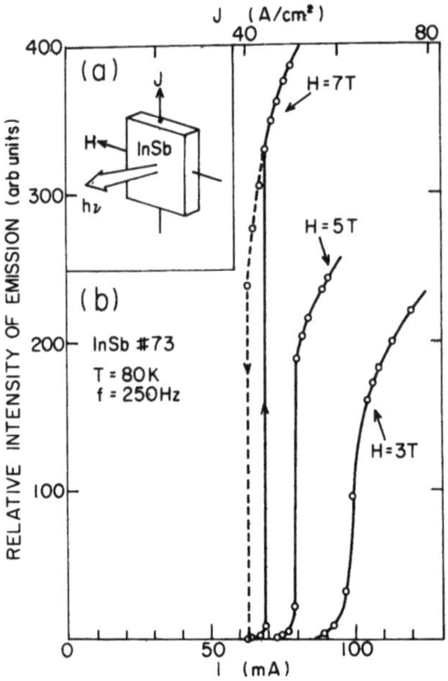

Fig.3. Current variation of the total output of emission at 80 K for a sample with cleaved surfaces constructing a Fabry-Pérot cavity. H = 3, 5 and 7 T.

cleavage to achieve a Fabry-Pérot cavity. The dimensions were 3.62 mm long, 0.63 mm wide and 0.25 mm thick. The carrier concentration n_0 and the Hall mobility μ were nominally 8×10^{14} cm^{-3} and 3×10^5 cm^2/Vs at 80 K, respectively. The current was chopped with a reference frequency 250 Hz, and the photosignal was synchronously detected for the configuration as in Fig.3(a). Owing to the high reflectivity of the emitting surface, the critical value of J_c for the stimulated emission is seen to shift to somewhat higher values than that of samples with etched emitting surfaces. However, the almost vertical rise-up of the emission observed as $H \to \infty$, which is much steeper compared with samples with etched surfaces (e.g., sample #M' as shown in Fig.2), indicates that the stimulated emission is enhanced through the feed-back of emission by the reflection at the surface. The amazingly low value of $J_c = 40$ A/cm^2 for $H = 7$T, is considered to arise from the extremely high value of the density of states for electrons at Landau levels, which is proportional to H [1,2]. Relatively, weak intensity of emission was detected when reversing the current direction, suggesting that the carrier population is concentrated to the front region of the sample for small current densities used here (for the carrier population refer to the case (b) in Fig.14 of Ref.5).

Figure 4 shows the voltage-current characteristics measured simultaneously at 80 K for constant current operation at given magnetic fields. From the comparison with Fig.3, one can clearly see that current saturation occurs in a way that $dJ/dE \to 0$ as $H \to \infty$ in the current region where the sharp stimulated emission is observed, indicating that a drastic "electric phase transition" takes place. The current saturation can be interpreted as arising from the rapid annihilation of electrons through the stimulated emission having an extremely high value of the probability of radiative recombination (> 50 %) [1,2]. It can also be seen that the saturation of the output of the emission occurs

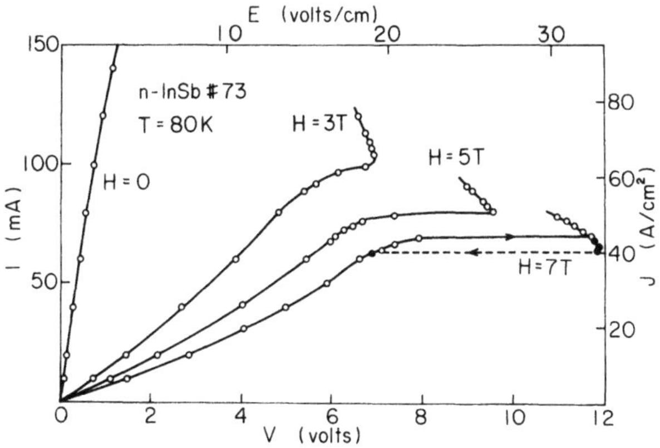

Fig.4. Voltage-current characteristics at 80 K for the sample with cleaved surfaces. $H = 0, 3, 5$ and 7 T.

accompanied by *S*-type negative resistance. It shows a rather stable hysteresis, as shown by the dotted line for H = 7T. The saturation of the output of the emission is interpreted to stem from the increase in the number of intrinsic carriers accompanied with the temperature rise [1,2]. These observations should have useful applications for tunable infrared diode lasers, and they are posing a new problem in quantum transport involving the excitation mechanism with finite electron lifetimes.

The authors are indebted to S.Ueda and Dr. S.Akai for preparation of samples and technical assistance. This work is partially supported by a Grant-in-Aid for Science Research from the Ministry of Education, Science and Culture.

1.T.Morimoto and M.Chiba, *Proc. 20th Int. Conf. Phys. Semiconductors*, Ed. E.M.Anastasskis and J.D.Joannopoulos (World Scientific, Singapore, 1990) p. 1847.
2. T.Morimoto and M.Chiba, to be published (1991).
3. For a review, see B.Lax, J. Magn. & Magn. Mater. **11**, 1 (1979).
4. C.R.Pidgeon and R.N.Brown, Phys. Rev. **146**, 575 (1966).
5. T.Morimoto and M.Chiba, Infrared Phys. **29**, 371 (1989).
6. A.G.Aronov, Soviet Phys.-Solid State **5**, 403 (1963).
7. For a review, see W.Zawdzkii, in *Physics of Solids in Intense Magnetic Fields,* Ed. E.D.Haidemenakis (Plenum, New York, 1969) p.311.

Part VII

Collective Effects, Magneto-Phonon Effect

Collective Excitations of Electrons in High Magnetic Fields

G. Meissner

Theoretische Physik, Universität des Saarlandes,
W-6600 Saarbrücken, Fed. Rep. of Germany

A simple but rigorous approach provides a study of the nature of intra–Landau–level collective excitations of interacting electrons within the fractionally filled lowest Landau level. A finite gap, due to correlations in the density fluctuations of the centers of cyclotron orbits, or no gap, because of broken magnetic translational invariance, are shown to exist rigorously in the long–wavelength limit of low–lying collective modes in two condensed phases, i.e., an incompressible quantum liquid and a two–dimensional electron quantum solid, respectively. Methods of applying our approach to examine current approximations such as the single–mode and the self–consistent–phonon approximation will be indicated. Implications to be drawn about the fractional quantum Hall effect and the two–dimensional Wigner crystal from a comparison with experimental findings and with numerical small–system calculations are briefly discussed.

1. INTRODUCTION

Interacting electrons of charge $-e$ being confined to a plane perpendicular to a magnetic field are a particularly interesting quantum many–body system, if the magnitude B of the field is sufficiently large that the number ν^{-1} of magnetic flux quanta ch/e per electron exceeds one [1], [2]. For a given area A of that plane this then corresponds to a fractional filling factor $\nu = 2\pi r_L^2 n_e < 1$, i.e., a fractional ratio of the mean areal density of electrons, $N_e/A = n_e$, and of flux quanta, $N_s/A = (2\pi r_L^2)^{-1}$, respectively, where $r_L = (c\hbar/eB)^{1/2}$ denotes the Larmor radius. The many–body ground state of such two–dimensional electron systems is expected to be a correlated incompressible quantum liquid (IQL) giving rise to a quantized Hall conductance $\sigma_{12} = \nu e^2/h$ for certain fractional filling factors $\nu = p/q < 1$, with p and q denoting mutual primes and q usually being odd. For a filling factor $\nu = 1/3$ and a typical electron density $n_e = 2 \cdot 10^{11} cm^{-2}$ [3], e.g., the required magnetic field $B = \nu^{-1} n_e(ch/e) \sim 24.84$ Tesla. Below a critical filling factor ν_c, another solid–like ground state with non–vanishing shear modulus μ is anticipated to exist. In the limit $\nu \to 0$ the triangular electron lattice formed by a classical two–dimensional (2D) Wigner crystal [4] may be identified with the ground state of this sort of a 2D electron quantum solid to be called quantum Hall crystal (QHC), since it results from broken magnetic translational invariance at zero temperature [5].

Early efforts to find experimentally the 2D Wigner crystal in high magnetic fields did not turn out to have been successful. Unexpectedly, however, they led to the discovery of the fractional quantum Hall effect (FQHE) in high–mobility samples of electron layers [6]. Since then, collective excitations of interacting electrons being confined to a plane perpendicular to high magnetic fields have been a subject of increasing interest, theoretically and experimentally. A striking difference in the collective excitation spectra of the two condensed phases exists, e.g., in the long–wavelength ($k \to 0$) limit. In the IQL, exhibiting a quantized Hall conductance σ_{12} being accompanied by minima in the

Springer Series in Solid-State Sciences, Vol. 101
High Magnetic Fields in Semiconductor Physics III Editor: G. Landwehr

longitudinal conductance σ_{11}, a finite gap is expected, whereas the QHC with a non-vanishing shear modulus μ should have a gapless collective mode. Experimentally, so far, neither the gap in the excitation spectrum of the incompressible quantum liquid nor the gapless collective mode of a magnetically induced 2D Wigner solid seems to have been observed directly. This is partly due to the fact that the range of accessible wave vectors is not wide enough, at present.

Therefore, an approach which provides a quantitative analysis of basic differences in the collective excitations of the two phases not resting on to restrictive approximations and assumptions is of great interest. For a fractionally filled lowest Landau level such an analysis is feasible, if the dispersion relation between frequencies ω and conserved wave vectors $\underset{\sim}{k}$ is derived from the response function for the density of the centers of cyclotron orbits, i.e., of the so-called guiding center coordinates. It is important in this context to notice, that the elementary excitations of our 2D interacting electron system in general have longitudinal as well as transverse components both in the IQL-phase and in the QHC-phase. In particular, for a charged electron fluid subjected to a perpendicular magnetic field it is possible to support transverse modes, too, because of the Lorentz force perpendicular to the particle velocity which can supply the necessary transverse restoring force. Thus, intra-Landau-level collective excitations of a fractionally filled lowest Landau level are quasi-transverse in nature. Their energies, $\hbar\omega(\underset{\sim}{k})$, being of the order of the Coulomb interaction between adjacent guiding centers, $e^2/\epsilon 2r_L$, in the limit of high magnetic fields ($r_L \to 0$) are clearly small in comparison to inter-Landau-level excitation energies being of the order $\hbar\omega_c = \hbar^2/m^* r_L{}^2$ and behaving quasi-longitudinal (ϵ: background dielectric constant, m*: effective electron mass, ω_c: cyclotron frequency).

Applying sum-rule techniques, a gap can be shown to exist exactly in the long-wavelength limit of these intra-Landau-level collective excitations of the IQL [7]. The concept of broken magnetic translational invariance may be employed in order to reveal the possibility of the incompressible quantum liquid of electrons to condense into a quantum Hall crystal below a critical filling factor ν_c. From the dispersion relation $\omega(\underset{\sim}{k}) \sim k^{3/2}$ obtained for the gapless symmetry-restoring Goldstone mode in the long-wavelength limit, this mode of the QHC can be identified with the magnetophonon of a 2D Wigner crystal [3]. It is worth noting that the difference between the $k \to 0$ magnetophonon dipersion of the QHC and the $k=0$ excitation gap of the IQL in such an approach is associated with the different long-wavelength behavior of the appropriate static susceptibilities in the two phases being proportional to k and k^4, respectively. Features of the massive degeneracy associated with 2D electron motion in a magnetic field, finally, are carried over in part into both condensed phases, by the non-commutivity of the guiding center coordinates and their density fluctuations, respectively.

The many-body Hamiltonian and the response function we consider in presenting this unified simple approach are described in the following Sect.2. Methods of applying our sum-rule techniques in both condensed phases to derive rigorous results and to examine current approximations, such as the single-mode approximation [8] for the IQL-phase and the renormalized harmonic approximation [9] for the QHC-phase, are presented in Sect.3. The possibility of going beyond such approximations in various respects is being indicated. Implications to be drawn from a comparison with experimental findings and with numerical small-system calculations are briefly discussed in Sect.4.

2. MODEL HAMILTONIAN AND RESPONSE FUNCTION

The many-body Hamiltonian

$$H = \sum_j (\underset{\sim}{p}_j - \frac{q}{c} \underset{\sim}{A}_j)^2/2m^* + V(\{\underset{\sim}{x}_j\}) \qquad (1)$$

with an interaction term $V = V_{e-e} + V_{e-b} + V_{b-b}$, containing the electron–electron interaction V_{e-e} and the interaction with a homogeneous charge–compensating background V_{e-b} together with the self–interaction of that background V_{b-b}, is considered to be sufficiently general to describe any phase, liquid or solid–like, of electrons of charge $q = -e$, being subjected to a perpendicular magnetic field $\underset{\sim}{B} = \underset{\sim}{\nabla}\times\underset{\sim}{A} = B\,\underset{\sim}{n}$. With the electron density n_e in the plane perpendicular to $\underset{\sim}{n} = (0,0,1)$ and the Fourier transformed Coulomb potential $v(\underset{\sim}{k}) = \int d^2r\, \exp(-i\underset{\sim}{k}\cdot\underset{\sim}{r})e^2/\epsilon r = 2\pi e^2/\epsilon k$ we have explicitly

$$V = \int\frac{d^2k}{(2\pi)^2}v(\underset{\sim}{k})\Big\{\frac{1}{2}\sum_{i\neq j} e^{i\underset{\sim}{k}\cdot(\underset{\sim}{x}_i-\underset{\sim}{x}_j)} - n_e\int d^2r\sum_i e^{i\underset{\sim}{k}\cdot(\underset{\sim}{x}_i-\underset{\sim}{r})} + \frac{1}{2}n_e^2\int d^2r\int d^2r' e^{i\underset{\sim}{k}\cdot(\underset{\sim}{r}-\underset{\sim}{r}')}\Big\}.$$

Cartesian components of position operators $\underset{\sim}{x}$ and canonical momentum operators $\underset{\sim}{p}$ of electrons obey canonical commutation relations, i.e., $[x_\alpha,p_\beta] = i\hbar\delta_{\alpha\beta}$, where $\alpha,\beta = 1,2$. Rewriting the kinetic energy part of the Hamiltonian

$$H - V = \sum_j (\underset{\sim}{p}_j - \frac{q}{c}\underset{\sim}{A}_j)^2/2m^* = \frac{1}{2}m^*\omega_c^2\sum_j \underset{\sim}{x}_L^2(j)$$

in terms of the coordinates, $\underset{\sim}{x}_L(j) = ((\underset{\sim}{p}_j - \frac{q}{c}\underset{\sim}{A}_j)\times\underset{\sim}{n})/m^*\omega_c$, of the cyclotron motion, eliminates any dependence on guiding center coordinates, $\underset{\sim}{X}(j) = \underset{\sim}{x}_j - \underset{\sim}{x}_L(j)$, which manifests the well–known massive degeneracy associated with 2D free electron motion perpendicular to a magnetic field. The single–electron wave function of the lowest Landau level, $<r\varphi|m0> = (2\pi r_L^2 m!)^{-1/2}(r/\sqrt{2}r_L)^m e^{-im\varphi}\exp(-r^2/4r_L^2)$, in polar coordinates (r,φ) reflects this degeneracy via the quantum numbers $m = 0,1,\ldots,N_s-1$ belonging all to the same energy $\hbar\omega_c/2 = \hbar^2/2m^*r_L^2$.

Using this sort of separation of variables, together with the canonical commutation relations, and with the linear spatial dependence of the vector potential for uniform magnetic fields, as, e.g., $\underset{\sim}{A}_j = (\underset{\sim}{B}\times\underset{\sim}{x}_j)/2$ in symmetric gauge, one finds the familiar commutation relations of the guiding center coordinates [10]

$$[X_\alpha(i),X_\beta(j)] = ir_L^2\delta_{ij}\epsilon_{\alpha\beta}\ , \tag{2}$$

where $\epsilon_{\alpha\beta}$ denotes components of the antisymmetric tensor in 2D with elements $\epsilon_{11} = \epsilon_{22} = 0$ and $\epsilon_{12} = -\epsilon_{21} = 1$. The non–commutivity of the guiding centers gives rise to a non–commutivity of their density fluctuations $\underset{\sim}{\Delta}(k) = \sum_j \exp(-i\underset{\sim}{k}\cdot\underset{\sim}{X}(j))$ as well, being reflected in the commutation relations

$$[\Delta(\underset{\sim}{k}),\Delta(\underset{\sim}{k}')] = -2i\sin(k_\alpha\epsilon_{\alpha\beta}k'_\beta r_L^2/2)\Delta(\underset{\sim}{k}+\underset{\sim}{k}')\ . \tag{3}$$

They in turn are a signature of the transverse nature of collective excitations being associated with correlations of these density fluctuations. With such commutation relations disposable, one now is in a position to obtain by means of sum–rule techniques the response function of the density fluctuations of the guiding centers being defined as

$$\chi_{\Delta\Delta}(\underset{\sim}{k},z = \omega + i\eta) = \frac{i}{\hbar}\int_0^{+\infty} dt\, e^{izt} <[\Delta(\underset{\sim}{k},t),\Delta(-\underset{\sim}{k},0)]> \tag{4}$$

and being of relevance for intra–Landau–level collective excitations. For the sake of theoretical simplicity, one uses to eliminate the fast degrees of freedom of the cyclotron motion already in the Hamiltonian in a mean–field–like fashion [11] which finally amounts to replacing the expressions of Eq.(1) by the many–body model Hamiltonian

$$\tilde{H} = (\tfrac{1}{2}\hbar\omega_c - \frac{e^2}{\epsilon r_L}\tfrac{1}{2}\sqrt{\pi/2})N_e + \tfrac{1}{2}\int\frac{d^2k}{(2\pi)^2}\, v(\underset{\sim}{k})e^{-k^2r_L^2/2}\Delta(\underset{\sim}{k})\Delta(-\underset{\sim}{k})\,.$$

3. COLLECTIVE EXCITATIONS

From zeros of the real part of the inverse of our density response function, i.e.,

$$\mathrm{Re}\,\chi^{-1}_{\Delta\Delta}(\underset{\sim}{k},\omega(\underset{\sim}{k})) \equiv 0, \tag{5}$$

the dispersion relation between frequencies ω and conserved wave vectors $\underset{\sim}{k}$ of the intra–Landau–level collective excitations, in principle, should be obtained not only in the IQL–phase but also in the QHC–phase. Physically, in the QHC–phase the expectation values $<\underset{\sim}{X}(j)>$ of the guiding center coordinates should form some kind of lattice at sufficiently high magnetic fields B or small Larmor radii $r_L = (c\hbar/eB)^{1/2}$, respectively, i.e., below a critical filling factor ν_c, in order to minimize the Coulomb interaction between guiding centers. First, however, we want to briefly review the present status of the theory of collective excitations in the liquid phase.

3.1. Incompressible quantum liquid (IQL)

Applying sum–rule techniques the following exact spectral representation for the inverse of our density response function

$$\overline{\omega}(\underset{\sim}{k})\chi^{-1}_{\Delta\Delta}(\underset{\sim}{k},z=\omega+i\eta) = -z^2 + \overline{\omega}(\underset{\sim}{k})\chi^{-1}_{\Delta\Delta}(\underset{\sim}{k},0) - z\,\overline{\omega}(\underset{\sim}{k})\int\frac{d\omega'}{\pi}\frac{\Gamma_{\Delta\Delta}(\underset{\sim}{k},\omega')}{\omega' - z} \tag{6}$$

was obtained [7]. The spectral width function $\Gamma_{\Delta\Delta}(\underset{\sim}{k},\omega)$ in (6) has the same symmetry properties as the spectral function $\chi''_{\Delta\Delta}(\underset{\sim}{k},\omega)$ being obtained from the Fourier transform of the commutator of the density–fluctuation operators. Therefore, the static $(t = 0)$ structure factor obeys a sum rule which at zero temperature is given as

$$S_{\Delta\Delta}(\underset{\sim}{k}) = <\Delta(\underset{\sim}{k})\Delta(-\underset{\sim}{k})>N_e^{-1} = (\hbar/n_e)\int_0^{\infty}\chi''_{\Delta\Delta}(\underset{\sim}{k},\omega)d\omega/\pi. \tag{7}$$

For the inverse of the static $(z = 0)$ susceptibility, $\chi^{-1}_{\Delta\Delta}(\underset{\sim}{k},0)$, the rigorous sum rule

$$\overline{\omega}(\underset{\sim}{k})\chi^{-1}_{\Delta\Delta}(\underset{\sim}{k},0) = \overline{\omega^3}(\underset{\sim}{k})/\overline{\omega}(\underset{\sim}{k}) - \overline{\omega}(\underset{\sim}{k})\int\frac{d\omega}{\pi}\Gamma_{\Delta\Delta}(\underset{\sim}{k},\omega) \tag{8}$$

holds. Evaluating the first and third frequency moment explicitly we found both of them

$$\overline{\omega^n}(\underset{\sim}{k}) = \int\omega^n\chi''_{\Delta\Delta}(\underset{\sim}{k},\omega)\,d\omega/\pi = C_n\,k^4 + \ldots \qquad (n = 1,3)$$

to behave like k^4 in the $k \to 0$ limit.The expression obtained for

$$C_1 = (r_L^2 n_e/16\hbar A)\sum_{\underset{\sim}{g}\neq 0} S_{\Delta\Delta}(\underset{\sim}{g})\,(\underset{\sim}{g}\cdot\partial_{\underset{\sim}{g}})\,(\underset{\sim}{g}\cdot\partial_{\underset{\sim}{g}})v(\underset{\sim}{g})\exp(-q^2 r_L{}^2/2)$$

turns out to depend on two–particle density correlations only, via the static structure factor $S_{\Delta\Delta}(\underset{\sim}{q})$, whereas the rather lengthy one for C_3 also contains four–particle correlations. From Eq. (8) we therefore conclude quite generally that $\chi_{\Delta\Delta}(\underset{\sim}{k},0) \sim k^4$ vanishes rapidly for small k, i.e., long–wavelength density fluctuations are strongly suppressed in our incompressible system. Using (5) and (6) one may conclude, moreover, a gap to exist at $k = 0$ in the collective excitation spectrum $\omega^2(\underset{\sim}{k})$ of the IQL. Mathematically, these results hold at least as long as the ω–integral over $\Gamma_{\Delta\Delta}$ in (8) for $k \to 0$ diverges slower than k^{-4}, if at all. Physically, rapidly varying parts, which may appear in $\Gamma_{\Delta\Delta}$ due to a coupling to other degrees of freedom, e.g., could cause the gap to vanish, if they led to a faster divergence of that ω–integral.

All of the foregoing results are exact. However, they can also be used for studying conveniently approximations of our many–body system. The single–mode–approximation (SMA) is readily obtained, e.g., if terms with $\Gamma_{\Delta\Delta}$ are disregarded in (6) and (8) for finite k, too. Omitting all unnecessary variables one then obtains in an obvious notation

$$\overline{\omega}\,\chi_0^{-1}(z) = -z^2 + \overline{\omega}\,\chi_0^{-1}(0) \ , \tag{6a}$$

$$\overline{\omega}\,\chi_0^{-1}(0) = \overline{\omega^3}/\,\overline{\omega} \ , \tag{8a}$$

and thus according to (5) for the collective excitation spectrum of the IQL–phase

$$\omega_0^2 = \overline{\omega}\,\chi_0^{-1}(0) = \overline{\omega^3}/\,\overline{\omega} \, . \tag{5a}$$

The gap of $\omega_0^2(\underset{\sim}{k})$ in Eq. (5a) at $k = 0$, results from the same k^4–behavior of both frequency–moments, $\overline{\omega^3}$ and $\overline{\omega}$, with their constant ratio being given as C_3/C_1. Using the spectral function $\chi_0''(\omega) = (\pi\overline{\omega}/2\omega_0)\{\delta(\omega-\omega_0) - \delta(\omega+\omega_0)\}$ in SMA one recovers then, via the sum rule (7) for the static structure factor, Feynman's relation

$$\omega_0(\underset{\sim}{k}) = \frac{\hbar\,\pi}{\nu}\, r_L^2\, \overline{\omega}(\underset{\sim}{k})/\ S_0(\underset{\sim}{k}) \ . \tag{7a}$$

With a gap in ω_0 and with the k^4–behavior of $\overline{\omega}$ in the $k \to 0$ limit, the k^4–behavior of the static structure factor now follows immediately from (7a) without having used the ground–state wave function at all. This k^4–behavior in our approach is actually associated with the constant ratio of $\overline{\omega^3}/\overline{\omega}$ in the $k \to 0$ limit which also provides the necessary quasi–transverse restoring force for the IQL–phase of the uniform system of interacting electrons in high magnetic fields. Another feature to be noticed quite generally from Feynman's relation in our approach is the magneto–roton minimum [8] in the excitation spectrum $\omega_0(\underset{\sim}{k})$ located at the peak of the static structure factor $S_0(\underset{\sim}{k})$ for wave vectors $\underset{\sim}{k}$ being reciprocal to nearest neighbor distances in the IQL–phase in agreement with small–system calculations and Monte Carlo studies [12].

3.2 Quantum Hall crystal (QHC)

For the question, whether there are reasons for the gap in the collective excitation spectrum $\omega(\underset{\sim}{k})$ of the IQL–phase to disappear, it is important to notice that there is a continuous symmetry in our correlated electron system, since the corresponding many–body Hamiltonian is invariant under infinitesimal magnetic translations [5].

Spontaneous symmetry breaking can thus give rise to a new ground state with crystalline–like correlations of displacements $u_\alpha(j) = X_\alpha(j) - R_\alpha(j)$ from the expectation values $R_\alpha(j) \equiv \langle X_\alpha(j) \rangle$ of guiding center coordinates. From Feynman's relation (7a) for the excitation spectrum $\omega_0(\underset{\sim}{k})$ of the IQL–phase one may expect the liquid–solid transition to occur again quite generally via softening of the magneto–roton minimum, with the peak of the static structure factor increasing and with the filling factor decreasing, respectively. The collapse of the gap at a critical value ν_c should take place for a certain finite wave vector $\underset{\sim}{k} = \underset{\sim}{G} \neq 0$ becoming a reciprocal lattice vector $\underset{\sim}{G}$ of the QHC–phase and therefore being close to a reciprocal nearest–neighbor distance in the IQL–phase, too. Due to lattice periodicity, the resulting gapless Goldstone mode restoring the broken magnetic translational invariance in the QHC, also develops at $\underset{\sim}{k} = 0$.

The dispersion relation in the QHC–phase can directly be determined from zeros of the 2×2 determinant of the inverse response function

$$|\mathrm{Re}\, \chi^{-1}_{\alpha\beta}(\underset{\sim}{k}, \omega(\underset{\sim}{k}))| \equiv 0 \tag{9}$$

for guiding center displacement fluctuations, $u_\alpha(\underset{\sim}{k}) = N_e^{-1/2} \sum_j u_\alpha(j) \exp(i\underset{\sim}{k} \cdot \underset{\sim}{R}(j))$, where the commutation relations

$$[u_\alpha(i), u_\beta(j)] = i r_L^2 \delta_{ij} \epsilon_{\alpha\beta} \tag{2a}$$

agree with those of (2) for the guiding center coordinates. The exact spectral representation of the inverse of the displacement response function replacing (6) now takes the form

$$\langle \omega^0_{\alpha\gamma} \rangle \chi^{-1}_{\gamma\beta}(z) = -z\delta_{\alpha\beta} + \langle \omega^0_{\alpha\gamma} \rangle \chi^{-1}_{\gamma\beta}(0) - z \langle \omega^0_{\alpha\gamma} \rangle \int \frac{d\omega}{\pi} \frac{\Gamma_{\gamma\beta}(\omega)}{\omega - z} \tag{10}$$

where all $\underset{\sim}{k}$–variables have been omitted. A rigorous sum rule relating components $\chi^{-1}_{\alpha\beta}(0)$ of the inverse of the static susceptibility and the ω–integral over components $\Gamma_{\alpha\beta}(\omega)$ of the spectral width function, i.e.,

$$\langle \omega^0_{\alpha\gamma} \rangle \chi^{-1}_{\gamma\beta}(0) = \langle \omega^0_{\alpha\gamma} \rangle^{-1} \langle \omega^1_{\gamma\beta} \rangle - \langle \omega^0_{\alpha\gamma} \rangle \int \frac{d\omega}{\pi} \Gamma_{\gamma\beta}(\omega) \tag{11}$$

is now determined by the zeroth and first frequency moment

$$\langle \omega^n_{\alpha\beta}(\underset{\sim}{k}) \rangle = \int \omega^n \chi''_{\alpha\beta}(\underset{\sim}{k}, \omega) d\omega/\pi \qquad (n = 0,1)$$

of the spectral function $\chi''_{\alpha\beta}(\underset{\sim}{k}, \omega)$. Making extensive use of the commutation relations (2a) we obtain again exact expressions for both of them. The zeroth frequency–moment $\langle \omega^0_{\alpha\beta}(\underset{\sim}{k}) \rangle = i\epsilon_{\alpha\beta}/m^*\omega_c$, being k–independent, is closely related to Kohn's theorem [13], [14]. The first frequency–moment $\langle \omega^1_{\alpha\beta}(\underset{\sim}{k}) \rangle = \epsilon_{\alpha\gamma}\epsilon_{\beta\delta}\, \phi^{(\infty)}_{\gamma\delta}(\underset{\sim}{k}) / m^*\omega_c^2$ is determined by the so–called non–dispersive part $\phi^{(\infty)}_{\alpha\beta}(\underset{\sim}{k})$ of the dynamical matrix [15] which in the $k \to 0$ limit behaves like k^2, only after subtraction of a non–analytic term, i.e.,

$$\phi^{(\infty)}_{\alpha\beta}(\underset{\sim}{k}) = 2\pi e^2 n_e k_\alpha k_\beta / (k\epsilon m^*) + Z^{(\infty)}_{\alpha\gamma\beta\delta} k_\gamma k_\delta + O_{\alpha\beta}(k^4) \, .$$

Combining (10) and (11) one thus obtains according to (9) a rigorous expression for the long-wavelength limit of the spectrum of collective excitations, being isotropic for a triangular lattice, i.e.,

$$\omega^2(\underset{\sim}{k}) = \frac{2\pi e^2}{m^{*2}\omega_c^2 \epsilon} \mu\, k^3 + O(k^4). \tag{12}$$

The isothermal shear modulus μ in (12) is rigorously related to the second derivative of the free energy with respect to displacement deformations acording to the generalized elastic sum rule [15].

These exact results may again be used for studying conveniently approximations. First it should be noticed that the magneto-phonon dispersion $\omega(\underset{\sim}{k}) \sim k^{3/2}$ of (12) is precisely the same as that found in harmonic approximation (HA) [9], and agrees also with that of the lower hybrid mode of the 2D Wigner solid [16]. Moreover, in complete analogy to the SMA in the IQL-phase, e.g., the self-consistent-phonon approximation (SCPA) is obtained by disregarding terms containing the spectral width function $\Gamma_{\alpha\beta}$ in (10) and (11) for finite wave vectors $\underset{\sim}{k}$, too, i.e.,

$$<\omega^0_{\alpha\gamma}> \chi^{-1}_{0\gamma\beta}(\underset{\sim}{k},z) = -z\delta_{\alpha\beta} + <\omega^0_{\alpha\gamma}>^{-1} <\omega^1_{\gamma\beta}(\underset{\sim}{k})>.$$

From this resulting equation it becomes then clear that the quasi-transverse restoring force in the crystalline-like phase is associated with the product of the inverse zeroth frequency-moment times the first frequency-moment being related to the dynamical matrix in conventional lattice dynamics. From $|\mathrm{Re}\chi^{-1}_{0\alpha\beta}(\underset{\sim}{k},z = \omega_0(\underset{\sim}{k}))| \equiv 0$ the dispersion relation $\omega_0^2(\underset{\sim}{k}) = \det|\phi^{(\infty)}_{\alpha\beta}(\underset{\sim}{k})|\,\omega_c^{-2}$ may readily be derived. Results as shown in Fig. 1, illustrate aspects of Fermi statistics in a modified HA [9] obtained practically from replacing $v(\underset{\sim}{k})$ by $v_{eff}(\underset{\sim}{k}) = v(\underset{\sim}{k})[1 - \sqrt{2\pi}(kr_L/2)\, I_0(k^2 r_L{}^2/4)\exp(k^2 r_L{}^2/4)]$ in the model Hamiltonian $\mathfrak{H}$ with $I_0(x) = (2\pi)^{-1}\int_0^{2\pi} d\varphi \exp(x\cos\varphi)$ denoting the modified Bessel function. Finally, it is worth noting that a consistent determination of coefficients in an expansion of the ground-state energy into powers of the filling factor up to terms of the order $\nu^{5/2}$, requires to include intermediate exchange of magneto-phonons, i.e., dispersive anharmonicities ($\Gamma_{\alpha\beta} \neq 0$) [9], having been disregarded in SCPA [17].

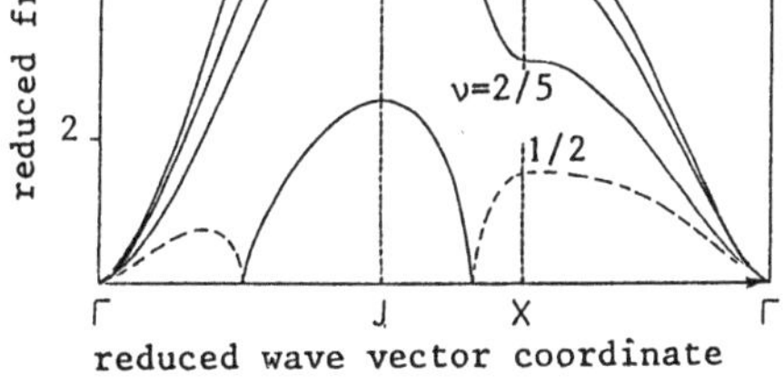

Fig. 1: Dispersion curves of magneto-phonon frequencies $\omega_0[\Omega^2/\omega_c]$ versus wave vectors $\underset{\sim}{k}[a/2\pi]$ in a particle-hole symmetric HA for a triangular lattice and for various filling factors ν using reduced units, where: $(r_L/a) \equiv (\sqrt{3}/4\pi)^{1/2}\,\nu^{1/2}$ and $\Omega \equiv (e^2/m^*a^3)^{1/2}$ with $a^2 = 2/(n_e\sqrt{3})$. Dashed lines denote negative values of $\omega_0^2(\underset{\sim}{k})$ indicating the non-existence of a solid phase at $\nu = 1/2$.

4. DISCUSSION AND CONCLUSIONS

From a fairly large amount of theoretical work, analytically and numerically [12], the nature of the low–lying collective modes within the lowest Landau level is considered to have been understood quite well. Experimentally, however, hardly any of the characteristic features of these collective excitations seems to have been observed directly, so far. This might partly be due to the fact that the range of accessible wave vectors needed for such experiments is not wide enough, at present. However, some of the explanations given for these features were incomplete, too. Therefore, a simple but rigorous approach providing a study of the nature of intra–Landau–level collective excitations should be of some interest.

The main advantages of the present theory of sum–rule techniques are, to provide simultaneously: (i) a clear and useful picture of the nature of the low–lying collective modes, (ii) a method to obtain rigorous results in limiting cases, and (iii) a convenient way to derive and analyse physically relevant approximations. Thus, differences and similarities in the incompressible quantum liquid and quantum solid phase of interacting electrons in high magnetic fields, e.g., can be associated quite generally with differences in their respective symmetries and various results are obtained without employing approximate wave functions.

This has been demonstrated rather explicitly in this paper for the finite–gap versus gapless behavior of collective excitations within the lowest Landau level in the long–wavelength limit. An important role in obtaining these results is played by the respective restoring forces. They are associated with those frequency moments in the spectral representations of the inverse response functions which enter into the non–dispersive, i.e., frequency–independent parts. These dissipationless parts of the inverse of the response functions are then determined by frequency moments which by the use of sum rules can be calculated independently from the nature of the single–particle excitations being fractionally charged or not.

For the IQL–phase of the uniform system of interacting electrons we have shown that the necessary quasi–transverse restoring force at k=0 is provided by the constant ratio of the third and first frequency–moment of the spectral function of guiding center densities. Therefore, the finite gap is a feature which can exist quite generally, becoming realized for fractional filling factors $\nu=p/q$, e.g., because of a novel broken symmetry [5], [18]. In the QHC–phase combinations of the first and zeroth frequency moment of the spectral function of guiding center displacements play a similar role for the quasi–transverse restoring force being related to the dynamical matrix of conventional lattice dynamics. A non–vanishing shear modulus $\mu > 0$, therefore, indicates the crystalline–like nature of that phase with broken magnetic translational invariance , as having been found, e.g., in a range of values $\nu < 0.4$ in an approach allowing for quantum statistical features and anharmonicities to one–loop order [7],[9]. A minimum discovered in the ν–dependence of μ in that approximation could possibly indicate a competition between the QHC–phase and the IQL–phase in the vicinity of fractional filling factors exhibiting quantized Hall conductance. This, finally, would be consistent with a diverging longitudinal resistivity having recently been observed experimentally in a narrow region around $\nu=1/5$ [19].

ACKNOWLEDGEMENTS

I would like to thank numerous colleagues for useful discussions on collective excitations in high magnetic fields at various occasions over many years, in particular, V.J. Emery, J. Hajdu, and G. Landwehr.

5. REFERENCES

1. R. B. Laughlin, Phys. Rev. Lett. 50,1395 (1983).
2. D. Arovas, J. R. Schrieffer, and F. Wilzcek, Phys. Rev. Lett. 53, 722 (1984).
3. T. Ando, A. B. Fowler, and F. Stern, Rev. Mod. Phys. 54, 437 (1982).
4. G. Meissner, H. Namaizawa, and M. Voss, Phys. Rev. B13, 1370 (1976); L. Bonsall and A. A. Maradudin, Phys. Rev. B15, 1959 (1977).

5. G. Meissner: In Recent Developments in Mathematical Physics, Eds. H. Mitter and L. Pittner (Springer, Berlin, Heidelberg, 1987), p. 275.
6. D. C. Tsui, H. L. Störmer, and A. C. Gossard, Phys. Rev. Lett. 48, 1559 (1982).
7. G. Meissner and U. Brockstieger: In Series in Solid State Sciences, Ed. G. Landwehr, 87 (Springer, Berlin, Heidelberg, 1989), p. 80.
8. S.M. Girvin, A.M. MacDonald, and P.M. Platzman, Phys. Rev. Lett. 54, 581 (1985).
9. G. Meissner and U. Brockstieger: In Series in Solid State Sciences, Ed. G. Landwehr, 71 (Springer, Berlin, Heidelberg, 1987), p. 85.
10. R. Kubo, S. J. Miyake, and N. Hashitsume: Solid State Physics, Ed. F. Seitz and D. Turnbull (Academic Press, 1965), 17, 269.
11. See, e.g., D. Yoshioka and H. Fukuyama, J. Phys. Soc. Jap. 47, 394 (1979).
12. T. Chakraborty and P. Pietiläinen, The Fractional Quantum Hall Effect (Springer, Berlin, Heidelberg, 1988).
13. W. Kohn, Phys. Rev. 123, 1242 (1961).
14. G. Meissner: In Lecture Notes in Physics, Ed. G. Landwehr, 177 (Springer, Berlin, Heidelberg, 1983), p. 70.
15. G. Meissner, Phys. Rev. B1, 1822 (1970).
16. G. Meissner, Z. Physik, B23, 173 (1976); A.V. Chaplik, Zh. Eksp. Teor. Fiz. 62, 746 (1972) [Sov. Phys.– JETP 35, 395 (1972)].
17. P.K. Lam and S.M. Girvin, Phys. Rev. B30, 473 (1984).
18. N. Read, Phys.Rev. Lett. 62, 86 (1989); P.W. Anderson, Phys.Rev. B28, 2264 (1983).
19. H.W. Jiang, R.L. Willett, H.L. Störmer, D.C. Tsui, L.N. Pfeiffer, and K.W. West, Phys. Rev. Lett. 65, 633 (1990).

Magnetoplasma Effects in Tunable Mesoscopic Systems on Si

J. Alsmeier

Sektion Physik, Universität München,
Geschwister-Scholl-Platz 1, W-8000 München 22, Fed. Rep. of Germany

A résumé of far-infrared spectroscopic studies on narrow two-dimensional as well as on one- and zero-dimensional inversion electron systems on Si is given. Stacked gate devices are described which are found to be best suited to study tunable low dimensional electron systems on Si. Experimental results are presented for narrow electron channels and quantum dots in high magnetic fields.

1. Introduction

Far-infrared (FIR) spectroscopy has been a successful and widely used tool to study magnetoplasma excitations in two-dimensional (2D) as well as in one-dimensional (1D) and zero-dimensional (0D) electron systems [1]. In laterally confined systems these excitations are observed as dimensional resonances which, in the classical regime, reflect the width of the confining potential [2]. In the special case of an external parabolic potential, which is experimentally approximated in most 1D or 0D structures, a single resonance is observed which is now understood to reflect the harmonic oscillator frequency of the center of mass motion of the electrons [3] and therefore gives a measure of the strength of the bare confining potential.

On Si 1D electron systems are up to now only realized in stacked gate devices [4,5] in contrast to low dimensional systems on GaAs heterostrucures, which for example can be achieved in split gate configurations [6]. The reason for this is the relatively high effective electron mass in Si and the low electron mobility compared to molecular beam epitaxy grown heterostructures. In our FIR measurements stacked gate devices not only allow the creation of 1D and 0D electron inversion systems but also prove to be highly tunable devices in which lateral potential strength, effective width of the electron system and, of course, the electron density can be changed continuously and nearly independently.

We first describe the sample preparation and the concept of our stacked gate device. For relatively wide channels we study the collective intraband excitations, i.e. several dimensional resonances, which were theoretically treated by Eliasson et al. [7]. In the transition regime from a density modulated two-dimensional (2D) electron gas to an array of narrow electron channels we elucidate the role of two different types of collective excitations that are dimensional resonances and plasmons, respectively. We then present experimental results on narrow 2D electron stripes as well as one-dimensional electron channels. From our results we can conclude that the external confinement of the electrons in the case of very narrow electron channels is essentially parabolic and only one single dimensional resonance is excited. The 1D subband spacings in these structures is estimated by comparison with samples prepared for DC-measurements as will be discussed later. Finally we present far-infrared spectra of electron dots containing 20 to 140 electrons per dot. In high magnetic fields the dot spectra show a splitting into a cyclotron resonance like and an edge like mode. Such splitting was initially identified in large electron disks on GaAs heterostructures [3]. We will demostrate that the edge mode is rather sensitive to the confinement potential.

Springer Series in Solid-State Sciences, Vol. 101
High Magnetic Fields in Semiconductor Physics III Editor: G. Landwehr

2. Stacked Gate Devices

The preparation starts with thermally oxidized (100)p-Si with oxide thickness d_1 between 50 and 100nm. Via holographic lithography, gratings with periods a in the range of 1 and 2µm but with high aspect ratios a/t between 4 and 5 are defined, where t is the geometrical width of the openings in the grating as shown in Fig. 1. Cross gratings with a period of a=400nm are fabricated by two consecutive exposures, the second after the sample is turned round by 90 degrees. To prepare the grating or mesh like bottom gate we evaporate about 10nm Tungsten or NiCr metal over an area of around $8mm^2$ followed by a lift-off step. After deposition of an SiO_2 isolator with thickness d_2=150nm in a PECVD (plasma enhanced chemical vapor deposition) reactor a second 5nm thick gate on top is evaporated. The gates have high sheet resistances (R≥1kΩ) and are therefore semi-transparent for FIR radiation.

Both gates are controlled independently with external voltage sources. Thus we can achieve widely tunable density modulated 2D electron systems as is sketched in the lower part of Fig. 1 marked by (a). Different channel widths are easily selected in a sample where the aspect ratio a/t of the bottom gate grid is different from unity either by inducing the electrons beneath the bottom gate stripes (b) or between the stripes (c). In the latter case, the so-called gap-confinement mode, the electrons are induced through the bottom gate openings via a positive voltage at the top gate. In configuration (c) electron dots can be formed with a mesh like bottom gate. From the infrared response we conclude that both electrons and holes can be induced side by side separated by a depletion zone thus forming a lateral nipi structure. The lateral depletion length w_{depl} is shorter than the bulk depletion length of the doped semiconductor material since the periodicity $a<w_{depl}$. For bias conditions, where electrons and holes are induced, the lateral potential modulation is of the order of the Si band gap.

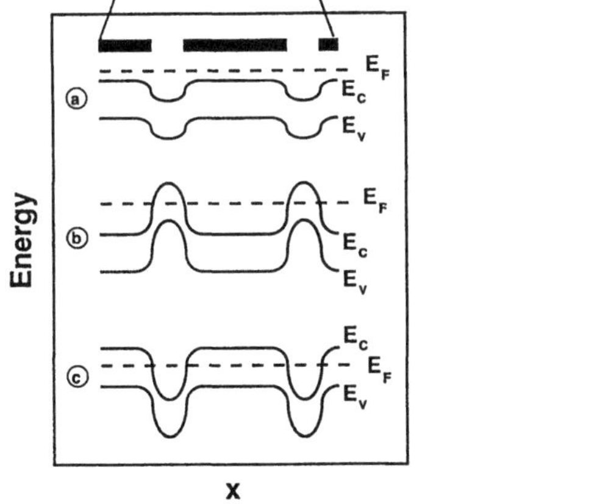

Fig. 1: A stacked gate device with a microstructured bottom gate embedded between two layers of SiO_2. In the lower part of the figure a density modulated system (a) and a system of isolated electron channels with different widths are sketched (b) and (c).

In the gap-confinement mode one- or zero-dimensional electron systems can be realized with a substrate bias technique that we use in the FIR measurements at low temperatures. For lack of ohmic contacts to the inversion electrons we charge the inversion layer by exciting electron hole pairs via a pulse of band gap radiation with appropriate voltages at both gates with respect to the substrate. After switching off the light an additional voltage, the substrate bias voltage, is applied to the top gate. This leads to a stable non-equilibrium condition for the Fermi energy, an increase of the depletion length in the bulk and to increased 2D subband separations [8]. The effect on the lateral potential is similar. The enhancement of the lateral depletion zone leads to a squeezing of the electron channels or dots which is directly seen in the far-infrared spectra as it will be shown in section 6.

3. Dimensional Resonances

We will start our discussion with electronic excitations in an array of relatively wide electron channels induced underneath the gate stripes. In all spectra that will be shown for arrays of electron channels the electrons are excited with an infrared electric field perpendicular to the channels. Excitations with the uniform far-infrared electric field along the channels only reflect the Drude conductivity. In Fig. 2 the relative change in transmission of an array of electron channels is shown for different linear densities $N_l=N_s \cdot a$ in the channel with a width around $W \cong a-t=1500$nm. Several distinct dimensional resonances are observed which are well described by [9]

$$\omega^2 = \frac{e^2 N_s}{\varepsilon_0 \varepsilon_{eff} m^* W} n, \qquad n = 1,3,5\ldots$$

with W the width of the channels and an effective dielectric constant ε_{eff} depending on a quasi wavevector $n\pi/W$ and obtained from the screening theory of plasmons in a 2DEG [10]. From the above equation the parameter W is determined and found to be slightly dependent on the

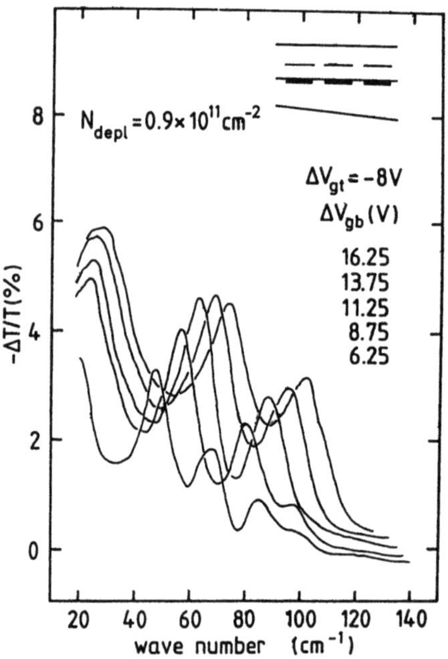

Fig. 2: FIR spectra of $W \cong 1500$nm wide electron channels. Several dimensional resonances can be observed in their dependence on the electron density in the channels. [from Ref. [9]]

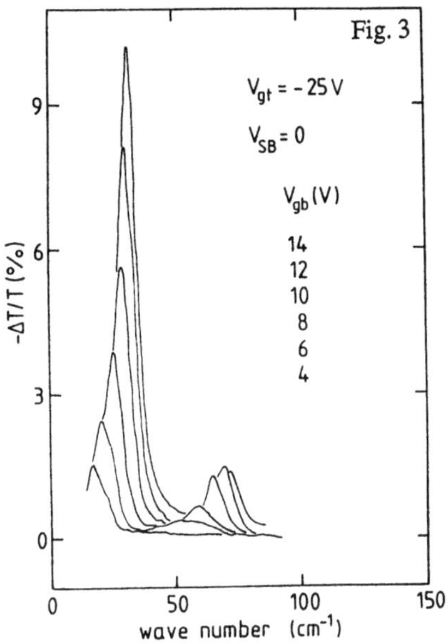

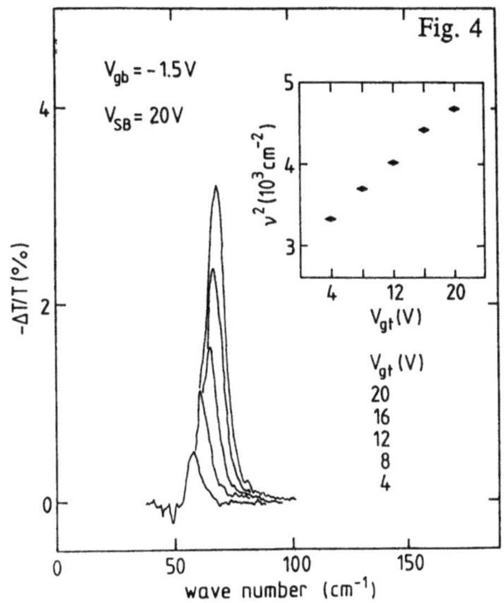

Fig. 3: Dimensional resonances for electron channels with widths of W≅a-t=800nm.

Fig. 4: In very narrow channels with W<200nm only one single dimensional resonance is observed.

electron density in the channels. This is explained as reflecting the increase of fringing fields with higher gate voltages. The width is found to vary between 1400nm and 1600nm from the lowest to the highest measured densities. The difficulty in determining ε_{eff} causes an uncertainty in the above equation where the factor n/W may be replaced by $(\pi/2)\cdot n/W$ and thus is identical with the result of Eliasson [7]. The exact factor will presumably depend strongly on the microscopic density distribution perpendicular to the channels [11]. The homogeneous FIR radiation couples to charge distributions, which accumulate at the boundaries of the channels and which have the highest dipole moments. So only dimensional resonances with odd n are excited. The observed magnetic field dependence of the dimensional resonances is

$$\omega^2 = \omega_{B=0}^2 + \omega_c^2$$

with ω_c the cyclotron frequency [9].

Dimensional resonances of higher order are only observable for sufficiently high negative bias at the top gate resulting in a lateral potential which is assumed to be essentially rectangular on a macroscopic scale. An interesting dependence on the width of the channel in exciting resonances of higher order is seen in Fig. 3 and 4. Fig 3 shows spectra with electron induced beneath the bottom gate stripes of width a-t=800nm whereas Fig. 4 shows spectra of electron channels induced in the gap confinement mode with W<100nm for the lowest densities. Comparing Fig.2, 3 and 4 we observe that higher order dimensional resonances can only be excited when the channel width W is well beyond 200nm. For the 800nm wide channel in Fig. 3 the external potential is likely still flat at the bottom. For low densities, i.e. low bottom gate voltages V_{gb}, this assumption seems no longer to be valid since only one resonance is observed. We think that the observation of only one dimensional resonance is caused by an essentially parabolic external confining potential for which a single resonance is expected, this is the case for Fig. 4, where a high substrate bias voltage is applied to the top gate and thus forming 1D channels. The effect of quantum confinement on the dimensional resonance is that the squared resonance frequencies of the excitations extrapolate to a finite value for vanishing electron density N_l in the channel. As judged from the oscillator strength of the dimensional resonances N_l is essentially proportional to the top gate voltage V_{gt}. This behavior is in contrast to the classical dependence of the squared resonance frequencies on N_l or N_s respectively. In recent theories /3,12/ a generalized Kohn theorem implies that the homogeneous FIR radiation couples to the center of mass motion of the electrons confined in a parabolic external potential. Thus the excitations give a direct measure of the strength of an external potential indicating very strong confinement achievable in our structures. We are able to estimate the 1D subband spacing for finite densities by comparison with measurements of magnetic depopulation of 1D hybrid subbands in a similar device to be about 1meV for densities $N_l \approx 1 \cdot 10^7 cm^{-1}$. This compares to a bare potential characterized by $\hbar\omega_0 \geq 7meV$. Thus screening seems to increase very effectively with increasing N_l.

4. Transition from a Density Modulated 2DEG to an Array of Narrow Electron Channels

The high tunability of the stacked gate device allows to study continuously the transition from a density modulated 2DEG to an array of narrow electron channels. The whole transition regime is shown in Fig. 5, where the bottom gate voltage V_{gb} is tuned from high electron density regions underneath the bottom gate stripes to isolated electron channels in the gap confinement mode. At V_{gb}=15V a splitting of the plasmon resonance is observed due to the periodic density modulation [13]. At V_{gb}=2V the system again becomes a density modulated 2DEG but now the

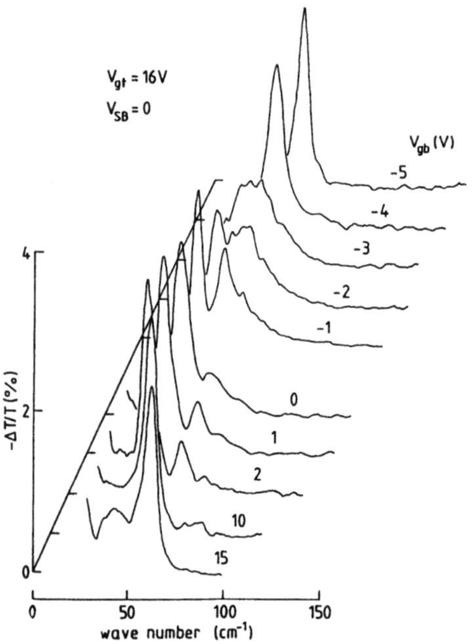

Fig. 5: With decreasing V_{gb} the electron system changes from a density modulated system with high density underneath the bottom gate stripes to one with the high density underneath the bottom gate opening. Finally at V_{gb}=-1V the system turns over to an array of isolated electron channels.

high density region is situated between the bottom gate stripes. In this case no splitting of the plasmon resonance is observed which may be due to the very low density of $N_s<5\cdot10^{11}cm^{-2}$ underneath the bottom gate stripes. At even lower gate voltages a quite complex behavior is observed for the transition to isolated narrow electron channels between V_{gb}=-1V and -4V. At V_{gb}=-5V well separated narrow electron channels are formed and only a single dimensional resonance is observed as discussed before. The transition regime is discussed in more detail in [14], where the transition is advantageously achieved with the substrate bias technique in which the average electron density $N_s=N_l/a$ remains constant. The resonance at lower frequencies in Fig. 5 proves to emerge out of the plasmon of the density modulated 2DEG with wavevector $q=2\pi/a$ and is excited by inhomogeneous electric fields due to some grating coupler efficiency of the structure or certainly by umklapp processes due to the periodicity of the electron system. The resonance at higher frequencies ends up in the dimensional resonance of the isolated electron channel and is determined by the width W of the electron channel. The dimensional resonance is excited by the homogeneous field of the FIR radiation. It reflects localisation of charges in a single channel and does not have significant dispersion perpendicularly to the channels. For a GaAs wire system this has recently been confirmed by Raman measurements [15].

5. Quantum Dots

Using a cross grating bottom gate we can induce electron dots below the gate openings in the gap-confinement mode. The lateral confinement of a quantum dot in both lateral directions leads to the excitation of two modes in the FIR spectra at finite magnetic fields which become degenerate at B=0. The single B=0 resonance is shown in Fig. 6a) for different top gate voltages V_{gt} and two different V_{SB}. The number of electrons does not depend linearly on the top gate voltage for V_{gb}=-2.25V as indicated by the non linear scale on the top of Fig. 6b). At a substrate-bias voltage V_{SB}=18V the resonance positions considerably shift to higher frequencies since the diameter of the electron dot is squeezed by the increased depletion potential. Also for lower bottom gate voltage the resonance positions shift to higher frequencies

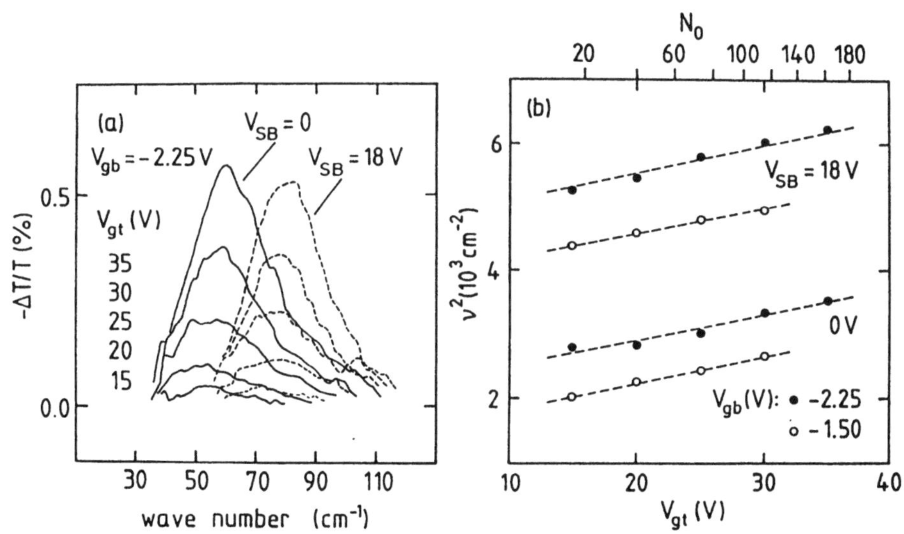

Fig. 6: Quantum dot spectra for two different substrate bias conditions V_{SB}(a). Squared resonance frequencies plotted against the top gate voltage V_{gt} (b). The electron numbers are only valid for V_{gb}=-2.25V. [from Ref. [17]]

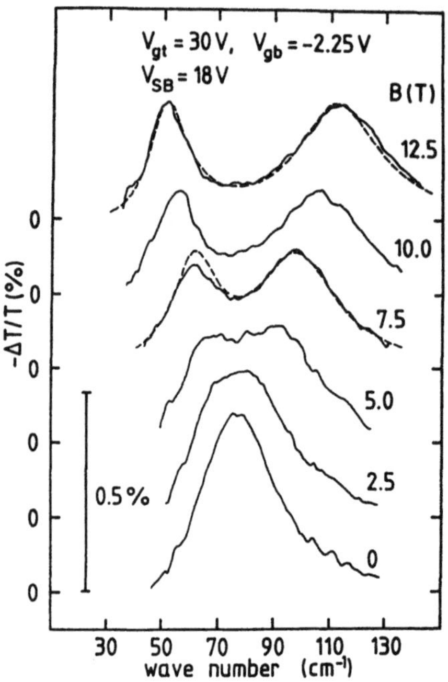

Fig. 7: Electron dot spectra at finite magnetic fields. [from Ref. [17]]

as can be seen in Fig. 6b) where the squared resonance frequencies are again plotted for different V_{gt}. The difference in resonance frequencies between V_{gb}=-1.5V and -2.25V is somewhat obscured since for V_{gb}=-1.5V the induced electron number that depends linearly on the top gate voltage is about a factor of 3 larger than for V_{gb}=-2.25V. At V_{gb}=-2.25V accumulation under the cross grating bottom gate is thought to be achieved. From this follows that the lateral potential roughly assumes the value of the Si band gap energy E_g resulting in extreme confinement conditions. For an external parabolic confinement the curvature k can be estimated to be

$$\frac{2E_g}{(a/2)^2} \le k \approx \frac{2E_g}{(t/2)^2}$$

giving $k \approx 4 \cdot 10^{10}$eV/cm^2 for t=150nm, which is comparable to the depletion potentials of highly doped semiconductors (see e. g. [16]).

In finite magnetic fields the single dimensional resonance at B=0 splits into two modes according to the well known dispersion relation for parabolic confinement in both lateral directions [e.g. 2]

$$\omega = \pm\frac{\omega_c}{2} + \sqrt{\left(\frac{\omega_c}{2}\right)^2 + \omega_{B=0}^2} \; .$$

Fig. 7 shows the spectra for an array of dots with $N_0 \cong 140$ electrons per dot. The dashed lines are a fit to the lineshapes according to a classical high-frequency conductivity of the electrons. The resonance positions follow the above dispersion exactly. However, the amplitude of the lower edge mode as well as its oscillator strength are smaller than expected from the classical theory in the low magnetic field regime. This behavior may be a consequence of the strong lateral 0D confinement since for high magnetic fields the lower edge mode behaves classically again when the cyclotron diameter is much smaller than the diameter of the dot. In our experiments a reduction of the amplitude or of the oscillator strength in electron disks with a few thousands and more electrons per disk is not observed[18].

6. Conclusion

Our experiments show that a stacked gate device is a versatile one with which not only low dimensional electron systems can be realized but which also allows to create density modulated electron systems with widely tunable modulation strength. The isolated 1D or 0D systems show a very strong confinement as can directly be seen in the far-infrared spectra. For relatively wide electron channels dimensional resonances of higher order become observable since we can create a steep confinement at the edges of the electron system with the stacked gate device. For narrow 2D as well as for 1D channels we observe only one dimensional resonance indicating essentially parabolic external potentials, which is also the case for the 0D spectra.

The dual gate device combines several advantages like high tunability, creation of steep confining potentials and very small dot sizes. In addition the stacked gate technology on Si is established and thus seems to compete with etched structures in the realization of strongly confined electron systems. Of course lateral superlattice effects such as those observed on GaAs [1] are harder to achieve on Si than on GaAs molecular beam epitaxy grown structures with a much larger elastic mean free path of the electrons.

I would like to thank E. Batke, J. P. Kotthaus, A. Lorke and J. Martinek for many discussions during the work summarized here and gratefully acknowledge financial support by the Stiftung Volkswagenwerk.

References

[1] W. Hansen, M. Horst, J. P. Kotthaus, U. Merkt, and Ch. Sikorski, Phys. Rev. Lett. 58, 2586 (1987); Ch. Sikorski and U. Merkt, Phys. Rev. Lett. 62, 2164 (1989); T. Demel, D. Heitmann, P. Grambow, and K. Ploog, Phys. Rev. Lett. 64, 788 (1990); A. Lorke, J. P. Kotthaus, and K. Ploog, Phys. Rev. Lett. 64, 2559 (1990);

[2] S. J. Allen, Jr., H. L. Störmer, and J. C. Hwang, Phys. Rev. B 28, 4875 (1983); S. J. Allen, Jr., F. De Rosa, G. J. Dolan, and C. W. Tu, Proc. of the 17th ICPS, San Francisco 1984, p.313, Springer.

[3] P. A. Maksym and T. Chakraborty, Phys. Rev. Lett. 65, 108 (1990).

[4] A. C. Warren, D. A. Antoniadis, and H. I. Smith, Phys. Rev. Lett. 56, 1858 (1986).

[5] J. H. F. Scott-Thomas, S. B. Field, M. A. Kastner, H. I. Smith, and D. A. Antoniadis, Phys. Rev. Lett. 62, 583 (1989).

[6] K.-F. Berggren, T. J. Thornton, D. J. Newson, and M. Pepper, Phys. Rev. Lett. 57, 1769 (1986)

[7] G. Eliasson, J.-W. Wu, P. Hawrylak, and J. J. Quinn, Solid State Commun. 60, 41 (1986)

[8] T. Ando, A. B. Fowler, and F. Stern, Rev. Mod. Phys. 54, 437 (1982).

[9] J. Alsmeier, E. Batke, and J. P. Kotthaus, Phys. Rev. B 40, 12574 (1989).

[10] A. V. Chaplik, Sov. Phys. JETP 35, 395 (1972) [Zh. Eksp. Teor. Fiz. 62, 746 (1972)].

[11] V. Cataudella and G. Iadonisi, Phys. Rev. B 35, 7443 (1987).

[12] L. Brey, N. F. Johnson, and B. I. Halperin, Phys. Rev. B 40, 10647 (1989).

[13] U. Mackens, D. Heitmann, L. Prager, J. P. Kotthaus, and W. Beinvogl, Phys. Rev. Lett. 53, 1485 (1984).

[14] J. Alsmeier, J. P. Kotthaus, T. M. Klapwijk, and S. Bakker, Proc. of the 20th ICPS, Thessaloniki 1990, Springer, to be published.

[15] Th. Egeler, G. Abstreiter, G. Weimann, T. Demel, D. Heitmann, and W. Schlapp, Proc. of the 20th ICPS, Thessaloniki 1990, Springer, to be published.
[16] M. A. Reed, J. N. Randall, R. J. Aggarwal, R. J. Matyi, T. M. Moore, and A. E. Wetsel, Phys. Rev. Lett. 60, 535 (1988).
[17] J. Alsmeier, E. Batke, and J. P. Kotthaus, Phys. Rev. B 41, 1699 (1990).
[18] J. Alsmeier, E. Batke, and J. P. Kotthaus, Surf. Sci. 228, 524 (1990).

Magnetophonon Resonance Amplitudes in GaAs-GaAlAs Heterojunctions

D.R. Leadley[1], *R.J. Nicholas*[1], *L. van Bockstal*[2], *F. Herlach*[2], *J.J. Harris*[3], *and C.T. Foxon*[3]

[1]Clarendon Laboratory, University of Oxford, Parks Road, Oxford, OX1 3PU, UK

[2]Department of Physics, KU Leuven, B-3030 Leuven, Belgium

[3]Philips Research Laboratories, Redhill, Surrey RH1 5HA, UK

Magnetophonon resonance is studied in GaAs–GaAlAs heterojunctions in steady and pulsed fields up to 36T. The resonance amplitudes are found to depend on the carrier density and a damping factor, γ, which is not constant. We find empirical expressions for the amplitude; and for γ, which contains terms proportional to T and $B^{3/2}$ and is dominated by remote impurity scattering.

Magnetophonon resonance (MPR) arises from absorption of optic phonons giving additional electron momentum relaxation whenever $\hbar\omega_0 = N\hbar\omega_c$, where ω_0 and ω_c are the phonon and cyclotron frequencies and N is an integer. It is seen as a series of maxima in the resistivity and has been extensively studied in bulk semiconductors. In 2–D systems there have been a number of studies where MPR has been observed [1–3], but few systematic studies of the amplitude. This paper represents an extention of our previous low field (<10T) studies [4,5] to include the N = 1 & 2 resonances by using steady magnetic fields up to 16T in the Oxford hybrid magnet and pulsed fields of up to 36T in Leuven.

The pulsed fields were generated by discharging a 40mF capacitor bank through a copper coil giving a pulse duration of ~30ms. A transient recorder was used to store ρ_{xx} and ρ_{xy} and a voltage $\propto {}^{dB}/_{dt}$, which was numerically integrated to find the field. Induced voltages in the sample leads were compensated for by additional single coils in the field. For low resistance samples there was good agreement between data taken on the rising and falling sides of the pulse and in steady fields. However, for resistances >200kΩ spurious effects were seen due to contact problems and RC phase lags, which made studying very low density material difficult.

The samples were all conventional GaAs–GaAlAs heterojunctions grown at Philips Research Laboratories, with undoped spacer layer thicknesses (L) between 3200Å and 17Å giving electron densities, n_s = 0.2 – 6 x 10^{11}cm^{-2}.

The magnetoresistance was measured between 100K and 300K, and the MPR extracted by subtracting a voltage linear in magnetic field (Fig. 1). To allow comparison between samples the resonance amplitudes, $\Delta\rho$, were normalised to the zero field resistivity $\rho(0)$ and are shown in Fig. 2 for the N = 1,2,3 resonances in the dark at 150K, where $\Delta\rho/\rho$ is largest due to competition between phonon population and Landau level broadening. $\Delta\rho/\rho$ shows a maximum of ~30% at a density of ~3 x10^{11}cm^{-2} for the N = 1 resonance, but at lower n_s for the higher harmonic numbers. At low concentrations the amplitude hardly changes with N, in some cases the N = 1 resonance is even smaller than the N = 2 peak.

Increasing n_s by illumination with a red LED increased $\Delta\rho/\rho$, but above 100K the photoconductivity was not persistent necessitating continuous illumination. This caused some parallel conduction in the bulk material, so only data taken in the dark will be used to compare absolute amplitudes.

Springer Series in Solid-State Sciences, Vol. 101
High Magnetic Fields in Semiconductor Physics III Editor: G. Landwehr

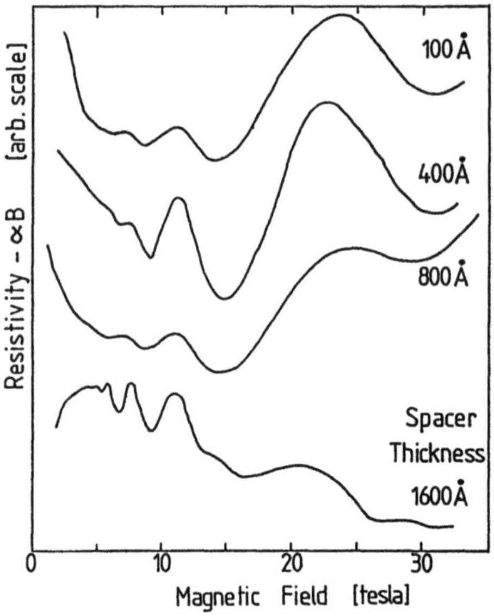

Figure 1 MPR in pulsed fields from a set of GaAs–GaAlAs heterojunctions.

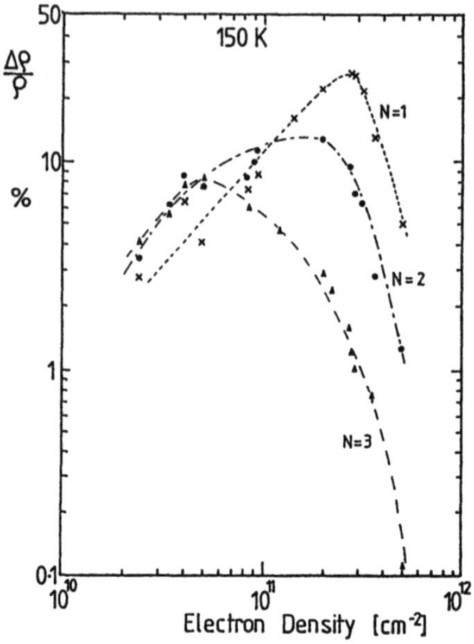

Figure 2 Normalised amplitudes for the first three resonances as a function of carrier concentration at 150K.

Before discussing the absolute amplitudes, we will study the relative sizes of the different harmonics. In 3–D magnetophonon resonances are well described by an exponentially damped series with a constant damping factor, γ [6]:–

$$\Delta\rho/\rho \propto \exp(-N\gamma). \tag{1}$$

For 2–D Mori *et al.* [7] found that for Lorentzian broadened Landau levels the oscillatory part of the transverse magnetoresistance is:

$$\frac{\Delta\rho_{xx}}{\rho_o} = 2\overline{\Delta\rho} \sum_{r=1}^{\infty} \exp(-2\pi r\gamma') \cos(2\pi r\omega_0/\omega_c) , \tag{2}$$

where γ' is the theoretical damping factor related to the Landau level width, Γ. For $2\pi\gamma > 1$ the r=1 term dominates allowing a correspondence to be drawn between the theoretical and experimental damping factors:

$$\gamma = 2\pi\gamma' \frac{\omega_c}{\omega_o} = \frac{2\pi\Gamma}{\hbar\omega_o} . \tag{3}$$

Thus the empirical damping factor will depend on magnetic field in the same way as the level width. Peeters *et al.*[8] have shown that variation of γ can have profound effects on the MPR amplitude and resonance position so we will attemplt to find an empirical expression for γ(L,T,B).

In practice one can only measure $\gamma_{ij} = \ln(\Delta\rho_i/\Delta\rho_j)$ – the damping factor between the ith and j^{th} peaks. Fig. 3 shows how γ_{ij} varies with spacer thickness for each pair of peaks – one could have shown γ_{ij} vs n_s, but n_s varied between samples of a given L while γ did not. On photoexcitation γ_{ij} usually decreased slightly, probably due to addition of some bulk MPR.

A number of points should be noticed in Fig. 3: (i) $\gamma_{ij} \propto -\log L$, showing the importance of remote impurity scattering; (ii) γ_{ij} is not constant but decreases

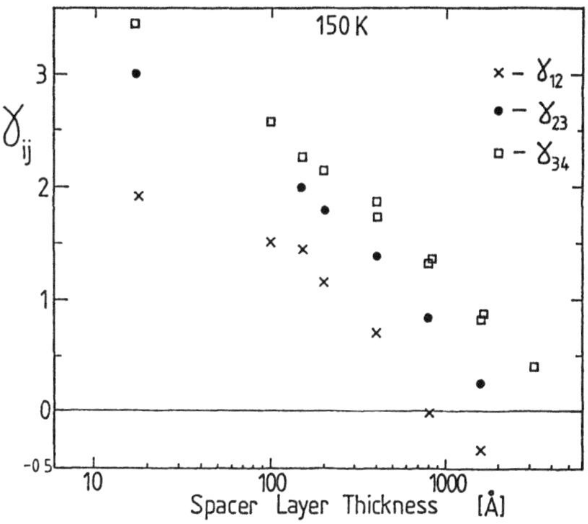

Figure 3 Measured damping factor from N = 1 to 4 peaks at 150K, shown as a function of spacer layer thickness.

at higher fields and (iii) for the widest spacer layers it actually becomes negative.

However, the measured γ_{ij} does not correspond to a particular magnetic field as it is a ratio of peaks at very different fields. So for behaviour like Eq. 1:

$$\gamma_{ij} = j\gamma_j - i\gamma_i . \tag{4}$$

If γ_{ij} does not depend on B Eq. 4 gives a constant γ_i as expected. However if γ_i increases with field Eq. 4 results in a relative damping γ_{ij} which decreases with field. Further, in order for γ_{ij} to be negative γ_i must have a superlinear increase with magnetic field.

If each contribution to Γ is Lorentzian then the total width is given as a sum of squares and so the damping factor can be writen as:

$$\gamma^2 = \gamma_*^2 + \gamma(B)^2 + \gamma(T)^2 . \tag{5}$$

Assuming that the B- and T-dependences of γ are mutually independent the temperature dependence can be readily extracted, since at a fixed magnetic field $\gamma_{ij}^2 = f(B,\gamma_*) + (j^2-i^2)\ \gamma(T)^2$. In Fig. 4 γ_{ij}^2 is plotted against T^2 for a 1600Å spacer sample and shows a linear dependence with $\gamma(T) \approx 0.8\ kT/\hbar\omega_0$ for both γ_{34} and γ_{45} – ie. a thermal contribution to Γ of ~kT/8. This same behaviour was seen in all the samples studied.

The true field dependence is more deeply embedded as it is first necessary to find the γ_i. In low n_s samples, where the damping factor is small and a large number of resonances are observed, γ_{ij} becomes constant at low field. We have equated this value to γ_7 (N = 7 is the resonance observed at lowest field) and calculated all the other γ_i from Eq. 4. In Fig. 5 γ_i^2 is plotted against logB for two 1600Å samples and clearly shows a B-dependent part at high fields and a constant term, γ_0, at lower fields. Subtraction of the constant term results in the same straight line for both samples, giving confidence in this approach, and gives the field dependent part of the damping factor as $(\pi\hbar\omega_c/\hbar\omega_0)^{3/2}$ – ie. $\Gamma \propto B^{3/2}$ – a very surprising result, since Γ is usually thought to increase as only $\sqrt{B}$ [7]! Inspection of Fig. 3 shows a similar difference between γ_{12} and γ_{34} for each sample which suggests this may be a general behaviour.

More problematic is extraction of the value of γ_*. Fig. 3' gives the qualitative result of a rapid decrease in γ_* with increasing spacer thickness, corresponding with the reduction in remote impurity scattering. Our major result is therefore that the damping factor is not constant in magnetic field as assumed in bulk but may be represented in the form:

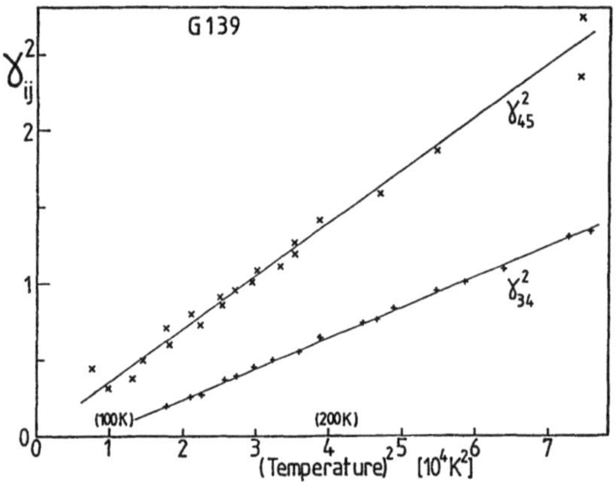

Figure 4 Temperature dependence of the measured damping factor in a 1600Å spacer layer sample.

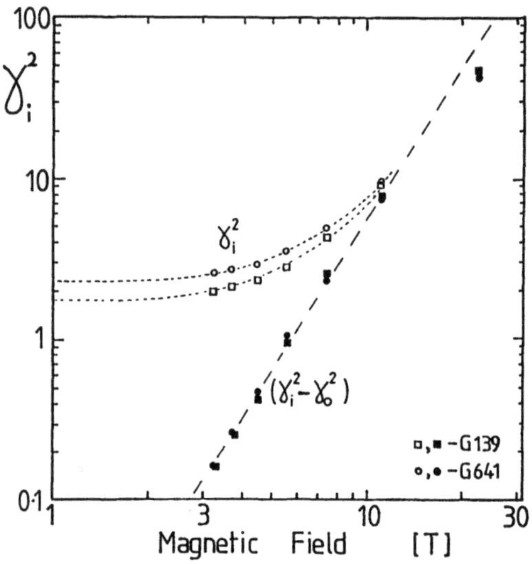

Figure 5 B–dependence of γ_i (open points) and after subtraction of the field independent term γ_0 (closed).

$$\gamma^2 \quad = \quad \gamma_*(L)^2 \quad + \quad \left[\frac{\pi \hbar\omega_c}{\hbar\omega_o}\right]^3 \quad + \quad \left[\frac{\pi}{4}\frac{kT}{\hbar\omega_o}\right]^2 \tag{6}$$

Returning to the magnetophonon amplitudes, it appears that two factors influence their absolute size: the damping of the oscillations; and the strength of the undamped coupling. From Fig. 2 this can be empirically represented by:–

$$\frac{\Delta\rho}{\rho_o} \quad = \quad A\ (n_s)^{3/2}\ \exp(-N\gamma_N) \tag{7}$$

Narrowing the spacer layer increases n_S, but can give more remote impurity scattering which broadens the Landau levels, thus γ increases with n_s and the net effect is a fall in $\Delta\rho/\rho$ for thin spacer layers. By contrast photoexcitation increases n_s without changing L, and always results in a greater resonance amplitude. Temperature variation of the amplitude will enter through both the prefactor, A, which will depend on phonon occupation number and give a reduction at low T; and on γ which accounts for the fall at high temperatures.

The reason for the underlying increase in $\Delta\rho/\rho$ with electron density is not clear, but may be related to changes in the strength of the electron-phonon coupling [5] or some amount of phonon focussing at the interface.

References

1. Nicholas R.J.; In *Landau Level Spectroscopy* Ed. G Landwehr & E.I. Rashba, North Holland in press (1990).
2. Englert Th., D.C. Tsui, J.C. Portal, J. Beerens and A.C. Gossard; *Solid State Commun.* **44**, 1301 (1982).
3. Kido G., N.Miura, H. Ohno & H. Sakaki; *J.Phys.Soc.Japan* **51**, 2168 (1982)
4. Brummell M.A., D.R. Leadley, R.J. Nicholas, J.J. Harris and C.T. Foxon; *Surf. Sci.* **196**, 451 (1988)
5. Nicholas R.J., D.J. Barnes, D.R. Leadley, C.J.G.M. Langerak, J. Singleton, P.J. van der Wel, J.A.A.J. Perenboom, J.J. Harris and C.T. Foxon; Plenum Physics Series B **206**, 451–470 (1989).
6. Stradling R.A. and R.A. Wood; *J. Phys. C* **1**, 1711 (1968).
7. Mori S, H.Murata, K.Taniguchi & C.Hamaguchi; *Phys.Rev.B* **38**, 7622 (1988)
8. F.M. Peeters, P. Warmenbol and J.T. Devreese; *Europhys. Lett.* **12**, 435 (1990)

Optically Detected Magnetophonon Resonance in GaAs-GaAlAs Heterojunctions

D.J. Barnes[1], *R.J. Nicholas*[1], *M. Watts*[1], *F.M. Peeters*[2], *X. Wu*[2], *J.T. Devreese*[2], *C.J.G.M. Langerak*[3], *J. Singleton*[3], *J.J. Harris*[4], *and C.T. Foxon*[4]

[1]Clarendon Laboratory, University of Oxford, Parks Road, Oxford, OX1 3PU, UK
[2]University of Antwerp, B-2610 Antwerpen, Belgium
[3]University of Nijmegen, 6525 ED Nijmegen, The Netherlands
[4]Philips Research Laboratories, Redhill, Surrey RH1 5HA, UK

Abstract. We present an experimental observation of the magnetophonon effect in the frequency dependent conductivity of a GaAl–GaAlAs heterojunction. The cyclotron resonance position and linewidth are found to oscillate as a function of magnetic field and show resonances for $\omega_{L.O.} \approx N\omega_c$. The resonances grow rapidly with increasing temperature. Good agreement is found with a one–electron theory of the cyclotron resonance absorption spectrum, which demonstrates that the observed oscillations have the same origin as the MPR oscillations in the resistivity.

Magnetophonon resonance (MPR) occurs when two Landau levels are a phonon energy apart, leading to an increased probability of the absorption or emission of (usually) optic phonons. We describe here a direct measurement of the resonant scattering through a study of the frequency dependent conductivity measured via the cyclotron resonance. This shows strong temperature dependent peaks in linewidth close to the classic magnetophonon resonance condition:

$$N\omega_c = \omega_{L.O.} \tag{1}$$

where ω_c is the cyclotron frequency (eB/m*) and $\omega_{L.O.}$ is the optic phonon frequency. The theoretical description of this effect uses the memory function approach. This predicts well the oscillatory linewidth, and shows a simultaneous shift in the cyclotron resonance position which is also observed experimentally.

The magnetophonon effect has been extensively studied in both bulk materials [1,2] and more recently in 2–D systems [2,3]. The oscillations are usually observed in the transport coefficients, such as the resistivity, which give a rather complicated average of the scattering processes. Our observation of the optically detected MPR (ODMPR) allows us to see coupling for specific transitons and to measure these quantitatively. The theory demonstrates the appearance of anticrossing behaviour in states previously unsplit at zero temperature.

Previous work on the cyclotron resonance linewidths at high temperatures [4] have shown that at relatively low fields (of order 6T) the linewidth is dominated by what seems to be non–resonant optic phonon scattering for temperatures above 60K. Fits to this data[5] require a relatively large level broadening, of order 0.1 $\hbar\omega_{L.O.}$, however for sufficiently well resolved levels this would lead us to expect that resonant broadening should occur.

Cyclotron resonance was observed in a series of conventional GaAs–GaAlAs heterojunctions over a range of temperatures (4.2–100K), using both a F.I.R. laser and a Fourier transform spectrometer. Fig. 1 shows a typical series of resonances taken at 80K, which shows a very obvious broadening, and decrease in resonance amplitude, at a frequency of around 140 cm^{-1}. Plots of the energy (and thus field) dependence of the linewidth taken at increasing temperature, are shown in fig. 2, again showing the strong peak appearing at an energy (field) of around 140 cm^{-1}

Springer Series in Solid-State Sciences, Vol. 101
High Magnetic Fields in Semiconductor Physics III Editor: G. Landwehr

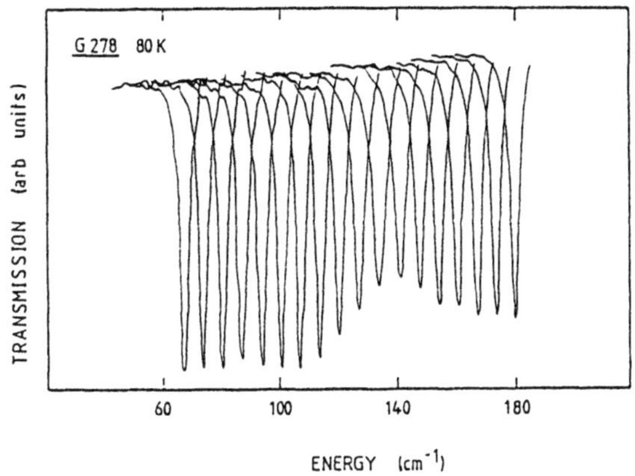

Fig. 1: A series of typical cyclotron resonance traces from 5 to 13.5 T.

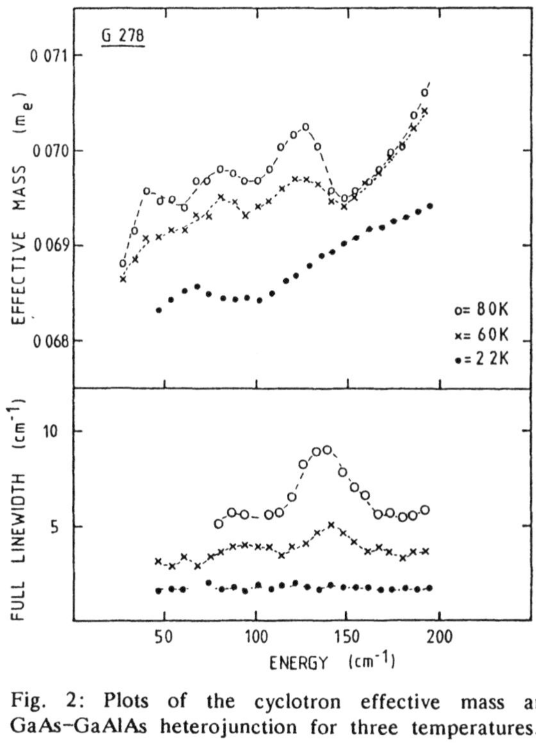

Fig. 2: Plots of the cyclotron effective mass and linewidth for a 200Å spacer GaAs–GaAlAs heterojunction for three temperatures.

($B \simeq 10.5T$) for high temperatures. This is close to the resonance condition N=2, known from conventional polaron coupling[6] and magnetophonon[7] measurements to occur at sub multiples of a field of 21–22T. Weaker features are also visible in some cases at N=3,4. The magnitude of the additional magnetophonon contribution to the linewidth appears to be almost independent of sample and carrier concentration, however the position of the peak appears to move up in energy as the temperature and carrier concentration increase. For the lower carrier concentrations the resonances appear to be rather lower in field than seen in

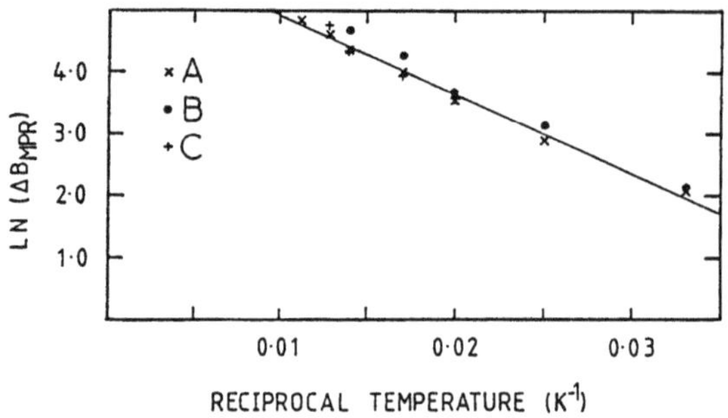

Fig. 3: An Arrhenius plot of the ODMPR amplitude of several heterojunctions.

equivalent MPR experiments in the resistivity [7], although the majority of these are performed at rather higher temperatures. Another noticeable difference is in the 'damping' of the oscillations, which measures the relative strength of the successively higher harmonic number resonances. This is again rather strong when compared with the equivalent oscillations in the resisitivity, and may be related to the shifts in resonance position[8]

The overall picture is thus one of a temperature dependent background, which is only weakly field dependent, with one or more resonant magnetophonon peaks superimposed on top of this. An Arrhenius plot of the resonant broadening (fig. 3) shows that the resonant term is sample independent, and has an exponential behaviour with an activation energy of 200K. This corresponds to half the optic phonon energy, as found for the non-resonant contribution to the linewidth[4] and in the calculations reported below.

The cyclotron effective masses, m*, also show oscillatory structure, and it is easier to distinguish the higher resonances in plots of the mass, also shown in fig. 2, than in the equivalent linewidth data. The resonances are, however, shifted in phase from those shown in the linewidth. This is due to the dispersive relationship in the field (and energy) dependence of the resonance positions and linewidths, which is the result of the Kramers-Kronig type of relationship between the two dependencies, as demonstrated recently by Nicholas et al [9]. The mass can be a more sensitive probe of the oscillations, due both to its derivative type relation to the linewidth, and the greater experimental accuracy in assigning the energy of an absorption peak as compared to its linewidth. The resonance positions deduced from the two measurements are nevertheles in good agreement. The magnetophonon resonances are essentially single particle effects, and will thus arise directly in a one polaron theory. Many particle effects[10] and other important factors such as finite thickness effects [11] will change the effect quantitatively, but will not be included in this first description of the theory.

The cyclotron resonance absorption spectrum is calculated from the complex conductivity $\sigma_{\mu\nu}(\omega)$, which is the time Fourier transform of the velocity-velocity relaxation function $(r_\mu(t),r_\nu(0))$ [12]. This can be written as

$$\sigma(\omega) = i \frac{e^2 n_e/m_b}{\omega - \omega_c - \Sigma(\omega)} \tag{2}$$

where n_e is the carrier concentration, ω_c is the 'bare' cyclotron frequency, and $\Sigma(\omega)$ is the memory function, which contains the effect of the electron-phonon interaction. The memory function is in essence a force-force correlation function, which it is sufficient to treat within second order perturbation theory for a weakly polar material such as GaAs. Following [12] we find

$$\Sigma(\omega) = \frac{1}{\omega} \int_0^\infty dt\ (1 - e^{i\omega t})\ Im F(t) \tag{3}$$

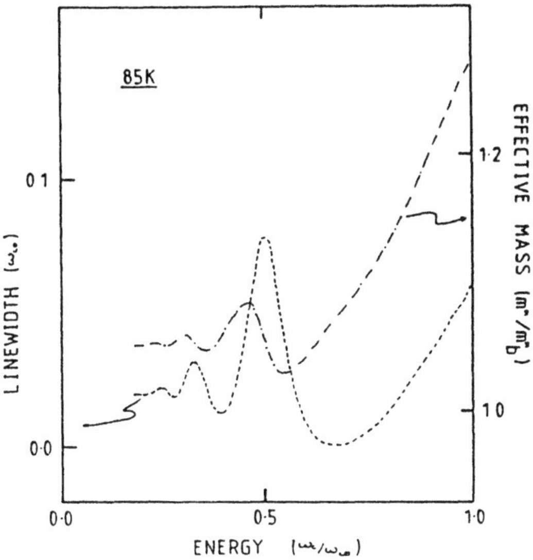

Fig. 4: Calculated values of the cyclotron linewidth and effective mass for a temperature of 90K and a broadening $\Gamma = 0.15\hbar\omega_{L.O.}$.

with

$$F(t) = -\sum \frac{k_{\|}^2}{m_b\hbar} |V_k|^2 \left[(1+n(\omega_{L.O.}))I(\mathbf{k},t) - n(\omega_{L.O.})I^*(-\mathbf{k},t)\right] e^{-i\omega_{L.O.}t} \tag{4}$$

where $n(\omega)$ is the L.O. phonon occupation number, and $k_{\|}^2 = k_x^2 + k_y^2$. The Fourier transform of the density-density correlation function is calculated for finite temperature

$$I(\mathbf{k},t) = \exp\left[-\frac{\hbar^2 k^2}{2m_b\omega_c}\left(1 - e^{i\omega_c t} + 4n(\omega_c)\sin^2(\omega_c t/2)\right) - \frac{\Gamma^2 t^2}{4\hbar^2}\right] \tag{5}$$

with Γ the Landau level broadening parameter. This broadening is introduced semi-empirically to remove the δ-function divergence of the Landau level densities of states, and will in practice depend on the magnetic field, temperature and Landau level index [13]. For simplicity we will take it to be constant.

The experimental results are simulated from equation (2), from which we can see that $Re\Sigma(\omega)$ is responsible for the shift in the cyclotron resonance frequency, which is governed by the reactive part of the interaction. $Im\Sigma(\omega)$ gives the width of the cyclotron resonance peak, representing the dissipative part of the interaction. These two components of the memory function are related, as mentioned above, by the Kramers-Kronig relation.

The zero frequency limit of Eq. 2 gives us directly the resisitivity by $\rho_{xx} = -Im\Sigma(\omega=0)$. This expression results in the conventional MPR seen in the resistivity, which was studied by Warmenbol et al[14] using similar approximations to those in the present paper. This comparison shows us the considerable interest and additional information given by the ODMPR. The oscillations are from a common origin, but the study of the frequency dependent conductivity now allows us to observe both components of the memory function, and in a direct way much less influenced by the other scattering processes taking place.

Numerical results were evaluated for the absorption spectra as a function of frequency using a broadening of $\Gamma=0.15\hbar\omega_{L.O.}$ and a temperature of $kT=0.21\hbar\omega_{L.O.}$

(≃90K). The cyclotron peak shows the expected broadening at the magnetophonon resonance conditions, as shown in fig. 4. Also shown is the cyclotron resonance mass, which also shows very clearly the appearance of oscillatory structure, with a derivative like relation between the mass and the linewidth as found in the experiment. When smaller values of the broadening are used in the simulations the resonances can split, showing an anticrossing behaviour, with the intensity switching from the lower to higher frequency components as the field goes through the resonance condition (1). At this temperature over 90% of the carriers are in the lowest Landau level, so that the anticrossing behaviour probably represents a splitting of the N=0 Landau level caused by the real phonon absorption.

The numerical results show that the amplitude of the oscillations increases rapidly with temperature, as expected, with a dependence proportional to $\surd n(\omega_{L.O.})$. This is also in agreement with the experimental results above, as seen for the non-resonant scattering[4,5], and is thought to be due to the internal self-consistency of the theory. The simplified theory overestimates the absolute magnitude of the oscillations by a factor of approximately 5, however this is not surprising, as the majority of the factors neglected will lead to a reduction of the effective interaction. The rather strong damping of the oscillations observed experimentally suggests that the damping may also need to be increased. This explains why no strong asymmetry or splitting of the resonance was observed close to the resonance conditions, but only the shift in position.

References

1. P.G. Harper, J.W. Hodby and R.A. Stradling, Rep. Prog. Phys. 36 1 (1973)
2. R.J. Nicholas, Prog. in Quantum Electron.
3. R.J. Nicholas, in *Landau level Spectroscopy* Ed. G. Landwehr & E.I. Rashba, (North Holland, Amsterdam) Mod. Probs. in Cond. Mat. Sci. 27.2 pp777 (1991)
4. M.A. Hopkins et al, Phys. Rev. B39 13302 (1989)
5. Xiaoguang Wu and F.M. Peeters, Phys. Rev. B41 3109 (1990)
6. C.J.G.M. Langerak et al, Phys. Rev. B38 13133 (1988)
7. D.R. Leadley et al, to be published, and D.Phil thesis, Univ. Oxford, (1989)
8. F.M. Peeters, P. Warmenbol, & J.T. Devreese, Europhys. Lett. 12 435 (1990)
9. R.J. Nicholas et al, Phys. Rev. B39 *** (1989)
10. P. Warmenbol,F.M. Peeters & J.T. Devreese, Sol. St. Electron 32 1545 (1989)
11. S. Das Sarma, Phys. Rev. B27 2590 (1983)
12. F.M. Peeters and J.T. Devreese, Phys. Rev. B34 8800 (1986)
13. N. Mori et al, Phys. Rev. B38, 7622 (1988)
14. P. Warmenbol, F.M. Peeters and J.T. Devreese, Phys. Rev. B37 4694 (1988)

Resonant Magneto-Polarons in InSb

P. Pfeffer and W. Zawadzki

Institute of Physics, Polish Academy of Sciences,
PL-02-668 Warsaw, Poland

Abstract. Experiments on magneto-polarons in bulk InSb are described theoretically using Green's function techniques. The narrow-gap band structure of InSb is taken into account not only in the energies but also in the electron wave functions. Resonant and nonresonant polaron corrections to the electron energies are included. Best description of different magneto-optical data is found for the phonon energy $\hbar\omega_L$=23.7meV and the polar coupling constant α=0.02.

Since the discovery of magneto-optical effects due to resonant interaction of electrons with optic phonons in InSb by Johnson and Larsen [1] the resonant polaron behavior has been the subject of substantial interest both in three-dimensional and two-dimensional semiconductor systems [2,3]. While many experiments have been performed on narrow-gap materials, where it is easy to reach the resonant condition $\hbar eB/m \approx \hbar\omega_L$, the existing theoretical work has used the standard one-band effective mass approximation. Swierkowski and Zawadzki [4] used the real structure of InSb-type materials to describe the resonant magneto-polaron behavior, but applied it to InAs in the region away from the resonance, where the band structure particularities are of little importance. In the present paper we describe the polaron behavior in InSb, which is a "model" narrow-gap semiconductor, consistently using the real band structure of this material (small energy gap and strong spin-orbit interaction) both in the energies and the wave functions. We apply the Green function formalism, necessary for a correct description of the upper polaron branch [5].

The initial Hamiltonian for our problem reads

$$H = P^2/2m_0 + V_0(\bar{r}) + H_{so} + H_{Fr} + H_{phn} \tag{1}$$

where m_0 is the free-electron mass, $\bar{P} = p + e\bar{A}$ is the kinetic momentum, V_0 is the periodic potential of the lattice, H_{so} is the spin-orbit interaction, H_{Fr} is the Fröhlich polar electron-phonon interaction (assumed to be weak), and H_{phn} is the free phonon field. The first three terms describe the band structure of a semiconductor in the presence of magnetic field. This electronic part is solved for the vicinity of the Γ point using the model of Pidgeon and Brown [6] in its spherical approximation. This takes exactly into account the $\bar{P}\cdot\bar{p}$ interaction between Γ_6, Γ_8 and Γ_7 levels (including explicitly the small energy gap ε_g and the spin-orbit energy Δ) and incorporating all other bands in the k^2 approximation. The resulting electron energies are obtained by diagonalizing numerically two 4x4 matrices for the spin-up and spin-down

Springer Series in Solid-State Sciences, Vol. 101
High Magnetic Fields in Semiconductor Physics III Editor: G. Landwehr

states. The resulting electron wave functions are

$$\Psi^{\pm}_{nk_z}(\bar{r}) = \sum_{l} c_l(n,k_z,\pm) f^n_l(\bar{r}) u_l(\bar{r}) \qquad (2)$$

where the summation is over the three levels (8 states including spin), f^n_l are the envelope functions of the harmonic-oscillator type, and u_l denote the Luttinger-Kohn periodic amplitudes. The envelope functions are given explicitly in ref. [7].

In our description of polarons the Fröhlich interaction is treated as a perturbation. We consider resonant case $\varepsilon(in)-\varepsilon(fi) \approx \hbar\omega_L$, in which the upper electron state $\Psi(in)$ can decay by a virtual (below the resonance) or real (above the resonance) emission of an optic phonon $\bar{q}$ to the final electron state $\Psi(fi)$. The polaron energies are given by maxima of the spectral function

$$A_n(E) = -\frac{1}{\pi} Im G_n(E) = -\frac{1}{\pi} Im\left[\frac{1}{E-(\varepsilon^0_n+\Sigma_n)}\right] \qquad (3)$$

where the selfenergy is

$$\Sigma_n(E) = \sum_{m=0} \sum_{q} \frac{|M_{nm}(\bar{q})|^2}{E-(\varepsilon^0_m+\delta_m+\hbar\omega_L)} \qquad (4)$$

in which E is the energy of interest, ε^0_m are the unperturbed Landau levels, and M_{nm} are the matrix elements of the interaction. In principle the energies in the denominators of Eq. (4) should be perturbed energies, which is included in the corrections δ_m. We correct all Landau levels by the same amount: $\delta_m = \Delta E(RS)$, equal to the second order Raleigh-Schrödinger perturbation cf. [8]. The selfenergy (4) has in general resonant and nonresonant parts, depending on the denominators. In the nonresonant parts the summation over Landau levels is converted to an integral.

The matrix elements of the Fröhlich interaction between the harmonic oscillator functions are (phonon emission, $m \geq n$, the phase factors are omitted),

$$\langle\phi_m|\exp(-i\bar{q}_\perp \bar{r}_\perp)|\phi_n\rangle = \exp(-a^2)(-a)^{m-n}(n!/m!)^{1/2} L^{m-n}_n(2a^2) \qquad (5)$$

where $a=q_\perp L/2$, $L^2=\hbar/eB$, and L^{m-n}_n are the associated Laguerre polynomials. The narrow-gap band structure of InSb, reflected in the form (2) of the wave functions results in a decrease of the total matrix element of the electron-phonon interaction, as compared to the case when only the leading terms in the wave functions are taken into account.

In various experiments performed on InSb one deals with different initial and final electron states coupled by the phonon emission. In the cyclotron-resonance experiments the initial state is 1^+ and the final state is 0^+, for the phonon emission. In this arrangement one can not follow well the resonant region, since the optical absorption at the resonance is dominated by the reststrahlen band. The upper polaron branch is much wider than the lower one. As a result, precision of the data above the resonance is not as good as that below. The results in Fig. 1 are described by the optic phonon energy $\hbar\omega_L$=23.7 meV (somewhat lower

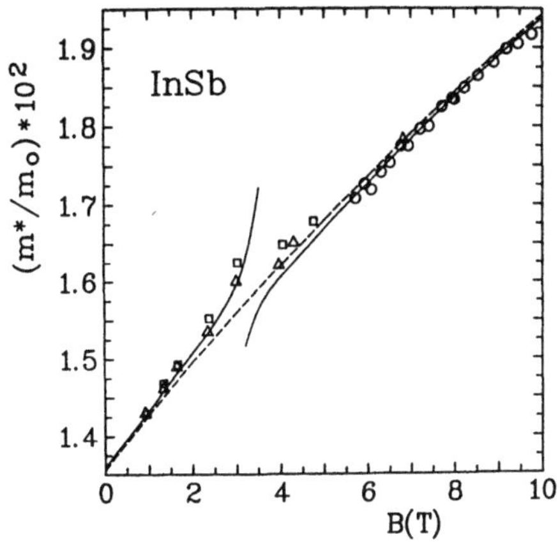

Fig.1. Cyclotron mass of conduction electrons in InSb versus magnetic field. Experimental data: triangles - Zawadzki et al. [9], squares - Huant et al. [10], circles - McCombe et al. [11]. The dashed line is calculated ignoring electron-phonon interaction, the solid lines are calculated including polaron effects.

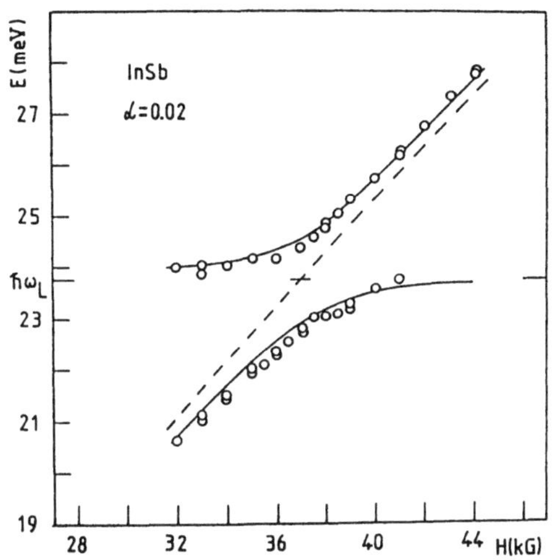

Fig.2. Cyclotron energy for conduction electrons in InSb (spin-down transitions) versus magnetic field. Circles - experimental data of McCombe and Kaplan [12]. The dashed line is calculated ignoring electron-phonon interaction, the solid lines are calculated including polaron effects.

than the generally accepted value of 24.4 meV) and the polar coupling constant α=0.02. The Pidgeon-Brown parameters used in the calculations are: ε_g=0.2352 eV, Δ=0.803 eV, $E_{Po}=P_0^2 2m_0/\hbar^2$= 21.92 eV, γ_1=3.25, $\bar{\gamma}$=0.35, κ=-1.3, N_1=-0.55, F=-0.2. This corresponds to the band-edge electron values $m_0^*=0.0136m_0$ and g_0^*=-51.0.

Fig.2 shows the data of McCombe and Kaplan [12], who investigated combined resonance transition $0^+ - 1^-$. This allows one to follow the 1^- Landau level through the resonant range of $\varepsilon_1^- - \varepsilon_0^- \approx \hbar\omega_L$ without the reststrahlen problems. The solid lines are calculated using the same phonon parameters, but employing the value of E_{Po}=21.92 eV. This corresponds to $m_0^*=0.01433m_0$ and g_0^*=-48.3.

Finally, Fig.3 shows the results of Koteles and Datars [13] at higher temperature of T=48K, which allowed the authors to observe four different CR transitions. The dominant polaron feature is seen for the $1^+ - 2^+$ transition, involving the resonant polaron in the range of $\varepsilon_2^+ - \varepsilon_0^+ \approx$

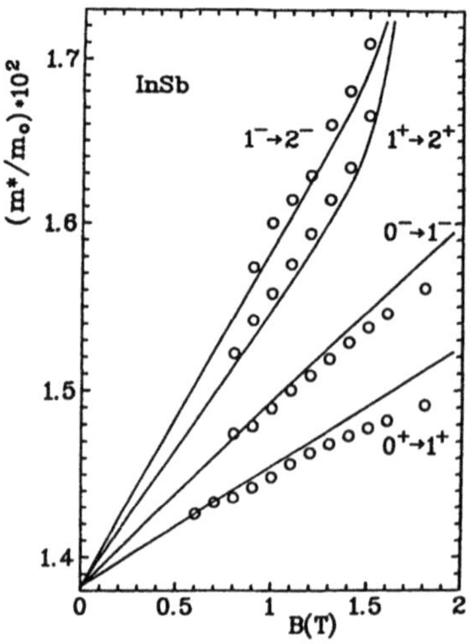

Fig.3. Cyclotron masses of conduction electrons in InSb at T=48K for four cyclotron resonance transitions versus magnetic field. Circles - experimental data of Koteles and Datars [13]. Solid lines are calculated including the polaron effects.

$\hbar\omega_L$. The solid lines are calculated using the same phonon parameters and E_{Po}=22.82 eV, corresponding to m_0^*=0.01382m_0 and g_0^*=-50.22.

We conclude that one has to account for the real band structure of InSb (both in the energies and the wave functions) in order to correctly describe resonant polaron effects in this semiconductor. Some inconsistencies between various experiments forced us to use slightly different electronic parameters in the best-fit procedures.

References

1. E.J. Johnson and D.M. Larsen, Phys. Rev. Lett. 16, 655 (1966).
2. R. Lassnig and W. Zawadzki, Surface Sci. 142, 388 (1984).
3. Wu Xiaoguang, F.M. Peeters, and J.T. Devreese, Phys. Rev. B36, 9760 (1987).
4. L. Swierkowski and W. Zawadzki, J. Phys. Soc. Japan 49, suppl.A, 767 (1980).
5. M. Nakayama, J. Phys. Soc. Japan 27, 636 (1969).
6. C.R. Pidgeon and R.N. Brown, Phys. Rev. 146, 575 (1966).
7. W. Zawadzki, in "Narrow Gap Semiconductors. Physics and Applications", Lecture Notes in Physics, Vol.133, Ed. W.Zawadzki (Springer, Berlin, 1980), p.85.
8. P. Pfeffer and W. Zawadzki, Phys. Rev. B37, 2695 (1988).
9. W. Zawadzki, S. Klahn, and U. Merkt, Phys. Rev. Lett. 55, 983 (1985)
10. S. Huant, L. Dmowski, M. Baj, and L.C. Brunel, Phys. Stat. Solidi (b), 125, 215 (1984).
11. B.D. McCombe, S.G. Bishop, and R. Kaplan, Phys. Rev. Lett. 18, 748 (1967).
12. B.D. McCombe and R. Kaplan, Phys. Rev. Lett. 21, 756 (1968).
13. E.S. Koteles and W.R. Datars, Phys. Rev. B14, 1571 (1976).

Hot Hole Magnetophonon Resonance in p-InSb in High Magnetic Fields up to 40T

N. Kamata[1], *K. Yamada*,[1], *and N. Miura*[2]

[1]Saitama University, Shimo-Ohkubo, Urawa, Saitama 338, Japan
[2]Institute for Solid State Physics, University of Tokyo, Roppongi, Minato-ku, Tokyo 106, Japan

Abstract. Hot hole magnetophonon resonance in p-InSb was measured in pulsed high magnetic fields up to 40T. In addition to Landau level to acceptor state transitions, a series of inter-Landau level transitions with spin splitting structures have been resolved for the first time. This enabled us to examine band parameters by observing an intrinsic intra-valence band process without extrinsic effects of impurities. The acceptor binding energy as a function of magnetic field $E_a(B)$ was determined by the Landau level to acceptor state transition. An excellent overall fit was obtained between theory and experiment.

1. Introduction

Investigation of valence bands in InSb under high magnetic fields has been one of the most fundamental problems in semiconductor physics. Pidgeon and Brown[1] determined band parameters from magnetoabsorption spectra. Grisar et al.[2] measured photoconductivity by CO_2 laser light (120meV) up to 20T. Ranvaud et al.[3] applied uniaxial stress in far-infrared resonance experiments by HCN laser light (3.68meV).

Magnetophonon resonance (MPR) in p-InSb has been studied by many authors since the first report by Amirkhanov et al.[4]. In hot hole MPR, holes at higher Landau levels lose energy by emitting LO phonons (24.3meV) and fall into either acceptor states or lower Landau levels as shown in Fig. 1. The resultant change in mobility can be observed by monitoring the magnetoresistance as a function of magnetic field. Therefore, in the hot hole regime, the heavy-hole Landau level to acceptor state (H-A) transition is observed, in contrast to the heavy-hole inter-Landau level (H-H) transition in the ohmic regime.

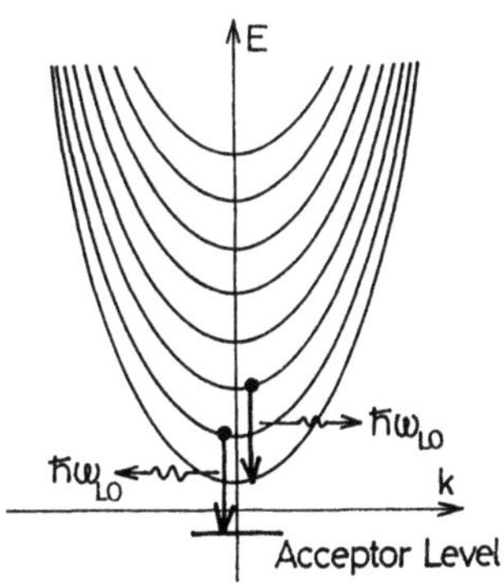

Fig. 1 Phonon emission processes in the hot-hole magnetophonon resonance. The heavy-hole Landau level to acceptor state (H-A) transition and the inter-Landau level transition (H-H) are shown. The sign of the energy has been inverted.

Springer Series in Solid-State Sciences, Vol. 101
High Magnetic Fields in Semiconductor Physics III Editor: G. Landwehr

Since the resonance is determined by an intra-valence band process, it is suitable for analyzing the valence band structure. Shimomae et al.[5] observed light hole transitions by the MPR and discussed the acceptor binding energy as a function of magnetic field $E_a(B)$ below 2T. We have reported hot-hole MPR up to 40T in a previous paper[6], showing distinct series of the H-H transition in addition to the H-A transition.

In this work an improved resolution revealed a twin-peaked structure of up- and down-spin components of the H-H transitions for B//<100>. The determination of band parameters became more accurate under high magnetic field conditions by observing an intrinsic intra-valence band process which includes no extrinsic effects of impurities, whereas $E_a(B)$ was determined from resonance peaks of the H-A transition. An excellent overall fit was obtained between theory and experiment.

2. Experimental

The samples used in this work were Ge doped p-InSb with $N_A-N_D=1.6\times10^{14}cm^{-3}$ and $\mu_p=57000cm^2/Vs$ at T=25K. Synchronized with the pulsed magnetic field with a half width of 5ms, an electric field of 20V/cm with 12ms duration was applied to the sample in order to avoid a lattice temperature rise. The MPR current and its derivative were recorded by a 16bit-resolution digital memory. The MPR spectrum was calculated by a computer.

MPR spectra for B//<111> and B//<100> at a lattice temperature of 4.2K are shown in Figs. 2 and 3. In this work the resolution of the measurements for B//<100> (Fig. 2(b)) was improved by accumulating and averaging the experimental data, which becomes clear when they are compared with previous data (Fig. 2(a)). Besides the major H-A transition, the H-H transition in the heavy hole band with a twin-peaked structure of the up- and down-spin split peaks was clearly resolved above B=10T in Fig. 2(b). For B//<111>, the same spin splittings were already observed as shown in Fig. 2[5]. The observation of the H-H transition was due to an increase in the density of states at each Landau level owing to the application of a high magnetic field.

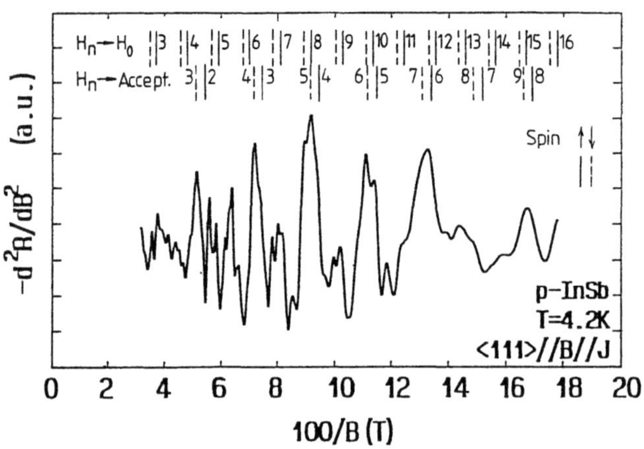

Fig. 2 MPR spectrum for B//<111>.

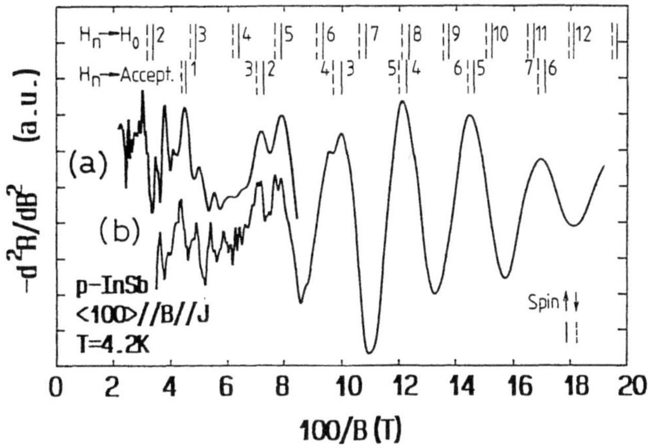

Fig. 3 MPR spectra for B//<100>. An improved resolution measurement (b) revealed the twin-peaked structure of the H-H transitions more clearly than the previous data (a).

3. Discussion

Each resonance peak in Figs. 2 and 3 corresponds to the energy relaxation of hot holes with LO-phonon (24.3meV) emission. Experimental data were compared with a theoretical energy band model.

3.1 Energy band model

Pidgeon and Brown[1] considered the interaction between 2 conduction and 6 valence bands within the framework of Luttingers theory[7]. Their parameters explained the magnetoabsorption spectra up to 10T well. Modifications of matrix elements and resultant different band parameters were reported by various authors[2,3].

For the heavy hole band, the parameters of Pidgeon and Brown[1] and those of Ranvaud et al.[3] resulted in the same Landau energy, which agreed well with the H-H transition in Figs. 2 and 3 including spin splitting structures. However, light hole Landau levels are sensitive to band parameters. Reasonable agreement between experimental data for light hole transition processes were reported for a Pidgeon and Brown model [1,5]. However for a comprehensive understanding it is necessary to take into account the different experimental situation.

3.2 Acceptor Binding Energy

Band parameters were examined by checking the resonance condition of the H-H transition within the heavy hole band. The major resonance of the H-A transition gives $E_a(B)$ as a function of magnetic field. In Fig. 4, $E_a(B)$ was plotted by using the MPR spectra of Figs. 2 and 3. The observed $E_a(B)$ showed a slightly quadratic behavior and the linearly extrapolated value was 9.8meV at B=0. The present $E_a(B)$ converges the result of Shimomae et al.[5] at low magnetic fields and agrees also well with the calculation by Lin-Chung and Henvis[8].

Following these procedures, all transitions were labeled in the upper part of Figs. 2 and 3. An excellent overall fit was obtained between

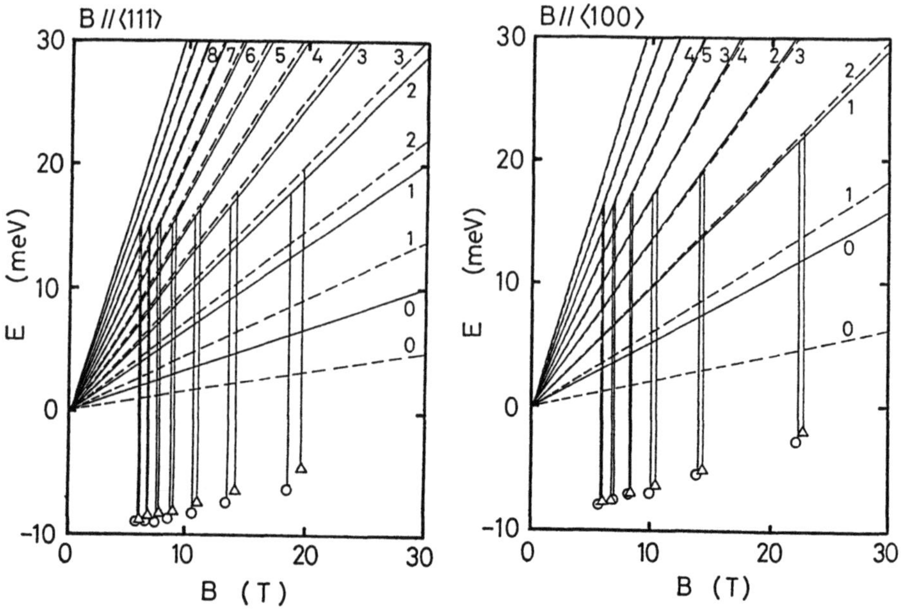

Fig. 4 Magnetic field dependent acceptor energies for B//<111> and B//<100>.

theory and experiment without significant change in band parameters from previous values[1,3].

3.3 Other Transitions

In addition to the H-A and H-H transitions, many fine structures of the MPR were determined much more precisely than in previous measurements. One possible mechanism is the light-hole to heavy-hole (L-H) transition since considerable mixing between original bands exists due to the narrow-gap nature of InSb and the mobility change by this transition is expected to be large. Assignment of the L-H transition becomes important since it gives a more precise means to investigate the light-hole band under high magnetic field.

4. Conclusions

We studied hot hole MPR of p-InSb under pulsed high magnetic fields up to 40T. In high magnetic fields, twin-peaked resonance structures due to up- and down-spin splittings were observed in the H-H transition for both B//<111> and B//<100>. The acceptor binding energy was determined as a function of the magnetic field and a reasonable fit between experiment and a theoretical model was obtained for both the H-H and H-A transitions. Further analysis of additional L-H transitions becomes important.

References

[1]C. R. Pidgeon and R. N. Brown, Phys. Rev., 146, 575(1966).
[2]R. Grisar, H. Wachernig, G. Bauer, J. Wlasak, J. Kowalski and W. Zawadzki, Phys. Rev., 18, 4355(1978).
[3]R. Ranvaud, H. R. Trebin, U. Rossler and F. H. Pollak, Phys. Rev. B, 20, 701(1979).
[4]K. I. Amirkhanov, R. I. Bashirov and Z. A. Ismailov, Sov. Phys. Semicond., 2, 356(1968).
[5]K. Shimomae, K. Kasai and C. Hamaguchi, J. Phys. Soc. Japan, 49, 1060 (1980).
[6]N. Kamata, K. Yamada and N. Miura, Proc. 3rd Int'l Conf. Phonon Physics, 805(1990).
[7]J. M. Luttinger, Phys. Rev., 102, 1030(1956).
[8]P. J. Lin-Chung and B. W. Henvis, Phys. Rev. B, 12, 630(1975).

Part VIII

Magneto-Tunneling, Magnetization in 2D Systems

Magnetotunnelling Spectroscopy to Measure the Electron and Hole $\varepsilon(k)$ Dispersion Curves in the Quantum Well of Resonant Tunnelling Structures

L. Eaves[1], *R.K. Hayden*[1], *D.K. Maude*[1], *M.L. Leadbeater*[1], *E.C. Valadares*[1], *M. Henini*[1], *O.H. Hughes*[1], *J.C. Portal*[2], *L. Cury*[2], *G. Hill*[3], *and M.A. Pate*[3]

[1]Department of Physics, University of Nottingham, Nottingham NG7 2RD, UK
[2]INSA-CNRS, F-31077 Toulouse Cedex and SNCI-CNRS, F-38042 Grenoble Cedex, France
[3]Department of Electronic Engineering, University of Sheffield, Sheffield S1 3JD, UK

We investigate the effect of a large magnetic field, B, applied parallel to the plane of the barriers on the resonant I(V) currents of n- and p-type double barrier structures. The shifts in the electron resonances with increasing B can be understood using the effective mass approximation. The results for the p-type structures allow us to probe the anomalous $\varepsilon(k_{\|})$ dispersion curves of the hole states in the quantum well.

1. Introduction

In bulk semiconductors, the dispersion curves of the light and heavy hole valence bands are parabolic at low hole energies. Their curvatures give the inverse masses of the light and heavy holes. However, when a quantising magnetic field ($\underline{B}\|\underline{z}$) is applied to a bulk semiconductor, the quantised hole energies or Landau levels, as a function of k_z, are highly complex [1,2]. The complex Landau level structure has been the subject of numerous investigations using magnetospectroscopic [3] and magnetotransport techniques [4].

A confining quantum well potential in, for example, an AlAs/GaAs/AlAs heterostructure, also gives rise to complex hole dispersion curves, as a function of the in-plane wavevector component, $k_{\|}$, even at zero magnetic field. An excellent review of this subject is given by Bastard [5]. Recently, Wessel and Altarelli [6] have calculated hole states in resonant tunnelling devices and have shown that the complex nature of the valence band strongly modifies the resonant tunnelling current-voltage characteristics, I(V). Mendez and co-workers [7] were, to our knowledge, the first to investigate p-type resonant tunnelling devices. Their structures show strong resonant features in I(V) and they were able to identify resonant tunnelling into both light and heavy hole states of the quantum well with some evidence of mixing.

Despite the considerable interest in hole states in quantum wells, there have been remarkably few direct experimental investigations of the anomalous hole dispersion curves, since spectroscopic techniques tend to average over k-space. In this article, we will describe a novel experimental method of determining the hole dispersion curves in the quantum well of resonant tunnelling devices using magnetotunnelling spectroscopy in which a strong magnetic field is applied parallel to the plane of the tunnel barriers. The experiments reveal directly the strong coupling between light and heavy hole states at finite $k_{\|}$ and show clearly that some of the states correspond to negative hole mass (hole energy *decreasing* with *increasing* $k_{\|}$).

We will preface our discussion of the holes by a description of the behaviour of electrons in n-type resonant tunnelling devices in the presence of a magnetic field applied parallel to the plane of the barriers [8,9,10]. These results will show that the shift in voltage of the peak current of the electron resonance with magnetic field is quadratic and essentially reflects the parabolic nature of the electronic $\varepsilon(k_{\|})$ curves. In contrast, we will see that the hole resonances shift non-quadratically, corresponding to the highly non-parabolic nature of the hole $\varepsilon(k_{\|})$ curves in the quantum well.

2. Sample characteristics

Both the n- and p-type resonant tunnelling devices used in this investigation were grown by molecular beam epitaxy on a Varian GEN II machine at Nottingham. The structures incorporate an undoped spacer layer between each barrier and the heavily-doped contact layer adjacent to it. This feature appears to increase the peak-to-valley ratio in the I(V) characteristics, probably by reducing ionised impurity scattering. Three structures are considered in this article. The n-type resonant tunnelling device (NU183) is shown schematically below.

NU183

0.5 μm GaAs, $n = 2 \times 10^{18}$ cm^{-3}, top contact
50 nm GaAs, $n = 10^{17}$ cm^{-3}
50 nm GaAs, $n = 10^{16}$ cm^{-3}
3.3 nm GaAs, undoped
11.1 nm $Al_{0.4}Ga_{0.6}As$, undoped (thick barrier)
5.8 nm GaAs, undoped (well)
8.3 nm $Al_{0.4}Ga_{0.6}As$, undoped (thin barrier)
3.3 nm GaAs, undoped
50 nm GaAs, $n = 10^{16}$ cm^{-3}
50 nm GaAs, $n = 10^{17}$ cm^{-3}
2 μm GaAs, $n = 2 \times 10^{18}$ cm^{-3}
n^{+}GaAs (001) substrate

Two p-type structures were examined; the first (NU448), with a well thickness of 4.2 nm, is shown schematically below. The second (NU490) has a well thickness of 6.8 nm, but with all of the other layer thicknesses the same as for structure NU448. All three structures were processed into circular mesas. For the p-type devices, the lower electrical contact was to the 3.0 μm heavily doped p-layer immediately above the n^{+} substrate.

NU448

0.6 μm GaAs, p = 2 x 10^{18} cm^{-3}, top contact
100 nm GaAs, p = 1 x 10^{18} cm^{-3}
100 nm GaAs, p = 5 x 10^{17} cm^{-3}
5.1 nm GaAs, undoped
5.1 nm AlAs, undoped (barrier)
4.2 nm GaAs, undoped (well)
5.1 nm AlAs, undoped (barrier)
5.1 nm GaAs, undoped
100 nm GaAs, p = 5 x 10^{17} cm^{-3}
100 nm GaAs, p = 1 x 10^{18} cm^{-3}
3.0 μm GaAs, p = 2 x 10^{18} cm^{-3}, lower contact
n^+ GaAs (001) substrate

3. **Electron resonant tunnelling in a transverse magnetic field**

The layer structure of the asymmetric n-type resonant tunnelling device (NU183) is given above. We consider the bias direction in which electrons are injected into the well through the thicker barrier (11.1 nm). They leave the well through the thinner (8.3 nm) barrier. Under these conditions there is little charge buildup in the well so that the resonance is not distorted by space-charge effects [11]. Tunnelling occurs from a two-dimensional electron gas (2DEG) which forms in the accumulation layer of the emitter contact when a bias is applied to the device. Figure 1 shows the current-voltage characteristics of a 200 μm diameter mesa at 4 K in magnetic fields B between 0 and 11.5 Tesla applied parallel to the plane of the barriers and perpendicular to the current flow $\underline{J}$. We define $\underline{B} \parallel \underline{z}$ and $\underline{J} \parallel \underline{x}$. The main effects are:

(a) There is a rapid decrease in the peak current with magnetic field; at 4 T the peak current is less than half the zero-field value. This trend is reversed at higher fields where there is a gradual increase.

(b) The peak position (V_p) moves to higher bias as the field increases. By 11 T, V_p has changed by more than 300 mV. The threshold voltage (V_{th}) above which the current increases rapidly to its resonant peak value also shifts to higher values of bias with increasing B, as does the cut-off voltage (V_v) beyond the resonant peak. Figure 2 plots the threshold, peak and cut-off voltages against B. The threshold and cut-off voltages were deduced by extrapolating the linear portions of I(V) on each of the resonant peaks to zero current.

(c) The resonance becomes very much broader. At B = 0, it is extremely sharp and has a full-width at half-maximum of less than 100 mV. This sharp resonance is characteristic of resonant tunnelling from a two-dimensional state with little charge buildup in the well. At 11 T the resonance is more than 400 mV wide.

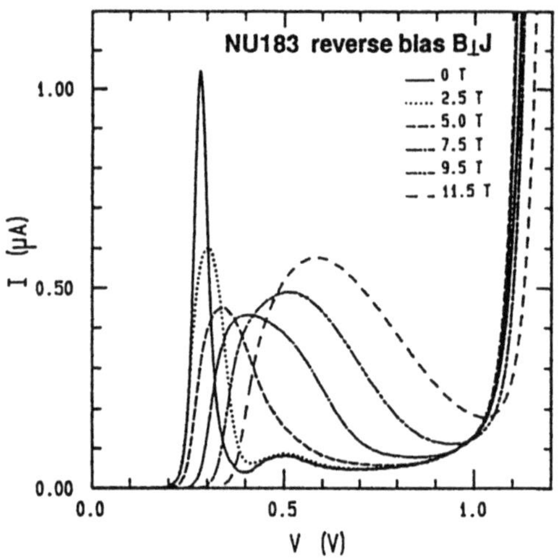

Figure 1: Current-voltage characteristics of a 200 μm diameter mesa of NU183 at the first resonance in reverse bias at 4 K in magnetic fields $\underline{B}\perp\underline{J}$ up to 12 Tesla.

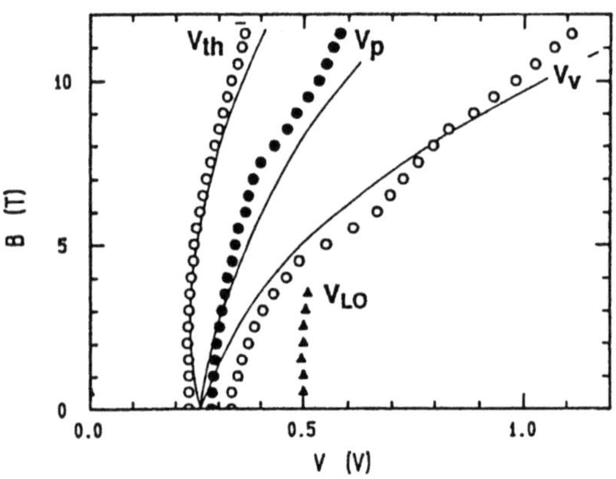

Figure 2: Plot of the threshold, peak and cut-off voltages for NU183 against magnetic field $\underline{B}\perp\underline{J}$, showing the broadening of the resonance and the diamagnetic shift in the peak position. The solid lines show the calculated field-dependence of the voltages discussed in the text. The position of the LO phonon peak (▲) does not change with field since this transition does not conserve transverse momentum.

(d) The peak-to-valley current ratio falls dramatically with field, from ~26:1 at zero field to 1.8:1 at 11 T.

(e) At B = 0 the resonant peak is almost symmetrical and the line-shape resembles the expected Lorentzian. However, at higher fields the peak in the current is markedly closer to the threshold voltage than to the cut-off.

(f) The LO phonon replica peak at V ~ 500 mV [12] does not shift with magnetic field and disappears for B > 5 T. The position of the LO phonon peak is also plotted in Figure 2, marked ▲.

Resonant tunnelling can only occur when there is an occupied state in the 2DEG of the emitter accumulation layer with the same values of total energy ε and components of canonical momentum k_y and k_z as an unoccupied state in the quantum well. Since both the accumulation layer potential and the quantum well are relatively narrow in this sample, the quantisation is predominantly electrical and we can describe the effect of a transverse field on the energy levels using a perturbation approach. The results of this calculation, with the zero of energy taken at the centre of the well, give the following values for the electron energy in the accumulation layer, ε_a, and well, ε_w:

$$\varepsilon_a = \varepsilon_0 + eE(b_1 + w/2) + \frac{3m^*\omega_c{}^2}{2a^2} + \frac{\hbar^2 k_z{}^2}{2m^*} + \frac{\hbar^2 (k_y - k_{y0})^2}{2m^*} \quad \text{and}$$

$$\varepsilon_w = \varepsilon_1 + \frac{m^*\omega_c{}^2 w^2}{24} + \frac{\hbar^2 k_z{}^2}{2m^*} + \frac{\hbar^2 k_y{}^2}{2m^*} ,$$

where ε_1 is the quantum confinement energy in the well, ε_0 is the quantum confinement energy in the accumulation layer, as shown in Figure 3. E is the applied electric field, a is the variational parameter of the Fang-Howard wavefunction, b_1 and w are the thicknesses of the emitter barrier and quantum well respectively, and $k_{y0} = eB\Delta s/\hbar = eB(3/a + b_1 + w/2)/\hbar$. This corresponds classically to the change in the wavevector due to the action of the Lorentz force on the tunnelling electron as it moves from its "quantum stand-off distance" 3/a in the emitter accumulation layer to the centre of the well. Note that k_{y0} corresponds to the acquired in-plane momentum in the quantum well. Resonant tunnelling occurs for $\varepsilon_a = \varepsilon_w$. If we assume that the effective masses are the same in the emitter and the well, the kinetic energies for motion in the z-direction are equal and so the conservation condition may be written as

$$\varepsilon_0 + eE(b_1 + w/2) + \Delta\varepsilon + \frac{\hbar^2 (k_y - k_{y0})^2}{2m^*} = \varepsilon_1 + \frac{\hbar^2 k_y{}^2}{2m^*} ,$$

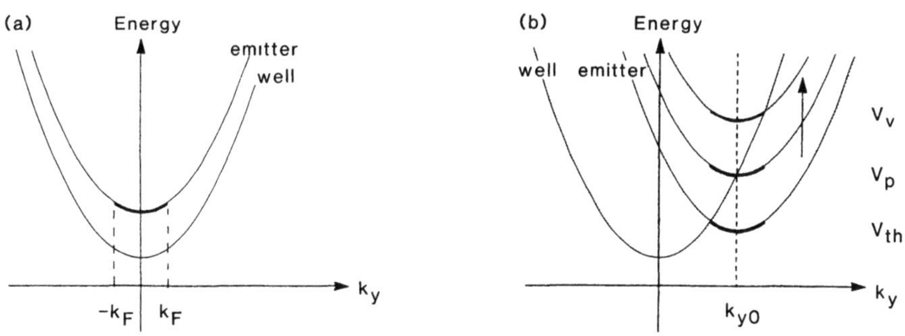

Figure 3: Schematic diagram illustrating the conservation conditions which govern resonant tunnelling in a transverse magnetic field. The occupied states in the emitter are represented by the portion of a parabola between $k_{y0} - k_F$ and $k_{y0} + k_F$. Resonant tunnelling occurs for biases where this curve intersects the parabola representing the electron states in the quantum well. (a) $B = 0$: the parabolae have a common axis and only coincide for one value of bias, producing a sharply peaked resonance, (b) $B > 0$: the origin of the emitter parabola is displaced by $k_{y0} = eB(3/a + b_1 + w/2)/\hbar$ and the two parabolae intersect for a range of bias voltages between V_{th} where $k_y = k_{y0} - k_F$ and V_v where $k_y = k_{y0} + k_F$. This produces a much broader resonance.

where $\Delta\varepsilon$ is the difference in the diamagnetic energy shifts, $m^*\omega_c{}^2(3/a^2 - w^2/12)/2$, which is small. At low temperatures, only states in the emitter with $|k_y - k_{y0}| < k_F$ are occupied where k_F is the Fermi wave-vector of the 2DEG in the emitter accumulation layer ($\varepsilon_F = \hbar^2 k_F{}^2/2m$). Therefore, filled states can be represented in ε - k_y space as a section of a parabola centred at $k_y = k_{y0}$, between the points $k_{y0} - k_F$ and $k_{y0} + k_F$. The well states lie on a parabola centred at $k_y = 0$. The conservation conditions can be interpreted graphically by looking for intersections of these two curves, as illustrated in Figure 3 [8,13,14]. At $B = 0$, $k_{y0} = 0$ so the two parabolae are centred about the same axis. At zero bias the emitter curve is below the energy level in the well and resonant tunnelling cannot occur. As the bias is swept, the emitter curve moves up relative to the well state and when the voltage drop across the emitter region and the first half of the quantum well is $eV_{e,th} = \varepsilon_1 + \varepsilon_F$, the device comes onto resonance. If we neglect the effects of space charge buildup in the well, this is the only applied bias for which resonant tunnelling can occur, so at zero magnetic field the threshold, peak and cut-off voltages are coincident, $V_{e,p} = V_{e,th} = V_{e,v} = V_0$.

The application of a transverse magnetic field shifts the centre of the emitter parabola to the right by k_{y0}. When $k_y = k_{y0} - k_F$, the threshold of the resonance is reached and electrons can tunnel into the well. As the bias is increased, the point of intersection moves to higher energy and k_y. There is a range of bias for which some of the states in the emitter are in resonance with states in the well. The cut-off occurs for the bias at which $k_y = k_{y0} + k_F$. If we neglect charge buildup in the quantum well and the variation in the transmission coefficient with k_y, the peak current occurs when there is the largest number of emitter states available to tunnel from, which is for $k_y = k_{y0}$. The corresponding voltage drops across the emitter are:

$$\text{Threshold} \quad eV_{e,th} = \varepsilon_1{}' + \frac{\hbar^2 (k_{y0} - k_F)^2}{2m^*} = eV_0 + \frac{\hbar^2 (k_{y0}{}^2 - 2k_{y0}k_F)}{2m^*} \; ; \quad (1)$$

$$\text{Peak} \quad eV_{e,p} = \varepsilon_1{}' + \frac{\hbar^2 (k_{y0}{}^2 + k_F{}^2)}{2m^*} = eV_0 + \frac{\hbar^2 k_{y0}{}^2}{2m^*} \; ; \quad (2)$$

$$\text{Cut-off} \quad eV_{e,v} = \varepsilon_1{}' + \frac{\hbar^2 (k_{y0} + k_F)^2}{2m^*} = eV_0 + \frac{\hbar^2 (k_{y0}{}^2 + 2k_{y0}k_F)}{2m^*} \; ; \quad (3)$$

$$\text{with} \quad k_{y0} = eB(3/a + b_1 + w/2)/\hbar \; ; \quad (4)$$

where $\varepsilon_1{}' = \varepsilon_1 - \Delta\varepsilon$. This predicts a quadratic shift in the peak position and a linear increase in the width of the resonance with magnetic field. However, the Fermi wavevector k_F of the emitter 2DEG is a function of the applied bias since it depends on the number density in the emitter n_a, $k_F = (2\pi n_a)^{1/2}$. This can be obtained from the periodicity of the oscillations in the tunnel current or differential capacitance when the magnetic field is applied perpendicular to the interfaces ($\underline{B}\|\underline{J}$) [15]. In addition, the voltage drop across the emitter region is not linearly related to the total applied bias. The distribution of the potential within the device varies due to the depletion of the collector contact; V_e can be approximately related to the total applied bias V by

$$V_e/V = (3/a + b_1 + w/2)/(3/a + \lambda_c + b_1 + b_2 + w + u)$$

where λ_c is the V-dependent collector screening length and $u = 3.3$ nm is the thickness of the undoped GaAs layer near the barriers. Since λ_c increases with bias whereas 3/a decreases, the proportion of the applied bias dropped across the emitter decreases as the voltage increases, so changing the dependence of V_p on B. Both of these factors tend to make the peak move more rapidly to higher bias than expected from equation (2). In particular, the increase in k_F with bias leads to the resonance being asymmetric about the peak, as observed, since the Fermi wavevector at cut-off is greater than that at threshold. Using this model, we can predict the variation with field of the total applied voltages (V) corresponding to $V_{e,th}$, $V_{e,p}$ and $V_{e,v}$. It demonstrates the usefulness of the effective mass-WKB approximation for treating the resonant tunnelling of conduction electrons [10].

To summarise this section: the quadratic shift in voltage of the electron resonance with increasing transverse magnetic field arises from the second term in equation (2) above, in which k_{y0} corresponds to the in-plane momentum, $k_{\parallel}$, acquired from the action of the magnetic field. The quadratic shift reflects the parabolic nature of the electron dispersion relation $\varepsilon(k_{\parallel})$ in the quantum well.

4. Hole resonant tunnelling in a transverse magnetic field

The valence band bending diagram of the structures is shown schematically in Figure 4 for an applied voltage (V), which gives rise to a hole accumulation layer adjacent to the emitter barrier. A two dimensional hole gas (2DHG) forms in the accumulation layer of the emitter with a quasi-bound hole state energy, ε_0. The quasi-bound hole states in the quantum well are given their conventional notation where LH and HH refer to light and heavy holes respectively. The states can be defined in this way only for $k_{\parallel} = 0$, since in this case there is no coupling between the light and heavy holes.

Figure 5 shows the I(V) characteristics obtained for structure NU448 at 4 K. Six distinct peaks are observed in I(V) over the measured voltage range up to 2.4 V. The figure shows the variation of the I(V) characteristics as a function of magnetic field applied parallel to the plane of the

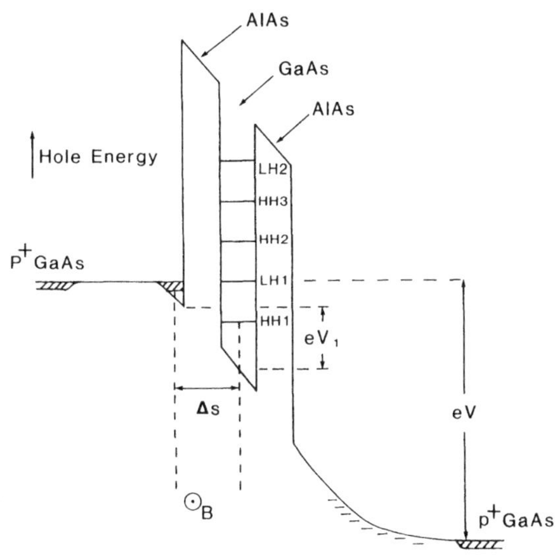

Figure 4: Schematic diagram of p-type resonant tunnelling device under an applied voltage V. Note that a two-dimensional hole gas forms in the accumulation layer adjacent to the emitter barrier. The applied magnetic field B causes the holes to acquire an additional in-plane momentum as shown schematically in the bottom part of the figure.

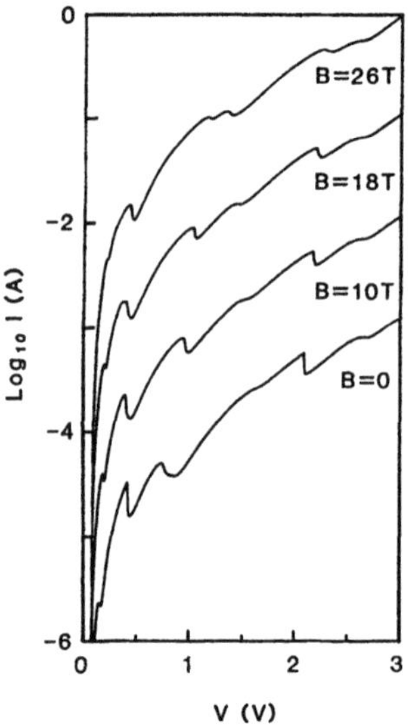

Figure 5: Logarithmic plots of the I (V) characteristics of NU448 (mesa diameter 100 μm) at various in-plane magnetic field values , T=4K. The logarithmic vertical axis gives the current for B=0. For clarity, the curves for 10, 18 and 26 T are displaced upwards by 1, 2 and 3 orders of magnitude respectively.

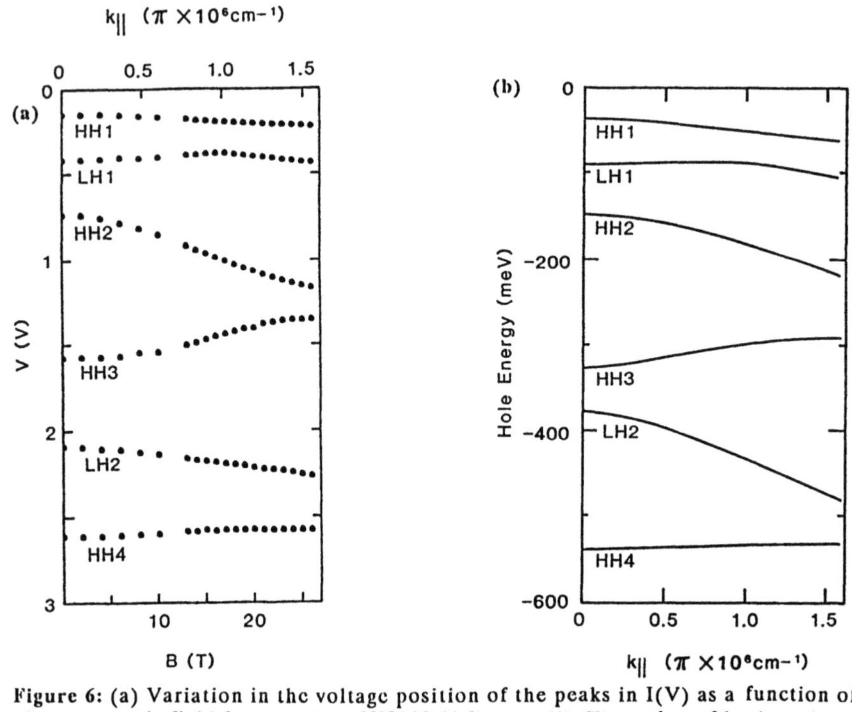

Figure 6: (a) Variation in the voltage position of the peaks in I(V) as a function of in-plane magnetic field for structure NU448 (4.2 nm well). The value of $k_{\parallel}$ is estimated using the procedure described in the text. (b) Calculated hole dispersion curves for an isolated AlAs/GaAs/AlAs quantum well of width 4.2 nm in zero electric and magnetic fields.

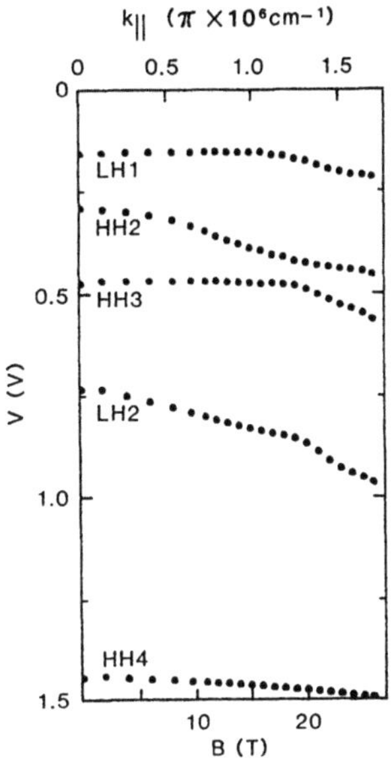

Figure 7: Variation in the voltage position of the peaks in I(V) as a function of in-plane magnetic field for structure NU490 (6.8 nm well). The value of $k_\parallel$ is estimated using the procedure described in the text.

barriers. The devices were examined in Nottingham up to fields of 12 T, and in Grenoble up to fields of 26 T, using the CNRS-MPI hybrid magnet system. It can be seen that the voltage positions of the peaks and thresholds of the resonances shift with increasing magnetic field, but at different rates and in different directions. The voltage positions of the peaks in I(V) are plotted in Figure 6(a) for structure NU448 (4.2 nm well). Figure 7 shows a similar plot for structure NU490 (6.8 nm well). These curves bear a remarkable resemblance to the ε versus $k_\parallel$ dispersion curves of holes in quantum wells.

As can be seen by reference to Figure 3 and equations 2 and 4 above, the applied magnetic field allows us to scan the occupied hole states in the 2DHG of the emitter accumulation layer through the $\varepsilon(k_\parallel)$ dispersion curves of the quantum well. The V versus B plots in Figures 6(a) and 7 therefore map out the anomalous dispersion curves of the hole states in the quantum well.

Using equation 4 we can estimate the $k_\parallel$ gained at a particular value of B. The corresponding $k_\parallel$ values are given on the upper axes of Figures 6(a) and 7. Using the Luttinger parameters for the valence band of GaAs [16], we estimate a hole quantum stand-off distance of 5 nm. Note that the energy of the tunnelling hole in the quantum well is considerably less than the total applied voltage and is given by $\varepsilon_{hw} = eE(3/a + b_1 + w/2)$ where E is the electric field in the region of the emitter barrier (b_1 is the thickness of the tunnel barrier).

A particularly interesting feature of Figure 6(a) is the weak dispersion of the HH1 and LH1 curves, due to their strong admixing with increasing $k_\parallel$. In contrast, the HH2 and HH3 curves are strongly dispersed in opposite senses, approaching each other at low B (low $k_\parallel$) and starting to repel each other at $B \simeq 26$ T ($k_\parallel \simeq 1.5\pi \times 10^6$ cm^{-1}). Such an interaction is expected [5,17] at this value of $k_\parallel$ ($\simeq \pi/w$) for a well width of 4.2 nm. Note from Figure 5 that the strengths of these two resonances in I(V) become comparable as the admixing increases. A similar repulsion between HH2 and HH3 is shown in Figure 7 for NU490. In this case, the repulsion occurs at lower values of B ($k_\parallel$) due to the wider well width ($w = 6.8$ nm for this structure). For NU490, the first HH1 resonance gives rise to too small a current to observe in the DC I(V) characteristics.

For a qualitative assessment of the data shown in Figure 6(a) we have derived the hole dispersion relations for an isolated AlAs/GaAs/AlAs quantum well of width 4.2 nm in zero electric and magnetic fields using a four component envelope function formalism [18]. This is shown in Figure 6(b). The calculated hole energies of the HH1, LH1, HH2 and HH3 states at $k_{\|} = 0$ correspond closely to the relative voltage values of the first four resonance peaks in the I(V) curve at B = 0 for NU448. Furthermore, the theoretical results reproduce accurately the main features of the experimental curves, in particular the weak dispersion of the HH1, LH1 and HH4 resonances and the much stronger dispersion of the HH2 and HH3 resonance. The only major qualitative difference between Figures 6(a) and (b) is the much larger dispersion obtained from theory for the LH2 resonance.

This remarkable correspondence shows that the applied magnetic field allows us to observe directly the strongly non-parabolic dispersion curves of holes confined in a quantum well potential. As Wessel and Altarelli [6] have shown, the dispersion curves for an isolated hole quantum well do not differ appreciably from those for a well with finite barriers provided the barriers are of sufficient width, as is the case here.

For a more quantitative comparison with experiment a full quantum mechanical treatment of hole resonant tunnelling in the presence of a magnetic field applied parallel to the barrier is required. This would need to consider the effect of the large electric field (> 10^7 V/m) on the hole states in the quantum well, and to describe the confined holes in the emitter accumulation layer. At fields B > 20 T, a non-perturbative approach is necessary since the magnetic length $(\hbar/eB)^{\frac{1}{2}}$ is comparable with the dimensions of electrical confinement. In addition, the finite spread of in-plane wave-vector ($-k_F < k_y < k_F$) in the emitter accumulation layer should be taken into account.

Finally, for a quantitative comparison between theory and the measured I(V) curves, it is necessary to model correctly the distribution of space charge throughout the device.

We have investigated the anisotropy of the resonances when the magnetic field is rotated in the plane of the barrier. Only very small differences are observed for B parallel to the [100] and [110] axes, indicating that anisotropy effects are relatively unimportant in the structures investigated here. In addition, we have not found any evidence for spin splitting of the hole state due to applied magnetic and electric fields.

Note that for sample NU448, the LH2 and HH4 resonances occur at high voltages, corresponding to resonant injection of holes near the top of the collector barrier.

5. Conclusion

We have investigated resonant tunnelling of electrons and holes in double barrier structures in the presence of a magnetic field applied parallel to the plane of the barriers. The magnetic field dependence of the electron resonances can be described using the effective mass approximation. Our measurements on the p-type structures have shown that the application of a magnetic field parallel to the plane of the barriers provides a new tunnelling spectroscopy technique for directly investigating the complex in-plane dispersion curves of holes in quantum well heterostructures.

Acknowledgements

This work is supported by SERC, CNRS and the European Community. We acknowledge invaluable discussions with Mr. R. Wessel, Professor M. Altarelli, Dr. F. W. Sheard and Dr. G. A. Toombs.

References

1. J. M. Luttinger and W. Kohn, Phys. Rev. 97, 869 (1955).
2. J. M. Luttinger, Phys. Rev. 102, 1030 (1956).
3. J. C. Hensel and K. Suzuki, Phys. Rev. B 9, 4219 (1974).
4. L. Eaves, R. A. Hoult, R. A. Stradling, S. Askenazy, R. Barbaste, G. Carrère, J. Leotin, J. C. Portal and J. P. Ulmet, J. Phys. C: Solid State Phys. 10, 2831 (1977).
5. G. Bastard, "Wave Mechanics Applied to Semiconductor Heterostructures", Les Éditions de Physique (1988).
6. R. Wessel and M. Altarelli, Phys. Rev. B 39, 12802 (1989).
7. E. E. Mendez, W. I. Wang, B. Ricco and L. Esaki, Appl. Phys. Lett. 47, 415 (1985).
8. M. L. Leadbeater, L. Eaves, P. E. Simmonds, G. A. Toombs, F. W. Sheard, P. A. Claxton, G. Hill and M. A. Pate, Solid State Electronics 31, 707 (1988).
9. E. S. Alves, M. L. Leadbeater, L. Eaves, M. Henini, O. H. Hughes, A. Celeste, J. C. Portal, G. Hill and M. A. Pate, Superlattices and Microstructures 5, 527 (1989).
10. M. L. Leadbeater, E. S. Alves, L. Eaves, M. Henini, O. H. Hughes, A. Celeste, J. C. Portal, G. Hill and M. A. Pate, J. Phys: Condens. Matter 1, 4865 (1989).
11. E. S. Alves, L. Eaves, M. Henini, O. H. Hughes, M. L. Leadbeater, F. W. Sheard, G. A. Toombs, G. Hill and M. A. Pate, Electronics Lett. 24, 1190 (1988).
12. M. L. Leadbeater, E. S. Alves, L. Eaves, M. Henini, O. H. Hughes, A. Celeste, J. C. Portal, G. Hill and M. A. Pate, Phys. Rev. B. 39, 3438 (1989).
13. L. Eaves, K. W. H. Stevens and F. W. Sheard, "The Physics and Fabrication of Microstructures and Microdevices", Springer Proceedings in Physics 13 (ed. M. J. Kelly and C. Weisbuch) (1986).
14. R. A. Davies, D. J. Newson, T. G. Powell, M. J. Kelly and H. W. Myron, Semicond. Sci. Technol. 2, 61 (1987).
15. M. L. Leadbeater, E. S. Alves, F. W. Sheard, L. Eaves, M. Henini, O. H. Hughes and G. A. Toombs, J. Phys.: Condens. Matter 1, 10605 (1989).
16. "Landolt-Börstein Numerical Data and Functional Relationships in Science and Technology", ed. V. Madelung (Springer, Berlin, 1982). Group III, Band 17.
17. R. Wessel, Diploma thesis at Technische Universität München, 1989.
18. L. C. Andreani, A. Pasquarello and F. Bassani, Phys. Rev. B 36, 5887 (1987).

Interband Tunneling in Semiconductor Inversion Layers in High Magnetic Fields

U. Kunze

Institut für Technische Physik, Universität Erlangen-Nürnberg,
W-8520 Erlangen und
Hochmagnetfeldanlage der Technischen Universität Braunschweig,
W-3300 Braunschweig, Fed. Rep. of Germany

Abstract. In electron inversion layers on degenerate p-type semiconductor substrates, tunneling from the electron subbands through the depletion layer into extended states of the valence band exhibits a resonance effect in the current-voltage characteristics. A review is presented of experiments under high magnetic fields, which illuminate the origin of the current resonance and yield some important parameters of interband tunneling in low-dimensional systems.

1. Introduction

Tunneling in semiconductor microstructures is one of the fundamental techniques for investigating the level structure of dimensionally confined electronic systems. In addition to basic research activities, there is currently a considerable interest in tunnel junctions that exhibit negative differential conductance (NDC), which is motivated by possible technical applications in ultrafast electronics. In junctions having three-dimensional (3D) electrodes, the occurence of NDC is well known from degenerate pn diodes [1], where interband tunneling is possible due to the high built-in electric field. The invention of bandgap engineering also makes it possible to obtain similar current-voltage (I-V) characteristics from a degenerate pn interband tunnel junction without electric field when the electrodes form a broken-gap configuration [2]. Even in intraband single-barrier junctions NDC has been observed, which arises from the energy dependence of the imaginary dispersion in the bandgap of the barrier [3-5].

However, if at least one electrode is two-dimensional (2D), transverse-wave-vector conservation in the tunneling process should lead to much higher NDC in interband [2,6-8] as well as in intraband [9] tunnel junctions. While intraband tunneling has been extensively studied in double-barrier resonant tunnel (DBRT) devices, interband tunneling in low-dimensional semiconductor structures is still on an early stage. First indications of interband tunneling in 2D systems were found in narrow pnp GaAs quantum well structures [10] and in reversely biased multiple-quantum well pin diodes [11]. Recent experiments on tunneling between the 3D valence band states and the electron subband of a heterostructure InAs quantum well [2] or of an InAs, InSb, or Si surface inversion layer [8,12-14] showed NDC with a partly very high current peak-to-valley (PTV) ratio. In the following we will briefly discuss the low-temperature I-V characteristics of electron subband-valence band tunnel (EVT) junctions in the absence of magnetic field. The rest of the paper deals with the effect of a high magnetic field on the tunneling characteristics of junctions prepared on degenerate p-type InAs [15-17]. These experiments yield the built-in electric field that determines the tunneling probability, confirm the effect of reduced dimensionality on the I-V characteristic, display the Landau-level structure of the nonparabolic ground subband, and reveal a large diamagnetic effect of the excited subband levels.

Springer Series in Solid-State Sciences, Vol. 101
High Magnetic Fields in Semiconductor Physics III Editor: G. Landwehr

2. Basic Junction Properties and Coherent-Tunneling Model

A number of common as well as different features makes it appear instructive to compare intraband DBRT devices with interband EVT junctions (Fig. 1). In both junctions two barriers separate the confined 2D layer from 3D electrodes. Whereas in the DBRT junction the barriers are formed by conduction-band offsets of lattice-matched semiconductor heterojunctions, the barriers in the EVT junction consist of a thin amorphous oxide layer and the bandgap barrier due to the surface band bending, according to a simple metal-oxide-semiconductor (MOS) arrangement (for details of the fabrication procedure see [8,13]). The oxide layer is thin enough to form a low-resistive contact of the metal layer to the inversion electrons in the occupied subbands. The band bending and the induction of surface electrons arise mainly from the low work function of the metal (Yb, Al, Ti) compared with the semiconductor electron affinity [8]. As will be shown below, the I-V characteristics are completely determined by the tunneling transitions between the 2D subband and 3D valence band.

Also in DBRT devices the sequential tunneling model [9] explains the I-V characteristics only from tunneling from the 3D emitter into empty subband states in the well, whereas the well-collector junction acts as a contact which introduces a series resistance. Here we disregard space-charge effects in the electrodes and in the well [18,19]. As shown in Fig. 2(a) transverse wave-vector conservation (coherent single-barrier tunneling) leads to an asymmetric triangular-shaped I-V characteristic, where the total width of the current peak is $2V_n$. Here $V_n=(E_{F3}-E_C)/e$ is given by the Fermi energy in the emitter, and the prefactor 2 reflects the fact that in symmetric DBRT devices half the applied bias drops across each barrier.

In EVT junctions (Fig. 2b) with equal effective masses in the subband and the light-hole valence band, the monotonous variation of the coherent tunneling interval results in a symmetric triangular I-V characteristic. The peak current occurs at a bias where the valence band Fermi level is aligned with the bottom of the subband. If the voltage drop across the oxide barrier is negligible the total peak width is given by $2V_p$, where $eV_p=E_V-E_{F3}$ is given by the Fermi energy in the valence band.

Fig. 2 makes clear that k_t-non-conserving transitions lead to an additional current component in DBRT devices beyond, and in EVT junctions below, the peak voltage. Thus the current PTV ratio decreases in DBRT and increases in EVT junctions. The result is that EVT junctions with a high scattering rate loose the resonance effect in the I-V characteristic which accordingly becomes similar to that of pn diodes with 3D electrodes but still exhibit particularly high PTV ratio [8,12].

The I-V curves at B=0 in Figs. 3 and 4 evidently show current peaks as given by the coherent-tunneling model. It should be noted that in junctions prepared on substrates with doping concentrations $N_A=1.9\times10^{18}cm^{-3}$ and $2.7\times10^{18}cm^{-3}$ (Fig. 3) only the ground subband is occupied, whereas in the junction used in Fig. 4 ($N_A=2.3\times10^{17}cm^{-3}$) the occurrence of two current peaks indicates two occupied subbands, which is due to the lower depletion field. Here we have interpreted

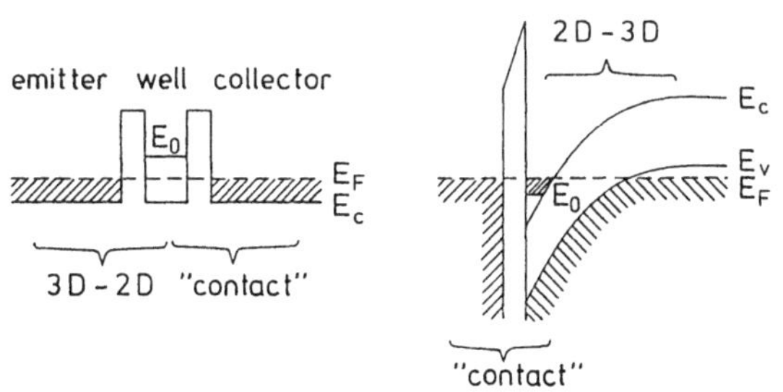

Fig. 1. Schematic energy-band diagrams of a DBRT device and an EVT junction. According to the sequential coherent tunneling model, the junctions of tunneling between the 2D and 3D electrodes which determine the occurrence of NDC and the junctions acting as contacts to the 2D electrode are indicated

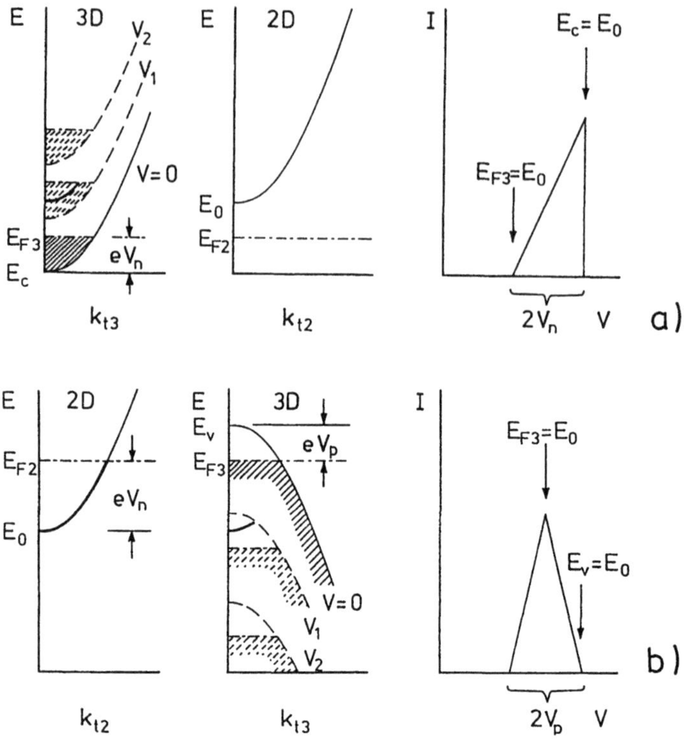

Fig. 2. Energy E with respect to the transverse wave vector k_{t2} in the 2D or k_{t3} in the 3D electrode, respectively, for (a) a DBRT device and (b) an EVT junction. Forward bias polarity ($V_2>V_1>0$) refers to a negative polarity at the emitter of the DBRT device or to positive polarity at the p-type substrate. Resonant states for coherent tunneling are indicated by the solid line in the E-vs-k_{t3} diagram at $V=V_1$. In the resulting I-V curve the biasing conditions of characteristic points are given by arrows

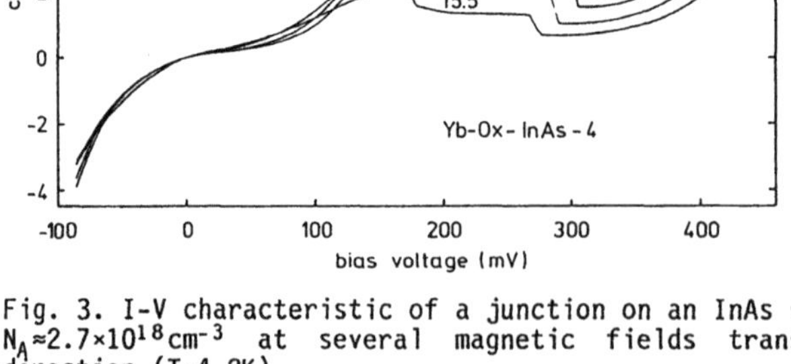

Fig. 3. I-V characteristic of a junction on an InAs substrate with doping level $N_A \approx 2.7\times10^{18}\,cm^{-3}$ at several magnetic fields transverse to the tunneling direction (T=4.2K)

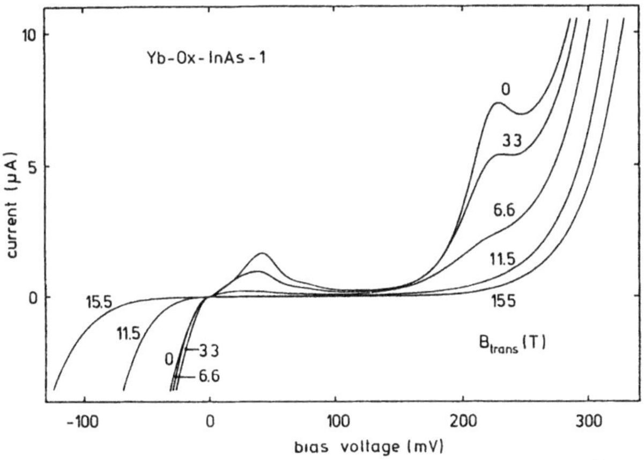

Fig. 4. As in Fig. 3, for a more lightly doped ($2.3\times10^{17}cm^{-3}$) sample. The two current peaks at 42 and 226mV reflect two populated subbands, according to a lower depletion field than that in the junction of Fig. 3 (T=4.2K)

the current peak position V_n as a direct measure of the subband energy E_n relative to the Fermi level, $eV_n=E_F-E_n$, according to the coherent-tunneling model, which, however, will be substantiated by the magneto-oscillatory effects discussed in Sect. 4.

3. Electric Field for Interband Tunneling

The maximum overlap of the localized subband and the valence band wave functions occurs in the depletion zone where the strong surface field is almost screened by the inversion charge. Since the depletion field associated with the ionized acceptors of density N_A decreases into the bulk the electric field F that determines the tunneling probability amounts to a fraction $\alpha<1$ of the depletion field $F_{depl}=[(E_g+eV_p+e\varphi_s-eV)2N_A/\varepsilon_S]^{1/2}$ at the InAs-oxide interface. Here the Fermi energy in the valence band eV_p is ranging from 3 to 17meV depending on N_A and φ_S is the total surface band bending. Since φ_S is not precisely known we can get a lower limit of F_{depl} if φ_S is replaced by the bias position of the current peak $V_0=(E_F-E_0)/e$. It schould be mentioned that in Yb-metallized junctions shown in the Figs. 3 and 4 V_0 decreases only a little as N_A increases, thus F_{depl} in both junctions varies $\propto N_A^{1/2}$. The difference in the peak current in these junctions is in accordnace with the uniform field model, where $I\propto\exp(-F_0/F)$. F denotes the tunneling electric field and F_0 is a material-dependent parameter [8].

In the presence of a magnetic field B transverse to the tunneling direction F has to be replaced by the effective field $F_{eff}=F(1-E_gB^2/2m^*F^2)^{1/2}$, where m* denotes the effective mass in a two-band model [19]. This leads to a reduction of the current as shown in Figs. 3 and 4. At a critical magnetic field $B_c=F(2m^*/E_g)^{1/2}$ F_{eff} decreases to zero and the current vanishes.

An accurate determination of the built-in electric field F can be gained from the measurement of the current as a function of transverse B [20]. In order to eliminate the influence of the B-dependent voltage drop across the series resistance we have evaluated the peak current (Fig. 5). At fit of the current decrease according to the uniform-field model yields the tunneling field F which is the only adjustable parameter. The same procedure can be carried out for the dependence of the current on longitudinal fields B (perpendicular to the interface), where the current decreases due to an increase of the effective energy gap between the subband and valence band Landau levels. The resulting electric fields are close to those obtained under

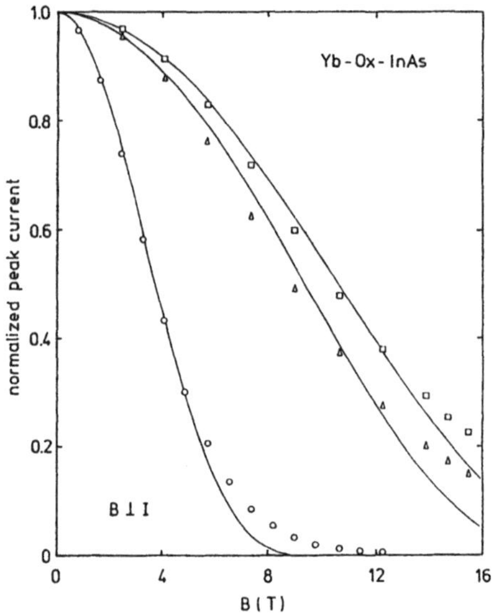

Fig. 5. Peak current at transverse magnetic field B, normalized to the peak current at B=0, as a function of B. Experimental points are measured from junctions having $N_A \approx 2.7\times10^{18}cm^{-3}$ (squares) and $1.9\times10^{18}cm^{-3}$ (triangles), and from the peak at 42mV of the junction having $N_A \approx 2.3\times10^{17}cm^{-3}$ shown in Fig. 4 (circles). Solid lines are fitted using $F=2.90\times10^7 V/m$, $2.65\cdot10^7 V/m$, and $1.43\times10^7 V/m$, respectively

transverse B. It is interesting to look at the resulting fractions α, which are 1.0 and 0.8 for the n=0 and n=1 subband in the junction having $N_A=2.3\times10^{17}cm^{-3}$ and decreases to 0.6 and 0.5 for $N_A=1.9\times10^{18}cm^{-3}$ and $2.7\times10^{18}cm^{-3}$, respectively. This can be understood from the extension of the subband wave function relative to the width of the depletion layer. The decrease of the width as N_A increases leads to a deeper penetration of the wave function into those regions where the depletion field is appreciably lower, as it is the case for excited subbands compared to the ground subband.

4. Subband Landau Levels

In high magnetic fields perpendicular to the interface the second derivative d^2I/dV^2, as a function of B, shows Shubnikov-de Haas type oscillations (Fig. 6). Measurements under tilted fields B revealed that the period of the oscillations only depends on the perpendicular component of B as expected for a 2D electronic system. The origin of the oscillations is the oscillatory density of states of the subband Landau levels which move across the valence band Fermi level when either the magnetic field or the bias voltage is swept. Apart from an unknown phase constant the bias position V of the oscillation minima represents the energy $E=E_F-eV$ of the subband Landau levels. Figure 7 shows a fan diagram gained from the two sets of oscillations that are visible in Fig. 6. Obviously the Landau-level lines of each set converge to a common voltage in the limit of vanishing B, and this voltage is close to one of the peak biases. This is the direct evidence that the current peak occurs at a bias V_n where the valence band Fermi level matches the n=0 or n=1 subband edge.

At constant B the separation of two adjacent lines in Fig. 7, which is equal to the cyclotron energy, is bias dependent according to the subband nonparabolicity. Another way of representing the subband dispersion law can be obtained from the magneto-quantum oscillations vs reciprocal field B: The period of oscillations $\Delta(1/B)$ measured at a bias V is related to an areal density N_k of subband states up to an energy $E=E_F-eV$, $\Delta(1/B)=2e/hN_k$, where N_k at V=0 is equal to the electron density N_n in the subbands n=0 and n=1 (Fig. 8). Additional weak oscillations measured at biases $V>V_n$ revealed periodicities close to those which correspond to the densities N_0 or N_1. These oscillations most likely reflect oscillations of the chemical potential in the inversion

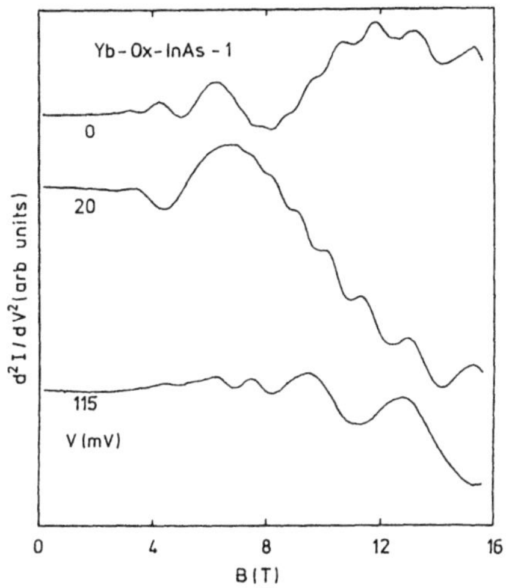

Fig. 6. d^2I/dV^2 with respect to the longitudinal magnetic field B at three different bias voltages V, measured from the same junction as in Fig. 4 (T=4.2K)

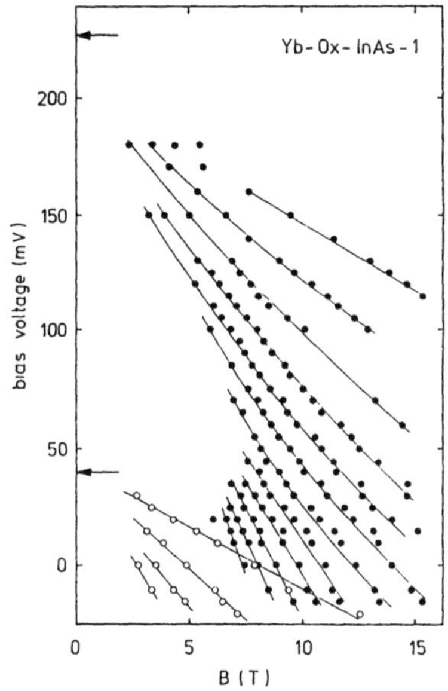

Fig. 7. Bias position of oscillation minima as a function of magnetic field B. Solid lines are guides to the eye. The arrows indicate the current peak positions

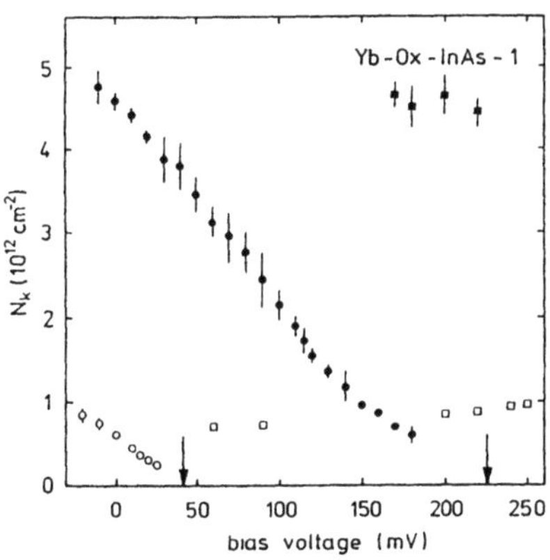

Fig. 8. Areal density of subband states up to an energy $E=E_F-eV$ (circles) and subband electron density (squares) as a function of bias V. Closed and open symbols correspond to the n=0 and n=1 subband, respectively. The arrows indicate the current peak positions

layer upon Landau levels crossing E_F. So we are able to obtain N_n as a function of V, which is needed for a quantitative analysis of the cyclotron energy and the density of states. The increase of N_1 with bias is due to the decrease of depletion charge density [8].

5. Excited Subbands

So far only the occupied subbands could be detected by current peaks in the I-V characteristic or by magneto-oscillations in d^2I/dV^2. At reverse biases structures of higher excited subbands are missing, probably because the signal-

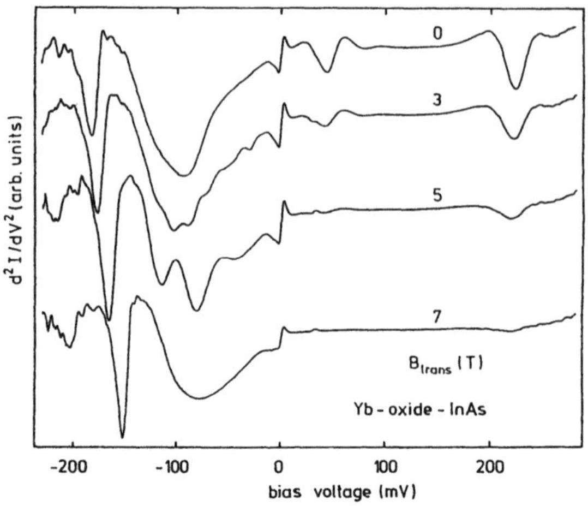

Fig. 9. d^2I/dV^2 with respect to bias V at different transverse magnetic fields, measured from the same junction as used in Fig. 4 (T=4.2K). New subband-edge-induced structures occur at small reverse biases, the origin of the large structure at V≈-150mV is not known

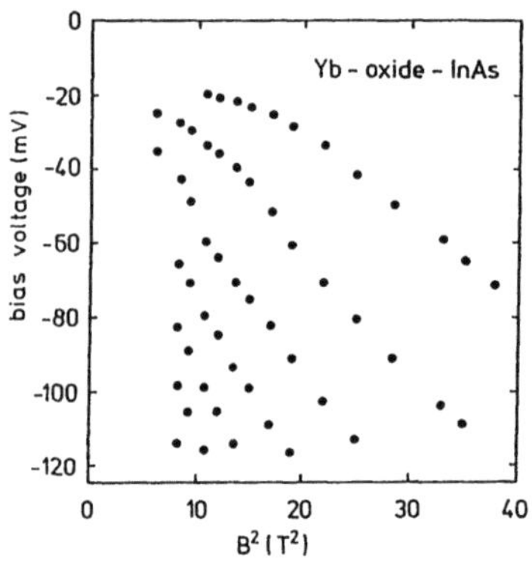

Fig. 10. Bias positions of the new dip structures in Fig. 9 as a function of B^2

to-noise ratio is worse in the low-resistive junction with high doping N_A, and because the subband separation is too small in the junctions having $N_A=2.3\times10^{17}cm^{-3}$. In the latter junctions the separation can be increased by the diamagnetic effect in a magnetic field parallel to the layer [17]. Figure 9 shows new structures arising at B≈3T, rapidly moving with increasing B towards larger reverse biases. These structures have amplitudes comparable to those of the n=0 and n=1 subbands, about two orders of magnitude larger than the magneto-oscillations in Fig. 6, and they vanish at $B>B_c$ as expected for subband-edge-induced structures. Figure 10 shows the bias positions of these diamagnetically shifted subband edges E_2 to E_8 as a function of square of magnetic field. Evaluating the slope of the approximately straight lines yields the spread of the wave functions perpendicular to the interface [21] ranging from 25nm (n=2) to 44nm (n=7), which is by a factor of about 3 larger than expected from the depletion field strength [17]. At present is is not clear whether the simple triangular-well model used in the calculation is inadequate or other reasons like peculiarities of tunneling in crossed electric and magnetic fields are responsible for the difference.

References

1. L. Esaki: Phys. Rev. 109, 603 (1958)
2. L.F. Luo, R. Beresford, and W.I. Wang: Appl. Phys. Lett. 55, 2023 (1989)
3. D.H. Chow and T.C. McGill: Appl. Phys. Lett. 48, 1485 (1986)
4. D.H. Chow, T.C. McGill, I.K. Sou, J.P. Faurie, and C.W. Nieh: Appl. Phys. Lett. 52, 54 (1988)
5. R. Beresford, L.F. Luo, and W.I. Wang: Appl. Phys. Lett. 54, 1899 (1989)
6. J.P. Leburton, J. Kolodzey, and S. Briggs: Appl. Phys. Lett. 52, 1608 (1988)
7. M. Sweeny, J. Xu: Appl. Phys. Lett. 54, 546 (1989)
8. U. Kunze: Z. Phys. B 76, 463 (1989)
9. S. Luryi: Appl. Phys. Lett. 47, 490 (1985)
10. Ch. Zeller, G. Abstreiter, and K. Ploog: Surf. Sci. 142, 456 (1984)
11. J. Allam, F. Beltram, F. Capasso, and A.Y. Cho: Appl. Phys. Lett. 51, 575 (1987)
12. U. Kunze and W. Kowalsky: Appl. Phys. Lett. 53, 367 (1988)
13. U. Kunze, U. Schumacher, and B. Föste: Superlattices and Microstructures 6, 115 (1989)
14. U. Kunze: Solid State Commun. 70, 573 (1989)
15. U. Kunze: In High Magnetic Fields in Semiconductor Physics II, G. Landwehr (ed.), Springer Series in Solid State Sciences Vol. 87, pp. 339-42 (Springer, Berlin 1989)
16. U. Kunze: Appl. Phys. Lett. 54, 2213 (1989)
17. U. Kunze: Phys. Rev. B 41, 1707 (1990)
18. V.J. Goldman, D.C. Tsui, J.E. Cunningham, and W.T. Tsang: J. Appl. Phys. 61, 2693 (1987)
19. M.H. Weiler, W. Zawadzki, and B. Lax: Phys. Rev. 163, 733 (1967)
20. A.R. Calawa, R.H. Rediker, B. Lax, and A.L. McWhorter: Phys. Rev. Lett. 5, 55 (1960)
21. U. Kunze: Surf. Sci. 170, 353 (1986)

Resonant Magnetotunneling Current Through Double Barriers: Coherent and Sequential Processes

G. Platero[1] *and C. Tejedor*[2]

[1]Instituto de Ciencia de Materiales (CSIC)
[2]Departamento de Fisica de la Materia Condensada, Universidad Autonoma, Cantoblanco, E-28049 Madrid, Spain

Abstract. Coherent and sequential contributions to the magnetotunneling current ($B \perp J$) are analyzed in the framework of the Transfer Hamiltonian formalism. The sequential tunneling calculated includes possible scattering processes in the well and is obtained invoking current conservation through the whole system. The characteristic curve I/V is qualitatively very different for coherent and sequential magnetotunneling and their relative intensity can be controlled by changing the external field and sample characteristics.

The aim of this work is the analysis of the effect of a magnetic field on the electronic tunneling current through semiconductor double barriers. We have considered a magnetic field applied in the plane of the barriers.In this configuration, the magnetic and barrier potential are superimposed in the growth direction and the electronic spectrum consists of dispersive bands rather than degenerate bulk Landau levels, whose associated eigenstates are localized in the growth direction.As an external bias is applied, electronic current flux takes place and we will show below that in this configuration, there is a discrete set of tunneling channels available to produce the current [1,2].

We have calculated the tunneling current in the scheme of the Transfer Hamiltonian techniques[3]. The experimental work [4-7]. on these systems reflects some clear features for the current as a function of the magnetic field and as a function of the external bias which cannot be completely understood by semiclassical arguments. There are two physically different processes which contribute to the total current : coherent and sequential magnetotunneling. The first one corresponds to the electronic tunneling from the left to the right which takes place through virtual transitions with the well states.This process occurs with energy and momentum conservation.The sequential tunneling describes the three step process in which the electrons cross sequentially the two barriers spending some time in the well. In order to calculate the coherent contribution to the current we have constructed a wave packet and we have analyzed its temporal evolution from the left to the right of the barrier system.For the analysis of the coherent processes a first order perturbation term (which allows an accurate description of the transmission probability coefficient for tunneling through a simple barrier) is not able to describe the virtual transitions with the well states and therefore we have extended the transfer Hamiltonian technique to infinite order in perturbation theory[8]. The transmission probability per unit of time from left to right through a double barrier system is given by :

$$P_{LR} = \frac{2\pi}{\hbar}\,\delta(E_L - E_R)\,|\,t_{LR}\,|^2 \qquad [1]$$

Where E_L and E_R are the eigenvalues of the auxiliary hamiltonians H_L and H_R respectively [1,2] (see fig. 1 a). We see from eq. 1 that the available coherent tunneling channels are those corresponding to the crossings E_{LR} as shown in fig. 1-b. The expression for the transmission t_{LR} is given in terms of the Green function of the system[8]. The transmission coming from the virtual processes involved in the coherent tunneling are represented by energy differences (E_L-E_C), where E_C is the energy of a state localized in the well, appearing in the denominators of the Green's function. These denominators give very important contributions to the transmission so that when a state localized in the well with energy E_C is close to E_L a peak appears in the current. From figure 1-b it is very simple to visualize that such a fact is going to occur several times when varying the bias. This gives a struc-

Springer Series in Solid-State Sciences, Vol. 101
High Magnetic Fields in Semiconductor Physics III Editor: G. Landwehr

1.a)

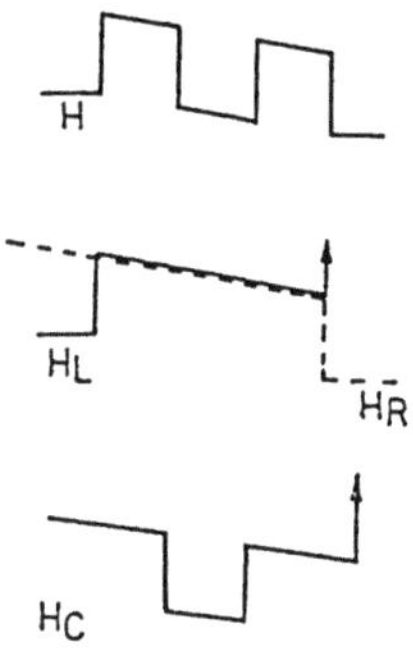

Fig. 1.a: Total (H), left (H_L), right (H_R) and Center (H_C) hamiltonians used in the GTD method.

1.b)

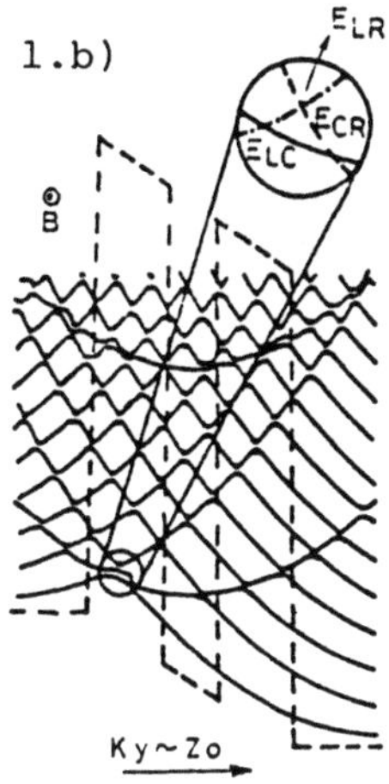

Fig. 1.b: Dispersion relation of the magnetic levels for a DB with B parallel to the interfaces.

ture of narrow peaks of j as a function of the bias. The sequential process is a three step process : the electrons cross the first barrier, then spend some time in the well where scattering processes can take place and finaly cross the second barrier. There is a pair of tunneling channels associated with each sequential process: lc (left-center) and cr (center-right) which are not at the same energy nor at the same momentum k_y (see fig. 1-b). In order to analyze the sequential contribution to the current we consider a macroscopic model which includes all the possible scattering processes in the well , which one would expect to be important when the well width is large, i. e. when the barriers are well separated[9].

In order to account also for these processes, a microscopic model for possible elastic or inelastic scattering processes is outside the scope of this work. However it is possible to calculate all the contributions to the sequential tunneling current in a general way by imposing electronic current conservation [10]. The double barrier system can be seen as two resistors in series, so that the current crossing the first barrier should be the same as the current crossing the second one [10]. It means that there is charge accumulated within the well in equilibrium, and the Fermi level in the well can be calculated from the condition described above : $J_L = J_R$ where J_L and J_R are the current through the first and second barriers respectively.

A double barrier where both coherent and sequential processes are possible, behaves like two channels in parallel. The resistance of each channel is proportional to the inverse of the corresponding transmission probabilities T_C or T_S respectively [10]. This gives for the total transmission probability: $T = T_C + T_S$ From the total transmission coefficient the total current can be obtained. Fig. 2 shows the total sequential contribution to the magnetotunneling current for a 100 Å AlGaAs double barrier separated by a 70 Å GaAs quantum well.For this sample, composed of thick barriers, the coherent contribution to the current is less than two orders of magnitude smaller than the sequential one which determines in this case the current features as a function of the external bias. Three different cases are calculated corresponding to three magnetic fields : B=6T, B=10 T and B=14 T. The current behaviour is similar to that observed experimentally : there is a threshold bias for the current which moves to higher bias as the magnetic field increases,the current peak broadens and also decreases with the field. As the barriers become thinner, the coherent tunneling current increases faster than the sequential one and is the first one which controls the total current. As it has been described before, narrow peaks appear in the

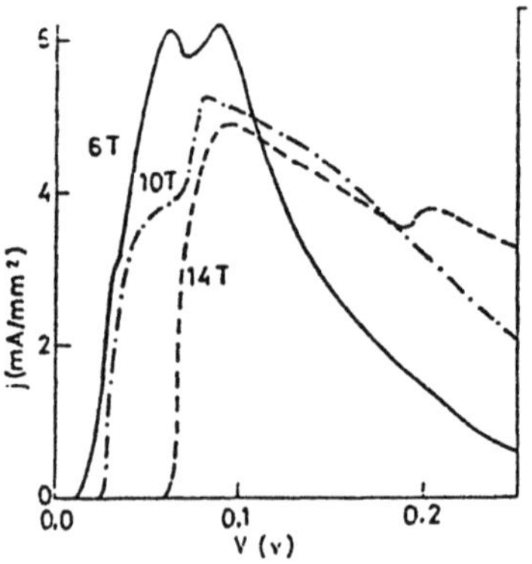

Fig. 2. Sequential tunneling current for a 100Å AlGaAs DB and a 70Å GaAs well as a function of the bias for B=14,10 and 6T.

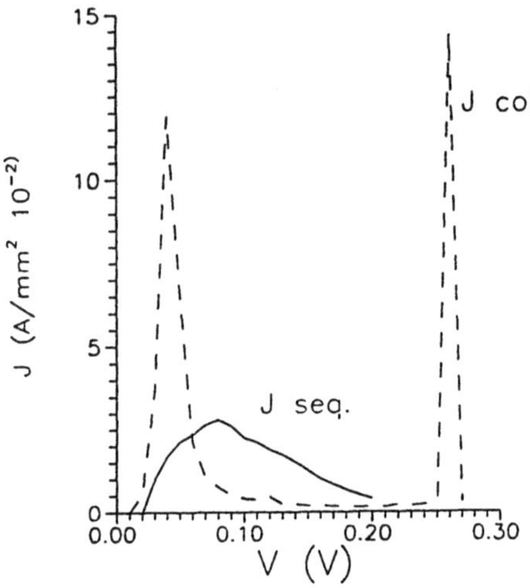

Fig. 3. Coherent (dashed line) and sequential (solid line) tunneling current for a 30 A AlGaAs DB and a 70 A GaAs well and a B = 10 T, as a function of the bias.

characteristic curve for the coherent contribution to the current. An example of this behaviour for narrow barriers is shown in fig. 3 where the I/V characteristic curve for a 30 Å AlGaAs double barrier separated by a 70 Å GaAs quantum well is represented, the Al barrier concentration is 40%.Here, the sequential contribution to the current is very small compared with the coherent one. The strong increase of the coherent tunneling contribution for thinner barriers comes from the fact that now the wavefunctions corresponding to the well states are not localized anymore in the well only but have non negligible weight outside of it, therefore the number of available coherent tunneling channels increases, giving a high contribution to the current. Unfortunately there is no experimental information to compare with for thin barriers in this configuration of the field.

We are indebted to CASA for the CRAY computing facilities. This work has been supported in part by the Comision Interministerial de Ciencia y Tecnologia of Spain under contract MAT88-0116-C02-01/02

REFERENCES

1. L.Brey, G.Platero and C.Tejedor, Phys. Rev. B, **38**, 9649 (1988)
2. G.Platero, L.Brey and C.Tejedor, Phys. Rev. B,**40**,8548 (1989)
3. C.B.Duke, "Tunneling in solids" Solid State Physics. Supplement 10. (Academic Press. New York, 1969)
4. M.L.Leadbeater, L.Eaves, P.E.Simmonds, G.A.Toombs, F.W.Sheard, P.A.Claxton, G.Hill and M.A.Pate, Solid State Electronics, **31**, 707 (1988)
5. M.L.Leadbeater, E.S.Alves, L.Eaves, M.Henini, O.H.Hughes, A.Celeste, J.C.Portal, G.Hill and M.A.Pate, Phys. Rev. B, **39**, 3438 (1989)
6. S.Ben Amor, K.P.Martin, J.J.L.Rascol, R.J.Higgins, A.Torabi, H.M.Harris and C.J.Summers, Appl. Phys. Lett. **53**, 2540 (1988)
7. P.Gueret, C.Rossel, E.Marclay and H.Meier, J.Appl.Phys,**66** ,278 (1989)
8. L.Brey, G.Platero and C.Tejedor, Phys. Rev. B, **38**, 10507 (1988)
9. M.Helm,F.M.Peeters,P.England,J.R.Hayes and E.Colas,Phys. Rev. B, **39**, 3427 (1989)
10. M.Büttiker, IBM J. Res. Dev. **32**, 63 (1988)

Magnetization of a Two-Dimensional Electron Gas with Broadened Landau Levels

K. Jauregui, W. Joss, V.I. Marchenko, S.V. Meshkov*, and I.D. Vagner*

Max-Planck-Institut für Festkörperforschung, Hochfeld-Magnetlabor, BP 166X, F-38042 Grenoble Cedex, France

*Permanent address: Institute of Solid State Physics, Academy of Sciences of the USSR, Chernogolovka, USSR

Abstract. We present an analytical study for the envelope of the magnetization of a two-dimensional electron gas considering finite temperature and a specific model of Landau level broadening due to inhomogeneities.

1. Introduction

The de Haas-van Alphen (dHvA) effect in strong magnetic fields provides important information about the properties of two-dimensional electron gases (2DEG). The information about the effective mass and the electron scattering is contained in the temperature and magnetic field dependence of the amplitude of the dHvA effect. The usual Lifshitz-Kosevich (LK) approach for the derivation of the magnetization using the Poisson summation formula is in the case of a 2DEG unsuitable because the entire discrete nature of Landau quantization causes bad convergence of the summation [1,2].

The difference for the dHvA effect in a three (3D) or two dimensional (2D) system is due to the fact that in a 2DEG the chemical potential is pinned only to the highest occupied Landau level (LL) and this at almost all values of the magnetic induction B, except for certain discrete values of B around which the chemical potential varies rapidly between adjacent LL. This results in strong quantum oscillations of the chemical potential [1,3], while in the 3D case the quantum oscillations of the Fermi energy are negligibly small.

In Shoenberg's expansion [4] of the theory of the dHvA effect in 2D, the Dingle temperature is easily included in an idealized case through a multiplicative factor. Finite relaxation time, finite temperature, etc. are treated as particular examples of *phase smearing* of the oscillations.

Shoenberg's elegant treatment has the disadvantage that it is valid only for low magnetic fields and high enough temperatures. In this situation the magnetization oscillations are smooth enough in order to take out the Poisson summation. In the following we are deriving a form for the magnetization which is valid for high fields and low temperatures.

Springer Series in Solid-State Sciences, Vol. 101
High Magnetic Fields in Semiconductor Physics III Editor: G. Landwehr

2. Formulation of the magnetization with Landau level broadening

We begin with the general formula for the thermodynamical potential of a 2DEG under tranverse magnetic field at finite temperature T.

$$\Omega = - g\, k_B T \sum_{n=0}^{\infty} \int_{-\infty}^{+\infty} \ln \left[1 + \exp\left(\frac{\mu - \epsilon}{k_B T} \right) \right] G(\epsilon - \epsilon_n)\, d\epsilon \ , \qquad (1)$$

where $g = 2B/\phi_0$ is the LL degeneracy, $\phi_0 = 2\pi\hbar/e$ is the quantum flux, μ is the chemical potential, and $G(\epsilon - \epsilon_n)$ is the density of states corresponding to the broadened LL n centered at the energy $\epsilon_n = \hbar\omega_c(n + 1/2)$, $\omega_c = eB/m_c$ with m_c being the cyclotron mass. The magnetization M can be presented in the form corresponding to Fermi filling of states:

$$M = - \left(\frac{\partial \Omega}{\partial B} \right)_\mu = \int_{-\infty}^{+\infty} f(\epsilon - \mu)\, m(\epsilon)\, d\epsilon \ , \qquad (2)$$

where $m(\epsilon)$ is the density of magnetic moment per energy and unit area. This density is given by a convolution of the broadening with the ideal density $m_0(\epsilon)$,

$$m(\epsilon) = \int_{-\infty}^{+\infty} G(\epsilon - \epsilon')\, m_0(\epsilon')\, d\epsilon' \ , \qquad (3)$$

with

$$m_0(\epsilon) = \frac{1}{\phi_0} \sum_{n=0}^{+\infty} [\Theta(\epsilon - \epsilon_n) - \epsilon_n\, \delta(\epsilon - \epsilon_n)] \ . \qquad (4)$$

$\Theta(\epsilon - \epsilon_n)$ is a step function being zero for $\epsilon < \epsilon_n$ and one otherwise; $\delta(\epsilon - \epsilon_n)$ is a Dirac function centered at ϵ_n . Note the finite value of $m_0(\epsilon)$ between LL describing the finite contribution of edge states into magnetization.

The case $T \neq 0$, without broadening was considered in the previous work [2], where a simple analytical formula for the magnetization amplitude has been found (Eq.6 in Ref.2):

$$M_{extr} \simeq \pm \frac{E_F}{\phi_0} \left[1 - \frac{1}{\alpha} \ln(2\alpha) - \frac{1}{\alpha} \right] \ . \qquad (5)$$

Here $E_F = \epsilon_n$ is the Fermi energy corresponding to the number of filled LL, $\alpha \equiv \hbar\omega_c / 2k_B T$, and $\omega_c = eB/m_c$.

3. Landau level broadening with a linear exponential decay

It is straightforward to find a similar result for broadened LL in the case T=0 for the envelope of the oscillation of the magnetization if one assumes a Landau level broadening which is described by the derivative of the Fermi-Dirac distribution function:

$$G(\epsilon - \epsilon_n) = \frac{1}{4\Gamma \cosh^2((\epsilon - \epsilon_n)/2\Gamma)} \quad . \tag{6}$$

We can apply directly this equivalence outlined by Shoenberg [4] to our analytical result for the envelope of the oscillation of the magnetization (Eq.5) with the substition of $k_B T$ by the width Γ. In the Fig.1 we compare the shape of the envelope of the magnetization defined by our (Eq.5) (Γ instead of $k_B T$) with the one given by the exact formula (Eq.3 in Ref.2) in taking only the extremal value of each oscillation.

For our model the density of magnetic moment (Eq.3) is given by

$$m(\epsilon) = \sum_{n=0}^{+\infty} [f(\epsilon_n - \epsilon) - \epsilon_n G(\epsilon_n - \epsilon)] \quad . \tag{7}$$

and is shown in Fig.2.

The asymptotics $k_B T \ll \hbar\omega_c$, $\Gamma \ll \hbar\omega_c$ can be considered at arbitrary case with the result similar to (Eq.5). Furthermore, it can been shown that both for $k_B T < \hbar\omega_c$, and $\Gamma < \hbar\omega_c$ except for very narrow region $\Gamma \simeq k_B T$ the expression (Eq.5) holds with corresponding definition of $\alpha = \hbar\omega_c / 2k_B T$ for $k_B T > \Gamma$ and $\alpha = \hbar\omega_c / 2\Gamma$ for $k_B T < \Gamma$.

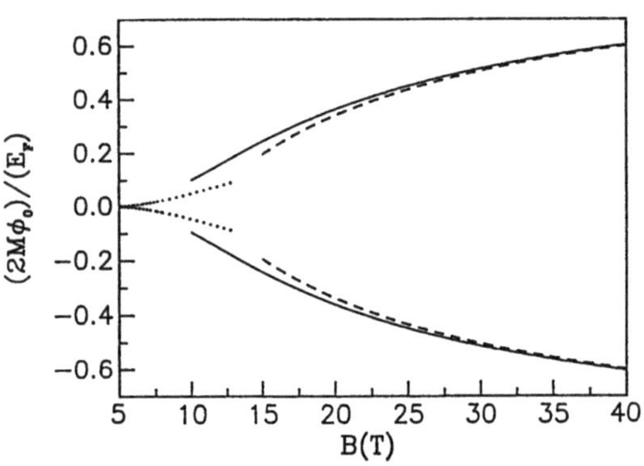

Fig.1. The low-field region (dotted line), as approximated by the LK formula, the high-field part (solid line) of the magnetization (Eq.3 in Ref.2), and the analytical expression (dashed line, Eq.5). For the three curves $k_B T$ is substitued by Γ (Γ=3.3 meV, m_c=0.067 m where m is the free-electron mass, and E_F=0.3 eV).

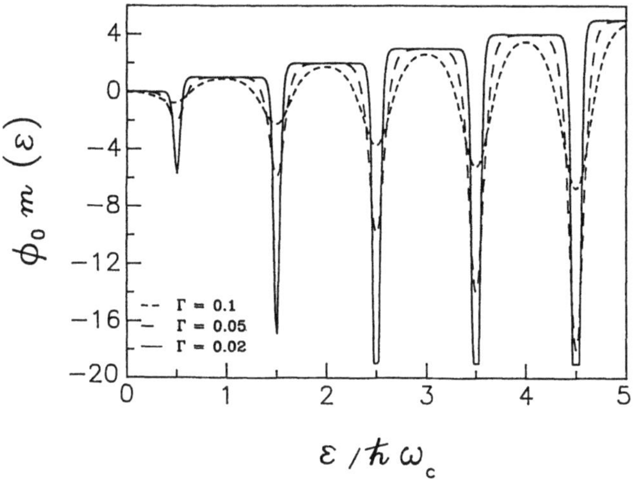

Fig.2. The density of magnetic moment defined by (Eq.7).

This research was supported by the G.I.F. for Scientific Research and Development, Grant No. *G*-112-279.7/88.

References

1 I.D. Vagner, T. Maniv and E. Ehrenfreund, Phys. Rev. Lett. **51**, 1700 (1983).
2 K. Jauregui, V.I. Marchenko, and I.D. Vagner, Phys. Rev. **B41**, 12922 (1990).
3 W. Zawadzki and R. Lassnig, Surf. Sci. **142**, 225 (1984).
4 D. Shoenberg, J. Low Temp. Phys. **56**, 417 (1984).

Part IX

Reports from High Magnetic Field Laboratories

Recent Progress of Semiconductor Physics at the Megagauss Laboratory of the University of Tokyo

N. Miura

Institute for Solid State Physics, University of Tokyo,
Roppongi, Minato-ku, Tokyo 106, Japan

Abstract. Recent progress of the high magnetic field technology has enabled precision measurements of solid state physics in very high pulsed magnetic fields up to a few megagauss. In this paper, a review is presented on the latest experimental results of semiconductor physics in high magnetic fields at the Megagauss Laboratory in Tokyo. The topics to be discussed include infrared cyclotron resonance in various semiconductors, magneto-optical spectra of layer type crystals and their superlattices, as well as of $(GaAs)_m/(Al_xGa_{1-x}As)_n$ short period superlattices, and magneto-tunneling phenomena in double barrier tunneling devices.

1. Introduction

In the last decade, a great deal of effort has been devoted to the generation of high magnetic fields as well as to their application to solid state physics in many high magnetic field facilities in the world[1]. Steady fields up to 30T are available by means of hybrid magnets in several laboratories, and the generation of higher fields is planned with enlarged power supplies. On the other hand, long pulse fields up to 50-60T have been generated by using wire-wound solenoids with especially reinforced strong wires containing Nb-Ti fibers[2] or fine Nb filaments in a Cu matrix[3]. However, there have been much less works in the megagauss range (>100T), as far as their application to solid state physics is concerned. This is mainly because of the technical difficulties in the generation of megagauss fields as well as in the data acquisition in short pulsed fields.

At the Megagauss Laboratory (MGL) of the University of Tokyo, we have developed techniques for generating magnetic fields higher than 100T by electromagnetic flux compression and the single-turn coil technique[4]. Megagauss fields have been successfully applied to various experiments[5]. Recently, the accuracy in the measurements has been greatly improved in various experiments such as magneto-optical and magnetic measurements or infrared laser spectroscopy. In the non-destructive field range, long pulsed fields up to about 40T have been conveniently utilized for various experiments.

In this paper, we report the latest results of semiconductor work in very high magnetic fields. In particular, an emphasis is put on the progress in the two years since the last Würzburg Conference[6].

2. Generation of Long Megagauss Pulse Fields

One of the progresses we have achieved concerning the very high magnetic field generation at the MGL is the success of the generation of megagauss

Springer Series in Solid-State Sciences, Vol. 101
High Magnetic Fields in Semiconductor Physics III Editor: G. Landwehr

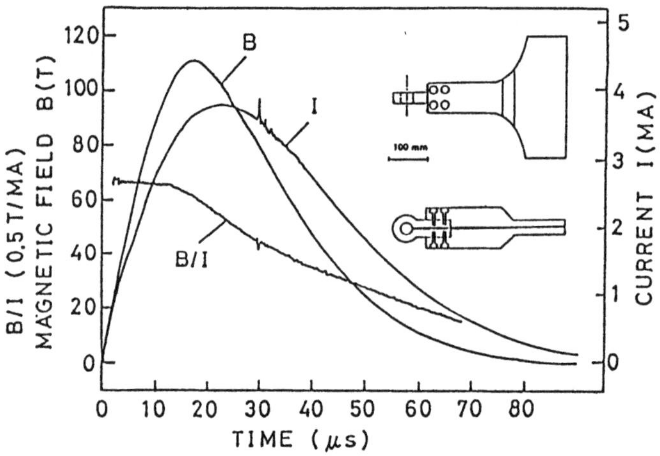

Fig. 1 Waveforms of the magnetic field B, the current I, and the field current ratio B/I for the thick walled single turn coil technique. The inset illustrates the design of the coil.

fields whose pulse duration is considerably longer than in previous methods. The duration of megagauss fields is usually very short of the order of a few μs. There is a tendency the higher the field, the shorter the duration. Fields higher than 1000T generated by an explosive method have a rise time of only about 2 μs[7]. The rise time of megagauss fields produced by electromagnetic flux compression and the single-turn coil technique at the MGL are about 8 μs and 2.5 μs, respectively. For some experiments, particularly for transport or magnetic measurements, it is desirable to have the longest possible rise time.

We have recently produced relatively long megagauss pulse fields by using thick-walled single turn coils, combined with a large condenser bank of 4MJ which has been installed for magnetic flux compression. The principle of the field generation is the same as the single turn coil technique using a fast condenser bank, but the period of the current pulse is an order of magnitude longer. We used a massive single turn coil made from steel, 20mm in thickness, 8mm in bore diameter, and 30mm in length. The coils fit to the coil clamping system which is normally used for the flux compression experiments. The profiles of the current and field are shown in Fig. 1 together with the coil design.

Owing to the slow current pulse and large coil mass, the pulse duration was as long as 50 μs and therefore an order of magnitude longer than ordinary megagauss field pulses. The maximum field of 110 T was generated when the maximum current of 3.8MA was supplied from the condenser of 5mF charged to 24kV. The discharge energy was 1.44 MJ. Supplying a larger current, even higher fields are expected to be generated. The long duration time is extremely useful for various experiments. Moreover, since a coil explosion occurs only in the outward direction, the samples and cryostats are not usually damaged. The experimental results obtained with these fields will be reported shortly.

3. Infrared Cyclotron Resonance

Employing the single turn coil technique with a fast condenser bank[4], cyclotron resonance was measured in various semiconductors in the infrared wavelength region. Typical examples of the experimental traces are shown in Fig. 2. In n-type PbTe, an anomalously large temperature dependence of the effective masses was observed in the CO_2-laser-wavelength region[6]. Namely, it was found that the effective mass increases almost 50% with increasing temperature from 10K to 300K. Experiments were extended to a longer wavelength of 16.9μm using an H_2O laser. A large temperature dependence similar to that observed around 10μm with a CO_2 laser was observed. This implies that the large temperature dependence is not merely a property in very high fields, but it is a property which should be observed also in the smaller photon energy region. A very similar temperature dependence was observed in n-PbSe. Cyclotron resonance was observed also for the narrow gap semimagnetic semiconductor HgMnSe:Fe[8].

In n-InP, cyclotron resonance and the impurity transition were observed simultaneously. However, spin splitting due to band non-parabolicity was not observed as in n-GaAs[9]. This is partly because of the broader linewidth in InP, but the main reason is that the splitting is much smaller in InP than in GaAs. The anisotropy of the resonance

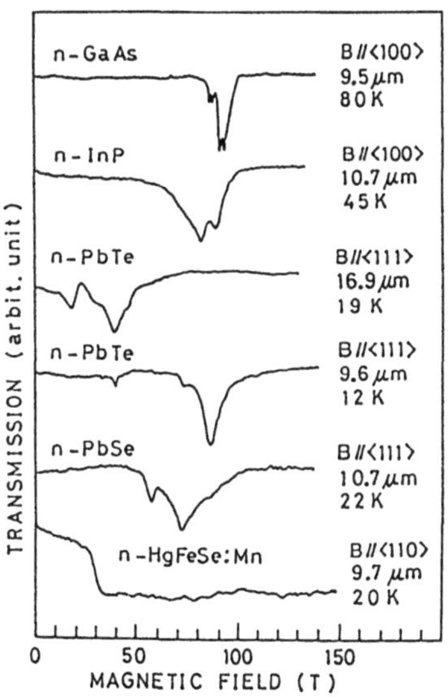

Fig. 2 Typical examples of the cyclotron resonance absorption in various semiconductors. Substance, wavelength of the incident radiation and temperature are indicated for each trace.

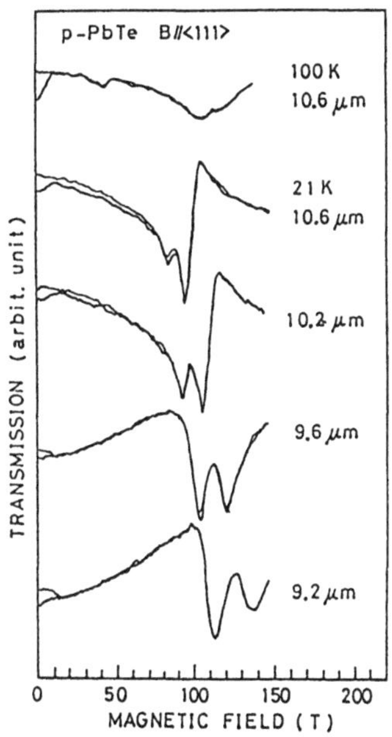

Fig. 3 Experimental recordings of the cyclotron resonance and the combined resonance in p-type PbTe. Except for the top trace which is for T=100K, the temperature is fixed at about 21K. The wavelength for the measurement is indicated for each figure.

with respect to the field direction against the crystal axes was also much smaller than in n-GaAs. The results indicate that the band structure in InP is more adequately expressed by the conventional 3 level model than in GaAs in which a 5 level model is necessary to explain the high field results.

Figure 3 shows the cyclotron resonance in p-PbTe at various photon energies at 21K together with a trace at 100K. In the normal cyclotron resonance at lower fields for $B // < 111 >$, two resonance peaks should be observed corresponding to the light and heavy mass holes located in the inequivalent L-points in the Brillouin zone.

In the present experiment, two peaks were resolved around the heavy mass peak. The peak at the higher field side is assigned to the cyclotron resonance and that at the lower field side to a combined resonance involving a spin-flip. This is the first observation of the combined resonance in the Faraday configuration, and it is striking that the resonance peak has an intensity as large as the cyclotron resonance. The combined resonance becomes allowed by the mixing of the two N=1 Landau levels with opposite spins due to the off-diagonal matrix elements in the $k \cdot p$ matrix. The mixing has been neglected in previous Landau level calculations for low fields, but it becomes important in high magnetic fields near 100T where the levels cross within the approximation neglecting the off-diagonal elements. By fitting the energy positions of the two resonance peaks to the Landau level calculation, a new set of band parameters for PbTe was determined[10]. The change of the relative magnitudes of the two peaks with photon energy is also consistent with the calculated transition probability.

4. Magneto-Optics of Excitons in Layer Type Crystals

There has been considerable controversy concerning the binding energy of the band edge exciton in PbI_2. The interval between the adjcent absorption lines is not consistent with the ideal Rydberg series. We have observed the diamagnetic shift and Zeeman splitting of the three main lines A_1, A_2 and A_3 for $B // c$ in the Faraday configuration, for the first time in high magnetic fields up to 150T[11]. The spectra were recorded by an image-converter camera in the megagauss range and by an OMA system in the non-destructive field range up to 40T. The conclusion was derived that the A_1 and A_2 lines correspond to the N=1 and N=2 levels of the Rydberg series[11]. The binding energy and the Bohr radius of the exciton were estimated to be 40meV and 21Å, respectively[10]. Thus the band edge exciton in PbI_2 is considered to be of the Wannier type.

We have observed magneto-absorption spectra of the ground state exciton also for $B \perp c$ in the Voigt configuration, as shown in Fig. 4[12]. In addition to the ground state line E_3, two lines E_1 and E_2 were observed in high fields. The lines exhibited non-linear shifts against field. According to the band calculations, the conduction and valence bands of PbI_2 consist mainly of the s- and p-orbital wave functions of Pb atoms, respectively. Taking account of the relatively small Bohr radius, therefore, one can imagine that the excitons may be described in terms of the cationic exciton model which deals with excitons as a local excitation

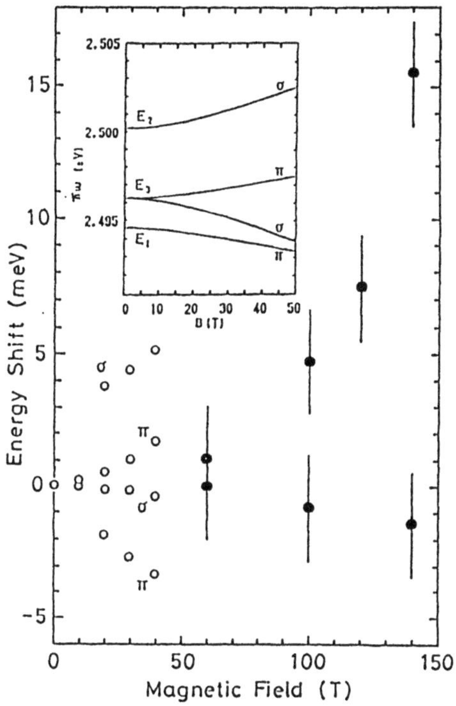

Fig. 4 Energy shift of the band edge exciton absorption lines for PbI_2 as a function of magnetic field. The inset shows the energy shift calculated on the basis of the cationic exciton model. E_1, E_2 and E_3 denote the pure triplet state, the z-like state and the xy-like states (doubly degenerate), respectively.

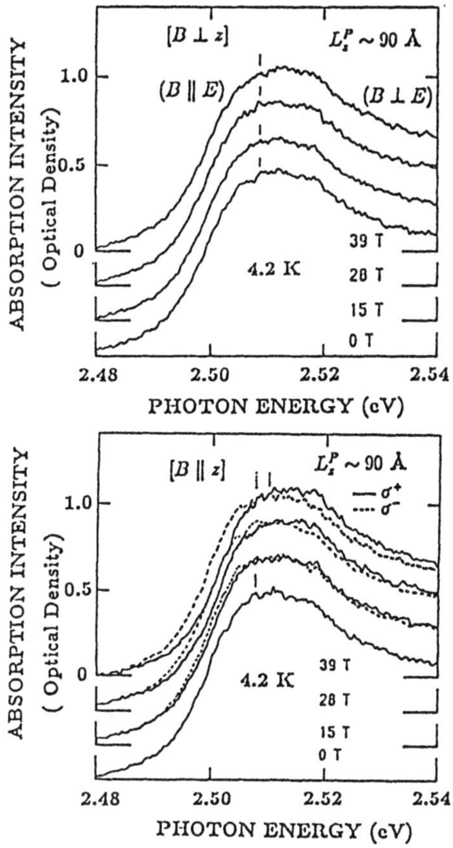

Fig. 5 Magneto-absorption spectra of excitons in the PbI_2 layers in a PbI_2/BiI_3 superlattice. The quantum well width is 93Å. The top and bottom figures show the spectra for B//layer (B⊥z) and those for B⊥layer (B//z), respectively. The σ_+ and σ_- denote the left- and right-handed circular polarizations, respectively.

within cations[13]. The inset of Fig. 4 shows the energy of the exciton levels calculated on the basis of the cationic exciton model. The curves are in reasonably good agreement with the experimental results including the selection rule with respect to the electric field direction relative to the magnetic field. This result implies that the excitons in PbI_2 are Wannier type excitons, but have a cationic exciton character at the same time by virtue of the small exciton wavefunction size.

PbI_2 is one of the layer type crystals in which each crystal layer is accumulated along the c-axis due to the Van der Waals force, so that the layer surfaces have no dangling bonds. Therefore, epitaxial growth in the c-direction on to a substrate of another crystal of a similar structure can easily be made[14]. We have grown PbI_2/BiI_3 superlattices by the hot-wall epitaxy technique[15]. High field magneto-absorption spectra have been measured on these superlattice samples. A quantum confinement effect of exciton absorption lines was observed only for the PbI_2 layers and not in the BiI_3 layers. In relatively wide PbI_2 layers ($L_z > 300\AA$), nearly the same shift and level splitting as those for bulk PbI_2 were observed for both field directions, as shown in Fig. 5. However, in narrower PbI_2 layers, the diamagnetic shift and level splitting were observed only for B⊥layer and those for B//layer were negligibly small. This can be explained in terms of the confinement effect of the exciton wavefunction in the z-direction; mixing of the E_3 states with the E_1 and E_2 states by the quantum well potential.

5. Magneto-Optics in Short Period Superlattices

Magneto-absorption and magneto-luminescence spectra were studied in $(GaAs)_m/(Al_xGa_{1-x}As)_n$ short period superlattices[16]. Samples of both type I($x \simeq 0.5$) and type II (x=1.0) were investigated. In the magneto-absorption for type I samples with (m,n) = (7,5) and (6,6), well resolved oscillatory structure was observed in high magnetic fields for both B⊥layers and B//layers as shown in Fig. 6[17]. For B//layers, a disappearance of the absorption peaks due to the broadening of the Landau levels was observed above the energy of the mini-band edge, as was previously reported by Belle *et al.* for a slightly longer period of (m,n)=(14,4)[18]. This is in contrast to the case of B⊥layers, where the oscillation was observed up to higher energy. The peak positions of the oscillation for B//layers are plotted in Fig. 7.

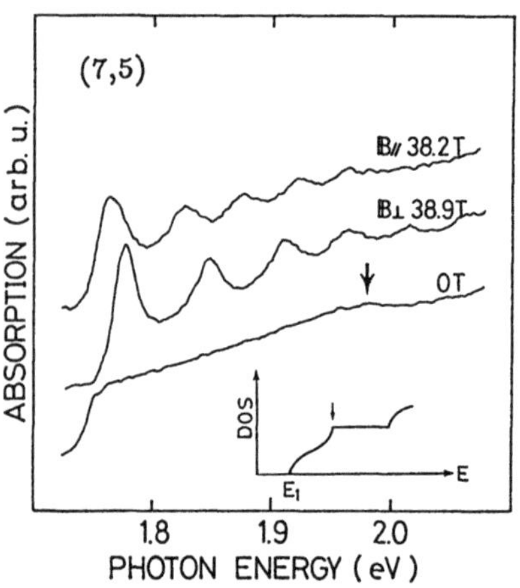

Fig. 6 Magneto-absorption spectra for a $(GaAs)_m/(Al_xGa_{1-x}As)_n$ short period superlattice with (m,n)=(7,5) for B//layer ($B_{//}$) and B⊥layer ($B_\perp$). The inset shows the density of states for the mini-band. At the mini-band edge, the oscillation disappears for $B_{//}$.

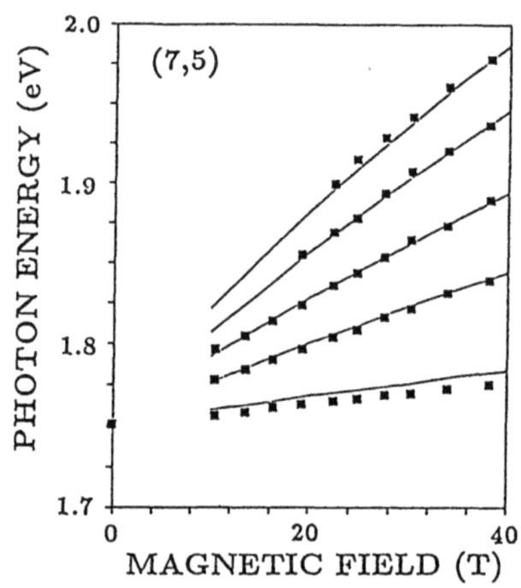

Fig. 7 Photon energies of the absorption peaks for a (m,n)=(7,5) sample as a function of magnetic field. The experimental data (closed points) are compared with the theoretical calculation (solid lines).

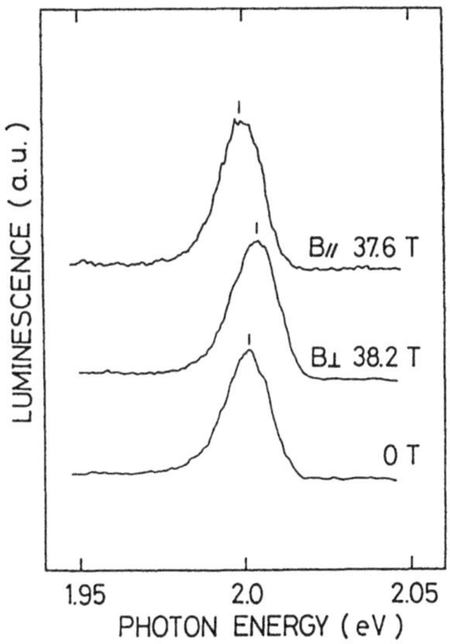

Fig. 8 The magneto-luminescence spectra of excitons for a type II GaAs/AlAs superlattice with (m,n)=(3,5). A considerable diamagnetic shift was observed for $B_{\perp}$ whereas the shift is negligibly small for $B_{//}$.

In order to test the validity of the effective mass theory for short period superlattices, we have numerically calculated the Landau level energies taking account of the periodic quantum well potential, on the basis of the Kronig-Penny model. The effect of the valence band degeneracy was appropriately treated by solving the $k \cdot p$ matrix directly. The broadening of the Landau levels was reproduced above the mini-band edge. The calculated curves of the transition energies between Landau levels are shown by solid lines in Fig. 7. The agreement between theory and experiment is excellent except for the lowest level where the excitonic effect is significant. This result demonstrates that the effective mass theory is still applicable for such short period superlattices. The experiment for samples with shorter periods is in progress. As for the lowest level, the magnetic field dependence of the exciton binding energy was obtained from the difference between the calculated energy and the experimental data. The binding energy thus obtained was found to be in reasonably good agreement with the theory of Makado and McGill[19].

For type II GaAs/AlAs superlattices, the oscillation in the magneto-absorption spectra was not so prominent, but a faint oscillation was obtained for B⊥layers in a sample with (m,n)=(5,6). Figure 8 shows the magneto-luminescence spectra in (m,n)=(3,5) for both field directions. It was found that there is a considerable diamagnetic shift for B⊥layers, but the shift is negligibly small for B//layers. This is consistent with a band calculation which predicts a very large anisotropy[20].

6. Magneto-Tunneling in Double Barrier Tunneling Devices

Magneto-transport phenomena have been measured in GaAs/AlGaAs double barrier tunneling devices in the non-destructive field range up to 40T in collaboration with the Nottingham University group[21]. Figure 9 shows typical magnetoresistance traces when the magnetic field

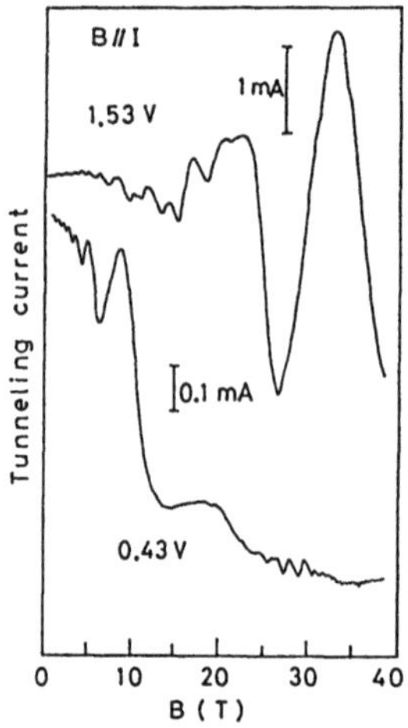

Fig. 9 Magneto-tunneling current against magnetic field at fixed bias voltages below (0.43V) and above (1.53V) the NDC region in a sample with L_z=50Å. Magnetic fields were applied perpendicular to the layers.

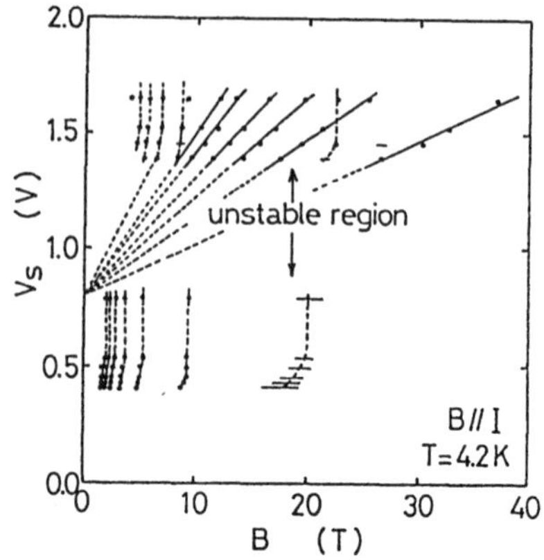

Fig. 10 Bias voltages for the current maxima as a function of magnetic field. The features are different in both sides of the unstable region due to NDC.

was swept at bias voltages below (0.43V) and above (1.53V) the negative differential conductivity (NDC) region for a sample with a well width of L_z=50Å. The fields were applied perpendicular to the barrier layers (B⊥layers). For low bias voltages, oscillatory structure was observed due to the magneto-quantum effect of the 2-dimensional electrons in the emitter with the same period as the Shubnikov-de Haas effect in magnetic fields lower than about 10T. At about 10T where the system enters the quantum limit regime, there is a large drop of the current. There is another structure at about 20T. At higher bias voltages above NDC, on the other hand, oscillations were found to consist of two series. The field positions for the current maxima are plotted in Fig. 10. The two series above the NDC were considered to be corresponding to the ocsillations due to 2D electrons and the LO phonon assisted resonant tunneling, respectively.

When we swept the bias voltage at constant magnetic fields, a remarkable hysteresis was observed between the up and down voltage sweeping in high magnetic fields in the quantum limit regime, as shown in Fig. 11. At 39T, the up sweeping trace shows a current peak at 0.92V and 1.1V. The first peak is only slightly shifted from the peak at 0.87V in zero magnetic field. In the down sweeping trace, peaks appeared at 0.68V and 1.1V. The origin of the hysteresis is not fully understood at this moment, but it may be caused by the charge built up in the quantum well[22]. The peak at 0.68V is vaguely observed also in the up sweeping trace as a shoulder, so that it must be related to some level structures. It may be a peak which arises from the lowest 2D or 3D Landau level at the emitter, whose density of states has a sharp peak in the high field.

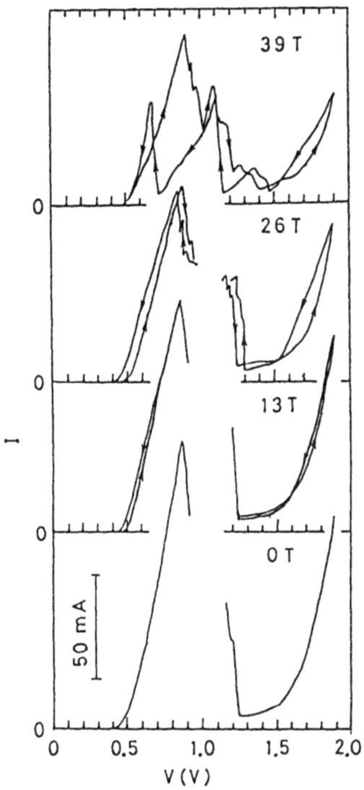

Fig. 11 The V-I characteristics at constant magnetic fields for a sample with L_z=50Å.

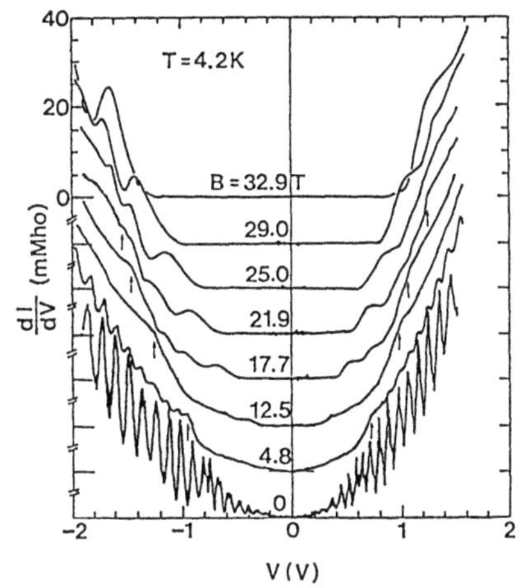

Fig. 12 First derivative of the magneto-tunneling current as a function of bias voltage at various constant fields in a wide well sample with L_z=600Å. Turning points were observed at voltages indicated by arrows.

When magnetic fields were applied parallel to the layer, a shift of the current threshold to the higher voltage side was observed. In order to explain the shift, the energy of the Landau levels was calculated as a function of the center coordinate of the cyclotron motion. The tunneling through the first barrier on the emitter side is possible if a level on the emitter side crosses a level originating from the well or from the collector below the Fermi level. From such a calculation, the shift was well explained in terms of the tunneling from skipping orbits on the emitter side to the traversing orbit in the well. These were recently investigated at lower fields up to 20T[23,24].

For magnetic fields parallel to the layers, very rich structures were observed in a sample with a wide well width of L_z=600Å[23]. Figure 12 shows a curve of the derivative of the current as a function of field at various fixed fields. In high magnetic fields, well-defined onsets of the current were observed in the $dI/dV - V$ curve. Moreover, many fine strutures were observed. At higher bias voltages, turning points were observed. These points are plotted in Fig. 13. As the interval between the levels is small in the wide well sample, and a relatively high voltage can be applied across the well, the tunneling current exhibits complicated structures. The peaks observed in a wide well sample can be classified

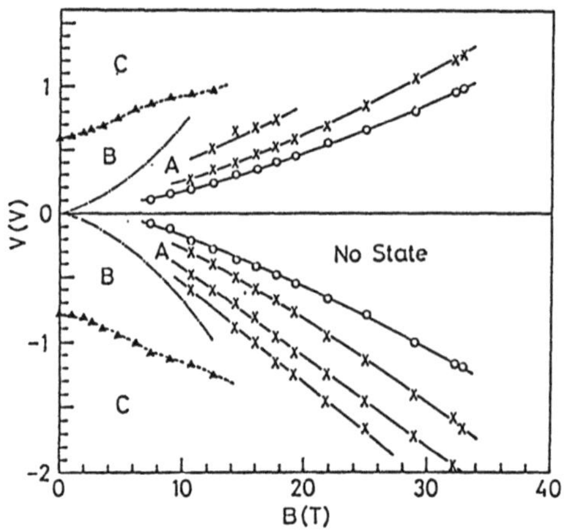

Fig. 13 Voltages for the structures in the V-I curve as a function of magnetic field for a sample with L_z=600Å. Regions A, B, C and No State designate different kinds of cyclotron orbits to which the tunneling takes place, as desctibed in the text.

according to the orbit to which the tunneling takes place from the orbit in the emitter: A. skipping orbits in the well, B. traversing orbits in the well, C. skipping orbits across the second barrier, and there is a region where no possible orbits exist for tunneling. The observed turning points form the boundary between the regions B and C. The boundary between A and B has been observed previously[23,24].

Acknowledgment

The author is indebted to the following colleagues and collaborators for the contribution to the works discussed in this paper: T. Takeyama, H. Yokoi, S. Sasaki, Y. Nagamune, T. Osada, K. Yamada, N. Kamata, Y. Horikoshi, S. Najda, G. Bauer, L. Eaves and M. von Ortenberg.

References

[1] For the recent progress of the high field magnet technology in high magnetic field facilities, see for example: Physica **B 155** (1989).

[2] N. Miura and S. Takeyama: In *Proc. 11th Int. Conf. Magnet Technology*, (To be published, 1990).

[3] S. Foner: Appl. Phys. Lett. **49** 982 (1986).

[4] N. Miura, T. Goto, K. Nakao, S. Takeyama, T. Sakakibara, T. Haruyama and T. Kikuchi: Physica **B 155** 23 (1989).

[5] N. Miura: Physica **B 164** 97 (1990).

[6] N. Miura: In *High Magnetic Fields in Semiconductor Physics*, ed. G. Landwehr (Springer-Verlag, 1989) p.618.

[7] A. I. Pavlovskii, N. P. Dolotenko, A. I. Bykov, A. A. Karpikov, V. I. Mamyshev, G. P. Spirov, O. M. Tatsenko, I. N. Markevtsev and P. V. Sosnin: In *Megagauss Fields and Pulsed Power Systems (Proc.*

5th Int. Conf. Megagauss Magnetic Field Generation and Related Topics (Novosibirsk, 1989), eds. V. M. Titov and G. A. Shvetsov (Nova Science Pub., New York, 1990) p.29.

[8] O. Portugall, H. Yokoi, S. Takeyama, L. van Bockstal, M. von Ortenberg, N. Miura, F. Herlach and W. Dobrowolski: To be presented at 20th Int. Conf. Phys. Semiconductors, Thessaloniki, 1990)

[9] S. P. Najda, S. Takeyama, N. Miura, P. Pfeffer and W. Zawadzki: Phys. Rev. **B 40** 6189 (1989).

[10] H. Yokoi, S. Takeyama, N. Miura and G. Bauer: To be presented at 20th Int. Conf. Phys. Semiconductors, (Thessaloniki, 1990).

[11] Y. Nagamune, S. Takeyama and N. Miura: Phys. Rev. **B 40** 8099 (1989).

[12] N. Miura, Y. Nagamune and S. Takeyama: To be presented at 20th Int. Conf. Phys. Semiconductors, (Thessaloniki, 1990).

[13] S. Takeyama, K. Watanabe, N. Miura, T. Kamatsu, K. Koike and Y. Kaifu: Phys. Rev. **B 41** 4512 (1990).

[14] A. Koma, K. Sunouchi and T. Miyajima: Microelectron. Eng. **2** 129 (1984).

[15] S. Takeyama, K. Watanabe, M. Ichihara, K. Suzuki and N. Miura: J. Appl. Phys. (To be published).

[16] N. Miura and S. Sasaki: To be published in Superlattices and Microstructures.

[17] S. Sasaki, N. Miura and Y. Horikoshi: J. Phys. Soc. Jpn. (To be published 1990).

[18] G. Belle, J. C. Maan and G. Weimann: Surf. Sci. **170** 611 (1986).

[19] P. C. Makado and N. C. McGill: J. Phys. C: **19** 873 (1986).

[20] T. Nakayama and H. Kamimura: J. Phys. Soc. Jpn. **54** 4726 (1985).

[21] N. Miura, K. Yamada, N. Kamata, T. Osada and L. Eaves: To be published in Superlattices and Microstructures.

[22] M. L. Leadbeater, E. S. Alves, F. W. Sheard, L. Eaves, M. Henini, O. H. Hughes and G. A. Toombs, J. Phys. Cond. Matter **1** 10605 (1989).

[23] E. S. Alves, M. L. Leadbeater, L. Eaves, M. Henini, O. H. Hughes, A. Celeste, J. C. Portal, G. Hill and M. A. Pate: Superlattices and Microstructures **5** 527 (1989).

[24] M. L. Leadbeater, E. S. Alves, L. Eaves, M. Henini, O. H. Hughes, A. Celeste, J. C. Portal, G. Hill and M. A. Pate: J. Phys. Cond. Matter **1** 4865 (1989).

Recent High-Field Investigations of Semiconductor Nanostructures and Other Systems at Nijmegen

J. Singleton, J.A.A.J. Perenboom, and J.-T. Janssen

High Field Magnet Laboratory and Research Institute for Materials, University of Nijmegen, 6525 ED Nijmegen, The Netherlands

Abstract. Some of the recent measurements carried out on semiconductor nanostructures using the far-infrared and high magnetic field facilities of the High Field Magnet Laboratory of the University of Nijmegen are reviewed. In addition, a brief summary of high field studies of other semiconductor systems is given.

The University of Nijmegen contains one of the world's foremost magnet facilities. Two hybrid magnets give fields of 25 and 30 T, and a variety of Bitter solenoids are used to provide up to 20 T [1]. Part of the laboratory's recent efforts have been directed towards magneto-transport and magneto-optics in semiconductor nanostructures. This work is supported by a European Community grant awarded jointly to the University of Nijmegen and the Cavendish Laboratory, Cambridge, which seeks to combine the complementary facilities of both laboratories. In addition, the laboratory is involved in a collaboration with the Delft Institute for Microelectronics and Submicron Technology (DIMES) on the physics of very small Si devices. In this paper we shall attempt to give a flavour of some of the recent measurements carried out within these collaborations.

1. 1D Subbands in Multiple Wires on Si MOSFET's

There has been much interest in transport and optics in quasi-one-dimensional (1D) semiconductor systems, in which the electrons are confined in two directions, whilst moving freely in the third (see *e.g.* [2]). One method of defining such system is to apply a negative voltage to gates a fraction of a μm apart on a semiconductor structure containing a two dimensional electron gas (2DEG); the electrons are depleted under the gates leaving a narrow channel between. The confinement leads to a series of 1D subbands, each of which introduces a peak in the density of states which can be detected in the conductivity of the device under certain conditions.

Most of the 1D subband studies reported have been carried out in devices based on GaAs-(Ga,Al)As heterostructures. In contrast, in Si-MOSFET's, universal conductance fluctuations (UCF) have tended to obscure the 1D subband effects in conductivity data [3]. We have overcome this problem by studying samples containing many parallel 1D channels [4]; the 1D subband effects are coherent for all the channels, but the UCF are averaged out. The devices

Springer Series in Solid-State Sciences, Vol. 101
High Magnetic Fields in Semiconductor Physics III Editor: G. Landwehr

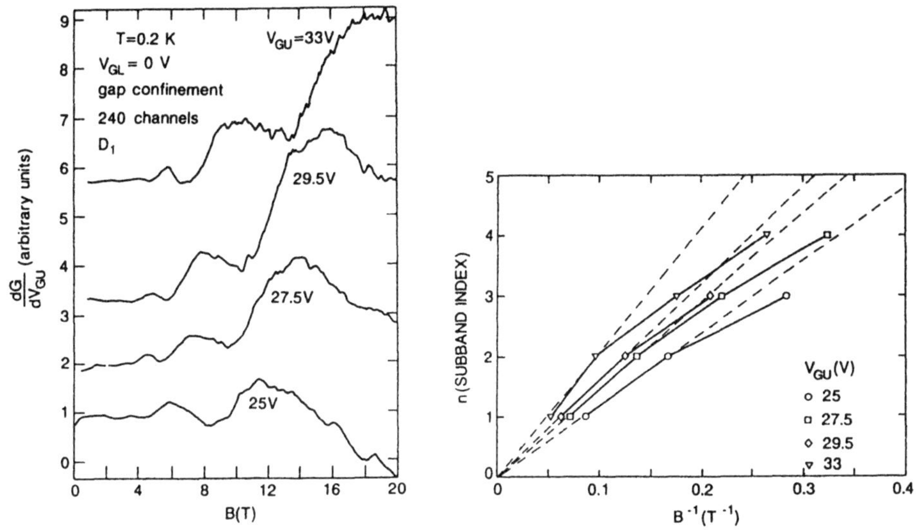

Fig.1: a) Oscillations in the conductivity G of the device for different values of the upper-gate voltage V_{GU}; b) Plot of the subband index n versus $1/B$ for different values of V_{GU}, deduced from the maxima of the oscillations in Fig.1a.

are dual-gate Si-MOSFET's fabricated on p-Si substrates at DIMES; after the growth of a 25 nm thick gate oxide and deposition of 80 nm n^+ polysilicon, 240 grating gates, of period 200 nm, with equal lines and spaces, were patterned in the polysilicon. Finally, 120 nm of CVD silicon nitride was deposited followed by an evaporated Al upper gate. The 1D channels are formed by applying a positive voltage V_{GU} to the upper Al gate to give inversion, whilst applying a negative voltage V_{GL} to the lower gate to form the confining potential; this is called the "gap-confinement" mode [4].

Fig. 1a shows $\mathrm{d}G/\mathrm{d}V_{\mathrm{GU}}$, where G is the wire conductivity, as a function of magnetic field B (perpendicular to the 2DEG plane), for different values of V_{GU}. Each curve exhibits three or four clear oscillations; the oscillation maxima are labelled with an index n and plotted as a function of $1/B$ in Fig. 1b. At low $1/B$ the points lie on lines through the origin, but for high $1/B$ they deviate towards the $1/B$-axis. This behaviour is due to the depopulation of 1D hybrid electric-magnetic subbands, which have an energy determined by contributions from the 1D confinement and the cyclotron energy [2, 4]; as B increases, the hybrid levels are pushed up through the Fermi energy and depopulate. When the cyclotron energy is much greater than the 1D confinement energy (or cyclotron radius $\ll$ channel width), the hybrid levels are 2D Landau levels, and the depopulation peaks are Shubnikov-de Haas oscilations; this is seen in the low $1/B$ region of Fig. 1b, where the maxima lie on a straight line through the origin. The deviation from linearity at high $1/B$ occurs when the confinement contribution to the hybrid subband spacing becomes comparable to the Landau level spacing; to our knowledge this is the first time that such 1D behaviour has been reported for a Si-MOSFET. The data can be used to estimate the 1D subband separations in zero field, assuming a parabolic potential, and values ≈ 3 meV are obtained for channel widths of ≈ 70 nm [4].

2. Photoconductive Response of a 1D Channel

Much of the physics of quasi-1D systems has been studied in so-called "ballistic channels" defined by a negative voltage applied to gates a fraction of a micron wide and a similar distance apart on top of a GaAs–(Ga,Al)As heterojunction containing a 2DEG [2]: the channel length is much shorter than the scattering lengths in the 2DEG, and so transport through it is ballistic. At $B = 0$ the resistance is quantised to a value $h/2ie^2$, where i is the number of occupied 1D subbands. There is much interest in the relationship between the gate voltage (V_g) and the potential between the gates, which determines the 1D subband spacing. In transport studies, the potential is determined indirectly by fitting theoretical predictions of the magnetic depopulation of 1D subbands to the data [5]. An alternative approach is to use a multiple quantum wire system, where an array of $\approx 10\,000$ channels ≈ 1 mm long is fabricated (see *e.g.* [6]). Although the large area of such samples means that the transmission of far-infrared (FIR) radiation can be used to detect transitions between 1D states, it also leads to poor transport properties, due to averaging effects.

We have chosen instead to study the photoconductive response of a single ballistic channel; our measurements indicate that the response may be used to detect transitions between 1D subbands [7]. The structure was fabricated at Cambridge on top of a GaAs-(Ga,Al)As heterojunction containing a 2DEG with carrier density $N_s \approx 2\times10^{11}$ cm^{-2} and 4 K mobility $\mu \approx 10^6$ cm^{2}V^{-1}s^{-1}. The gate is a metal line, 0.5 μm wide, bisecting the mesa between source and drain; the split in this gate defining the 1D channel is 0.5 μm wide. The photoresponse experiments were carried out at a temperature $T \approx 400$ mK, with an optically-pumped laser as FIR source [7].

The resistance of the sample showed ≈ 14 plateaux as a function of V_g (see inset to Fig. 2). The photoresponse (*i.e.* change in resistance due to the FIR) at $B = 0$ as a function of V_g shows clear oscillations each time the resistance increases to the next plateau; this is shown in Fig. 2 for illumination with FIR of wavelength $\lambda = 699$ μm. The oscillations tend to get stronger for more negative V_g and the amplitude of the response exhibits a quasi-linear increase with laser intensity; however, no clear dependence on FIR wavelength was found. This non-resonant part of the photoresponse is due to the heating of the electrons by the FIR [7], and can be simulated by subtracting resistance $R(T)$ versus V_g data for two different T, T_1 and T_2 ($T_1 < T_2$). The rise in T broadens the region between resistance plateaux so that $R(T_2) - R(T_1)$ yields a sharp negative peak followed by a broader peak of the opposite sign in this region; this is of the same qualitative phase and form as the photoresponse.

However, a *resonant* photoresponse was found for magnetic fields around the cyclotron resonance (CR) field for electrons in the 2DEG [7]. Fig. 3a shows the photoresponse signal as a function of V_g for a laser wavelength of

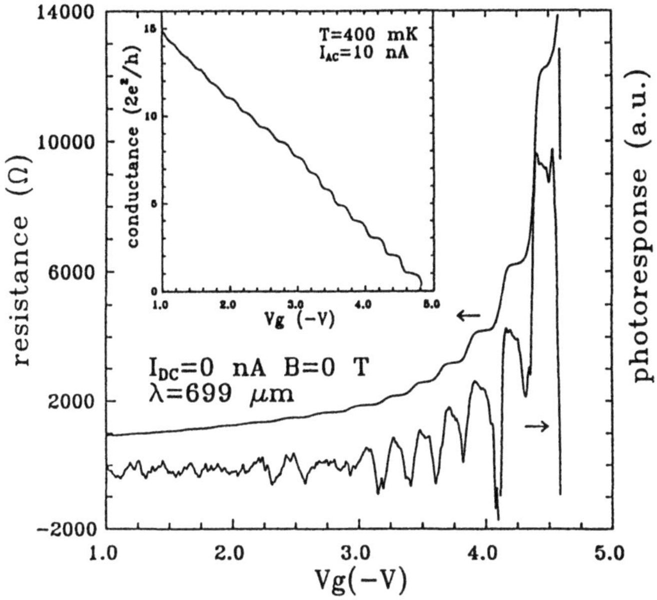

Fig.2: Photoresponse (right-hand scale) and resistance (left-hand scale) observed in a 1D channel as fuction of the gate voltage V_g. The conductance (see inset) shows well-resolved plateaux at $2ie^2/h$ for i occupied 1D subbands.

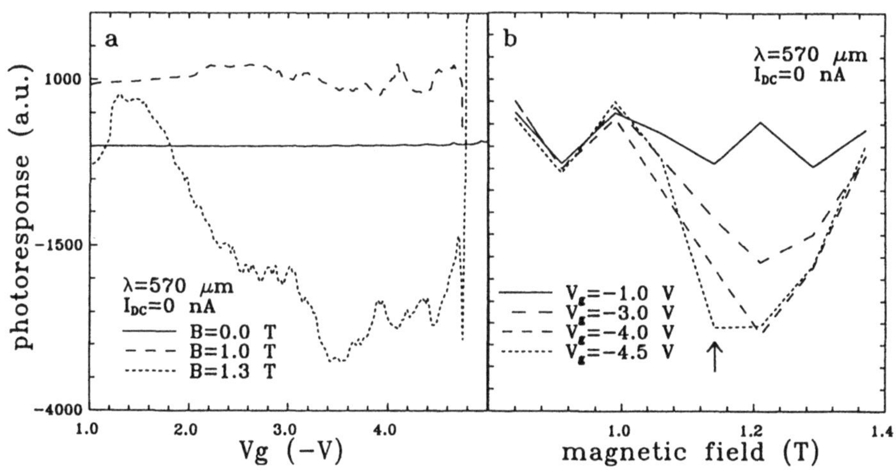

Fig.3: a) Photoresponse at FIR wavelength of 570 μm, measured as a function of gate voltage V_g for different values of the magnetic field. b) The magnetic field dependence of the photoresponse deduced from data as in Fig.3a reveals a resonant photoresponse.

$\lambda = 570\ \mu m$; the signal changes drastically in size and form as the CR-field (1.32 T for $\lambda = 570\ \mu m$) is approached. The effect can be seen more clearly by taking a number of V_g-sweeps at different B and plotting the photoresponse signal as a function of B for different values of V_g (Fig. 3b). A very clear dip in the photoresponse can be observed which deepens and shifts to lower B as V_g increases; at all V_g where this feature is present it is well below the field at which CR in the 2DEG occurs. This behaviour is typical of a transition between hybrid electric-magnetic subbands in the 1D channel [6]; these have separations containing contributions from both electric and magnetic quantisations (*e.g.* for a parabolic confining potential, the hybrid subbands will have spacings given by $\omega_{res}^2 = \omega_0^2 + \omega_c^2$, where $\hbar\omega_c = e\hbar B/m^*$ is the cyclotron energy and $\hbar\omega_0$ the zero-field 1D subband spacing [2]). The hybrid subband separation will thus be greater than either the Landau level or zero-field 1D subband spacing, so that for fixed energy a transition between hybrid levels will occur at a lower B than the CR. As V_g becomes more negative, the confinement increases, raising the hybrid level separation, and moving the transition to lower fields. The resonance observed in Fig. 3b is broad, due to the fact that the 1D subband spacing varies throughout the channel, from zero at the edge to a maximum value at the channel centre. Taking the point indicated by the arrow to be the transition between the channel centre states, and assuming a parabolic potential, we obtain $\hbar\omega_0 = 1.1$ meV for $V_g = -4.5$ V [7]. This value is similar to, but somewhat smaller than, the values found in magneto-transport studies on narrower structures (typically of width $\approx 0.3\ \mu m$ as opposed to $\approx 0.5\ \mu m$ in this case), which are of the order of 2–3 meV [5]. The narrower structures will have an increased subband spacing close to pinch-off [5].

The change in dc resistance of the 1D channel due to the FIR was also measured for dc currents up to 500 nA [7]. A consequence of the dc bias is observed directly in the resistance, where new plateaux appear at half integer values $h/(2i+1)e^2$ in between those seen at zero bias. This effect has been predicted but there have been no previous clear experimental observations [8]. The intermediate plateaux originate from the fact that the number of populated 1D subbands is no longer equal for both velocity directions; a description of the mechanism is given in [9], and details of our observations are in [10].

Under dc bias there is a very strong photoresponse when the FIR is resonant with the Landau level spacing outside the channel (*i.e.* at the CR field) [7]. The photo-excited electrons populate the first empty Landau level above the Fermi energy, from where they can tunnel through the channel region. This "photon assisted tunnelling", driven by the dc bias dropped across the channel, can also be seen as the injection of hot electrons into the 1D channel, and thus gives rise to a large oscillatory photoresponse as a function of V_g.

3. Magneto-optics and Magneto-capacitance of Quantum Dots

We have performed simultaneous magneto-capacitance and FIR magneto-transmission studies of an array of GaAs–(Ga,Al)As quantum dots [11]. There are

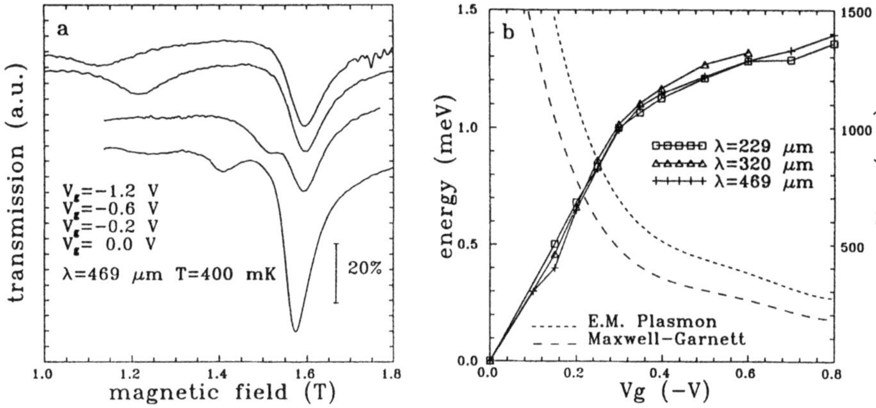

Fig.4: a) FIR transmission of dots versus magnetic field. b) Data: plasmon energy of the dots (left-hand scale), and dashed lines: radius of the dots (right hand scale).

considerable advantages in measuring both at once, as the SdH-like oscillations in the capacitance as B is swept allow one to determine the carrier density within the dots, thus removing one variable parameter from the analysis of the optical measurements (see *e.g.* [12]). A lattice (unit cell 2.25×2.5 μm; area 4 mm^2) of pillars of high-resolution negative resist 0.5 μm high and diameter ≈ 1.0 μm was fabricated on a GaAs-(Ga,Al)As heterojunction containing a 2DEG close to the interface and a Si δ-doped layer 200 nm below; this acts as a transparent back-contact for the capacitance measurements. Finally, a 5 nm thick NiCr gate was evaporated over part of the lithographed area. The gate is not continuous over the sides of the pillars, so that no gate voltage (V_g) will appear on the pillars. Therefore, on applying negative V_g, the 2DEG between the pillars will be depleted, leaving "dots" of 2DEG underneath; the capacitance measurements indicate that this occurs at $V_g \approx -0.3$ V [11].

The transmission of the dots is shown as a function of B in Fig. 4a for various V_g. With zero V_g applied, the dominant features of the FIR magneto-transmission spectra are the cyclotron resonance (CR) of the 2DEG, and at slightly lower field, a 2D magnetoplasmon which couples to the FIR via the periodic gate. The experimental field position of the 2D magnetoplasmon is in good agreement with theoretical calculations including the screening due to the gate [13], and this allows the screening functions for the calculations described in the next paragraph to be chosen with a degree of confidence [11].

On applying a negative V_g, the 2D magnetoplasmon disappears, and then a resonance associated with the dots forms close to the CR position, and moves to lower B as V_g becomes more negative (Fig. 4a). For fixed V_g, the resonance has the energy dependence $\omega = [(\omega_c/2)^2 + \omega_0^2]^{0.5} + \omega_c/2$, where ω_0 is a frequency associated with transitions between states of the dot [11]. Values of ω_0 are shown in Fig. 4b for three FIR energies. No energy dependence was observed, indicating that the potential in the dot is roughly field-independent.

A knowledge of the electron density within the dots allows their response to be modelled using the Maxwell-Garnett (MG) [14] or edge magnetoplasmon (EM plasmon) [13] models with the dot radius as an adjustable parameter. The two models give values of the radius within $\approx 30\%$ of each other for $V_g < -0.3$ V (*i.e.* isolated dots) and the results are shown in Fig. 4b. With the MG model a value is found close to the physical size of the dots at the point at which they are first isolated at $V_g \approx -0.35$ V. In this regime, it appears that the radius can be varied by $\approx 50\%$ by the action of V_g [11].

4. Thermal Transport in Free-Standing GaAs Wires

We have studied heating effects in free-standing n-GaAs wires in magnetic fields up to 20 T [15]. Our aim is to detect 1D, laterally-confined phonons using the thermal conductivity κ of the wires; the phonon dimensionality determines the T-dependence of κ (*i.e.* 1D: $\kappa = \eta T$; 3D: $\kappa = \Lambda T^3$, with Λ and η constants [16]). The wires have a triangular cross-section (sides ≈ 0.5 μm), and below 4.2 K exhibit 1D weak localisation and 3D electron–electron interaction corrections to their electronic conductivity σ; the conducting width of each wire is ≈ 0.125 μm [17]. Electron heating experiments performed at $B = 0$ gave $\kappa = \eta T$, but a Lorentz number $\kappa/\sigma T$ close to the theoretical value for electrons, indicating that most of the heat generated in the wires is carried away by electrons; no appreciable 1D phonon contribution was observed [16].

High magnetic fields were therefore used in the electron heating experiments to suppress the electronic contribution to κ. The measurements were performed at $T = 0.4$–1.2 K with the sample *in vacuo* [15]; the latter condition avoids thermal short-circuiting of the wires, and simplifies the heat-flow models to 1D equations. Initially, the field B quenches the weak localisation and then induces a strong positive magnetoresistance for $B > 6$ T, due to confinement of the impurity wavefunctions (Fig. 5a). For $B > 10$ T, σ decreases by several orders of magnitude, indicating a suppression of the electronic thermal conductivity by a comparable factor; in this region three separate regimes in the behaviour of κ are observed, characterised by its temperature dependence: *i)* for $B \leq 13$ T, $\kappa = \Lambda T^3$ appears to be a much better fit to the data (*cf.* Fig. 5b); *ii)* for 14 T $\leq B \leq$ 15 T, the behaviour changes, and $\kappa = \eta T$ is the better fit; *iii)* for $B \geq 16$ T, neither ηT nor ΛT^3 adequately fits the data. In all cases ($B \geq 10$ T) $\kappa/\sigma T$ considerably exceeds the theoretical Lorentz number; *i.e.* the thermal conductivity is primarily due to phonons.

It seems that the aim of the experiment, to suppress the electronic contribution to κ, has succeeded. However, the results are not simple to interpret and the behaviour for $B \geq 14$ T may indicate that the assumption of metallic diffusive electronic conductivity [16] is breaking down. Estimates of the phonon mean free paths l may be obtained from the values of Λ in the field range 10–13 T, indicating that if the phonons are 3D then $l \approx 5$ nm. The high quality of the T^3 (3D) fits in this region implies that the phonon temperature is well defined at all points in the wire, consistent with the short l, and this can be interpreted as evidence for phonon localisation, as there is no other obvious

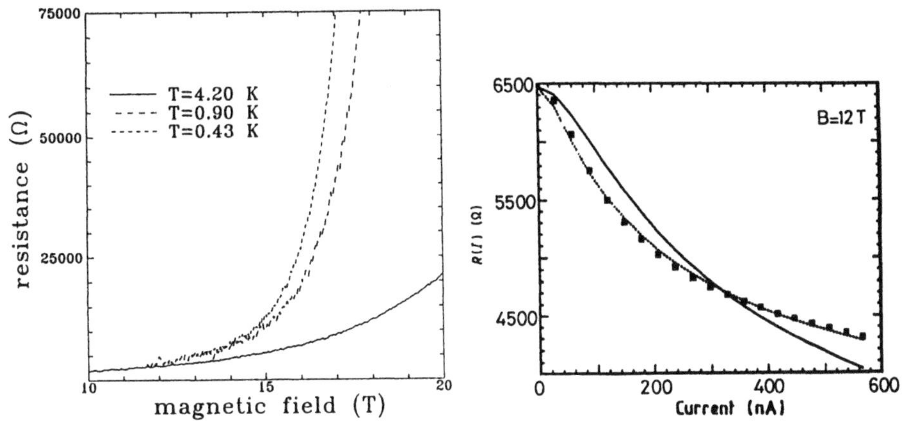

Fig.5: a) Magnetic field dependence of the resistance of 27 wires in parallel. b) Resistance of 27 wires in parallel as a function of the heating current, $T = 595$ mK and $B = 12$ T; the ΛT^3-law (dotted line) fits better to the data than the ηT-law (solid line).

reason why 3D phonons should be so strongly scattered at $T \leq 1$ K. Phonon localisation is an oft-predicted consequence of low-dimensionality (see*e.g.* [16]), and so more experiments are under way to probe this regime.

5. Other Highlights

We shall conclude with a very brief summary of a few of the other activities in our laboratory. The resonant polaron effects in the $1s$–$2p^+$ transitions of well- and barrier-centre donors in a GaAs-(Ga,Al)As quantum well (QW) have been studied, using $B \leq 25$ T [18]. The resonant polaron effect occurs in both cases close to the GaAs LO phonon energy, and the strength of the polaron coupling was found to be greatest for the barrier centre donors, due to their lower effective dimensionality. Very recently we have made a direct comparison of the resonant polaron effects in the D^- to $N = 1$ Landau level transition for shallow impurities in bulk n-GaAs and GaAs-(Ga,Al)As QW's, to directly assess the effects of dimensionality reduction [18, 19].

Another interesting discovery made in work on GaAs-(Ga,Al)As heterojunctions is the observation of a "magnetophonon effect" in the CR linewidth, which oscillates as a function of energy at temperatures greater than 20 K, with maxima occurring close to sub-harmonics of the LO phonon energy [20].

The CR of the 2DEG in the layered semiconductor InSe has been investigated for $B \leq 30$ T (where $\hbar\omega_c \approx 75\%$ of the LO phonon energy, 220 cm^{-1}). At these energies, the cyclotron effective mass is enhanced by 20% over the value expected from band non-parabolicity alone (Fröhlich constant $\alpha = 0.3$), showing the strongest polaron effects yet seen in a 2DEG system (see Fig. 6). In addition, a resonant polaron effect due to 2D electron-homopolar phonon coupling was observed for the first time [21].

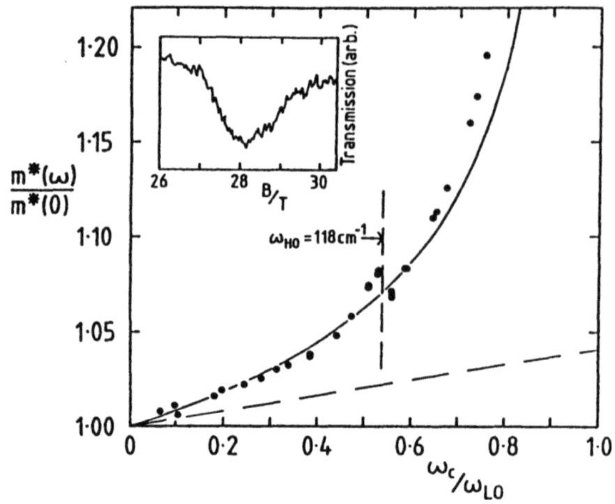

Fig.6: Polaron effect of the cyclotron resonance in the layered semiconductor InSe.

In magneto-transport, we have extended our studies to $\delta-$doped GaAs [22], InSb-CdTe heterojunctions [23] and systems with multiple occupied subbands; the latter data have been modelled using a phenomenological quasi-self-consistent approach to reveal details of the Landau level density of states [24]. Magneto-transport has also been performed for heavily-doped n-GaAs under hydrostatic pressure up to 20 kbar, to reveal the importance of the spatial correlation of charged donors [25].

Acknowledgements. This work is supported by the European Community, and FOM and NWO (the Netherlands). We are very grateful to our many colleagues cited in this work, for fruitful and stimulating collaborations.

References

[1] J.A.A.J. Perenboom and K. van Hulst, Physica B **155**, 74 (1989).
[2] H. van Houten, C.W.J. Beenakker and B.J. van Wees, in: *Semiconductors and Semimetals*, M.A. Reed ed. (Academic Press, New York, 1990).
[3] J.R. Gao, J. Caro, A.H. Verbruggen, S. Radelaar and J. Middelhoek, Phys. Rev. B **40**, 11676 (1989).
[4] J.R. Gao, C. de Graaf, J. Caro, S. Radelaar, M. Offenberg, V. Lauer, J. Singleton, T.J.B.M. Janssen and J.A.A.J. Perenboom, Phys. Rev. B **41**, 12315 (1990); and these proceedings.
[5] D.A. Wharam, U. Ekenberg, M. Pepper, D.G. Hasko, H. Ahmed, J.E.F. Frost, D.A. Ritchie, D.C. Peacock and G.A.C. Jones, Phys. Rev. B **39**, 6283 (1989).
[6] J. Alsmeier, E. Batke and J.P. Kotthaus, Phys. Rev. B **40**, 12574 (1989).
[7] N.K. Patel, T.J.B.M. Janssen, J. Singleton, M. Pepper, H. Ahmed, D.G. Hasko, R.J. Brown, J.A.A.J. Perenboom, G.A.C. Jones, J.E.F. Frost, D.C. Peacock and D.A. Ritchie, Proc. 20th Int. Conf. on the Physics of Semiconductors, ICPS20, Thessaloniki, 1990 (to be published).

[8] L.P. Kouwenhoven, B.J. van Wees, C.J.P.M. Harmans, J.G. Williamson, H. van Houten, C.W.J. Beenakker, C.T. Foxon and J.J. Harris, Phys. Rev. B **39**, 8040 (1989).
[9] L.I. Glatzmann and A.V. Khaetski, Europhys. Lett. **9**, 263 (1989).
[10] N.K. Patel, L. Martin-Moreno, M. Pepper, H. Ahmed, D.G. Hasko, R.J. Brown, J.A.A.J. Perenboom, G.A.C. Jones, J.E.F. Frost, D.C. Peacock, D.A. Ritchie, T.J.B.M. Janssen and J. Singleton, J. Phys.: Condens. Matter (to be published).
[11] N.K. Patel, T.J.B.M. Janssen, J. Singleton, M. Pepper, H. Ahmed, D.G. Hasko, R.J. Brown, J.A.A.J. Perenboom, G.A.C. Jones, J.E.F. Frost, D.C. Peacock and D.A. Jones, these proceedings.
[12] D. Demel, D. Heitmann, P.G. Grambow and K. Ploog, Phys. Rev. Lett. **64**, 788 (1990).
[13] A.L. Fetter, Phys. Rev. B **32**, 7676 (1985).
[14] S.J. Allen Jr., H.L. Störmer and J.C.M. Hwang, Phys. Rev. B **28**, 4875 (1983).
[15] A. Potts, J. Singleton, T.J.B.M. Janssen, M.J. Kelly, C.G. Smith, D.G. Hasko, D.C. Peacock, J.E.F. Frost, D.A. Ritchie, G.A.C. Jones, J.R. Cleaver and H. Ahmed, these proceedings.
[16] A. Potts, M.J. Kelly, C.G. Smith, D.G. Hasko, J.R.A. Cleaver, H. Ahmed, D.C. Peacock, D.A. Ritchie, J.E.F. Frost and G.A.C. Jones, J. Phys.: Condens. Matter **2**, 1817 (1990).
[17] A. Potts, D.G. Hasko, J.R.A. Cleaver, C.G. Smith, H. Ahmed, M.J. Kelly, J.E.F. Frost, G.A.C. Jones, D.C. Peacock and D.A. Ritchie, J. Phys.: Condens. Matter **2**, 1807 (1990).
[18] S. Huant, C.J.G.M. Langerak, J. Singleton, R.A.J. Thomeer, G. Hai, F.M. Peeters, J.T. Devreese and B. Etienne, Proc. 20th Int. Conf. on the Physics of Semiconductors, ICPS20, Thessaloniki, 1990 (to be published).
[19] M.B. Stanaway, J.M. Chamberlain, J. Singleton, C.J.G.M. Langerak and C. Stanley, Proc. 17th Int. Symp. on GaAs and Related Compounds, Jersey, 1990 (to be published by IOPP Ltd, Bristol).
[20] D.J. Barnes, R.J. Nicholas, M. Watts, F.M. Peeters, J.T. Devreese, C.J.G.M. Langerak, J. Singleton, J.J. Harris and C.T. Foxon, these proceedings.
[21] D.F. Howell, T.J.B.M. Janssen, R.J. Nicholas, C.J.G.M. Langerak, J. Singleton and A. Chevy, Surf. Science. **229**, 496 (1990)
[22] P.M. Koenraad, A.P.J. Voncken, J. Singleton, F.A.P. Blom, C.J.G.M. Langerak, M.R. Leys, J.A.A.J. Perenboom, S.J.R.M. Spermon, W.C. van der Vleuten and J.H. Wolter, Surf. Science **228**, 538 (1990).
[23] S.K. Greene, J. Singleton, T.D. Golding, M. Pepper, C.J.G.M. Langerak and J. Dinan, Surf. Science **228**, 542 (1990).
[24] R.M. Kusters, J. Singleton, G. Gobsch, G. Paasch, D. Schulze, F. Wittekamp, G.A.C. Jones, J.E.F. Frost, D.C. Peacock and D.A. Ritchie, Proc. 5th Int. Conf. on Superlattices and Microstructures, ICSM90, Berlin, 1990 (to be published).
[25] P. Wisniewski, L. Dmowski, C. Skierbiszewski, T. Suski, P.J. van der Wel, J. Singleton, K. Ploog and J.J. Harris, Proc. 20th Int. Conf. on the Physics of Semiconductors, ICPS20, Thessaloniki, 1990 (to be published).

Index of Contributors

Springer Series in Solid-State Sciences

Editors: M. Cardona P. Fulde K. von Klitzing H.-J. Queisser

1 **Principles of Magnetic Resonance**
3rd Edition By C. P. Slichter
2 **Introduction to Solid-State Theory**
2nd Printing By O. Madelung
3 **Dynamical Scattering of X-Rays in Crystals** By Z. G. Pinsker
4 **Inelastic Electron Tunneling Spectroscopy**
Editor: T. Wolfram
5 **Fundamentals of Crystal Growth I**
Macroscopic Equilibrium and Transport Concepts. 2nd Printing
By F. Rosenberger
6 **Magnetic Flux Structures in Superconductors** By R. P. Huebener
7 **Green's Functions in Quantum Physics**
2nd Edition By E. N. Economou
8 **Solitons and Condensed Matter Physics**
2nd Printing
Editors: A. R. Bishop and T. Schneider
9 **Photoferroelectrics** By V. M. Fridkin
10 **Phonon Dispersion Relations in Insulators** By H. Bilz and W. Kress
11 **Electron Transport in Compound Semiconductors** By B. R. Nag
12 **The Physics of Elementary Excitations**
By S. Nakajima, Y. Toyozawa, and R. Abe
13 **The Physics of Selenium and Tellurium**
Editors: E. Gerlach and P. Grosse
14 **Magnetic Bubble Technology** 2nd Edition
By A. H. Eschenfelder
15 **Modern Crystallography I**
Symmetry of Crystals.
Methods of Structural Crystallography
By B. K. Vainshtein
16 **Organic Molecular Crystals**
Their Electronic States. By E. A. Silinsh
17 **The Theory of Magnetism I**
Statics and Dynamics. 2nd Printing
By D. C. Mattis
18 **Relaxation of Elementary Excitations**
Editors: R. Kubo and E. Hanamura
19 **Solitons.** Mathematical Methods for Physicists. 2nd Printing
By G. Eilenberger
20 **Theory of Nonlinear Lattices**
2nd Edition By M. Toda
21 **Modern Crystallography II**
Structure of Crystals
By B. K. Vainshtein, V. M. Fridkin, and V. L. Indenbom
22 **Point Defects in Semiconductors I**
Theoretical Aspects
By M. Lannoo and J. Bourgoin
23 **Physics in One Dimension**
Editors: J. Bernasconi, T. Schneider
24 **Physics in High Magnetic Fields**
Editors: S. Chikazumi and N. Miura
25 **Fundamental Physics of Amorphous Semiconductors** Editor: F. Yonezawa
26 **Elastic Media with Microstructure I**
One-Dimensional Models. By I. A. Kunin
27 **Superconductivity of Transition Metals**
Their Alloys and Compounds
By S. V. Vonsovsky, Yu. A. Izyumov, and E. Z. Kurmaev
28 **The Structure and Properties of Matter**
Editor: T. Matsubara
29 **Electron Correlation and Magnetism in Narrow-Band Systems** Editor: T. Moriya
30 **Statistical Physics I**
Equilibrium Statistical Mechanics
By M. Toda, R. Kubo, N. Saito
31 **Statistical Physics II**
Nonequilibrium Statistical Mechanics
By R. Kubo, M. Toda, N. Hashitsume
32 **Quantum Theory of Magnetism**
2nd Edition By R. M. White
33 **Mixed Crystals** By A. I. Kitaigorodsky
34 **Phonons: Theory and Experiments I**
Lattice Dynamics and Models of Interatomic Forces. By P. Brüesch
35 **Point Defects in Semiconductors II**
Experimental Aspects
By J. Bourgoin and M. Lannoo
36 **Modern Crystallography III**
Crystal Growth 2nd Edition
By A. A. Chernov
37 **Modern Crystallography IV**
Physical Properties of Crystals
Editor: L. A. Shuvalov
38 **Physics of Intercalation Compounds**
Editors: L. Pietronero and E. Tosatti
39 **Anderson Localization**
Editors: Y. Nagaoka and H. Fukuyama
40 **Semiconductor Physics** An Introduction
5th Edition By K. Seeger
41 **The LMTO Method**
Muffin-Tin Orbitals and Electronic Structure
By H. L. Skriver
42 **Crystal Optics with Spatial Dispersion, and Excitons** 2nd Edition
By V. M. Agranovich and V. L. Ginzburg
43 **Structure Analysis of Point Defects in Solids by Multiple Magnetic Resonance Spectroscopy**
By J.-M. Spaeth, J. R. Niklas, and R. H. Bartram
44 **Elastic Media with Microstructure II**
Three-Dimensional Models By I. A. Kunin
45 **Electronic Properties of Doped Semiconductors**
By B. I. Shklovskii and A. L. Efros
46 **Topological Disorder in Condensed Matter**
Editors: F. Yonezawa and T. Ninomiya
47 **Statics and Dynamics of Nonlinear Systems**
Editors: G. Benedek, H. Bilz, and R. Zeyher
48 **Magnetic Phase Transitions**
Editors: M. Ausloos and R. J. Elliott
49 **Organic Molecular Aggregates** Electronic Excitation and Interaction Processes
Editors: P. Reineker, H. Haken, and H. C. Wolf

Springer Series in Solid-State Sciences

Editors: M. Cardona P. Fulde K. von Klitzing H.-J. Queisser

50 **Multiple Diffraction of X-Rays in Crystals**
By Shih-Lin Chang

51 **Phonon Scattering in Condensed Matter**
Editors: W. Eisenmenger, K. Laßmann, and S. Döttinger

52 **Superconductivity in Magnetic and Exotic Materials** Editors: T. Matsubara and A. Kotani

53 **Two-Dimensional Systems, Heterostructures, and Superlattices**
Editors: G. Bauer, F. Kuchar, and H. Heinrich

54 **Magnetic Excitations and Fluctuations**
Editors: S. Lovesey, U. Balucani, F. Borsa, and V. Tognetti

55 **The Theory of Magnetism II** Thermodynamics and Statistical Mechanics By D. C. Mattis

56 **Spin Fluctuations in Itinerant Electron Magnetism** By T. Moriya

57 **Polycrystalline Semiconductors,**
Physical Properties and Applications
Editor: G. Harbeke

58 **The Recursion Method and Its Applications**
Editors: D. Pettifor and D. Weaire

59 **Dynamical Processes and Ordering on Solid Surfaces** Editors: A. Yoshimori and M. Tsukada

60 **Excitonic Processes in Solids**
By M. Ueta, H. Kanzaki, K. Kobayashi, Y. Toyozawa, and E. Hanamura

61 **Localization, Interaction, and Transport Phenomena** Editors: B. Kramer, G. Bergmann, and Y. Bruynseraede

62 **Theory of Heavy Fermions and Valence Fluctuations** Editors: T. Kasuya and T. Saso

63 **Electronic Properties of Polymers and Related Compounds**
Editors: H. Kuzmany, M. Mehring, and S. Roth

64 **Symmetries in Physics** Group Theory Applied to Physical Problems
By W. Ludwig and C. Falter

65 **Phonons: Theory and Experiments II**
Experiments and Interpretation of Experimental Results By P. Brüesch

66 **Phonons: Theory and Experiments III**
Phenomena Related to Phonons
By P. Brüesch

67 **Two-Dimensional Systems: Physics and New Devices**
Editors: G. Bauer, F. Kuchar, and H. Heinrich

68 **Phonon Scattering in Condensed Matter V**
Editors: A. C. Anderson and J. P. Wolfe

69 **Nonlinearity in Condensed Matter**
Editors: A. R. Bishop, D. K. Campbell, P. Kumar, and S. E. Trullinger

70 **From Hamiltonians to Phase Diagrams** The Electronic and Statistical-Mechanical Theory of sp-Bonded Metals and Alloys By J. Hafner

71 **High Magnetic Fields in Semiconductor Physics**
Editor: G. Landwehr

72 **One-Dimensional Conductors**
By S. Kagoshima, H. Nagasawa, and T. Sambongi

73 **Quantum Solid-State Physics**
Editors: S. V. Vonsovsky and M. I. Katsnelson

74 **Quantum Monte Carlo Methods** in Equilibrium and Nonequilibrium Systems Editor: M. Suzuki

75 **Electronic Structure and Optical Properties of Semiconductors** 2nd Edition
By M. L. Cohen and J. R. Chelikowsky

76 **Electronic Properties of Conjugated Polymers**
Editors: H. Kuzmany, M. Mehring, and S. Roth

77 **Fermi Surface Effects**
Editors: J. Kondo and A. Yoshimori

78 **Group Theory and Its Applications in Physics**
By T. Inui, Y. Tanabe, and Y. Onodera

79 **Elementary Excitations in Quantum Fluids**
Editors: K. Ohbayashi and M. Watabe

80 **Monte Carlo Simulation in Statistical Physics**
An Introduction
By K. Binder and D. W. Heermann

81 **Core-Level Spectroscopy in Condensed Systems**
Editors: J. Kanamori and A. Kotani

82 **Introduction to Photoemission Spectroscopy**
By S. Hüfner

83 **Physics and Technology of Submicron Structures**
Editors: H. Heinrich, G. Bauer, and F. Kuchar

84 **Beyond the Crystalline State** An Emerging Perspective By G. Venkataraman, D. Sahoo, and V. Balakrishnan

85 **The Fractional Quantum Hall Effect**
Properties of an Incompressible Quantum Fluid
By T. Chakraborty and P. Pietiläinen

86 **The Quantum Statistics of Dynamic Processes**
By E. Fick and G. Sauermann

87 **High Magnetic Fields in Semiconductor Physics II**
Transport and Optics Editor: G. Landwehr

88 **Organic Superconductors**
By T. Ishiguro and K. Yamaji

89 **Strong Correlation and Superconductivity**
Editors: H. Fukuyama, S. Maekawa, and A. P. Malozemoff